Experimental Methods in the Physical Sciences

Volume 49

Neutron Scattering - Applications in Biology, Chemistry, and Materials Science

Experimental Methods in the Physical Sciences

Thomas Lucatorto, John T. Yates,
and Kenneth Baldwin
Editors in Chief

Experimental Methods in the Physical Sciences

Volume 49

Neutron Scattering - Applications in Biology, Chemistry, and Materials Science

Edited by

Felix Fernandez-Alonso
ISIS Facility
Rutherford Appleton Laboratory
Didcot, United Kingdom

and

Department of Physics & Astronomy
University College London
London, United Kingdom

David L. Price
Synchrotron Radiation and Neutron Research Group, CEMHTI
Centre National de la Recherche Scientifique
Orléans, France

ELSEVIER

ACADEMIC PRESS
An imprint of Elsevier

Academic Press is an imprint of Elsevier
50 Hampshire Street, 5th Floor, Cambridge, MA 02139, United States
525 B Street, Suite 1800, San Diego, CA 92101–4495, United States
The Boulevard, Langford Lane, Kidlington, Oxford OX5 1GB, United Kingdom
125 London Wall, London, EC2Y 5AS, United Kingdom

ISBN: 978-0-12-805324-9
ISSN: 1079-4042

Publisher: John Fedor
Acquisition Editor: Anita A. Koch
Editorial Project Manager: Amy Clark
Production Project Manager: Maria Bernard
Cover Designer: Greg Harris

Typeset by SPi Global, India

Contents

Volume 43. Radiometric Temperature Measurements. II. Applications
Edited by Z. M. Zhang, B. K. Tsai, and G. Machin

Volume 44. Neutron Scattering – Fundamentals
Edited by Felix Fernandez-Alonso and David L. Price

Volume 45. Single-Photon Generation and Detection
Edited by Alan Migdall, Sergey Polyakov, Jingyun Fan, and Joshua Bienfang

Volume 46. Spectrophotometry: Accurate Measurement of Optical Properties of Materials
Edited by Thomas A. Germer, Joanne C. Zwinkels, and Benjamin K. Tsai

Volume 47. Optical Radiometry for Ocean Climate Measurements
Edited by Giuseppe Zibordi, Craig J. Donlon, and Albert C. Parr

Volume 48. Neutron Scattering - Magnetic and Quantum Phenomena
Edited by Felix Fernandez-Alonso and David L. Price

Volume 49. Neutron Scattering - Applications in Biology, Chemistry, and Materials Science
Edited by Felix Fernandez-Alonso and David L. Price

Preface

Just over 80 years ago, a brief letter from James Chadwick to *Nature* [1,2] presented conclusive experimental evidence unveiling the existence of a neutral particle (nearly) isobaric with the proton. The discovery of the henceforth-to-be-known-as "neutron" had profound consequences for both scientific research and the destiny of humankind, as it led to the unleashing of the might of nuclear power in less than a decade [3].

The first use of these "neutral protons" to probe the microscopic underpinnings of the materials world around us also dates back to those early years, with pioneering neutron-diffraction experiments at Oak Ridge National Laboratory (USA) in the mid 1940s, and the subsequent development of neutron spectroscopy at Chalk River (Canada) in the 1950s. Since then, neutron-scattering techniques have matured into a robust and increasingly versatile toolkit for physicists, chemists, biologists, materials scientists, engineers, and technologists. At the turn of the last century, the 1994 Nobel Prize in Physics awarded to C.G. Shull and B.N. Brockhouse recognized their ground-breaking efforts towards the development and consolidation of neutron science as a discipline in its own right [4]. This milestone also served to define neutron scattering as the technique *par excellence* to investigate *where atoms are* (structure) and *what atoms do* (dynamics), a popular motto across generations of neutron-scattering practitioners.

Sustained and continued developments in experimental methods over the past few decades have greatly increased the sensitivity and range of applications of neutron scattering. While early measurements probed distances on the order of interatomic spacings (fractions of a nm) and characteristic times associated with lattice vibrations (picoseconds), contemporary neutron-scattering experiments can cover length scales from less than 0.01 to 1000 s of nanometers, and time scales from the attosecond to the microsecond. These advances have been made possible via a significant expansion of the range of neutron energies available to the experimenter, from micro-electron-volts (particularly at cold sources in research reactors) to hundreds of electron-volts (at pulsed spallation sources), as well as by unabated progress in the implementation of a variety of novel and ingenious ideas such as position- and polarization-sensitive detection or back-scattering and spin-labeling methods. As a result, neutron science has grown beyond traditional research areas, from the conventional determination of crystal structures and lattice dynamics of half-a-century ago (not to forget their magnetic analogs), to high-resolution structural studies of disordered thin films, liquid interfaces, biological structures, macromolecular and supramolecular architectures and devices, or the unraveling of the dynamics and energy-level structure of complex molecular solids, nanostructured materials and surfaces, or magnetic clusters and novel superconductors. Along with these scientific and technical developments, the community of neutron scientists has also expanded and diversified beyond recognition. Whereas the early stages of neutron scattering had its roots in condensed-matter physics and crystallography, present-day users of central neutron-scattering facilities include chemists, biologists, ceramicists, and metallurgists, to name a few, as well as physicists with an increasingly diverse range of transdisciplinary interests, from the foundations of quantum mechanics to soft matter, food science, biology, geology, or archeometry.

This book series seeks to cover in some detail the production and use of neutrons across the aforementioned disciplines, with a particular emphasis on technical and scientific developments over the past two decades. As such, it necessarily builds upon an earlier and very successful three-volume set edited by K. Sköld and D.L. Price, published in the 1980s by *Academic Press* as part of *Methods of Experimental Physics* (currently *Experimental Methods in the Physical Sciences*). Furthermore, new third-generation spallation sources have either been constructed in the United States and Japan, or are in the advanced construction stage in China. Likewise, the development of long-pulse, ultra-intense spallation neutron sources is well underway in Europe, including the construction of the European Spallation Source. These global efforts have been accompanied by an increasing interest in time-of-flight and broadband neutron-scattering techniques. Correspondingly, the improved performance of cold moderators at both reactors and spallation sources has extended long-wavelength capabilities to such an extent that a sharp distinction between fission- and accelerator-driven neutron sources may no longer be of relevance to the future of the discipline.

On a more practical front, the chapters that follow are meant to enable you to identify aspects of your work in which neutron-scattering techniques might contribute, conceive the important experiments to be done, assess what is required to carry them out, write a successful proposal to a user facility, and perform these experiments under the guidance and support of the appropriate facility-based scientist. The presentation is aimed at professionals at all levels, from early-career researchers to mature scientists who may be insufficiently aware or up-to-date with the breadth of opportunities provided by neutron techniques in their area of specialty. In this spirit, it does not aim to present a systematic and detailed development of the underlying theory, which may be found in superbly written texts such as those of Lovesey [5] or Squires [6]. Likewise, it is not a detailed hands-on manual of experimental methods, which in our opinion is best obtained directly from experienced practitioners or, alternatively, by attending practical training courses at the neutron facilities. As an intermediate (and highly advisable) step, we also note the existence of neutron-focused thematic schools, particularly those at Grenoble [7] and Oxford [8], both of which have been running on a regular basis since the 1990s. With these primary objectives in mind, each chapter focuses on well-defined areas of neutron science and has been written by a leading practitioner or practitioners of the application of neutron methods in that particular field.

The first volume of this series *Neutron Scattering – Fundamentals* [9] gave a self-contained survey of the theoretical concepts and formalism of the technique and established the notation used throughout the series. Subsequent chapters in this first volume reviewed neutron production and instrumentation, respectively, areas that have profited enormously from recent developments in accelerator physics, materials research and engineering, or computing, to name a few. The remaining chapters treated several basic applications of neutron scattering including the structure of complex materials, large-scale structures, and dynamics of atoms and molecules. The appendix went back to some requisite fundamentals linked to neutron-matter interactions, along with a detailed compilation of neutron scattering lengths and cross sections across the periodic table.

The second volume *Neutron Scattering – Magnetic and Quantum Phenomena* [10] placed the focus on traditional (and still highly popular) areas of neutron scattering. These included a detailed presentation of neutron optics and spin-labeling methods along with a broad introduction to the interaction of neutrons with electronic spins in condensed matter, a subject also of direct relevance to the further development of novel instrumentation at present and in the foreseeable future. The following chapters discussed recent developments in the use of neutron scattering to investigate quantum phase transitions, high-temperature superconductors, magnetic structures, multiferroics, nanomagnetism, and nuclear magnetism.

The present and final volume of the series is dedicated to the application of neutron scattering techniques across contemporary biology, chemistry, and materials science. As such, our presentation has been necessarily restricted to a handful of topics, in order to highlight and bring to the fore specific areas of activity that continue to gain importance and relevance. We start with the structure and dynamics of biological systems, followed by what one may consider as "more chemical" applications including water and aqueous systems, ionic liquids, catalysis on surfaces and nanostructured media, ionic and proton conductors, and soft matter. In some cases, we have chosen to place the emphasis on a detailed description of specific techniques and their recent use across disciplines. These include high-temperature and high-pressure techniques, or the (quite unique) use of epithermal neutrons to interrogate the quantum behavior of atoms in materials, and how it relates to function and performance. Our presentation is wrapped-up by an up-to-date overview of applications in engineering, an area which certainly links the technique to an everyday context quite close to our lives. Relative to the previous two volumes, we have found it much more difficult to keep the overall size of this third and final volume within reasonable bounds. In our view, this challenge is a testament to the growth and success witnessed by the field over the past two decades. The future of neutron scattering as a discipline, therefore, looks very bright!

In closing this preface, we wish to thank all authors for taking time out of their busy schedules to be part of this venture, Drs. T. Lucatorto, J.T. Yates, and K. Baldwin for inviting us to undertake this work, and the staff of *Academic Press* for their encouragement, diligence, and forbearance along the way.

Felix Fernandez-Alonso
David L. Price

REFERENCES

[1] J. Chadwick, Nature 129 (1932) 312.
[2] www.nobelprize.org/nobel_prizes/physics/laureates/1935/ (last accessed on 12 March 2017).
[3] A. MacKay, The Making of the Atomic Age, Oxford University Press, Oxford, 1984.
[4] www.nobelprize.org/nobel_prizes/physics/laureates/1994/ (last accessed on 12 March 2017).
[5] S.W. Lovesey, Theory of Neutron Scattering from Condensed Matter, vols. I and II, Oxford University Press, Oxford, 1986.
[6] G.L. Squires, Introduction to the Theory of Thermal Neutron Scattering, third ed., Cambridge University Press, Cambridge, 2012.
[7] www.hercules-school.eu/ (last accessed on 12 March 2017).
[8] www.oxfordneutronschool.org/ (last accessed on 12 March 2017).
[9] F. Fernandez-Alonso, D.L. Price (Eds.), Neutron Scattering – Fundamentals, Experimental Methods in the Physical Sciences, vol. 44, Academic Press, Amsterdam, 2013.
[10] F. Fernandez-Alonso, D.L. Price (Eds.), Neutron Scattering – Magnetic and Quantum Phenomena, Experimental Methods in the Physical Sciences, vol. 48, Academic Press, Amsterdam, 2015.

Symbols

Note: Other symbols may be used that are unique to the chapter where they occur.

a	scattering length in centre-of-mass frame
b	bound scattering length
A	atomic-mass (or nucleon) number
$\bar{b}$	coherent scattering length
b^+	scattering length for $I+\frac{1}{2}$ state
b^-	scattering length for $I-\frac{1}{2}$ state
b_i	incoherent scattering length
b_N	spin-dependent scattering length
c	speed of light in vacuum $=299{,}792{,}458$ m s^{-1} [1]
$\mathbf{D}$	dynamical matrix
$\mathbf{D}_\perp(\mathbf{Q})$	magnetic interaction operator
d	mass density
$\mathbf{d}$	equilibrium position of atom in unit cell
$d\sigma/d\Omega$	differential scattering cross section
$d^2\sigma/d\Omega\,dE_f$	double differential scattering cross section
E	neutron energy transfer $(E_i - E_f)$
E_n	neutron energy
E_i, E_f	initial, final (scattered) neutron energy
e	elementary charge $= 1.6021766208\times10^{-19}$ C
$\boldsymbol{e}^k(\mathbf{q})$	polarization vector of normal mode k [$\boldsymbol{e}_d^k(\mathbf{q})$ for a non-Bravais crystal]
$F(\boldsymbol{\tau})$	unit-cell structure factor
f	force constant
$f(\mathbf{Q})$	form factor
$G(\mathbf{r},t)$	space-time (van Hove) correlation function $[G_d(\mathbf{r},t)+G_s(\mathbf{r},t)]$
$G_d(\mathbf{r},t)$	"distinct" space-time correlation function
$G_s(\mathbf{r},t)$	"self" space-time correlation function
$g(\mathbf{r})$	pair distribution function
g_e	electron g factor $= -2.00231930436182$
g_n	neutron g factor $= -3.82608545$
h	Planck constant $= 6.626070040\times10^{-34}$ J s
$\hbar$	Planck constant over $2\pi (h/2\pi) = 1.054571800\times10^{-34}$ J s
$\mathbf{I}$	angular-momentum operator for nucleus
$I(\mathbf{Q},t)$	intermediate scattering function

$I_s(\mathbf{Q},t)$	self intermediate scattering function
k_B	Boltzmann constant = $1.38064852 \times 10^{-23}$ J K^{-1}
$\boldsymbol{k}_i, \boldsymbol{k}_f$	initial, final (scattered) neutron wave vector
$\boldsymbol{k}_n$	neutron wave vector ($k_n = 2\pi/\lambda_n$)
L_i, L_f	initial (primary), final (secondary) flight path
$\mathbf{l}$	unit-cell position vector (lowercase "L")
M	atom mass
$\boldsymbol{M}$	magnetization operator
m_e	electron mass = $9.10938356 \times 10^{-31}$ kg
m_n	neutron mass = $1.674927471 \times 10^{-27}$ kg
m_p	proton mass = $1.672621898 \times 10^{-27}$ kg
N	number of atoms in sample
N_A	Avogadro constant = $6.022140857 \times 10^{23}$ mol^{-1}
N_c	number of unit cells in crystal
$\mathbf{R}_j$	position vector for particle j
$\mathbf{Q}$	scattering vector ($\boldsymbol{k}_i - \boldsymbol{k}_f$)
$\mathbf{q}$	reduced wave vector ($\mathbf{Q} - \boldsymbol{\tau}$)
r_0	classical electron radius ($e^2/4\pi\varepsilon_0 m_e c^2$) = $2.8179403227 \times 10^{-15}$ m
$\mathbf{S}$	atomic-spin operator
$S(\mathbf{Q})$	static structure factor $I(\mathbf{Q},0)$
$S(\mathbf{Q},E)$	dynamic structure factor or scattering function
$S_c(\mathbf{Q},E)$	coherent dynamic structure factor or scattering function
$S_i(\mathbf{Q},E)$	incoherent dynamic structure factor or scattering function
$\mathbf{s}$	electron-spin operator
T	temperature
u	unified atomic-mass unit = $1.660539040 \times 10^{-27}$ kg
$\mathbf{u}^l$	vibrational amplitude ($\mathbf{u}_d^l$ for non-Bravais crystals)
V	volume of sample
v_i, v_f	initial, final (scattered) neutron velocity
v_n	neutron velocity
v_0	unit-cell volume
$2W$	exponential argument in the Debye-Waller factor
Z	atomic (or proton) number
$Z(E)$	phonon density of states
Γ	spectral linewidth
γ	neutron magnetic moment to nuclear magneton ratio (μ_n/μ_N) = -1.91304273
γ_n	neutron gyromagnetic ratio = 1.83247172×10^8 s^{-1} T^{-1}
Θ	Debye temperature
θ	Bragg angle
λ_i, λ_f	initial, final (scattered) neutron wavelength
$\boldsymbol{\mu}$	atomic magnetic moment
μ_N	nuclear magneton ($e\hbar/2m_p$) = $5.050783699 \times 10^{-27}$ J T^{-1}

μ_B	Bohr magneton ($e\hbar/2m_e$)= $927.4009994 \times 10^{-26}$ J T^{-1}
μ_n	neutron magnetic moment = $-0.96623650 \times 10^{-26}$ J T^{-1}
ν	frequency ($\omega/2\pi$)
$\rho(\mathbf{r}, t)$	particle-density operator
ρ_0	average number density
Σ	macroscopic cross section associated with a given cross section σ (see below)
σ_a	absorption cross section
σ_t	bound total cross section ($\sigma_s + \sigma_a$)
σ_c	bound coherent scattering cross section
σ_i	bound incoherent scattering cross section
σ_s	bound scattering cross section ($\sigma_c + \sigma_i$)
$\frac{1}{2}\boldsymbol{\sigma}$	neutron-spin (Pauli) operator
$\boldsymbol{\tau}$	reciprocal lattice vector $\{2\pi[(h/a), (k/b), (l/c)]\}$
Φ	neutron flux (typically defined as neutrons crossing per unit area per unit time)
ϕ	scattering angle ($=2\theta$)
χ	susceptibility
$\chi(\mathbf{Q}, E)$	generalized susceptibility
Ω	solid angle
ω	radian frequency associated with neutron energy transfer $E = \hbar\omega$
$\omega_k(\mathbf{q})$	frequency of normal mode k

REFERENCES

[1] All numerical values have been taken from the CODATA Internationally Recommended Values of the Fundamental Physical Constants, as detailed in P.J. Mohr, D.B. Newell, B.N. Taylor, Rev. Mod. Phys. 88 (2016) 035009 and P.J. Mohr, D.B. Newell, B.N. Taylor, J. Phys. Chem. Ref. Data 45 (2016) 043102. Updates can be found online at physics.nist.gov/cuu/Constants/ (last accessed on 12 March 2017).

Chapter 1

Biological Structures

Zoe Fisher*,†, Andrew Jackson*,‡, Andrey Kovalevsky§, Esko Oksanen*,¶ and Hanna Wacklin*,‡,1

*European Spallation Source, Lund, Sweden; †Department of Biology, Lund University, Lund, Sweden; ‡Department of Chemistry, Lund University, Lund, Sweden; §Biology and Soft Matter Division, Oak Ridge National Laboratory, Oak Ridge, TN, United States; ¶Department of Biochemistry and Structural Biology, Lund University, Lund, Sweden

1Corresponding author: e-mail: hanna.wacklin@esss.se

Chapter Outline

Experimental Methods in the Physical Sciences, Vol. 49. http://dx.doi.org/10.1016/B978-0-12-805324-9.00001-7

simply the availability of more instruments, there has also been a push by several facilities to offer users access to support laboratories for sample preparation, in particular deuteration.

Recent developments in crystallographic refinement software have also played a major role in improving the throughput of neutron protein structures. Since ~50% of the atoms in a biological crystal are hydrogens, crystallographic refinement of the hydrogen positions effectively doubles the number of parameters, while the number of observed Bragg reflections remains the same for a given resolution. This unfavorable effect on the data-to-parameter ratio can be mitigated by using also X-ray data in the refinement to provide information about the positions of heavier atoms. This so-called joint refinement has been implemented in the software packages CNS and PHENIX [16].

1.2.1.1 Instruments

Neutron beams are made in two ways for research applications [17], at reactor-based sources that are usually continuous and at spallation sources that are usually pulsed. Globally, there are a handful of instruments dedicated to macromolecular crystallography available at both types of sources. Their properties are listed in Tables 2 and 3. The different neutron facilities that host these instruments offer a range of user support for sample preparation, data collection, and analysis. Despite the seemingly large variety in data collection methods (time of flight and Laue vs quasi-Laue vs monochromatic), instrument geometries, available neutron energies, beam delivery systems, beam divergence, detector technologies, etc., the final diffraction data measured at any of these instruments are very similar. Yet, there are advantages and disadvantages in using one over the other for a given sample. Things to consider include the crystallographic unit cell, crystal volume, and whether it is partially or perdeuterated.

When comparing instrument features, there are some indications as to which instrument may be better suited for a particular sample. Most instruments are well suited to unit cells that are no bigger than ~100 Å on each edge, but this is heavily dependent on the crystal size and mosaicity. Depending on the source and detector type, some instruments offer higher gains for perdeuterated crystals over those partially deuterated by solvent H/D exchange.

In general, the advantage of perdeuteration is larger at reactor-based Laue instruments, where it significantly pushes the volume limit down. For example at LADI-III, data were successfully collected from perdeuterated crystals smaller than $0.3\ mm^3$ [9]. Beam time at the facilities hosting the instruments is allocated based on scientific merit after review of a beam time proposal. As the allocation of beam time is incredibly competitive, users tend to use all the available instruments. In addition to the instruments currently available for users, there are plans for new instruments at both existing and future

TABLE 2 Reactor-Based Instruments for Neutron Protein Crystallography

	LADI-III [18]	Biodiff [19]	Imagine [20]	BIX-3 [21]	BIX-4 [22]
Facility/source	ILL [23] 58 MW	MLZ, FRM-II [24] 20 MW	ORNL, HFIR [25] 84 MW	JRR-3M [26] 20 MW	JRR-3M [26] 20 MW
Laue versus monochromatic	Quasi-Laue	Monochromatic	Quasi-Laue	Monochromatic	Monochromatic
Wavelengths min and max (Å)	2.6–3.4, 3.2–4.2, 3.8–5.0, 4.1–5.5	2.4–5.6	2.0, 2.8, 3.3 3.0, 4.0, 4.5	2.4	2.6
Detector type	Cylindrical image plate	Cylindrical image plate	Cylindrical image plate	Cylindrical image plate	Cylindrical image plate
Sample environment	Ambient	Ambient, 90–550 K, 3.5–325 K	Ambient, 100 K cryogenic	Ambient, 100 K cryogenic	Ambient, 100 K cryogenic
Crystallographic resolution (Å)	~1.5	~1.5	~1.5	~1.5	~1.5
Unit cell (Å)	~150 Å each edge	~100 Å each edge	~100 Å each edge	~100 Å each edge	~100 Å each edge
Sample volume	>0.05 mm^3	>1 mm^3	>1 mm^3	>1 mm^3	>1 mm^3

TABLE 3 Spallation-Source-Based Instruments for Neutron Protein Crystallography

	PCS [27]	MaNDi [28]	iBIX [29]	NMX (future) [30]
Facility/source	LANSCE [31], 0.1 MW, 20 Hz	ORNL, SNS [32], 2 MW, 60/30 Hz	J-PARC, MLF [33], 1 MW, 25 Hz	ESS [34], 5 MW, 14 Hz
Laue versus monochromatic	Laue	Laue	Laue	Laue
Wavelengths min and max (Å)	0.6–7.0	1–10 ($\Delta\lambda$ 2.15 or 4.3)	0.7–7.8 ($\Delta\lambda\sim4$)	1.8–10 ($\Delta\lambda\sim1.75$)
Detector type	^{3}He	Anger camera	λ-Shifting fiber detectors	Gd-coated GEM
Sample environment	Ambient, 100 K	80–400 K	Ambient, 100 K	Ambient, 100 K, controlled humidity
Crystallographic resolution (Å)	>1.0	~1.5	~1.5	~1.5
Unit cell (Å)	~100 Å one up to 200 Å	~150 Å, 300 Å for large crystals (>1 mm^3)	~100 Å one up to 200 Å	~300 Å all edges
Sample volume	>0.5 mm^3	>0.1 mm^3	>1 mm^3	>0.01 mm^3

neutron sources. The plans at existing facilities for new macromolecular diffractometers are not yet very advanced, but the NMX diffractometer at the European Spallation Source (ESS)—currently under construction in Lund, Sweden—is in the detailed engineering design phase and is likely to be available for users ca. 2023. The ESS source is ideally suited for macromolecular crystallography and large gains are expected (Table 2).

The data reduction for different types of instruments is also rather different. With monochromatic instruments, the software packages used for monochromatic X-ray diffraction can be used with minor modification. For reactor-based Laue instruments, software has been adapted from X-ray Laue software, but for the time-of-flight Laue method, new, dedicated software has been developed. At present, each facility has its own solution, and a common approach is yet to emerge.

1.2.1.2 Sample Considerations for Neutron Protein Crystallography

Instrumentation and software advances have certainly improved the throughput of neutron protein crystallography beamlines, but crystallization remains a major bottleneck. Macromolecular crystals are typically of large systems that diffract weakly and have a lot of solvent (30–70%), making them particularly intractable for inherently flux-limited neutron experiments. Even for X-ray applications where orders of magnitude smaller crystals are needed, crystal growth is difficult to study and control. This is because it is expensive and time-consuming to prepare large quantities of labeled (and unlabeled) sample, and they are often not stable enough to enable exhaustive screening for optimal crystallization conditions. For neutron experiments, there is the added burden of needing crystals in the 1 mm^3 range, and this consumes enormous amounts of sample, typically hundreds of milligrams of pure protein. For most crystallographers, this is very difficult, even impossible, to achieve. However, in the recent years, there have been many reports of large crystal growth for neutron experiments, and some techniques seem to work better than others. As for X-ray crystallography, proteins should be >95% pure and as concentrated as the protein solubility permits. It is beneficial to centrifuge the concentrated protein solution prior to setting up crystallization drops as this removes any particulate or insoluble matter that may interfere with nucleation and crystal growth.

If enough sample is available, it is very useful to investigate the solubility of the protein as a function of precipitant/pH/temperature (or some other parameter like additive concentration) through a phase diagram. If done properly and carefully, a well-conceived phase diagram will indicate a range of parameters from which to further optimize on large scale. In an idealized phase diagram, there are four zones: solution, metastable, labile, and precipitation zones. Nucleation occurs in the labile zone, followed by crystal growth in the metastable zone. In the solution zone, the drops are clear with no nucleation occurring. In the precipitation zone, there is an irreversible step where the protein comes out of solution. A phase diagram can be set up in a 24- or 96-well plate format with relatively little sample. Many of the crystallization robots (e.g., Oryx8 from Douglas Instruments) can be programmed to set these up, covering an extensive solubility matrix for systematic study of protein behavior. To maximize the amount of soluble protein available for crystal growth, it is necessary to scale up the drop sizes. The conditions that produced the best crystals in small volume have to be further optimized once scale-up has occurred. For most successful neutron crystal preparations, the drops should be 100–1000 μL in volume. This implies that the usual sitting-drop setups that work well for X-ray crystallography are not sufficient or practically possible for neutron applications. Due to the large volumes required, the approaches that can routinely produce large crystals are sitting-drop vapor diffusion, batch, and dialysis.

Various commercial supports are available to ease the growth of large crystals for neutron crystallography. For sitting-drop vapor diffusion, the sandwich box setup from Hampton Research utilizes a large reservoir and a nine-well glass plate, each well can hold up to 1000 μL. This has been used with huge success for a large number of different proteins that crystallize with sitting-drop vapor diffusion [35–40]. The nine-well glass plate can be used for batch crystallization also, as long as the plates are well sealed with tape so that no evaporation occurs. Large, single crystals of pyrophosphatase were grown on the vapor diffusion concept but in capillaries [41].

Crystallization by dialysis is a powerful variation of the vapor diffusion technique. Here, the concentrations of the protein and precipitant increase in the dialysis button due to slow diffusion of the water out of the sample into the mother liquor and counterdiffusion of the precipitant through a semipermeable membrane with a defined molecular weight cutoff. Dialysis is the only approach used for crystallization that works by lowering, rather than increasing, the precipitant concentration. For some proteins, crystals can be grown by dialysis against H_2O or D_2O due to a salting-in effect. Dialysis button and setup kits are commercially available from Hampton Research and can be bought for volumes up to 350 μL. Care has to be taken during setup and crystal harvest as the buttons are quite small and it is easy to lose a crystal when the button is opened up. This method has been used with success for growing neutron quality crystals of a handful of proteins [42,43]. There has also been a recent development to automate dialysis crystallization with temperature control and real-time visualization. This device is called the Opticrys and is commercially sold by NatX-ray. This approach is different from previous work since it allows the user to set up parallel experiments with the same protein sample. It is a nondestructive method, and the conditions can be altered or reversed in the same vessel while testing multiple crystallization conditions in a rational, controlled way [44]. The current maximum sample volume is 90 μL, but the system is suitably flexible so that larger volume can be used with a custom made button.

Most neutron diffraction experiments with macromolecules are performed at ambient temperature and pressure. Neutrons, in the energy spectrum most suitable for macromolecular crystallography, are not destructive to biological samples, and they do not cause the typical radiation damage seen with X-rays. This makes it very simple to collect data at room temperature with the samples mounted in thick-walled quartz capillaries, with the ends sealed airtight with beeswax. These capillaries are easy to work with, and quartz scatters thermal or cold neutrons very weakly.

Using capillaries also allows the user to perform H/D exchange through the vapor in a sealed capillary prior to data collection. This technique is quite gentle on the sample compared with growing the crystal in deuterated solutions or soaking the crystals after they have formed. It also leads to equivalent levels of H/D exchange provided at least ~3 weeks are allowed for H/D exchange prior to data collection. The exchange in the capillary occurs from the D_2O vapor that comes from the deuterated mother liquor plug that is introduced during crystal mounting. For most large, stable crystals, capillary mounting and room-temperature data collection work well but are clearly not a good solution for transient complexes or for crystals that will not survive at room temperature for the days/weeks it can take to collect neutron data. Every macromolecular beamline is equipped with standard cryosystems that allow data collection at 100 K, but freezing large crystals at home before data collection can be very challenging or even impossible to do without causing catastrophic damage to the crystal lattice. Long, serial soaking in increasing amounts of cryoprotectants has been demonstrated to help with this, but either many systems are not amenable to this due to a lack of sufficient numbers of large crystals to experiment on or that the crystal simply does not tolerate this [45]. ADC offers a stand-alone high-pressure crystal freezing device that may present a practical and technically feasible solution to freezing large crystals without the use of cryoprotectants. These devices probably have a very good place at most neutron protein crystallography beamlines where crystals bigger than 1 mm^3 are still required. For crystals 0.5 mm^3 and smaller, it is usually sufficient to use standard procedures as is routinely done for X-ray work [46].

1.2.2 Neutron Fiber Diffraction

Neutron fiber diffraction has been applied to study the structure of many different kinds of biologically important polymers, such as DNA [47], troponin C, cellulose [48], chitin [49], and amyloid fibers [50,51] For this technique, samples have to be aligned fibers and H/D solvent exchanged to improve the signal-to-noise ratio for neutron scattering experiments. The alignment has to be such that the long axes are parallel to the fiber axes. The highly ordered crystalline regions of fibers are usually interspersed with less ordered, amorphous regions that complicate the diffraction patterns. In addition, how the fibers are packed and aligned with respect to the individual chains being in register will largely determine the quality of the measured diffraction pattern. As Figure 2 illustrates, the variation can be quite extreme, from a continuous molecular transform to sharp, well-defined Bragg peaks.

With a well-prepared sample and other favorable conditions, the resolution can be extended to atomic resolution. For most biological polymers that occur in a fibrous state, it is the atomic structure, arrangement of monomers, and the higher-order interactions between chains that are crucial to how they function. Fiber diffraction is also amenable to measuring conformational changes in biological polymers as the fibers can often tolerate changes in environment better than single crystals. This makes it possible to subject a sample to an array of different humidities, ionic strengths, pH, and so on. Neutron fiber diffraction is highly complementary to X-ray methods as it

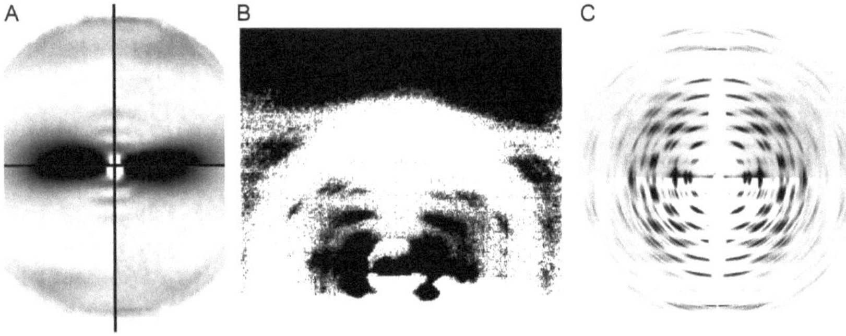

FIGURE 2 Neutron fiber diffraction data recorded on the D19 diffractometer at ILL from a variety of samples indicate the wide variation in details of observed diffraction patterns. (A) Troponin C, (B) DNA, and (C) cellulose. *Reproduced with permissions from [48,51,52].*

can provide high-resolution atomic information in addition to enabling water location and H/D atom visualization [4].

While different forms of cellulose appear naturally as crystalline microfibers, they are not sufficiently well ordered to be suitable for diffraction experiments without some additional preparation. First, the thin microcrystalline sticks are purified, and the suspension is then cast as films in which the microcrystals were oriented parallel to one another. These films can then be used for "as is" for X-ray diffraction experiments. The samples are subsequently subjected to H/D exchange to make them suitable for neutron diffraction experiments [48]. Also, the samples have to be much bigger for neutron experiments than for X-rays, and this means the films or fibers have to be made thicker by stacking or some other manual manipulation. For work on troponin C, the samples were prepared as a solution in heavy water (D_2O) and sealed in 3 mm diameter capillaries. The polymers were then oriented by placing the capillaries in an 18 T magnetic field [51]. Samples of amyloid fibers were prepared by drying solutions of the peptide in between two glass rods. The resulting fibers were then stacked in a parallel array to match the neutron beam size. Fibers are often subjected to H/D exchange in a humidity chamber prior to and during data collection. The humidity can also be controlled to drive crystalline transitions for functional studies [50].

Most neutron fiber diffraction experiments in the last decade have been conducted at beamline D19 at the ILL (Grenoble, France). A thorough description of the D19 beamline and instrument capabilities can be found in Teixeira et al. [53].

1.3 SMALL-ANGLE NEUTRON SCATTERING (SANS)

Small-angle neutron scattering [1] provides structural information on nanometer to micrometer length scales, covering the sizes of interest for biological systems from lipid bilayers, to proteins and protein aggregates, to liposomes and viruses. Some of the earliest SANS measurements were made on biological systems [54–56] and summarized by Jacrot [57]. Using contrast variation and matching, SANS is a powerful tool for the study of biological systems in solution, particularly the study of multidomain proteins [58–60], protein-lipid complexes [61] [62], and nucleic acids [63]. Use of deuterated proteins and lipids allows such complex systems to be studied and details of intermolecular arrangements to be determined.

Small-angle scattering data are recorded as scattered intensity as a function of wave-vector transfer, q, which is defined in Eq. (1):

$$q = \frac{4\pi \sin\theta}{\lambda} \quad (1)$$

where 2θ is the scattering angle and λ is the neutron wavelength. The method takes its name from the fact that the scattering angles used are small compared with diffraction measurements, with the upper limit widely considered to be $2\theta = 45°$. The size of the scattering object has an inverse relationship with the q vector, so measuring at low angles means measuring at low q, which leads to studying scattering from larger structures than those probed by diffraction measurements. Thus, whereas diffraction methods probe interatomic distances, small-angle scattering allows for studies of molecular conformation, association, and arrangement. The measured scattering has a coherent part, which contains structural information and an incoherent part that contains no structural information and is a source of background in a SANS measurement [64].

1.3.1 SANS Instrumentation

SANS is measured in transmission geometry (Figure 3), with the neutron beam being directed onto a platelike sample through a pair of collimating apertures that determine the angular divergence of the beam and hence both the minimum accessible wave-vector transfer, q, and the angular resolution

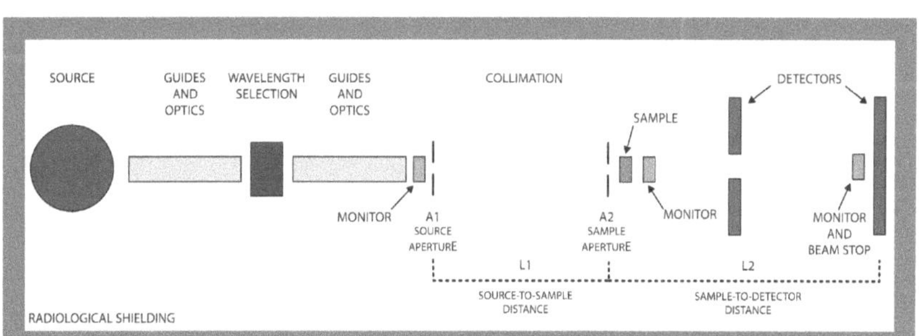

FIGURE 3 Schematic of a pinhole SANS instrument showing the major components and terminology. Not to scale.

of the instrument, dθ/θ. In order to reach lower values of scattering vector and hence study larger structures, SANS instruments tend to have long collimation distances and long sample-to-detector distances, usually in the range of 10–20 m each, and use long wavelengths of neutrons, up to 20 Å. The scattered neutrons are then counted on one or more area detectors in order to measure at multiple angles simultaneously. Access to lower values of scattering vector can be achieved via either measuring at a smaller angle or through the use of longer wavelength neutrons. Two approaches exist for the design of pinhole SANS instruments, continuous/monochromatic or pulsed/time of flight. In the continuous approach, a beam of neutrons with a narrow band of wavelengths, usually with 10–15% $\Delta\lambda/\lambda$, is directed onto the sample. In this case, the simultaneously measurable q-range is determined solely by the angular coverage of the detectors. In the time-of-flight approach, a pulsed beam of neutrons is required, and the time between the neutron pulse and the time of measurement of a signal on the detector is used to determine the neutron velocity and therefore the neutron wavelength. The pulse of neutrons will contain a broad range of wavelengths, so a wider q-range is measurable at one time since wavelength and scattering angle are varied simultaneously.

In addition to these "pinhole geometry" SANS instruments (Figure 3), which are the most common, other methods exist to reduce the minimum accessible scattering vector. USANS, "ultra-" or "ultra-high-resolution" small-angle neutron scattering [65,66], is a term usually reserved for two-axis monochromator-analyzer crystal instruments. These instruments make use of diffraction from crystals to define the incoming wavelength and also provide a very narrow angular resolution and hence much lower scattering vectors. USANS instruments generally allow access to scattering from structures between 0.5 and 10 micrometers in size. VSANS, "very" small-angle neutron scattering, uses systems of converging collimation slits or pinholes to overlap the size range of SANS and USANS [67,68].

The neutron has a spin, and as such, a beam of neutrons can be polarized such that it contains only neutrons of one spin state [69]. The polarization of the beam is affected by the interactions of the neutron with materials and magnetic fields and is most commonly used to probe the magnetic structure of nanomaterials. This can apply in some biologically related studies, such as therapeutic nanoparticles, but can also be used as a means of quantifying the incoherent scattering background that comes primarily from the water in a sample [70–72].

Polarized neutron beams are also used in spin-echo SANS [73,74], which is a less common approach to measuring scattering at small angles. Here, the polarized beam of neutrons passes through a magnetic field, which causes their spin to precess. A second field after the sample exactly matches the first but in the reverse direction. If there is no interaction with the sample, then there is no change to the polarization of the neutron beam. If there is scattering, however, the path length through the second field will be different from that through the first, and the spin precession will not exactly match that on the incoming side. By measuring the polarization of the beam at the end of the instrument as a function of magnetic field, the scattering can be determined. Here, the result is given as a function of "spin-echo length," which is a real-space quantity, unlike q. Similar data analysis methods can be employed for calculating the scattering function and comparing it to the measured data. The SESANS instrument at the Reactor Institute Delft [75] employs this method, as does the LARMOR instrument at the ISIS facility, Rutherford Appleton Laboratory [76]. LARMOR can also use a modified version of the SESANS method, known as spin-echo modulated SANS (SEMSANS) [77] to measure SESANS and pinhole SANS simultaneously.

1.3.1.1 Currently Available Instruments

Every neutron scattering facility has at least one dedicated SANS instrument or another instrument that is used to make SANS measurements. Table 4 gives a list of SANS instruments available at major facilities worldwide, along with whether they can use polarized neutron beams, whether they are time-of-flight instruments, whether they can measure USANS, and whether they are capable of SESANS measurements.

The choice of where to perform an experiment is driven by the specifications of the instrument (e.g., q-range), the availability of the required sample environment equipment (e.g., temperature-controlled sample stages), and availability of instrument time, which is usually awarded on a peer review basis.

1.3.1.2 Future Instruments

At the ESS ERIC, two SANS instruments are currently under construction, to be available to user operations ca. 2023–2024. Loki [123], a broad wavelength band nonpolarized SANS instrument, will cater for biological small samples and measurements with a broad simultaneous q-range. SKADI [124,125], a longer polarized SANS instrument, will be applicable to all types of SANS measurements, with an optional grazing-incidence geometry (GISANS; see Section 1.4 for thin film samples).

1.3.2 Sample Preparation

The smallest sample size that is commonly measurable on the highest flux SANS instruments today is around 100 μL for a sample with good contrast and a reasonable concentration (~1 wt%). In practice, to manage the sample in the quartz cuvettes normally used, around 500 μL of sample solution is required. The usual approach in SANS measurements is to place the sample solution into quartz cuvettes, often referred to as "sample cells." Quartz is transparent to neutrons and has weak scattering so that it contributes little to

TABLE 4 SANS Instruments at Neutron Facilities

Facility	Country	Instrument	Pol	TOF	USANS	Comments
ANSTO [78]	Australia	Quokka [79]	Yes	No	No	
ANSTO [78]	Australia	Bilby [80]	No	Yes	No	
ANSTO [78]	Australia	Kookaburra [81]	No	No	Yes	
HZB BERII [82]	Germany	V4 SANS [83]	No	No	No	
HZB BERII [82]	Germany	V16 VSANS [84]	No	Yes	No	VSANS
HZB BERII [82]	Germany	V12 USANS [85]	No	No	Yes	
BRR KFKI [86]	Hungary	SANS [87]	No	No	No	
FRM-II [24]	Germany	KWS-1 [88]	Yes	No	No	
FRM-II [24]	Germany	KWS-2 [89]	Yes	No	No	
FRM-II [24]	Germany	KWS-3 [90]	No	No	No	Focusing mirror VSANS
FRM-II [24]	Germany	SANS-1 [91]	Yes	No	No	
HANARO [92]	South Korea	40 m SANS [93]	Yes	No	No	
HANARO [92]	South Korea	18 m SANS [93]	No	No	No	
HANARO [92]	South Korea	USANS	No	No	Yes	
ILL [23]	France	D11 [94]	No	No	Yes	Pinhole USANS Option
ILL [23]	France	D22 [95]	Yes	No	No	
ILL [23]	France	D33 [96]	Yes	Yes	No	TOF Option

RI Delft [97]	The Netherlands	SANS [98]	No	No	No	In commissioning
RI Delft [97]	The Netherlands	SESANS [99]	No	No	Yes	Spin-echo
ISIS [100]	The United Kingdom	LOQ [101]	No	Yes	No	
ISIS [100]	The United Kingdom	SANS2D [102]	No	Yes	No	
ISIS [100]	The United Kingdom	LARMOR [76]	Yes	Yes	Yes	Pinhole + spin-echo
ISIS [100]	The United Kingdom	ZOOM [103]	Yes	Yes	No	VSANS
J-PARC [33]	Japan	TAIKAN [104]	Yes	Yes	No	
JRR-3 [26]	Japan	SANS-U	No	No	No	
JRR-3 [26]	Japan	SANS-JII	No	No	No	
JRR-3 [26]	Japan	ULS USANS	No	No	Yes	
LANSCE [31]	The United States	LQD [105]	No	Yes	No	
LENS [106]	The United States	NBL1 [107]	No	Yes	No	
LLB [108]	France	PACE [109]	No	No	No	
LLB [108]	France	PAXY [109]	No	Yes	No	TOF option
LLB [108]	France	PA20 [109]	Yes	No	No	
LLB [108]	France	TPA [109]	No	No	No	VSANS
NIST [110]	The United States	NGB 10 m [111]	No	No	No	
NIST [110]	The United States	NGB 30 m [112]	No	No	No	
NIST [110]	The United States	NG7 SANS [113]	Yes	No	No	

Continued

TABLE 4 SANS Instruments at Neutron Facilities—cont'd

Facility	Country	Instrument	Pol	TOF	USANS	Comments
NIST [110]	The United States	BT-5 [114]	No	No	Yes	USANS
NIST [110]	The United States	NG3 [115]	No	No	No	VSANS
ORNL SNS [32]	The United States	EQ-SANS [116]	No	Yes	No	
ORNL SNS [32]	The United States	USANS [117]	No	Yes	Yes	
ORNL HFIR [25]	The United States	BIO-SANS [118]	No	No	No	
ORNL HFIR [25]	The United States	GP-SANS [119]	Yes	No	No	
SINQ PSI [120]	Switzerland	SANS-1 [121]	Yes	No	No	
SINQ PSI [120]	Switzerland	SANS-2 [122]	Yes	No	No	

the background signal. It is also transparent to light that enables visual inspection of the state of the sample.

In the case of determination of the solution structure of a protein, it is important to ensure that the sample is dilute, noninteracting and not aggregated, so that it has a monodisperse size distribution. Thus, sample conditions (pH, ionic strength, and temperature) need to be chosen such that the sample will not aggregate during measurement. Characterization of the sample before measurement using UV-vis spectroscopy and dynamic light scattering is necessary to determine monodispersity, as is filtration of the sample when placing into the sample cell. If solvent contrast variation is being used, the sample must be dialyzed against a large volume of solvent of the required scattering length density and the same pH and ionic strength as the sample. Depending on the volume ratio of sample to solvent reservoir, the dialysis will need to be repeated multiple times.

1.3.3 Data Processing and Analysis

The data collected by the instrument are not immediately amenable to analysis and must be processed before analysis in terms of the structure of the sample can proceed. This processing or "data reduction," is instrument-specific and is done with software provided at the instrument.

The value of interest from a scattering experiment is the macroscopic differential cross section, which is related to the number of scattered neutrons per incoming neutron and is often referred to as I(q). It is given by the amplitude of the Fourier transform of the scattering length density distribution in the sample as a function of q (Eq. 2):

$$\frac{d\Sigma}{d\Omega}(q) = \frac{1}{V}\left|\int \rho(r)e^{iqr}dr\right|^2 \quad (2)$$

and contains the structural information we seek about how matter is arranged in the sample.

However, the actual measured intensity distribution is only proportional to this value and so must then be corrected for instrumental effects and sample transmission to place the result on "absolute scale," giving the macroscopic differential cross section as a function of q, convoluted with the resolution function of the instrument. After processing, the result can then be analyzed.

In order to be able to obtain this result, measurements of the solvent alone (to measure the background scattering from the solvent) and the transmission of the solvent and sample must be made in addition to the scattering measurement from the sample of interest. There also need measurements for detector sensitivity, q calibration, and intensity calibration.

Analyzing the scattering data once it is corrected should proceed from simple approaches of data inspection, to simple shape modeling, to more complex computational approaches including shape reconstruction and comparison with crystal structures and molecular dynamic simulation.

1.3.3.1 Model Independent Data Analysis

The Guinier approximation [126,127] gives Eq. (3):

$$I(q) = I(0)e^{\frac{-q^2R_g^2}{3}} \tag{3}$$

and thus, a linear fit of $\ln(I(q))$ against q^2 gives R_g and $I(0)$. R_g, the radius of gyration of the particle, is the RMS distances within the particle from the center of mass, with differently shaped particles of the same volume having different values of R_g, and gives information about the mass distribution in the particle. $I(0)$ is related to the molecular weight, MW, of the particle through Eq. (4):

$$I(0) = N(\Delta\rho V)^2 = \frac{C\Delta\rho^2 v^2 MW}{N_A} \tag{4}$$

where N is the number density of particles in the sample, V is the particle volume, C is the mass per unit volume, $\Delta\rho$ is the scattering contrast, and N_A is Avogadro's number.

In the case of monodisperse, dilute, particulate systems, such as proteins in solution, the intensity of scattering can be related to the distance distribution function of the particle, P(r), as shown in Eq. (5):

$$I(q) = 4\pi \int_0^\infty P(r) \frac{\sin(qr)}{qr} dr \tag{5}$$

Ideally, therefore, a Fourier transform could be applied to the scattering data to directly obtain the distance distribution function. However, due to the limited q-range of the measurement and the loss of phase information in the scattering process and instrumental effects, it is not possible to directly transform the data in this way. Thus, the indirect Fourier transform method of Glatter [128–131] is used to extract P(r). This method is available in a number of software packages, including the ATSAS suite and SasView (see Table 5). Before applying such methods, it is important to understand whether the collected data meet the prerequisites of being monodisperse and dilute, that is, noninteracting. Jacques and Trewhella [5] provide a thorough approach to the initial stages of data inspection, making use of linearized plots and P(r) to determine the radius of gyration of a protein and to identify the presence of interparticle interactions and aggregation.

1.3.3.2 Geometric Modeling

A common modeling approach is to represent the particle structure as a geometric form, calculate the scattering from such an object [141], and compare it to the data using a nonlinear least-square method. In applying this method, care must be taken to account for the instrumental resolution [142,143] and understand the relationship between the model parameters; for example, the

TABLE 5 Software Packages for the Analysis of SANS Data

Package name	Features and address
SasView [132,133]	Geometric modeling and indirect Fourier transform. 2-D scattering from oriented systems. Magnetic scattering
SasFit [134]	Geometric modeling. Magnetic scattering
NCNR Igor [135]	Geometric modeling
ATSAS [136]	Suite of software tools aimed at biological small-angle scattering. Includes indirect Fourier transform and shape reconstruction tools
Scatter [137]	Geometric modeling
Scatter [138]	Data treatment and analysis for scattering from biological molecules
SASSIE [139,140]	Atomistic and coarse-grained simulation of biomolecules and calculation of scattering curves from simulation results

contrast and volume fraction of particles are usually directly correlated and so cannot both be fitted as free parameters. A number of software packages exist to make this approach accessible to nonexperts, and the most widely used are summarized in Table 5.

1.3.3.3 Shape Reconstruction

This approach to modeling is widely employed in biological small-angle scattering and builds up a scattering particle from subresolution spheres randomly distributed within a set of constraints and compares the calculated scattering with the measured data [144]. A family of possible structures is thus generated and merged to give a most probable particle shape. This can then be compared with crystal structures to see if the solution structure is the same or how it is different. Care must be taken in applying this approach to ensure that the data are of adequate quality and that sufficiently many structures have been tested.

1.3.3.4 Molecular Dynamics Simulations and Conformational Searches

Another approach for determining biomolecular structure from scattering results is to compare the scattering with scattering calculated from a real-space chemical model of the molecule or molecular assembly. Molecular dynamic (MD) simulations of large biomolecules are usually performed using coarse-grained techniques in order to improve computational efficiency [145], since the resolution of a small-angle scattering measurement is such that atomic level structure is not needed. Where a molecule has a flexible region or is intrinsically disordered, the large conformation space means that a purely MD approach would be prohibitively computationally expensive. An alternative, therefore, combines MD with conformational search approaches where the shape of the linker is varied randomly and the nonflexible portions, which may have been determined from MD simulations, remain unchanged [139,146]. The scattering will be the result of averaging over all orientations of the molecule, and so many structures can give similar scattering patterns. Using this approach, families of structures will emerge that have similar scattering patterns. These can be compared with experimental data, and a best fit family can be determined. This, for instance, allows determination of whether a protein is compact or extended in solution and what is the most probable arrangement of subgroups.

1.4 NEUTRON REFLECTOMETRY

Neutron reflection [64] probes the structure of biological and biomimetic surfaces and interfaces on the angstrom-nanometer scale at which most of the functional interactions between proteins and membranes take place. In contrast to synchrotron X-ray beams or high-intensity lasers, the low neutron energies used (meV) give rise to no physical damage to the samples and can easily penetrate through bulk materials to access biological or biomimetic surfaces under solution conditions. The scattering of neutrons only depends on the nuclear composition, and thus, neutrons are applicable to samples that exhibit optical anisotropy, such as lipid membranes below the gel-to-fluid transition, which can be difficult to study using ellipsometry or dual-polarization interferometry [147]. The neutron wavelengths employed (1–30 Å) are well matched to the length scale of biological membranes and proteins, in addition to which the samples are aligned on a planar surface giving rise to better resolution (2–3 Å) than in isotropic solution small-angle scattering (5–10 Å). Depending on the instrument, specular neutron reflection allows surface structures of up to $\sim$1 μm thick to be investigated, offering highly complementary information about the structures across biological interfaces to the lateral structures accessed by modern fluorescence and scanning probe microscopy techniques. Off-specular reflectometry [148] and grazing-incidence small-angle neutron scattering GISANS [149] in principle offer the possibility to probe 2-D organization of biological surfaces at micrometer and nanometer length scales, respectively, but their use has so far been rather limited due to the weakly scattering nature of thin lipid and protein films. It is however expected that these techniques will emerge as powerful characterization tools to complement grazing-incidence small/wide and X-ray scattering (GISAXS/GIWAXS) and surface diffraction (GIXD), which are more widely employed to characterize lipid membranes.

1.4.1 Specular and Off-Specular Reflection

Specular neutron reflection [64] (in which the angle of reflection is equal to the angle of incidence) measures the structure of thin films at surfaces or interfaces between two materials (Figure 4). The films are characterized by their thickness and scattering length density (or neutron refractive index) profile of a film along the direction normal to its surface. The scattering length density ρ of a material is the sum of the coherent nuclear scattering lengths b_i of its constituent atoms and their number density n_i (Eq. 6):

$$\rho = \sum_i b_i n_i \tag{6}$$

Neutron reflectivity data are measured as the reflected intensity relative to the incident beam ($R = I/I_0$) and recorded as a function of the momentum transfer vector q, defined in Eq. (1).

In specular reflection, only momentum transfer in the direction perpendicular to the surface is measured along the z-axis, while the signal from the whole illuminated area is averaged. Thus, specular reflectivity can be measured in two ways: either by using a single wavelength (monochromatic mode) and varying the angle of incidence or by using a polychromatic beam (time-of-flight mode) and recording reflectivity at a few fixed angles of incidence to cover the desired range of q-values. The resolution of the experiment is determined by the wavelength resolution and the angular resolution of the incident beam, with $\Delta q/q = 0.01$–0.1 typically employed.

When the off-specular reflections arise from structures that are much smaller than the coherence length of the neutrons impinging on the surface (~10–100 μm along the beam), an average of the lateral scattering length density profile is observed by the specular reflectivity. If the lateral structure has a characteristic length scale much larger than the neutron coherence length, the

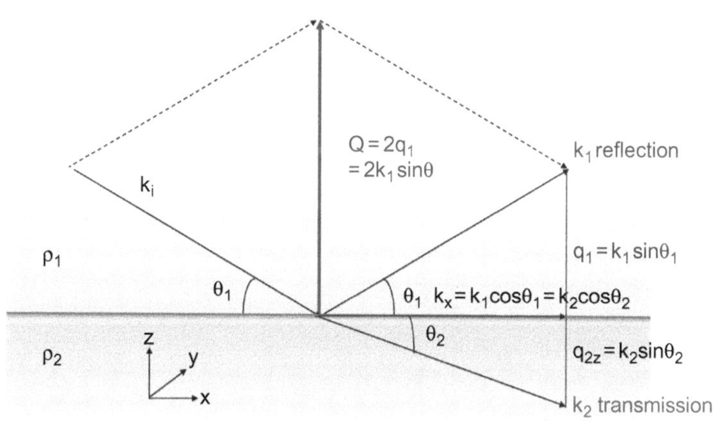

FIGURE 4 Scattering geometry for specular and off-specular reflection.

reflections arising from it are no longer coherent and have to be treated as the sum of reflections from the independent structures. In the intermediate size range, structures with characteristic dimensions approaching the neutron coherence length lead to significant off-specular scattering, which decreases the specular reflection signal. Off-specular reflection could in principle be used to detect protein or membrane domains, but in practice, the membrane domains may be so small (from 100 nm to 1 μm) that the off-specular signal, which is also very weak from 1 to 10 nm membranes/protein films, is not observed but manifests itself as an "interfacial roughness," which slightly decreases the specular intensity. Larger domains lead to observable off-specular signal, but the analysis of off-specular reflectivity in general is still in its infancy.

1.4.2 Grazing-Incidence SANS (GISANS)

In addition to the specularly reflected neutrons, it is possible that a laterally structured sample gives rise to off-specular reflection, that is, there are momentum transfer components parallel to the surface (x and y) plane (Figure 5). Grazing-incidence small-angle neutron scattering (GISANS) can in principle detect lateral structures perpendicular to the neutron beam (y) with nm resolution, using a laterally collimated beam (1–5 mm and collimation/detector distances similar to transmission SANS, Section 1.3).

Using time-of-flight measurements [150], it is also possible to probe lateral structures as function of distance from the interface in thicker systems, as the neutron penetration depth is wavelength-dependent. GISANS is typically done on SANS instruments, which have the collimation distances required, but experiments have yet to be realized on biological membrane or protein systems, which often suffer from incoherent background arising from the bulk solvent phase. It is also possible to encode the off-specular scattering/GISANS signal by employing neutron spin interaction with a magnetic field, but the emerging technique of spin-echo grazing-incidence scattering (SERGIS) [151] has not yet been applied to biological surfaces.

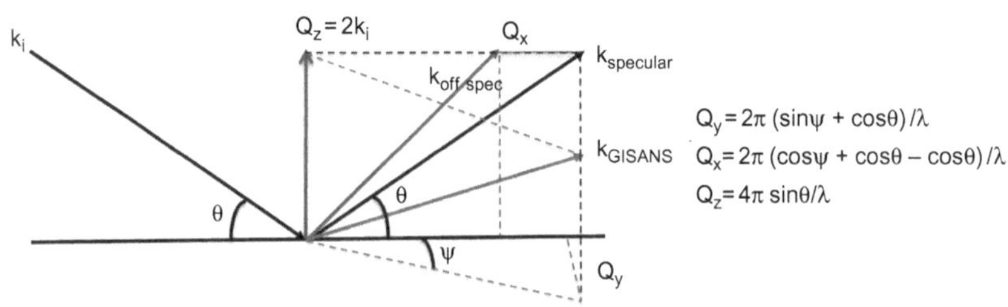

FIGURE 5 Reflection geometry for off-specular reflection and GISANS.

1.4.3 Instruments

Neutron reflectometers are popular instruments for investigating surfaces in a wide range of sciences, and one or more exist at most neutron scattering facilities (Table 6). At the ESS ERIC [34], two new reflectometers are currently under construction. FREIA [189], a high-flux horizontal sample reflectometer for liquid surfaces, is optimized to use the long pulse of the ESS source for kinetic studies using a broad simultaneous wavelength band, and ESTIA [190], a polarized focusing reflectometer with a vertical sample geometry, is designed to work with very small samples that are not accessible on low-flux instruments. Elsewhere, CANDOR, the new reflectometer at the NIST Centre for Neutron Scattering is under construction, and the plans for the second target station of the Spallation Neutron Source at Oak Ridge National Laboratory include three reflectometers [191].

1.4.4 Experimental Considerations

Neutron reflection experiments can be done at liquid and solid surfaces, which determines the required instrument configuration—solid supports and solid-liquid chambers can be mounted either horizontally or vertically on a reflectometer, but liquid surfaces require a horizontal sample geometry, preferably with minimal sample movement during the measurements to avoid disturbing the fragile samples. In all cases, the key prerequisite for neutron reflection is that the sample surface is both very flat and smooth (<20 Å rms) over the entire illuminated area, as irregularities reduce the signal-to-noise ratio rapidly by off-specular scattering. The solid supports used can be mechanically polished to the desired roughness, but for liquid surfaces, this implies that the samples need to be vibration isolated to minimize capillary waves on the surface. In an ideal case, the sample surface will have less than 5 Å roughness. Neutrons are strongly scattered by liquids, but many solid materials such as silicon, sapphire, and quartz are nearly transparent, making them suitable as the incident medium at solid-liquid interfaces. To avoid incoherent scattering background, it is often desirable to choose the medium with less hydrogen as the incident medium at liquid-liquid interfaces (aided further by deuteration); therefore, sometimes, an inverted beam geometry is beneficial if this means that the interface has to be illuminated from below the horizon. Only the FIGARO [192] reflectometer at ILL has this option available today.

1.4.4.1 Resolution

Biological samples in a noncrystalline state often form relatively thin films on surfaces, and for this reason, biological experiments typically do not require very high q-resolution. Most instruments have suitable resolution options ($dq/q = 1–10\%$) available, although it can be beneficial to choose a lower resolution in order to increase the flux of neutrons that can be used, especially on

TABLE 6 Neutron Reflectometers at Major Reactor and Spallation Sources

Facility	Country	Instrument	Interface	Pol[a]	TOF	Comments
ANSTO [78]	Australia	Platypus [152]	Horizontal	Yes	Yes	
ANSTO [78]	Australia	Spatz-BioRef [153,154]	Vertical	No	Yes	
HZB BERII [82]	Germany	V6 [155]	Horizontal	Yes	No	
HZB BERII [82]	Germany	V14 [156]	Vertical	Yes	No	
BRR KFKI [86]	Hungary	GINA [157]	Vertical	Yes	No	
BRR KFKI [86]	Hungary	REF [158]	Vertical	Yes	No	
FRM-II [24]	Germany	MARIA [159]	Vertical	Yes	No	
FRM-II [24]	Germany	MIRA [160]	Vertical	Yes	No	
FRM-II [24]	Germany	N-REX$^+$ [161]	Horizontal	Yes	No	
FRM-II [24]	Germany	REFSANS [162]	Horizontal	Yes	Yes	GISANS
HANARO [92]	South Korea	REF-V [93]	Vertical	No	No	
HANARO [92]	South Korea	REF-H [92]	Horizontal	No	No	
ILL [23]	France	SuperADAM [163]	Vertical	Yes	No	Swedish CRG
ILL [23]	France	D17 [164]	Vertical	Yes	Yes	Also monochromatic
ILL [23]	France	D50 [165]	Vertical	No	No	Angle dispersive mode
ILL [23]	France	Figaro [166]	Horizontal	No	Yes	Inverted beam available
RI Delft [97]	The Netherlands	ROG [167]	Horizontal	Yes	Yes	

ISIS [100]	The United Kingdom	CRISP [168]	Horizontal	Yes	Yes	
ISIS [100]	The United Kingdom	SURF [169]	Horizontal	No	Yes	
ISIS [100]	The United Kingdom	Inter [170]	Horizontal	No	Yes	
ISIS [100]	The United Kingdom	PolRef [171]	Horizontal	Yes	Yes	
ISIS [100]	The United Kingdom	OffSpec [172]	Horizontal	Yes	Yes	Also vertical, SERGIS
J-PARC [33]	Japan	SOFIA [173]	Horizontal	No	Yes	
J-PARC [33]	Japan	SHARAKU [174]	Vertical	Yes	Yes	
JINR [175]	Russia	REFLEX [176]	Vertical	Yes	Yes	
JINR [175]	Russia	REMUR [177]	Vertical	Yes	Yes	
JINR [175]	Russia	GRAINS [178]	Horizontal	Yes	Yes	
LANSCE [31]	The United States	ASTERIX [179]	Vertical	Yes	Yes	
LANSCE [31]	The United States	SPEAR	Horizontal	No	Yes	
LLB [108]	France	DESIR [109]	Horizontal	No	Yes	
LLB [108]	France	EROS [109]	Horizontal	No	Yes	
LLB [108]	France	PRISM [109]	Vertical	Yes	No	
NIST [110]	The United States	CANDOR [180]	Vertical	No	No	Under construction
NIST [110]	The United States	MAGIK [181]	Vertical	No	No	
NIST [110]	The United States	NG1-PBR [182]	Vertical	Yes	No	
NIST [110]	The United States	NG7 [183]	Horizontal	No	No	

Continued

TABLE 6 Neutron Reflectometers at Major Reactor and Spallation Sources—cont'd

Facility	Country	Instrument	Interface	Pol	TOF	Comments
ORNL SNS [32]	The United States	BL-4B liquid reflectometer [184]	Horizontal	No	Yes	
ORNL SNS [32]	The United States	BL-4A magnetism reflectometer [185]	Vertical	Yes	Yes	
SINQ PSI [120]	Switzerland	AMOR [186]	Horizontal	Yes	Yes	Wide divergence mode also monochromatic
SINQ PSI [120]	Switzerland	Morpheus [187]	Vertical	Yes	No	
SINQ PSI [120]	Switzerland	Narziss [188]	Vertical	Yes	No	

[a] *Pol = polarized reflectometry available.*

small biological samples. The minima in a reflectivity curve are separated by $\delta q = 2\pi/d$, where d is the thickness of the sample to be measured in Å. Therefore, a protein film with a thickness of 100 Å will have the first minimum at $q = 0.063$ Å^{-1}, and an absolute q-resolution (separation of data points) of 0.0035 Å^{-1} would correspond to a relative q-resolution ~5%.

1.4.4.2 Sample Preparation and Sample Environments

Modern neutron reflectometers typically offer a range of sample environments for user experiments that are custom made or adapted to the instrument's geometry and particular types of samples. Table 7 lists some of the most common types of sample environment and their purpose.

TABLE 7 Sample Environments

Environment	Function	Experiments	Samples
Air-liquid troughs	Temperature-controlled sealed environment	Adsorption at gas-water interface	Soluble proteins, biosurfactant nucleic acids, biopolymers
Solid-liquid cells	Temperature-controlled flow cell	Adsorption and interactions in solid-supported layers	Lipid bilayers, membrane-protein interactions, biomaterial surfaces, biomedical devices
Langmuir film balance	Control of surface area and pressure	Insoluble monolayers, protein and drug interactions	lipid monolayers, protein/drug interactions, lung surfactant
HPLC pump (+column)	Automated solvent dispensing-controlled oligomerization	Solid-liquid experiments in flow cells	Contrast variation experiments
Syringe pump	Automated liquid dispensing/sample injection	Solid-liquid experiments	Contrast variation experiments, protein/drug interactions
Electromagnet	Sample magnetization for magnetic contrast variation	Solid-liquid experiments	Protein arrays, sensor
Electrochemical cell	*In situ* electrochemistry	Solid-liquid experiments	Impedance spectroscopy, electrodes, sensors

In practice, the nature of biological samples means that the surface films for reflectometry experiments are most often prepared *in situ* on the instrument by, for example, adsorbing vesicles to form a lipid bilayer and surface adsorption of proteins from solution.

1.4.4.3 Contrast Variation

In biological studies, the large water content of the materials gives access to a number of chemically similar systems by varying the H/D content of the water or by isotopic substitution of the biomolecules themselves. By measuring neutron reflectivity from a number of such contrasts, a unique solution to the structure is usually found from simultaneous fitting of these, aided by basic chemical knowledge of the materials involved. The scattering length densities of membrane lipids can be calculated, for example, from volume elements obtained from molecular dynamic simulations [193,194], which have been shown to be in good agreement with experimental data from density [195] and X-ray/neutron diffraction measurements [196]. Tables 8 and 9 give a summary of some typical lipid scattering lengths, volumes, and scattering length densities used in data analysis. The values of protein scattering lengths can be calculated from the protein sequence assuming that all labile

TABLE 8 Phospholipid Component Volumes from Molecular Dynamic Simulations [194]

Group	V (DPPC)[a] (Å^3)	V (DOPC)[b] (Å^3)	V (POPC)[c] (Å^3)
CH_3	53.57±0.91	52.79±0.4	50.41±0.77
CH_2	27.93±0.1	28.13±0.09	28.24±0.17
C=C		45.91±0.69	42.1±2.31
COO carbonyl	43.58±1.05	37.4±1.28	38.43±1.47
C_3H_5 glycerol backbone	65.47±2.11	81.62±2.96	72.48±2.28
PO_4 phosphate	66.36±2.95	51.05±6.31	52.12±2.23
$C_5H_{13}N$ choline	107.62±2.49	129.68±5.25	120.68±2.98
H_2O	29.5±0.5	29.5±0.5	29.5±0.5
Total	1215.79±14.27	1322.19±20.52	1255.78±17.12
Experimental [195–198]	1232	1295	1267

[a]*1,2-Di-palmitoyl-phosphatidylcholine ($C_{40}H_{80}NO_8P$).*
[b]*1,2-Dioleylphosphatidylcholine ($C_{44}H_{84}NO_8P$).*
[c]*1-Palmitoyl, 1-oleylphosphatidylcholine ($C_{42}H_{82}NO_8P$).*

TABLE 9 Scattering Length Densities and Volumes of Common Phospholipids

	h-DPPC	d_{62}- DPPC	DOPC	POPC
$b_{headgroup}$[a] (10^{-5} Å^{-1})	60.0432	60.043	60.0432	60.0432
$V_{headgroupg}$[a] (Å^3)	326.25	326.25	337.15	322.14
$\rho_{headgroupg}$[a] (10^{-6} Å^{-2})	1.840	1.840	1.781	1.864
b_{chains}[b] (10^{-5} Å^{-1})	-32.5372	612.98	-20.916	-26.726
V_{chains}[b] (Å^3)	889.18	889.18	985.04	933.68
ρ_{chains}[b] (10^{-6} Å^{-2})	-0.3659	6.894	-0.2123	-0.286

[a]*The headgroup is assumed to contain all polar groups including the choline, phosphate, glycerol backbone, and carbonyl groups.*
[b]*The chain volume is calculated from C2 onward, excluding the carbonyl carbon.*

hydrogens [57] of the residues will exchange with the solvent. If the crystal structure of a protein is not known, its molecular volume can be estimated from the sum of the solution volumes of amino acid residues [199].

Table 10 shows the neutron scattering length b of amino acids and the number of labile hydrogens that need to be considered when calculating the value of b in different solvent contrasts. Estimating the molecular or partial molecular scattering length densities allows contrast matching to be used to selectively highlight or hide parts of a complex system.

In neutron reflectivity from supported films (lipid bilayers and protein arrays), the solid material is chosen to be highly transparent to neutrons so that the membrane-water interface can be approached through the solid to avoid scattering in the bulk liquid phase. The reflectivity contribution from the solid material (most commonly silicon, quartz, or sapphire) can be eliminated by solvent contrast matching, but as reflectivity is most sensitive to differences in scattering length density, this is often used as a means to boost the signal from the sample. Interfacial roughness is the main limiting factor in reflectivity studies, and the solid substrates need to be polished to a high degree of smoothness (rms < 5 Å) and flatness.

1.4.5 Data Analysis and Refinement

The most common way to analyze reflectivity data is to compare measured profiles with simulated profiles generated by an optical matrix calculation [200] in which the sample is described by homogeneous flat layers with a characteristic thickness, scattering length density, and interfacial roughness, which are varied until a fit to the experimental data is obtained. Several programs (Table 11) exist that allow both least squares and other types of

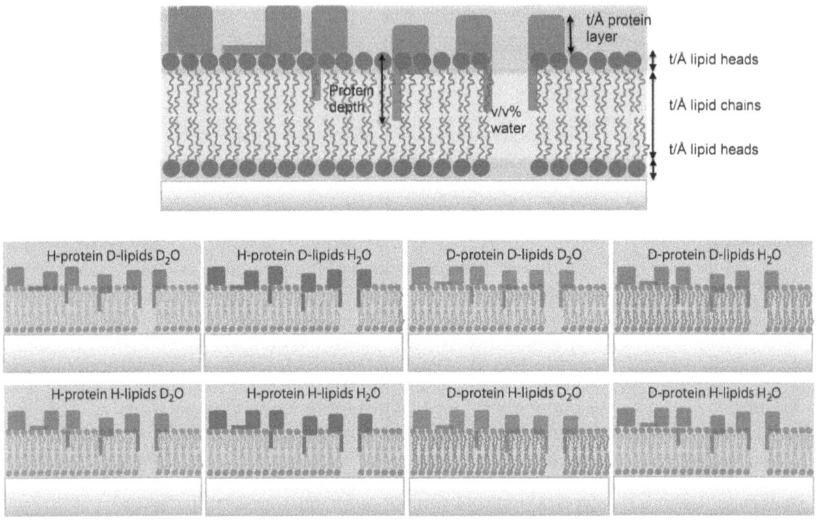

FIGURE 6 Modeling neutron reflectivity data from a protein-membrane system using contrast variation by deuterating lipids and proteins in turn. *Adapted with permissions from [207].*

If more than one chemical species is present, then ρ_s is determined by the volume fraction and scattering length density of each molecular species in the layer. In practice, programs like Motofit allow fitting of ρ_s and ϕ_s as separate variables, so that if the molecular scattering length density is known, only the volume fraction is fitted. When multiple contrast data are fitted simultaneously, the number of fitting variables is greatly reduced by constraining the volume fractions, layer thicknesses, and interfacial roughness values to be the same in all contrasts. For the molecular scattering length densities of, for example, proteins, the exchange of labile protons with the solvent needs to be taken into account in the different contrasts. The scattering length density of a layer can also be expressed as shown in Eq. (8):

$$\rho = \frac{\sum_i b_i n \cdot b_w}{V_a} \tag{8}$$

with $\Sigma_i b_i$ corresponding to neutron scattering lengths of the nuclei and b_w to the water. If A is the area per molecule, the volume per molecule and the water associated with it are given by Eq. (9):

$$V_a = A \cdot t_{layer} = V_m + n \cdot V_{aw} \tag{9}$$

where V_m is the dry molecular volume of the protein, related to the fitted volume fraction and layer thickness through Eq. (10):

$$V_m = \phi \cdot A \cdot t_{layer} \tag{10}$$

If V_m is known from, for example, a crystal or solution structure, then the area (A) available for the protein and the number of associated water molecules (n) can be calculated using Eqs. (11) and (12):

$$A = \frac{n \cdot V_w}{t_{layer} \cdot (1 - \phi)} \tag{11}$$

$$n_w = \frac{(1 - \phi)\sum_i b_i}{\rho_{layer} V_w - (1 - \phi) b_w} \tag{12}$$

It is also common to refer to the surface excess Γ (μmol m^{-2}), obtained from Eq. (13), if A is given in $Å^2$ from which the number of proteins per unit area can be derived:

$$\Gamma = \frac{10^{26}}{A \cdot N_A} \tag{13}$$

In a supported lipid membrane, for example, the scattering length density profile (sld) normal to the interface ρ(z) is modeled in terms of the contributions from the solid surface, the lipid chains, the headgroups, and the hydration present, which depends on the thickness and packing density of the lipids. In addition, proteins or other additive components may contribute to the sld. For example, in a multicomponent membrane containing lipids, protein, and water, the total scattering length density is described in terms of the volume fraction and molecular sld of the components, for example, as in Eq. (14):

$$\rho_{layer} = \phi_{lipid}\rho_{lipid} + \phi_{protein}\rho_{protein} + \phi_{water}\rho_{water} \tag{14}$$

where the volume fractions ϕ_i for the protein, lipids, and water should add to one. Using the equation above, the volume fractions of individual components in a multicomponent system can be determined by recording reflectivity profiles in several solvent contrasts (H_2O/D_2O mixtures). The resolution in scattering length density is typically ± 0.2 $Å^{-2}$ if data are recorded in multiple contrasts, which determines the detection limit of components depending on the difference in their molecular scattering length densities. In more complex systems, for example, membranes supported on a polymer cushion or a multilayered substrate, the structure of the cushion/substrate is typically determined *in situ* by measuring its reflectivity before membrane formation and is typically chosen so that it can be easily distinguished from, for example, a lipid membrane due to high solvent content or by selective deuteration. In protein interaction experiments, the membrane structure is ideally determined in several contrasts before and after protein addition to have a reference structure. Smaller components in the membrane, such as sterols and membrane-binding drugs, can be detected to relatively low volume fractions deuterated, the so-called selective deuteration or segmental labeling. The sections below focus on current methods for deuteration of biomolecules, including information on user-accessible deuteration facilities.

1.6.1 Protein Deuteration

There are several well-established protocols for protein deuteration, specifically for neutron experiments. Depending on the method chosen, different levels of D incorporation can be achieved, from H/D exchanged to full (per) deuteration. This means from ~30% of H atoms exchanged with D to ~85% and >99% D incorporation. Protein deuteration by overexpression in microbial cell cultures however requires the adaptation of the microorganism to the deuterated growth conditions and other necessary modifications in, for example, carbon source, temperature, and antibiotic resistance. While the microbiology and biochemistry are standard, neutron experiments often require scale-up of the methods in order to produce large (10–100 mg) quantities of purified proteins. The vast majority of proteins deuterated are soluble proteins that can be expressed in bacterial cultures, but proteins with post-translational modifications require higher host organisms, such as yeast, that are more challenging, as the isotopic effects of deuteration have a larger impact on their more complex cell biochemistry. Membrane proteins pose a special challenge due to their limited solubility and need for maintaining a membrane-like environment.

The easiest and most effective way to perdeuterate proteins with D incorporation >99% is to grow the protein expression strain of *E. coli* in a variant of M9 minimal media, with all stock solutions and components either deuterated or dissolved in D_2O. This includes the use of a labeled carbon source such as deuterated glycerol or glucose. It is beneficial to add ferrous sulfate and a generic cell-culture vitamin mix to the media to promote bacterial growth in D_2O [214]. Deuterated media is toxic to bacteria and causes the cells to grow very slowly and often leads to decreased protein yields, especially compared with what can be achieved in rich media made with water. However, most expression protocols in bacteria that users are already using for making unlabeled proteins are readily amenable to repeating under deuterated conditions.

Using minimal media in D_2O with a labeled carbon source is the most expensive way to make perdeuterated proteins, and with lower yields, it becomes necessary to do many repeat preparations to get enough material. However, it is very effective for experiments where D occupancy should be maximized, such a single-crystal diffraction methods where a low signal-to-noise dominates what is practically achievable.

A different way to make perdeuterated rich media is to use an algal extract. Algae are cultivated in deuterated minimal media-specific for the strain of algae grown, under ~2% CO_2 in a dedicated incubator, equipped with programmable lighting to provide the photosynthetic organisms with light/dark cycles. A variety of algal strains can be grown like this and harvested by centrifugation [215]. The cells are freeze dried and then subject to deuterated hydrochloric acid hydrolysis in an acid-reflux setup under an inert gas, such as argon. After neutralization and clarification, this results in a concentrated solution of perdeuterated algal cell extract, ready to dilute in pure D_2O and used as you normally would a rich cell-culture media. Some algal strains produce a lot of different fatty acids and lipids, and these can also be extracted for neutron experiments depending on what types the user is interested in. The bacterial cells grow very well in this algal-derived media, and protein yields are much higher than in traditional M9 minimal media. Growing the algae to prepare this rich media can take a lot of time as the cells grow slower in heavy water, but it is not resource-intensive. Also, it is cost effective to recycle the growth media D_2O through rotary evaporation, and this can be reused many times without a measureable loss in growth or deuteration.

In many cases, it is sufficient to have ~85% D incorporation. In these cases, the same modified M9 minimal media can be prepared in D_2O, but an unlabeled (hydrogenous) carbon source is fed to the cells, leading to better cell health and higher protein yields. Of course, this does not give full (per) deuteration, but for many techniques, this is sufficient and is quite inexpensive and trivial to do in the lab.

Pichia pastoris yeast can be used to express eukaryotic proteins that are difficult to express in *E. coli*, as it is capable of some posttranslational modifications including disulfide-bond formation and yeast-specific glycosylation and proteolytic processing [216]. *Pichia pastoris* can be grown in a simple medium consisting of yeast nitrogen base (YNB), basal salt media (BSM) supplemented with *Pichia* trace metals (PTM1) in either flask cultures or fermenters with high protein expression levels. For perdeuteration, deuterated glycerol is used as carbon source until protein expression is induced with methanol. *Pichia pastoris* has also been used to express deuterated membrane proteins [217] and can also be used as an economical source of deuterated yeast extract for insect cell cultures [218].

Insect cells [219] transfected with a baculovirus vector can be used to express membrane proteins, multiprotein complexes, secreted proteins, or other proteins of human origin. Insect cells are able to correctly fold a larger fraction of human proteins and produce a well-defined glycosylation pattern. Insect cells however grow much more slowly than bacteria and require a medium supplemented with sugars and amino acids, making deuterium labeling more challenging. Therefore, deuterated extracts from yeast [220] or algae are typically used to supplement insect cell-culture media. Even so, because deuterated glycerol cannot be added to insect cell cultures and the cells demand a complex cell culture medium, deuteration levels of only up to 76% are achieved and are generally less homogeneous than in *E. coli* and yeast, and the protein yields are reduced to 30–60% of that in unlabeled cultures. Thus far, insect cells have been used to produce a small number of deuterated proteins for NMR studies, such as the human Abelson kinase domain [218,220] and beta-1 adrenergic receptor β_1AR [218].

For neutron reflectometry and SANS, it is often desirable to deuterate only specific residues in a protein to serve as an indication of its orientation or location with respect to another protein or an interface such as a biological membrane or implant/sensor surface. This can be achieved by feeding perdeuterated amino acids to the cell culture [221] or by solid-phase peptide synthesis, if the sequence to be deuterated is not very long. In other cases, partial labeling of specific sequences or subunits is desirable. Unless the deuterated/nondeuterated subunits can be expressed and reconstituted separately [222], segmental labeling by coupling deuterated segments with unlabeled parts of the protein is required [223,224], but this has not yet used been for deuterium labeling for neutron scattering experiments.

Due to numerous factors, there are few examples of neutron protein crystal structures determined from perdeuterated crystals. There are multiple reasons for this: perdeuteration required costly reagents and specialized equipment in some cases, the lack of know-how from structural biologists, the resulting protein may have slightly different physical characteristics (solubility and activity), and in most cases, the crystallization parameters—such as pH—are affected [225]. It is often observed that perdeuterated proteins will crystallize but that crystals do not grow as large as their hydrogenous counterparts. Crystallizing perdeuterated proteins will require some additional screening and optimization of conditions, including protein concentration.

Despite the obvious advantages to using perdeuterated samples for neutron scattering experiments, the vast majority of protein crystal structures have been determined from H/D exchanged samples. In many cases, this is sufficient as the scientific questions being asked involve hydrophilic residues and/or waters, and these are amenable to H/D exchange. However, for many other situations, it is more desirable to have perdeuterated samples. In addition to unambiguous neutron density maps for amino acid residues and solvent, the overall quality of the data and resulting maps tend to be much better compared with data from H/D exchanged crystals. There are different approaches to H/D exchange that include dialyzing the protein solution prior to crystallization against a deuterated buffer or soaking crystals in deuterated mother liquor prior to data collection. A very effective and gentle way to exchange crystals is to mount the crystal in a quartz capillary and introduce a deuterated mother liquor plug away from the crystal prior to sealing the capillary. This allows H/D exchange to occur through the vapor phase and causes the least amount of damage to the crystal.

For future, high scientific impact, neutron protein crystallography, there is a need to expand the kinds of proteins that are studied. A search of the Protein

Data Bank will reveal that many deposited neutron crystal structures have been determined for proteins that are relatively small (5–100 kDa), soluble, easy to express in bacteria, or commercially available. There is a need to expand the use of neutron protein crystallography for larger, more complex, and membrane-associated proteins, and this will require the use of yeast and/or insect cell expression systems in deuterated media, which are still at an early development stage, and require significant future investments.

1.6.2 Biopolymer Deuteration

Bacterial cellulose is an interesting candidate for biocompatible coatings in a range of industries and as a feedstock for biofuel production. Deuterated extracellular cellulose has been successfully produced in acetic acid bacteria in both partially and fully labeled from and shown to have an identical structure to nondeuterate cellulose [226,227]. Bacteria also produce other types of biopolymers mainly for intracellular energy storage. Bacterial polyesters or polyhydroxyalkanoates (PHAs) [228] have been perdeuterated by feeding deuterated fatty acids (synthesized chemically) to *Pseudomonas oleovorans* for use as a contrast agent in infrared microspectroscopy to study the polymer phase separation. Chitosan has been deuterated and purified from *P. pastoris* yeast [229].

Higher organisms typically withstand deuterium only in limited quantities, but recent efforts have increased the fraction of deuteration up to 50% in some plants [230], with the aim to use deuterium substitution not only as a means to study biosynthetic pathways but also as a source of plant-based deuterated materials.

1.6.3 Biomembrane and Small Molecule Deuteration

Smaller biological molecules, such as membrane phospholipids, sterols, short peptides, and biological detergents, can be either produced in biological expression systems or synthesized chemically. The method chosen is often determined by the requirements on yield, purity, and cost, which are mainly determined by the deuterated raw materials needed. Few commercial sources exist for perdeuterated starting materials beyond simple solvents, small organic molecules, and bulk chemicals, although some manufacturers of speciality chemicals provide a selection of aliphatic fatty acids, phospholipids, and other fully or partially labeled molecules, such as glycerol and some offer custom synthesis of related compounds.

The synthesis of deuterated organic compounds typically starts by catalytic hydrogen-deuterium exchange in a hydrothermal reaction using heavy water and a metal catalyst, in which all hydrogens are replaced by deuterium [231]. The process is typically repeated 2–3 times to achieve up to 98%

deuteration in the best cases, but the degree of deuteration depends on the molecule deuterated. The process is harsh and only applicable to molecules whose structure can withstand the conditions without degradation or isomerization, typically aliphatic hydrocarbons, but these often require the presence of a functional group such as a carboxyl or alcohol, and hydrogens in very long chains are more difficult to exchange catalytically. The same applies to aromatic compounds, in which lower deuteration degrees are often achieved even with optimized catalysts [232]. In most cases, the deuterated product requires further synthesis into the target compound; for example, unsaturated fatty acids and lipids are synthesized from separately deuterated fragments [233]. As many biological molecules have a well-defined structure, many reactions need to be stereospecific, for example, to create, for example, isomerically pure cis-double bonds, as few living organisms have trans fats. Often, it is desirable to partially deuterate a molecule to enhance neutron contrast to specific parts, such as the lipid acyl chains or polar headgroup. Chiral molecules, such as amino acids, pose a particular synthetic challenge as they do not withstand the H/D exchange conditions and require enantioselective synthesis or separation methods. For this reason, few deuterated amino acids have been synthesized chemically and often in racemic DL-form [234]. For membrane phospholipids with a chiral glycerol-phosphate headgroup, a total synthesis including perdeuterated glycerol backbone was recently developed, which allows any part of asymmetrical chain, unsaturated phospholipids such as 1-palmitoyl, 2-oleyl, and 3-sn-phosphatidylcholine to be selectively deuterated [235].

The main advantage of chemical synthesis is the ability to produce large (mg-g) quantities of chemically pure compounds but can be costly and very labor-intensive due to the often complex synthetic routes required. For this reason, biological deuteration using microbial cultures is becoming an attractive alternative for producing lipids, sterols, and peptides when suitable purification and analysis methods exist to obtain sufficiently pure compounds. Deuterated amino acids can be fed to cell cultures for partial labeling of proteins or peptides, which was used to create the deuterated lipopeptide surfactin with the hydrophobic leucine residues deuterated [234]. Sterols can be only partially deuterated by chemical synthesis, whereas they can be perdeuterated by yeast cultures, which naturally produce ergosterol, and have been genetically modified to produce deuterated cholesterol [236,237].

The main challenge in producing lipids and sterols in cell cultures is to analyze the composition of the membranes, which can contain hundreds of different molecular species, and to separate the desired compounds. Lipids can be separated into nonpolar and polar lipid classes with relative ease using either thin layer or column chromatography and analyzed for their total fatty-acid composition using gas chromatography [238]. By analyzing the lipid composition, it is also possible to discover any isotopic effects of growing the cell cultures in a deuterated medium. It was recently determined that the model yeast *P. pastoris* produces the same deuterated phospholipid classes as in normal media, but their fatty-acid chain composition is altered, with considerably less polyunsaturated lipids produced as a consequence of reduced enzymatic activity [239]. Bacterial cultures such as *E. coli* are a facile source of bacterial lipids [240] and have been used to generate specifically labeled lipids for contrast matching in neutron experiments [241].

1.6.4 Current and Future Deuteration Facilities

Several neutron user facilities in the world have in the last decade established support labs for labeling biomaterials. This is an important commitment that all facilities should aim to do as proper sample preparation is key to optimal use of neutron beam time and especially so for biological systems.

1.6.4.1 D-Lab, ILL-EMBL, Grenoble (France)

The ILL Deuteration laboratory (D-Lab) [242] is hosted at the Partnership for Structural Biology, not only primarily supports the ILL neutron user program but also potentially offers access to other users. The ILL has pioneered large-scale protein deuteration by overexpression in high-density *E. coli* cultures for neutron protein crystallography and small-angle scattering experiments. The D-Lab also has the capability to express proteins in yeast and produce cell paste for other deuterated molecules such as membrane lipids and DNA. The D-Lab also has supporting facilities for protein crystallization.

1.6.4.2 Biodeuteration Lab (BDL), Oak Ridge National Laboratory, Oak Ridge TN (USA)

The Biodeuteration Lab at Oak Ridge National Laboratory (ORNL) [243] was established to support ORNL user's research using neutrons. The lab facilities are most commonly used by the structural biology community. The BDL supports the culturing of bacteria in deuterated media for the production of not only labeled proteins but also other biomolecules, such as bacterial cellulose. They support selective, partial, and full deuteration of biomolecules. They also offer the use of fermenters so that more materials can be prepared while using less D_2O. User can submit a proposal along with their neutron beam time proposal to either one of the on-site neutron facilities—the Spallation Neutron Source (SNS) or high-flux isotope reactor (HFIR). The proposals are then screened for feasibility and scientific impact, and access to lab resources may then be allocated. Labeled samples prepared through the BDL have been used in a wide array of neutron scattering methods, such as SANS, QENS, and NPX.

is only able to make immature, noninfectious virus particles. PR serves as a good example of successful structure-based drug design using X-ray crystallography. Since the first reported structure of PR, numerous clinical drugs (PR inhibitors or PI) have been developed based on these data [12]. Structural analysis of PR/PI complexes can reveal unfilled pockets in the active site, the role of water-mediated binding, and areas where H-bonding can be introduced/removed by chemical modification of ligands.

The PR active site contains a conserved triad, Asp25-Thr26-Gly27 and Asp25′-Thr26′-Gly27′ from the adjacent monomer. In this configuration, the two Asp residues face each other, and substrate peptides/proteins and PIs bind between the catalytic Asp residues and the flexible flap regions. Recent work by Weber et al. [12] reported the first neutron structure of HIV PR in complex with a clinical drug, amprenavir (APV), and these data were compared with another HIV PR neutron structure in complex with a nonclinical inhibitor KNI-272. The crystals were $0.2\ mm^3$ and deuterated by expression of the PR in perdeuterated M9 minimal media, but using hydrogenous (unlabeled) glucose as carbon source. The resulting protein was ~85% deuterated and was confirmed by mass spectrometry. Neutron diffraction data to 2 Å were collected at the quasi-Laue instrument LADI-III (ILL, Grenoble, France). Interestingly, the authors note that fully deuterated (>99% D incorporation) HIV PR did not yield crystals in a feasible size range for neutron diffraction studies. The 85% deuterated PR crystallized very well and provided all the benefits of working with a fully deuterated protein (data collection from smaller crystals, good diffraction, and high-quality nuclear density maps).

The neutron data confirmed previous ^{13}C NMR measurement that indicated mismatched pKa for catalytic Asp residues, Asp25 and Asp25′. In this structure, Asp25 is protonated and donating an H-bond to the hydroxyl group of APV, while Asp25′ accepts an H-bond from the same hydroxyl. Overall, the distribution of H-bonds within the active site varied considerably compared with the PR/KNI-272 structure and showed that the exact location of H/D on Asp side chains and the identity of H-bond donors/acceptors are very much dependent on the chemistry of the ligand. Previous studies by the same group focused on ultra-high-resolution (0.84 Å) X-ray structures and identified a number of potential H-bonds between the ligand and PR that may enhance binding. Surprisingly, the neutron data showed the details of these interactions and contradicted some of the inferred or assumed H-bonds and their relative importance in determining binding. In fact, the neutron data show only two very strong H-bonds between APV and catalytic Asp residues and between tetrahydrofuran moiety and amides of Asp29 and Asp30. There is also a water-mediated H-bond to the flaps and Asp30′. This results in three total H-bonds and not the four that were deduced from the X-ray data alone. The strongest H-bond to the Asp 25 and Asp25′ is asymmetrical due to pKa mismatch. The neutron data support the idea that these interactions can be improved by matching the pKa; this can be achieved by introducing a strong electron-accepting group in the APV molecules. It could also be advantageous to incorporate groups to directly H-bond to the PR; getting rid of water-mediated interactions will improve the enthalpy of binding.

In general, these studies highlight several issues: X-Ray data alone can be misleading for rational drug design as only a partial picture is seen, even at ultra high resolution; neutrons give information on H-bonds and make it possible to estimate their importance based on distance and angle between donors/acceptors; the physical basis to why some drugs bind better than others may be very subtle. For these reasons, it is important to always try and use more than one method or approach to study binding interactions. Neutrons clearly have an important future role in structure-based drug design, and this may be one of the best applications of the technique in structural biology. With new sources and improved instruments, data collection in a day on small(er) crystal will transform the practical feasibility of NPX as applied to drug design efforts.

1.7.1.4 Neutron Studies of an Engineered Photo-Switching Chromoprotein, Dathail

Fluorescent protein (FP) occurs naturally with many different fluorescent characteristics. Protein engineering, mutagenesis, and directed evolution have dramatically improved the variety and properties of FPs. As such, they are widely used in cell biology, molecular biology, and biotechnology research as probes and reporters for different experiments. Green fluorescent protein (GFP) was first isolated from jellyfish, and later, X-ray studies showed it to have a β-barrel structure made of 11 β strands with the chromophore located in the middle. The chromophore is formed after expression by the cyclization of three amino acids to form p-hydroxybenzylidenimidazolinone. The pocket around the chromophore is lined with amino acid side chains and ordered water molecules. It has been shown that altering the pocket can have dramatic effects and can change spectral properties of the resulting protein. Some of these proteins have the useful property of being able to photoswitch between fluorescent and nonfluorescent states by illumination of a specific wavelength of light. X-ray work has shown that this switching may be related to the chromophore being in the cis-conformation during the ground state and in a trans state when not fluorescent. It has been suggested that protonation and chemical environment of the chromophore play a role [247]. Some FPs have a very low quantum yield, and when it is below 0.001, they are classified as chromoproteins. Chromoproteins that can convert reversibly between the fluorescent and nonfluorescent state and that have very low quantum yield are called reversibly switchable chromoproteins (RSCP). They have application in high-resolution microscopy and in photochromic Förster resonance energy transfer (FRET). The use of RSCP is limited due to several characteristics, having a narrow spectral range, poor spectral overlap, high background

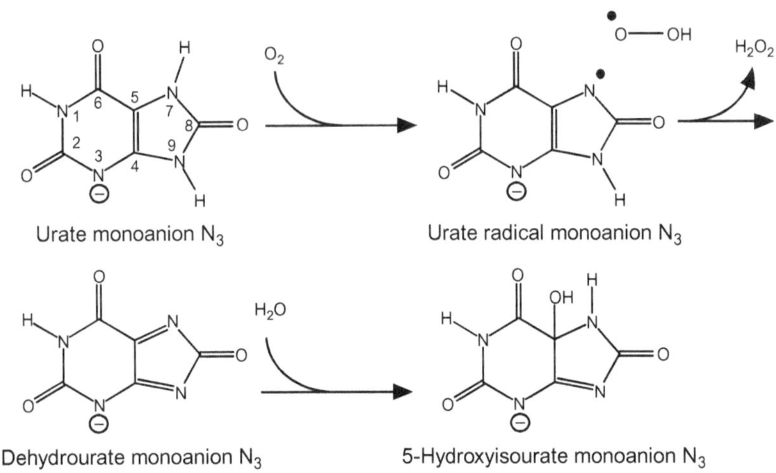

FIGURE 7 The reaction catalyzed by urate oxidase. *Reproduced with permissions from [249].*

The neutron structure at 2.35 Å resolution (Figure 8) [249] showed that in fact, the species in the active site was the enol tautomer of urate, 8-hydroxyxanthine. It was shown by quantum chemical calculations that despite the enol tautomer being unstable in solution, the active site stabilizes it, so the deprotonation occurs on the oxygen instead of the nitrogen. This allowed a more plausible mechanism to be postulated and demonstrates how the unique information from a neutron structure can give completely new insights.

1.7.2 Fibre Biomaterials and Structural Biopolymers

Cellulose is the most abundant material in the biosphere, making up half of all wood mass. The detailed structure of cellulosic materials and its function to the properties is however still not well understood. Native cellulose structures are only solved crystallographically if the cellulose is exceptionally ordered and well oriented and if the microfibrils are thick enough. Neutron and X-ray small-angle scattering have therefore contributed important insights into the structure and of cellulose microfibrils in wood, in particular the diameter of the fibrils (2.9–3.0 nm) from which the number of chains in each could be derived to be 24 in spruce wood [250]. Celery collenchyma is typically very ordered, which allowed both WANS data to be collected to resolve the disordered microfiber components [251]. The advantage of using neutrons here is that more contrast can be created by soaking the fibers in heavy water, which both inserts D_2O between the fibrils and deuterates their surface by H-D exchange. The partially deuterated cellulose was in the latter case also used in polarized FTIR spectroscopy measurements to identify the solvent accessible hydroxyl groups.

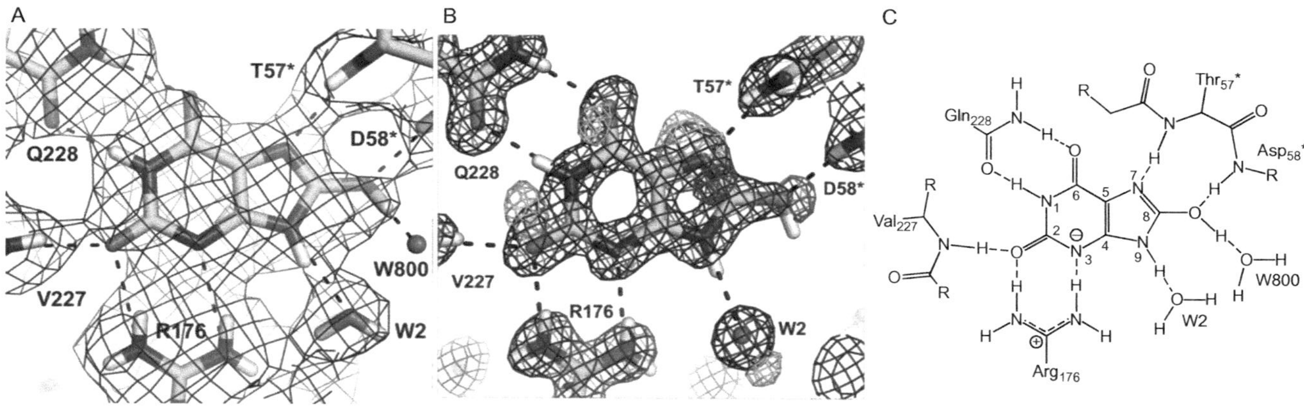

FIGURE 8 The 2.35 Å neutron map of urate oxidase (A) compared with the 1.05 Å X-ray map (B) and a schematic representation of the surroundings of the urate substrate (C).

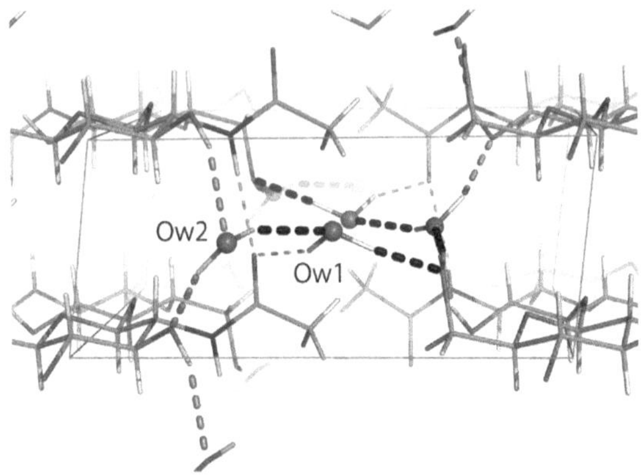

FIGURE 9 Final structure model and hydrogen bond networks of dihydrate β-chitin. It is seen along the chain direction and ab projection. *Reproduced with permissions from [49].*

Chitin is the second most common structural biopolymer found in living organisms after cellulose and forms fibrous structures in cuticles, tendons, and extracellular filaments. Neutron fiber diffraction was used to elucidate the incorporation of water in β-chitin (Figure 9) in which the hydrogen positions revealed a different arrangement of the waters from that predicted from the oxygen atom positions obtained from X-ray fiber diffraction and prevents intramolecular hydrogen bonding [49].

Many structural studies have been carried out on amyloid fibers, but major questions still remain about how exactly these protein deposits are associated with a range of pathologies from Alzheimer's disease, Parkinson's disease, Huntington's disease, type 2 diabetes mellitus, and transmissible spongiform encephalopathies. The quality of amyloid fiber diffraction data is usually not very good compared with other filamentous systems due to poor alignment. As consequence, many studies have focused instead on amyloid fibers made from synthetic peptides. Most studies of amyloid fibers have been carried out using X-rays, but neutron fiber diffraction offers the possibility to use H_2O/D_2O exchange to probe the role of water in amyloids. So far, only one neutron fiber diffraction study has been published, but the results show clearly the incorporation of D_2O into the fibers [50].

1.7.3 Protein Solution Structures

Small-angle neutron scattering is a powerful tool for investigating protein complexes in solution by contrast variation. Cardiac myosin-binding protein C (cMyBP-C) is vital for maintaining regular heart function, with its four N-terminal regulatory domains, C0-C1-m-C2 (C0C2), influencing actin and myosin interactions. Using deuterated C0C2, it was shown by Whitten et al. [58] that C0C2 forms a repeating assembly with filamentous actin, in which the C0 and C1 domains attach near the DNase I-binding loop and

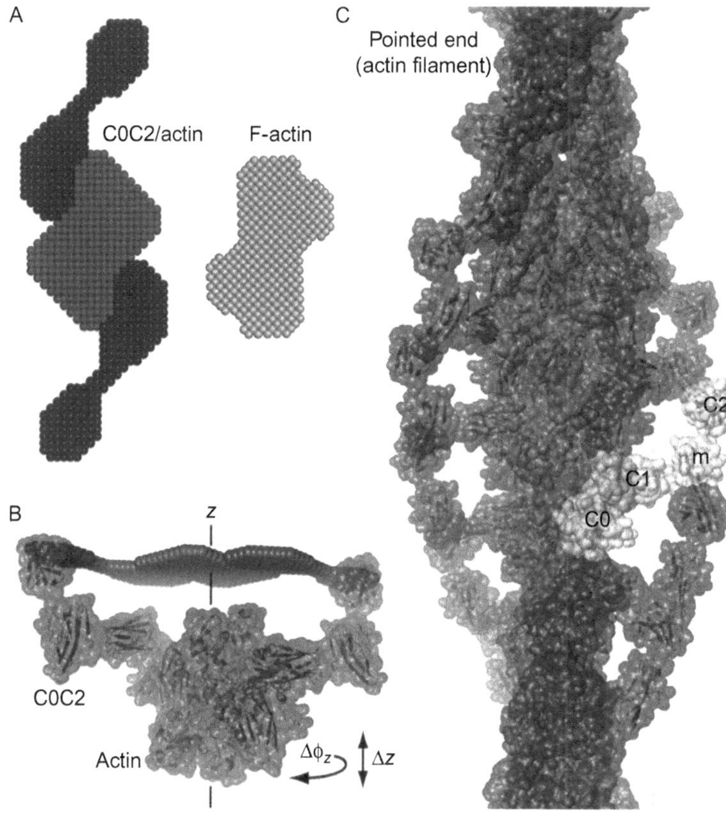

FIGURE 10 Models derived from SANS data, with actin shown in red and C0C2 in blue. (A) Cross-sectional dummy atom model of the C0C2-cardiac actin assembly obtained from the contrast variation series. The cross-sectional model for cardiac F-actin (light magenta) was obtained from optimization against the three contrast points measured for F-actin. (B) Relationship between the dummy atom model and the 3-D model of the complex. (C) Final 3-D model of the C0C2-actin complex. *Reproduced with permissions from [58].*

subdomain 1 of adjacent actin monomers (Figure 10). This confirmed that direct interactions between the N terminus of cMyBP-C and actin provide a mechanism to modulate muscle contraction by affecting the regulatory state of the thin filament and its ability to interact with myosin.

Membrane proteins need to be solubilized in a detergent of lipid carrier for solution scattering experiments. Neutrons are a good tool for characterizing protein-lipid complexes due to the possibility to contrast-match the lipids with the solvent. For this purpose, so-called stealth nanodiscs were developed, in which the lipid deuteration level is contrast matched to D_2O [252], leaving only the protein inside visible to neutrons. This was recently used to obtain the solution structures of cytochrome P450 [253] (Figure 11) and monomeric bacteriorhodopsin [254] embedded in nanodiscs.

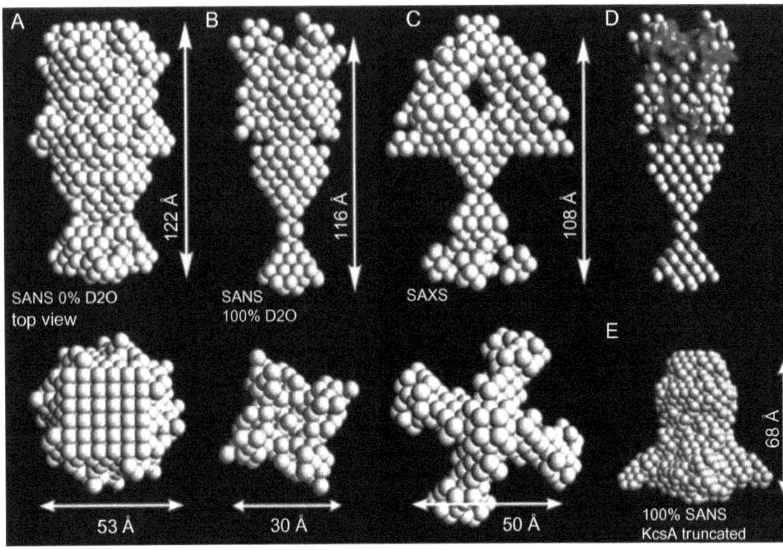

FIGURE 13 Dummy atom model of full-length KcsA derived from SANS and SAXS data at pH 7.5. Shape models derived from 0 (A) and 100% (B) D_2O SANS data are compared with the model derived from X-ray scattering (C). (D) A surface representation of the C-terminally truncated KcsA channel (red surface, pdb code 1BL8) aligned with the head domain of the full-length KcsA model derived from the 100% D_2O SANS data. (E) Shape reconstruction from 100% D_2O SANS of truncated KcsA at pH 7.5. *Reprinted with permission from Elsevier from [257].*

flip-flop upon heating the membrane to a fluid state [259]. By using deuterated DMPC vesicles, Gerelli et al. were able to determine the exchange kinetics of lipids with a nondeuterated supported DMPC bilayer [260].

Similar experiments using SANS revealed that the exchange kinetics between deuterated and nondeuterated DMPC vesicles are considerably slower than in supported bilayers [261,262], which is thought to be due to there always being some defects in supported membranes. In vesicles, the rate-limiting step was found to be lipid flip-flop from the inner to the outer monolayer of the vesicles. With careful contrast matching, it is also possible to detect the size and number of nanoscopic membrane rafts in unilamellar vesicles using SANS [263], and the same method can also be used to study their mechanical properties [264], as shown in Figure 14.

Membrane diffraction from lamellar stacks is another method for probing the internal structure of membranes [196], particularly useful for determining the location of specific groups, such as the terminal methyl groups [265] using selective deuteration. This has been used to determine the disposition of ceramide [266], beta-amyloid peptides [267], and coenzyme Q_{10} in lamellar membranes [268]. Cholesterol has been observed to orient parallel to the lipids in saturated lipid membranes [269] but to locate horizontally in the

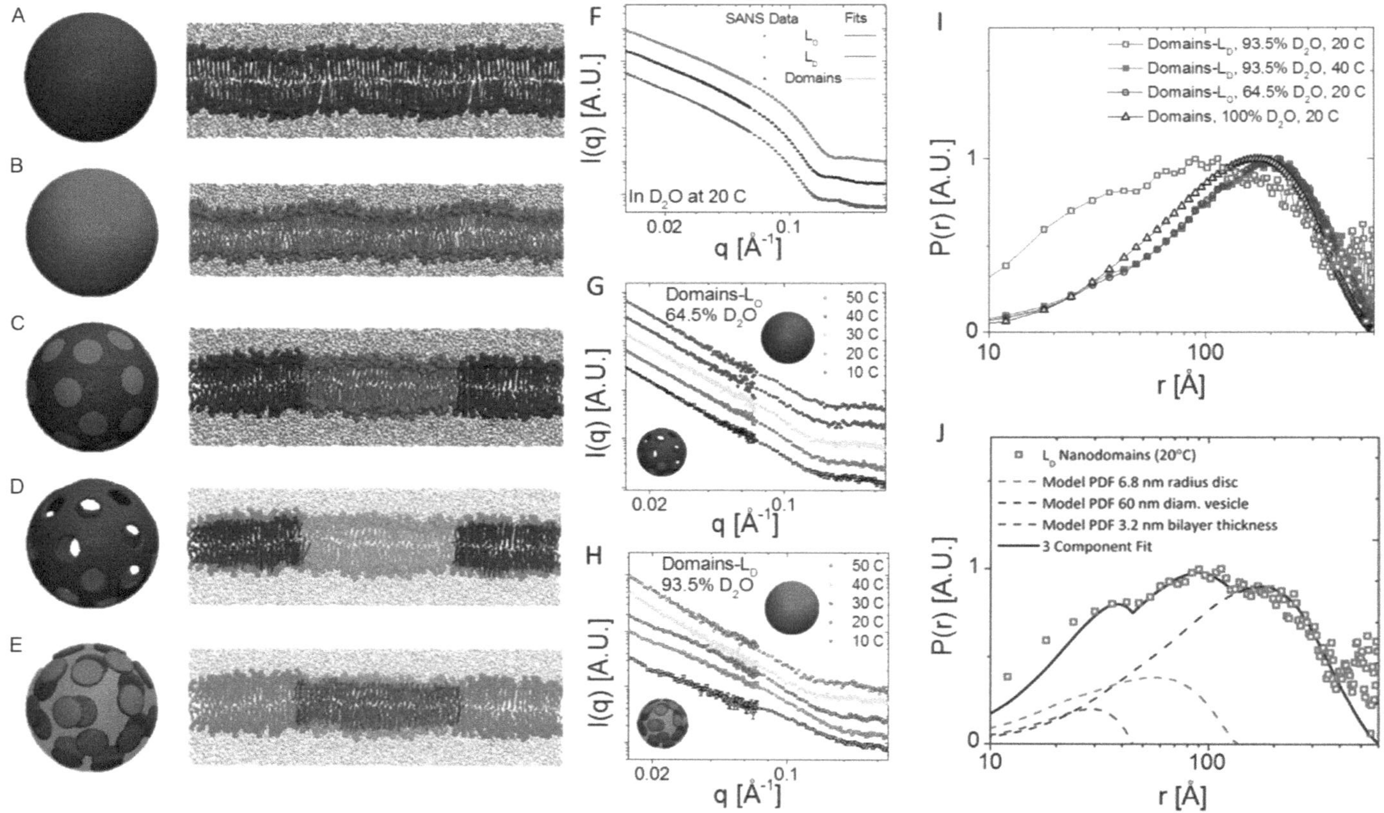

FIGURE 14 (A–E) SANS data and neutron contrast matching schemes to detect membrane rafts in vesicles. (F) SANS data and fits for A–C. (G) SANS data for D and (H) E as a function of temperature. (I) Pair-distribution functions calculated from the SANS data emphasizing the structural changes in the domains-LD (nanodomains) sample above and below the domain miscibility temperature of 40°C. (J) Pair-distribution function of the domains-LD (nanodomains) sample at 20°C compared with a model pair-distribution function. *Reprinted with permissions from [264]. Copyright (2015) American Chemical Society.*

center of polyunsaturated membranes [270]. Diffraction from nonlamellar membranes [271] can also be performed, and it has been used to determine lipid demixing due to the distortion of the inverse hexagonal phase using deuterated lipids [272]. Neutron diffraction is also the only technique capable of detecting peptide pores in fluid membranes [273].

Neutrons are also a great tool for investigating the structure of proteins in membranes and their interaction with lipids [274]. Reflectometry experiments can be conducted in an aqueous environment, and interactions can be followed *in situ* with time-of-flight measurements that capture a broad q-range using a polychromatic neutron beam. Neutron reflection has been used to investigate the membrane-binding mechanism of many different types of peptides and proteins with the view to understand how lipid composition regulates the mechanism. Contrast variation allows the composition of even complex systems to be determined, which, for example, allows the detection of aqueous membrane pores. For the eukaryotic cytolysin equinatoxin II, deuteration of the lipids, proteins, and solvent in turn allowed the observation of a new membrane-bound intermediate in sphingomyelin and cholesterol-containing membranes before the formation of an aqueous pore [207], as illustrated in Figure 15.

Neutron reflectometry allows membrane proteins to be investigated in a physiologically relevant noncrystalline environment, important for understanding their function, but only few examples have so far been published due to the complexity of forming suitable samples by reconstituting membrane proteins into lipid bilayers. This can in principle be done either by the

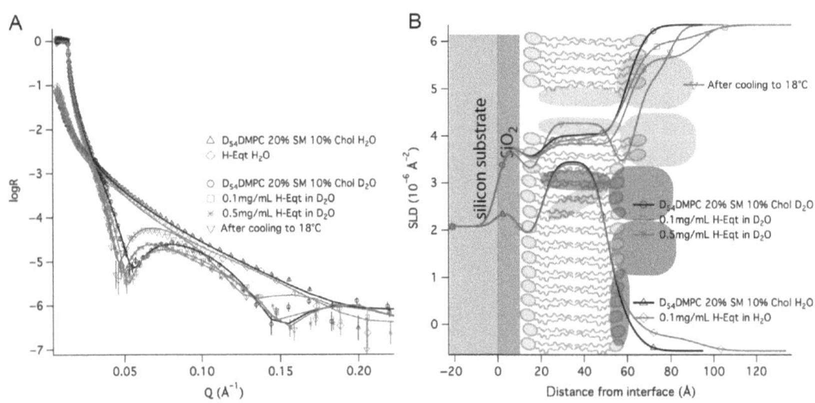

FIGURE 15 Interaction of EqtII with DMPC and d_{54}-DMPC membranes containing SM and Chol. (A) Reflectivity profiles and scattering length density profiles and model structures of d_{54}-DMPC membranes with 10 mol% cholesterol and 20 mol% sphingomyelin before and after EqtII injection at 0.1 mg/ml in H_2O and D_2O. (B) Illustration of EqtII location and possible orientations of the embedded α-helix. The green EqtII cartoon in (B) illustrates the membrane thickening and aqueous pore formation that occurs upon cooling the membrane-EqtII assembly to 18°C, crossing the gel-phase boundary. *Reproduced with permissions from [207].*

adsorption of proteoliposomes or from detergent-solubilized proteins, but the challenge is typically in ensuring correct orientation of the proteins and avoiding interactions with the supporting solid material. The first integral membrane-protein structure solved by NR was for α-hemolysin [275] in a tethered hybrid bilayer and since then structures of the voltage sensor domain of the voltage-gated potassium channel [276], cytochrome P450 in lipid nanodiscs [277], and the leucine transporter LeuT [278]. For all these studies, it was essential to be able to study the protein conformation *in situ*.

It is also possible to follow the kinetics of biological reactions using neutron reflectometry. The mechanism of phospholipase A_2 hydrolysis of lipids was elucidated in different lipid membranes [279], which allowed the location of the enzyme to be determined during the lag phase and the active phase, and partial deuteration of only the sn-1 fatty acyl chain of POPC allowed the distribution of the reaction products, a (nondeuterated) fatty acid and a deuterated lysophospholipid to be monitored *in situ* [280]. This showed for the first time that only the more water-soluble lysolipid leaves the reacting membrane, while accumulation of the fatty acid leads to further enzyme adsorption on the membrane in an autocatalytic fashion. In contrast, in membranes in which the lysolipid cannot leave, the enzyme activity is rapidly inhibited.

The oxidative damage caused to lung surfactant proteins [281] and lipids by ozone [282] has also been investigated using time-resolved neutron reflection. In these studies, synthesis of partially deuterated lipids and the fast time resolution (5 s) enabled by the high-flux reflectometers Inter at ISIS and Figaro at ILL were crucial for determining the reaction mechanism and kinetics.

In many cases, for example, for the PTEN tumor suppressor [283], the experimental data are complemented by molecular dynamic simulations [284], which can be used to compute the neutron scattering length density profile of the simulated membrane-protein system. Simulation tools are also helpful to determine the molecular volumes of lipids in mixtures and for comparing experimental and simulated structures [205].

Much of the neutron scattering work on membranes has been carried out using model systems composed of a limited number of synthetic lipids due to the necessity of deuterium labeling, but recently, efforts at deuteration facilities have made it possible to purify deuterated biological lipid mixtures from cell cultures. Phospholipids from *P. pastoris* yeast were used to reconstitute membranes for investigating the mechanism of the antifungal agent amphotericin B, which showed that the drug both inserts in membranes and forms a sterol-extracting sponge layer [285] (Figure 16) and that its activity is partly regulated by the degree of lipid polyunsaturation, which was determined by analysis of the lipid composition by gas chromatography. In this study, perdeuterated membranes were a prerequisite for creating sufficient contrast to the polyene drug that would have been impossible to synthesize in deuterated form. Similarly, deuterated phospholipids from *E. coli* have

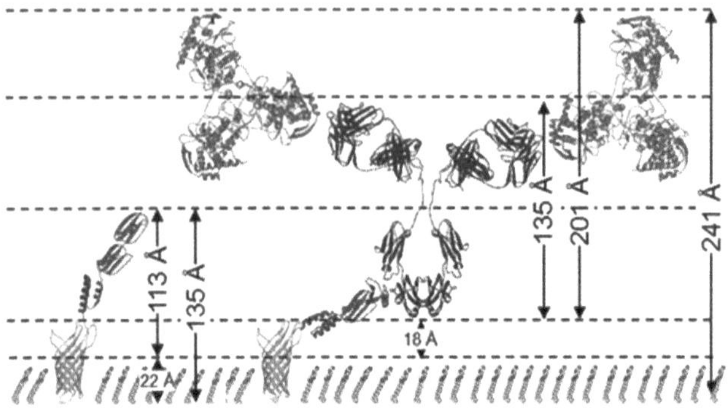

FIGURE 18 The dimensions of each component of a GGZctOmpA array before and after IgG and rNP binding. ctOmpA is shown in light green and is bound to the gold surface and is surrounded by thioPEG (the molecular packing is not represented accurately) with a protein A Z domain linker (dark green) and a tandem pair of B domains from protein G (red) to which there is an IgG (blue) bound at the Fc region. Bound to the IgG Fab regions, there is NP (orange) to which a secondary IgG can be bound (not shown). *Reproduced without changes under the Creative Commons Attribution 4.0 International License [191] from [287].*

essential to determine the binding orientation and dimensions without disturbing the protein assembly by solvent exchange.

Zhao et al. [288] used NR to show that a mouse monoclonal antibody laid flat on the a sensor surface at low surface density with a thickness around 40 Å, equivalent to the short axial length of the antibody molecule and that the loose packing of the antibody within this range resulted in better human prostate-cancer-specific antigen-binding efficiency, while an increase in the surface-adsorbed amount reduced the antigen binding due to steric hindrance. The antigen inserted into the antibody layer rather than forming an additional layer on the top and remained stable and active over 4 months.

McIntosh et al. used neutron reflectometry to detect the structure of deuterated/nondeuterated fibronectin layers adsorbed on titania surfaces to prevent cell adhesion on bioimplants and found that the proteins showed a transition from monolayer to bilayer structured at high concentrations with the protein longitudinal orientation moving away from the plane of the TiO_2 surface.

Surfactants produced by bacteria are of significant interest as biodegradable and sustainable detergents and neutron reflectivity and SANS have contributed much to understanding their solution behavior and structures at the air-water interface. Most studied are rhamnolipids from *P. aeruginosa* and sophorolipids from *Candida bombicola*. Obtaining deuterated biosurfactants was essential in investigating their adsorption at the air-water interface [289] and interaction with synthetic surfactants by SANS [290].

Cellulose films are of technological interest for coating applications. NR was useful for determining the swelling and water update of ultrathin cellulose films, which are of interest for their hydrogel properties [291]. Improving the efficiency of enzymatic digestion of cellulose is one of the key challenges in producing ethanol and other biofuels from lignocellulosic material. A better understanding of how soluble cellulose enzymes interact with insoluble cellulose will aid in the design of more efficient enzyme systems. NR has been used to detect the mechanism of cellulose film digestion by different cellulases [292] and their activation by a carbohydrate-binding module [293]. In these thick films, the location of the enzyme is not resolved, but NR was critical in detecting water penetration into the film and the roughness caused by enzyme activity, which are indicative of how the enzyme accesses the insoluble substrate.

1.8 FUTURE PERSPECTIVES

In this chapter, we have shown that neutron scattering techniques can be used to solve biological structures in many different types of systems and yield information from the arrangement of atoms in protein and enzymes to supramolecular structure of fibers, membranes, and solution aggregates at the nanometer and micrometer length scales. In most cases, the ability to use deuterium labeling to highlight components gives unique information not accessible using other techniques and allows investigation of complex system all the way up to the cellular level in some cases. The last decade has seen a strong expansion in the use of neutron scattering techniques for biological structure determination, largely due to the availability of new high-flux instruments and supporting facilities for deuteration, which has meant that obtaining samples that are large enough is less of a problem, and the throughput of the instruments is beginning to allow biologically relevant questions to be probed with multiple contrast variation. The neutron facilities worldwide are also recognizing that more and more investigators are using neutrons as one technique among many methods needed to elucidate intricate biological questions and are offering increased support for also nonexpert users. With the advent of the ESS in Lund and the second target station at the SNS in Oak Ridge on the horizon, we can expect new instrument capabilities to probe smaller samples on faster timescales that will allow this trend to continue and make neutron techniques a mainstay in structural biology.

REFERENCES

[1] D.L. Price, F. Fernandez-Alonso, An introduction to neutron scattering, in: F. Fernandez-Alonso, D.L. Price (Eds.), Experimental Methods in the Physical Sciences, Neutron Scattering, vol. 44, Academic Press, San Diego, CA, 2013, pp. 1–136.

[2] J. Fitter, T. Gutberlet, J. Katsaras, Neutron Scattering in Biology – Techniques and Applications, Springer Berlin Heidelberg, Heidelberg, 2006.

[3] M.P. Blakeley, P. Langan, N. Niimura, A. Podjarny, Curr. Opin. Struct. Biol. 18 (2008) 593–600.
[4] V.T. Forsyth, I.M. Parrot, High-angle neutron fiber diffraction in the study of biological systems, in: Neutron Scattering in Biology: Techniques and Applications, Springer Berlin Heidelberg, Berlin, Heidelberg, 2006, pp. 85–105.
[5] D.A. Jacques, J. Trewhella, Protein Sci. 19 (2010) 642–657.
[6] H.P. Wacklin, Curr. Opin. Colloid Interface Sci. 15 (2010) 445–454.
[7] G. Fragneto, Eur. Phys. J. Spec. Top. 213 (2012) 327–342.
[8] https://www.ncnr.nist.gov/resources/n-lengths/ (last accessed on 01/01/017).
[9] M.P. Blakeley, S.S. Hasnain, S.V. Antonyuk, IUCrJ 2 (2015) 464–474.
[10] J. Dawidowski, J.R. Granada, J.R. Santisteban, F. Cantargi, L.A.R. Palomino, Neutron scattering lengths and cross sections, in: F. Fernandez-Alonso, D.L. Price (Eds.), Experimental Methods in the Physical Sciences, vol. 44, Academic Press, New York, 2013, pp. 471–528 (Appendix).
[11] T. Yokoyama, M. Mizuguchi, A. Ostermann, K. Kusaka, N. Niimura, T.E. Schrader, I. Tanaka, J. Med. Chem. 58 (2015) 7549–7556.
[12] I.T. Weber, M.J. Waltman, M. Mustyakimov, M.P. Blakeley, D.A. Keen, A.K. Ghosh, P. Langan, A.Y. Kovalevsky, J. Med. Chem. 56 (2013) 5631–5635.
[13] R. Michalczyk, C.J. Unkefer, J.P. Bacik, T.E. Schrader, A. Ostermann, A.Y. Kovalevsky, R. McKenna, S.Z. Fisher, Proc. Natl. Acad. Sci. U. S. A. 112 (2015) 5673–5678.
[14] S.C. Capelli, Structure of complex materials, in: F. Fernandez-Alonso, D.L. Price (Eds.), Experimental Methods in the Physical Sciences, vol. 44, Academic Press, San Diego, CA, 2013, pp. 321–352.
[15] B.P. Schoenborn, Acta Crystallogr. D Biol. Crystallogr. 66 (2010) 1262–1268.
[16] P.V. Afonine, M. Mustyakimov, R.W. Grosse-Kunstleve, N.W. Moriarty, P. Langan, P.D. Adams, Acta Crystallogr. D Biol. Crystallogr. 66 (2010) 1153–1163.
[17] F.J. Bermejo, F. Sordo, Neutron sources, in: F. Fernandez-Alonso, D.L. Price (Eds.), Experimental Methods in the Physical Sciences, vol. 44, Academic Press, New York, 2013, pp. 137–243.
[18] https://www.ill.eu/instruments-support/instruments-groups/instruments/ladi-iii/description/instrument-layout/ (last accessed on 01/01/2017).
[19] http://www.mlz-garching.de/biodiff (last accessed 01/01/2017).
[20] https://neutrons.ornl.gov/imagine (last accessed on 01/01/2017).
[21] I. Tanaka, K. Kurihara, T. Chatake, N. Niimura, J. Appl. Cryst. 35 (2002) 34–40.
[22] K. Kurihara, I. Tanaka, N. Niimura, M. Refai Muslih, A. Ostermann, J. Synchrotron Radiat. 11 (2004) 68–71.
[23] https://www.ill.eu/ (last accessed on 01/01/2017).
[24] http://www.frm2.tum.de/en/home/ (last accessed on 01/01/2017).
[25] https://neutrons.ornl.gov/hfir (last accessed on 01/01/2017).
[26] http://jrr3.jaea.go.jp/jrr3e/index.htm (last accessed on 01/01/2017).
[27] http://lansce.lanl.gov/lujan/instruments/PCS.shtml (last accessed on 01/01/2017).
[28] https://neutrons.ornl.gov/mandi (last accessed on 01/01/2017).
[29] https://j-parc.jp/researcher/MatLife/en/instrumentation/ns_spec.html#bl03 (last accessed on 01/01/2017).
[30] https://europeanspallationsource.se/macromolecular-diffractometer-0 (last accessed on 01/01/2017).
[31] http://lansce.lanl.gov (last accessed on 01/01/2017).

[32] https://neutrons.ornl.gov/sns (last accessed on 01/01/2017).
[33] https://j-parc.jp/researcher/MatLife/en/index.html (last accessed on 01/01/2017).
[34] https://europeanspallationsource.se/ (last accessed on 01/01/2017).
[35] K.H. Sippel, J. Bacik, F.A. Quiocho, S.Z. Fisher, Acta Crystallogr. Sect. F Struct. Biol. Cryst. Commun. 70 (2014) 819–822.
[36] J.C. Chen, B.L. Hanson, S.Z. Fisher, P. Langan, A.Y. Kovalevsky, Proc. Natl. Acad. Sci. U. S. A. 109 (2012) 15301–15306.
[37] T. Bryan, J.M. Gonzalez, J.P. Bacik, N.J. Denunzio, C.J. Unkefer, T.E. Schrader, A. Ostermann, D. Dunaway-Mariano, K.N. Allen, S.Z. Fisher, Acta Crystallogr. Sect. F Struct. Biol. Cryst. Commun. 69 (2013) 1015–1019.
[38] Z. Fisher, A.Y. Kovalevsky, M. Mustyakimov, D.N. Silverman, R. McKenna, P. Langan, Biochemistry 50 (2011) 9421–9423.
[39] A.Y. Kovalevsky, B.L. Hanson, S.A. Mason, T. Yoshida, S.Z. Fisher, M. Mustyakimov, V.T. Forsyth, M.P. Blakeley, D.A. Keen, P. Langan, Angew. Chem. Int. Ed. Engl. 50 (2011) 7520–7523.
[40] M. Ohlin, L. Von Schantz, T.E. Schrader, A. Ostermann, D.T. Logan, S.Z. Fisher, Acta Crystallogr. Sect. F Struct. Biol. Cryst. Commun. 71 (2015) 1072–1077.
[41] R.C. Hughes, L. Coates, M.P. Blakeley, S.J. Tomanicek, P. Langan, A.Y. Kovalevsky, J.M. Garcia-Ruiz, J.D. Ng, Acta Crystallogr. Sect. F Struct. Biol. Cryst. Commun. 68 (2012) 1482–1487.
[42] B.L. Hanson, P. Langan, A.K. Katz, X. Li, J.M. Harp, J.P. Glusker, B.P. Schoenborn, G.J. Bunick, Acta Crystallogr. D Biol. Crystallogr. 60 (2004) 241–249.
[43] T. Ishikawa, T. Chatake, Y. Morimoto, M. Maeda, K. Kurihara, I. Tanaka, N. Niimura, Biochem. Biophys. Res. Commun. 376 (2008) 32–35.
[44] N. Junius, E. Oksanen, M. Terrien, C. Berzin, J.L. Ferrer, M. Budayova-Spano, J. Appl. Cryst. 49 (2016) 806–813.
[45] L. Coates, S. Tomanicek, T.E. Schrader, K.L. Weiss, J.D. Ng, P. Juttner, A. Ostermann, J. Appl. Cryst. 47 (2014) 1431–1434.
[46] V.G. Vandavasi, K.L. Weiss, J.B. Cooper, P.T. Erskine, S.J. Tomanicek, A. Ostermann, T.E. Schrader, S.L. Ginell, L. Coates, J. Med. Chem. 59 (2016) 474–479.
[47] M.W. Shotton, L.H. Pope, T. Forsyth, P. Langan, R.C. Denny, U. Giesen, M.T. Dauvergne, W. Fuller, Biophys. Chem. 69 (1997) 85–96.
[48] Y. Nishiyama, J. Sugiyama, H. Chanzy, P. Langan, J. Am. Chem. Soc. 125 (2003) 14300–14306.
[49] D. Sawada, Y. Nishiyama, P. Langan, V.T. Forsyth, S. Kimura, M. Wada, PLoS One 7 (2012) e39376.
[50] S.M. Tiggelaar, E. Mossou, P. Callow, S. Callow, S.C. Teixeira, E.P. Mitchell, A. Mitraki, V.T. Forsyth, Acta Crystallogr. Sect. F Struct. Biol. Cryst. Commun. 67 (2011) 332–335.
[51] S. Fujiwara, F. Matsumoto, J. Mol. Biol. 367 (2007) 16–24.
[52] W. Fuller, T. Forsyth, A. Mahendrasingam, Philos. Trans. R. Soc. Lond. B Biol. Sci. 359 (2004) 1237–1247. discussion 1247–1238.
[53] S.C. Teixeira, J. Ankner, M.C. Bellissent-Funel, R. Bewley, M.P. Blakeley, L. Coates, R. Dahint, R. Dalgliesh, N. Dencher, J. Dhont, P. Fischer, V.T. Forsyth, G. Fragneto, B. Frick, T. Geue, R. Gilles, T. Gutberlet, M. Haertlein, T. Hauss, W. Haussler, W.T. Heller, K. Herwig, O. Holderer, F. Juranyi, R. Kampmann, R. Knott, J. Kohlbrecher, S. Kreuger, P. Langan, R. Lechner, G. Lynn, C. Majkrzak, R. May, F. Meilleur, Y. Mo, K. Mortensen, D.A. Myles, F. Natali, C. Neylon, N. Niimura,

[189] https://europeanspallationsource.se/article/revealing-change-over-time-freia-brings-fast-kinetic-studies-reflectometry (last accessed on 01/01/2017).
[190] https://europeanspallationsource.se/estia-truly-focusing-reflectometer (last accessed on 01/01/2017).
[191] http://neutrons.ornl.gov/sts (last accessed on 01/01/2017).
[192] R. Campbell, H. Wacklin, I. Sutton, R. Cubitt, G. Fragneto, Eur. Phys. J. Plus 126 (2011) 1–22.
[193] R.S. Armen, O.D. Uitto, S.E. Feller, Biophys. J. 75 (1998) 734–744.
[194] H.I. Petrache, S.E. Feller, J.F. Nagle, Biophys. J. 72 (1997) 2237–2242.
[195] S.C. Costigan, P.J. Booth, R.H. Templer, Biochim. Biophys. Acta Biomembr. 1468 (2000) 41–54.
[196] M.C. Wiener, S.H. White, Biophys. J. 61 (1992) 434–447.
[197] D.M. Small, The Physics and Chemistry of Lipids, Plenum Press, New York, 1986.
[198] J.F. Nagle, M.C. Wiener, Biochim. Biophys. Acta 942 (1988) 1–10.
[199] A.A. Zamyatin, Prog. Biophys. Mol. Biol. 24 (1972) 107–123.
[200] O.S. Heavens, Optical Properties of Thin Solid Films, Butterworths, London, 1955.
[201] A. Nelson, J. Appl. Cryst. 39 (2006) 273–276.
[202] P.A. Kienzle, K.V. O'Donovan, J.F. Ankner, N.F. Berk, C.F. Majkrzak, ReflPak, 2000-2006.
[203] Y. Gerelli, J. Appl. Cryst. 49 (2016) 330–339.
[204] M. Bjorck, G. Andersson, J. Appl. Cryst. 40 (2007) 1174–1178.
[205] L. Darré, J. Iglesias-Fernandez, A. Kohlmeyer, H. Wacklin, C. Domene, J. Chem. Theory Comput. 11 (2015) 4875–4884.
[206] T.K. Lind, L. Darré, C. Domene, Z. Urbanczyk-Lipkowska, M. Cárdenas, H.P. Wacklin, Biochim. Biophys. Acta Biomembr. 1848 (2015) 2075–2084.
[207] H.P. Wacklin, B.B. Bremec, M. Moulin, N. Rojko, M. Haertlein, T. Forsyth, G. Anderluh, R.S. Norton, Biochim. Biophys. Acta 1858 (2016) 640–652.
[208] J.K. Blasie, L. Herbette, J. Pachence, J. Membr. Biol. 86 (1985) 1–7.
[209] https://www.ill.eu/instruments-support/instruments-groups/instruments/d16/description/instrument-layout/ (last accessed on 01/01/2017).
[210] S.A. Tristram-Nagle, Methods Mol. Biol. 400 (2007) 63–75.
[211] U. Mennicke, T. Salditt, Langmuir 18 (2002) 8172–8177.
[212] T.A. Harroun, H. Fritzsche, M.J. Watson, K.G. Yager, O.M. Tanchak, C.J. Barrett, J. Katsaras, Rev. Sci. Instrum. 76 (2005) 065101.
[213] J.F. Nagle, J. Katsaras, Phys. Rev. E Stat. Phys. Plasmas Fluids Relat. Interdiscip. Top. 59 (1999) 7018–7024.
[214] D.M. Lemaster, F.M. Richards, Anal. Biochem. 122 (1982) 238–247.
[215] H.F. Daboll, H.L. Crespi, J.J. Katz, Biotechnol. Bioeng. 4 (1962) 281–297.
[216] A.R. Pickford, J.M. O'Leary, Isotopic labeling of recombinant proteins from the methylotrophic yeast *Pichia pastoris*, in: A.K. Downing (Ed.), Protein NMR Techniques, Humana Press, Totowa, NJ, 2004, pp. 17–33.
[217] S. Massou, V. Puech, F. Talmont, P. Demange, N.D. Lindley, M. Tropis, A. Milon, J. Biomol. NMR 14 (1999) 231–239.
[218] C. Opitz, S. Isogai, S. Grzesiek, J. Biomol. NMR 62 (2015) 373–385.
[219] L. Skora, B. Shrestha, A.D. Gossert, Isotope labeling of proteins in insect cells, in: K. Zvi (Ed.), Methods in Enzymology, vol. 565, Academic Press, San Diego, CA, 2015, pp. 245–288 (Chapter 11).

[220] A. Sitarska, L. Skora, J. Klopp, S. Roest, C. Fernández, B. Shrestha, A.D. Gossert, J. Biomol. NMR 62 (2015) 191–197.

[221] H.-H. Shen, R.K. Thomas, P. Taylor, Langmuir 26 (2009) 320–327.

[222] P. Callow, A. Sukhodub, J.E. Taylor, G.G. Kneale, J. Mol. Biol. 369 (2007) 177–185.

[223] D. Liu, R. Xu, D. Cowburn, Methods Enzymol. 462 (2009) 151–175.

[224] P.E. Dawson, T.W. Muir, I. Clark-Lewis, S.B. Kent, Science 266 (1994) 776–779.

[225] F. Meilleur, J. Contzen, D.A. Myles, C. Jung, Biochemistry 43 (2004) 8744–8753.

[226] G. Bali, M.B. Foston, H.M. O'Neill, B.R. Evans, J. He, A.J. Ragauskas, Carbohydr. Res. 374 (2013) 82–88.

[227] H. O'Neill, R. Shah, B.R. Evans, J. He, S.V. Pingali, S.P. Chundawat, A.D. Jones, P. Langan, B.H. Davison, V. Urban, Methods Enzymol. 565 (2015) 123–146.

[228] R.A. Russell, T.A. Darwish, L. Puskar, D.E. Martin, P.J. Holden, L.J.R. Foster, Biomacromolecules 15 (2014) 644–649.

[229] R.A. Russell, C.J. Garvey, T.A. Darwish, L.J.R. Foster, P.J. Holden, Biopolymer deuteration for neutron scattering and other isotope-sensitive techniques, in: K. Zvi (Ed.), Methods in Enzymology, vol. 565, Academic Press, San Diego, CA, 2015, pp. 97–121 (Chapter 5).

[230] B.R. Evans, R. Shah, Development of approaches for deuterium incorporation in plants, in: K. Zvi (Ed.), Methods in Enzymology, vol. 565, Academic Press, San Diego, CA, 2015, pp. 213–243 (Chapter 10).

[231] N.R. Yepuri, S.A. Jamieson, T.A. Darwish, A. Rawal, J.M. Hook, P. Thordarson, P.J. Holden, M. James, Tetrahedron Lett. 54 (2013) 2538–2541.

[232] T.A. Darwish, A.R.G. Smith, I.R. Gentle, P.L. Burn, E. Luks, G. Moraes, M. Gillon, P.J. Holden, M. James, Tetrahedron Lett. 53 (2012) 931–935.

[233] T.A. Darwish, E. Luks, G. Moraes, N.R. Yepuri, P.J. Holden, M. James, J. Label. Compd. Radiopharm. 56 (2013) 520–529.

[234] H.-H. Shen, R.K. Thomas, C.-Y. Chen, R.C. Darton, S.C. Baker, J. Penfold, Langmuir 25 (2009) 4211–4218.

[235] N.R. Yepuri, T.A. Darwish, A.M. Krause-Heuer, A.E. Leung, R. Delhom, H.P. Wacklin, P.J. Holden, ChemPlusChem 81 (2016) 315–321.

[236] C.M. Souza, T.M.E. Schwabe, H. Pichler, B. Ploier, E. Leitner, X.L. Guan, M.R. Wenk, I. Riezman, H. Riezman, Metab. Eng. 13 (2011) 555–569.

[237] A. Alfonso-García, S.G. Pfisterer, H. Riezman, E. Ikonen, E.O. Potma, J. Biomed. Opt. 21 (2015) 061003.

[238] W.W. Christie, X. Han, Chromatographic analysis of lipids: general principles, in: Lipid Analysis, fourth ed., Woodhead Publishing, Cambridge, UK, 2012, pp. 21–54 (Chapter 2).

[239] A. De Ghellinck, H. Schaller, V. Laux, M. Haertlein, M. Sferrazza, E. Marechal, H. Wacklin, J. Jouhet, G. Fragneto, PLoS One 9 (2014) 9.

[240] T.K. Lind, H. Wacklin, J. Schiller, M. Moulin, M. Haertlein, T.G. Pomorski, M. Cárdenas, PLoS One 10 (2015) e0144671.

[241] S. Maric, M.B. Thygesen, J. Schiller, M. Marek, M. Moulin, M. Haertlein, V.T. Forsyth, M. Bogdanov, W. Dowhan, L. Arleth, T.G. Pomorski, Appl. Microbiol. Biotechnol. 99 (2015) 241–254.

[242] https://www.ill.eu/sites/deuteration/ (last accessed on 01/01/2017).

[243] http://csmb.ornl.gov/bdl/ (last accessed on 01/01/2017).

[244] http://www.ansto.gov.au/ResearchHub/Bragg/Facilities/NationalDeuterationFacility/ (last accessed on 01/01/2017).

[245] http://www.isis.stfc.ac.uk/support-laboratories/isis-isotope-facility/isis-deuteration-facility 15371.html (last accessed on 01/01/2017).

[246] G.M. Sheldrick, Acta Crystallogr. A 64 (2008) 112–122.

[247] P.S. Langan, D.W. Close, L. Coates, R.C. Rocha, K. Ghosh, C. Kiss, G. Waldo, J. Freyer, A. Kovalevsky, A.R. Bradbury, J. Mol. Biol. 428 (2016) 1776–1789.

[248] C.M. Casadei, A. Gumiero, C.L. Metcalfe, E.J. Murphy, J. Basran, M.G. Concilio, S.C. Teixeira, T.E. Schrader, A.J. Fielding, A. Ostermann, M.P. Blakeley, E.L. Raven, P.C. Moody, Science 345 (2014) 193–197.

[249] E. Oksanen, M.P. Blakeley, M. El-Hajji, U. Ryde, M. Budayova-Spano, PLoS One 9 (2014) e86651.

[250] A.N. Fernandes, L.H. Thomas, C.M. Altaner, P. Callow, V.T. Forsyth, D.C. Apperley, C.J. Kennedy, M.C. Jarvis, Proc. Natl. Acad. Sci. U. S. A. 108 (2011) E1195–E1203.

[251] L.H. Thomas, V.T. Forsyth, A. Sturcova, C.J. Kennedy, R.P. May, C.M. Altaner, D.C. Apperley, T.J. Wess, M.C. Jarvis, Plant Physiol. 161 (2013) 465–476.

[252] S. Maric, N. Skar-Gislinge, S. Midtgaard, M.B. Thygesen, J. Schiller, H. Frielinghaus, M. Moulin, M. Haertlein, V.T. Forsyth, T.G. Pomorski, L. Arleth, Acta Crystallogr. Sect. D 70 (2014) 317–328.

[253] N. Skar-Gislinge, S.A.R. Kynde, I.G. Denisov, X. Ye, I. Lenov, S.G. Sligar, L. Arleth, Acta Crystallogr. D Biol. Crystallogr. D71 (2015) 2412–2421. International Union of Crystallography, 2015, 1–10.

[254] S.A.R. Kynde, N. Skar-Gislinge, M.C. Pedersen, S.R. Midtgaard, J.B. Simonsen, R. Schweins, K. Mortensen, L. Arleth, Acta Crystallogr. D Biol. Crystallogr. 70 (2014) 371–383.

[255] C. Breyton, F. Gabel, M. Lethier, A. Flayhan, G. Durand, J.M. Jault, C. Juillan-Binard, L. Imbert, M. Moulin, S. Ravaud, M. Hartlein, C. Ebel, Eur. Phys. J. E Soft Matter 36 (2013) 71.

[256] K.A. Rubinson, C. Pokalsky, S. Krueger, L.J. Prochaska, Protein J. 32 (2013) 27–38.

[257] J. Zimmer, D.A. Doyle, J.G. Grossmann, Biophys. J. 90 (2006) 1752–1766.

[258] H.P. Wacklin, Langmuir 27 (2011) 7698–7707.

[259] Y. Gerelli, L. Porcar, G. Fragneto, Langmuir 28 (2012) 15922–15928.

[260] Y. Gerelli, L. Porcar, L. Lombardi, G. Fragneto, Langmuir 29 (2013) 12762–12769.

[261] M. Nakano, M. Fukuda, T. Kudo, H. Endo, T. Handa, Phys. Rev. Lett. 98 (2007) 238101.

[262] M. Nakano, M. Fukuda, T. Kudo, N. Matsuzaki, T. Azuma, K. Sekine, H. Endo, T. Handa, J. Phys. Chem. B 113 (2009) 6745–6748.

[263] J. Pencer, T. Mills, V. Anghel, S. Krueger, R. Epand, J. Katsaras, Eur. Phys. J. E 18 (2005) 447–458.

[264] J.D. Nickels, X. Cheng, B. Mostofian, C. Stanley, B. Lindner, F.A. Heberle, S. Perticaroli, M. Feygenson, T. Egami, R.F. Standaert, J.C. Smith, D.A.A. Myles, M. Ohl, J. Katsaras, J. Am. Chem. Soc. 137 (2015) 15772–15780.

[265] M. Mihailescu, R.G. Vaswani, E. Jardón-Valadez, F. Castro-Román, J.A. Freites, D.L. Worcester, A.R. Chamberlin, D.J. Tobias, S.H. White, Biophys. J. 100 (2011) 1455–1462.

[266] D. Groen, G.S. Gooris, D.J. Barlow, M.J. Lawrence, J.B. Van Mechelen, B. Demé, J.A. Bouwstra, Biophys. J. 100 (2011) 1481–1489.

[267] S. Dante, T. Hauss, N.A. Dencher, Biophys. J. 83 (2002) 2610–2616.

[268] T. Hauß, S. Dante, T.H. Haines, N.A. Dencher, Biochim. Biophys. Acta Bioenerg. 1710 (2005) 57–62.

[269] E.H. Mojumdar, D. Groen, G.S. Gooris, D.J. Barlow, M.J. Lawrence, B. Deme, J.A. Bouwstra, Biophys. J. 105 (2013) 911–918.

[270] T. Harroun, J. Katsaras, S. Wassall, Biochemistry 45 (2006) 1227–1233.

[271] L. Ding, W. Liu, W. Wang, C.J. Glinka, D.L. Worcester, L. Yang, H.W. Huang, Langmuir 20 (2004) 9262–9269.

[272] L. Ding, T. Weiss, G. Fragneto, W. Liu, L. Yang, H. Huang, Langmuir 21 (2005) 203–210.

[273] L. Yang, T. Weiss, R. Lehrer, H. Huang, Biophys. J. 79 (2000) 2002–2009.

[274] L.A. Clifton, C. Neylon, J.H. Lakey, Methods Mol. Biol. 974 (2013) 119–150.

[275] D.J. Mcgillivray, G. Valincius, F. Heinrich, J.W.F. Robertson, D.J. Vanderah, W. Febo-Ayala, I. Ignatjev, M. Losche, J.J. Kasianowicz, Biophys. J. 96 (2009) 1547–1553.

[276] S. Gupta, J.A. Dura, J.A. Freites, D.J. Tobias, J.K. Blasie, Langmuir 28 (2012) 10504–10520.

[277] M. Wadsäter, T. Laursen, A. Singha, N.S. Hatzakis, D. Stamou, R. Barker, K. Mortensen, R. Feidenhans'l, B.L. Møller, M. Cárdenas, J. Biol. Chem. 287 (2012) 34596–34603.

[278] V. Jagalski, R.D. Barker, M.B. Thygesen, K. Gotfryd, M.B. Kruger, L. Shi, S. Maric, N. Bovet, M. Moulin, M. Haertlein, T.G. Pomorski, C.J. Loland, M. Cardenas, Soft Matter 11 (2015) 7707–7711.

[279] H.P. Vacklin, F. Tiberg, G. Fragneto, R.K. Thomas, Biochemistry 44 (2005) 2811–2821.

[280] H.P. Wacklin, F. Tiberg, G. Fragneto, R.K. Thomas, Biochim. Biophys. Acta Biomembr. 1768 (2007) 1036–1049.

[281] J.M. Hemming, B.R. Hughes, A.R. Rennie, S. Tomas, R.A. Campbell, A.V. Hughes, T. Arnold, S.W. Botchway, K.C. Thompson, Biochemistry 54 (2015) 5185–5197.

[282] K.C. Thompson, S.H. Jones, A.R. Rennie, M.D. King, A.D. Ward, B.R. Hughes, C.O.M. Lucas, R.A. Campbell, A.V. Hughes, Langmuir 29 (2013) 4594–4602.

[283] S. Shenoy, P. Shekhar, F. Heinrich, M.-C. Daou, A. Gericke, A.H. Ross, M. Loesche, PLoS One 7 (2012) e32591.

[284] S.S. Shenoy, H. Nanda, M. Loesche, J. Struct. Biol. 180 (2012) 394–408.

[285] A. De Ghellinck, G. Fragneto, V. Laux, M. Haertlein, J. Jouhet, M. Sferrazza, H. Wacklin, Biochim. Biophys. Acta Biomembr. 1848 (2015) 2317–2325.

[286] A. Junghans, M.J. Waltman, H.L. Smith, L. Pocivavsek, N. Zebda, K. Birukov, M. Viapiano, J. Majewski, Mod. Phys. Lett. B 28 (2014) 1430015.

[287] A.P. Le Brun, A. Soliakov, D.S.H. Shah, S.A. Holt, A. Mcgill, J.H. Lakey, Biomed. Microdevices 17 (2015) 9951.

[288] X. Zhao, F. Pan, B. Cowsill, J.R. Lu, Langmuir 27 (2011) 7654–7662.

[289] M. Chen, C. Dong, J. Penfold, R.K. Thomas, T.J.P. Smyth, A. Perfumo, R. Marchant, I.M. Banat, P. Stevenson, A. Parry, I. Tucker, R.A. Campbell, Langmuir 27 (2011) 8854–8866.

[290] M. Chen, C. Dong, J. Penfold, R.K. Thomas, T.J.P. Smyth, A. Perfumo, R. Marchant, I.M. Banat, P. Stevenson, A. Parry, I. Tucker, I. Grillo, Langmuir 29 (2013) 3912–3923.

[291] E. Kontturi, M. Suchy, P. Penttila, B. Jean, K. Pirkkalainen, M. Torkkeli, R. Serimaa, Biomacromolecules 12 (2011) 770–777.

[292] G. Cheng, Z. Liu, J.K. Murton, M. Jablin, M. Dubey, J. Majewski, C. Halbert, J. Browning, J. Ankner, B. Akgun, C. Wang, A.R. Esker, K.L. Sale, B.A. Simmons, M.S. Kent, Biomacromolecules 12 (2011) 2216–2224.

[293] V. Reyes-Ortiz, R.A. Heins, G. Cheng, E.Y. Kim, B.C. Vernon, R.B. Elandt, P.D. Adams, K.L. Sale, M.Z. Hadi, B.A. Simmons, M.S. Kent, D. Tullman-Ercek, Biotechnol. Biofuels 6 (2013) 93.

Chapter 2

Dynamics of Biological Systems

Tilo Seydel[1]
Institut Max von Laue – Paul Langevin, Grenoble, France
[1]*Corresponding author: e-mail: seydel@ill.eu*

Chapter Outline

Experimental Methods in the Physical Sciences, Vol. 49. http://dx.doi.org/10.1016/B978-0-12-805324-9.00002-9

2.1 INTRODUCTION

Through recent advances in instrumentation, neutrons access the hierarchy of molecular dynamics of biological macromolecules with unprecedented precision. It has for instance become possible to separate and quantitatively describe the superposition of the translational and rotational center-of-mass diffusion of proteins in aqueous solutions as well as the internal diffusive motions of protein side-chains. An increasing number of questions concerning the dynamics of biological systems can be addressed by neutron spectroscopy. Current topics include, among others, the diffusive dynamics in crowded media, the formation of both static and dynamic protein clusters, the fluctuations of biological membranes, the effect of pressure and temperature on living systems, as well as aggregation and self-assembly phenomena, and biological fibers. Along with the progress in the experimental methods, new data analysis frameworks and more accurate models are being employed, such as those derived from colloid physics to describe the diffusion of proteins in aqueous solutions.

Living cells can generally be considered as soft matter and, unlike single crystals, are mostly characterized by the absence of long-range order. Biological macromolecules rather display local order only. The degree of ordering may strongly depend on the type of molecule and its biological role, considering for instance that membranes constitute comparatively highly ordered entities, while in so-called intrinsically disordered proteins a stable tertiary structure can be nearly completely absent. The absence of long-range order in soft matter and biological systems makes the case for experimental methods that can probe local order on a molecular level. Moreover, since biological function is governed by the underlying molecular motions, the experimental method must be sensitive to these motions on a time scale ranging from less than a picosecond to several tens of nanoseconds. Spectroscopy using the scattering of cold and thermal neutrons responds to these requirements and in addition carries the advantages of not being invasive or destructive to the samples and of being highly sensitive to the hydrogen atoms. Relaxational and diffusive molecular dynamics is of extraordinary importance in biology. The concept of quasi-elastic neutron scattering [1] constitutes a key framework to explore this type of dynamics.

The ensemble averaging and the superposition of signals from a hierarchy of different types of molecular motions pose a particular challenge in analyzing and interpreting neutron spectroscopy data from biological systems. Historically, in an approach that is to some extent analog to previous research in polymer physics, biological macromolecules and, notably, proteins have in the past been explored mainly in a (hydrated) "powder" state [2]. In this way, the internal molecular motion of these macromolecules could be investigated as a function of for instance the hydration level and the temperature. The technical advantage of such studies on powder systems as opposed to macromolecules suspended in aqueous solutions has been and still is the suppression of any contribution of a center-of-mass motion of the molecules. Since recently, however, biological macromolecules suspended in aqueous solutions have moved in the focus of the research interest. Such samples much closer to the *in vivo* intracellular environment can be successfully explored subsequent to significant advances in both neutron sources and instrumentation as well as improved models.

The neutron spectroscopy techniques used to explore the dynamics of biological systems mainly consist in time-of-flight spectroscopy (TOF), backscattering spectroscopy (NBS), and spin-echo spectroscopy (NSE). Various different instrument design concepts for TOF and NBS spectrometers have been employed at different neutron sources. Nevertheless, experiments using TOF and NBS are conceptually very similar and mainly differ in the accessed energy transfer and scattering vector ranges and resolutions. Most investigations of a particular biological system combine at least two different types of such instruments, typically a TOF and an NBS spectrometer, in order to access a sufficiently large dynamic range. By contrast, NSE is peculiar as it uses the neutron spin to encode dynamic information accessed by the scattering process. NSE can be seen as being to some extent analogous to dynamic light scattering (DLS), while accessing significantly larger scattering vectors.

The present chapter is devoted to provide a tutorial introduction to the neutron experimental methods and data analysis addressing the molecular dynamics in biological samples. It complements the preceding chapter in this volume on biological structures. The tutorial attempts to provide selected textbook examples for different types of experiments, but does not aim for a comprehensive review of the field. Rather, by the help of this chapter, the reader shall obtain "recipes" to be able to design, carry out, and analyze neutron spectroscopy experiments in the field of biology. The adjacent topics of spectroscopy on soft condensed matter and on biologically inspired materials closely interact and provide mutual inspiration. The present chapter also highlights some of these links. Moreover, the chapter illustrates examples where predictions from theoretical physics have been quantitatively confirmed in biological systems by employing large series of systematic neutron spectroscopy experiments.

2.2 FUNDAMENTAL ASPECTS AND OVERVIEW

2.2.1 Neutron Spectroscopy Fundamentals to Study Biological Systems

The rather obvious assets of spectroscopy using thermal or cold neutrons, i.e., neutrons with kinetic energies of about 3meV or less, consist in the accessible length scales and energies of atomic motion commensurate with the biomolecular structure and dynamics. Thus the explored length scale typically ranges from a fraction of a nanometer up to on the order of 100 nanometers, and the accessible energy transfer translates into time scales of molecular motion from less than a picosecond up to several 100 nanoseconds. Importantly, the employed neutron energy is several orders of magnitude below the typically required ionization energy of soft (e.g., polymeric) or biological macromolecules. Neutron scattering therefore constitutes a very gentle method for biophysics and in addition does not require any "tagging" of molecules other than in some cases an isotopic substitution. Moreover, the large scattering cross-sections of the hydrogen isotopes ^{1}H and ^{2}H=D, which additionally differ drastically in their incoherent part, provide a very strong scattering signal from the hydrogen-rich biological samples.

Incoherence may have a somewhat negative connotation in common speech and hopefully does not characterize the reasoning that follows, but incoherent scattering probably constitutes one of the most important and least trivial assets of neutron spectroscopy on soft and biological materials. Confusion may arise, since scattering experiments—be it using X-rays or neutrons or visible light—that are designed to probe static structure generally require a spatial coherence volume of the impinging radiation that is larger than the structure to be explored, i.e., larger than several times the distance of neighboring atoms. A spatial and temporal coherence volume is defined by the design of any scattering experiment "through the very process of propagation" [3, 4]. By contrast, spatial incoherence in a neutron scattering experiment enters via the nuclear spin statistics [5]. Thus a neutron scattering experiment which is by way of the spin-statistics spatially incoherent nevertheless does require coherence in time—that is, an energy resolution—in order to access time-correlations in the sample. It is then the time-correlations of the motions of single atoms during this coherence time that, subsequent to the ensemble-averaging over the illuminated sample volume, are the observable quantity in the incoherent neutron scattering experiment. This single-particle self-correlation function is of tremendous importance. Since Chapter 1 and in particular Section 1.2.5 in Volume 44 of this book series [6] has been devoted to these foundations of neutron scattering, we will only briefly repeat the van Hove correlation functions relevant for all what follows in Section 2.2.3.

2.2.2 Disambiguation of Symbols

Wherever possible, the present chapter follows the general assignment of symbols for the entire book series. Due to the frequent superposition of several spectral widths in measured signals from biological samples, however, both the symbols Γ and γ will be used to denote spectral widths. Γ will generally denote a width arising from internal molecular motions, while γ is assigned to the signal from a global center-of-mass motion of a macromolecule. Also, the neutron spectra from biological samples are mostly recorded in a "powder" scattering geometry where only the absolute value $q := \|\mathbf{Q}\|$ of the scattering vector $\mathbf{Q}$ is relevant. As is common in the literature on neutron spectroscopy of biological systems, ω in $E = \hbar\omega$ will be used synonymously for the energy E. The symbols $f(x)$ and $g(x)$ will be used to denote scalar functions of x. We note the abbreviations for the frequently cited neutron scattering facilities where example instruments for this chapter are located: ILL for the international Institut Max von Laue - Paul Langevin [7] in Grenoble, France, and SNS for the Spallation Neutron Source (SNS) [8] in Oak Ridge, Tennessee. It is emphasized that several other instruments similar to those chosen as examples exist worldwide—reviewing these would exceed the scope and space allocated to this chapter.

2.2.3 Collective and Self-Correlation Functions

The classical van Hove correlation functions [10] constitute the foundation of the access to biomolecular dynamics using neutrons:

$$G(\mathbf{r},t) = \frac{1}{N}\sum_{i,j=1}^{N}\left\langle \delta\left(\mathbf{r}-\mathbf{R}_j(t)+\mathbf{R}_i(0)\right)\right\rangle \tag{1}$$

$$G_s(\mathbf{r},t) = \frac{1}{N}\sum_{j=1}^{N}\left\langle \delta\left(\mathbf{r}-\mathbf{R}_j(t)+\mathbf{R}_j(0)\right)\right\rangle, \tag{2}$$

where t denotes the time, $\mathbf{r}$ the instantaneous position of the neutron, and $\mathbf{R}_{i,\,j}$ the time-dependent positions of all N nuclei in the sample. The correlation function $G(\mathbf{r}, t)$, equation (1), contains the sum of pairs of distinct nuclei $\mathbf{R}_{i,\,j}$. It therefore includes both temporal and spatial correlations. By contrast, the self-correlation function $G_s(\mathbf{r}, t)$, equation (2), includes the time-correlation of nuclei with identical index j only. It therefore constitutes an ensemble-averaged single-particle self-correlation function. The two different correlation functions, equations (1) and (2), are schematically illustrated in figure 1 for the case of randomly moving particles being observed at two arbitrary moments in time. The importance of equation (2) arises from the possibility to unambiguously access self-diffusion, as will be seen in section 2.3. No readily available method other than neutrons can access the self-diffusion on similarly short time and length scales in a nondestructive manner on biological samples.

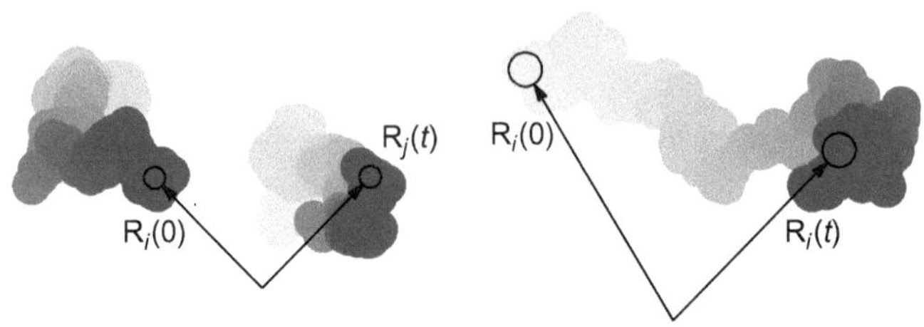

FIGURE 1 Schematic illustration for the interpretation of the van Hove correlation functions for coherent scattering (left, equation (1)) and incoherent scattering (right, equation (2)), respectively: By coherent scattering, the ensemble-average over pairs of particles with coordinates $\mathbf{R}_{i,j}(t)$ is observed, marked in the schematic by the time-dependent coordinates $\mathbf{R}_{i,j}(t)$ and shadows indicating the random traces followed by a pair of particles on the observation time scale. By incoherent scattering, the ensemble-average over single-tagged particles is observed, indicated in the schematic by the trace followed by a single particle and being observed at the moments 0 and t in time. (Figure rendered in MATHEMATICA [9].)

2.2.4 Observable Quantities

The scattering function $S(q, \omega)$ depending on the scattering vector q and energy transfer ω observable in neutron scattering experiments has been introduced in previous chapters, notably in Chapter 1 of Volume 44 [6]. In the case that we carry out an incoherent neutron spectroscopy experiment where the accessed range in ω is small enough to not reach vibrational or phonon excitations and centered at 0 (i.e., typically $|\omega| \leq 2\text{meV}$), assuming that vibrations and diffusive motions are uncoupled, we can write the scattering function separated in two factors,

$$S(q,\omega) = \exp\left(-\frac{1}{d} q^2 \langle u_{\text{vib}}^2 \rangle\right) \cdot S_{\text{diff}}(q,\omega), \tag{3}$$

where $S_{\text{diff}}(q, \omega)$ contains all contributions from diffusive or relaxational molecular motions which are typically present in soft matter and biological materials. $S_{\text{diff}}(q, \omega)$ is, at least at physiological temperatures, a symmetric function centered at $\omega = 0$ and denoted the quasi-elastic scattering function, since it experimentally manifests itself in a symmetric line broadening relative to the spectrometer resolution function (cf. section 2.4, figure 7). Here we note that we do not specifically address the vibrational spectrum that is recorded using inelastic (as opposed to quasi-elastic) neutron scattering in the present chapter, but will nevertheless later mention a few examples where inelastic neutron scattering is applied to biological systems. For the theory behind vibrational spectroscopy, we refer to the dedicated chapters of this book series, notably Chapters 1 and 3 in Volume 44 [6, 11].

In equation (3), $\langle u_{\text{vib}}^2 \rangle$ denotes the ensemble-averaged mean-squared displacement (MSD) of the atomic vibrations, and the term $\exp\left(-\frac{1}{d} q^2 \langle u_{\text{vib}}^2 \rangle\right)$

is known as the vibrational Debye-Waller factor of the system with d dimensions (typically $d = 3$). This Debye-Waller factor therefore results in a q-dependent reduction of the elastic scattering signal. The scattering function due to the diffusive motions can mostly be written as [12]

$$S_{\text{diff}}(q,\omega) = \mathscr{L}(\gamma(q),\omega) \otimes \{A(q)\delta(\omega) + [1 - A(q)]\mathscr{L}_\alpha(\Gamma(q),\omega)\}, \tag{4}$$

where $\mathscr{L}(\gamma(q),\omega)$ is a Lorentzian function with the linewidth $\gamma(q)$ accounting for the global translational and rotational diffusive motions of a macromolecule in a liquid solution. The symbol $\otimes$ represents a convolution, and $\mathscr{L}_\alpha(\Gamma(q),\omega)$ is the function describing the ensemble-averaged internal diffusive molecular motions of the sample macromolecule(s). In a rather general form we can write $\mathscr{L}_\alpha(\Gamma(q),\omega)$ in terms of a Kohlrausch-Williams-Watts (KWW) function [12, 13]

$$\mathscr{L}_\alpha(\Gamma(q),\omega) = \int_{-\infty}^{\infty} \frac{dt}{2\pi} e^{-i\omega t} \exp\left(-|t\Gamma(q)|^\alpha\right) \tag{5}$$

with the phenomenological parameter $0 \leq \alpha \leq 1$. The amplitude factor $A(q)$ is known as the elastic incoherent structure factor [14].

Equation (3) containing the vibrational MSD inspires the definition of the q-dependent generalized mean-squared displacement $\langle u^2 \rangle_q$ [12]

$$\exp\left[-\frac{1}{d}\langle u^2 \rangle_q q^2\right] := S(q, |\omega| < \Delta\omega), \tag{6}$$

where $\Delta\omega$ is the energy resolution of the spectrometer. This definition can be further simplified in the limit of small scattering vectors

$$\langle u^2 \rangle := -d \lim_{q \to 0} \left\{ \frac{\log([S(q, |\omega| < \Delta\omega)])}{q^2} \right\}. \tag{7}$$

The definitions of the mean squared displacement according to equations (6) and (7) are of extraordinary importance in the study of biological systems using neutron spectroscopy, as they are the key to the interpretation of so-called "elastic fixed-window" scans on high-resolution neutron spectrometers explained in section 2.6.2. The generalized and vibrational MSD are not to be confused with the diffusive mean-squared displacement which will be defined later in equation (14).

2.2.5 Past and Present Topics in the Dynamics of Biological Systems

One of the main interests in experiments on biomolecular dynamics using neutron spectroscopy quite generally consists in the establishment of structure-dynamics-function relationships. Notwithstanding the general complexity of living cells, neutron spectroscopy experiments pursue a test of relatively simple physical models, that, remarkably, often turn out to be successful, such as the finding that relatively simple colloid models can predict the short-time diffusion of proteins [15]. The scope of neutron spectroscopy studies of biological systems ranges from highly simplified model systems that consist of one type of macromolecule, such as a hydrated protein or a protein dissolved in water, to entire living cells (e.g., Refs. [16, 17]). The investigation of molecular dynamics in biological systems using neutron spectroscopy has for instance addressed the internal relaxational and diffusive motions for a very large selection of different proteins. For these studies, the samples consisted of protein powders with different levels of hydration. The hydration was typically adjusted by a storage in a humid atmosphere, and among the advantages of these samples was the absence of any center-of-mass diffusion of the entire proteins inside the samples and the possibility to freeze them to low temperatures without damage. The absence of the center-of-mass diffusion in the hydrated powder samples used in the past has greatly simplified the data analysis. Specific examples will be provided in section 2.8, and for a review we refer to Ref. [2]. The studies on hydrated protein powders started in the late 1980s [18], and approximately ten years later, along with progress in neutron instrumentation, the exploration of liquid solutions of proteins began [19]. Subsequent studies of proteins in aqueous solution addressed for instance the effect of macromolecular crowding on the protein diffusion [15, 20, 21]. Even more recent is the interest in the neutron spectroscopic investigation of the formation of protein clusters, which may be transient or static in time, and which may be caused by temperature, concentration, or added salts [22–24]. These studies on the cluster formation also point toward the future perspectives for a controlled self-assembly of biological macromolecules from liquid solutions, and more generally toward controlling structure, dynamics, and transport in biological soft matter.

Membranes constitute another topic addressed by neutron spectroscopy of biological systems, which are studied both in bulk, i.e., in a powder scattering geometry, as well as in the form of planar lipid membranes [25–27]. Also, biological fibers such as collagen or silk are being explored regarding their molecular dynamics [28–30]. Recently, another topic given by intrinsically disordered proteins has moved in the focus of interest, where methods exploring structure alone are particularly limited and require the dynamic information as an unavoidable complement [31–33]. Finally, the influence of pressure on biological molecular dynamics is being investigated since recently [34–36]. Specific examples of some of the above topics will be presented in section 2.8, but it is emphasized again that this brief overview is not to be understood as a review of the field.

2.3 CONCEPTS OF DIFFUSION

2.3.1 Diffusion Fundamentals

Diffusive motion is ubiquitous [37–39]. In biological systems, diffusion crucially influences transport and signal transduction on the molecular level

[40]. High-resolution neutron spectroscopy is ideally suited to study diffusion on molecular length scales and biologically relevant time scales. Due to the existence of spin and isotope incoherent and coherent neutron scattering, neutrons can probe both self-diffusion and collective diffusion phenomena in many-particle systems (cf. equations (1), (2)). In the present section only the most prominent terms and definitions regarding diffusion will be reviewed, mostly restricting the discussion to simple Brownian motion and extensions for confined geometries [14], fractional diffusion, and crowded colloidal suspensions [41]. The focus will be on the molecular diffusive properties in biological samples accessible to neutron scattering. For more detailed considerations we refer to dedicated textbooks and lecture notes such as Refs. [42–44].

Brownian diffusion can be seen as overdamped motion driven by thermal fluctuations in a solvent medium. In modern physics, the concepts of Brownian diffusion have been extended to include so-called anomalous diffusion processes such as subdiffusion and superdiffusion [39]. Recent research has also addressed the "hydrodynamic memory" [45] arising from a diffusing particle acting back on its solvent. Historically, the first mathematical picture of record describing diffusion goes back to A. Fick [46]. According to Fick's second law, diffusion can be written analogously to the heat equation as a concentration field which changes with time as

$$\frac{\partial \rho}{\partial t} = D \triangle^2 \rho, \tag{8}$$

where ρ is the particle concentration and D the diffusion coefficient. This equation can be written in a generalized form known as Smolochowski or Fokker-Planck equation for a many-particle system [42, 47]

$$\frac{\partial P(\mathbf{r},t)}{\partial t} = \hat{\mathbf{L}}_S P(\mathbf{r},t) \tag{9}$$

with the probability density function $P(\mathbf{r}, t)$ and the Smoluchowski operator $\hat{\mathbf{L}}_S$ acting on the function $\mathcal{F}$ through

$$\hat{\mathbf{L}}_S(\mathcal{F}) = \triangle \cdot D(\mathbf{r}) \cdot [\beta[\triangle U(\mathbf{r})](\mathcal{F}) + \triangle(\mathcal{F})]. \tag{10}$$

Therein, $\beta = k_B T$, $\mathbf{r}$ is the vector of all particle coordinates, $U = U(\mathbf{r})$ is the potential energy of all molecules, and $D(\mathbf{r})$ is the diffusion tensor.

Diffusion can also be described in terms of a generalization of the classical Newton equation of motion to represent a many-particle system by introducing a time-dependent stochastic force term $F_{\text{fluct}}(t)$ [48]. The Newton equation for a particle with velocity v then becomes the Langevin equation

$$\dot{v} + \alpha v = \tilde{F}(t) \tag{11}$$

with $\alpha = -\kappa/m$ and $m\,\tilde{F}(t) = F_{\text{fluct}}(t)$, where $\kappa = 6\pi\eta_0 a > 0$ is a constant accounting for the drag experienced by the diffusing particle with radius a

and mass m in the system (solvent) with viscosity η_0. κ is also known as hydrodynamic friction coefficient (cf. section 2.3.2). The stochastic force $F_{\mathrm{fluct}}(t)$, denoted Langevin force, can in most cases be written as a δ-correlated function with zero mean and the fluctuation strength ξ

$$\left\langle \widetilde{F}(t) \right\rangle = 0, \ \left\langle \widetilde{F}(t)\, \widetilde{F}(t') \right\rangle = \xi\, \delta(t - t'). \tag{12}$$

Importantly, the solution of the Langevin equation (11),

$$v(t) = v_0 \exp(-\alpha t) + \int_0^t dt' \exp(-\alpha(t - t'))\, \widetilde{F}(t') \tag{13}$$

with the initial condition $v_0 = v(0)$, gives rise to the definition of the so-called mean-squared displacement of the diffusing particle (cf. figure 3)

$$\begin{aligned} W(t) &= \frac{1}{2d}\left\langle (r(t) - r(0))^2 \right\rangle = \frac{1}{2d}\int_0^t \int_0^t dt_1\, dt_2\, \langle v(t_1) v(t_2) \rangle \\ &= \frac{1}{2d}\frac{\xi}{\alpha^2} t \left(1 - \frac{(1 - \exp(-\alpha t))}{\alpha t} + \frac{(1 - \exp(-\alpha t))^2}{2\alpha t} \left(\frac{2\alpha}{\xi} v_0^2 - 1 \right) \right). \end{aligned} \tag{14}$$

Therein, r is the particle position and d the dimension of the system. For $t \to \infty$, eq. (14) reduces to $W(t) = D_0 t$ with the Brownian diffusion coefficient

$$D_0 = \frac{\xi}{2d\alpha^2} = \frac{k_B T}{\kappa}. \tag{15}$$

In equation (15) we recognize the well-known Stokes-Einstein relation, assuming that the relation between ξ and κ is given by the equipartition theorem of classical mechanics [43].

For reasons that will become clear later, it is convenient to rewrite the diffusion equation (8) in terms of the probability density function $G(r, t)$ depending on the spatial r and temporal t coordinates

$$\frac{\partial G}{\partial t} = D \frac{\partial^2}{\partial r^2} G(r,t), \tag{16}$$

which is solved by the Gaussian function

$$G(r,t) = \frac{1}{\sqrt{4Dt}} \exp\left(-\frac{r^2}{4Dt} \right), \tag{17}$$

and note its spatial Fourier transform in terms of the scattering vector q

$$\frac{\partial G}{\partial t} = -Dq^2 G(q,t), \tag{18}$$

which is solved by the exponential or Debye relaxation function

$$G(q,t) = \exp(-Dq^2 t). \tag{19}$$

In equation (19) we recognize the product Dq^2 linking the observed decay rate to the scattering vector q in a scattering experiment on a suspension of particles (such as proteins dissolved in water) obeying Brownian diffusion. That is, in a scattering experiment, the width γ of the Lorentzian $\mathscr{L}(\gamma,\omega)$ describing the center-of-mass diffusion of such tracer particles reads

$$\gamma(q) = D\,q^2. \tag{20}$$

An example of proteins in aqueous solution following this relationship is shown in figure 9, which will be further explained in section 2.4. When a single large tracer particle is suspended in a liquid solution, it generally follows this Brownian diffusion, equation (20). However, the molecules of the liquid themselves giving host to the tracer particle—such as the solvent water molecules— usually follow a more complex diffusion. One example is the so-called jump diffusion which constitutes a generalization of the simple Brownian diffusion by introducing a residence time τ between molecular diffusive jumps. In this case the Lorentzian linewidth associated with the solvent diffusive motion reads

$$\gamma(q) = \frac{D\,q^2}{1 + D\,q^2\,\tau}. \tag{21}$$

This relationship has been confirmed for the case of the diffusion of water molecules in pure bulk water [49].

As soon as more than one large particle (e.g., a protein) suspended in a liquid solution of small solvent (e.g., water) molecules is being observed, the terms of collective diffusion and self-diffusion, respectively, need to be distinguished (cf. section 2.2.3 and figure 1). Moreover, the concentration or volume fraction φ occupied by the particles in the liquid suspension crucially influences the diffusion of the individual particles. This leads us to the following section on the diffusion in colloidal suspensions.

2.3.2 Diffusion in Colloidal Suspensions

As we shall see later in this chapter, an aqueous solution of biological macromolecules such as proteins can to some extent be understood in terms of the physics of a colloidal suspension [15, 21, 52, 53]. Here we summarize the most important terms required to understand the diffusion in colloidal suspensions, and we refer to the textbooks and seminal articles [15, 42, 44, 54, 55], as well as to the references given therein, for further reading. In a dense colloidal suspension, different diffusive time scales occur [54]. At very short times $t \ll \tau_B \approx m/(6\pi\eta R)$ the diffusion of a large sphere with the radius R surrounded by small solvent molecules with the solvent viscosity η is not yet influenced by the solvent molecules and therefore can be described as undergoing a "free flight" with its position $\mathbf{r}$ evolving in time t according to

generalized by a scalar function $f(\varphi, t)$ depending on the volume fraction φ occupied by the colloidal particles in the suspension and on the observation time t:

$$D = \frac{k_B T}{6\pi\eta R} f(\varphi, t) \tag{25}$$

where $\kappa = 6\pi\eta R$ is denoted hydrodynamic friction coefficient with the solvent viscosity η and the colloidal particle hydrodynamic radius R. In both the short-time and long-time limits, the dependence of f on the time t vanishes (cf. figure 3), i.e., $f = f(\varphi)$, and therefore, these limiting cases are particularly interesting for a systematic comparison of experiment and theory. It is therefore particularly useful that neutron spectroscopy can access the short-time limit of the diffusion of proteins, where hydrodynamic interactions prevail and are not "obstructed" by direct interactions [15]. For both the short-time and long-time limits, theoretical predictions for $f(\varphi)$ have been derived for hard-sphere model colloids, for both the case of non-charged and homogeneously charged spheres [41, 50], and for both the case of translational and rotational diffusion. Figure 4 shows the resulting theoretical dependencies $D(\varphi) = D_0 f(\varphi)$ for several cases of colloidal self-diffusion. With increasing φ, the colloidal suspensions become increasingly "crowded," resulting in a significant reduction of the diffusion compared with the dilute limit. This

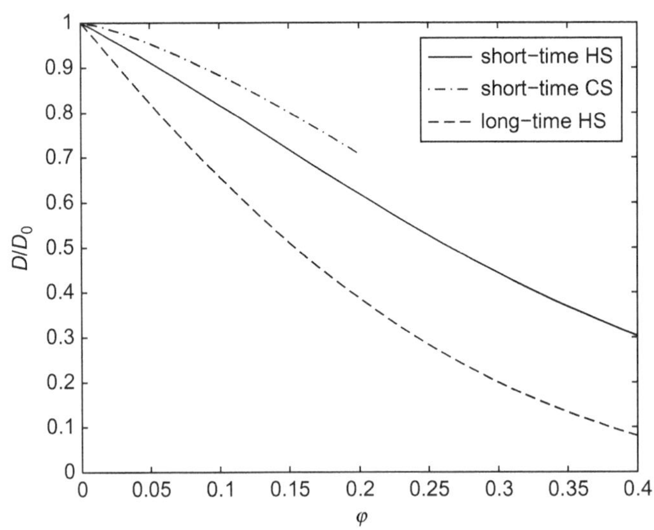

FIGURE 4 Normalized theoretical translational self-diffusion coefficients $D(\varphi)/D_0$ of colloidal hard-sphere suspensions as a function of the colloidal volume fraction φ. The different curves denote the diffusion for hard spheres (HS) for the short-time and long-time limits, respectively, according to Ref. [50], and for homogeneously charged spheres (CS) in the short-time limit according to Ref. [41]. D_0 denotes the diffusion coefficient of the colloid particle in the limit of infinite dilution. (Figure rendered in MATLAB [51].)

reduction of the diffusion due to crowding is of great importance in living cells, where the intracellular fluid constitutes a crowded suspension of macromolecules [57, 58], and where diffusion is one of the main transport processes [40] (see also section 2.8.2). Figure 4 shows the situation of self-diffusion, and it is noted that also the collective diffusion is subject to theoretical modeling, which converges to self-diffusion in the short-time limit at large q, but differs in the long-time limit. By essentially being governed exclusively by hydrodynamic interactions, the regime of short-time self-diffusion as explored by incoherent neutron spectroscopy can be assumed to be a key to the understanding of intracellular communication. The hydrodynamic interactions can be considered as quasi-instantaneous on time scales of biological function.

While the theory of the diffusion in colloidal suspensions applies to the global diffusion of colloidal objects in liquid solutions, the internal molecular diffusive motions within soft macromolecules such as biological polymers or proteins require an additional very different level of modeling. This topic of molecular dynamics in geometrical confinement is the subject of the next section.

2.3.3 Diffusion in Confined Geometries

The diffusion in geometrical confinement contributes to the atomic motion inside biological macromolecules. Remarkably, as indicated in the introduction of this chapter, incoherent neutron spectroscopy provides information on local order in soft materials including biological materials. It does so by providing information on the geometrical confinement of diffusive atomic displacements via the elastic incoherent structure factor (EISF) [14, 59, 60]. Quite generally, the EISF is the amplitude $A(q)$ in the part of the scattering function describing the localized diffusive motions, such as in equation (4). Here we note that equation (4) describes the very general case where a global motion of a macromolecule in solution is superimposed on the internal motions. In most cases studied using neutron spectroscopy on biological samples, the global center-of-mass motion contribution does not occur, i.e., $\mathscr{L}_\gamma(q) \equiv 1$, because (hydrated) powder samples are studied.

Historically, models to describe how such diffusion manifests itself in elastic and quasi-elastic neutron spectra have been developed for various geometries such as localized jumps of atoms between different neighbor sites. One of the most famous examples is the diffusion inside a sphere [61]:

$$A_{\text{sphere}}(q) = \left| \frac{3j_1(qR)}{qR} \right|. \tag{26}$$

This can for instance be imagined as the motion of a protein side chain. A jump of an atom between two or three equivalent sites with distance a is also frequently observed, which can for instance occur for the atoms of methyl or ammonium groups of a protein. For a two-site jump the EISF reads

$$A_{2-\text{jump}}(q) = 1 - 2p[1 - j_0(qa)], \tag{27}$$

where p is the probability of locating the diffusing particle at one of the two sites. The two-site jump may be generalized to a larger number of sites (e.g., $A_{3-\text{jump}}(q) = \frac{1}{3}[1 + 2j_0(qa)]$ for a jump between three equivalent sites with equal probability), and the discussion of various geometries and their corresponding EISF can be found in the textbook by M. Bee [14]. In the earlier equations, $j_l(x)$ is the spherical Bessel function of the first kind and order l.

A relatively simple example of a more complex EISF typically observed in proteins is provided by the superposition of a two-site jump diffusion, a diffusion inside a sphere, the relative weight of which is given by the factor c_1, as well as an elastic (static) contribution with the constant contribution c_0 (cf. e.g., Ref. [28]):

$$A_0(q) = (1 - c_0)\left[c_1 A_{\text{jump}}(q) + (1 - c_1)A_{\text{sphere}}(q)\right] + c_0. \tag{28}$$

Here we have tacitly assumed that the investigated model system is dynamically heterogeneous such that the individual dynamic components can simply be added [62]. Such an EISF is illustrated schematically in figure 5. Please note that here we have neglected the fact that the elastic (static) contribution c_0 depends on q by way of the thermal Debye-Waller factor $\exp\left(-q^2\langle u_{\text{vib}}^2\rangle\right)$ (cf. equation (3)).

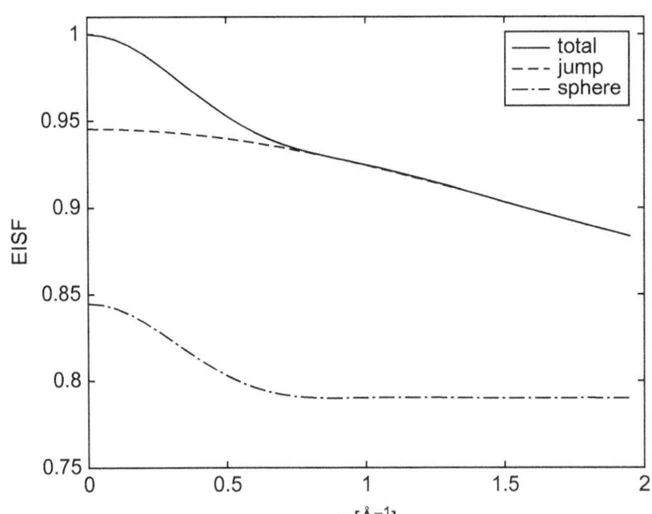

FIGURE 5 Calculated example of an elastic incoherent structure factor (EISF) $A(q)$ depending on the absolute value q of the scattering vector (solid line), composed of the superposition of a two-site jump diffusion with jump distance 1.33Å (dashed line), the diffusion inside a sphere with radius 5.07Å (dotted line), and an elastic contribution resulting in a constant offset of all lines (equation (28)). (Figure rendered in MATLAB [51].)

Geometrically confined diffusion processes not only result in a rather complex q-dependence of the EISF, but also of the quasi-elastic line broadening that goes along with it. For the example EISF defined by equation (28), the corresponding QENS contribution reads [14, 28]

$$S_{\text{diff}}(q,\omega) = (1-c_2)\left\{(1-c_1)\left[\left(\frac{3j_1(qR)}{qR}\right)^2\delta(\omega)+\left(1-\left(\frac{3j_1(qR)}{qR}\right)^2\right)\frac{1}{\pi}\frac{\Gamma}{\Gamma^2+\omega^2}\right]\right.$$
$$\left.+c_1\left[\frac{1}{2}(1+j_0(qa))\,\delta(\omega)+\left(\frac{1}{2}(1-j_0(qa))\,\frac{1}{\pi}\,\frac{2\tau}{4+\tau^2\omega^2}\right)\right]\right\}$$
$$+c_2\left(\frac{1}{\pi}\frac{\Gamma_u(q)}{\Gamma_u^2(q)+\omega^2}\right), \tag{29}$$

where Γ is the linewidth due to the diffusion inside a sphere and τ is the residence time on one of the sites between two-site jumps. Equation (29) in addition contains a contribution due to unrestricted jump diffusion with the width $\Gamma_u(q)$ defined by equation (21). Such unrestricted jump diffusion can for instance be associated with weakly bound or unbound water in a humid protein sample and would not contribute to the EISF. The corresponding component is therefore added to the QENS model, equation (29), as an additional component with its contribution being weighted by the scalar fit parameter c_2 [28]. For further reading, in particular regarding a very large choice of established models of confinement geometries, we refer to the textbook by M. Bee [14].

2.3.4 Fractional Generalization of the Diffusion Equation

In biological macromolecules such as proteins, internal molecular motions are rarely of a simple Brownian type. Complex relaxation modes rather manifest themselves in decay functions that deviate from the simple Debye-relaxation, equation (19). Possible generalizations and extensions are given by the Kohlrausch-Williams-Watts stretched exponential function [13, 63] (cf. Equation (5) and Volume 44, Section 6.8.1, of this book series [11])

$$G(t) \propto \exp(-t^{\beta}) \tag{30}$$

with $0 \leq \beta \leq 1$, and by the inverse power law

$$G(t) \propto \frac{1}{(1+t)^{\alpha}}. \tag{31}$$

Beyond these rather empirical models, an extension of the models describing diffusion has been pursued. Brownian diffusion models are based on Markovian random walk models, which assume that the diffusive process at any moment in time contains no memory of its preceding states. Later, a systematic generalization of the models describing the "random walks" of the diffusing particles in space and time has been achieved. This has led to so-called continuous time random walk models [64]. To this effect, the diffusion equation (eq. (8)) has been generalized to contain noninteger derivatives in time [65],

$$^{C}D^{\alpha}G(r,t) = K_{\alpha}\nabla^{2}G(r,t), \tag{32}$$

where $G(r, t)$ is the van Hove autocorrelation function and

$$^{C}D^{\alpha}[\Phi](t) = \frac{1}{\Gamma(1-\alpha)}\int_{0}^{t}\frac{\Phi'(\xi)}{(t-\xi)^{\alpha}}d\xi \tag{33}$$

(where ^{C} denotes "Caputo"-type) is the operator defining the generalized fractional derivative of the function $\Phi = \Phi(t)$. Therein, Φ' denotes the first-order integer derivative of Φ and $\Gamma(x)$ is the gamma function (not to be confused with the Lorentzian spectral linewidth also being assigned the symbol Γ in different contexts).

$$K_{\alpha} = \frac{\sigma_{c}^{2}}{\tau_{c}^{\alpha}} \tag{34}$$

is the generalized fractional diffusion coefficient. Therein, σ_c and τ_c are the characteristic spatial and time scales, respectively, and the scalar $0 < \alpha \leq 1$ is a real number interpreted as the memory parameter of the system. For $\alpha = 1$, the diffusion would be of the classical Fickian type (see equation (8)). In an analog way, the diffusion equation may be further generalized to contain noninteger derivatives in space, but we will not discuss this possibility here. The model of fractional diffusion has been successfully applied for instance to describe internal molecular relaxations in proteins [66, 67] as well as to describe the intermediate-time diffusion of proteins in polymeric solutions [68].

The spatial Fourier transform of equation (32) is solved by the Mittag-Leffler function and thus gives the intermediate scattering function due to fractional diffusion as

$$I(q,t) \propto E_{\alpha,1}\left(-\left|\frac{t}{\tau}\right|^{\alpha}\right) \quad \text{with} \quad \tau = \left(K_{\alpha}q^{2}\right)^{-1/\alpha} \tag{35}$$

with the fractional relaxation time τ and the Mittag-Leffler function

$$E_{\alpha,\beta}(z) = \sum_{n=0}^{\infty}\frac{z^{n}}{\Gamma(\alpha n + \beta)}. \tag{36}$$

The Mittag-Leffler function (cf. figure 6) constitutes the generalization of the exponential function $\exp(z) = E_{1,1}(z)$, as it can be easily seen that equation (36) for $\alpha = 1$, $\beta = 1$ is the Taylor series of the exponential function.

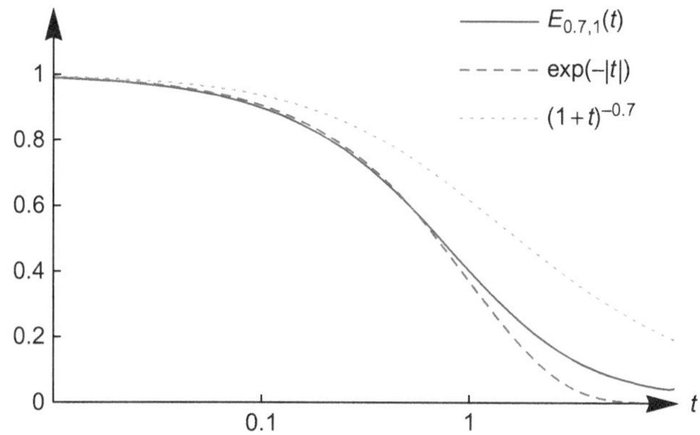

FIGURE 6 The Mittag-Leffler function $E_{\alpha,\beta}(t)$ (solid line for $\alpha = 0.7$, $\beta = 1$), equation (36), interpolates between the (stretched) exponential decay (equation (30), dashed line) at small times and the inverse power law (equation (31), dotted line) at large times. (Figure rendered in MATHEMATICA [9].)

Remarkably, the relation $\tau = (K_\alpha q^2)^{-1/\alpha}$ between the generalized diffusion coefficient and the fractional relaxation time fully describes the dependence of the measured spectra on the scattering vector and therefore links the spatial and temporal components of the model. It can be seen as a generalization of the simple Brownian diffusion $\tau = (Dq^2)^{-1}$ and has been verified for the first time in neutron spectra from silk fibers [30]. In equation (5), the Kohlrausch-Williams-Watts model is consequently replaced by the Fourier transform of the Mittag-Leffler function to establish the fractional generalization of the scattering function. The approach of describing diffusion using the fractional calculus is mathematically elegant. Moreover, this approach can describe the dynamics in suitable complex systems by just two characteristic quantities, namely K_α and α. It is the dependence of these quantities on system parameters such as the tensile stress on a biological fiber that then provides very useful insights. Moreover, recent molecular dynamics simulations of several different prototypical proteins support the view that the internal molecular motions of individual proteins can quite generally be interpreted in terms of models of fractional diffusion over a very wide range of time scales [69]. However, as a general criticism to the fractional dynamics approach it may be noted that it does not convey direct specific information on the confinement geometry of the molecular diffusive processes, as opposed to the approach described in section 2.3.3. Concluding this paragraph, we remark that the terms "fractional" and "fractal" are not to be confused. Nevertheless, while "fractional" essentially refers to the derivatives of noninteger order given by equation (32), it is often an underlying "fractal" geometry that causes "fractional" diffusion.

2.4 MACROMOLECULES IN AQUEOUS SOLUTIONS

2.4.1 Center-of-Mass Diffusion

Unlike colloidal hard spheres, proteins and other biological macromolecules are soft, "floppy" materials with a generally nonspherical shape. Nevertheless, in the short-time limit accessible by neutron spectroscopy, the global diffusion of proteins in aqueous solutions can be understood in terms of a picture of the diffusion of rigid colloidal particles [15, 21]. Moreover, aqueous solutions of purified proteins can be considered as highly monodisperse colloidal suspensions. However, on the observation length and time scale of neutron spectroscopy experiments, different types of motions are typically superimposed in the observed spectroscopic signal. In this way, a neutron spectroscopy experiment on biological macromolecules dispersed in a liquid solution is in general sensitive to both the global center-of-mass motion of the molecules as well as their internal molecular motions. This situation may under certain conditions be simplified depending on the resolution and energy range accessed by the spectrometer as well as on the size of the macromolecules. Here we consider the general case, in which the spectra from the macromolecules—without restriction of the generality we here think of proteins—in solution can be modeled by the resolution-convoluted (incoherent) scattering function

$$\begin{aligned} S^{(\mathrm{exp})}(q,\omega) = \mathscr{R}(q,\omega) \otimes \{\beta(q)[A(q)\mathscr{L}(\gamma(q),\omega)\ldots \\ +(1-A(q))\mathscr{L}(\gamma(q)+\Gamma(q),\omega)]\ldots \\ +\beta_{D_2O}(q)\mathscr{L}(\gamma_{D_2O}(q),\omega)\}. \end{aligned} \tag{37}$$

Therein, $\mathscr{R}(q,\omega)$ denotes the instrumental resolution function, $\otimes$ a convolution, $\beta(q)$ a scaling factor, and $\mathscr{L}$ a Lorentzian function

$$\mathscr{L}(\sigma(q),\omega) = \frac{1}{\pi}\frac{\sigma(q)}{\omega^2 + \sigma(q)^2}. \tag{38}$$

The scalar function $A(q)$ is an amplitude which can be identified with the Elastic Incoherent Structure Factor (EISF) (cf. section 2.3.3). The term $\beta_{D_2O}(q)\ \mathscr{L}(\gamma_{D_2O}(q),\omega)$ describes the Lorentzian-shaped solvent (D_2O) water contribution to the signal with the amplitude $\beta_{D_2O}(q)$. $\mathscr{L}(\gamma(q),\omega)$ is associated with the global motion of the protein in the solution, consisting of contributions from both rotational and translational motions [19]. $\mathscr{L}(\gamma(q)+\Gamma(q),\omega)$ models the internal diffusive motions of the proteins, and we note that this Lorentzian may be replaced by a more complicated function in some cases (cf. equation (5) and section 2.3.4). We also note that this Lorentzian depends on the sum $\gamma(q) + \Gamma(q)$ for technical reasons only, because in this way the numerical fit procedure becomes more stable (cf. section 2.5). An example spectrum (filled symbols) of a protein solution at $T = 295$K measured on a medium-resolution backscattering spectrometer (energy resolution $\Delta\omega \approx 3.5\ \mu$eV

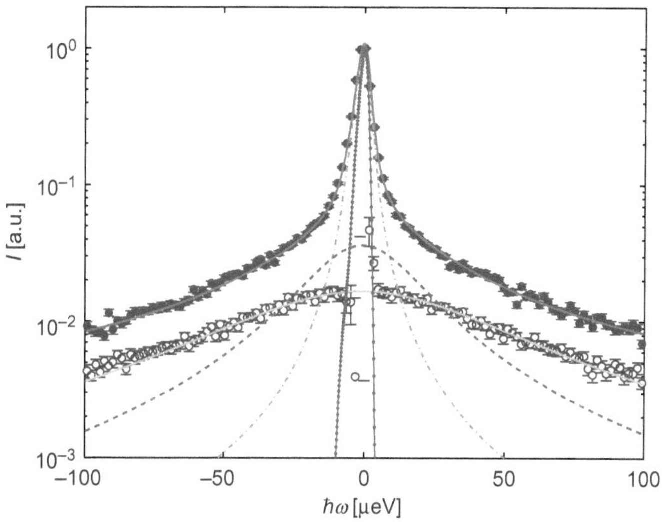

FIGURE 7 Example spectrum (filled symbols) from an aqueous (D_2O) solution of *Bovine Serum Albumin* proteins at the concentration of 200mg/ml and the sample temperature $T = 295$ K recorded on the backscattering spectrometer BASIS (SNS, Oak Ridge) at the scattering vector $q = 0.65 Å^{-1}$. The solid line superimposed on the data (filled symbols) denotes the fit according to equation (37) consisting of the contributions from the protein center-of-mass diffusion associated with the Lorentzian linewidth γ (narrow dash-dotted line), the internal molecular relaxations associated with the Lorentzian linewidth $\gamma + \Gamma$ (broader dashed line), and the contribution from the solvent water (D_2O). The latter has been experimentally determined from a pure D_2O sample (open symbols) and fitted with a Lorentzian with the linewidth γ_{D_2O} (solid line superimposed on the open symbols). The narrow line with the dot symbols marks the instrumental resolution. The three Lorentzian contributions are displayed subsequent to the convolution with this experimental resolution. See text for further explanations. (Figure reprinted with permission from Ref. [70]. Copyright 2015 owned by the authors.)

FWHM) and the associated fit with equation (37) (solid line) is displayed in figure 7. The figure additionally depicts the separately measured pure solvent (D_2O) spectrum (open symbols). On the observation energy range of this backscattering spectrometer, all dynamic components modeled by equation (37) become well visible: The global motion of the proteins leads to a narrow Lorentzian that is significantly broader than the resolution linewidth. The internal molecular motions inside the protein cause a significantly broader Lorentzian that is nevertheless narrower than the third Lorentzian associated with the solvent water. Section 2.5 will in detail address the different steps in the data analysis and the numerical methods leading to fits as exemplified in figure 7. Here we note that the q-dependence of the width of the narrow Lorentzian $\mathscr{L}(\gamma(q),\omega)$ in equation (37) of a typical protein solution such as that of bovine serum albumin or γ-globulin is consistent with Brownian diffusion (equation (20), see figure 9, right), i.e., $\gamma(q) = D\, q^2$, in perfect agreement with similar observations on colloidal hard sphere suspensions. Importantly, this result is

return confidence limits on the obtained fit parameters. For an efficient use of these algorithms, it is useful to program a user interface that can translate regular string expressions entered on the interface level into the code that is interpreted by the fit algorithm. In this way, the model function can easily be modified, and various models can be tested. Moreover, although in most cases it is more interesting to not impose any known q-dependence on a fit of neutron spectra, in some cases it is useful if the fit algorithm nevertheless allows for the option to include a known q-dependence (e.g., for implementing the two-state switching diffusion model briefly mentioned in section 2.4.2). In this case, the entire spectral matrix $S(q, \omega)$ needs to be fitted at once as opposed to a sequential fit of the individual spectra $S(q = q_i, \omega)$ for all given q_i. If the q-dependence is not imposed in the model *a priori*, for instance the simple Brownian diffusion for the global motion of proteins in solution is confirmed without any bias, as shown in figure 9. By contrast, if the Brownian global diffusion can be reasonably assumed to be confirmed, it can be fixed in a model including the q-dependence, and in this way the number of free fit parameters can be reduced to obtain more accurate information on the internal motions of a macromolecule in solution. In some cases, a model to describe complex neutron spectra is not *a priori* known. In this case, in addition to the "classical" spectral fitting by least-squares minimization, the methods of inverse Laplace transform analysis known as CONTIN analysis [83] and Bayesian inference are being employed [84]. In other cases, an analytical implementation of the model function in terms of the scattering function $S(q, \omega)$ may be significantly more difficult than an implementation in terms of the intermediate scattering function $I(q, t)$ (e.g., for equation (35)). In this situation, a numerical Fourier transform needs to be provided by the data analysis software when fitting a model to a measured $S(q, \omega)$.

Prior to the fits, the raw neutron data need to be reduced, which is in general done by software provided by the neutron scattering facility. These steps include the weighting of the measured intensities by the individual detector efficiencies, the subtraction of the empty can and sample environment contributions to the scattering signal, the grouping of data, the calibration of the energy axis, and in some cases absorption and self-shielding corrections. The latter corrections take into account that the empty sample container and empty sample environments scatter more neutrons than the same containers and environments when loaded with a sample, because the sample already scatters part of the beam that is then no longer available to illuminate the container or environment. These corrections also take into account that the scattering and absorption by a sample may depend on the direction of travel of a scattered neutron. Finally, multiple scattering effects may need to be taken into account, but ideally a neutron scattering environment is designed to reduce the effect of multiple scattering as far as possible by using samples with a scattering probability of less than 10 %, i.e., a double-scattering probability of less than 1 %. The data file formats and software for treatment are increasingly becoming standardized across different neutron scattering facilities using *Nexus* / *hdf* file formats and modern treatment tools such as *Mantid* [85]. The possibility to record data in the so-called event mode storing detailed information on the time dependence of a data acquisition rather than just integral detector counts will greatly help in particular experiments on fragile biological samples. Besides "classical" fit methods, some modern neutron data treatment tools including *Mantid* [85] already offer the above-mentioned Bayesian inference methods for advanced data analysis [84].

2.5.2 Modeling of the Solvent Water Contribution

Even if heavy water (D_2O) is used as a solvent for "native" (i.e., protonated) biological macromolecular solutions, this solvent water nevertheless in general adds a nonnegligible q-dependent contribution to the measured scattering signal (cf. figures 7 and 9, right). When a backscattering spectrometer such as IN16B at the ILL (Grenoble) [86] or BASIS at the SNS (Oak Ridge) [87] is used to record the comparatively slow motions of the biological molecules of interest, the energy range of these backscattering spectrometers may be too narrow to fully and accurately access the spectral contribution from the fast solvent molecules at physiological or even higher temperatures. In this case it has been proven useful to only determine the amplitude of the water solvent contribution from a measurement of pure water (D_2O) samples on these backscattering spectrometers at the respective sample temperatures. This amplitude information may then be combined with the temperature-dependent spectral linewidth information obtained from measurements of pure water (D_2O) on a time-of-flight spectrometer such as IN5 [88] or IN6 at the ILL. In the fit of equation (37), $\beta_{D_2O}(q)$ is therefore fixed from fit results of the amplitude information obtained from pure water measurements on the respective backscattering spectrometer [70]. This amplitude is scaled according to the excluded solvent volume occupied by the biological macromolecules. The second parameter, namely the width of the water contribution $\mathcal{L}(\gamma_{D_2O}(q),\omega)$ in equation (37), is fixed from the fits to the pure D_2O neutron time-of-flight spectral data [70]. Here it is noted that the time-of-flight spectra from D_2O of course differ significantly from the well-studied H_2O time-of-flight spectra [49]. Importantly, the amplitude and width of the D_2O spectra are strongly influenced by coherent scattering effects near the q-position of the water nearest-neighbor correlation (about 2Å^{-1} near room temperature), resulting in a de Gennes-narrowing of their width at this position [89].

2.5.3 Separation of the Rotational and Translational Diffusions

In this section we explain the separation of the rotational D_r and translational D_t diffusions of proteins or other macromolecules in liquid solutions based on recently developed analysis frameworks based on an implicit function

$D = D(D_r, D_t)$ linking D_r and D_t to the experimentally observable quantity D [15, 19, 21, 70], i.e., the apparent diffusion coefficient explained in section 2.4. For further reading we in particular refer to Grimaldo et al. [70], which is an open-access article. Because of being experimentally observable, D is often also denoted apparent diffusion coefficient. We stress that D is the quantity that is directly obtained from the width γ in equation (37). The incoherent scattering function due to translational diffusion can be written as

$$S_{\mathrm{inc}}^{\mathrm{T}}(q,\omega) = \mathscr{L}(D_t\, q^2,\omega), \tag{44}$$

and the scattering due to rotational diffusion reads

$$S_{\mathrm{inc}}^{\mathrm{R}}(q,\omega) = \sum_{l=0}^{\infty} B_l(q)\; \mathscr{L}[l\,(l+1)\, D_r,\omega], \tag{45}$$

where

$$B_l(q) = (2\,l+1) \int \rho_H(r)\, j_l^2(q\,r)\, \mathrm{d}r \tag{46}$$

with the radial hydrogen density distribution function $\rho_H(r)$ and the spherical Bessel function j_l of the first kind and order l. In equation (44) we have imposed that in the case of Brownian diffusion the associated Lorentzian line-width is proportional to q^2. The convolution of the translational and rotational contributions then reads

$$\begin{aligned} S_{\mathrm{inc}}^{\mathrm{TR}}(q,\omega) &= S_{\mathrm{inc}}^{\mathrm{R}}(q,\omega) \otimes S_{\mathrm{inc}}^{\mathrm{T}}(q,\omega) \\ &= \sum_{l=0}^{\infty} B_l(q) \mathscr{L}(\Gamma_l,\omega), \end{aligned} \tag{47}$$

where $\Gamma_l = l(l + 1)D_r + q^2 D_t$. The fit of the single Lorentzian with the width $\gamma = \gamma(q)$ describing the apparent diffusion coefficient D to the experimental spectra is equivalent to writing

$$\min_{\alpha,\gamma} \left\{ \int \left[S_{\mathrm{inc}}^{\mathrm{TR}}(q,\omega) - \alpha\, \mathscr{L}(\gamma,\omega)\right]^2 \mathrm{d}\omega \right\}, \tag{48}$$

that is, the minimization of a norm with respect to the parameters α and γ. Here we note that also α depends on q. Solving equation (48) leads to the conditions [70]

$$\alpha = 2\gamma \sum_{l=0}^{\infty} \frac{B_l}{\Gamma_l + \gamma}, \quad \text{and} \tag{49}$$

$$0 = \sum_{l=0}^{\infty} \frac{B_l}{\Gamma_l + \gamma} \left(\frac{1}{\Gamma_l + \gamma} - \frac{1}{2\,\gamma} \right). \tag{50}$$

By assuming that $\gamma = Dq^2$, i.e., a Brownian-type global diffusion, and rearranging equation (50), we obtain [70]

$$\sum_{l=0}^{\infty} B_l(q) \frac{l(l+1)D_r + q^2(D_t - D)}{[l(l+1)D_r + q^2(D_t + D)]^2} = 0. \tag{51}$$

In practice, we approximate equation (51) by a finite sum

$$\sum_{l=0}^{n} B_l(q) \frac{l(l+1)D_r + q^2\left(D_t - D^{(n)}\right)}{[l(l+1)D_r + q^2(D_t + D^{(n)})]^2} = 0, \tag{52}$$

where $D^{(n)} = D^{(n)}(q, D_r, D_t)$ is the approximate apparent diffusion coefficient. $D^{(n)}$ in general reasonably quickly converges to D with increasing n, because $B_l(q)$ strongly decays for $l > q\tilde{r}$, where $\tilde{r}$ is the largest radius with nonvanishing hydrogen proton density of the diffusing macromolecule. In practice, with n on the order of 100 for a globular protein such as BSA, a good convergence is achieved. We remark that neutron backscattering explores rather large scattering vectors, where $D^{(n)}$ only very weakly depends on q. Other types of experiments may access a q-range where the situation becomes more complicated. Associated illustrations for the situation at small q-values for incoherent scattering can be found in Grimaldo et al. [70], and the situation in the case of neutron spin-echo spectroscopy is explained by Biehl et al. [72]. Equation (52) can be used to obtain either of the three quantities D, D_t, D_r given that the other two are known. When only D is experimentally accessible, the usual approach is to assume the theoretical short-time diffusion for charged colloidal spheres $D_r = D_r(0)\,(1 - 1.3\varphi^2)$ [41] with the dilute limit rotational diffusion $D_r(0)$ to be valid at least for a globular protein and use this relation in equation (52) [15]. Finally, for proteins the radial hydrogen density distribution $\rho_H(r)$ entering equation (46) can be calculated rather easily from the associated protein data bank (pdb) file.

2.5.4 Molecular Dynamics Simulations

Neutron spectroscopy experiments addressing biomolecular dynamics are often complemented by molecular dynamics simulations, and here we refer to Chapter 6 in Volume 44 of this book series [11] for a dedicated introduction to the field. A particular strength of neutron spectroscopy in this context consists in the unique access to the single-particle dynamics as a function of reciprocal spatial coordinates, permitting a direct testing of predictions from simulations on the single-particle level. We note that with software packages such as nMOLDYN [90, 91], extremely useful tools are readily available to extract the dynamic scattering function $S(q, \omega)$ from the simulated atom trajectories. One general advantage of simulations is that the impact of particular "components" of a biomolecule such as a chosen side chain of a protein can in principle be selectively "switched off" in the simulation, and the resulting impact on the overall scattering function can be observed. Such type of "reverse engineering" can help to better understand and improve the suitable model scattering function for the experimental data. Moreover, simulations can easily distinguish between collective- and self-dynamic effects that in the experiment would require to distinguish coherent and incoherent scattering, which is not often easily possible. Molecular dynamics simulations of biological systems are, however, particularly limited by the large size of the "simulation box" that is required to describe the generally large and complex biomolecules. When performing fully atomistic simulations as opposed to so-called coarse-grained simulations, and when using typical simulation time steps on the order of a femtosecond, obviously, the larger the required total time described by the simulation, the larger the computational cost. It is therefore the slow nanosecond motions explored by high-resolution neutron spectrometers that are most difficult to simulate. Present state-of-the-art simulations can successfully describe protein motions from the atomic level of internal fluctuations up to the fluctuation of entire subdomains [92]. The latter protein subdomain fluctuations on nanosecond time scales and beyond, which are important for example in the enzymatic activity, can be experimentally tested using neutron spin-echo spectroscopy [71].

2.6 SPECTROMETERS TO STUDY BIOLOGICAL DYNAMICS

2.6.1 Types of Spectrometers for Biological Dynamics

For systematic studies of the dynamics of biological systems, very often, experiments are carried out on a combination of several different types of neutron spectrometers to cover a large dynamic range. Suitable instruments have been presented in Chapter 6, Section 6.4, of Volume 44 within the present book series [11]. The most frequently used spectrometers comprise neutron time-of-flight spectrometers, neutron backscattering spectrometers, and neutron spin-echo spectrometers, all of which have been introduced to some extent in the preceding chapters of this book series, and notably in Chapter 3 of Volume 44 [81]. Sometimes, but rather rarely, polarization analysis is added to the time-of-flight technique in order to discriminate coherent from incoherent scattering, and one of the suitable spectrometers for that purpose is D7 at the ILL. In the absence of polarization analysis, both backscattering and time-of-flight spectrometers record the total, i.e., coherent and incoherent, scattering from the sample. In the case of nondeuterated biological samples, the incoherent scattering from the hydrogen ^1H usually largely dominates the signal. A large uninterrupted energy range can be accessed by combining exact backscattering spectrometers such as IN16B at the ILL with an energy resolution of better than 1 μeV FWHM and a maximum energy transfer of approximately 35 μeV, near-backscattering spectrometers such as BASIS at the SNS with an energy resolution of about 3.5 μeV and a maximum energy range of 100 μeV, and time-of-flight spectrometers with an elastic energy resolution of typically about 90 μeV and an in principle unlimited energy range on the neutron energy gain side (with, however, decreasing resolution with increasing energy). While neutron backscattering and time-of-flight experiments are conceptually similar by recording data on an energy axis and accessing a similar q-range, the neutron spin-echo technique [93] is different in that it accesses the intermediate scattering function $I(q, t)$. Moreover, spin-echo spectrometers such as IN15 [94] at the ILL usually explore a q-range commensurate with the range accessed by small-angle scattering instruments, and the spin-echo technique is essentially a coherent scattering technique based on polarization analysis. On protein solutions, the spin-echo experiment often permits to measure on a q-range from below to above the colloidal structure correlation peak, and to identify the peak in the coherent scattering. It is mostly assumed that at q-values far above the q-position of the structure factor maximum, the observed diffusive dynamics to a good approximation can be considered as self-diffusion. By contrast, on the correlation peak position, the spin-echo technique accesses the collective diffusion.

Backscattering and time-of-flight spectrometers are mainly used to study quasi-elastic scattering from biological samples, thus exploring relaxational and diffusive dynamics. Time-of-flight spectrometers may in addition be used to study inelastic scattering arising from vibrational dynamics. Such vibrational dynamics in biological samples is not in the main focus of the present chapter, because the biophysical aspects of vibrational excitations here are less different from aspects of general chemical spectroscopy, but we note that neutron spectroscopy can very helpfully complement infrared and Raman spectroscopy techniques, inelastic X-ray scattering as well as Terahertz spectroscopy with electromagnetic radiation. In all cases, neutron spectroscopy profits from the absence of selection rules and the sensitivity to the hydrogen atom. Moreover, of course, neutron vibrational spectroscopy also does not cause beam damage as opposed to, e.g., Raman and X-ray spectroscopy. Vibrational dynamics can be explored using neutrons in an energy range from on the order of 100 μeV using cold neutron time-of-flight spectrometers and up to on the order of 1eV using vibrational spectrometers on thermal and hot neutron sources. For further reading on neutron spectrometers, we also refer to Chapter 3 in Volume 44 of this book series [81].

2.6.2 Types of Measurements

The standard acquisition mode on a time-of-flight or backscattering spectrometer is to record a full spectrum $S(q, \omega)$ depending on the energy ω and scattering vector q, where it is noted that of course technically, the raw data constitute intensities depending on the scattering angle 2θ and on the neutron flight time in the case of time-of-flight spectrometers or a related parameter such as the Doppler shift in the case of an exact backscattering spectrometer. $S(q, \omega)$ is subsequently obtained by applying the suitable reduction and interpolation steps. On time-of-flight spectrometers, very often the incident wavelength can be freely tuned over a wide range, such as IN5 at the ILL, which can access wavelengths from 2.8 to 20Å. The q-range of course varies accordingly, since the range in the accessible scattering angle 2θ remains fixed. Moreover, the incident flux strongly depends on the chosen wavelength due to the Maxwellian spectrum of the neutron source, the wavelength-dependent neutron guide filling, and the setting-dependent chopper transmission. On backscattering spectrometers, the elastic and inelastic fixed-window scans provide additional types of data acquisition that were mainly developed for saving neutron beam time. In the elastic fixed-window method, $S(q, |\omega| \leq \Delta\omega)$ (cf. equation (6)) with the energy resolution $\Delta\omega$ of the spectrometer is measured by not scanning the incident energy and instead illuminating the sample with a static monochromatic beam. On an exact backscattering spectrometer located at a continuous neutron source, this type of scan dramatically increases the effective flux at the elastic energy, i.e., at identical incident and analyzed energies, because the entire continuous incoming flux is applied. By contrast, on backscattering spectrometers located at pulsed neutron sources, there is no equivalent gain in efficiency from such elastic fixed-window scans, because the suitable incident energy only arrives "once per pulse" at the sample. At backscattering spectrometers with monochromators that are tunable either by a mechanical Doppler-shifting or by a temperature-controlled variable crystal lattice spacing, so-called inelastic fixed-window scans [95] may in addition be applied. In this case, the spectrum $S(q, |\omega - \omega_0| \leq \Delta\omega)$ from the sample is measured at a fixed offset energy $\omega_0 \neq 0$. Technically, this offset scan is achieved in the case of a mechanical Doppler shift by moving the monochromator crystal at a constant speed along the crystal normal axis for most of its mechanical path and then "turn it around quickly" into its return path, equally moving at constant speed for most of the path length. This type of "chain saw" velocity profile replaces the standard sinus velocity profile applied when continuously scanning the incident energy. This inelastic fixed-window technique can be used to efficiently access general information on the level of confinement of a diffusive process [96]. In both the elastic and inelastic fixed-window methods, the data are typically recorded while continuously scanning the sample temperature. The varied sample property of interest may of course also be a different parameter such as the pressure or time (if the sample slowly evolves) or (rarely in biological systems) the magnetic field.

2.6.3 Recent Advances in Neutron Optics

Details on progress regarding neutron sources and instruments can be found in other chapters of this book series. Notably, neutron sources are addressed in

Chapter 2 of Volume 44 [98], and instruments for spectroscopy are presented in Chapter 6, Section 6.4, of Volume 44 [11]. The present section is dedicated to some innovations that specifically help the investigation of (biological) molecular dynamics using spectrometers with a high energy resolution. Remarkably, recently commissioned neutron spectrometers such as for instance IN16B [86, 99] at the ILL and BASIS [86, 100] at the SNS provide a neutron flux at the sample that is by at least a factor of 10 higher than at conceptually similar predecessor instruments. This increase in performance will contribute significantly to the field of the dynamics of biological systems. IN16B and BASIS represent examples of spectrometers optimized for continuous neutron sources (ILL) and pulsed neutron sources (SNS), respectively, and complement each other in the accessible dynamic ranges. Neutron spectrometers are characterized by moving neutron optical elements that operate at the technological limit for mechanical load, including fast spinning chopper disks, velocity selectors, and oscillating Doppler drives. Some of the recent progress is owed in part to improvements of these devices, and also importantly to the improvements of neutron guide coatings, guide layout, larger and better monochromator crystals, and a complete in-vacuum operation of the spectrometers.

An important general observation regarding cold or thermal neutron spectrometers is one of scale. Since it takes a length on the order of 10 cm to moderate a fission or spallation neutron from its initial energy on the order of 2MeV in the case of fission to the required thermal energy on the order 2meV [101], the length of 10 cm can be seen as setting the typical source size, and thus initial beam size which is transported to the spectrometer. As a result, a typical neutron guide has an initial cross-section of 12 cm height by 6 cm width, where the optimal use of angular space of a neutron facility in the horizontal laboratory plane superimposes on the otherwise isotropic shape of the neutron source. At a wavelength of 6Å, the neutron beam transported through a typical neutron guide (with a coating corresponding to twice the critical angle of total reflection of Nickel) has an angular divergence of $\pm$ 1.2°. For incoherent neutron scattering experiments, a beam divergence of $\approx\pm 4°$ is acceptable, since no distinct structural features need to be resolved through the q-dependence of the scattering signal. This allows to implement focus optics with a beam compression ratio of about 3 to 4, such that the beam size at the sample is approximately 3 cm high by 3 cm wide. To let a sample of that size appear nearly point-like with respect to the secondary spectrometer and to achieve a good resolution of the neutron flight time, the typical neutron flight path length in a secondary spectrometer (i.e., the part of the instrument behind the sample) has to be at least approximately 2 m (in the case of a backscattering spectrometer this value applies for each direction). The resulting large size of the primary spectrometer (needing space for the focus optics, choppers and other items) and of the secondary spectrometer results in a large space requirement of the spectrometer within the neutron scattering facility.

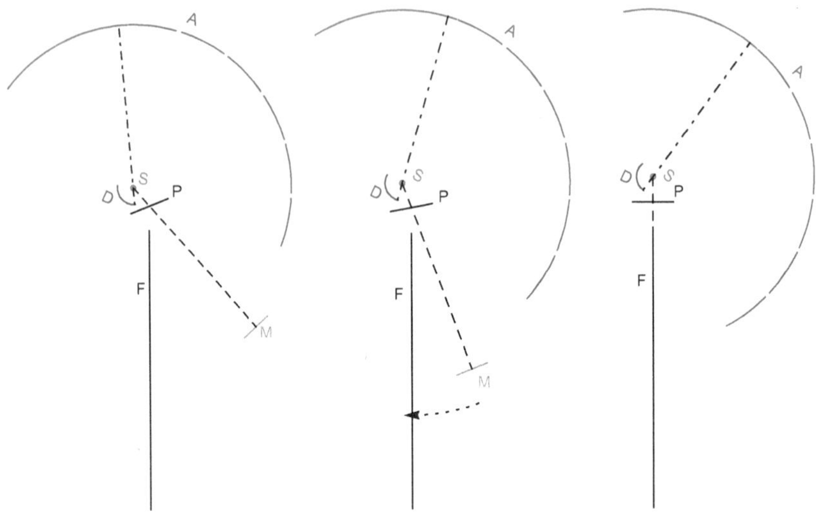

FIGURE 10 Schematic top view of the "BATS-transformation" [97] illustrating the change from a backscattering spectrometer with a monochromator in exact backscattering toward a time-of-flight-backscattering spectrometer ("BATS" abbreviates "Backscattering And Time-of-flight Spectroscopy"). A typical backscattering spectrometer at a continuous neutron source such as a reactor defines the incident neutron energy by a single-crystal monochromator M in exact backscattering (left image). The neutrons travel through the focus guide F via a phase-space transformation (PST) chopper P toward M, and the monochromatic neutrons return along the incident path (dashed line) toward the sample S. Neutrons are scattered by S toward the analyzers A, and neutrons with the analyzed wavelength may return from A to the detectors D. When transforming to a backscattering spectrometer using a pulsed neutron source, which can be a spallation source or a reactor source combined with pulse choppers, the spectrometer is rotated around the PST chopper pivot axis (center image), and the monochromator M is removed. The resulting "indirect-TOF" setup (right image) uses an incident beam where the incident wavelength toward S varies as a function of time due to the spectral broadening with travel distance from the source of initially short neutron beam pulses. In this "indirect-TOF" setup, the PST chopper P becomes a mere pulse-shaping chopper and no longer reflects neutrons. (Figure rendered in MATHEMATICA [9].)

In some cases the secondary spectrometer also needs to be able to move relative to some parts of the primary spectrometer to allow for different instrument configuration options such as different analyser setups to be implemented, thus further adding to the space requirements, cf. figure 10. The space requirement as well as the wavelength setting through flight time from a pulsed (spallation) source or pulse choppers for some instruments [87, 97] results in a large distance of the spectrometer from the neutron source. This distance requirement of typically on the order of 50 to 100m from the neutron source has resulted in the implementation of so-called ballistic neutron guides which reduce the neutron transport losses over the distance [102, 103]. The large distance from the source of modern high-resolution spectrometers also helps to reduce the background induced by fast neutrons and γ-radiation from the neutron source.

Focus optics was implemented in so-called second-generation spectrometers [104] such as the backscattering instruments IN16 (ILL) or HFBS (NIST) as well as most modern time-of-flight spectrometers. This focus optics has allowed in particular to match the vertical beam divergence at the sample to the requirements of an efficient trade-off of flux and angular resolution. In highest resolution neutron backscattering, the horizontal beam divergence has been fully increased toward this efficiency requirement only recently by using the technologically extremely challenging phase-space transformation (PST) technique [105, 106] implemented by the reflection of the incident neutron beam from a fast moving mosaic crystal mounted on a chopper disk with a circumferential speed of approximately 230m/s. The PST increases the flux at the sample by achieving a "compression" of the incident neutron beam in energy at the expense of the above-mentioned increase of the beam divergence. The PST is therefore superior to a mere focus technique which does not affect the energy distribution of the neutron beam. All of the above considerations regarding the source size and reasonable beam divergence at the sample position for a spectrometer with a high energy resolution imply that the sample for this type of experiment should have a cross-section of approximately $3 \cdot 3\,\mathrm{cm}^2$. This requirement obviously poses a challenge to experiments on biological samples, which are often available in very limited quantities only. More strict beam divergence criteria apply to spin-echo spectrometers, which in general require an access to smaller scattering angles as opposed to backscattering and time-of-flight spectrometers as well as an accurate definition of the neutron beam paths through the magnetic fields. Therefore, focus optics is in general not applied for these instruments, but the beam size delivered to the sample is similar.

Reactor-based backscattering spectrometers (figure 10, left) typically use a Doppler monochromator crystal moving according to a periodic velocity profile to set and tune the incident wavelength. The maximum energy transfer range in this case is limited by the maximum speed of the crystal, which is nowadays about 4.7m/s for a typical monochromator mass including its carbon fiber support of 1kg. When using a Si(111) crystal as monochromator (wavelength 6.27Å), this speed translates into a maximum energy transfer range of about 35 μeV. The maximum energy range on backscattering spectrometers using a pulsed neutron source is limited by the condition of no overlap between the scattering from the individual source pulses in the detectors of the spectrometer. For instance, BASIS is routinely operated with a maximum energy transfer of 100 μeV. Along with the different designs of continuous-time source and pulsed-source backscattering spectrometers (figure 10), different energy resolutions are achieved. The exact backscattering geometry of a continuous-source backscattering spectrometer allows for the higher resolution. Using Si(111) crystals as monochromator and analyzers, a standard Gaussian resolution of 0.75 μeV FWHM is obtained on IN16B. This resolution can be further improved by not bending the crystals to follow their spherical supports. In this case a resolution of approximately 0.4 μeV FWHM is obtained, which, however, in this case contains a noticeable Lorentzian component in its line shape. IN16B can also be configured to use Si(311) crystals corresponding to a center incident wavelength of 3.28Å. In this configuration, the maximum energy transfer range, resolution linewidth, and q-range approximately double compared with the standard Si(111) configuration. Finally, a configuration using GaAs(200) single crystals (wavelength 5.65Å) is foreseen for the more distant future on IN16B, which is expected to dramatically enhance the energy resolution (by up to a factor of 10, obviously at the expense of neutron flux). The BASIS spectrometer achieves a resolution of approximately 3.5 μeV FWHM [87] using Si(111) crystals, which significantly differs from a single Gaussian line shape. This deviation of the resolution from the Gaussian line shape can be seen with a closer look at figure 7. However, its resolution function can be well described analytically by a sum of several Gaussians.

2.6.4 Practical Aspects of Experiments on Biological Samples

Neutron spectroscopy experiments on biological samples rarely require extreme temperatures, neither below 1K nor above 400K. They also do not normally require extreme magnetic fields, if any. They do occasionally require very high pressures. However, the most important requirement generally is the stability of the environmental parameters near ambient conditions, i.e., near ambient temperature, pressure, and for instance humidity. These requirements are not necessarily trivial to fulfill, since it is often near ambient conditions that proportional-integral-differential (PID) feedback algorithms are most difficult to adjust.

In most cases, sample containers made of aluminum (Al) are suitable for neutron spectroscopy experiments when the neutron wavelength is long enough to not excite the Al Bragg peaks. The most frequently used wavelengths $\lambda \geq 5$Å therefore do not pose any problems. Pure aluminum is, however, extremely difficult to be machined, and the ideal alloy should contain few elements that become strongly radioactive in the neutron beam. A good experience has been made with an alloy that contains silicon as the main additive element and nearly no other impurities besides Si. Another advantage of Al containers is that they can be well sealed with Indium wire and used at cryogenic temperatures. When the samples are corrosive, which can also occur in biological samples (e.g., in lysozyme protein solutions), the aluminum containers may be coated with gold, which increases the absorption and background scattering, but is by experience better than an anodization of the Al surfaces. The latter creates an even higher background scattering. Alternatively, a thin Teflon coating of Al sample holders may also be considered. Al sample holders are particularly suitable for large scattering vectors while quartz cells are ideal for very low scattering vectors. Such small scattering vectors are often explored in neutron spin-echo experiments. By contrast, quartz scatters strongly at large scattering vectors and is therefore not useful in backscattering or time-of-flight spectroscopy experiments. Al sample holders can be easily sealed with Indium wire and are therefore compatible with a surrounding vacuum, which is required for most standard sample environments on neutron spectrometers. Moreover, in this case the sample holders are in addition compatible with extremely low temperatures, which is occasionally required also in biological studies.

An advantage of neutron spectroscopy experiments consists in the possibility to measure on opaque samples as opposed to dynamic light scattering experiments. Moreover, unlike in light scattering, neutron scattering is in general not severely affected by small impurities of dust in liquid sample solutions of biological macromolecules. A disadvantage, however, is the large amount of sample material required. For instance, a pure protein sample should have a mass of 50 to 100mg for a neutron spectroscopy experiment. If a protonated, i.e., "native" protein is studied in an aqueous (D_2O) solution, at least 50mg of the protein should be available per sample for a dissolution in on the order of 1ml of D_2O. Here it is noted that NMR experiments on standard laboratory devices need similar amounts of sample material. Another advantage of neutron spectroscopy experiments over some other types of spectroscopy experiments is a relatively easy access to the sample environment for a combination with other *in situ* characterization or manipulation methods, allowing for instance in principle to carry out dielectric spectroscopy or optical spectroscopy or optical stimulation *in situ* during a neutron scattering experiment [107]. Of course, the external access option can also be used to apply pressure to samples, which is of significant current interest in biology [34]. Moreover, samples may be *in situ* humidified in a quantitatively controlled way or be exposed to a tensile strain, which is useful for biological fiber samples (section 2.8.5).

A backscattering spectrum on a new instrument such as IN16B or BASIS requires a counting time of typically two to four hours. This constitutes a significant reduction compared to spectrometers from the preceding generation and allows to explore a larger parameter space during a typical experiment with a duration of three to five beam days. A time-of-flight spectrum on IN5 at an incident wavelength of 5Å on a typical proton-rich biological sample may require even less counting time on the order of one hour. On spin-echo spectrometers of the type of IN15, $I(q, t)$ is measured sequentially at each q, which is required for the accuracy of the magnetic fields of the spectrometer, thus allowing for the highest possible accessible echo times. This requirement of course affects the counting time per sample on this type of instrument.

2.7 NEUTRON SPECTROSCOPY IN THE CONTEXT OF COMPLEMENTARY METHODS

In this section a few very brief notes have been compiled to provide an overview in which way neutron spectroscopy complements or is complemented by other methods frequently applied to biological macromolecules. It is not the purpose of these brief notes to reach any level of comprehensiveness. Moreover, there is no room to even mention all complementary methods, including for example dielectric spectroscopy and Terahertz spectroscopy. Remarkably, some of the complementary methods can in principle be combined *in situ* with neutron spectroscopy, such as for instance Raman scattering [108] and dielectric spectroscopy. In some cases, as briefly mentioned before, it is possible to combine complementary methods and the control of several sample parameters at once *in situ* with a neutron spectroscopy experiment. For instance, neutron spectroscopy can in principle be performed simultaneously *in situ* with a combination of high pressure and temperature control and dielectric spectroscopy, or with a tensile strain apparatus and humidity control.

2.7.1 Dynamic Light Scattering (Visible and X-Ray Photons) and Other Photon Methods

Dynamic light scattering (DLS), be it using visible or X-ray photons, is a coherent scattering technique that requires the coherent illumination of a macroscopic sample volume commensurate with many particle distances of a colloidal suspension such as a protein solution. DLS is based on the fluctuation of the speckle pattern arising from the interference of light undergoing different phase shifts while traveling along different paths within the sample. From the scattered light intensity recorded at a given detector position as a function of time, the time-autocorrelation function is calculated to obtain information on the sample dynamics. Therefore, DLS is also known as photon correlation spectroscopy (PCS), or XPCS in the case of an X-ray photon source. Both XPCS (for geometrical reasons of the speckle size [109]) and even more so visible-light DLS are restricted to small scattering vectors compared to neutron spectroscopy. Moreover, photon correlation techniques are by their very nature "photon-hungry" techniques, because a large number of photons per unit time needs to be detected to build a correlation function. This photon statistics requirement sets the fastest detectable time scale. When a synchrotron or free electron laser X-ray source is used for XPCS, the time structure of the source sets another limit (which is given by the electron bunch spacing typically on the order of a nanosecond in the case of a synchrotron and may in principle be overcome by so-called delay-line techniques [110]). It further results from the nature of photon correlation techniques that they cannot directly access incoherent scattering and therefore cannot provide information on self-diffusion. Nevertheless, macroscopically coherent photon scattering techniques can in principle access a type of information related to the information on local order in long-range disordered materials. This type of information arising from the concept of the elastic incoherent structure factor in neutron spectroscopy is obtained via higher-order correlation functions of series of "snapshots of speckles" in the case of coherent X-rays [111, 112]. Notably, photon correlation spectroscopy techniques display some distant conceptual resemblance to neutron spin-echo spectroscopy. In the latter method, the transverse coherence of a photon beam is replaced by the polarization analysis of the neutron beam.

Vibrational spectroscopy using inelastic neutron scattering can be complemented by inelastic X-ray scattering as well as by infrared and Raman spectroscopy using visible or infrared light. For these photon-based methods, the associated charge fluctuations and polarization effects have to be taken into account, i.e., unlike in the case of neutrons, the access to the dynamics of the atoms is not direct [113].

It is noted that obviously, all X-ray photon scattering experiments on biological samples are concerned by the issue of radiation damage inflicted on the samples.

2.7.2 Nuclear Magnetic Resonance

Nuclear Magnetic Resonance (NMR) is not a scattering technique and accesses angular rather than spatial correlation functions, but displays some formal resemblance to incoherent neutron scattering [114]. From a practical perspective, NMR differs strongly from incoherent neutron spectroscopy in terms of the accessible length and time scales. The accessible time scales by NMR in general are governed by the Larmor frequency which is proportional to the applied magnetic field. For the strongest readily available magnetic fields (on the order of 10 Tesla), the Larmor frequency of ^{1}H nuclei is of the order of 1 Gigahertz. NMR has been enhanced by so-called diffusion-ordered spectrometry (DOSY) [115], and DOSY-NMR is for instance applied to filter out the solvent water contribution by making use of the resonance fine splitting. NMR displays the general advantage of being sensitive to several other nuclear isotope contrasts of biological interest beyond the H/D-contrast mostly exploited in neutron spectroscopy.

The center-of-mass diffusion of macromolecules in solution can be explored by the pulsed-field gradient NMR (PFG-NMR) technique which displays some remarkable conceptual resemblance to the neutron spin-echo spectroscopy technique (and was actually invented before NSE [93, 116]). PFG-NMR typically accesses time scales from milliseconds to seconds (related to the spin-lattice relaxation time) and length scales down to micrometers [117]. Due to the large observation length scale, the method is sensitive to turbulences in the sample. PFG-NMR is the method of choice to observe the long-time self-diffusion of proteins in solution.

2.7.3 Fluorescence Correlation Spectroscopy

Fluorescence correlation spectroscopy (FCS) needs to be mentioned as a technique that can access the self-diffusion, or—synonymously and in this case quite literally—the tracer diffusion of proteins, where in terms of colloidal diffusive time scales the observation scale corresponds to the long-time limit. FCS is a classical tracer technique that can be combined with extremely high-resolution optical microscopy. However, due to the requirement to place fluorescent labels to the proteins or other macromolecules of interest, it is an invasive method as opposed to neutron spectroscopy [118].

2.8 APPLICATIONS

Most of the example sample systems that will be discussed in terms of neutron spectroscopy in the following sections have already been addressed from the perspective of structural investigations in the preceding chapter on biological structures in the present volume of this book series. Some more general aspects of membrane vesicles and thin membrane films have also been addressed in the context of large-scale structures in Chapter 5 of Volume 44 [119]. Notably, protein solutions as well as membrane vesicle solutions have been subject to intense investigations using neutron small-angle scattering. For protein solutions, the static small-angle scattering experiments contribute important information on the protein-protein interaction potentials as well as on the shape of the proteins. This information is often essential to interpret the dynamic information from the neutron spectroscopy experiments. Planar lipid membranes have been studied using neutron reflectivity experiments, and biological fibers have been subject to fiber diffraction experiments, all of which have been presented in the previous chapter. Here we note that wide-angle diffraction can with some limitations due to the generally larger beam divergence sometimes be recorded *in situ* along with the dynamic information on neutron spectrometers, which can be very useful when exploring oriented membranes or fibers [26, 28].

The technique of the elastic fixed-window scans to be discussed in the following section in the context of hydrated protein powder samples has already been mentioned in a more general context in Chapter 6, Section 6.8.4, of Volume 44 [11] in this book series.

2.8.1 Proteins as Hydrated Powders

The roles of the proteins and their hydration water are closely interconnected and have been subject to experimental studies using neutron, nuclear magnetic resonance, and Terahertz spectroscopy [120]. In particular, the molecular dynamics of hydrated protein powder samples as a function of temperature has been studied rather comprehensively using neutron spectroscopy. The first studies of this kind date back to the late 1980s [18] and were inspired by earlier work on polymeric glass formers (e.g., Frick et al. [121]). These samples were generally prepared by storing lyophilized powders of proteins in atmospheres with a well-defined humidity level to adjust an accurate humidity of the samples (typically on the order of 0.4g water per 1g of protein). An important question concerned the role of this hydration water for the biological function of the proteins as manifested in the hydration- and temperature-dependent molecular mobility. In these studies, elastic fixed-window scans as a function of the sample temperature were employed as an important tool on backscattering spectrometers (cf. section 2.6.2) in order to obtain the apparent mean-squared displacement (MSD) (equation (7)) of the prevailing hydrogen atoms in these samples. As a general observation, a temperature-dependent so-called dynamical transition [18] with the onset of nonharmonic motions was found for hydrated proteins typically at a transition temperature near 200 K, which is at present still subject to intense discussions as to its fundamental meaning [122, 123]. A difficulty for the interpretation of these data arises from the superposition of vibrational and diffusive contributions in this observed apparent MSD $\langle u^2 \rangle$ (cf. section 2.2.4), but nevertheless the elastic fixed-window method has been proven useful for a direct, efficient, and accurate comparison of different proteins and different hydration levels. Moreover, the effect of a partial deuteration of the samples has been systematically explored. A comprehensive review can be found in Ref. [2].

The interpretation of the MSD data has been carried further by defining the resilience of a protein to the onset of diffusive motions with rising temperature. This resilience can be described in terms of an effective force constant F introduced by G. Zaccai [124] motivated by the dimensional equivalence of the inverse slope of the mean-squared displacement with temperature to a force and given as follows [124]:

$$\langle F \rangle \propto \left(\frac{d\langle u^2 \rangle}{dT} \right)^{-1}, \tag{53}$$

where $\langle u^2 \rangle$ is the observed apparent mean-squared displacement as defined by equation (7). $\langle u^2 \rangle$ in equation (7) is also known as Gaussian approximation (i.e., $S(q,|\omega| < \Delta\omega) \propto \exp\left(-1/3 \langle u^2 \rangle q^2\right)$) and has been used in most studies so far discussing the apparent mean-squared displacement as obtained from backscattering fixed elastic window scans on (hydrated) protein powders. It can be generalized toward a q-dependent MSD $\langle u^2 \rangle_q$ defined by equation (6), which can be written as [12]

$$\langle u^2 \rangle_q = \langle u^2_{\text{vib}} \rangle - \frac{3}{q^2} \log\left(\xi(q)\right), \tag{54}$$

where

$$\xi(q) = A(q)F_1\left(\frac{\gamma}{\Delta\omega},0\right) + [1 - A(q)]F_\beta\left(\frac{\gamma}{\Delta\omega},\frac{\Gamma}{\Delta\omega}\right) \tag{55}$$

is obtained by evaluating equation (3) using equations (4) and (5), and writing the integral expression [12]

$$F_\beta(x,y) = \sqrt{\frac{2}{\pi}}\exp\left(\frac{x^2}{2}\right)\int_0^\infty \exp\left[-\frac{1}{2}(\zeta + x)^2 - |y\zeta|^\beta\right]d\zeta. \tag{56}$$

It is most convenient to fit a Taylor series expansion of equation (54) to experimental results, and further considerations for interpreting the resulting Taylor coefficients can be found in Ref. [12].

Lately, both the extraction of the mean-squared displacement from backscattering fixed elastic window scans as well as the discussion of the internal molecular motions in terms of effective force constants have been carried to aqueous solutions of proteins, where the data analysis becomes significantly more involved [12]. The study by Hennig et al. [12] notably also discusses elevated temperatures and the effect of thermal denaturing of the proteins.

The role of the hydration water for the biological function of proteins remains a current question [32] being presently addressed by both backscattering fixed-window scans and full QENS scans. Hydrated (as well as dry) protein powder samples have also been subject to comprehensive inelastic neutron scattering studies addressing their vibrational dynamics. Among these studies, a low-energy excitation known as the Boson peak already from the field of glass physics has been found and explored in proteins (see, e.g., Ref. [125] and references therein). Such studies are typically carried out on time-of-flight spectrometers such as IN5 and IN6 at the ILL. Deep-inelastic scattering experiments on biological samples profit from new dedicated high-flux vibrational spectrometers such as IN1/Lagrange at the ILL or VISION at the SNS, where first studies on hydrated protein samples have recently been carried out (e.g., Ref. [113]). These inelastic neutron scattering experiments notably complement infrared and Raman scattering experiments.

2.8.2 Proteins in Solution, Crowding, and Cluster Formation

Since inside living cells proteins occur in the aqueous intracellular fluid, it is of tremendous importance to systematically address the dynamics of proteins in aqueous solutions. On the molecular level, both global center-of-mass motions as well as internal molecular motions are superimposed, adding to the difficulty of the analysis of the resulting spectra (cf. section 2.4). Since some years ago, the effects of macromolecular crowding, temperature, pressure, and salt-induced charges on the dynamics of aqueous protein solutions have moved in the focus of interest (e.g., Refs. [15, 21, 24, 34, 75] and references therein). It has been started to very systematically explore the effect of these parameters on *in vitro* model protein solutions employing the recently commissioned most advanced neutron spectrometers.

The first of this set of parameters, macromolecular crowding [57, 58], not only affects the time-averaged structure of a living cell, but importantly also the transport of molecules and signals. Using high-resolution neutron backscattering spectroscopy it has been found that the diffusion of model proteins in aqueous solutions drastically decreases with increasing protein concentration in the solution already in the short-time limit [15, 21], reaching a decrease by a factor of approximately 5 compared with the diffusion in the dilute limit near physiological volume fractions (figure 11). In these studies, the protonated proteins in aqueous (D_2O) solutions served both as tracer

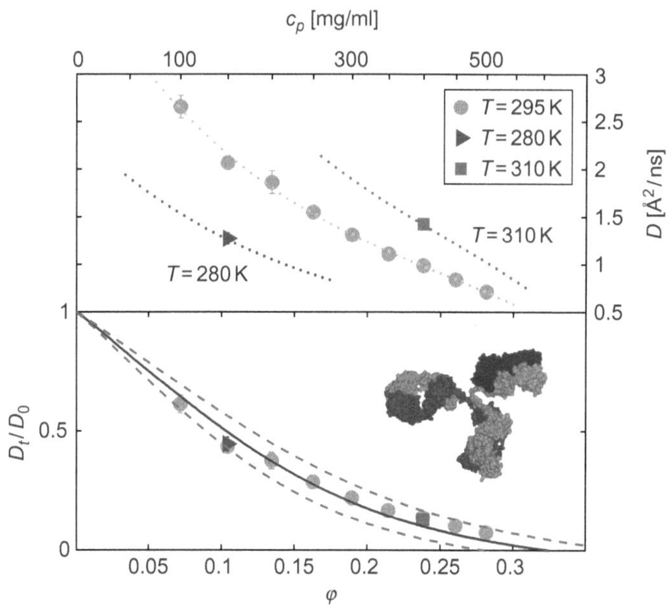

FIGURE 11 Top: Apparent center-of-mass diffusion coefficients D of bovine gamma-globulin proteins in aqueous (D_2O) solutions for different temperatures T and protein concentrations c_p (symbols) obtained from QENS spectra recorded at the backscattering spectrometer IN16B [99] at the ILL. The dotted lines are guides to the eye. Bottom: Translational components D_t of the above center-of-mass diffusion coefficients D separated according to the framework described in section 2.5.3 after normalization to the temperature-dependent dilute-limit diffusion coefficient $D_0 = D_0(T)$ (symbols). The bottom horizontal axis denotes the dry protein volume fraction φ calculated from the protein concentration c_p. The solid line marks the theoretical self-diffusion $D_s^{\text{theo}}(\varphi) = D_t(0) f^{\text{theo}}(\varphi_t)$ of colloidal hard spheres (cf. section 2.3.2), where f^{theo} is the normalized theoretical short-time translational diffusion of hard spheres and φ_t is the theoretical hydrated volume fraction of equivalent effective spheres calculated from the dry volume fraction φ by making assumptions on the hydration shell [21]. (Figure reprinted with permission from Ref. [21]. Copyright 2014 American Chemical Society.)

particles and as crowding agent by tuning the protein concentration c_p (or volume fraction φ) of the solution. Moreover, it has been found that the short-time self-diffusion of the investigated model proteins as a function of the protein volume fraction φ can be understood in terms of models from colloid physics (cf. section 2.3.2) [15, 21]. Hence, the translational short-time self-diffusion $D_t(\varphi, T)$ of these model proteins follows a master curve

$$D_t(\varphi,T) = D_{t,0}(T) f(\varphi), \tag{57}$$

which separates into the product of the dilute-limit translational Stokes-Einstein diffusion coefficient carrying the dependence on the temperature T, and the scalar function f depending solely on φ and understandable in terms of colloid models (cf. equation (25)). Remarkably, the experimentally found master curve also scales as expected with temperature by a Stokes-Einstein prefactor, equation (15), as long as the proteins are in their native (i.e., not denatured) state [75] (cf. figure 11). It is noted that for an accurate comparison with theory, the calibration of the effective hydrodynamic volume fraction occupied by the proteins in the solution including their hydration shell is crucial, and a considerable effort has been made toward the accuracy of the determination of this volume fraction [15]. This determination of the volume fraction works reliably for globular proteins such as *Bovine serum albumin* (BSA). Figure 11 also illustrates the high accuracy with which diffusion coefficients of proteins in solution can be determined on new neutron spectrometers such as IN16B [99] at the ILL.

In the presence of suitable salts, protein solutions can display rather complex phase diagrams [127, 131, 132] comprising regimes where the proteins are fully dissolved as well as precipitate regimes (figure 12). The types of salts inducing such complex phase diagrams are mostly trivalent, such as YCl_3. Inside the precipitate phase region, another closed region displaying a liquid-liquid phase separation (LLPS) [127, 131, 132] can often be observed, where micrometer-sized droplets of dense protein phases are surrounded by a rather dilute solution. When traversing the phase diagram (figure 12) vertically, i.e., along the c_s-axis for a fixed $c_p >> 0$, a protein solution that is initially transparent and fully dissolved (regime 1) with increasing salt concentration becomes turbid in the precipitate regime 2. Remarkably, when the salt concentration further increases toward regime 3, the protein solution reenters a transparent regime. This phenomenon is known as reentrant dissolution and ascribed to the inversion of the protein surface charge by the presence of the salt ions [126, 128, 133, 134].

It is very interesting to explore the phase diagram schematically shown in Figure 12 from the perspective of the global diffusion of the proteins. Remarkably, when approaching the separation line c^* for $c_s < c^*$, the self-diffusion observed by high-resolution neutron spectroscopy follows a universal scaling behavior for aqueous YCl_3 salt solutions of the model protein BSA [24]

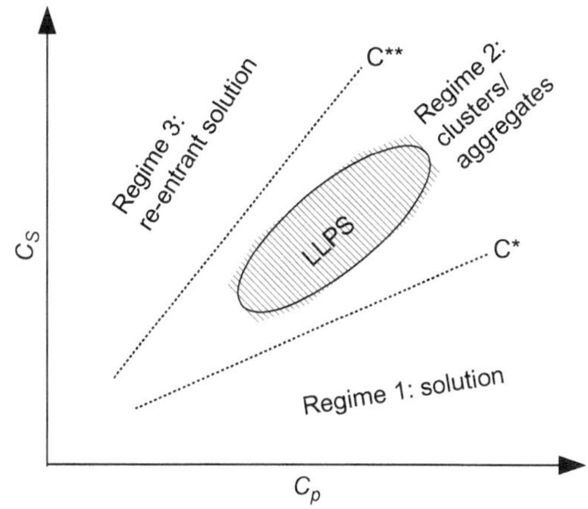

FIGURE 12 Approximate sketch of a typical phase diagram of a protein solution with the protein concentration c_p and tri-valent salt concentration c_s, drawn based on observations from several studies on different globular proteins [24, 126–129] (where the c_s and c_p axes are to be imagined as logarithmic and where the temperature-dependence of the phases is not shown). Depending on (c_p, c_s), different regimes can be observed: In regime 1, the proteins are fully dissolved. In regime 2 beyond the separation line $c^* = c^*(c_s, c_p, T)$, aggregates occur that may precipitate. Within regime 2, a closed phase region marks the regime of the liquid-liquid phase separation (LLPS), where protein-rich microdroplets are surrounded by a protein-depleted solution. In regime 3 beyond the separation line $c^{**} = c^{**}(c_s, c_p, T)$, the proteins undergo a reentrant dissolution to form a monomeric solution as in regime 1. (Figure rendered in LibreOffice [130].)

$$D(c_s, c_p) = D(c_s = 0, c_p)\, g(c_s/c_p) \tag{58}$$

with a scalar function g that only depends on the ratio c_s/c_p of the salt and protein concentration. $g(c_s/c_p)$ describes a universal slowing-down of the global diffusion when approaching c^* and can be linked to the formation of protein clusters that are static on the observation time on the subnanosecond to nanosecond scale [24]. Moreover, $g(c_s/c_p)$ can be quantitatively understood in terms of a picture of patchy colloidal particles [24, 55, 129]. The concept of patchy colloids is inspired by the observation that proteins generally display an inhomogeneous surface charge distribution. More specifically, in these terms $g(c_s/c_p)$ can be directly linked to a universal scaling of the probability of a bond formation between ion-activated patches [24]. Cluster phases in different protein systems have also been explored using neutron spin-echo spectroscopy [22, 23, 135], underlining the importance of information on the diffusive dynamics to be able to discern for instance static from dynamic or transient protein cluster phases, and thus complementing static small-angle scattering data [136].

Finally, it is also of interest to systematically explore the effect of temperature on protein solutions even for temperatures above the point where proteins start to thermally denature [12, 75]. This allows for a better understanding of the protein backbone and side chain motions by this systematic temperature variation (cf. figure 8, right). Moreover, the global protein diffusion can be blocked by the cross-linking of fully unfolded, denatured proteins.

2.8.3 Membrane Vesicles

The previous sections have mainly been concerned with proteins in aqueous solutions where a complex superposition of contributions from internal motions and the global center-of-mass motions is generally contained in the dynamic scattering function. By contrast, membrane vesicles in solution built from lipid molecules are bio-macromolecular assemblies that constitute much larger objects with typical diameters of several 100 nm. Given this large size, their center-of-mass diffusion is generally negligible even on nanosecond time scales at physiological temperatures. A membrane vesicle can for instance consist of a lipid bilayer forming a hollow spherical object (filled with a fluid) in an aqueous solution. It can also consist of several concentric bilayers forming a so-called multilamellar vesicle. Lipid membrane vesicles are also known by the name liposome in living systems. The molecular diffusive dynamics of membranes accessible by neutron spectroscopy is of fundamental interest and differs from observations that can be made using macroscopic measurements [137], because, as explained in section 2.3.2, observation length and time scales are fundamentally important when investigating diffusive processes.

Membrane vesicles in aqueous solutions can be considered as useful model systems for biological cell membranes that display complex self-assembly patterns and lipid phase diagrams and are sensitive to the influence of additives such as peptides, proteins, anesthetics, or cholesterol embedded in the membranes. An example is given by the effect of antimicrobial peptides on the subnanosecond vesicle membrane fluidity [138]. The motion of proteins embedded in membranes displays a remarkable complexity, and their anomalous diffusion within the membrane depends on the situation of "protein crowding" of the membrane [139]. There is a considerable practical interest in vesicles as "vehicles" for drug delivery, including vaccines, and present research for instance addresses the stability and phase behavior of phospholipid vesicle formation [140]. Here, high-resolution neutron spectroscopy can explore the lateral and internal diffusive motions of surfactant molecules depending on the vesicle phase [140]. Besides membrane vesicles, it is worth to mention the much smaller lipid micelles (cf. Figure 5.20 in Volume 44 of this book series), since along with the structural interest of micellar solutions explained in Chapter 5 of Volume 44 [119], there is an equivalent interest in exploring their dynamics, for instance as a function of the micellar concentration and ionic strength of the micelle solutions.

While membrane vesicles in liquid solutions can be considered a good model system for membranes of living cells in a well-controlled environment close to *in vivo* conditions, the planar lipid membranes addressed in the following section permit the access to nonisotropic diffusive motion. Studies of membrane fluctuations in vesicles in liquid solutions and of fluctuations of planar lipid membranes that are supported by solid surfaces can therefore be considered complementary.

2.8.4 Planar Lipid Membranes

The present chapter has mainly been focused on a "powder" scattering geometry by discussing samples of biological macromolecules that occur as dry or hydrated powders or as liquid suspensions. In all these cases, the structural information – if not suppressed anyway by the fact that the scattering experiment is incoherent—only occurs as a powder-averaged scattering signal. By contrast, planar quasi 2-dimensional samples constitute an exciting new geometry for biological dynamics studies where the observed dynamic scattering function becomes dependent on the orientation of the sample with respect to the incident beam. Lipid membranes can be configured in such a geometry by spreading them on a solid support or a liquid surface, and neutron reflectivity studies on such samples have been addressed in the previous chapter on biological structures.

Spectroscopy on lipid membrane samples is more challenging due to the much smaller quasi-elastic and inelastic scattering signals compared to the elastic scattering sufficient for the structural studies. Rather than using a single lipid bilayer, neutron spectroscopy experiments have been carried out on stacks of hundreds or even thousands of such bilayers [25–27, 141, 142]. Since these stacks have a very small mosaicity of better than 0.1°, they can be considered mostly equivalent to a single layer. Such membrane stacks are typically supported by Silicon wafers [26] or glass surfaces [143] and enclosed in sealed sample holders to ensure a constant well-defined humidity level. Thereby, each Si wafer or glass surface carries on the order of 100 bilayers. In most experiments using neutron spectrometers, these membrane stacks have been mounted such that the membrane surface plane was vertical and perpendicular to the horizontal scattering plane of the spectrometer [26, 27, 143]. In this case, the angle between the membrane normal axis and the incident beam axis defines the scattering vector, consisting of components parallel and perpendicular to the membrane surface. The experiments have routinely been carried out at different sample orientations with predominant scattering vector components in the membrane plane and perpendicular to the membrane plane, respectively. Moreover, membranes being oriented parallel to the scattering plane of the spectrometer have also been measured [142]. Concluding from the different spectra obtained at the different membrane surface orientation relative to the incident beam, the nonisotropic diffusive dynamics within the membranes can be explored on intermolecular length scales [142], and the effect of additives such as cholesterol or peptides on the membrane fluctuations can be accessed [143, 144]. In this context it is noted that the geometries of confinement of the molecular diffusive processes as reported in section 2.3.3 exclusively for isotropic geometries become non-isotropic for the highly oriented membrane samples. Moreover, it has become possible to study the dynamics of fully deuterated membranes hydrated with heavy water. In this way, significant coherent scattering effects appear, and the link of the sample structure and dynamics becomes observable in a rather direct way, associated with a distinct de Gennes-narrowing along with correlated coherent motions [27, 89]. This access to coherent dynamics has for instance permitted the exploration of collective overdamped lateral motions in lipid membranes on nanosecond time scales and nanometer length scales [27]. In earlier work it has also been demonstrated that neutron spin-echo spectroscopy is capable of accessing the fluctuations in single free-standing polymer membranes [145] and complement the time scales accessible by X-ray photon correlation spectroscopy toward faster motions.

The investigation of fully deuterated lipid membranes using neutron spectroscopy is related to the entirely static studies using neutron reflectivity as introduced in the preceding chapter. It may in this context be envisaged to explore the membrane surface dynamics unambiguously by accessing a so-called evanescent wave scattering regime, which up to now has only been applied statically on reflectometers or SANS instruments and which is based on illuminating a sample surface at grazing angles of incidence below the critical angle of total external reflection [146]. By accessing the evanescent wave scattering regime using neutron spectroscopy, the lateral membrane dynamics may in principle be observed over much longer lateral correlation lengths, but the demonstration of the feasibility is at present still pending.

2.8.5 Biological Fibers

Besides the experiments on oriented lipid membranes introduced in the preceding section, the scattering on oriented, aligned fibers represents another scattering geometry deviating from the usual "powder scattering." Some biological fibers are known for their extraordinary mechanical properties combining elasticity and strength. These fibers notably include spider dragline silk [147–149], which is a hierarchically structured nanocomposite protein polymer fiber consisting of small polypeptide crystallites interconnected to microfibrils by polymeric regions with smectic order embedded in a fully amorphous polymer matrix [150, 151]. These fibers emerge biologically from a protein aggregation process taking place in aqueous solutions at a physiological temperature and ambient pressure. Nevertheless, the so-called toughness of spider dragline silk fibers per unit mass, i.e., their capability to absorb kinetic energy by deformation, exceeds that of steel. Adsorbed water drastically influences the mechanical properties notably of spider silk but also of other biological fibers such as silkworm silk [152]. It is therefore assumed that the molecular mobility of both the adsorbed water and the water-accessible amorphous polymer matrix are important for the mechanical properties. Systematic neutron spectroscopy experiments combined with *in situ* tensile testing have therefore been carried out to address the diffusive dynamics of the adsorbed water and amorphous polymer matrix of silkworm silk and collagen fibers on the molecular level [28–30, 153]. In these experiments, bundles of well aligned fibers with a total mass of 100–200 mg served as samples, thus providing a scattering geometry depending on the sample orientation. In this way, the change of the geometrical confinement of the diffusive motions as a function of the externally applied tensile strain has been explored in terms of the models for the quasi-elastic neutron scattering given in section 2.3.3. Moreover, an analysis in terms of fractional diffusion (section 2.3.4) has been established for humid silk fibers [30]. In both pictures, the diffusive dynamics of the amorphous polymer matrix as a function of the tensile stress supports a model of entropy elasticity [152], where the tensile deformation reduces the entropy. In the picture of fractional diffusive dynamics, this reduction of entropy is reflected in a loss of memory, thus increasing the memory parameter α in equation (32) toward $\alpha = 1$ with increasing stress, consistent with simple Brownian diffusion in this limit. Remarkably, the observed picosecond dynamics within the amorphous fractions increases with rising stress applied to the humid fibers [28, 153]. This faster diffusive dynamics on nanometer length scales is reflected in the model-free observation of an increased quasi-elastic line broadening in the neutron spectra as well as in an increased generalized fractional diffusion coefficient K_α describing the generalized Brownian diffusion $\Gamma(q) = K_\alpha q^2$ with the generalized spectral linewidth $\Gamma(q)$ entering the Fourier transform of equation (35).

The experiments on the aligned silk fibers also illustrate the possibility to use rather sophisticated "Taylor-made" sample environments in neutron spectroscopy experiments, where in the present example a humidity control within a closed humidity chamber as well as the computer-controlled tensile testing has been combined *in situ* with the neutron experiment [28]. It is noted that the required tensile forces generated by the sample environment are quite considerable and reach the order of 1kN to enter the regime of plastic deformation of a dry silkworm silk fiber bundle [153]. Further challenges in the design of this sample environment arise from the requirement to minimize the neutron flight path length through the humid air inside and to interface the humidify chamber with the vacuum chamber of the neutron spectrometer [28]. In the case of spider silk, it is obviously not trivial to obtain the required large sample [151]. Nevertheless, *in situ* tensile test experiments combined with neutron time-of-flight spectroscopy have recently been carried out on a bundle of aligned humid spider dragline silk fibers [154], confirming the picture of strain-dependent fractional diffusion previously seen for silkworm silk [30]. For spider silk, smaller values for α and K_α, but larger values for the fractional relaxation time τ (equation (35)) were found compared with silkworm silk [154]. These findings are consistent with a higher degree of "fractionality" and more pronounced memory effects in spider silk compared with silkworm silk. Due to the alignment of the fibers, structural information can be obtained simultaneously with the spectroscopic data, although the larger beam divergence of spectrometers as opposed to dedicated fiber diffractometers obviously somewhat smears out the structure information. By using heavy water for humidification, a selective H/D-exchange can be achieved in the amorphous fraction of silk fibers, since the crystalline parts are not accessible to water [151].

The protein folding process in the presence of water is often denoted "hydrophobic collapse." This hydrophobic collapse is known by the term "supercontraction" in the case of spider dragline silk [155]. The tensile test experiments combined with neutron spectroscopy on humid collagen and silk fibers, as illustrated by the above examples, constitute a restraining—by the external tensile force—of the hydrophobic collapse of proteins within these fibers that would otherwise be caused by the humidification [29, 154]. Such tensile test experiments may therefore provide further fundamental insight on protein folding.

In passing, it is further noted that besides the quasi-elastic scattering, also the inelastic signal carries information on the geometrical confinement of molecular motions, for example by way of the so-called libration motions of confined water molecules [156, 157].

As a further example, deoxyribonucleic acid (DNA) also forms fibers, and large samples of oriented DNA fibers have already been used to explore acoustic phonons using inelastic neutron spectroscopy at different tensile strains applied to such fibers [158]. A direct access to the elasticity, stiffness and force constants on the molecular level has been obtained from these experiments in combination with simulations [158].

2.8.6 Other Systems

Other biological systems presently being in the focus of experimental studies regarding their molecular dynamics comprise for instance the so-called intrinsically disordered proteins (see, e.g., Refs. [33, 159] and references therein). As its name suggests, this type of protein lacks a stable structure and can be considered as essentially being defined by its dynamic behavior. Experimental methods that only explore static, time-averaged properties are in particular need of being complemented by spectroscopic methods in order to further understand intrinsically disordered proteins.

Additional topics that have not been particularly addressed in the present textbook chapter further comprise the entire field of biological deuteration. Partial deuteration can be used to reduce the incoherent scattering signal from a particular biological macromolecule or molecular subgroup to better distinguish different dynamic contributions. Deuteration can further be used in combination with polarization analysis [160], which can help to distinguish between coherent and incoherent scattering. Deuterated macromolecules in aqueous (D_2O) solution may also be used to establish the situation of "external crowding" of protonated tracer proteins. It is noted that unlike in the case of small-angle neutron scattering, the benefit from "contrast matching" experiments is limited in QENS experiments, because dynamic contributions from relaxational or diffusive motions do not add in the same straight-forward way as in SANS due to "dynamic window" effects. Finally, the study of biological systems also comprises entire, even living cells (e.g., Refs. [16, 17]), including partially or nearly fully deuterated living cells exposed to high pressure [36].

Regarding simplified systems, it may be worth to mention artificial gels, which may in the future serve as drastically reduced model environments with a lower complexity than entire living cells, but displaying some of the relevant properties [161, 162]. Notably, gels can macroscopically immobilize their solvent water while leaving the picosecond diffusion of the water molecules on nanometer observation length scales inside the gels largely unaffected, i.e., identical to the bulk water diffusion [161]. In this way, suitable gels resemble to living cells, which at ambient conditions also display a picosecond intracellular water dynamics that is largely unaffected compared to pure bulk water [163].

Finally, amino acids and peptides, the building blocks of proteins, are also subject to neutron spectroscopic investigations regarding their diffusive and vibrational dynamics as a function of various conditions and external parameters. Two examples from the vast literature, addressing the water dynamics in peptide solutions and vibrational spectroscopy on polypeptides, respectively, are given by Refs. [164, 165].

2.9 FUTURE PERSPECTIVES

Neutron spectroscopy has proven to contribute to answering important biological questions by exploring molecular dynamics on a wide range of time scales from below picoseconds up to microseconds, by being nondestructive and uniquely sensitive to the prevailing hydrogen atoms, by providing a unique access to the ensemble-averaged single-atom self-dynamics, as well as to different hierarchical levels of molecular dynamics in both fully disordered materials such as intrinsically disordered proteins, as well as in materials with different levels of local or long-range order, such as membranes and fibers. It will be able to do so in an even better way in the future due to the considerable recent and present progress on neutron instrumentation owed to improved neutron optics and spectrometer layout. Combined with future, brighter neutron sources, neutron spectroscopy will increasingly be a standard tool for biophysics. The higher available neutron flux will allow for more systematic studies and a substantially improved accuracy of the results. With the new capabilities, smaller, more dilute samples will be explored with smaller statistical errors and an increased "throughput." For example, the diffusion of proteins with concentrations as low as 10 mg/ml has been investigated using the most advanced backscattering spectrometers [166].

The present textbook chapter has provided an attempt of a tutorial introduction to biomolecular dynamics studies using neutrons with a particular focus on the separation of the different hierarchically superimposed dynamic contributions to the measured scattering signal from rather complex biological samples. Moreover, the present tutorial chapter has attempted to provide examples of novel results that have for the first time become accessible due to the increased flux of novel neutron spectrometers and progress in data analysis. These examples comprise the access to the center-of-mass diffusion of proteins in dilute aqueous solutions and the simultaneous access to the internal molecular diffusive motions of proteins in such aqueous solutions. The spectroscopic investigation of these internal motions permits to test molecular dynamics simulations and to determine conformational entropies and activation energies. The list of examples also includes the exploration of minute scattering signals arising from the directional molecular fluctuations in thin lipid membrane films and from dilute membrane vesicle solutions as well as from the molecular motions of polymers inside biological fibers. With the novel instruments and methods at hand, it is the systematic and quantitative information that will further connect biological dynamics to a fundamental physical understanding and ultimately to a rational control of parameters that govern biological function. Both novel theoretical approaches and further improved molecular dynamics simulations will be accurately tested with advanced neutron spectrometers and data analysis software. For instance, a systematic understanding of the underlying mechanisms of protein cluster formation may open up the perspective for the rational choice of pathways for a controlled self-assembly of biological macromolecules [24] from liquid solution into clusters, aggregates, fibers, or crystals. The possible applications of knowledge gained using neutron spectroscopy regarding the biomolecular dynamics include the fundamental understanding of the molecular processes of life, the direct testing of predictions from theoretical physics, the calibration of force fields for molecular dynamics simulations, the design of new macromolecules for pharmacy, and drug delivery.

REFERENCES

[1] G.H. Vineyard, Phys. Rev. 110 (5) (1958) 999.

[2] F. Gabel, D. Bicout, U. Lehnert, M. Tehei, M. Weik, G. Zaccai, Q. Rev. Biophys. 35 (4) (2003) 327–367.

[3] A.S. Marathay, Elements of Optical Coherence Theory, John Wiley & Sons, Chichester, 1982.

[4] M. Köppe, M. Bleuel, R. Gähler, R. Golub, P. Hank, T. Keller, S. Longeville, U. Rauch, J. Wuttke, Phys. B Condens. Matter 266 (1) (1999) 75–86.

[5] G.L. Squires, Introduction to the Theory of Thermal Neutron Scattering, Dover, Mineola, NY, 1978.

[6] D.L. Price, F. Fernandez-Alonso, An introduction to neutron scattering, in: F. Fernandez-Alonso, D.L. Price (Eds.), Experimental Methods in the Physical Sciences, Neutron Scattering, vol. 44, 2013, pp. 1–136.
[7] Institut Max von Laue – Paul Langevin (ILL), Available from: https://www.ill.eu (accessed 3.12.16).
[8] Spallation Neutron Source (SNS), Available from: http://neutrons.ornl.gov/sns (accessed 3.12.16).
[9] Wolfram Mathematica, Wolfram Research, Available from: http://www.wolfram.com/mathematica/ (accessed 11.07.16).
[10] L. Van Hove, Phys. Rev. 95 (1) (1954) 249–262.
[11] M.R. Johnson, G.J. Kearley, Dynamics of atoms and molecules, in: F. Fernandez-Alonso, D.L. Price (Eds.), Experimental Methods in the Physical Sciences, Neutron Scattering, vol. 44, Academic Press, 2013, pp. 415–470.
[12] M. Hennig, F. Roosen-Runge, F. Zhang, S. Zorn, M.W.A. Skoda, R.M.J. Jacobs, T. Seydel, F. Schreiber, Soft Matter 8 (5) (2012) 1628–1633.
[13] G. Williams, D.C. Watts, Trans. Faraday Soc. 66 (1970) 80–85.
[14] M. Bée, Quasielastic Neutron Scattering, Adam Hilger, Bristol, 1988.
[15] F. Roosen-Runge, M. Hennig, F. Zhang, R.M.J. Jacobs, M. Sztucki, H. Schober, T. Seydel, F. Schreiber, Proc. Natl Acad. Sci. USA 108 (29) (2011) 11815–11820.
[16] M. Jasnin, M. Moulin, M. Haertlein, G. Zaccai, M. Tehei, Biophys. J. 95 (2) (2008) 857–864.
[17] A. Stadler, I. Digel, G.M. Artmann, J.P. Embs, G. Zaccai, G. Büldt, Biophys. J. 95 (11) (2008) 5449–5461.
[18] W. Doster, S. Cusack, W. Petry, Nature (London, UK) 337 (1989) 754–756.
[19] J. Perez, J. Zanotti, D. Durand, Biophys. J. 77 (1) (1999) 454–469.
[20] S. Longeville, W. Doster, G. Kali, Chem. Phys. 292 (2–3) (2003) 413–424.
[21] M. Grimaldo, F. Roosen-Runge, F. Zhang, T. Seydel, F. Schreiber, J. Phys. Chem. B 118 (25) (2014) 7203–7209.
[22] L. Porcar, P. Falus, W.-R. Chen, A. Faraone, E. Fratini, K. Hong, P. Baglioni, Y. Liu, J. Phys. Chem. Lett. 1 (1) (2009) 126–129.
[23] E.J. Yearley, P.D. Godfrin, T. Perevozchikova, H. Zhang, P. Falus, L. Porcar, M. Nagao, J.E. Curtis, P. Gawande, R. Taing, I.E. Zarraga, N.J. Wagner, Y. Liu, Biophys. J. 106 (8) (2014) 1763–1770.
[24] M. Grimaldo, F. Roosen-Runge, M. Hennig, F. Zanini, F. Zhang, M. Zamponi, N. Jalarvo, F. Schreiber, T. Seydel, J. Phys. Chem. Lett. 6 (2015) 2577–2582.
[25] M.C. Rheinstädter, W. Häußler, T. Salditt, Phys. Rev. Lett. 97 (2006) 048103.
[26] M.C. Rheinstädter, T. Seydel, T. Salditt, Phys. Rev. E 75 (2007) 011907.
[27] M.C. Rheinstädter, J. Das, E.J. Flenner, B. Brüning, T. Seydel, I. Kosztin, Phys. Rev. Lett. 101 (2008) 248106.
[28] T. Seydel, W. Knoll, I. Greving, C. Dicko, M.M. Koza, I. Krasnov, M. Müller, Phys. Rev. E 83 (2011) 016104.
[29] S. Perticaroli, G. Ehlers, N. Jalarvo, J. Katsaras, J.D. Nickels, J. Phys. Chem. Lett. 6 (20) (2015) 4018–4025.
[30] I. Krasnov, T. Seydel, M. Müller, Phys. Rev. E 91 (2015) 042716.
[31] Y. Fichou, G. Schirò, F.-X. Gallat, C. Laguri, M. Moulin, J. Combet, M. Zamponi, M. Härtlein, C. Picart, E. Mossou, L.-J. Hugues, J.-P. Colletier, D.T. Tobias, M. Weik, Proc. Natl Acad. Sci. USA 112 (20) (2015) 6365–6370.

[32] G. Schirò, Y. Fichou, F.-X. Gallat, K. Wood, F. Gabel, M. Moulin, M. Härtlein, M. Heyden, J.-P. Colletier, A. Orecchini, A. Paciaroni, J. Wuttke, D.J. Tobias, M. Weik, Nat. Commun. 6 (2015) 6490.
[33] S. Lenton, T. Seydel, T. Nylander, C. Holt, M. Härtlein, S. Teixeira, G. Zaccai, J. R. Soc. Interface 12 (110) (2015) 20150506.
[34] M. Erlkamp, J. Marion, N. Martinez, C. Czeslik, J. Peters, R. Winter, J. Phys. Chem. B 119 (14) (2015) 4842–4848.
[35] J. Marion, M. Trovaslet, N. Martinez, P. Masson, R. Schweins, F. Nachon, M. Trapp, J. Peters, Phys. Chem. Chem. Phys. 17 (5) (2015) 3157–3163.
[36] F. Foglia, R. Hazael, G.G. Simeoni, M.-S. Appavou, M. Moulin, M. Haertlein, V.T. Forsyth, T. Seydel, I. Daniel, F. Meersman, P.F. McMillan, Sci. Rep. 6 (2016) 18862.
[37] R. Brown, Philos. Mag. Ser. 2 4 (21) (1828) 161–173.
[38] A. Einstein, Ann. Phys. (Leipzig) 17 (1905) 549–560.
[39] J. Klafter, I.M. Sokolov, Phys. World 18 (2005) 29–32.
[40] E.M. Purcell, Am. J. Phys. 45 (1) (1977) 3–11.
[41] A.J. Banchio, G. Nägele, J. Chem. Phys. 128 (10) (2008) 104903.
[42] J.K.G. Dhont, An Introduction to the Dynamics of Colloids, Elsevier Science B.V., Amsterdam, 1996.
[43] H. Risken, The Fokker-Planck-Equation. Methods of Solution and Applications, second ed., Springer-Verlag, Berlin, Heidelberg, 1989.
[44] G. Nägele, The Physics of Colloidal Soft Matter, Centre of Excellence for Advanced Materials and Structures, Warsaw, 2004.
[45] T. Franosch, M. Grimm, M. Belushkin, F.M. Mor, G. Foffi, L. Forró, S. Jeney, Nature (London, UK) 478 (2011) 85–88.
[46] A. Fick, Philos. Mag. Ser. 4 10 (1855) 30–39.
[47] M. von Smoluchowski, Ann. Phys. (Leipzig) 353 (1915) 1103–1112.
[48] P. Langevin, C. R. Acad. Sci. 146 (1908) 530–532.
[49] J. Qvist, H. Schober, B. Halle, J. Chem. Phys. 134 (14) (2011) 144508.
[50] M. Tokuyama, I. Oppenheim, Phys. Rev. E 50 (1) (1994) 16–19.
[51] MATLAB, The MathWorks, Inc., Available from: http://www.mathworks.com/products/matlab/ (accessed 11.07.16).
[52] S. Longeville, W. Doster, G. Kali, Chem. Phys. 292 (2–3) (2003) 413–424.
[53] W. Doster, S. Longeville, Biophys. J. 93 (4) (2007) 1360–1368.
[54] G. Nägele, Phys. Rep. 272 (5–6) (1996) 215–372.
[55] F. Roosen-Runge, F. Zhang, F. Schreiber, R. Roth, Sci. Rep. 4 (2014) 7016.
[56] F. Roosen-Runge, M. Hennig, T. Seydel, F. Zhang, M.W.A. Skoda, S. Zorn, R.M.J. Jacobs, M. Maccarini, P. Fouquet, F. Schreiber, BBA Proteins Proteom. 1804 (1) (2010) 68–75.
[57] R.J. Ellis, Curr. Opin. Struct. Biol. 11 (1) (2001) 114–119.
[58] S.B. Zimmerman, A.P. Minton, Annu. Rev. Biophys. Biomol. Struct. 22 (1) (1993) 27–65.
[59] M. Bée, Phys. B 182 (4) (1992) 323–336.
[60] D.J. Bicout, Phys. Rev. E 64 (1) (2001) 011910.
[61] F. Volino, A.J. Dianoux, Mol. Phys. 41 (2) (1980) 271–279.
[62] D.J. Bicout, Incoherent neutron scattering functions for combined dynamics, in: Proceedings of the ILL Millennium Symposium and European User Meeting, vol. 1, 2001, pp. 60–62.
[63] R. Kohlrausch, Ann. Phys. Chem. 91 (1854) 56–82.
[64] E.W. Montroll, G.H. Weiss, J. Math. Phys. 6 (2) (1965) 167–181.
[65] R. Metzler, J. Klafter, Phys. Rep. 339 (2000) 1.

[66] G.R. Kneller, K. Hinsen, J. Chem. Phys. 121 (20) (2004) 10278–10283.

[67] G.R. Kneller, Phys. Chem. Chem. Phys. 7 (13) (2005) 2641–2655.

[68] S. Gupta, R. Biehl, C. Sill, J. Allgaier, M. Sharp, M. Ohl, D. Richter, Macromolecules 49 (2016) 1941–1949.

[69] X. Hu, L. Hong, M.D. Smith, T. Neusius, X. Cheng, J.C. Smith, Nat. Phys. 12 (2016) 171–174.

[70] M. Grimaldo, F. Roosen-Runge, N. Jalarvo, M. Zamponi, F. Zanini, M. Hennig, F. Zhang, F. Schreiber, T. Seydel, EPJ Web Conf. 83 (2015) 02005.

[71] R. Biehl, B. Hoffmann, M. Monkenbusch, P. Falus, S. Préost, R. Merkel, D. Richter, Phys. Rev. Lett. 101 (13) (2008) 138102.

[72] R. Biehl, M. Monkenbusch, D. Richter, Soft Matter 7 (2011) 1299–1307.

[73] L. Stagg, S.-Q. Zhang, M. Cheung, P. Wittung-Stafshede, Proc. Natl Acad. Sci. USA 104 (48) (2007) 18976–18981.

[74] F. Roosen-Runge, D.J. Bicout, J.-L. Barrat, J. Chem. Phys. 144 (2016) 204109.

[75] M. Grimaldo, F. Roosen-Runge, M. Hennig, F. Zanini, F. Zhang, N. Jalarvo, M. Zamponi, F. Schreiber, T. Seydel, Phys. Chem. Chem. Phys. 17 (2015) 4645–4655.

[76] J. Fitter, Biophys. J. 84 (6) (2003) 3924–3930.

[77] A.M. Stadler, F. Demmel, J. Ollivier, T. Seydel, Phys. Chem. Chem. Phys. (2016).

[78] L. Harris, S. Larson, K. Hasel, A. McPherson, Biochemistry 36 (7) (1997) 1581–1597.

[79] O.G. Randl, H. Franz, T. Gerstendörfer, W. Petry, G. Vogl, A. Magerl, Phys. B Condens. Matter 234–236 (1997) 1064–1065.

[80] J. Humlíček, J. Quant. Spectrosc. Radiat. Transf. 21 (4) (1979) 309–313.

[81] M. Arai, Experimental techniques, in: F. Fernandez-Alonso, D.L. Price (Eds.), Experimental Methods in the Physical Sciences, Neutron Scattering, vol. 44, Academic Press, 2013, pp. 245–320.

[82] Python Software Foundation, Available from: https://www.python.org (accessed 11.07.16).

[83] S.W. Provencher, Comput. Phys. Commun. 27 (3) (1982) 229–242.

[84] D. Monserrat, A. Vispa, L.C. Pardo, R. Tolchenov, S. Mukhopadhyay, F. Fernandez-Alonso, J. Phys. Conf. Ser. 663 (1) (2015) 012009. http://stacks.iop.org/1742-6596/663/i=1/a=012009.

[85] MANTID; Manipulation and Analysis Toolkit for Instrument Data; Mantid Project, Available from: http://www.mantidproject.org (accessed 11.07.16).

[86] B. Frick, E. Mamontov, L. van Eijck, T. Seydel, Z. Phys. Chem. 224 (1–2) (2010) 33–60.

[87] E. Mamontov, K.W. Herwig, Rev. Sci. Instrum. 82 (8) (2011) 085109.

[88] The IN5 spectrometer at the ILL, Available from: https://www.ill.eu/instruments-support/instruments-groups/instruments/in5/description/instrument-layout/ (accessed 3.12.16).

[89] P.G. De Gennes, Physica 25 (7) (1959) 825–839.

[90] T. Róg, K. Murzyn, K. Hinsen, G.R. Kneller, J. Comput. Chem. 24 (5) (2003) 657–667.

[91] nMOLDYN, Available from: http://dirac.cnrs-orleans.fr/plone/software/nmoldyn/ (accessed 11.07.16).

[92] L. Hong, N. Smolin, J.C. Smith, Phys. Rev. Lett. 112 (2014) 158102.

[93] F. Mezei, Z. Phys. 255 (1972) 146–160.

[94] The IN15 spectrometer at the ILL, Available from: https://www.ill.eu/instruments-support/instruments-groups/instruments/in15/description/instrument-layout/ (accessed 3.12.16).

[95] B. Frick, J. Combet, L. Van Eijck, Nucl. Instrum. Methods Phys. Res. Sect. A 669 (2012) 7–13.

[96] F. Roosen-Runge, T. Seydel, A generalized mean-squared displacement from inelastic fixed window scans of incoherent neutron scattering as a model-free indicator of anomalous diffusion confinement, in: EPJ Web of Conferences, vol. 83, EDP Sciences, 2015.

[97] L. van Eijck, L. Gérard, B. Frick, T. Seydel, H. Schober, Nucl. Instrum. Methods Phys. Res. Sect. A 672 (2012) 64–68.

[98] F.J. Bermejo, F. Sordo, Neutron sources, in: F. Fernandez-Alonso, D.L. Price (Eds.), Experimental Methods in the Physical Sciences, Neutron Scattering, vol. 44, Academic Press, 2013, pp. 137–244.

[99] The IN16B spectrometer at the ILL, Available from: https://www.ill.eu/instruments-support/instruments-groups/instruments/in16b/description/instrument-layout/ (accessed 3.12.16).

[100] The BASIS spectrometer at the SNS, Available from: http://neutrons.ornl.gov/basis (accessed 3.12.16).

[101] F. Mezei, J. Neutron Res. 6 (1–3) (1997) 3–32.

[102] H. Abele, D. Dubbers, H. Häse, M. Klein, A. Knöpfler, M. Kreuz, T. Lauer, B. Märkisch, D. Mund, V. Nesvizhevsky, A. Petoukhov, C. Schmidt, M. Schumann, T. Soldner, Nucl. Instrum. Methods Phys. Res. Sect. A 562 (1) (2006) 407–417.

[103] H.N. Bordallo, B. Frick, H. Schober, T. Seydel, J. Neutron Res. 16 (1) (2008) 39–54.

[104] B. Frick, A. Magerl, Y. Blanc, R. Rebesco, Phys. B Condens. Matter 234–236 (1997) 1177–1179.

[105] J. Schelten, B. Alefeld, Backscattering spectrometer with adapted Q-resolution at the pulsed neutron source, in: R. Scherm, H.H. Stiller (Eds.), Proceedings of the Workshop on Neutron Scattering Instrumentation for SNQ, vol. Jül-1954, Kernforschungsanlage, Jülich, 1984, p. 378.

[106] M. Hennig, B. Frick, T. Seydel, J. Appl. Crystallogr. 44 (3) (2011) 467–472.

[107] S. Combet, J. Pieper, F. Coneggo, J.-P. Ambroise, M.-C. Bellissent-Funel, J.-M. Zanotti, Eur. Biophys. J. 37 (5) (2008) 693–700.

[108] M.A. Adams, S.F. Parker, F. Fernandez-Alonso, D.J. Cutler, C. Hodges, A. King, Appl. Spectrosc. 63 (2009) 727–732.

[109] A. Madsen, T. Seydel, M. Tolan, G. Grübel, J. Synchrotron Radiat. 12 (6) (2005) 786–794.

[110] G. Grübel, G.B. Stephenson, C. Gutt, H. Sinn, Th. Tschentscher, Nucl. Instrum. Methods Phys. Res. Sect. B 262 (2) (2007) 357–367.

[111] P. Wochner, C. Gutt, T. Autenrieth, T. Demmer, V. Bugaev, A.D. Ortiz, A. Duri, F. Zontone, G. Grübel, H. Dosch, Proc. Natl Acad. Sci. USA 106 (28) (2009) 11511–11514.

[112] F. Lehmkühler, G. Grübel, C. Gutt, J. Appl. Crystallogr. 47 (4) (2014) 1315–1323.

[113] C.A. Crain, N.A. Strange, J.Z. Larese, Inelastic neutron scattering studies of natural silk-worm proteins, in: MRS Proceedings, vol. 1793, Cambridge University Press, 2015, pp. 41–46.

[114] G. Fleischer, F. Fujara, NMR as a generalized incoherent scattering experiment, in: P. Diehl, E. Fluck, H. Günther, R. Kosfeld, J. Seelig (Eds.), NMR—Basic Principles and Progress, vol. 30, Springer, Berlin, 1994, pp. 159–207.

[115] C.S. Johnson, Prog. Nucl. Magn. Reson. Spectrosc. 34 (3) (1999) 203–256.

[116] E.O. Stejskal, J.E. Tanner, J. Chem. Phys. 42 (1) (1965) 288–292.

[117] J. Kärger, G. Fleischer, Trends Anal. Chem. 13 (1994) 145.

[118] M.K. Quinn, N. Gnan, S. James, A. Ninarello, F. Sciortino, E. Zaccarelli, J.J. McManus, Phys. Chem. Chem. Phys. 17 (46) (2015) 31177–31187.

[119] J. Penfold, I.M. Tucker, Large-scale structures, in: F. Fernandez-Alonso, D.L. Price (Eds.), Experimental Methods in the Physical Sciences, Neutron Scattering, vol. 44, Academic Press, 2013, pp. 353–414.

[120] M.-C. Bellissent-Funel, A. Hassanali, M. Havenith, R. Henchman, P. Pohl, F. Sterpone, D. van der Spoel, Y. Xu, A.E. Garcia, Chem. Rev. (2016).

[121] B. Frick, D. Richter, W. Petry, U. Buchenau, Z. Phys. B Condens. Matter 70 (1) (1988) 73–79.

[122] W. Doster, Eur. Biophys. J. 37 (5) (2008) 591–602.

[123] S. Magazu, F. Migliardo, A. Benedetto, J. Phys. Chem. B 115 (24) (2011) 7736–7743.

[124] G. Zaccai, Science 288 (5471) (2000) 1604–1607.

[125] S. Perticaroli, J.D. Nickels, G. Ehlers, A.P. Sokolov, Biophys. J. 106 (12) (2014) 2667–2674.

[126] F. Zhang, M.W.A. Skoda, R.M.J. Jacobs, S. Zorn, R.A. Martin, C.M. Martin, G.F. Clark, S. Weggler, A. Hildebrandt, O. Kohlbacher, F. Schreiber, Phys. Rev. Lett. 101 (14) (2008) 148101.

[127] F. Zhang, R. Roth, M. Wolf, F. Roosen-Runge, M.W.A. Skoda, R.M.J. Jacobs, M. Sztucki, F. Schreiber, Soft Matter 8 (5) (2012) 1313–1316.

[128] E. Jordan, F. Roosen-Runge, S. Leibfarth, F. Zhang, M. Sztucki, A. Hildebrandt, O. Kohlbacher, F. Schreiber, J. Phys. Chem. B 118 (38) (2014) 11365–11374.

[129] D. Soraruf, F. Roosen-Runge, M. Grimaldo, F. Zanini, R. Schweins, T. Seydel, F. Zhang, R. Roth, M. Oettel, F. Schreiber, Soft Matter 10 (6) (2014) 894–902.

[130] LibreOffice, The Document Foundation, Available from: http://www.libreoffice.org (accessed 11.07.16).

[131] F. Zhang, F. Roosen-Runge, A. Sauter, R. Roth, M.W.A. Skoda, R.M.J. Jacobs, M. Sztucki, F. Schreiber, Faraday Discuss. 159 (1) (2012) 313–325.

[132] F. Zhang, F. Roosen-Runge, A. Sauter, M. Wolf, R.M.J. Jacobs, F. Schreiber, Pure Appl. Chem. 86 (2) (2014) 191–202.

[133] F. Zhang, S. Weggler, M.J. Ziller, L. Ianeselli, B.S. Heck, A. Hildebrandt, O. Kohlbacher, M.W.A. Skoda, R.M.J. Jacobs, F. Schreiber, Proteins Struct. Funct. Bioinf. 78 (16) (2010) 3450–3457.

[134] F. Roosen-Runge, B.S. Heck, F. Zhang, O. Kohlbacher, F. Schreiber, J. Phys. Chem. B 117 (18) (2013) 5777–5787.

[135] P. Falus, L. Porcar, E. Fratini, W.-R. Chen, A. Faraone, K. Hong, P. Baglioni, Y. Liu, J. Phys. Condens. Matter 24 (6) (2012) 064114.

[136] A. Stradner, H. Sedgwick, F. Cardinaux, W.C.K. Poon, S.U. Egelhaaf, P. Schurtenberger, Nature (London, UK) 432 (7016) (2004) 492–495.

[137] S. Busch, C. Smuda, L.C. Pardo, T. Unruh, J. Am. Chem. Soc. 132 (10) (2010) 3232–3233.

[138] V.K. Sharma, E. Mamontov, D.B. Anunciado, H. O'Neill, V.S. Urban, Soft Matter 11 (34) (2015) 6755–6767.

[139] R. Metzler, J.-H. Jeon, A.G. Cherstvy, Biochim. Biophys. Acta 1858 (2016) 2451–2467.

[140] S. Mitra, V.K. Sharma, V. Garcia-Sakai, A. Orecchini, T. Seydel, M. Johnson, R. Mukhopadhyay, J. Phys. Chem. B 120 (2016) 3777–3784.

[141] M. Rheinstädter, T. Seydel, F. Demmel, T. Salditt, Phys. Rev. E 71 (2005) 061908.

[142] L. Toppozini, F. Roosen-Runge, R.I. Bewley, R.M. Dalgliesh, T. Perring, T. Seydel, H.R. Glyde, V. García-Sakai, M.C. Rheinstädter, Soft Matter 11 (42) (2015) 8354–8371.

[143] M.A. Barrett, M. Trapp, W. Lohstroh, T. Seydel, J. Ollivier, M. Ballauff, N.A. Dencher, T. Hauß, Soft Matter 12 (2015) 1444–1451.

[144] C.L. Armstrong, W. Häußler, T. Seydel, J. Katsaras, M.C. Rheinstädter, Soft Matter 10 (15) (2014) 2600–2611.

[145] I. Sikharulidze, B. Farago, I.P. Dolbnya, A. Madsen, W.H. de Jeu, Phys. Rev. Lett. 91 (16) (2003) 165504.

[146] H. Dosch, K. Al Usta, A. Lied, W. Drexel, J. Peisl, Rev. Sci. Instrum. 63 (12) (1992) 5533–5542.

[147] J.M. Gosline, M.E. DeMont, M.W. Denny, Endeavour 10 (1) (1986) 37–43.

[148] F. Vollrath, D. Porter, Soft Matter 2 (5) (2006) 377–385.

[149] D. Porter, J. Guan, F. Vollrath, Adv. Mater. (Weinheim, Ger.) 25 (9) (2013) 1275–1279.

[150] C. Riekel, M.C.G. Gutierrez, A. Gourrier, S. Roth, Anal. Bioanal. Chem. 376 (5) (2003) 594–601.

[151] D. Sapede, T. Seydel, V.T. Forsyth, M.A. Koza, R. Schweins, F. Vollrath, C. Riekel, Macromolecules 38 (20) (2005) 8447–8453.

[152] J.M. Gosline, M.W. Denny, M.E. Demont, Nature (London, UK) 309 (5968) (1984) 551–552.

[153] T. Seydel, K. Kölln, I. Krasnov, I. Diddens, N. Hauptmann, G. Helms, M. Ogurreck, S.-G. Kang, M.M. Koza, M. Müller, Macromolecules 40 (2007) 1035–1042.

[154] I. Krasnov, T. Seydel, I. Greving, M. Blankenburg, F. Vollrath, M. Müller, J. R. Soc. Interface 13 (122) (2016) 20160506.

[155] R.W. Work, J. Arachnol. 9 (3) (1981) 299–308.

[156] N.W. Ockwig, R.T. Cygan, M.A. Hartl, L.L. Daemen, T.M. Nenoff, J. Phys. Chem. C 112 (35) (2008) 13629–13634.

[157] V. Crupi, D. Majolino, P. Migliardo, V. Venuti, J. Phys. Chem. A 104 (47) (2000) 11000–11012.

[158] L. van Eijck, F. Merzel, S. Rols, J. Ollivier, V.T. Forsyth, M.R. Johnson, Phys. Rev. Lett. 107 (2011) 088102.

[159] A.M. Gaspar, M.-S. Appavou, S. Busch, T. Unruh, W. Doster, Eur. Biophys. J. 37 (5) (2008) 573–582.

[160] A.M. Gaspar, S. Busch, M.-S. Appavou, W. Haeussler, R. Georgii, Y. Su, W. Doster, BBA Proteins Proteom. 1804 (1) (2010) 76–82.

[161] T. Seydel, L. Wiegart, F. Juranyi, B. Struth, H. Schober, Phys. Rev. E 78 (6) (2008) 061403.

[162] J.T. Trevors, G.H. Pollack, Prog. Biophys. Mol. Biol. 89 (1) (2005) 1–8.

[163] M. Jasnin, M. Moulin, M. Haertlein, G. Zaccai, M. Tehei, EMBO Rep. 9 (2008) 543–547.

[164] C. Malardier-Jugroot, D.T. Bowron, A.K. Soper, M.E. Johnson, T. Head-Gordon, Phys. Chem. Chem. Phys. 12 (2) (2010) 382–392.

[165] H.D. Middendorf, R.L. Hayward, S.F. Parker, J. Bradshaw, A. Miller, Biophys. J. 69 (2) (1995) 660.

[166] S. Fujiwara, K. Araki, T. Matsuo, H. Yagi, T. Yamada, K. Shibata, H. Mochizuki, PLoS ONE 11 (4) (2016) e0151447.

Chapter 3

The Structure of Water and Aqueous Systems

Alan K. Soper[1]
Rutherford Appleton Laboratory, Chilton, United Kingdom
[1]Corresponding author: e-mail: alan.soper@stfc.ac.uk

Chapter Outline

Experimental Methods in the Physical Sciences, Vol. 49. http://dx.doi.org/10.1016/B978-0-12-805324-9.00003-0

3.1 INTRODUCTION

Water, water, every where, And all the boards did shrink; Water, water, every where, Nor any drop to drink.

Samual Taylor Coleridge (1798)

The water in Coleridge's famous poem, *The Rime of the Ancient Mariner*, was of course seawater and therefore undrinkable. At this point in the story the ship is becalmed in steamy, tropical, waters, and one by one, the ship's crew are dying of thirst.

The sense of frustration felt by the sailors in that poem—surrounded by water but unable to drink any of it—in many ways parallels the dilemmas often faced by scientists today who decide to investigate the science of water and its interactions with other substances. Many decades of research have been devoted to the study of pure water and its interactions with salts, solutes, and surfaces, yet clear, definitive answers to simple questions, like "What is water? Why is water like it is?" still elude us.

How is this? We are indeed surrounded by water: it constitutes around 70% of our bodies and most of that water is "salty", like seawater. We wash in it, drink it, and put it to countless uses. It is a primary component of most industries and is a major driving force behind the weather systems that affect us every day. Moreover it appears to be widespread beyond planet Earth as well. Any search for extra-terrestrial life usually begins as a search for water.

Yet how well do we really understand water, and why do we *want* to understand it? Factually of course we know a lot about water—perhaps more so than any other material—but that knowledge does not, unfortunately, always translate into understanding. Water by itself, and in its interactions with dissolved substances and interfaces, has properties that enable life to happen and be sustained under a range of conditions from low to high temperature, at low and high salt content, at low humidity, and at high pressure, but what causes these properties is still largely a matter for debate. For example, if only we could understand why a protein molecule folds one way and not another in a particular aqueous environment, then we would have come a long way in understanding the causes of, and perhaps identified cures for, numerous diseases. Because our understanding of water is still inadequate sharp controversies exist, right up to the present, about the true nature of water substance, and at this point in time there does not appear to be an end to those controversies in sight. Hence a scientist who has been studying water for a long time might be forgiven for developing a sense of desolation not dissimilar to that felt by the sailors in Coleridge's poem!

Therefore, before getting into the details of the role that neutron scattering can play in the study of water and aqueous systems, it is worth taking a look at a few of the more pertinent controversies that have affected our thinking about water and aqueous solutions in recent decades. The question then arises: how can neutron scattering help? Attempting to answer that question will form the main thrust of this chapter.

3.1.1 Water Controversies

3.1.1.1 A "Theory" of Water?

It is well-known that crystalline water, ice Ih, is less dense than the liquid, so ice floats on water. As you cool water at ambient pressure its density reaches a maximum at about 4°C, then it starts to fall again. If you cool water below 0°C the density continues to fall, but you can't keep supercooling water indefinitely: sooner or later water will spontaneously crystallize. This is called homogeneous nucleation of ice. Austen Angell and Robin Speedy and others [1–3] studied these properties extensively in the 1970s and 1980s. Not only did they observe the density decrease, but they also observed that the isothermal compressibility, κ_T, and the specific heat at constant pressure, C_p, both *increased* markedly on cooling. This was strange behavior indeed. Because specific heat is a measure of how much thermal energy a material can absorb per unit temperature rise, normally one would expect the specific heat to decrease on cooling, since, as you cool them down, the atoms should have become more restricted in motion, and therefore less able to accept thermal agitation when heat was applied. Somehow, water was different. Somehow, as you cooled them down, water molecules were becoming more able to break any bonds and so absorb thermal agitation than at higher temperature. Moreover, the more Angell and Speedy supercooled the water the more the compressibility and specific heat increased. Since such an increase could not go on indefinitely, Speedy [4] surmised that this behavior was indicative of some sort of fundamental breakdown in the stability of water as a liquid that would occur if it were possible to go to even lower temperatures. But of course they could not go lower in temperature, because of homogeneous nucleation, and so the question remained unresolved.

In 1991, Peter Poole and colleagues [5] made a bold proposal, based on computer simulations of a particular water model, ST2 [6], that there is a hidden, second, critical point in the phase diagram of water. This critical point is probably below the homogeneous nucleation temperature, and therefore unobservable in the real liquid, but it would imply that if you could hypothetically get to these low temperatures (below ~220 K) and bypass the formation of ice, then you would observe two states of water, a high-density phase (HDL) and a low-density phase (LDL), with a sharp change in density as you go between them. This proposal fit quite nicely with the known fact that at much lower temperatures, below 100 K, water can exist in at least two glassy forms, namely low-density (LDA) and high-density (HDA) amorphous ice, with an apparently first order-like transition between them [7]. The same second (liquid-liquid) critical point would explain the apparent divergence in compressibility and specific heat, and the decrease in density in the supercooled region: thermodynamic behaviors at a critical point can extend well away from the critical point itself, and into the metastable and stable regions above the homogeneous nucleation temperature. Indeed, much more recently Holten and Anisimov [8] used a simple two state model, in which water molecules are free to transfer between the two states, low density (LDL) and high density (HDL), to fit the density of water quite accurately over a wide range of temperature and pressure, and showed that this model also has a second, liquid-liquid, critical point, with very realistic behavior for the isothermal compressibility and specific heat. The model cannot tell us what these two states look like, but since they must both be made up of only water molecules, the differences between them must be related to different local molecular arrangements. Fundamentally, the transition from LDL to HDL in this model is seen as an entropy-driven transition. There have been several attempts to generate an order parameter that could be used to define whether a water molecule is in one state or the other [9–11], but these are really only applicable if computer simulation analysis is available.

Despite the apparent success of these ideas, their speculative nature means that there has been strong opposition from some authorities. Limmer and Chandler in particular have written several critical theses [12–15], claiming that the transition seen in computer simulations of ST2 water is in fact a liquid-solid transition. Although Liu, Palmer et al. [16–18] strongly refute this, they do recognize that both high- and low-density liquids are metastable with respect to crystalline ice, and so may not exist for very long in reality. Whilst it is formally correct to note that the existence of a second critical point is impossible to observe, for the finite sized systems that are typically used in computer simulation and in a region where the liquid is metastable with respect to the solid [19], this does not stop a particular model of water from being a useful mechanism for understanding what might be going on in the stable and supercooled liquids. After all, Newton's theory of gravitation is known to be only approximate on some length- and time-scales, but that does not stop it from being extremely useful for understanding, indeed predicting, the motion of celestial objects. Hence the second critical point model of water appears to fit into the category of a model that may be useful for understanding, but formally incorrect for, the real material. With this point of view one might say therefore that water behaves *as if* there were a second critical point below the homogeneous nucleation temperature.

At the time of writing there does not appear to be any resolution to this controversy. It has to be said however that the so-called "mW" model of water [20] does a good job at describing *qualitatively* the thermodynamic properties of supercooled water [21]: in this case no liquid-liquid transition is observed with this model because the relaxation time for ice formation at low temperature becomes shorter than the relaxation time for the transition from one liquid to the other. Although this model gives only qualitative agreement with experiment, the lack so far of any clear evidence for a structural transition all the way down to *below* the homogeneous nucleation temperature [22] makes it tempting to conclude that the mW model is correct in this regard.

3.1.1.2 Icebergs in Water

Another widely cited controversy which is still unresolved emanated from the original work of Frank and Evans [23] who studied the free energy and entropy of mixing of a large number of solutes in water. One particular result, still referred to, was their claim that around hydrophobic entities (which are unable to form hydrogen bonds with water) in solution, there exists a structured region, which they loosely referred to as a "microscopic iceberg" (p. 520). This description came with the following caveat: "Throughout the rest of this paper we shall use the word iceberg without quotation marks or apology, to represent a microscopic region, either of pure water or surrounding a solute molecule or ion, in which water molecules are tied together in some sort of quasi-solid structure. It is not implied that the structure is exactly ice-like, nor is it necessarily the same in every case where the word iceberg is used." Their conclusion was arrived at because of the loss of entropy when a number of hydrophobic gases were dissolved in water, although there was at that time no independent experimental corroborating evidence for the so-called "icebergs" in these systems. This was later paralleled by the notion that hydrophobic effects were the primary driving forces in protein folding [24], an idea that has now apparently been largely debunked [25], also partly as a result of neutron scattering research, to be mentioned below. However further work with light scattering has clarified some of the dynamics in these systems [26, 27]. In particular it appears the OH intramolecular stretching frequency can be significantly red-shifted for water molecules close to small hydrophobic entities in aqueous solution compared to the bulk liquid, but this effect diminishes as the temperature increases or the hydrophobic chain length increases. These shifts are interpreted as resulting from "enhanced water structure with greater tetrahedral order" in the immediate vicinity of the solute. As with all such statements in the literature, however, the reader is left to imagine what the authors actually mean by "enhanced" in this context. Do they mean the disordered "icebergs" as envisaged by Frank and Evans, or do they mean something else? By how much is the tetrahedral order increased? On these matters the spectroscopy remains silent.

3.1.1.3 Water Chains and Rings

A 21st century controversy in water science that arose in 2004 was the claim [28] that water was primarily a chain-like liquid, rather than the tetrahedral-like network that has been traditionally envisaged for many decades [29]. The basis for this claim was some new data, namely the near-edge oxygen X-ray absorption spectrum for water (XAS) which had originally been investigated and commented on only relatively recently [30]. In the later work the authors compared the XAS spectrum from water with that from crystalline ice Ih and noticed a profound difference between the two. Water has a marked main edge peak at ~537 eV and significant pre-edge feature at ~535 eV, while in ice the main edge and pre-edge peaks are much weaker, and the spectrum is dominated by a larger post-edge feature at ~540 eV. Knowing that in ice each water molecule is tetrahedrally coordinated with exactly 4 neighbors, the authors concluded, by means of a (perhaps contrived?) method of extracting particular configurations of water molecules from computer simulations of water, that water structure is quite different from ice, with only two strong hydrogen bonds per water molecule (and perhaps two weak ones), so that the liquid was comprised primarily of chains of water molecules compared to the tetrahedral network of ice. Later work showed that the water oxygen XAS spectrum was sensitive to temperature and pressure and to the concentration and nature of dissolved species. [31–33].

On the other hand total X-ray scattering from water had shown for a long time that the first shell coordination number of water is >4, about 4.5 in fact. This number arises from the higher density of water compared to ice, and was also confirmed by several computer simulations of water. The traditional view was that four of these molecules were in approximately ice-like positions, but that a fifth molecule, not hydrogen bonded to the central molecule, could approach any given molecule quite closely [34]: it was this extra molecule that broke the symmetry and forced the liquid state, and so gave the liquid its rather unique structure. There was no way a chain-like structure could occur in such a situation.

To date a comprehensive understanding of the XAS spectrum of water and solutions is lacking, but it seems likely the initial interpretation of these spectra was too primitive, and that full account has to be taken of the quantum nuclear motion as well as inhomogeneous screening effects [35, 36]. The apparent asymmetry observed between water and ice in the XAS spectrum then arises from the very high time resolution (femtoseconds) of the XAS measurements, which disappears at longer times [37]. A consequence of this analysis is that longer-time view of water structure as it is traditionally understood does not change, but we are now much more aware of the highly dynamic nature of that structure and significant asymmetries in the hydrogen-bonding electron distributions that can occur at the very short time-scale.

3.1.2 Water: The Role of Neutron Scattering

A salient feature of all three water controversies summarized above (plus others not listed) is the lack of decisive information about what is going on at the molecular level in these materials. It can be seen that spectroscopic methods can be highly sensitive to changes in the local interactions between molecules, but discerning what those changes imply in terms of the way molecules are arranged or move is not so straightforward. Electromagnetic radiation interacts with the electron distribution of the water molecule, which we know is readily polarized by its neighbors. The scattering and absorption of electromagnetic radiation therefore involves a complex interaction mechanism, which, despite its sensitivity to local conditions, is difficult to emulate in reconstructions of the local molecular dynamics and arrangements. Neutrons on the other hand interact primarily with the nuclei of the molecule. (None of the systems considered in this chapter involve magnetic atoms, so the neutron magnetic interaction can be ignored here.) At the low neutron energies that are typically employed for this work there is no nuclear angular momentum change so the scattering is billiard ball-like: the neutron is scattered equally in any direction from the stationary single nucleus, and so it samples the nuclear positions and dynamics directly.

The following gives a somewhat abbreviated account of how the neutron experiment applies to liquids and disordered materials, but other information is available from other volumes in this series, particularly Chapter 1 (Introduction to Neutron Scattering), Chapter 3 (Instrumentation), Chapter 6 (Dynamics of Atoms and Molecules), and Appendix (Cross Sections) of the first volume [38].

3.1.2.1 The Total Neutron Scattering Cross Section

In order to understand how this works in practice it is necessary to write down the expression for the total scattering of a material by a neutron. Otherwise known as the differential scattering cross-section (DCS), this can be described by the following equation:

$$\frac{d\sigma}{d\Omega}(k_e, 2\theta) = \sum c_\alpha \langle b_\alpha^2 \rangle S_\alpha^{(s)}(k_e, 2\theta) + \sum c_\alpha c_\beta (2 - \delta_{\alpha\beta}) \langle b_\alpha \rangle \langle b_\beta \rangle H_{\alpha\beta}(Q_e) \quad (1)$$

where k_e is the elastic neutron wavevector (see Section 3.2.3 for more details), 2θ the scattering angle, and c_α and b_α the atomic fraction and neutron scattering length of species α. $S_\alpha^{(s)}(k_e, 2\theta)$ is an integral of the single atom van Hove dynamic structure factor for atom type α over energy transfers for a given k_e and 2θ—see for example [39, 40]—and so is subject to inelasticity (so-called Placzek) corrections, while $H_{\alpha\beta}(Q_e)$ is the site-site partial structure factor for atom pair $\alpha\beta$ as a function of elastic wavevector change $Q_e = 2k_e \sin\theta$, which is a similar integral of the van Hove distinct dynamic structure factor. As discussed in Section 3.2.3, this latter term, as far as anyone can tell, is largely immune from inelasticity corrections because the first moment of the distinct dynamic structure factor is identically zero, i.e., the average change in energy from distinct atom scattering (scattering involving pairs of different atoms) is zero [41]. The form (1) therefore highlights the fact that the single atom scattering in a total scattering experiment is likely to be subject to inelasticity effects, whilst for the interference (distinct) scattering, these effects are probably small. In fact the $H_{\alpha\beta}(Q)$ are related directly by Fourier transform to the site-site pair distribution functions (PDF) which form the basis for the statistical mechanical theory of liquids [42]:

$$\begin{aligned} H_{\alpha\beta}(Q) &= 4\pi\rho\int\left[g_{\alpha\beta}\left(\vec{r}\right)-1\right]\exp\left(i\vec{Q}\cdot\vec{r}\right)d\vec{r} \\ &= 4\pi\rho\int r^2\left[g_{\alpha\beta}(r)-1\right]\frac{\sin Qr}{Qr}dr \end{aligned} \tag{2}$$

where ρ is the total atomic number density, and the lower form of Eq. (2) (the most common form) arises in most liquids, glasses, and powdered crystals when the orientation of the $\vec{Q}$ vector is isotropic with respect to $\vec{r}$ on average. Note that slight variations of the definitions (1) and (2) occur in the literature. The definitions used here are those of Faber and Ziman [43], and are probably the most commonly used in experimental work.

For neutrons, an important aspect of the DCS (1) is the angle brackets around the neutron scattering lengths, which represent averages over the spin and isotope states of the respective nuclei. These averages can have a profound impact on what is observed. Light hydrogen, H, has a large spin-incoherence, much larger than any other element. This means that the scattering from materials, which contain H are dominated by the single atom scattering ($\langle b_H^2\rangle \gg \langle b_H\rangle^2$), whilst for deuterium, D, we find $\langle b_D^2\rangle \sim \langle b_D\rangle^2$ so the scattering from materials containing D is a roughly equal mixture of both single atom and distinct atom (interference) scattering. $\langle b_\alpha\rangle$ is referred to as the "coherent" neutron scattering length of an atom. A consequence of this distinction between H and D for neutrons is that whenever possible the deuteriated version of a material is usually advocated for structural work, while the protiated version is used for spectroscopic studies of the hydrogen atomic motion.

3.1.2.2 First Order Difference Method

An idea, which emerged originally in the 1970s and before there were any significant attempts to measure the structure of water itself with neutrons, was to use the change in neutron scattering length with isotope to look at the local order around specific ions in water [44]. Some of the ions studied included Ni and Cl—these atoms have large contrasts between ^{nat}Ni and ^{62}Ni, and between ^{35}Cl and ^{37}Cl respectively. In these cases two otherwise identical solutions are made up (usually in D_2O to avoid the large single atom backgrounds associated with H_2O) with different Ni or Cl isotopes. The two solutions are measured by neutron total scattering under as close to identical conditions as is practically possible. A simple "first-order" difference between these two solution datasets then yields only those correlations associated with the ions that have been isotopically substituted. In practice these correlations are dominated by the ion-water correlations, namely ion to water-oxygen and ion to water-hydrogen. The first-order difference approach has been applied successfully to a number of ions in solution, both cations and anions, and the reviews by Herdman and Neilson, and Mason et al. summarize much of the available data [45, 46]. Other ions that have been studied in this way include nitrate, carbonate, phosphate, guanadinium, and thiocyanate [47–49].

There has also been an application in the recent past to use this first order difference method on water, by exploiting the small difference in coherent scattering length between ^{nat}O and ^{18}O ($\langle b \rangle = 5.80$ fm, $\langle b \rangle = 6.01$ fm) [50]. This gives rise to a small change in the DCS between the two isotopic samples. As a result, the method has been applied to date to look at the difference between the two differential cross sections, from $H_2{}^{nat}O$ and $H_2{}^{18}O$, and from $D_2{}^{nat}O$ and $D_2{}^{18}O$, yielding a mix of the O–O and O–H or OD partial structure factors respectively, but it did help to resolve an earlier controversy about the quantum difference between the O–H and O–D intramolecular bond lengths in liquid water [51].

3.1.2.3 *Second Order Difference Method*

In principle, by measuring a third sample with another isotopic contrast on the same ion, the ion-ion correlation function can be extracted directly. This possibility is sometimes called the "second order difference" method, because it involves the difference between two first-order difference experiments. The second order difference method has been successfully exploited in the case of binary alloys [52], molten salts [53], and binary network glasses [54], and, on some occasions, for aqueous solutions [55], but in most aqueous solution instances of interest, the signal from the ion-ion correlation has a statistical weighting of around 1% in the total scattering pattern, which is about the same as the underlying systematic uncertainty in the data, making its determination by this method unreliable. Such uncertainty arises from the requirement that all three samples must have identical hydrogen/deuterium ratios, which is difficult to achieve in practice, and because of small detection instabilities in the neutron scattering experiment, which can be hard to eliminate reliably at this level.

One situation where the second order difference method has been successfully applied is in the structure of the water itself, both pure and in solution. Hydrogen (H) and deuterium (D) have markedly different coherent neutron scattering lengths, $\langle b_H \rangle = -3.74$ fm, while $\langle b_D \rangle = +6.67$ fm, while the scattering length for oxygen is $\langle b_O \rangle = +5.80$ fm. Since there are only three site-site partial structure factors for pure water, namely O–O, O–H, and H–H, this means that by measuring three different mixtures of heavy and light water, and making the reasonable assumption that heavy and light water have closely similar structures [56], one has in principle the possibility of extracting all the partial structure factors for water. In particular the negative scattering length of H means that the OH partial structure factor has a negative weighting in the DCS for H_2O compared to the positive contributions from O–O and

H–H. For D_2O, on the other hand, all three partial structure factors make a positive contribution to the DCS. As a result the neutron DCS for H_2O, D_2O, and a mixture of these two liquids, are markedly different in shape and magnitude.

The downside of this second order difference method is of course that H_2O has a large, indeed dominant, contribution from the single atom scattering (1). This contribution is exacerbated by the fact that due to inelasticity effects, which in turn arise from the nearly equal mass of the neutron and the hydrogen atom, the single atom scattering has a significant shape which depends on both k_e and 2θ [40]. As a result in the past a great deal of uncertainty has surrounded the precise determination of the site-site partial structure factors for water.

Fortunately there is a circumstance that can lead to an almost error free determination of the H–H function in water. If a mixture of x parts H_2O and $(1 - x)$ parts D_2O is measured and compared to the DCS from the pure liquids (three samples in total), then, assuming that the single atom scattering for H in a mixture of H_2O and D_2O is the same as the single atom scattering for H in pure H_2O, likewise for D in the mixture compared to pure D_2O, it can be shown that the single atom scattering will cancel in this case [56], with a residue related to the H–H partial structure factor alone. However this method does not give the O–O or O–H terms directly, so that nowadays empirical methods are used to remove the inelasticity effects and these are leading to much more consistent outcomes [57].

3.1.2.4 Inelastic and Quasielastic Scattering From Water

So far the discussion has been about structure determination in water and aqueous systems, but there is of course another dimension to neutron scattering in that the energy transfer that occurs in the scattering process in principle tells us about the dynamics of the atoms and molecules in the system. However, in order to detect energy change in the scattering experiment one has to introduce an additional element in the form either of a chopper—to select a particular incident neutron energy and study the distribution of scattered energies by time-of-flight—or else by using a monochromator, such as a single crystal, in the scattered beam to fix the final neutron energy. There is a dramatic drop in counting efficiency compared to the total scattering experiment because only those neutrons that meet a particular combination of incident and scattered energies can be detected at any one time. Fortunately modern techniques, such as focussing neutron guides and monochromators, can help to alleviate some of these deficiencies. For very high neutron energy transfers, greater than ~2 eV, nuclear resonances can be used as the monochromator in either the incident or scattered beams.

Quasielastic neutron scattering (QENS) is a variant of the inelastic method in that it focuses on the small Q region and with energy transfers close to zero

(<1 meV)—hence "quasi" elastic. This region is dominated by diffusive motions, if present, in the scattering system, so that the data can be related to structural and dynamical relaxation times. Generally high energy resolution is required for these measurements.

Full range dynamical studies of water and aqueous systems using inelastic neutron scattering are surprisingly few and far between, probably because of the large amount of beam time and subsequent data analysis that are required. There were a number of early studies by Springer [58] and Safford [59, 60] which were of a qualitative nature, where the possibilities of the technique were first being explored. Although there have been numerous individual studies on particular aspects of water dynamics since then, it is probably correct to assume there are no comprehensive studies covering the full energy transfer and Q scale in existence at this time, certainly not to the same extent that the structure of water has been studied and determined. Fortunately there have been some recent and significant efforts to redress this imbalance [61, 62].

Perhaps the most controversial studies in this regard relate to the purported existence of a fragile to strong transition in confined supercooled water. In this case, quasielastic scattering was used to infer the relaxation time of confined water or water on the surface of a protein, as a function of temperature or pressure [63, 64]. A relatively sharp change in dynamics is observed as a function of temperature at lower pressures, but above about 1.6 kbar for confined water, the dynamical transition becomes continuous, suggesting that above this pressure the system is in a single phase region of the phase diagram, whilst below it a transition from a higher density structure to a lower density structure is occurring. The pressure of 1.6 kbar is then associated with a liquid-liquid critical point in the phase diagram of supercooled water, by inference from computer simulations of bulk supercooled water.

This hypothesis was challenged [65] on the basis of the technical limitations of the neutron spectrometers where this work was performed. In essence the width of the QENS at low temperature becomes comparable to the resolution function of the spectrometer, rendering information on the relaxation time of the diffusing species unreliable, so the true status of this topic is unknown at the present time.

Note that the spectroscopy of nuclear quantum effects in water are discussed in this volume by Andreani et al. (Chapter 7), while water dynamics in biological systems are discussed by Seydel (Chapter 2).

3.1.3 Scope of This Chapter: What Is Included and What Is Not Included

Hopefully the above discussions give a flavor of how neutron scattering can contribute to the science of water and aqueous systems. Large volumes could be written about the structure and dynamics of water, indeed the earlier books by Felix Franks attempted to do just that [66]. Such information rapidly becomes out of date however, so here we will attempt to give the principles of what is currently possible, rather than concentrating excessively on the latest results. In particular, there is no space here to adequately discuss the dynamical aspects of water and aqueous systems as determined by neutron scattering, except to the extent that such dynamics affect the structural experiment. Hence the present account will focus primarily on the structure of water and aqueous systems, how it is determined using total neutron scattering methods, and some of the limitations associated with the method. This will be followed by some of the better established examples of its application.

3.2 THE STRUCTURE OF A LIQUID AND HOW IT IS MEASURED

3.2.1 The Structure of a Liquid

Equation (2) already preempts a fundamental thesis of this chapter, namely that the structure of *any* material is characterized by a hierarchy of correlation functions, ranging from the single body correlation function (in essence the local density at any point in the material) to the many body correlation function which involves all the atoms in the material. For a crystal the one-body correlation function is a well defined quantity, varying from place to place in a regular, periodic manner, but for a liquid or glass the one-body correlation has no characteristic structure and, for a liquid in particular, will be flat, since the atoms and molecules of the liquid can move anywhere.

The first nontrivial correlation for a liquid (or a glass) is the two-body correlation function. In this case the position of one atom or molecule is correlated with that of another as a function of their separation. This correlation will never be flat, even for a dilute gas, since two atoms cannot occupy the same space. Therefore at short enough distances the two-body correlation is always zero, but becomes nonzero further out where atoms do not overlap. If pairs of atoms are strongly correlated, as occurs in a molecule, then the pair correlation function has sharp peaks at the different bonding distances, whereas if two atoms repel each other, as occurs when they have the same Coulomb charge for example, then the pair correlation function will tend to be weak and vary slowly with distance. It is this pair correlation function that is probed directly by the neutron scattering experiment [67].

This simple picture only works, however, in very low-density systems. As the density is increased to the sorts of values that occur in actual liquids and solids, then higher-order correlation functions come into play, in particular the three-body correlation function, which correlates the position of *three* atoms or molecules. These higher order correlation functions can cause the pair correlation to have strong correlation peaks, even when the attraction between atoms is weak. Consider the case of three atoms, A, B, and C, Figure 1. These

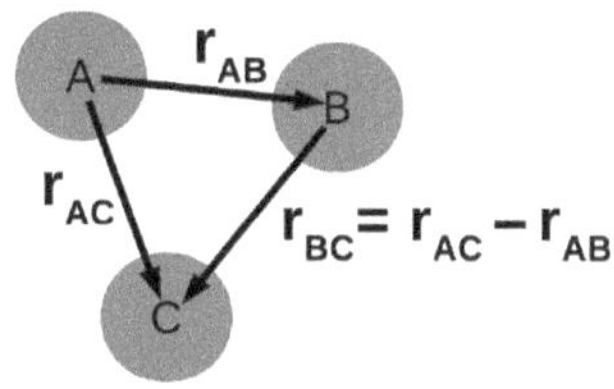

FIGURE 1 Representation of a three-body correlation and possible three-body force. Each of atoms B and C interact with atom A by means of their "direct" interatomic separations, $\mathbf{r}_{AB}$ and $\mathbf{r}_{AC}$. However atom C cannot go anywhere around A since it will also interact directly with atom B via their separation, $\mathbf{r}_{BC}$. This gives rise to a *three-body correlation*, which in turn will affect the measurable pair correlation. Similarly, the presence of atom C in the vicinity of atoms A and B could affect the electron distributions in those atoms, and so modify the interaction between those atoms. Hence the interaction between A and B may not depend *only* on the value of $\mathbf{r}_{AB}$, but also on $\mathbf{r}_{AC}$ and $\mathbf{r}_{BC}$ as well. This is called a *three-body force*. Analogous many body correlations and forces can be defined.

three atoms repel each other strongly out to some distance σ but interact only weakly at larger distances. As the density increases B will get closer to A as will C. However C cannot go just anywhere around A since if it did so it might overlap with B. Hence the presence of B in the vicinity of A places a restriction on where C can go. This restriction occurs whatever the nature of the interaction forces between A, B, and C. As the density is increased further, additional atoms will attempt to get close to A, but they also will be affected by the other atoms that are already around A. Hence there is a need for a hierarchy of correlation functions to adequately describe the structure of a liquid at each particular density, temperature, and pressure.

That observation immediately poses the question: how do you measure the many-body correlation function in a liquid? The short answer is you cannot! The only correlation function we have access to is the pair correlation function, via the neutron (or X ray or electron) scattering experiment (Eqs. 1 and 2). The theory of extended X-ray absorption fine structure (EXAFS) includes many-body effects, due to the way the electron is scattered after it is ejected from an atom, but even here determining those many-body effects from the data, without resort to some modeling technique, can be a major challenge. In addition it should be noted that the pressure derivative of the liquid structure factor does, in principle, give you access to an integral of the three-body correlation [68], although there are very few applications of this result in practice due to the difficulty of measuring pressure derivatives reliably.

The theory of liquids [41] was invented to address the question of how to go from a known force of interaction between pairs of atoms to the fully fledged many-body correlation function of a liquid. However it really only works for simple force fields, and even then requires certain simplifying assumptions to be made. To circumvent this restriction, computer simulation has become the normal approach [69], even though this method also invokes approximations. The simulation has to be performed with a finite number of atoms or molecules placed in a simulation "box" on the computer. To capture the full three-dimensional nature of the liquid, periodic boundary conditions are imposed, which means that, in effect, each simulation box is surrounded by exact replicas of itself, which proceed to infinity in all directions. Provided the correlation functions are only calculated out to distances no more than half the dimensions of the box, then the approximation of using periodic boundary conditions is not a major handicap, but it always has to be borne in mind, especially when attempting to simulate systems with large-scale heterogeneities. A lot will be said shortly about how computer simulation can be used to model the structure of a glass or liquid based on total scattering data.

3.2.2 The Molecular Pair Correlation Function

Molecular materials such as water, namely those where groups of atoms are joined by covalent bonds and move around as single entities, present a significant degree of extra complexity for the total scattering experiment. The formulae (1) and (2) still apply in this case, but now the quantities of interest regarding the structure are different.

First of all the radial distribution functions can be split into correlations involving atoms in the same molecule—the intramolecular structure—and correlations between atoms on distinct molecules, the so-called intermolecular structure. This distinction is usually quite important since the intramolecular correlations will normally be short ranged and sharp because most molecules have a very well defined short range structure. On the other hand the intermolecular correlations will generally be weaker than the intramolecular correlations—often *much* weaker—and hence otherwise hidden under the much stronger intramolecular structure. Yet it is the intermolecular structure that tells us how the molecules are correlated with each other.

In addition, as can be seen in Figure 2, each molecule will have an orientation, defined by three Euler angles $(\alpha\beta\gamma)$, so that now the pair correlation becomes a function of nine variables, namely the three Euler angles needed to define the orientation of the *central* molecule at the origin, the three Euler angles needed to define the orientation of the *second* molecule away from the origin, and the three coordinates (xyz) or $(r\theta\varphi)$ needed to define the *position* of the second molecule relative to the first. (The definitions and conventions used here follow those described by Gray and Gubbins [70].) In practice only six of these coordinates are independent, since the laboratory coordinate axes can always be rotated to coincide with the coordinate axes of each central molecule for which the correlation is required without loss of generality. In that case $(\alpha\beta\gamma)$ becomes the *relative* orientation of the second molecule with respect to the central molecule at the origin. Even so attempting to visualize a quantity that is a function of six independent variables is not easy, and this

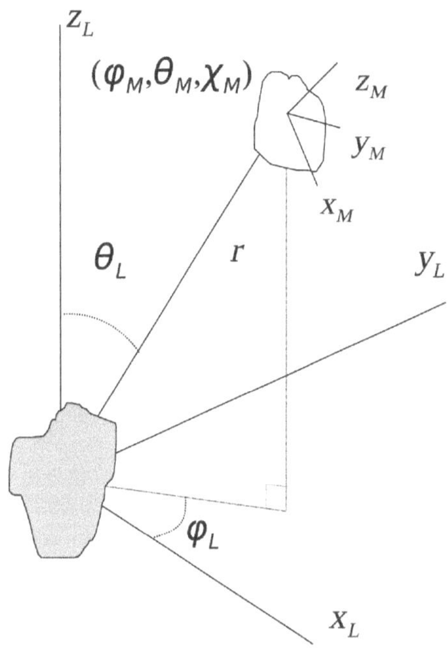

FIGURE 2 The molecular pair correlation function. A molecule (shaded) sits at the coordinate origin. The density of the second molecule is then mapped out as a function of *relative* position, (r,θ_L,φ_L), and *relative* orientation, $(\varphi_M,\theta_M,\chi_M)$.

gets even harder if a molecule has internal degrees of freedom such as dihedral variability, as can occur in long or chain-like molecules.

For simplicity we refer to the position of the second molecule with respect to the first as $(r\omega_L)$, where $\omega_L = (\theta_L\varphi_L)$, and the relative orientation of the second molecule with respect to the first as $\omega_M = (\varphi_M,\theta_M,\chi_M)$, so that the molecular pair correlation function, also sometimes called the orientational pair correlation function, is represented by $g(r, \omega_L, \omega_M)$. Based on this definition, two subsidiary definitions are frequently used to simplify the process of visualizing this function of six independent variables. One of these, the spatial density function, which represents the distribution of second molecules with respect to the first, irrespective of their relative orientation, is achieved by averaging the full molecular pair correlation function over the orientations of the second molecule:

$$g_{SD}(r, \omega_L) = \langle g(r, \omega_L, \omega_M) \rangle_{\omega_M} \tag{3}$$

This function can be useful, for example, for showing the local structure around particular side chains in solution. A further simplification can be achieved by averaging (3) over the directions ω_L to yield the molecular centers pair distribution function:

$$g_c(r) = \langle g_{SD}(r, \omega_L) \rangle_{\omega L} \tag{4}$$

It is important to stress here that unlike the site-site radial distribution functions that appear in the scattering equations (1 and 2), the molecular pair correlation function, $g(r, \omega_L, \omega_M)$, and their derived spatial density and centers correlation functions, cannot be determined directly by experiment, even in principle, so that inevitably there is a lack of information in the scattering experiment on the relative orientations and positions of neighboring molecules. However Gray and Gubbins show that the full molecular pair correlation function can be expanded as a series in generalized spherical harmonic functions. The site-site pair distribution functions can also be expanded as a subset of the same series, so that there *is* a relationship between the site-site terms and the full orientational pair correlation function. The relationship is not complete because not all the terms that appear in the molecular pair correlation function also appear in the site-site functions. As a result, attempts to reconstruct the full molecular pair correlation function from total scattering data are likely to contain ambiguities and uncertainties about the degree and nature of intermolecular correlations. In that case any information about structure in the molecular material should ideally be accompanied by information about which aspects of the structure are well defined by the data, and which aspects are more ambiguous.

3.2.3 The Neutron Total Scattering Experiment

Figure 3 shows the basic geometry of the experiment. For any given material the double differential scattering cross section per unit solid angle and per unit energy transfer into a detector at angle 2θ and with energy transfer E is given by $\frac{d^2\sigma}{d\Omega dE} = \frac{k_f}{k_i} F_n(Q, E)$ where k_i, k_f are the incident and scattered wavevectors, and

$$F_n(Q, E) = \sum c_\alpha \langle b_\alpha^2 \rangle S_\alpha^{(s)}(Q, E) + \sum c_\alpha c_\beta (2 - \delta_{\alpha\beta}) \langle b_\alpha \rangle \langle b_\beta \rangle S_{\alpha\beta}(Q, E). \tag{5}$$

Here $S_\alpha^{(s)}(Q, E)$ is the van Hove single atom dynamic structure factor for atom type α, and $S_{\alpha\beta}(Q, E)$ is the distinct dynamic structure factor for pairs

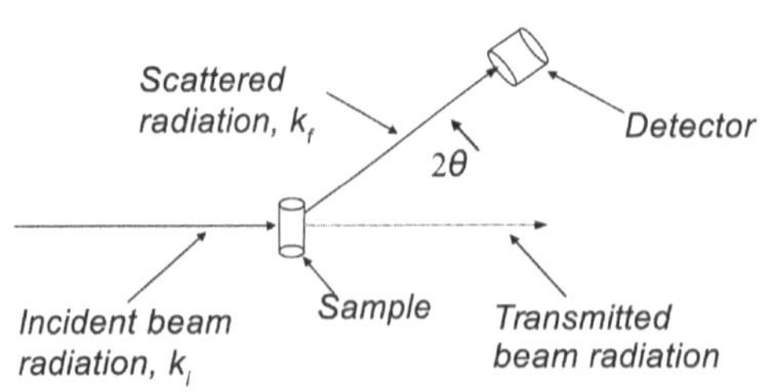

FIGURE 3 Geometry of total scattering experiment. A beam of radiation enters from left and is scattered through angle 2θ to a detector. The sample scattering is integrated over the allowed energy transfers.

of atom types, $\alpha\beta$. As already discussed in Section 3.1.2.1, the angle brackets in (5) signify averages over the spin and isotope states of the respective nuclei.

The neutron scattering event is subject to three constraints, namely conservation of momentum:

$$Q = \sqrt{k_i^2 + k_f^2 - 2k_i k_f \cos 2\theta}, \tag{6}$$

conservation of energy:

$$E = \frac{\hbar^2}{2m}\left(k_i^2 - k_f^2\right) \tag{7}$$

and conservation of time-of-flight:

$$\frac{(1+R)}{k_e} = \frac{1}{K_i} + \frac{R}{K_f} \tag{8}$$

where k_e is the elastic wave vector, i.e., the wave vector for which $k_i = k_f = k_e$ and $E = 0$, and R is the ratio of the distance between sample and detector to the distance between the neutron source and sample. The case where the incident wave vector is fixed by a monochromator is obtained by setting $R = 0$ in (8), since then $k_i = k_e$ for all k_f.

The total differential scattering cross section is defined in terms of its dynamic counterpart, Eq. (5), by integrating over energy transfers:

$$\frac{d\sigma}{d\Omega}(k_e, 2\theta) = \int \frac{\Phi(k_i)}{\Phi(k_e)} \left(\frac{\partial k_i}{\partial k_e}\right)_E \frac{F(k_f)}{F(k_e)} \frac{k_f}{k_i} F_n(Q, E) dE. \tag{9}$$

For any given detector position and 2θ this integral over energy transfers is achieved at each of a series of k_e values. The path of the integral through (Q, E) space is defined by Eqs. (6)–(8), but a Jacobian is required to describe the amount the incident neutron spectrum, $\Phi(k_i)$, is sampled for each value of (Q, E):

$$\left(\frac{\partial k_i}{\partial k_e}\right)_E = \frac{k_i^2}{k_e^2} \frac{(1+R)}{\left[1 + R\left(k_i/k_f\right)^3\right]} \tag{10}$$

In the limit $R \to 0$, $k_i = k_e$ for all k_f, and the Jacobian ≈ 1.0 for energy transfers up to $E = \frac{\hbar}{2m} k_i^2$ and zero beyond. For this reason Egelstaff called this Jacobian the "sampling factor" [71]. The normalization to the elastic incident spectrum and detector efficiency shown in (9) is achieved in the experiment using the scattering from a standard sample such as vanadium to calibrate the incident neutron spectrum, detector efficiency, and detector solid angle at each detector position, as discussed later.

3.2.3.1 Impact of Inelastic Scattering and How It Can Be Corrected

The integral in (9) is performed over the same (Q,E) path for each of the self and distinct terms in the summations in (5), but the result depends significantly on the nature of the scattering atom involved, and whether the single atom scattering or the distinct scattering is being considered. Placzek (1952) showed that the mean energy transfer (first moment) associated with the single atom dynamic scattering law is the nuclear recoil energy, $\frac{\hbar^2 Q^2}{2M}$, where M is the mass of the scattering atom, so that if the (Q,E) path is highly curved (as occurs at low 2θ in a time-of-flight experiment) or the scattering is cut-off prematurely by the sampling factor (as occurs at high 2θ in a fixed incident wavelength experiment) then significant impact on the measured total single atom scattering would occur. He also showed that the first moment of the distinct scattering is identically zero, so that, to a good approximation, inelasticity effects in the integral over energy transfers for the distinct terms should be small. Hence traditional practice is to ignore the likely small distortion to the distinct total scattering caused by inelasticity effects. As a consequence, Eq. (1) shows the elastic wave vector and angle dependence to the single atom scattering, but not to the distinct scattering.

The general conclusion from a careful analysis [41] suggests that for the majority of elements the effects of the inelastic scattering are small and Placzek proposed a simple approximation based on a Taylor expansion around the first and second moments of the dynamic scattering law which would work for most elements and which could be evaluated more or less exactly. However how this expansion should be implemented for different experimental geometries and materials became a matter of some debate. Nowadays it is far easier and just as quick to evaluate the required integral numerically using an appropriate simple model of the dynamic scattering law, such as the ideal gas scattering law, for an atom of specified mass. This integral naturally incorporates all the known effects such as incident neutron spectral shape, detector efficiency, and detector flight path ratio R as required.

For light elements such as hydrogen and deuterium, neither of these methods work, even when quite sophisticated models of the dynamic scattering law are employed [40] and over the years several methods have been tested with varying degrees of success. The most recent method, which will only work when energy dispersive neutron scattering is used in combination with an array of detectors which covers a wide range of scattering angles, involves an empirical method whereby the wavelength dependence of the total differential cross section is first estimated, by averaging over scattering angles for each elastic neutron wavevector. This is then used to subtract the energy dependence of the scattering prior to averaging over neutron wavevectors and scattering angles at fixed Q_e values, to give the Q_e dependence of the total differential scattering. The latter is then used to make the next estimate of the angle-averaged wavevector dependence of the total scattering, and so the process can be repeated until a satisfactory subtraction is achieved.

The advantage of this method is that it does not require prior knowledge of the dynamic scattering function in order to make the inelasticity correction, which in turn means it can be applied to any material. It does however require a broad range of neutron energies and scattering angles to be available, and therefore would not work at a fixed incident wavelength neutron source, such as a reactor source. Also it is still only approximate and may not work satisfactorily in every case. The maximum benefit is achieved if further constraints on the extracted total differential scattering cross section are applied, such as it satisfies certain known criteria at large Q_e and when Fourier transformed to r-space [57].

3.2.3.2 Fixed Wavevector Versus Time-of-Flight Options

Traditional total neutron scattering is performed at a reactor neutron source where the incident wavevector is fixed using a monochromator. Scanning as a function of Q_e is achieved by varying the detector angle or having many detectors over a range of scattering angles. Since it is mechanically quite difficult to have detectors at every possible scattering angle, such detector arrays need some degree of angle scanning. The D4C and D20 diffractometers at the Institut Laue Langevin, Grenoble, France are prime examples of such diffractometers. They deliver extremely high count rates and because reactor neutron sources generate very uniform fluxes of neutrons over a period of time, they tend to be exceptionally stable, making it possible to detect small changes in scattering between samples or between different state conditions of the same sample. In addition the counting statistics are very uniform across the entire Q_e range. They generally work using a few predefined incident neutron wavelengths. These in turn lead to well defined minimum and maximum Q_e that are achievable with these machines. Their weakness from the point of view of studying aqueous systems is the strong inelasticity effects that show up with both H and D at large scattering angles and hence at the large Q_e values, as discussed in the previous section. Even so, some of the most highly cited work on the structure of water as determined with neutrons was in fact performed on a reactor neutron source [72].

Newer neutron sources are tending to be accelerator based, partly because of the problems of getting new reactors licensed. Most of these newer sources are pulsed, so that a pulse of neutrons with a range of energies start from the neutron target at one point in time, then travel down the beamline to the sample position. The traveling time (time-of-flight, TOF) from source to detector is used to sort the neutrons into energies, so that total scattering of these neutrons is an energy dispersive method of measuring the DCS. This has the advantage that a wide range of energies are available, from sub-meV to several tens of eV, covering maybe two orders of magnitude in neutron wavelength. In addition, because the method is energy dispersive, the detectors do not need to move. This means large arrays of detectors can be built to maximize the count rate, without the problem of devising some scheme to allow them to scan as a function of scattering angle. Furthermore, by concentrating the detector solid angle at low scattering angles, large angle inelasticity effects can be avoided. It also means that a very wide range of Q_e values is accessible: in the case of the NIMROD diffractometer on the ISIS Second Target Station this range covers >3 orders of magnitude in Q_e. Some of the drawbacks of the method are that pulsed sources are intrinsically less stable than reactor sources, which can make the detection of small signals difficult. In addition the incident neutron intensity varies dramatically with neutron energy, being extremely intense at some, mostly lower, energies, and much weaker at higher energies. As a result the measured DCS usually has a wide variation of counting statistics as a function of Q_e.

To be effective over a wide range of Q_e values, the time-of-flight total scattering method requires the initial neutron pulse to be very short, of order of a few μs. The European Spallation Source being built at Lund, Sweden, is also pulsed, but has much longer neutron pulses, of order >1 ms. By themselves such pulses are unsuitable for total scattering studies, and ideas to correct this by chopping the incident pulse into a sequence of much sharper pulses would mean removing a large fraction of the incident intensity. At this stage it is not clear whether there would be any advantage to doing this, compared to conventional short pulse neutron sources or reactor-based fixed wavelength sources.

3.2.3.3 Samples That Are Required or Desirable

A concept fundamental to the total scattering technique is that the data need to be put onto an absolute scale of differential scattering cross section. This is because they can be compared directly with simulations of the same structure, and also so that quantities such as $g(r)$ can be calculated reliably. Data analysis codes such as Gudrun [73] can do this in a relatively seamless manner, but they do require that certain sets of data, other than simply the sample of interest, are available.

Perhaps the most important of these is the normalization sample. This is typically a rod or slab of vanadium of similar dimensions to the sample, placed in the same position as the sample. Vanadium is rather unique material for neutrons since $\langle b_V \rangle \approx 0$, so that, although this material scatters neutrons quite strongly, the scattering is almost incoherent, with only very small Bragg peaks from the coherent scattering. The scattering can be further rendered perfectly incoherent by adding about 8 at.% Nb, which has the effect of canceling out the slightly negative scattering length of the vanadium: Nb apparently mixes in a random manner with V. Since we can calculate what the DCS from V or V/Nb should be, the scattering measured from this sample can be used as a calibration for the detector efficiency and solid angle, and, in the case of

TOF scattering, the variation of the incident spectrum as a function of neutron energy.

Other measurements that are required are for the empty sample container, if there is one, for the empty instrument, to check the magnitude of the neutron background, and, if a particular sample environment is being used such as a furnace, cryostat, or pressure cell, then for the empty sample environment.

Normalizing the data to the vanadium sample works if the sample fills the container completely. However this does not work if the material is in the form of a powder since it is the impossible to fill all the space of the container with such a sample. Hence an additional "packing fraction" needs to be estimated to correct for packing effects. This can be measured by weighing the container empty and with the sample, or else by choosing a multiplying factor that leads to the expected limit at high Q_e.

3.2.3.4 Corrections That Are Needed to the Data

Scattering equations such as (1) and (9) imply that neutron enters the sample, scatters once, and is then detected. In practice of course, the finite thickness of the sample means it may self-shield the detector, either by absorbing the neutron before or after it is scattered (attenuation), or by scattering the neutron a second or third or more times (multiple scattering) before it reaches the detector. The sample containment will also be subject to both effects. Provided sample geometries are kept simple, such as cylindrical, flat plate, or spherical, the calculation of corrections for these effects, using the known scattering cross sections of the relevant materials, is straightforward. For more complicated geometries, the corrections are more difficult to evaluate, probably requiring some form of Monte Carlo simulation to generate a realistic replica of the sample scattering environment. However to ignore these corrections for neutron scattering introduces a degree of uncertainty in the final DCS that may render it unreliable for subsequent interpretation in terms of atomistic structure. This is because the corrections will be angle dependent and neutron energy dependent—the latter being particularly important for TOF scattering, where neutron cross sections can vary quite dramatically with neutron energy, particularly if hydrogen is present or one component has a pronounced neutron capture cross section.

The transmitted neutron intensity can be used in principle to estimate the neutron cross section of the sample and its containment as a function of neutron energy. This requires monitor detectors to be placed in the incident and transmitted neutron beams, and ensures the cross sections used to estimate the attenuation and multiple scattering corrections is precisely that of the current sample—such cross sections can depend rather sensitively on the composition and chemical state of the sample. If such transmission data is not

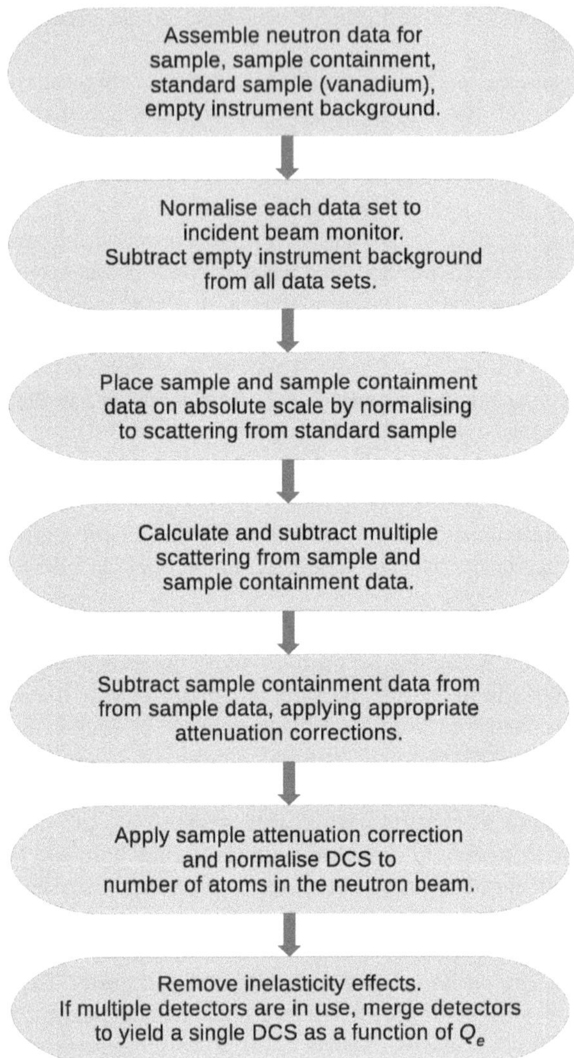

FIGURE 4 Typical sequence of data analysis steps.

available the cross sections as a function of neutron energy can be obtained from tabulations such as the ENDF/B compilations [74].

The method of applying such corrections is routine and well documented in places such as the Gudrun Manual [73], and so will not be repeated here, but the underlying sequence of operations is given in Figure 4, which shows a typical example—particular cases might be more or less complicated depending on the details of the sample containment and environment.

3.2.3.5 *Comparison With Total X-Ray Scattering*

Increasingly, X-ray scattering is being combined with neutron scattering in the investigation of liquids and disordered materials. X-rays provide a powerful contrast to neutrons because the scattering amplitude of each element depends on the atomic number of that element. This means X-rays are generally sensitive to heavier elements in a sample, while neutrons are more sensitive to the lighter elements such as hydrogen and oxygen. The equations and methods that apply to X-ray scattering data are closely analogous to those given here for neutrons, with one major difference being that the scattering lengths, usually referred to as X-ray form factors, are Q-dependent, which arises because X-rays are scattered by the atomic electron clouds, which have a similar dimension to the wavelength of the radiation. Typically, the form of the X-ray form factor is determined for isolated atoms—the independent atom form factors. For heavier atoms, which are dominated by scattering from the core electrons, this works well, but for lighter atoms, particularly where hydrogen-bonding is involved, it is not clear that use of the independent atom form factors is appropriate. A fully satisfactory resolution to this conundrum does not exist, although there have been some attempts [75, 76].

3.2.4 How the Scattering Data Are Interpreted

Until the advent of Reverse Monte Carlo computer simulation (RMC) in 1988 [77] the experimenter was usually left with the only option to try to invert the Fourier transform (2) intrinsic to the scattering experiment, and, depending on the real space structure that this produced, attempt to infer something about the local order of the liquid or disordered material in question. At best they might attempt some simple molecular modeling to remove the dominant structure which arises from intramolecular correlations, so that the more subtle intermolecular correlations could be revealed. If isotope substitution had been performed, this could very often eliminate some of the site-site distribution functions contained in the total scattering data, so giving a more direct view of individual site-site correlations. However there was nothing equivalent to the Rietveld method used routinely in crystallography since 1967 [78] in which a model of the crystal structure was refined against a set of diffraction data, often yielding unprecedented insight and detail. The problem here was that for a liquid or glass there is no long range order or crystal symmetry that can be applied as in the case of crystal structure refinement. Essentially one was dealing with a unit cell of infinite extent and only very basic symmetry.

The RMC approach [77], and its subsequent development, empirical potential structure refinement (EPSR) [79] changed the situation quite profoundly. Computer simulation as a tool to produce liquid-like structures had been around for a number of decades. The transformative step that was made with RMC, and later EPSR, was to refine the simulated structure of the liquid against a set of total scattering data obtained by experiment. With RMC this was achieved by checking the fit to the data after each atom move using a statistical measure. If the fit improved (fitting parameter got smaller), the move was always accepted. If the fit got worse (fitting parameter got larger) the move was only accepted with some probability, which got progressively smaller depending on by how much the fitting parameter got worse. In other words it followed the standard Monte Carlo method, but with energy replaced by the fitting parameter. Constraints on the possible moves were introduced so that pairs of atoms did not get too close and that specified coordination constraints would be satisfied.

EPSR built on this idea to perform simulations involving molecules. In fact water was the first system attempted [79]. As already described in Section 3.2.2, molecular systems present a particular challenge to the total scattering experiment because the required underlying molecular correlation functions are not derivable from the scattering data, even in principle. To try to alleviate this difficulty, EPSR went much further than RMC by introducing the idea of a "reference potential" that would attempt to encapsulate known information about the system, such as the structure of the molecules, whether or not they would be likely to form hydrogen bonds, and whether they were charged or neutral. Hydrogen bonding would be modeled by introducing partial charges on particular atoms, as had been done in conventional computer simulation for many decades, although there was no reason, in principle, why other representations of hydrogen bonding could not be used. Typically the intermolecular part of the reference potential uses a Lennard-Jones term to prevent atomic overlap and to include attractive dispersion forces, a Coulomb term to capture possible charge-charge interactions, and an additional exponential repulsive term designed to keep specific pairs of atoms outside a specified minimum separation:

$$U_{\alpha\beta}^{(Ref)}\left(r_{ij}\right)=4\varepsilon_{\alpha\beta}\left[\left(\frac{\sigma_{\alpha\beta}}{r_{ij}}\right)^{12}-\left(\frac{\sigma_{\alpha\beta}}{r_{ij}}\right)^{6}\right]+\frac{q_\alpha q_\beta}{4\pi\varepsilon_0 r_{ij}}+C_{\alpha\beta}\exp\left[\frac{\left(r_{\alpha\beta}-r_{ij}\right)}{w}\right] \quad (11)$$

where $\epsilon_{\alpha\beta}$ and $\sigma_{\alpha\beta}$ are the Lennard-Jones well-depth and range parameters respectively, q_α is the electronic charge on atomic type α, w controls the hardness of the additional repulsive potential, and r_{ij} is the separation of atoms i of type α and j of type β. The constant $C_{\alpha\beta}$ is adjusted automatically by the program to maintain the required minimum separation, $r_{\alpha\beta}$, of this pair of atom types.

Although the form (11) is commonly used in EPSR simulations there is no reason in principle why other forms for the reference potential cannot be used. For example the reference potential could include a many body potential if required.

3.2.4.1 Dealing With Underdetermined Scattering Matrices

The question then arises: how do you introduce the scattering data into EPSR? To understand this we need to delve a little further into the consequences of the scattering Eq. (1). In order to make progress we will assume we have successfully removed the single atom scattering and its problematic inelasticity corrections, to leave the distinct (or *interference*) differential scattering cross section:

$$F^{(d)}(Q) = \sum c_\alpha c_\beta (2 - \delta_{\alpha\beta}) \langle b_\alpha \rangle \langle b_\beta \rangle H_{\alpha\beta}(Q) \tag{12}$$

as a function of Q. It will be noted that if there are N atomic components in the material, there will be $N(N+1)/2$ terms on the right-hand side of (11). Each term in $H_{\alpha\beta}(Q)$ is weighted by coefficient $w_{\alpha\beta} = c_\alpha c_\beta (2 - \delta_{\alpha\beta}) \langle b_\alpha \rangle \langle b_\beta \rangle$ in the total function. Hence to have any chance of inverting this equation to extract the $H_{\alpha\beta}(Q)$ we will need at least $N(N+1)/2$ measurements of (12), and the different coefficients $w_{\alpha\beta}$ in these measurements must have a nonvanishing determinant. If N is 2, as in the case of water (atoms O and H), then there are three partial structure factors in (12) and this condition can be met provided we measure three samples with different O or H isotopes, and the third sample is a mixture of the first two samples. As N increases above 2, this condition becomes progressively more difficult to satisfy since many common atoms like carbon, silicon, nitrogen, sulfur, etc., do not have suitable isotopes that can be exploited, or if they do, can be prohibitively expensive, if not practically impossible, for use in this way. Hence simply relying on isotope substitution to extract the partial structure factors becomes increasingly infeasible in practice as the systems investigated become more complex.

For RMC the problem of insufficient data is, at first sight, not a problem, since the only criterion that is used to refine the structure is how well the simulated structure fits the available data. However simply fitting a set of scattering data is no guarantee of success, so that if there are not enough datasets to adequately define all the site-site distributions, any proposed solution that apparently fits the data may not be the only structure to do so and hence could be ambiguous. In other words the problem of insufficient data sets can affect the outcome of RMC just as much as with other methods.

To circumvent this difficulty in EPSR two factors come into play. Firstly, the assumed molecular structure and the reference potential, even though it is not necessarily the exact potential for the system in question, has the effect of driving the simulation box into a realistic region of phase space associated with the atoms and molecules of the material in question. They help to ensure the simulation does not sample nonphysical regions of phase space and can be regarded as additional constraints that the structure has to satisfy as well as the scattering data. If a fit is not possible with these constraints, it means either the assumed molecular structure or reference potential is wrong or there is something wrong with the neutron data themselves. Since in general the user will have some choice over the assumed molecular structure and reference potential that is employed, these devices gives the user the opportunity to develop a confidence in which features of the simulated structure are supported by the data, and which features are more ambiguous.

Secondly, a set of fictitious weights, based on the simulated partial structure factors, are added to the weights derived from the scattering lengths and atomic fractions. The neutron scattering weights are multiplied by a "feedback" factor, f, and the simulated distributions are given a weighting proportional to $(1-f)$. Typically $f = 0.9$, so that the simulated distributions are given a weighting typically around 10% of the weighting given to the neutron data, with each simulated site-site distribution being treated as a separate dataset. With this device the scattering equations become overdetermined (more equations than unknowns) so that an inversion of the weights matrix can be performed, by least squares analysis, to produce the most probable inversion matrix based on the neutron weights and computer simulation weights combined.

Having found an inversion to the matrix of neutron weights, in EPSR this inversion is applied to the difference between data and fit to generate a perturbation, called the empirical potential, to the reference potential. This empirical potential is combined with the reference potential to revise the simulated structure over a large number of atom and molecular moves. The process is iterated repeatedly until a satisfactory fit to the scattering data is achieved. Full details are available in Refs. [79–81].

3.2.4.2 *Increasing Importance of X-Ray Scattering Data*

Although this chapter is focussed on neutron scattering, it would be incorrect to convey the impression that neutron scattering, when combined with data modeling tools such as RMC or EPSR, can solve all the problems associated with lack of determinacy in the scattering weights matrix. In this regard, in the past 7 years or so X-ray scattering has proved increasingly powerful at providing an independent look at local order for modeling studies of disordered materials. The scattering equation for X-rays is in essence closely analogous to (1), with the principle change being that neutron scattering lengths are replaced by independent atom form factors (IAFF) which represent the scattering of a single atom on its own. In addition, of course, there is no averaging over the spin and isotope states of each nucleus. These form factors are Q_e dependent because of the finite size and shape of the electron cloud around each atom: this cloud is assumed to be independent of the chemical environment of each atom, an approximation that generally works well because it is the core electrons that do the majority of the X-ray scattering.

The magnitude of the atomic form factors depends on the atomic number of the scattering atom. Hence hydrogen atoms scatter X-rays very weakly compared to other atoms such as carbon and oxygen. This means the total

X-ray scattering pattern from a material is highly complementary to that using neutrons, and the corresponding X-ray scattering weights are therefore quite different to the neutron weights. Hence the determinacy of the scattering weights is strongly enhanced when X-ray data is included alongside the neutron data. In the case of water, for example, the total scattering X-ray experiment is weighted strongly by the O–O partial structure factor, and only much weaker by the O–H term. Meanwhile the H–H structure factor makes a negligible contribution to the X-ray experiment, whilst it strongly dominates the neutron data from both heavy and light water. Hence it is advisable to include X-ray data when attempting to refine the structure of a disordered material, particularly when hydrogen atoms are present.

3.2.4.3 *When Is a Fit a Fit?*

In a typical situation, the fit of a model structure to a set of data can be described by a χ^2 statistic where

$$\chi^2 = \frac{1}{N_p}\sum\frac{(data - model)^2}{\sigma^2} \tag{13}$$

where N_p is the number of data points, σ is the statistical uncertainty in the data, and the sum is over all the measured data points—in the present case all the measured Q_e values and all the measured data sets. If the model were a good fit to the data then we would expect $\chi^2 \approx 1$. Unfortunately, for disordered materials, there are only very broad peaks in the scattering pattern, and, as we have already seen, there may be uncertainties in the data that we do not know about, such as the inelasticity corrections or problems in the absolute normalization and correction of the data. Hence for these kinds of data, the statistical uncertainty may not be particularly helpful in assessing the quality of fit, since the unknown systematic uncertainties could be much larger than the known statistical uncertainties. What is probably true however is that the systematic uncertainties will not oscillate with Q_e like true structural data, but have a much more gradual dependence instead. Hence there is still a good chance of assessing the quality of fit with a χ^2-like statistic, the difference being simply that we might observe some systematic difference between data and model which varies slowly with Q_e and in a manner not like the oscillations seen in the data. This of course introduces a degree of subjectivity into the decision as to whether a particular model is or is not a good representation of the data. To this end, for disordered materials data σ is set to 1.0, so that χ^2 measures the mean standard deviation between data and model. As will be seen in the next section, this still allows for possible ambiguities in the interpretation of a set of data, so that on the basis of achieving a particular χ^2 different models of a set of scattering data can give equally good representations of the data. Viewed subjectively, this means the model that we might propose for a material is not necessarily the one with lowest χ^2.

Note that associated with (12) is its Fourier counterpart in real space:

$$f^{(d)}(r) = \frac{1}{2\pi^2\rho}\int Q^2 F^{(d)}(Q)\frac{\sin Qr}{Qr}dQ = \sum c_\alpha c_\beta (2-\delta_{\alpha\beta})\langle b_\alpha\rangle\langle b_\beta\rangle(g_{\alpha\beta}(r)-1) \tag{14}$$

which can be calculated for both the data and simulation. Because the data have statistical noise and possible systematic uncertainties, and because only a finite range Q is available for both simulation and data, the calculated $f^{(d)}(r)$ are likely to have truncation oscillations and other artifacts which have nothing to do with the structure of the material. Ideally our model structure should not be biased by these truncation effects. Application of the inverse weights matrix to $f^{(d)}(r)$ gives rise to estimates of the $g_{\alpha\beta}(r)$ based on the data and simulation, $g_{\alpha\beta}^{(\mathrm{estd})}(r)$, which can be compared with the same quantities derived directly from the simulation box, $g_{\alpha\beta}^{(\mathrm{sim})}(r)$. Note that because the inverse weights matrix depends on the relative weighting of each site-site term in the scattering data and the feedback factor described in Section 3.2.3.1, the $g_{\alpha\beta}^{(\mathrm{estd})}(r)$ will depend to some extent on the simulated distributions and on this feedback factor. The extent of that dependence is related to the weighting of a particular site-site distribution in the data. A very strongly weighted pair distribution will be only slightly influenced by the simulation, whereas a very weakly weighted distribution will be strongly influenced by the simulated function. A feedback factor of 0.0 would mean no feedback from the data, so that $g_{\alpha\beta}^{(\mathrm{estd})}(r) \equiv g_{\alpha\beta}^{(\mathrm{sim})}(r)$ in all cases.

Another useful guide to how well a particular fit is doing is the local coordination number for particular site-site distributions,

$$N_{\alpha\beta}(r_{\min}, r_{\max}) = 4\pi\rho c_\beta \int r^2 g_{\alpha\beta}(r)dr, \tag{15}$$

which represents the number of β atoms around an α atom in the distance range $r_{\min}$ to $r_{\max}$. Note that whereas $g_{\alpha\beta}(r) \equiv g_{\beta\alpha}(r)$ it does not necessarily follow that $N_{\alpha\beta}(r_{\min}, r_{\max}) = N_{\beta\alpha}(r_{\min}, r_{\max})$ For example in amorphous silica we would expect to find $N_{\mathrm{SiO}}(0,2) \approx 4$ (where the distances are specified in Å) in which case we would also find $N_{\mathrm{OSi}}(0,2) \approx 2$.

In order to control the leveling of fit in an EPSR simulation, the empirical potential (derived from the comparison of the simulated with the measured total scattering data), $U_{\alpha\beta}^{(\mathrm{ep})}(r)$, is characterized by an amplitude:

$$A^{(\mathrm{ep})} = 4\pi\rho \sum c_\alpha c_\beta \int r^2 \left| U_{\alpha\beta}^{(\mathrm{ep})}(r) \right| dr. \tag{16}$$

As $A^{(\mathrm{ep})}$ increases from zero, so the fit should improve. However this improvement cannot proceed indefinitely, otherwise the simulation will start to reproduce the truncation and other unphysical artifacts in $g_{\alpha\beta}^{(\mathrm{estd})}(r)$. Hence, although some methods for automatically adjusting $A^{(\mathrm{ep})}$ have been developed in the latest versions of EPSR, the final decision of what value of this amplitude to use is a subjective decision left to the user. If the errors in the scattering data were purely statistical in nature, then it would be simple to adjust $A^{(ep)}$ until the required value of χ^2 was achieved, but since the most likely scenario is that systematic uncertainties are often larger than the statistical uncertainties, this is not an option in practice. Moreover, although the problem will be couched in different terms for RMC or other types of total scattering data interpretation, exactly the same outcome will occur, namely the final structural model that is presented will be subject to a decision on what constitutes a reasonable fit on the part of the experimenter. Presently there does not appear to be a rigorous solution to this dilemma.

3.3 PURE WATER SUBSTANCE

The most recent attempts to extract the structure of water from a set of total scattering experiments using both neutrons and X-rays were described in [57]. Purely by chance, this paper was published almost simultaneously with an independent analysis of a series of X-ray-only experiments [82]. Remarkably, the results for the O–O radial distribution function for water presented in these two papers were almost indistinguishable, suggesting that a consensus was, after >40 years of trying, at last beginning to emerge about the heights, shapes, and positions of peaks in this distribution function. Up to that time, such consensus had not existed, with some very big variations in the reported radial distribution functions of water.

Why was that? We have already alluded to the fact that total scattering data can have uncertainties that are not purely statistical in origin. Moreover these uncertainties are not always well understood or well-characterized. This is equally true of both X-ray and neutron data. In the case of X-rays, uncertainties can arise from the sample geometry used (cylindrical versus flat plate), the extent to which polarization corrections are required, and the ever present Compton scattering (X-ray equivalent of inelastic scattering with neutrons), the amount of which is included in the data is strongly affected by the detector characteristics, which may or may not be known. Attempts to remove the Compton scattering by energy analysis of the scattered X-rays introduce further uncertainty in the final outcome, since however the energy analysis is performed there will always be a region of Q where both elastically scattered and inelastically scattered X-rays will contribute to the total scattering. Furthermore, it must be remembered that X-rays are scattered by the electron cloud around an atom, and not directly from the atomic nucleus. To go from X-ray scattering pattern to a nucleus-nucleus correlation function requires an assumption of how the electrons are distributed around the nuclei, and in the case of water, the hydrogen bonding might affect that distribution in a manner, which is currently difficult to quantify.

In the case of neutron data, we have already seen that inelasticity associated with scattering from hydrogen and deuterium produces a separate set of uncertainties, plus the effects of sample dependent background can be significant, depending on the geometry of the neutron experiment.

As a result it is highly unlikely that different sets of data measured on different X-ray and neutron sources will give identical results. The most recent attempts with X-rays, where several datasets were measured and compared, are therefore to be welcomed [82]. There is however an even more fundamental reason why different estimates of the radial distribution functions of water have been so variable. These derive from the way the data are interpreted. Bearing in mind that we cannot directly measure these radial distribution functions, the only way to access them is via some manipulation of the data, using the known neutron or X-ray weightings, to yield the individual partial structure factors, and then attempt to Fourier transform the results to real space. This was done in the early days before data modeling was implemented and gave wildly discrepant results—compare [83] with [56, 72]. Later RMC and EPSR methods became available, in which a computer simulation model of water was refined against a set of scattering data. Even so the outcomes were not overly impressive—compare [84] with [85, 86].

3.3.1 A Case Study: Ambient Water

The answer to this conundrum lies fairly and squarely on the different uncertainties and errors in different data sets, as well as the ever present finite range of Q that blights every experiment. Even the Q-scale itself can be the subject of some uncertainty because the low resolution of a typical disordered materials diffractometer means the Bragg peaks from a crystalline calibrator are broad, making their precise position determination uncertain. To illustrate the consequences of these considerations we consider here a set of neutron and X-ray total scattering data measured at the ISIS pulsed neutron source and at a laboratory (fixed wavelength) X-ray diffractometer (XRD) located at the same facility. These data are included in the compilation already presented in [57].

For the neutron experiments, data were recorded for four samples, namely pure H_2O, pure D_2O, a 50:50 mole ratio mixture of H_2O and D_2O, referred to here as "HDO", and 64:36 mole ratio mixture of H_2O and D_2O, referred to as "null" because at this composition the mean neutron scattering length of the hydrogen atoms is zero. This quality of the "null" sample means it is sensitive to only the O–O partial structure factor.

To interpret these data, two EPSR models were set up, described in Table 1: Model A, has a negative charge on the oxygen atom and neutralizing positive charges on each hydrogen atom, while Model B has no charges on any atoms. The Lennard-Jones parameters for each model were chosen so that without any structure refinement the O–O distance was close to 2.8 Å, as indicated from the X-ray data and the "null" sample data. In addition, to prevent unphysical overlaps, a minimum approach distance of 1.5 Å for the intermolecular H–H separation was specified. The simulation was run including the

TABLE 1 Parameters for the Intermolecular Reference Potential for EPSR Simulations of Pure Water Described Here

Parameter	Model A		Model B	
σ_O [Å]	3.5		2.9	
σ_H [Å]	0.0		0.0	
ε_O [kJ/mol]	0.3		0.3	
ε_H [kJ/mol]	0.0		0.0	
q_O [e]	−1.0		0.0	
q_H [e]	+0.5		0.0	
r_{HH} [Å]	1.5		1.5	
C_{HH} [kJ/mol] (approx.)	15.6		27.3	
w [Å]	0.3		0.3	
Results	**With X-ray**	**Without X-ray**	**With X-ray**	**Without X-ray**
Configurational energy [kJ/mol]	−45	−40	+10	+33
Pressure [GPa]	+2.2	+2.5	+2.0	+3.5
χ^2 [$\times 10^4$]	7.3	8.1	7.3	11.0
Empirical amplitude [kJ/mol]	50	50	110	110
N_{OO}(0,3.39 Å)	4.8	4.7	4.8	4.8
N_{OH}(0,2.40 Å)	1.8	1.8	1.7	1.8
N_{HH}(0,3.00 Å)	5.5	5.6	5.6	5.5

The Lennard-Jones parameters are determined from the Lorentz-Berthelot mixing rules, namely $\sigma_{\alpha\beta} = \frac{1}{2}(\sigma_\alpha + \sigma_\beta)$ and $\varepsilon_{\alpha\beta} = \sqrt{\varepsilon_\alpha \varepsilon_\beta}$. The values of C_{HH} are only indicative of the typical values because they vary continuously as the simulation proceeds. Other values (configurational energies, pressures, χ^2) represent averages over ~1000 configurations.

X-ray data in the structure refinement and also using only the neutron data in the structure refinement. In the latter case the simulated X-ray scattering pattern was still calculated even though the simulation was not refined against the X-ray data. Also calculated are the simulated nearest neighbor coordination numbers.

Looking at Table 1, one is struck immediately that although all four models do a reasonable job at fitting the data ($\chi^2 \approx 10^{-3}$), the energies and pressures of the simulations differ quite widely. In particular it is found with Model B (no charges) that the configuration energy is positive, implying the system is unbound and would disintegrate if allowed to do so. For Model A, the configurational energy is negative and consistent with the known latent heat of evaporation of water (~44 kJ/mol) at 298 K. This immediately tells us that a model with charges is likely to be more realistic than one without charges. At the same time the amplitude of the empirical potential required to obtain a reasonable fit is more than a factor of 2 smaller for this model. In none of these cases however is the pressure of the simulation particularly good: at these pressures (>2 GPa) in real water and at 298 K water occurs as Ice VII! The most likely cause of these high pressures is the hard repulsive edge of the Lennard Jones potential, which rises too quickly at short distances with decreasing atomic separation compared with the more physically realistic exponential rise that is expected in practice. Lennard-Jones is used here in these examples to counteract the divergence of the Coulomb potential that would otherwise occur. In practice, more realistic core potentials can be employed, based on the ideas of Buckingham [87], but in that case steps have to be taken at small r to prevent the Coulomb $1/r$ divergence.

Note that these considerations do not occur in traditional RMC simulations, since there is no energy or pressure in such simulations. The latest variants of RMC in fact do include the equivalent of a reference potential [86] so it would seem using energies and pressures to assess the validity of a particular structure refinement does have an intrinsic value in data interpretations of these kinds.

With regard to the fits in Q space, Figure 5, it can be seen that within some uncertainty, each fit is equivalent to the others, irrespective of the energy or pressure, and this is borne out by the values of χ^2: only for the simulation with Model B and no-X-ray data is this χ^2 value slightly larger than the other simulations. Based on the visual quality of fit shown in Figure 5, it is not clear whether this difference is significant. At the same time it is observed that there remain, in all cases, differences between the EPSR simulation and the data. The method is apparently not able to produce a "perfect" fit to the data, i.e., one that goes uniformly through the measured data points. So the question arises as to how reliable these EPSR simulations are?

We can partly address this question by looking at the nearest neighbor coordination numbers, Table 1. It can be seen that for all four simulations the results are same for each of the O–O, O–H, and H–H site-site radial distribution functions within 0.1 atoms. These values are also closely similar to those found in numerous other computer simulations of water. The O–O coordination number is almost universally found to be >4, but <5, while the O—H coordination number is always <2. If we assume the first intermolecular peak in the O—H distribution represents hydrogen bonding between water molecules, a coordination number of ~1.8 means that in water every water

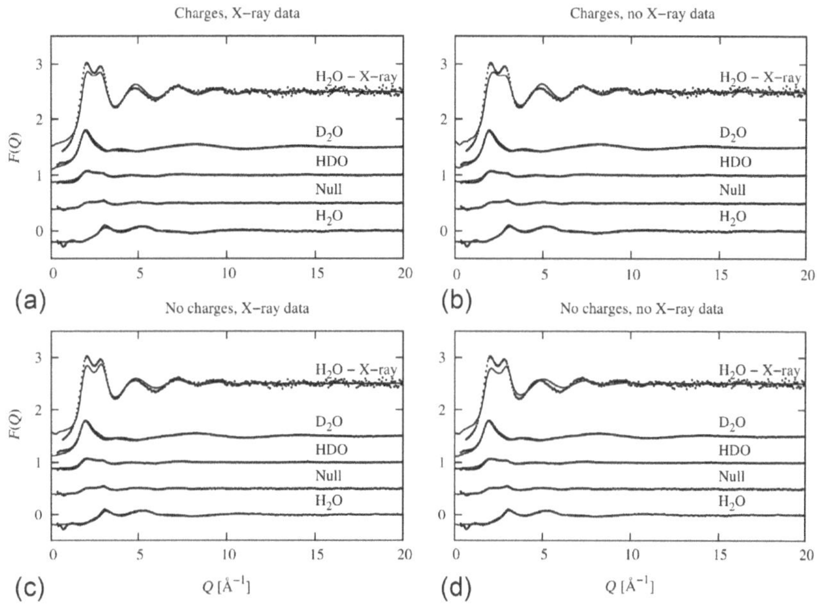

FIGURE 5 Measured (dots) and EPSR simulated fits to (lines) total scattering X-ray and neutron interference scattering functions, $F^{(d)}(Q)$, for, in order from top of each figure, X-ray scattering, neutron D_2O, neutron HDO, neutron "null" water, and neutron H_2O. The top row, (a) and (b), shows the simulations with charges in the reference potential for the oxygen and hydrogen atoms (Model A), and the bottom row, (c) and (d), shows the simulations with no charges (Model B). The left hand column, (a) and (c), shows the results when X-ray data are included in the structure refinement, while the right-hand column, (b) and (d), shows the results where the X-ray data are simulated, but not used in the structure refinement.

molecule will be hydrogen bonded to ~3.6, other water molecules on average, meaning roughly 10% of the possible hydrogen bonds in water are broken.

So far so good: there doesn't appear to be any doubt on these assessments, and if someone proposes a different scenario for the local bonding in water, as occurred in 2004 [28], where it was proposed that water is a chain structure and not a network, it can be shown that this is impossible for a material with a coordination number >4 [88].

Much more contentious, however, is the *shape* of the associated $g_{\alpha\beta}(r)$ functions. Here there are far more divergent views, yet those computer simulators who aim to derive better and better molecular potentials for water which capture the correct energetics, structure, and dynamics from the gas phase to the crystalline phases, seek highly accurate values for these functions in order to assess and refine their models. Figure 6 gives an indication of why this might be a difficult, if not impossible, assignment to achieve in practice.

For example, Model A, the model with atomic charges, gives quite a pronounced first peak in the O–O radial distribution function, going above 3 at its highest point. On the other hand in Model B with no atomic charges, the same

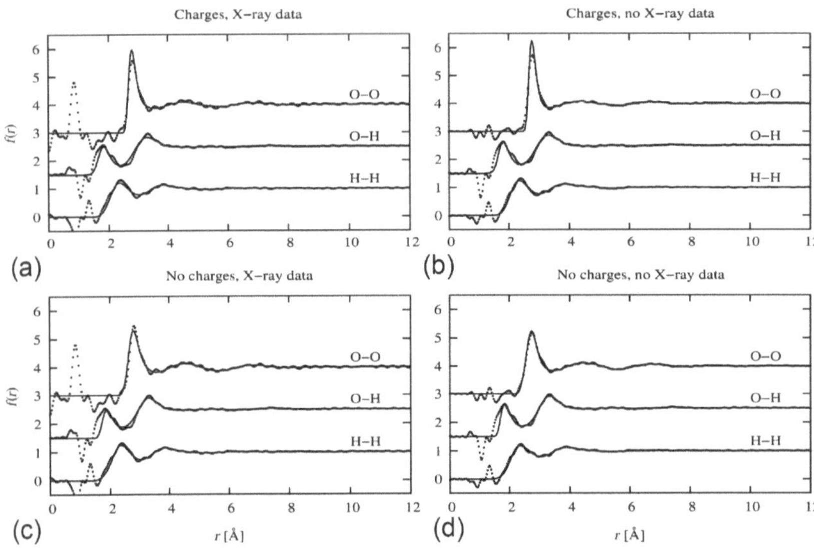

FIGURE 6 Comparison of $g_{\alpha\beta}^{(\mathrm{estd})}(r)$ (r) (dots) with $g_{\alpha\beta}^{(\mathrm{sim})}(r)$ (lines) for each of the EPSR simulations described in Table 1 and Section 3.3.1. The top row, (a) and (b), shows the simulations with charges on the oxygen and hydrogen atoms in the reference potential, and the bottom row, (c) and (d), shows the simulations with no charges in the reference potential. The left hand column, (a) and (c), shows the results when X-ray data are simulated and included in the structure refinement, while the right-hand column, (b) and (d), shows the results where the X-ray data are simulated, but not used in the structure refinement.

peak is broader and lower, with maximum height around 2.3. Based on the supplied X-ray and neutron scattering data, there simply is no reliable way of saying which is better. Obviously the energy for Model A is much more realistic, but without extensive further testing, there is nothing to say a different version of Model A still with atomic charges but with softer core potential would not also produce a softer first peak in $g_{OO}(r)$. Generally it is assumed that the true peak height for water is somewhere between these extremes.

The role of X-ray scattering data in finding water structure is also apparent in Figure 6. The left hand graphs in this figure show the results when X-ray scattering data are included. Although these data do not apparently help the debate about the height and sharpness of the first peak in $g_{OO}(r)$, they do stabilize the second peak in this function. The right hand graphs show the fits when the X-ray data are not included, and it can be seen that the shape and position of this second peak varies with the choice of reference potential. On the other hand for the left-hand graphs, the second and subsequent peaks in this function are much more similar to each other, irrespective of which reference potential is employed. At the same time, comparing the left- and right-hand sides shows the results where the X-ray data are simulated, but not used in the structure refinement of this figure, the O–H and H–H distributions are much less affected by whether the X-ray data are included or not: as expected, these functions appear well characterized by the neutron data and are not strongly affected by which reference potential is used, or whether X-ray data are included or not.

The point of the above analysis is to highlight the fact that even when water molecules are defined by realistic (in this case harmonic) forces, and atomic overlap is controlled by means of realistic repulsive and attractive (dispersion) forces, the outcome of any given structure refinement of a set of total scattering data, still depends to an extent on the starting assumptions. If repulsive forces had been represented by hard-sphere type interactions, as usually occurs in RMC simulations, or molecules are represented by simple coordination constraints, even more divergent solutions might have been found. Total scattering data, even when the matrix of scattering weights is well behaved as here, and even when statistical errors are small as here, do not sufficiently characterize the local atomic and molecular structure by themselves. Sadly, this fact has often been poorly recognized in the literature with claims and counter-claims based on different datasets, when the real problem is not the data themselves but the different methods used to interpret those data.

3.3.2 Beyond the Site-Site Radial Distribution Functions

Traditional wisdom has it that water is a hydrogen-bonded, tetrahedral structure. If that was strictly true, then any given water molecule would be surrounded by four other water molecules, and the angle between any three of them would be the tetrahedral angle, namely $\cos^{-1}\left(-\frac{1}{3}\right) = 109.47^{\circ}$. Moreover the Bernal-Fowler ice rules would require that two of those surrounding water molecules would have one of their hydrogen atoms pointing toward the central water molecule, and two of them would have both their hydrogen atoms pointing away from the central molecule. Is that what happens in the real liquid? We have seen that the nearest neighbor coordination number for water is >4, so if there is tetrahedral-like local order, it will be more complicated than a simple tetrahedron. To answer these questions we consider two quantities that are derivable from the EPSR simulation of water, but which cannot be extracted directly from the scattering data. The first of these is the triangle distribution function—aimed at answering our first question.

3.3.2.1 Triangle Distribution Functions

Within the simulation box, two water molecules can be regarded as bonded if their separation is less than some specified distance. Since the nature of the "bond" in this case is undefined, one possibility, in keeping with our aim of determining the extent to which water is a tetrahedral structure, is to choose that distance at which the nearest neighbor coordination number reaches 4.

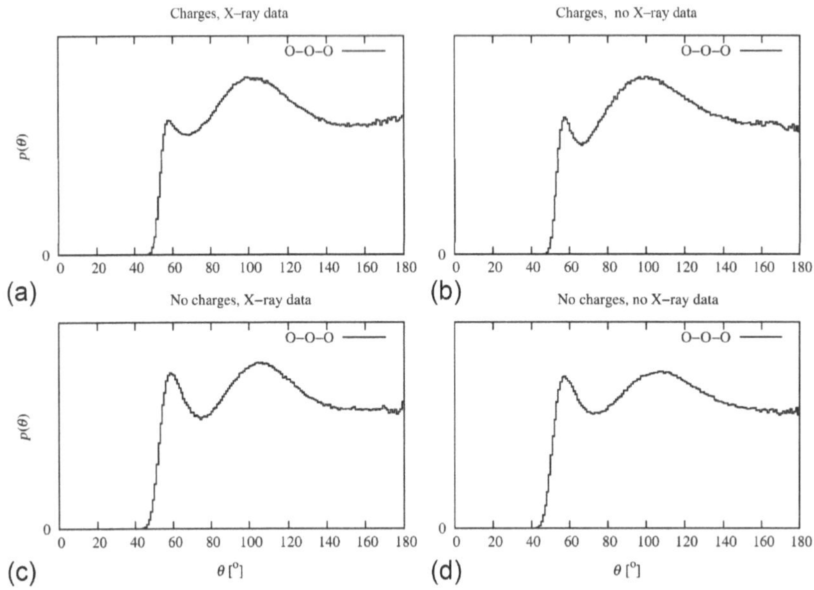

FIGURE 7 Triangle distribution functions for the four EPSR simulations shown in Table 1. Water molecules are regarded as bonded if their separation is 3.2 Å or less. θ is the included angle between three bonded water molecules. The top row, (a) and (b), shows the simulations where charges are included in the reference potentials for O and H, while the bottom row, (c) and (d), shows the simulations where the charges are set to zero. The left hand graphs, (a) and (c), include X-ray data in the structure refinement, while the right hand graphs, (b) and (d), shows the results where the X-ray data are simulated, but not used in the structure refinement.

For the simulations shown above this is $r \sim 3.2$ Å. The results of calculating the distribution of triangles of water molecules that satisfy this distance criterion for the same four models already described are shown in Figure 7.

Now we discover a problem with our starting assumption. Firstly the distribution of angles, far from being well-defined around 109.47 degree is in fact rather broad and extends over all angles from ≤ 60 degree up to 180 degree. Secondly there are not one but *two* peaks in this distribution, one very broad one at ~ 102 degree, the other more pronounced near 60 degree. The three water molecules involved in the latter peak clearly cannot all be hydrogen bonded in the traditional sense, unless one adopts an extremely bent hydrogen bond definition for one or other of the water molecules. Clearly, therefore, the simple picture generated by the statement that water is "tetrahedral" doesn't work in practice: the real structure is much more complicated and, unlike a typical crystal structure, it is described by multivariate distribution functions, not by specific angles and positions. Furthermore, we can also comment on the idea that all water molecules must be hydrogen bonded. To have a peak at 60 degree in the triangle distribution means that even in the nearest neighbor shell there will be some water molecules that are not hydrogen bonded, according to any reasonable definition of a hydrogen bond that might be assumed. In fact this result tallies with the fact that the O–H coordination number is around 1.8, i.e., not 2 as would be expected for a fully hydrogen bonded system. In water some hydrogen-bonding possibilities are always present that are not satisfied.

The other thing to be noted about these triangle distribution functions is that they are more sensitive than the site-site distribution functions to the chosen reference potential. Where the reference potential has Coulomb charges, so the 102 degree peak is stronger and the 60 degree peak is weaker. When charges are not included, the 60 degree peak is more pronounced.

3.3.2.2 The Spatial Density Function

A more graphical way to visualize the structure of water is to calculate the spatial density function described in Section 3.2.3. This is a three-dimensional representation of the pair correlation function after averaging over orientations of the second molecule, $g_{SD}(r,\omega_L)=\langle g(r,\omega_L,\omega_M)\rangle_{\omega M}$. Since this is a function of three dimensions, the best one can do is present a surface contour of the spatial distribution, with the contour level chosen to capture a representative view of the local density over some specified distance range. In the case of water it is necessary to specify several distance ranges because the structure changes quite characteristically as the separation of the first and second molecules is increased.

There are different ways to calculate these spatial density functions, but in EPSR the method involves expanding the full orientational pair correlation function as a series of generalized spherical harmonic functions, then reconstructing the spatial density function from the calculated coefficients of that expansion. The full details of how this is done are given elsewhere [89] and so will not be repeated: here we show only the results of this analysis, Figure 8. We will do this for Model A where the X-ray data are included in the structure refinement.

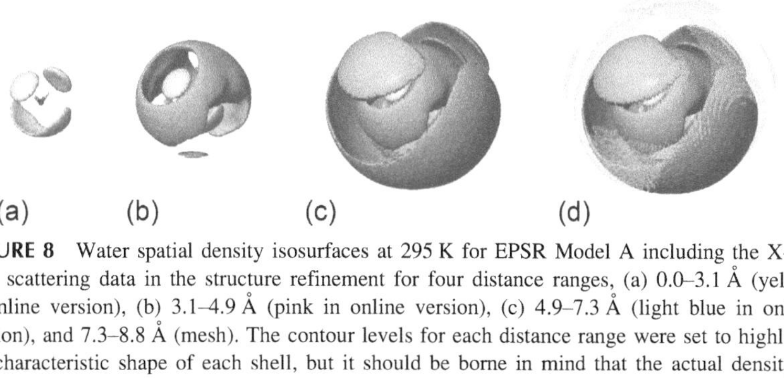

FIGURE 8 Water spatial density isosurfaces at 295 K for EPSR Model A including the X-ray total scattering data in the structure refinement for four distance ranges, (a) 0.0–3.1 Å (yellow in online version), (b) 3.1–4.9 Å (pink in online version), (c) 4.9–7.3 Å (light blue in online version), and 7.3–8.8 Å (mesh). The contour levels for each distance range were set to highlight the characteristic shape of each shell, but it should be borne in mind that the actual density is continuous: these plots are simply highlighting the regions of highest density within the specified distance range. In each case the inner shell or shells are shown to expose how the current shell fits into the previous shell. Horizontal dimension of each plot is approximately 20 Å.

On inspection of these functions, one is struck by the characteristic motif of water structure, namely (a) an inner core of molecules which form acceptor hydrogen bonds with the two hydrogen atoms on the central molecule. This is accompanied by a band of molecules underneath the central molecule, which would traditionally be regarded as donor hydrogen bonding to the lone-pair electrons on the central molecule. The important point is that this first shell does not follow exact tetrahedral symmetry: the top two lobes are always distinct and separated, whatever contour level is chosen, but the bottom lobes become single at a sufficiently low contour level as shown here.

By contrast, the second shell (b) is in almost perfect antiphase to the first shell. Presumably these are molecules that are bonded to the first shell, but not bonded to the central molecule. What is perhaps not expected, however, is that the third shell (c) is a more disordered version of the first shell, with distinct lobes opposite each hydrogen atom on the central molecule, just at greater separation, while the fourth shell (d) is also closely analogous to the second shell (b) in shape. It is possible that this pattern repeats itself even further away from the central water molecule. It is clear therefore that although the spatially averaged radial distribution function of water apparently becomes nearly uniform after about $r \approx 7.5$ Å (Figure 6), there is an underlying pattern of characteristic 3-dimensional density correlations that persist beyond this. How far away from the central molecule these density correlations actually proceed is so far undetermined. This pattern of alternating shells in the spatial density function was first noticed by [90].

3.3.2.3 *Orientational Correlations*

It is possible to take the analysis described here one step further to explore how the orientations of the first, second, third, etc., neighbors correlate with the central molecule. This introduces a further 3 dimensions to the visualization problem, so will not be pursued further here, although the example of water around glutamine in aqueous solution [90a] does show these orientational correlations.

3.3.3 Water Structure: Effect of Temperature and Pressure

Given that the tools described above now exist to explore water structure in a variety of environments, the obvious next question is how is water structure altered by changes in temperature and pressure.

3.3.3.1 *Effect of Temperature at Ambient Pressure*

The effect of temperature on water structure at ambient pressure has been explored several times with X-rays [22, 91, 92] but rarely with neutron scattering. As can be judged from Section 3.1.1.1 there is much interest in the structure of water in the supercooled region, but this is difficult with neutrons because of the problems of crystallization in the typically large (~1 mL) samples that are required for these experiments. There have been some successful attempts with microemulsions [93] but such measurements will include contributions from the water-surfactant and water-oil correlations, so that reliable interpretation of the scattering patterns is not possible, other than some rather general observations.

For temperatures above 273 K there have been some recent neutron measurements published [94], which can be compared with new X-ray measurements [92]. The work of Skinner focuses on the fact that the isothermal compressibility of water has a *minimum* at 319 K: at lower temperatures the compressibility *increases* with decreasing temperature, making water rather unusual in this regard compared to most other liquids. They observe that up to the compressibility minimum the second peak in the O–O radial distribution function is nearly stationary in position with temperature, but at higher temperatures the peak moves out to larger distances with increasing temperature, and it is speculated this represents a transition from anomalous to normal liquid behavior. The same peak trends are seen in the Varenna data [94], so it is interesting to calculate the spatial density functions based on these latter data as a function of temperature. This was achieved using EPSR simulations in exactly the same way as described previously and with the same reference potential. The results for the first and second coordination shells at 280 K, 313 K, and 365 K are shown in Figure 9.

Comparing these with Figure 8, one is struck by the similarity of all these SDFs, which appear almost invariant in shape with temperature. There is some blurring of the distributions at the highest temperature, but, instead of the second shell increasing to larger distances as the temperature increases, as stated by Skinner et al., there is in fact a slight but noticeable contraction of this shell onto the first shell with increasing temperature. Presumably this is a consequence of the slightly weakened and hence distorted hydrogen bonds at the higher temperatures. The outwards movement of the second peak in

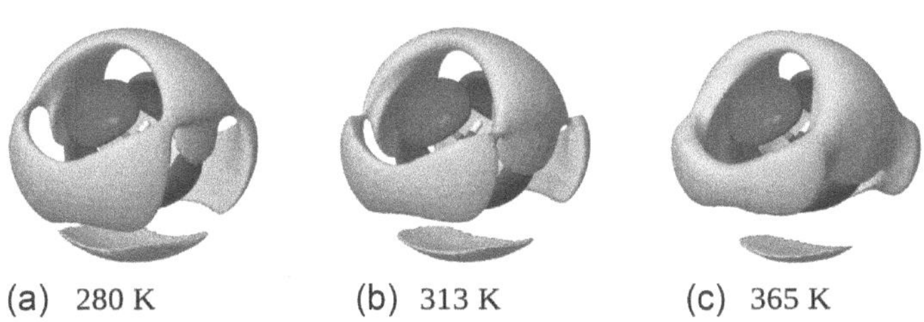

FIGURE 9 EPSR simulated spatial density functions for water at three temperatures, namely (a) 280 K, (b) 313 K, and (c) 365 K. The reference potential used in these simulations is the same as for Figure 8. The first and second coordination shells are shown. The spatial density functions are shown in the distance ranges 0.0–3.1 Å (yellow in online version) and 3.1–4.9 Å (pink in online version). The horizontal dimension of each plot is approximately 10 Å.

$g_{OO}(r)$ has to be seen as a rather subtle consequence of the increase in temperature and not as some fundamental change in local order above the isothermal compressibility minimum. Indeed it is likely that there is a remaining signature of the underlying tetrahedral-like nearest-neighbor coordination even in the dense gas above the water liquid-vapor critical point at 647 K [95]. Therefore, although water is a highly mobile, dense liquid in the stable condensed phase, it does nonetheless retain a very characteristic local order that renders it rather unique amongst most known liquid materials. These results also highlight the fallacy of attempting to interpret features in the radial distribution functions of water in terms of structure. As has been alluded to several times in this article, these radial functions hide a lot of the true nature of water structure.

3.3.3.2 *Effect of Increased Pressure*

As soon as one goes away from ambient pressure (0.1 MPa) the difficulties for the neutron (and X-ray) experiment grow considerably. The main issue is that to make a significant change in water structure you need a change in pressure of around 0.1 GPa. This translates to an increase in thickness of the walls of the containing vessel by a factor of at least 6—much larger if much higher pressures are needed. Some of the first experiments were performed by Neilson et al. [96], which was followed much later by an experiment on supercooled water under pressure by Bellisent-Funel [97]. In the latter case, the experiment was performed partly in an emulsion to prevent water crystallization. The effect of the emulsion on the observed water structure is unknown at this point in time, and, furthermore, there was no attempt to analyze the data in terms of the three site-site radial distribution functions or spatial density functions. Later these data were analyzed to indicate a possible two-state model for water [98] in which any particular density and temperature is regarded as a linear mixture of two end-point states, namely a low-density liquid and a high-density liquid.

For X-ray scattering, the most definitive study was performed by Okhulkov et al. [99] who looked at water in the pressure range 0.1–770 MPa and at 293 K. Their primary observation, Figure 10, was that the nearest neighbor peak in $g_{OO}(r)$ moves very little with increased pressure, but the second peak at ~4.5 Å changes considerably, being eventually replaced by a pronounced shoulder on the large r side of the first peak near $r = 3.3$ Å and a broad minimum near 4.8 Å. The subsequent oscillations in $g_{OO}(r)$ also compress inwards with increased pressure.

To understand in greater detail what was going on, Soper and Ricci [100] published a neutron total scattering study of water under pressure at pressures of 26, 209, and 400 MPa, and at 268 K, i.e., just below the normal water freezing point. Hydrogen-deuterium substitution was performed in order to extract the three site-site partial structure factors, namely O–O, O–H and H–H. The data at all three pressures were subjected to EPSR simulations in

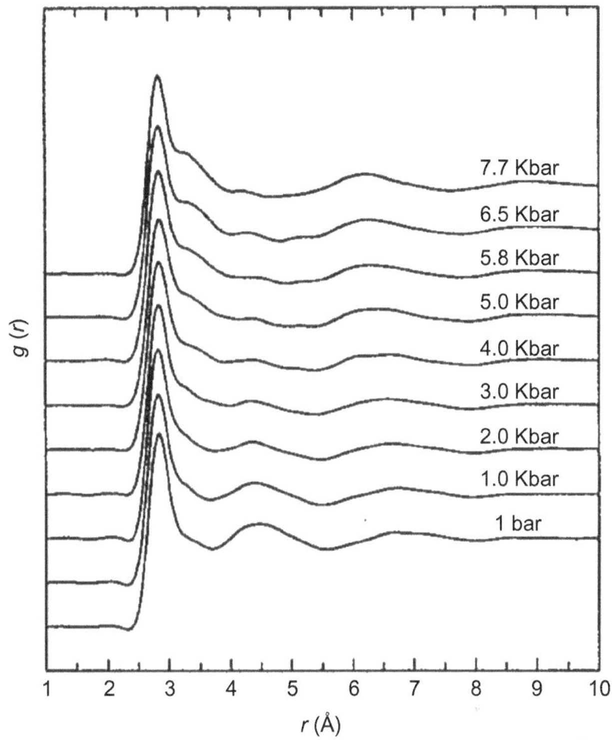

FIGURE 10 X-ray radial distribution functions for water as a function of increasing pressure. *Reprinted with permission from Ref. [99].*

the same way as has already been described, and the results for the site-site radial distribution functions are shown in Figure 11 from a modern version of those simulations, in which the Model A reference potential (Table 1) was used.

It can be seen that the pressure dependence of the extracted O–O function was very similar to that shown previously by Okhulkov et al. using X-ray scattering, with the peak at ~4.5 Å moving to lower *r* with increased pressure. Another indication of the change of structure is the triangle distribution functions for the O⋯O⋯O and O⋯O–H included angles, Figure 12. For the latter case the H atom is on the same molecule as the middle O atom, so the latter distribution is the distribution of angles the O–H bond makes with the intermolecular O⋯O axis. This is peaked near 0 degree as might be expected for linear hydrogen bonds, but becomes notably broader with increased pressure. Equally the O⋯O⋯O triangle distribution starts out with a broad peak near 100 degree at low pressure as for ambient water, but moves to progressively smaller angles with increased pressure, signaling again the increased degree of hydrogen-bond bending at high pressure.

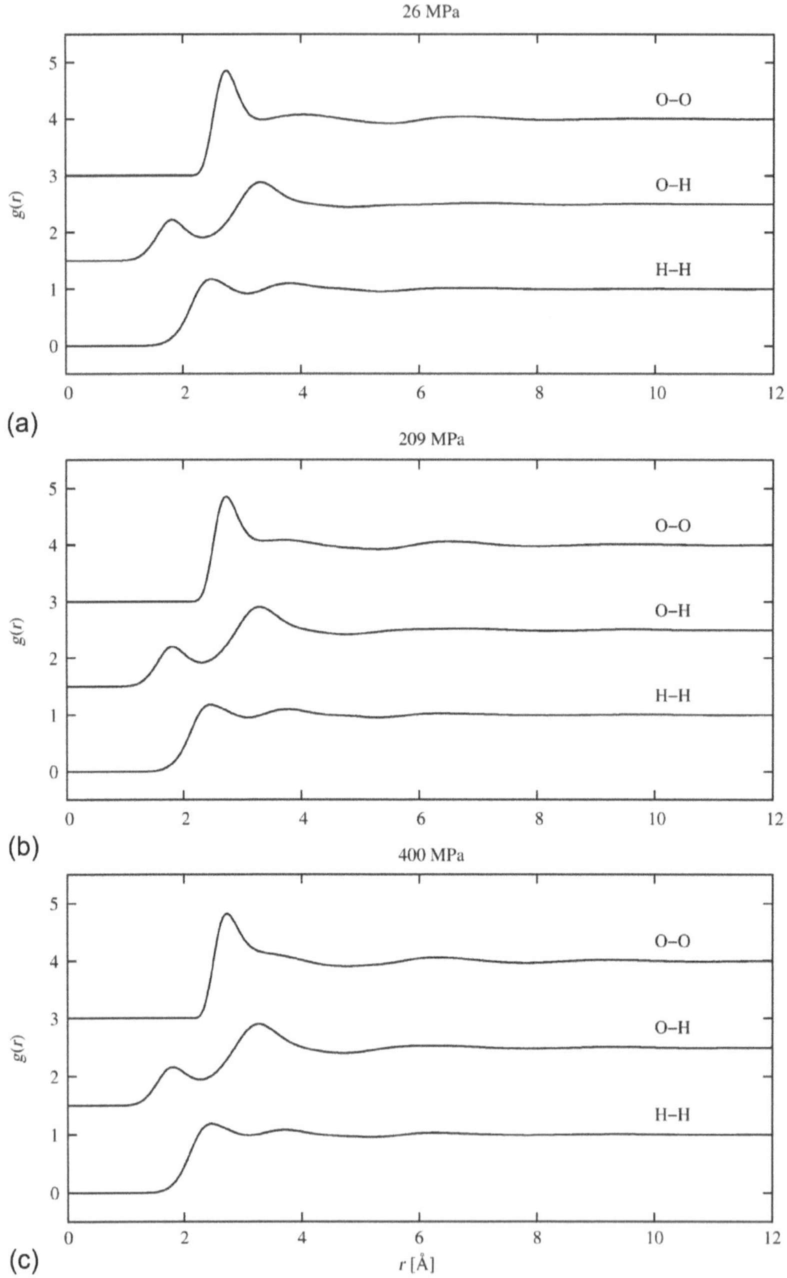

FIGURE 11 Radial distribution functions for water at 268 K as a function of increasing pressure, namely 26 MPa (a), 209 MPa (b), and 400 MPa (c), as determined by neutron scattering with isotope substitution [100].

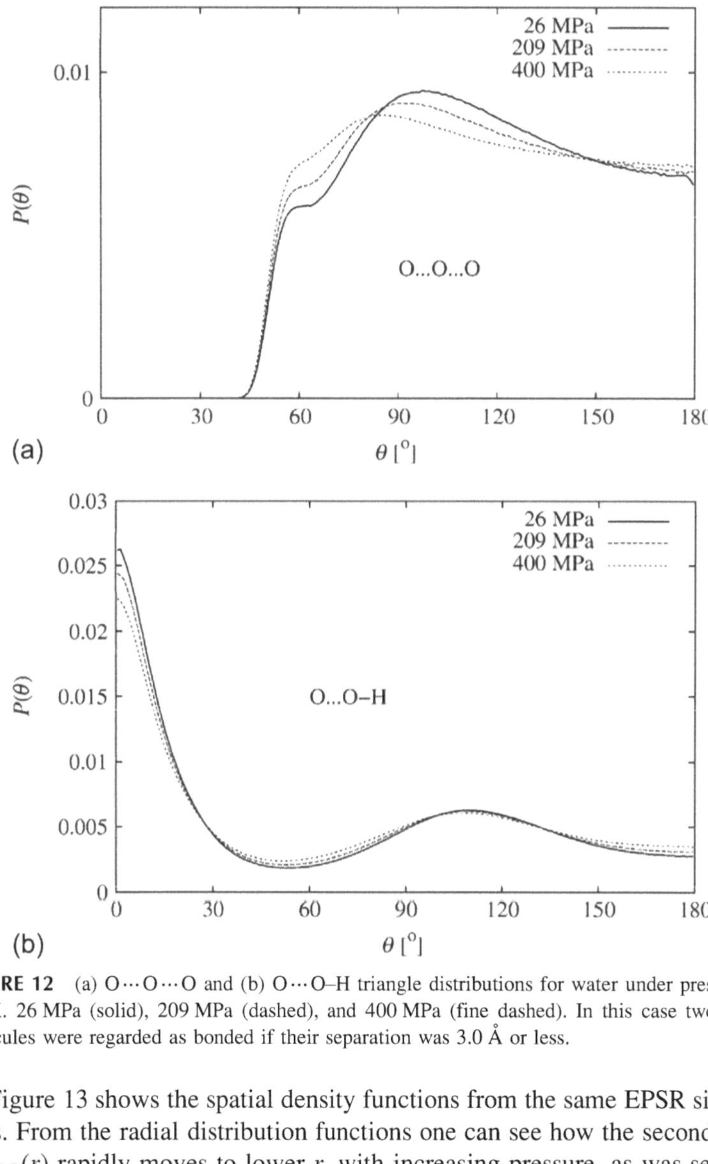

FIGURE 12 (a) O···O···O and (b) O···O–H triangle distributions for water under pressure at 268 K. 26 MPa (solid), 209 MPa (dashed), and 400 MPa (fine dashed). In this case two water molecules were regarded as bonded if their separation was 3.0 Å or less.

Figure 13 shows the spatial density functions from the same EPSR simulations. From the radial distribution functions one can see how the second peak in $g_{OO}(r)$ rapidly moves to lower r, with increasing pressure, as was seen by Okhulkov et al. On the other hand, from the spatial density distributions it can be seen that, as with the effect of changing the temperature, the first coordination shell is largely unaffected in this range of pressures, while the second coordination shell, which is in antiphase to the first shell, contracts slightly but noticeably as the density is increased, and starts to merge with the first shell. The temperature and pressure effects are not directly comparable, however, because the increase in temperature over the range studied, 280 K to

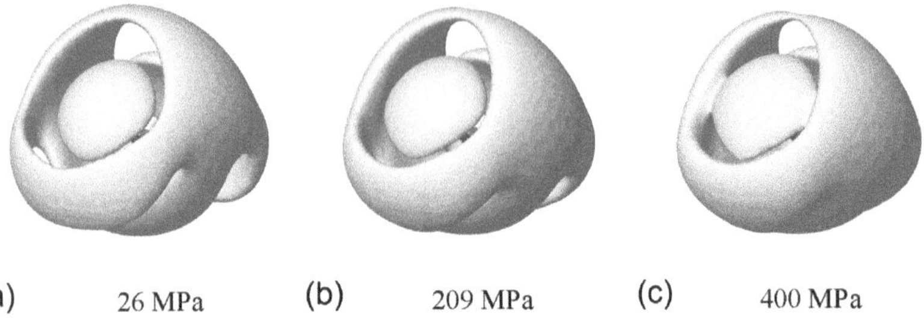

FIGURE 13 EPSR simulated spatial density functions for water at three pressures, namely (a) 26 MPa (density 0.0338 mols/Å^3), (b) 209 MPa (density 0.0362 mols/Å^3), and (c) 400 MPa (density 0.0381 mols/Å^3), all at 268 K. The reference potential used in these simulations is the same as for Figure 8. The first and second coordination shells are shown together in the distance range 0.0–5 Å. The horizontal dimension of each plot is approximately 10 Å.

365 K at ambient pressure, is associated with a *decrease* in density, whereas density *increases* with increasing pressure. Moreover with increasing temperature the position of the second peak in $g_{OO}(r)$ moves to larger r, while with increasing pressure the same peak moves to lower r. One is struck here once again by the resilience of the three-dimensional spatial water structure to relatively large changes in pressure, while the second and subsequent peaks in the radial distribution functions are more volatile to changes in state conditions. What is clear, therefore, is that changes to these peak positions cannot be simply related to changes to the three-dimensional structure of water. The real liquid is far more subtle than that!

3.3.3.3 *Spatial Structure of Ice Ih and the Amorphous Ices*

In addition to the above results there is a recent new study of the spatial structure in crystalline ice Ih [101] using the same approach with the total scattering technique and EPSR simulations that has been used here for the liquid. In the case of ice it is observed that whereas the first coordination shell is even more structured than in the liquid, the second shell, although now well separated from the first shell, is actually *more* disordered than the liquid. It is believed that this counter-intuitive result arises from the fact that in ice each water molecule is held in place by four more-or-less equal strength hydrogen bonds which gives the water some orientational freedom to experience its local environment, whereas in the liquid, the reduced symmetry in the first shell means only one hydrogen of a given water molecule forms a strong, more linear, hydrogen bond, leaving the molecule with actually less orientational freedom than in the solid form. The orientational freedom is more restricted because there are nonbonded molecules in the first shell—these are not present in the case of ice.

Further back there have also been several studies of the 3-dimensional structure in amorphous ice [90, 102, 103]. The trends seen in those studies exactly parallel those shown here. In particular, the second shell for low density amorphous ice (LDA) is closely similar to that found in crystalline ice, whereas that for high-density amorphous ice (HDA) is closer to that found in bulk water.

3.4 IONIC SOLUTIONS

We all know that when you add salt to ice two things happen simultaneously: the ice starts to melt and the resulting ice/water/salt mixture gets colder. An explanation of this effect is usually rushed off by saying that salt lowers the freezing point of water, causing it to melt (depending on how much salt is actually present of course) and the melting of ice involves breaking hydrogen bonds which will absorb energy, causing the mixture to cool. The strangeness of this explanation is that when salt is first added to the ice, the ice contains no salt, so why should it start melting? We have already seen that water structure is resilient to quite large increases in pressure, and ice is a hard unyielding substance, which is even more resilient to changes in pressure. How can a few specks of salt melt it, apparently without any effort?

The truth is of course that this explanation hides a feature of atomic and molecular interactions that frequently gets ignored. The forces that hold matter together are ostensibly strong and immovable, yet if something comes along that can override them at the molecular level, that matter can change phase quite readily. We mostly have no concept of the size or nature of the change that simple materials like sodium chloride can have on the local order of something like ice. We nonchalantly talk about salt "dissolving" in water as if it just disappears, whereas in fact what it is doing is more like a wrecking ball to a tall building: it is bending and drastically weakening hydrogen bonds between water molecules and grabbing them for itself. It is doing to the water structure what would otherwise require the equivalent of several thousand atmospheres of applied pressure to the pure liquid. Hence a simple speck of salt can do far more "damage" to water structure than any number of Titanics crashing into icebergs!

Neutron scattering has been able to map out some of these effects. We now have a pretty good idea of how water structure is affected by dissolved salts, and how individual ions might coordinate water molecules. All the smaller ions have a very well defined "hydration shell" in which water molecules are strongly oriented relative to the ion [45, 104]. Beyond the first shell, it is less clear how far the interaction might proceed, but certainly some effects have been detected some distance away from any given ion [105]. A lesser known issue is how the dissolved ions themselves are distributed. This was the subject of endless debate in the past which remains unsolved, but the problem for the neutron experiment is that even when a saturated solution is considered, the contribution of the ion-ion correlation functions to the total scattering pattern for most salts remains quite small—typically around 1%—making reliable experimental determination of these correlation functions close to impossible. Of course, with the advent of computer simulation methods such as EPSR or RMC to interpret scattering data, one can estimate the ion-ion correlation functions, but these will only ever be estimates that are consistent with the data: they will not have been determined directly.

3.4.1 The Dissimilar Pair, NaCl and KCl

Sodium and potassium ions perform a critical role in the function of our nerve cells. In a neuron's resting state there is a higher concentration of Na ions outside the cell membrane and higher concentration of potassium ions inside the neuron, a status quo maintained by ion-specific channels and pumps in the cell membrane. If the nerve cell is stimulated, Na^+ ions flow into the neuron and K^+ ions flow out, changing the potential of the neuron. It is this process that causes the nerve signal to be transmitted along the neuron. Until the sodium and potassium ion channel proteins were identified and extracted there was very little understanding how the nerve was able to distinguish between the two ions that chemically appear very similar. Even so, there has to be some physical way these channel proteins can identify and either block sodium and transmit potassium, or vice versa. Since the charge on the ions is the same, this ability must relate to the different radii of the ions. For sodium the nearest water molecules are about 2.3 Å away, while for potassium the same distance is ~2.7 Å. This should mean water molecules are more tightly bound to sodium compared to potassium. As a consequence, we should be able to observe some difference in the local structure surrounding these ions, even when observed at the same molar concentration.

An early experiment on SANDALS attempted to observe this difference in local order. As it happens the results have never been published, so they are given here as an illustration of the neutron scattering method in its current form as applied to ionic solutions, and to show whether there are indeed measurable differences between these two ions. It should be pointed out of course that chlorine isotope substitution had previously been applied to both salts in solution, using the first order difference method to identify the nature of the chloride hydration shell. Equally potassium isotope substitution had been performed to yield some information on the potassium hydration shell. The equivalent experiment on the sodium cation is not possible because there is only one stable sodium isotope. A summary of these results can be found in [45], but it should be noted that these ion isotope substitution experiments tell us little about how the water structure itself is affected by the presence of the ions. To do this you need hydrogen isotope substitution on the water hydrogen atoms, as in the case of pure water.

The SANDALS experiment consisted, for each ion pair and at each molar concentration, of measuring three samples, one made from pure H_2O, one from pure D_2O, and one from an equimolar mixture of H_2O and D_2O. These three samples were corrected for attenuation, multiple scattering, container scattering, and inelastic scattering, in the usual way, and combined to produce three adjunct data sets, XX, XH, and HH, where X represents all the atoms except

TABLE 2 List of Lennard-Jones Parameters and Coulomb Charges Used in the EPSR Simulations of NaCl and KCl Aqueous Solutions

Atom	σ [Å]	ε [kJ/mol]	q [e]
O	3.166	0.6500	−0.8476
H	0.000	0.0000	+0.4238
Na	2.290	0.5144	+1.0000
K	2.940	0.5144	+1.0000
Cl	4.191	0.5660	−1.0000

the hydrogen atoms, so X represents Na, Cl, and O in the case of NaCl, and K, Cl and O in the case of KCl. This method—which is now no longer used—is described in detail in [106]. The three data sets were then analyzed with EPSR in the same way as described above for pure water. For each NaCl salt four concentrations were studied 0.6, 1.3, 3.0, and 5.2 mol of salt per 50 mol of water. For KCl the four concentrations were 0.6, 1.3, 3.0, and 4.0 mol of salt per 50 mol of water, the last concentration being limited by the saturation concentration of KCl in D_2O. Each simulation box had 1000 water molecules with the numbers of each ion corresponding to the stated molar concentration. In this case the SPC/E potential parameters [107] were used for the reference potential for water, and a full list of these parameters for each ion and water molecule can be found in Table 2. At the time of the experiment there were no readily accessible X-ray facilities to produce a further X-ray data set for each solution to be included in the structure refinement. Table 3 lists some of the ion-water coordination numbers coming out of these EPSR simulations.

3.4.1.1 Water Structure

The simplest way to demonstrate the water structure is via the site-site radial distribution functions. The primary effect is in the OO function, Figure 14, where the second peak at ~4.5 Å is successively replaced by a shoulder on the large r side of the first peak as the salt concentration is increased. This trend is slightly less marked in the case of KCl compared to NaCl, although the highest concentration for KCl is lower than that for NaCl. The trends seen in the OO functions are mirrored by similar changes as with pressure in the OH and HH functions (not shown here).

Further indications of the impact of the ions on water structure is to calculate the O···O···O (Figure 15) and O···O–H (Figure 16) triplet distribution functions for nearest neighbor water oxygen atoms, and to calculate the water-water spatial density function as a function of concentration (Figure 17), as was done previously for pure ambient water, and water under pressure.

TABLE 3 Ion-Water Coordination Numbers at Different Concentrations

Concentration [Salt:Water Mole Ratio]	N_{NaO} (0,3.00 Å)	N_{NaCl} (0,3.30 Å)	N_{ClH} (0,2.85 Å)	N_{ClO} (0,3.81 Å)
		(a) NaCl		
6:500	5.1	0.2	5.9	7.0
13:500	4.9	0.4	5.6	6.9
30:500	4.7	0.7	5.4	7.0
52:500	4.2	1.1	4.8	6.6
Concentration [Salt:Water Mole Ratio]	N_{KO} **(0,3.42 Å) (117)**	N_{KCl} **(0,3.54 Å) (121)**	N_{ClH} **(0,2.82 Å) (97)**	N_{ClO} **(0,3.78 Å) (129)**
		(b) KCl		
6:500	6.3	0.2	6.0	6.8
13:500	5.9	0.4	5.6	6.5
30:500	5.6	0.9	5.2	6.2
40:500	5.3	1.0	5.0	6.1

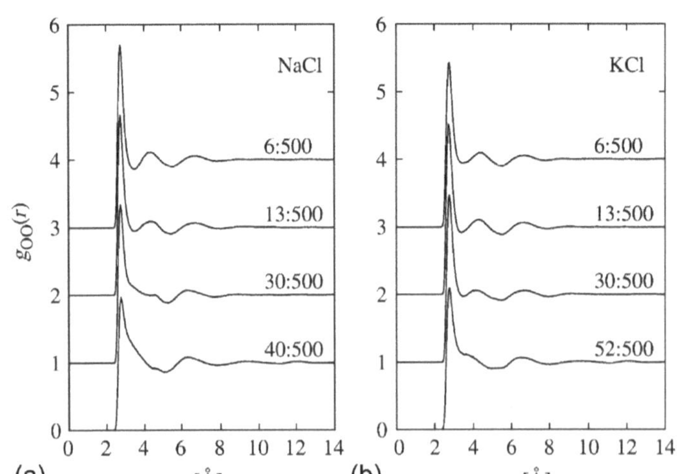

FIGURE 14 Oxygen-oxygen radial distribution functions in (a) NaCl and (b) KCl aqueous solutions, as a function of concentration, determined from EPSR simulations of neutron total scattering data with hydrogen/deuterium substitution.

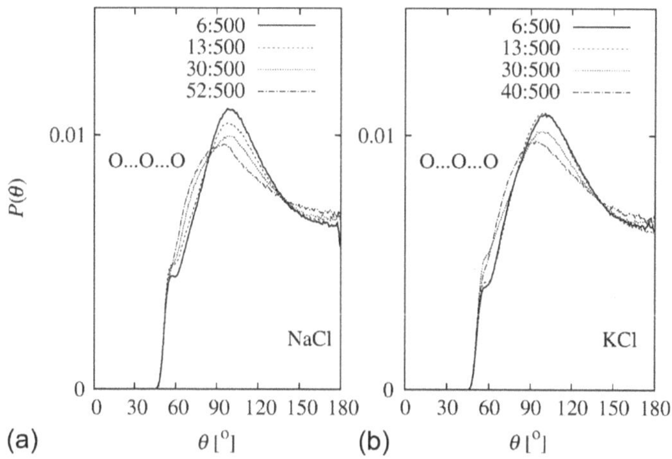

FIGURE 15 O···O···O triplet distribution functions from EPSR simulations of (a) NaCl and (b) KCl aqueous solutions and over a range of concentrations at mole ratios 6:500 (solid), 13:500 (dashed), 30:500 (dotted), and 52:500 (NaCl) or 40:500 (KCl) (dot-dash). Two water molecules were regarded as bonded if their oxygen atom separation was 3.2 Å or less.

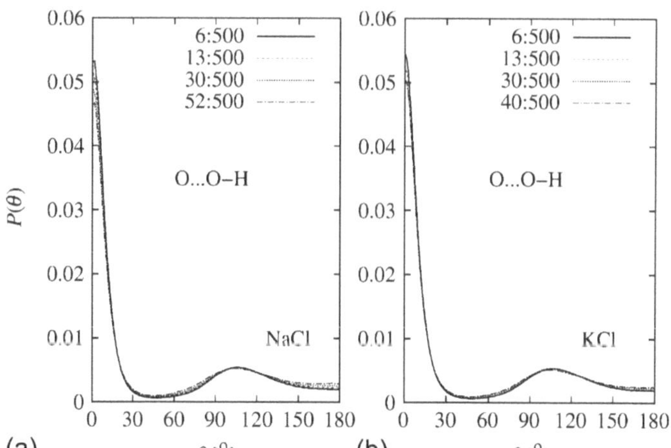

FIGURE 16 O···O–H triplet distribution functions from EPSR simulations of (a) NaCl and (b) KCl aqueous solutions and over a range of concentrations at mole ratios 6:500 (solid), 13:500 (dashed), 30:500 (dotted), and 52:500 (NaCl) or 40:500 (KCl) (dot-dash). Two water molecules were regarded as bonded if their oxygen atom separation was 3.2 Å or less.

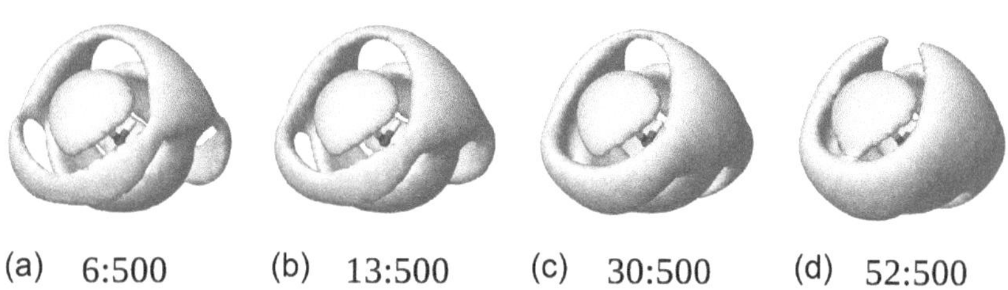

FIGURE 17 Water-water spatial density function in NaCl aqueous solutions as a function of concentration at mole ratios (a) 6:500, (b) 13:500, (c) 30:500, and (d) 52:500.

Comparing these figures with Figures 11–13 for the pure liquid, it becomes clear that the primary effect on the structure of water by adding salt is equivalent to raising the pressure of pure water by several 100 MPa, the amount depending on the concentration of salt. This has sometimes been referred to as the "effective pressure" of an aqueous solution, although it has to be made quite clear that the actual pressure of the solutions remains at 0.1 MPa. The equivalent pressure simply refers to that pressure that would be required on the pure liquid to induce a similar structural change as observed in solution.

3.4.1.2 Ion-Water Structure

Figures 18 and 19 show the radial distribution functions associated with the ions. These are characterized by sharp first peaks for both anions and cations, and for both water oxygen and water hydrogen atoms, signaling a well defined ion hydration shell for all the ions, in which the water molecules are strongly oriented by the ionic electric field. For the cations, the water oxygen atom is the closest to the ion, as might be expected, with the water hydrogen atoms repelled by the ionic field. For the anions the reverse is true: now the water hydrogen atoms are closest, while the water oxygen atoms are repelled. In all cases these basic trends are only slightly affected by the change in concentration.

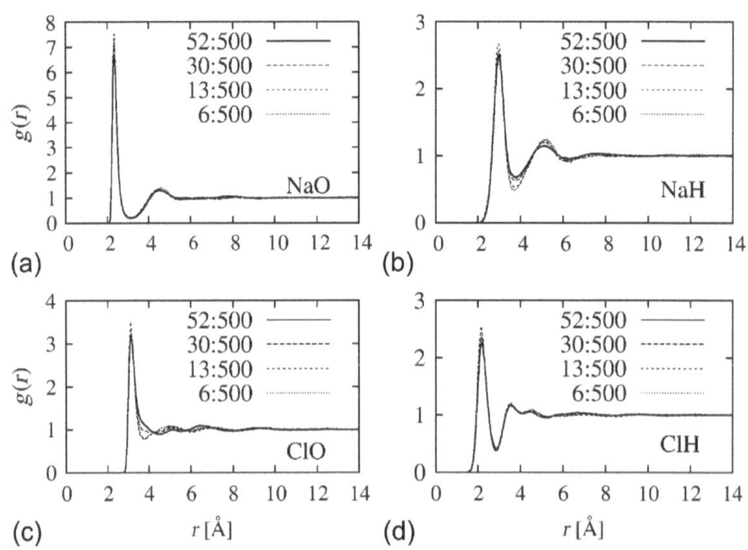

FIGURE 18 Anion- and cation-water radial distribution functions, determined by EPSR simulation, as a function of salt concentration for NaCl at mole ratios 6:500 (dotted), 13:500 (fine dashed), 30:500 (dashed), and:500 (solid) for (a) NaO, (b) NaH, (c) ClO and (d) ClH pairs.

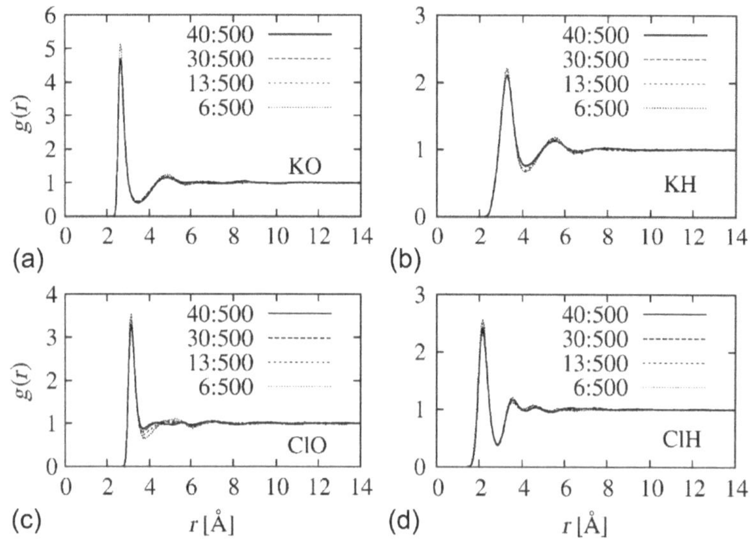

FIGURE 19 Anion- and cation-water radial distribution functions, determined by EPSR simulation, as a function of salt concentration for KCl at mole ratios 6:500 (dotted), 13:500 (fine dashed), 30:500 (dashed), and 40:500 (solid) for (a) KO, (b) KH, (c) ClO and (d) ClH pairs.

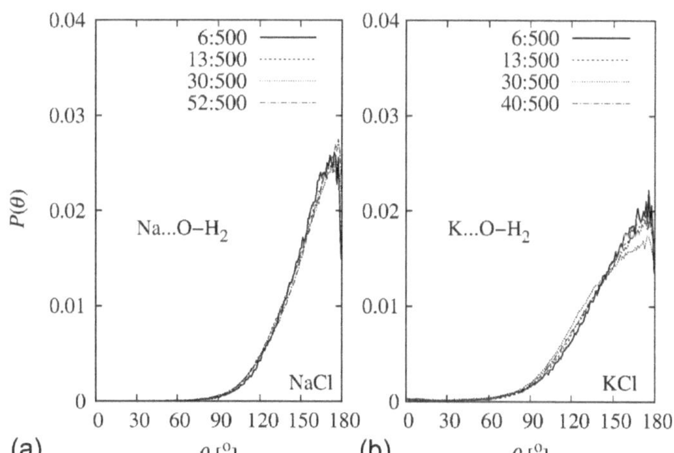

FIGURE 20 Cation-water triangle distribution functions for (a) NaCl and (b) KCl at mole ratios 6:500 (solid), 13:500 (dashed), 30:500 (dotted), and 52:500 (dot-dash). In this case the angle is the included angle between the cation-water oxygen axis, and the axis from the water oxygen to the bisector of the same water molecule's two hydrogen atoms, in other words the water molecule's dipole moment vector.

These ideas are perhaps better illustrated by the cation and anion to water angle plots, Figures 20 and 21. For the cations, which repel the water molecule hydrogen atoms, the useful quantity is the angle the water molecule dipole moment makes with the cation—water oxygen axis, while for the

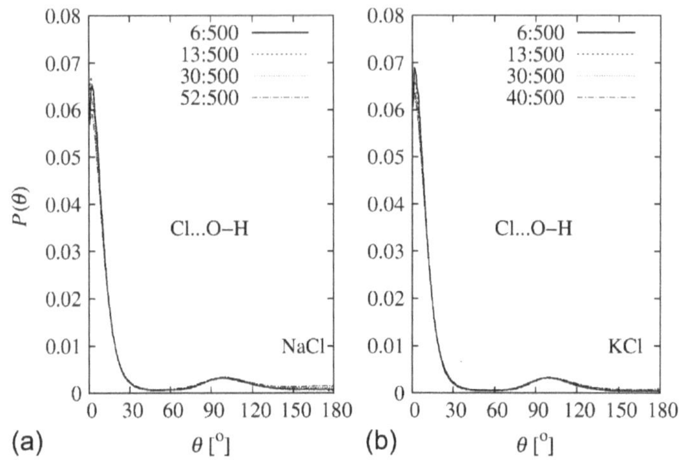

FIGURE 21 Anion-water triangle distribution functions for (a) NaCl and (b) KCl, at mole ratios 6:500 (solid), 13:500 (dashed), 30:500 (dotted) and 52:500 (NaCl) or 40:500 (KCl) (dot-dash). The angle is defined as the angle between the anion-water oxygen axis and the axis from the water oxygen to one of the same water molecule's hydrogen atoms.

anions, which attract the water molecule hydrogen atoms, the angle is between the anion-water oxygen axis and the water molecule's O–H vector. The latter distributions have two peaks, corresponding to the two hydrogen atoms with an internal angle of $\sim$105 degree.

For the cations, Figure 20, it is found that the most probable anion-dipole moment angle is 180 degree, i.e., the water dipole moment points directly away from the cation. However there is some spread in this angle, $\sim\pm 40$ degree for NaCl and $\sim\pm 50$ degree for KCl, and since $\theta > 180$ degree has no meaning in this context because it would be replaced by its conjugate $(180 - \theta)$ in any calculation of angles, it means the average angle of the dipole moment with respect to the ion-water axis would always be reported as less than 180 degree. It is notable here how the cation-water dipole angle distribution is notably broader for KCl and more variable with salt concentration than for NaCl, where the same angle distribution is narrower and less concentration dependent. In addition for KCl the angle distribution gets slightly broader as the concentration increases. These two factors signal that the K^+ hydration shell is weaker and more easily perturbed than Na^+ shell, which appears robust to composition changes.

In contrast, for the anions, Cl^-, Figure 21, the hydration shell is robust to both changes in composition and nature of the counterion. Here the O–H vector is seen to line up strongly with the Cl$\cdots$O axis, with a marked second peak corresponding to the second hydrogen atom on the coordinating water molecule. This second peak samples rotations of the water molecule about the O–H axis, as well as rotations perpendicular to this axis, and so should be broader than the peak at $\theta = 0$ degree, which it is, but not a lot so, signaling a well defined orientation of the water molecule with respect to the Cl^+ ion.

The trends seen here in the EPSR simulations of simple ionic solutions are in good accord with what has been observed previously using isotope substitution on the ions themselves [45].

3.4.1.3 Ion-Ion Structure

For completeness the ion-ion radial distribution functions associated with these simulations are shown in Figures 22 and 23. These are very noisy at the lower concentrations, so only the results at the highest concentrations for NaCl and KCl are shown here. As already indicated it must be emphasized that these cannot be extracted directly from the experimental neutron scattering data: they are simply a consequence of running a structure refinement simulation on the scattering data and using that simulation to estimate the corresponding distribution functions. Hence they will be consistent with the data, but that does not mean they are necessarily the only functions that satisfy that requirement.

Assuming the calculated functions are sensible, then it can be seen from the NaCl and KCl radial distributions that cation-anion contact pairs occur

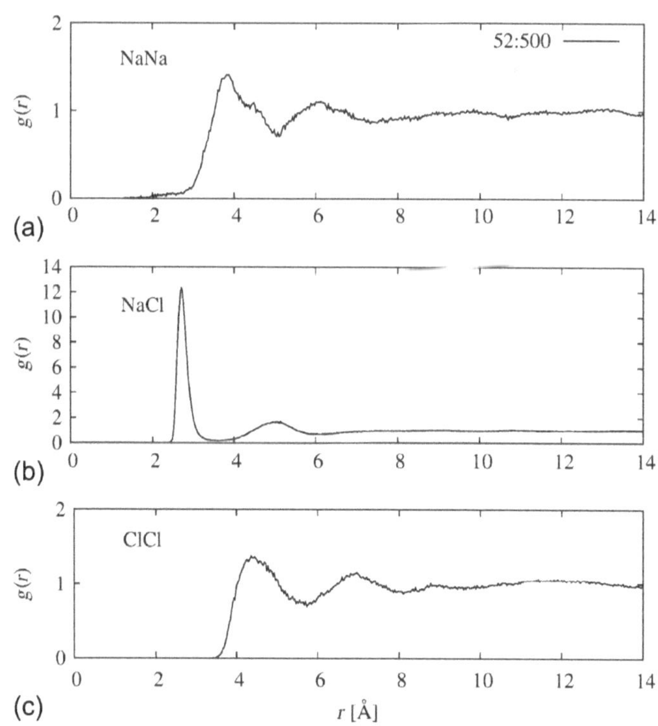

FIGURE 22 Ion-ion radial distribution functions for NaCl in water at a mole ratio of 52:500 for (a) NaNa, (b) NaCl and (c) ClCl pairs.

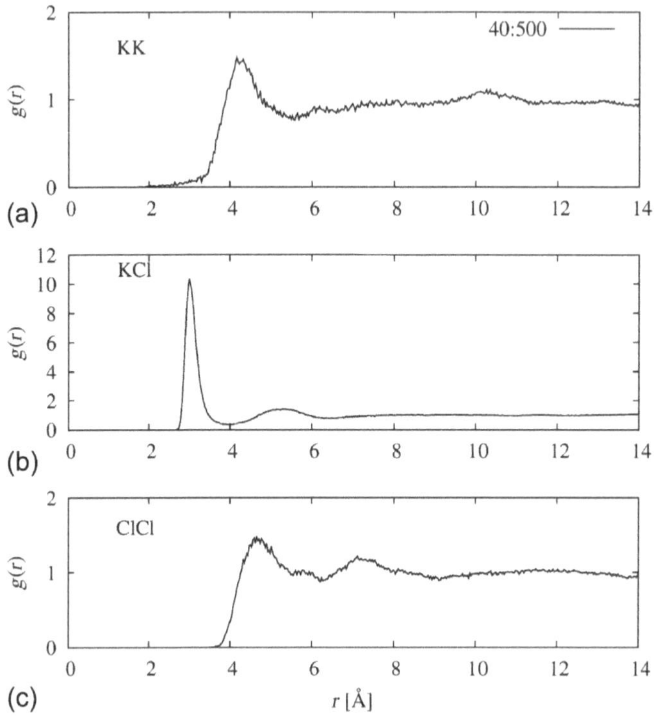

FIGURE 23 Ion-ion radial distribution functions for KCl in water at a mole ratio of 40:500 for (a) KK, (b) KCl and (c) ClCl pairs.

readily in these simulations. For Na^+ the Cl^- ion is beyond the water oxygen atoms of the hydration shell, but very close in at $r \sim 2.7$ Å nonetheless. For K^+ the counterion is at about $r \sim 3.0$ Å. In both cases the coordination number of this peak grows steadily with concentration and almost exactly compensates for the loss of water as the concentration increases, Table 3. Because of the weakness of the contribution that the ion-ion correlations make to the total scattering pattern, this cation-anion correlation close in has been mostly overlooked in the earlier work using chlorine and potassium isotope substitution work. Although well defined, the ion hydration shells are apparently not strong enough to exclude the counterions in these cases.

For the like-like ion-ion distribution functions, we observe a broad peak near $r \sim 4$ Å for the cation-cation distributions, and a similar peak near $r \sim 4.3$ Å for the anion-anion distributions. The like-like and like-unlike distances seen here seem to mirror those observed in pure molten NaCl in the early work on this molten salt by Edwards and Enderby [53], suggesting a degree of charge ordering in the aqueous solution not dissimilar to that found in the melt.

3.4.1.4 Discussion

Combining the results from the previous three sections, it has been seen that the primary effect of dissolved 1:1 salts is to cause a compression of the water structure equivalent to pressurizing the pure liquid by several hundred MPa (100 MPa = 1 kbar). This is an effect that has been identified and discussed in several publications [108]. Sometimes the effect of ions on water is described as "electrostriction", implying the ions actually draw the water molecules closer together. However this is not really possible in the present instance. For example if the Na–O distance is ~2.3 Å and the water molecules are arranged octahedrally around the sodium, the water-water distance in the Na^+ shell is ~3.25 Å, which is larger than the O–O nearest neighbor distance in pure water (~2.8 Å). If the arrangement around the sodium were tetrahedral this O–O distance would be even larger. Therefore simply dragging the water molecules closer together by the ions does not happen. What does happen of course is that the hydration water molecules are strongly oriented by the ions, as seen in Figures 20 and 21. For the chlorine ions this is perhaps not so serious since only one of the water molecule hydrogen atoms is involved in bonding to the anion, leaving the other hydrogen atom still free to bond to other water molecules or other ions, while the water oxygen atom is also still available to form acceptor hydrogen bonds. For cations however, both hydrogen atoms are oriented, pointing mostly away from the ion. This creates a deficiency in hydrogen bonding acceptor sites in the solution and a surplus of hydrogen bond donor sites. This can only result in a severe disruption to the water network and produces a pattern similar to what happens when pure water is pressurized.

A further issue is that it appears quite common for counterions to exist in the neighborhood of any given ion: the ion hydration shells in these cases are not strong enough to repel counterions. As a result ion-water coordination numbers (Table 3) do not adopt well defined values like 4 or 6 or something else. Instead they depend on the concentration of the solution and the nature of the counterion.

3.4.2 Other Studies of Ions in Water

The above account is intended of course to give only a flavor of the type of information that can be extracted from ion hydration experiments using total neutron scattering. Although ion isotope substitution experiments are rarely performed nowadays due to the prohibitive cost of the required isotopes, there has been a recent resurgence of this work, combining neutron scattering with hydrogen isotope substitution, X-ray scattering, and EXAFS data, with computer modeling. This has resulted in some intriguing new insights into the ways much more complex ions are coordinated when dissolved in water [109–114].

3.5 MICROHETEROGENEITY IN AQUEOUS MIXTURES

If you mix alcohol or sugar syrup or glycerol with water in a glass you might initially see some "ripples" in the transmitted light caused by the refractive index varying from place to place in the solution. These are due to the concentration of alcohol or sugar or glycerol not being uniform at the beginning of mixing. Very quickly, however, the light ripples disappear and the solution becomes fully mixed and clear. At that point we might think the concentration of solute is uniform through the liquid and without obvious density or concentration fluctuations. Strictly speaking however this statement is only true for the wavelength of the radiation we are using to observe the solution, namely ~550 nm in the case of visible light: we cannot see fluctuations that are much smaller than this length scale with this radiation. Yet interatomic distances are of order 0.3 nm, that is ~1000 times smaller than the wavelength of visible light, so is it possible there are still fluctuations at this much smaller length scale? In a sense the answer to this question will always be "yes" because different molecules cannot overlap each other so there would naturally be some stochastic fluctuation in numbers and concentrations of molecules when a solute is dissolved in water.

So the question arises whether there are fluctuations in density and concentration which are much smaller than those visible by ordinary light, but larger than those due to the random motion of individual atoms or molecules—in other words are there *correlated* motions or structures which involve groups of atoms or molecules? Until fairly recently we didn't have a clear answer to this question at the molecular length scale, even though truly heterogeneous systems such as micelles, liquid crystals, colloidal systems, and so on, had been known about for a long time. In the latter cases the density and concentration fluctuations could extend over 1–1000 nm and are typically characterized by dynamic light scattering and small-angle neutron or X-ray scattering. It is well known that the excess entropy of mixing in these solutions is often negative, even for simple alcohol-water solutions where longer range correlations are not expected. This implied that there was some nonideal restructuring taking place when the solution was made, and led to speculation that so-called "icebergs" might be forming around hydrophobic headgroups in solution [23], as has already been alluded in Section 3.1.1.2.

One of the first hints of a system not being fully mixed at the molecular length scale came from some little known computer simulations of Palinkas, Hawlicka, and Heinzinger [115]. Their statement to have seen some "water aggregation" in computer simulations of concentrated methanol water solutions largely went unnoticed, perhaps being regarded as a possible artifact of the computer simulation procedure. Yet their observations were pretty dramatic: in 0.9:0.1 mole ratio methanol:water solution, they observed that the statistical probability of a water molecule being bonded to another water molecule was around 40% even though the statistical likelihood of this in a random system would be closer to approximately 1% at this concentration. Other experimental observations seemed to hint at the same conclusion [116, 117]. Then in 2002, a neutron total scattering study of methanol:water at the mole ratio of 7:3 was published in which hydrogen isotope substitution was performed on both the methanol and the water atoms, leading to 7 distinct sets of data being interpreted using a single computer simulation box [118]. These data were compared with earlier data on pure methanol. It was shown that in solution the methyl groups got pushed slightly closer together compared to the pure liquid, while the water tended to congregate around the OH groups on the methanol, leading to interpercolating patches of water and patches of methyl groups. Despite this partial segregation of the solution, the structuring did not appear to induce any undue excess small Q scattering, so might not have been detectable in a small-angle scattering experiment.

The effect was studied at different concentrations, temperatures, and pressures [119] and found to be largely unchanged at ambient temperatures and below. In particular, lowering the temperature and raising the pressure both appeared to increase the degree of clustering observed, although at no point was there any sign of the solution actually phase separating in the simulation: such a phase separation would have invariably induced a large rise in the small Q simulated functions which would have been immediately visible if it had occurred. So was this partial water segregation a feature of methanol: water only, or did it occur in other systems?

Several companion studies to the methanol:water work suggest partial segregation of water in aqueous solution probably has a widespread occurrence. Examples of other systems where it has been observed experimentally include tertiary-butanol in water [120], ethanol in water [121], glycerol in water [122], sorbitol in water [123], and a cryopreservative in water [124]. A fundamental idea that emerges in this work is that the solute can actively limit the extent of the water network that would otherwise occur if ice formed when these solutions are cooled down, and so inhibit crystallization.

3.5.1 A Case Study of Alcohol:Water Mixtures

To illustrate this microsegregation we consider the case of ethanol in water, which was measured with neutrons, using isotope substitution on both the methyl hydrogen and water hydrogen atoms, at three alcohol:water mole ratios, namely 1:9, 2:8, and 4:6 [121]. The simplest way to demonstrate microsegregation is to plot the box of atoms, highlighting only the water oxygen and ethanol carbon atoms, Figure 24.

Another idea is to look only at the water distribution and draw surface contour around all regions where the water density goes above a predefined level. To achieve this surface contour plot each water oxygen is "smeared" in 3-dimensions using a Gaussian function of specified width [125] to produce a water oxygen density map, which is then used to produce the surface

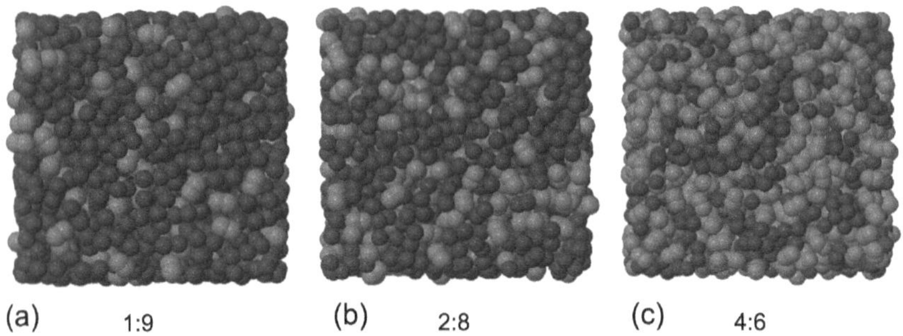

FIGURE 24 EPSR simulated boxes of ethanol:water mixtures at mole ratios (a) 1:9, (b) 2:8, and (c) 4:6. The numbers under each box correspond to the ethanol:water mole ratios. Only the ethanol carbon atoms (gray in online version) and water oxygen atoms (red in online version) are shown.

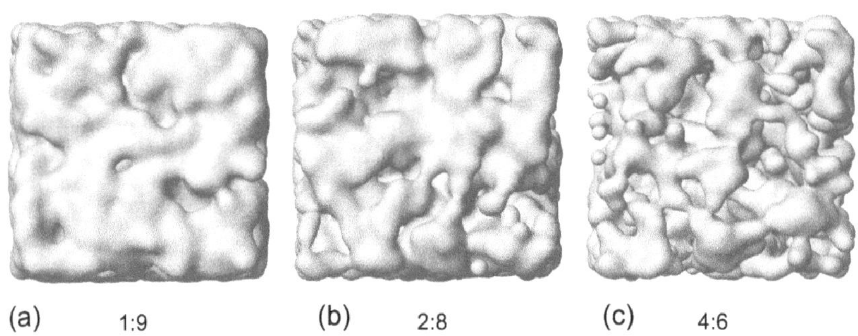

FIGURE 25 Water density plots at mole ratios (a) 1:9, (b) 2:8, and (c) 4:6 showing surface contours around regions where the water density is above 0.01 molecules/Å^3. On this scale bulk water has a density of 0.0334 molecules/Å^3.

contour. The precise value of the RMS deviation for the Gaussian does not affect the overall result very significantly, so it is chosen to correspond approximately to the radius of a water molecule, that is 1.6 Å.

The surface contours obtained for the three concentrations of ethanol in water are shown in Figure 25. It is seen that the water does indeed appear to aggregate in clusters, many of which are joined to neighboring clusters, forming a percolating network. The effect becomes more pronounced as the alcohol concentration increases, as might be expected.

A natural question to ask at this juncture is: how do we know these distributions are any different to what would occur with random packing? Could they simply be due to steric effects from the shapes of the molecules? To address this question Towey et al. [122] used EPSR simulations, in parallel to those used to interpret the scattering data, in which all the Coulomb charges on the molecules were set to zero and no structure refinement was allowed (empirical potential set to zero), while maintaining exactly the same density, molecular geometry, and repulsive and dispersion forces as for the simulations with structure refinement. In this way all the hydrogen bonding between molecules was switched off. This is probably the closest you can get to a truly random system which includes only the packing effects. They then investigated the nature of water and glycerol clustering, and showed there was a marked reduction in clustering when the hydrogen bonding was switched off.

To show the effect of switching off the hydrogen bonding, a similar set of EPSR simulations were performed for the ethanol:water mixtures discussed here, in which all the Coulomb charges were set to zero and structure refinement switched off. Unfortunately a simple view of the simulation box when all the hydrogen bonding has been switched off is not particularly informative since the equivalent of figures such as in Figure 24 or 25 do not appear much different to those figures: at this level the with- and without-hydrogen bonding simulations look quite similar, even though from the point of view of the site-site radial distribution functions they clearly are very different, Figure 26. In that figure we can see the drastic effect of switching off the hydrogen bonding, leading to rather structureless local order, particularly in those correlations where hydrogen-bonding is expected to dominate, such as those between O and OW and H and HW.

To illustrate the effect of hydrogen bonding and structure refinement for the present case of ethanol in water, the water-water and ethanol-ethanol cluster distributions were calculated in the same way as done by Towey et al. For this purpose two water molecules were regarded as in the same cluster if their oxygen atoms were separated by 3.5 Å or less, this distance corresponding to the first minimum in the water O–O radial distribution function. Two ethanol molecules were regarded as being in the same cluster if their terminal carbon atoms (CM) were separated by 4.7 Å or less, this being the distance of the first minimum of the corresponding site-site radial distribution function. These definitions are somewhat arbitrary but the precise values are not important for the present purpose of demonstrating how the clustering is affected by the hydrogen bonding. Results are shown for the mole ratio 4:6 in Figure 27 with and without hydrogen bonding.

These figures indicate that there are two types of clusters in these simulations. On the left are isolated clusters involving a few, or perhaps 10s of molecules. On the right are relatively sharp peaks, which involve up to 1000 molecules or more. (The current simulations at a mole ratio of 4:6 were performed with 2000 molecules in total, 1200 water molecules and 800 ethanol molecules.) The latter clusters are above the percolation threshold (power law, dashed line, $n_c(s) \sim s^{-2}$ [126]) and so must represent percolating clusters—clusters which effectively span the simulation box and would be infinite for an infinitely large simulation. For simulations without hydrogen bonding, it is seen that the water clusters in particular have become far less percolating. The sharp peak near 1000 molecules has disappeared in this case, with maximum cluster size around 800 molecules and at a probability one

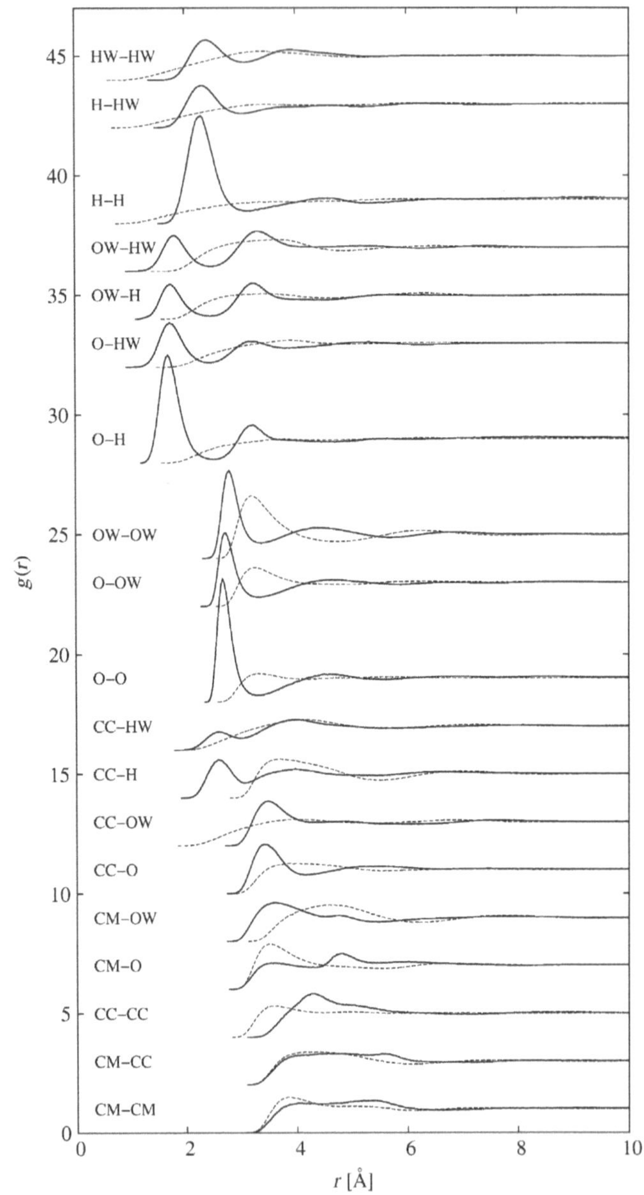

FIGURE 26 Selected site-site radial distribution functions for ethanol:water solutions at a mole ratio of 2:8. Atoms shown are, respectively, CM and CC, central and methyl carbon atoms on ethanol, O and H, oxygen and hydroxyl hydrogen atoms on ethanol, OW and HW, and water oxygen and water hydrogen atoms. The solid lines are from the fit to the scattering data with Coulomb charges to emulate hydrogen bonds. The dashed lines are with no fit to the scattering data and with all the Coulomb charges set to zero.

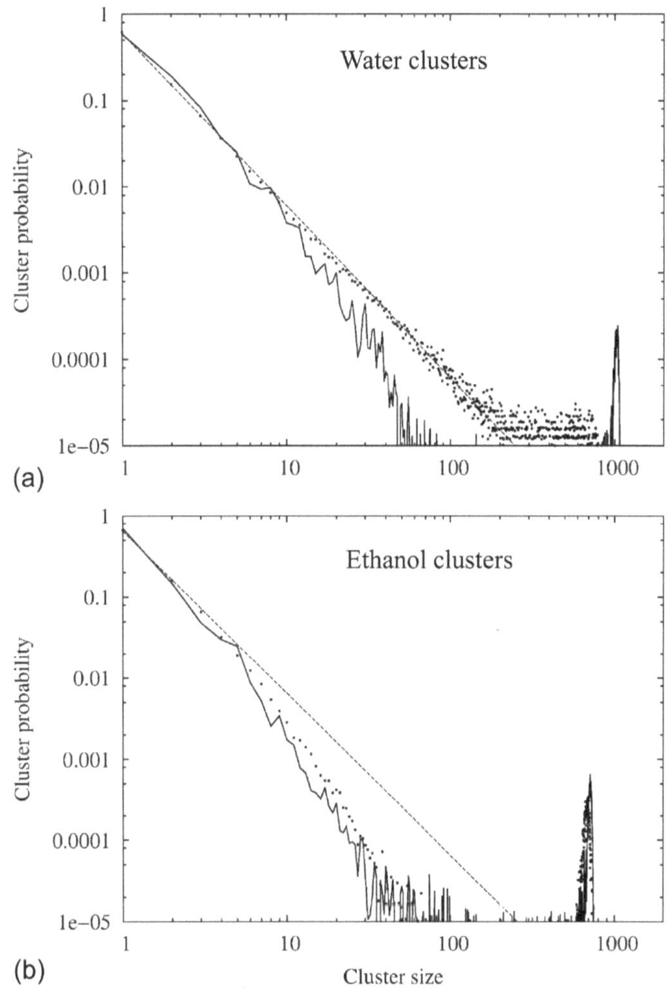

FIGURE 27 Distribution of cluster sizes for (a) water clusters, and (b) ethanol clusters at an ethanol:water mole ratio of 4:6. For (a) two water molecules are regarded as bonded if their oxygen atoms are separated by 3.5 Å or less. For (b) two ethanol molecules are regarded as bonded if their terminal carbon atoms (CM) are separated by 4.7 Å or less. The lines show the distribution when charges are included in the reference potential and full structure refinement is performed with EPSR. The dots are when the empirical potential is set to zero, and the electronic charges used to generate hydrogen bonding are switched off. The dashed lines show the percolation threshold, as described by Stauffer [126].

order of magnitude smaller than with hydrogen bonding. Also there are far more nonpercolating clusters. Hence large clusters can occur in the nonhydrogen bonding system, but with a far lower probability than when hydrogen bonding is present. At the same time there are now many more smaller and nonpercolating clusters. The effect of hydrogen bonding on the ethanol clusters is much less pronounced than on the water clusters, although there is analogous behavior, with slightly enhanced clustering of ethanol molecules when hydrogen bonding is present.

The cluster size distribution is most sensitive for indicating the microheterogeneity effect if one of the clusters is close to the percolation threshold, as in case (a) of Figure 27. Another way to illustrate the excess clustering induced by hydrogen bonding is to divide the simulation box into a large number of volume pixels and then, using a suitable distance criterion, map out all the pixels that are "occupied" by water molecules, namely those pixels which lie within the specified distance from *any* water molecule. The distance is chosen such that the fraction of occupied pixels corresponds approximately to the water volume fraction in the solution. In the present case the distance from a water molecule was set at 2.1 Å. A pair distribution function is calculated for the occupied pixels—this will be called the "water volume correlation function" here—and the slope of this correlation function at low r is related to the "confining length" of the water—see [125, 127, 128]. This can of course be done for both the simulations with hydrogen bonding and structure refinement, and those without. The results for ethanol:water at the three concentrations are shown in Figure 28.

Here one discovers three features are to be noted. Firstly for $0 \leqslant r \leqslant 5$ Å the correlation functions are essentially the same shape, but invariably the simulation without hydrogen bonding shows a weaker correlation at all concentrations. This fits with the fact that the fraction of occupied pixels is smaller by ~10% when hydrogen bonding is present compared to when it is absent, so hydrogen bonding has the effect of drawing the water molecules closer together on average, as might be expected. In addition the water volume correlation function grows in this region with increasing ethanol

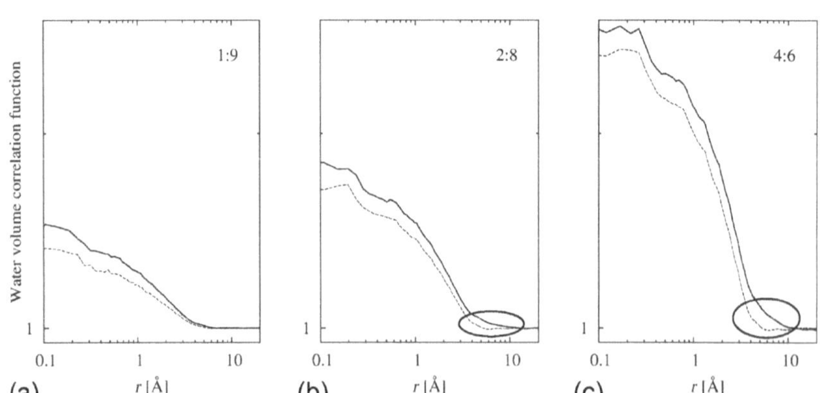

FIGURE 28 Water volume correlation functions for ethanol-water solutions at three concentrations, as described in the text both with (solid) and without (dashed) hydrogen bonding at mole ratios (a) 1:9, (b) 2:8, and (c) 4:6.

concentration. Because these correlation functions are normalized to the average density of occupied pixels, the value at low r relates inversely to the fraction of occupied pixels that are deemed occupied. Therefore as the ethanol concentration increases, the fraction of occupied pixels decreases, and the correlation function at short r increases accordingly.

A second feature is that for the simulations without hydrogen bonds and structure refinement the correlation functions drop quickly to 1.0 at $r \approx 5$ Å and actually go slightly below 1.0 in the region $5 \leqslant r \leqslant \sim 10$ Å, signaling, if anything, a slightly *lower* than average water density in this region. On the other hand, for the simulations with hydrogen bonds and structure refinement, there is a positive tail to the correlation functions in this region, particularly for the 2:8 and 4:6 mole ratios.

Finally from $r > 10$ Å to the maximum value of around 20 Å (set by the size of the simulation box) the correlation function is close to unity in both cases, signaling no longer-ranged correlations. It is the excess correlation in the water volume in the region $0 \leqslant r \leqslant \sim 10$ Å that is the primary signal of water microsegregation caused by water hydrogen bonding to other water molecules and to the hydroxyl group on the alcohol. Since the confining length is quite short in these cases—around 5–6 Å, it means that water is pulled by the hydrogen bonding into the kind of strings and sheets that are seen in Figure 25. This segregation is over and above what can be caused by random packing effects, but is of too short a length scale for the system to be classified as colloidal. It is a perfect example of the nonideal mixing at the molecular level that can take place when amphiphilic solutes are dissolved in water.

3.5.2 Other Studies

Beyond the studies listed here, there are relatively few experimental studies that refer specifically to microheterogeneity in aqueous systems. Part of the problem is that the effect is often not directly discernible in the scattering data: microheterogeneity can occur without an obvious signal in the low Q region because as seen in Figures 25 and 28 the effect is relatively subtle, involves a fairly fractal surface, and may only occur at distances <10 Å, that is outside the distance range from which small-angle scattering signals typically emanate.

Perera et al. [129, 130] have performed extensive computer simulations of systems that exhibit microheterogeneity, and identified some of the associated pitfalls. Without experimental data to back them up, questions will always be raised as to whether any observed microheterogeneity is not an artifact of the particular computer simulation protocol used. These authors go so far as to calculate the Kirkwood-Buff integrals [131], which can be related to thermodynamic properties of the solution. Whether such integrals would be helpful for constraining the EPSR simulations—as an extra "data point" remains to be seen. Equally Hawlicka has continued to investigate microheterogeneity via computer simulation [132], particularly with regard to the effect of dissolved ions on the structure of the mixture.

The main point however is that the hydrogen bonding between water molecules and between water and other molecules can produce some fairly subtle effects which nonetheless are crucial to understanding complex aqueous environments of the kind that occur in every day life. It is likely there are many more such effects waiting to be discovered.

3.6 WATER AT INTERFACES

Without an interface, probably water would lose much of its interest: it is at the water interface that interesting things happen—structures are formed, chemical reactions occur, changes in physical state occur, and materials are dissolved. Traditionally such interfaces were not the realm of total scattering techniques. This is because in a bulk system the interfaces normally make only a tiny contribution to the total scattering pattern, which is dominated by local interactions between atoms and molecules. As a result much more surface sensitive techniques are usually employed, such as small-angle X-ray and neutron scattering, neutron and X-ray reflectometry, direct scattering of X-rays from surfaces, second harmonic generation, electron microscopy, and so on.

The situation changes if we can somehow enhance the surface compared to the bulk, such as can occur in mesoporous materials like Vycor glass for example [133]. Vycor is a form of amorphous silica which has a network of cavities, not dissimilar in shape to the water network shown in Figure 25 in ethanol:water solutions. In addition, a critical aspect is being able to model the surface, since in a microporous material there will be a mixture of bulk and interfacial scattering, which, even with isotope substitution or combining X-ray data with neutron data, cannot uniquely be separated. Interfacial total scattering experiments are always going to be harder than for the bulk fluid, with corresponding increased ambiguity in any interpretations that are made. In this regard Vycor may not be a good substrate for total scattering experiments because the surface is generally of quite a complex shape.

One of the earliest attempts to look at water structure near a surface using total scattering techniques investigated the way counterions and water were distributed between vermiculite clay surfaces [134]. Such clays are nearly ideal substrates for this work because they present well defined, very flat, atomically smooth, nearly parallel, and crystalline surfaces of lateral dimensions at least 1 μm but it can be much larger than this. Meanwhile the separation of the surfaces is around 1 nm. Depending on the amount of charge on the surfaces and the type of counterion between them, their spacing can be controlled with the amount of hydration, enabling a range of interfaces to be generated. In the earlier work, the density profile of water sandwiched between the clay layers was determined by modeling total scattering data in which hydrogen isotope substitution had been performed on the water.

Laponite is a synthetic version of clay, which has closely similar characteristics: in this case the separation between the layers can be increased indefinitely, depending on the water content, producing a series of gel and liquid crystal-like states. In both of these cases the structure of the substrate is well known, and this enables detailed modeling at the atomic and molecular level [135].

A significant disadvantage of clay systems is that the surfaces are necessarily, charged, so that counterions will always occur between the layers. This is useful if you are interested in the way charged ions are arranged between surfaces emersed in water, but is a disadvantage if you are interested in the properties of the solvent alone in confinement.

3.6.1 MCM-41 Substrate

Another system that has attracted much attention in recent years for surface studies is the porous silica material called MCM-41 [136]. This is made from aqueous aluminosilicate gels in the presence of surfactants, where the surfactant forms cylindrical micelles, which arrange themselves on a regular hexagonal lattice. The surfactant and water are removed by a process of calcination, leaving an array of cylindrical pores embedded in an amorphous silica matrix, which however also contains some silanol (–OH) groups. As a result the process of calcining the material tends to make it strongly hydrophilic, so that removing all the water from it after manufacture can be difficult. The diameter and spacing of the pores can be controlled by the length of chain on the surfactant and the physical and chemical treatments prior to calcining. It is believed the pore diameter can be controlled down to ~1.5 nm or 15 Å. The material has no net electrical charge, unlike the clay surface, and will readily absorb a variety of fluids such as nitrogen and water, without the need for counterions or other entities. Indeed nitrogen absorption is often used to estimate the size of the pores, by observing the amount of nitrogen required to complete a layer of fluid on the surface. This layering is characterized by one or more steps in the absorption isotherm as a function of applied pressure. Figure 29 gives a computer simulated version of what this material might look like at the atomic level.

MCM-41 probably presents us with the best chance to date for studying water in confinement. One of the properties that makes this potentially scientifically interesting is that when absorbed in MCM-41 with 20 Å diameter pores, water can readily be supercooled down to around 200 K without any apparent ice crystallization. This observation is based on differential scanning calorimetry (DSC) data [137]: with larger pore diameters crystallization of ice occurs, with the freezing temperature depending sensitively on the diameter of the pores. Whether crystallization really does not occur in 20 Å pores is yet to be determined. DSC has to scan as a function of temperature over a period of time, so that if the equilibration time is

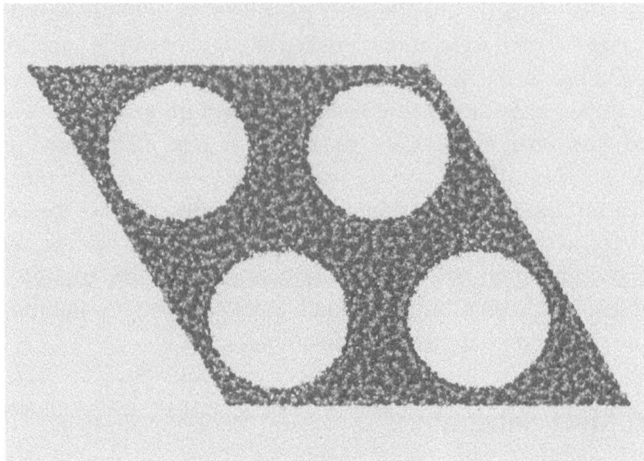

FIGURE 29 Computer simulated image of MCM-41, showing a 2 × 2 unit cell section of the hexagonal lattice. Oxygen atoms are colored red (in online version) and silicon atoms pink (in online version). Some white hydrogen atoms can be seen near the surface corresponding to silanol groups buried in the substrate and near the surface of the pores. The pores are approximately 25 Å in diameter. This is a view along the cylindrical (**c**-) axis of the pore.

sufficiently slowed down by the pore confinement, crystallization might become undetectable by this technique.

Notwithstanding this caveat, assuming water does not crystallize in MCM-41, it becomes an interesting candidate to explore water properties in a region of the phase diagram, which would otherwise be inaccessible for the bulk liquid. To this end, Sow-Hsin Chen and colleagues have performed a series of studies of water confined in this medium, using small-angle neutron scattering [138]. This is in addition to the previously mentioned work on this system using quasielastic neutron scattering, Section 3.1.2.4. In the small-angle scattering work they concluded water was being supercooled to a significant extent and that its density continued fall as the temperature was lowered through 200 K. Later work looking at the same system under pressure showed hysteresis in the temperature dependence of the small-angle scattering, apparently implying water was undergoing some sort of transition across a boundary between two phases. This transition was interpreted as a high-density to low-density transition [139].

To understand how this second experiment worked it is necessary to understand how a material like MCM-41 scatters neutrons.

3.6.2 Scattering From MCM-41

Figure 30 shows the hexagonal layout of cylindrical pores in MCM-41. The intensity of any given Bragg peak is proportional to the multiplicity of that peak times the modulus squared of a crystallographic structure factor $C(Q)$, $I(Q) = |C(Q)|^2$, where

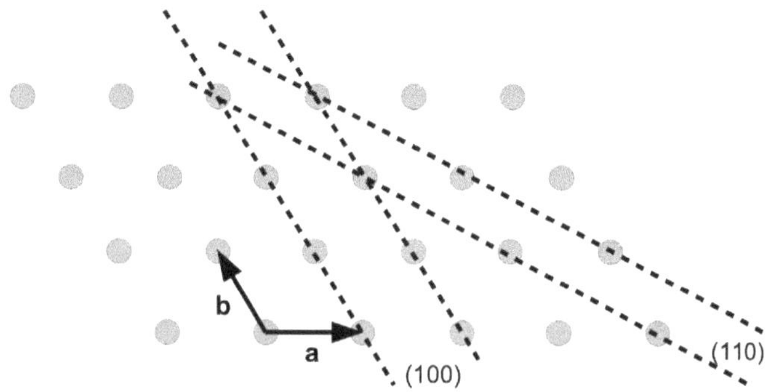

FIGURE 30 Designation of crystallographic axes for MCM-41. The blue (in online version) circles (not drawn to scale) represent cylindrical pores arranged on an hexagonal lattice. The planes corresponding to the first two Bragg reflections are shown, namely (100) and (110). The crystallographic **c**-axis is vertically out of the page.

$$C(Q) = \int dxdydz(\rho(x, y, z) - \rho_0) \exp[i(Q \cdot r)], \tag{17}$$

and $\rho(x,y,z)$ is the scattering length density at position (x,y,z) in the unit cell while ρ_0 is the average scattering length density for the substrate, so $C(Q)$ is related to the distribution in scattering length density as a function of position compared to what it would be like if the silica had no pores. The integral in (17) is performed over the hexagonal unit cell, and so includes both the amorphous silica substrate and whatever material is in the pores. It is possible to approximate $\rho(x,y,z)$ as a product of two terms, one which describes the radial dependence of the scattering length density, $\rho(r)$, where r is measured from the central axis of the pore, and one which measures the longitudinal variation in density (along the **c** axis or z axis), $M(z)$. Provided the longitudinal variations in density are much longer ranged than the diameter of the pore, as seems likely unless the pores are very poorly constructed, then at the Q values at which the Bragg peaks occur they will be insensitive to this longitudinal variation. (However the longitudinal variation may give rise to part of the pronounced Porod scattering seen in this material.) Hence the Bragg peaks arising from the hexagonal lattice will be dominated by the radial component of the density from the center of the pore. We write

$$C(Q) \sim \int (\rho(r) - \rho_0) J_0(Qr) r dr \tag{18}$$

where $J_0(Qr)$ is the zeroth order cylindrical Bessel function of the first kind. Hence the intensities of the Bragg peaks are determined by the distribution of scattering length density across the pore, and across the substrate as a whole. However the substrate density distribution is not expected to change when fluids are absorbed in the pore.

Figure 31 shows a typical diffraction pattern from MCM-41 containing different amounts of nitrogen absorbed in the pore. Three primary regions

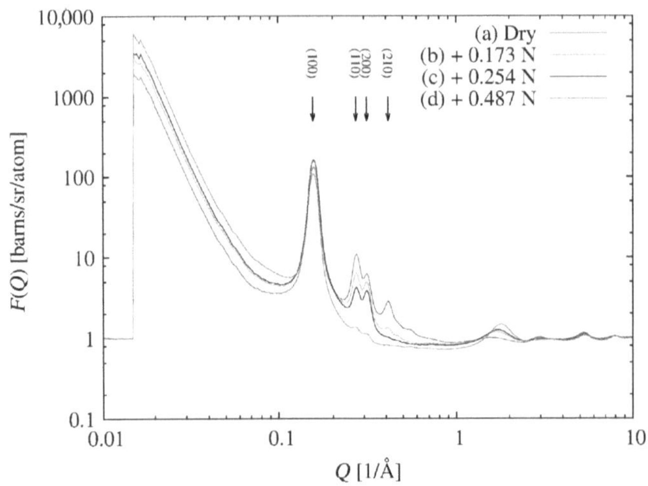

FIGURE 31 Neutron total scattering pattern for MCM-41, measured on the NIMROD diffractometer at ISIS, as a function of increasing nitrogen adsorption. The numbers opposite each curve describe the mole fraction of N relative to a mole of substrate (SiO_2 plus silanols) [139a].

in the diffraction pattern can be identified. At low Q, <0.1 Å^{-1}, there is a sharp rise in the scattering—this typically takes on a power law dependence, ($1/Q^{\nu}$), and signals the presence of interfaces over a range of length scales in the material: this "Porod" scattering is a very common feature of scattering patterns from heterogeneous materials. For intermediate Q, $0.1 \leq Q \leq 1.0$ Å^{-1}, Bragg peaks appear which are related to the hexagonal arrangement of the pores. Beyond this, the scattering pattern is dominated by atom-scale correlations, primarily (for the dry MCM-41) from the amorphous silica network, but increasingly dominated by nitrogen correlations as the amount of nitrogen increases. What is very clear is that both the Porod scattering and the intermediate Q Bragg peak heights relating to the hexagonal lattice of pores change dramatically as fluid is added to the pores: this is a direct indication that the nitrogen (in this case) is indeed being absorbed into the cylindrical pores. (It may of course be absorbed at other sites as well, but in the present case the nitrogen pressure was kept below the saturated vapor pressure of nitrogen at 77 K, and the temperature of the MCM-41 held at 100 K, so the chances of surface absorption outside of the cylindrical pores in the present case are highly unlikely.) Since the nitrogen clearly cannot completely fill the pores for the initial stages of nitrogen absorption, the distribution of density, $\rho(r)$, across the pore will not be uniform and will change as nitrogen is added. Changes to this radial distribution—otherwise known as the density profile—as well as the total amount absorbed, therefore give rise to marked changes in the Bragg peak intensities.

3.6.3 Using the (100) to Characterize Water in Confinement

The second set of experiments by the Chen group referred to above [138] concentrates on the behavior of just *one* of these peaks, namely the (100) peak. Assuming the radial dependence of the density profile, $\rho(r)$, is invariant to changes of state condition, they map the changes of intensity of the (100) peak with temperature and pressure, and relate these changes of intensity directly to the average density of water in the pores. Furthermore, by observing the hysteresis that occurs in these intensities between cooling and heating scans, they conclude they are witnessing a high-density to low-density liquid transition [139].

The issues raised by this work of the Chen group have been the subject of several criticisms [140–143]. In particular there is no evidence to date that the density distribution across the pore is unchanging with temperature. In addition, as discussed above, it is claimed that water does not crystallize if the pore diameter is small enough. This is not proven however. Firstly, in every likelihood, as indicated in the cited references, and, as already pointed out in much earlier work [144], the pore diameters in MCM-41 are likely significantly larger—by as much as 10 Å—than those given by traditional gas adsorption analysis. Secondly it turns out that it is perfectly feasible, in theory, to fit a cylinder of ice into a cylindrical pore of this diameter, as shown in Figure 32.

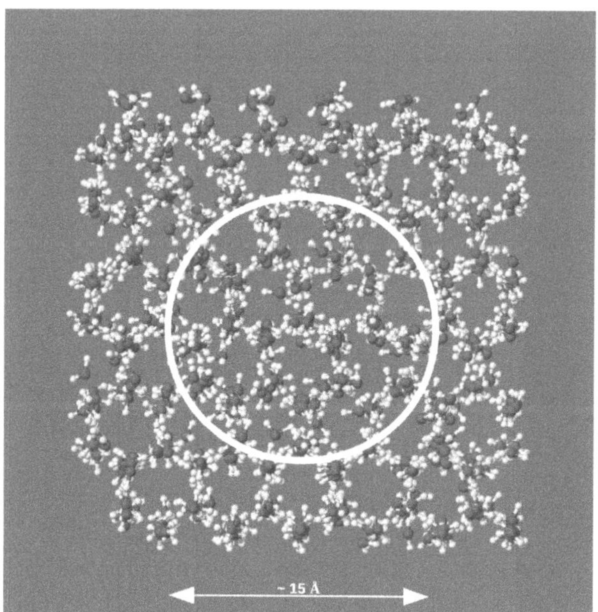

FIGURE 32 View of a disordered ice Ih lattice looking down along the **c**-axis. A circle of diameter 15 Å (approximate) is drawn to show how such a crystal structure might fit into a cylindrical pore of the same diameter in MCM-41. *The structure is adapted from [145].*

Of course if this were to happen in practice the structure would need to be distorted by the interactions of water molecules with the silica atoms on the surface of the pore, but it illustrates the point that there is in principle no reason why ice cannot form in small pores.

Further EPSR simulations, in which the whole region of the scattering pattern from $Q = 0.1$ Å^{-1} to $Q = 20$ Å^{-1} was fit, were presented in [128]. Figure 33 shows the density profiles that were obtained in those simulations, while Figure 34 compares the confined water radial distribution functions with those found in bulk water. The density profiles do not agree in detail with those shown by Mancinelli et al., but this is because the assumed pore diameter for the former simulations was 15 Å, while for the later work it was 25 Å. However they both record the significant penetration of water into the amorphous silica substrate. Equally the water radial distribution functions

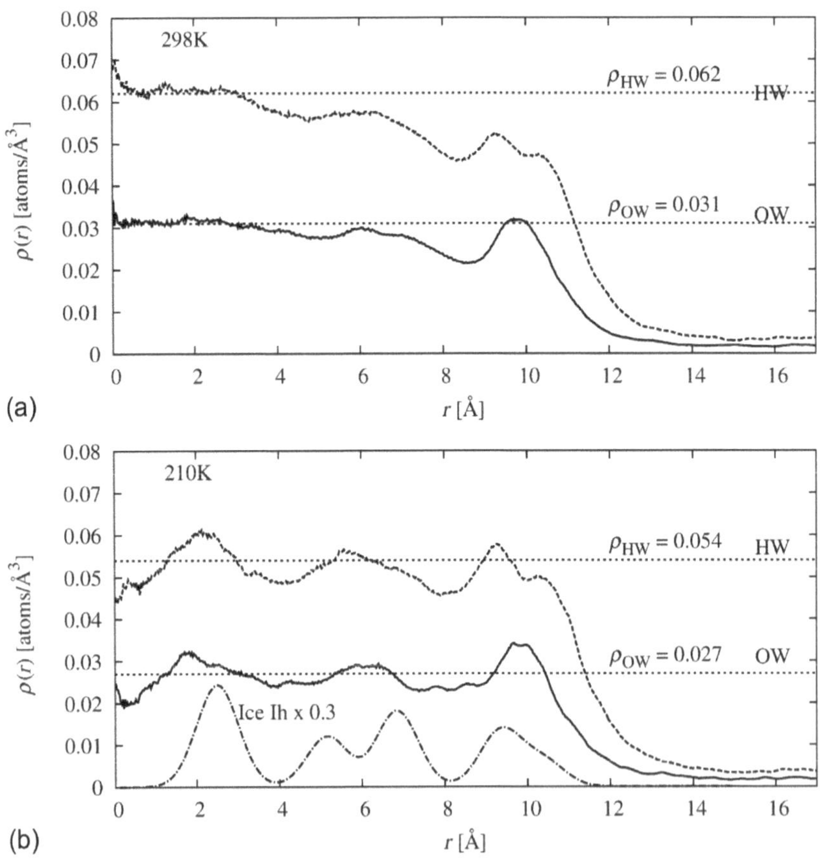

FIGURE 33 Density profile, measured from the center of the pore, of water adsorbed in MCM-41, at ambient temperature (a), and at 210 K (b). As originally shown in [128].

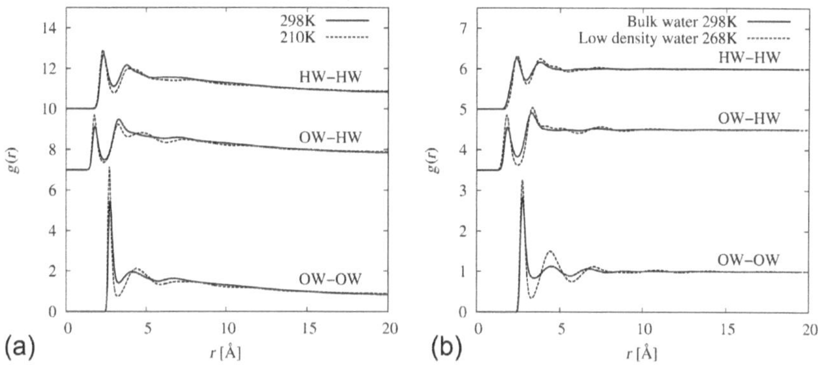

FIGURE 34 (a) Water-water radial distribution functions for water adsorbed in MCM-41, at ambient temperature (line) and at 210 K (dashed). (b) Same functions for water at ambient (line) and as projected to occur at low density at 268 K [100]. For (a) the pronounced slope on the distribution functions arises from the confinement of the water, and is called an "excluded volume effect". As originally shown in [120].

show the increase in tetrahedrality of the water which occurs at the lower temperature compared to ambient conditions. This is shown by the marked sharpening of the 4.5 Å peak in the $g_{OO}(r)$ function on lowering the temperature.

None of these simulations actually answers the question of what is the state of the water in these pores at low temperature. Is it a (disordered) crystal, is it a glass, or is it, as some would claim, a low-density form of water? There is nothing in the data presented so far that is convincing in any of these regards, but given the likely complex interaction of water with the silica surface, plus the complete lack of conclusive evidence to the contrary, it would appear that a simple interpretation based on a thermodynamic model of water which ignores the effect of the surface is untenable.

3.7 TOWARD EVER GREATER COMPLEXITY—SOME CONCLUDING REMARKS

The foregoing account has hopefully given a feel for the scope of modern neutron total scattering experiments on aqueous systems. Combination of accurate total scattering data, placed on an absolute scale of neutron or X-ray differential scattering cross section, with the ability to model those data using computer simulation in which the force field is modified *by* the data to give the best possible fit, whilst at the same time obeying known physical and chemical constraints, such as atomic overlap forces, dispersion forces, Coulomb forces, molecular structure, and so on, prove to be extremely powerful for understanding these systems at the molecular level.

Because this method is so highly developed and quite specialized, the present account naturally has big gaps in it. For example there is only scant reference to the important role that inelastic and quasielastic neutron scattering have in these systems in telling us about the dynamics. Indeed compared to what has been achieved with total scattering, these dynamical matters are still open to serious questions, a case in point being the controversial claims that supercooled water in confinement undergoes a fragile to strong transition (Section 3.1.2.4). Was this real, or was it an artifact of the instrumental set up? If I could use inelastic neutron scattering to investigate dynamics, how would I distinguish between solute and solvent? Is inelastic scattering amenable to the kind of detailed computer refinement performed here on the structure of the liquid? These questions remain still largely unanswered. Nonetheless the importance of the dynamics of these systems cannot be overstated: after all the dynamics is what make them liquids!

In this regard there are a number of pulsed optical experiments on aqueous systems of the kinds explored here [146] which can apparently sample the short time (ps) dynamical properties of water molecules in solution. However the interpretation of these experiments tends to be heuristic, based on trends observed with temperature and concentration, rather than via the kind of mathematical modeling on a computer described here. The interaction of light with molecules is via the polarizability tensor and this makes a simple transition from model to data much more complicated than for the cases of neutron and X-ray scattering, which both probe directly the nucleus-nucleus correlation functions. The future of these and other probes in this field has still to be quantified.

Since 2009 at ISIS the Near and Intermediate Range Order Diffractometer, NIMROD, has come on stream. This permits total scattering measurements in the Q range $\sim$0.01 to $\sim$30 Å^{-1}. The data from this new machine give atomic scale resolution over distances up to $\sim$600 Å, which is 1 order of magnitude larger than had been available previously on equivalent diffractometers at ISIS (SANDALS) or the ILL (D4C) or SNS (NOMAD). Other recent developments include the NOVA diffractometer at J-PARC in Japan, which has a Q-range equivalent to NIMROD, and there are now several total scattering X-ray diffractometers around the world, including at the APS, Argonne, SPRING-8, Japan, and the new XPDF machine at Diamond in the UK, This means studies of simple structures such as ionic and nonionic aqueous solutions are becoming superseded by studies of much more complex objects, a case in point being micellar systems. Equally the complex environment of biologically relevant molecules, such as drug molecules and amino acids and peptide mimics are being explored in molecular detail. In some cases the neutron data are combined with NMR data which can give specific coordination constraints on particular atom pairs.

All this however comes with an additional serious challenge, namely how to interpret the new data? The current EPSR method is good up to about 150,000 atoms, but this gives a simulation box size which is still an order of magnitude too small compared to the Q range of the data. Simply increasing the box size to the required value is possible in principle but would require $>10^7$ atoms and lead to unacceptably slow execution times, even with massive parallelization. The problem here is that because the data give long range information, the empirical potential used in the structure refinement would also need to be correspondingly long ranged, so that many of the tricks to speed up execution that are typically used in large scale computer simulations could not be applied.

The increasing complexity also brings with it another challenge. We have already seen that even at short length scales, interpretation of a given set of scattering data is not necessarily unambiguous. Because our data are necessarily incomplete—not all the required site-site functions can be extracted independently or are weakly weighted in the total scattering data—it may be possible to fit a given set of data equally well to more than one structural model. This problem will only get larger with longer length scales. For example is a micelle a well defined spherical object with sharp boundaries between water, surfactant and oil, as it is often portrayed in textbooks, or is it a rougher, less spherical object, with some water penetrating the oil and surfactant, as found in recent total scattering studies [147]? A full answer to this question, and many other similar types of question, is not available at present. Currently the most likely way around the scale problem, and to keep the structure refinement on a practical time and size scale, is to employ some form of coarse-graining to deal with the long range part of the data. How this develops will hopefully emerge in the coming few years.

REFERENCES

[1] C.A. Angell, J. Shuppert, J.C. Tucker, J. Phys. Chem. 77 (1973) 3092–3099.
[2] C.A. Angell, J.C. Tucker, Science 181 (1973) 342–344.
[3] R.J. Speedy, C.A. Angell, J. Chem. Phys. 65 (1976) 851.
[4] R.J. Speedy, J. Phys. Chem. 86 (1982) 982–991.
[5] P.H. Poole, F. Sciortino, U. Essmann, H.E. Stanley, Nature 360 (1992) 324–328.
[6] F.H. Stillinger, A. Rahman, J. Chem. Phys. 60 (1974) 1545.
[7] O. Mishima, L.D. Calvert, E. Whalley, Nature 314 (1985) 76–78.
[8] V. Holten, M.A. Anisimov, Sci. Rep. 2 (2012) 713.
[9] E.B. Moore, V. Molinero, J. Chem. Phys. 130 (2009) 244505.
[10] J. Russo, H. Tanaka, Nat. Commun. 5 (2014) 3556.
[11] L.G.M. Pettersson, A. Nilsson, J. Non-Cryst. Solids 407 (2015) 399–417.
[12] D.T. Limmer, D. Chandler, J. Chem. Phys. 135 (2011) 134503.
[13] D.T. Limmer, D. Chandler, J. Chem. Phys. 138 (2013) 214504.
[14] D.T. Limmer, D. Chandler, Mol. Phys. 113 (2015) 2799–2804.
[15] D.T. Limmer, D. Chandler, Phys. Rev. E 91 (2015) 16301.
[16] Y. Liu, J.C. Palmer, A.Z. Panagiotopoulos, P.G. Debenedetti, J. Chem. Phys. 137 (2012) 214505.
[17] J.C. Palmer, R. Car, P.G. Debenedetti, Faraday Discuss. 167 (2013) 77.

[18] J.C. Palmer, F. Martelli, Y. Liu, R. Car, A.Z. Panagiotopoulos, P.G. Debenedetti, Nature 510 (2014) 385–388.
[19] K. Binder, Proc. Natl. Acad. Sci. 111 (2014) 9374–9375.
[20] V. Molinero, E.B. Moore, J. Phys. Chem. B 113 (2009) 4008–4016.
[21] E.B. Moore, V. Molinero, Nature 479 (2011) 506–508.
[22] J.A. Sellberg, C. Huang, T.A. McQueen, N.D. Loh, H. Laksmono, D. Schlesinger, R.G. Sierra, D. Nordlund, C.Y. Hampton, D. Starodub, D.P. DePonte, M. Beye, C. Chen, A.V. Martin, A. Barty, K.T. Wikfeldt, T.M. Weiss, C. Caronna, J. Feldkamp, L.B. Skinner, M.M. Seibert, M. Messerschmidt, G.J. Williams, S. Boutet, L.G.M. Pettersson, M.J. Bogan, A. Nilsson, Nature 510 (2014) 381–384.
[23] H.S. Frank, M.W. Evans, J. Chem. Phys. 13 (1945) 507.
[24] W. Kauzmann, Adv. Protein Chem. 14 (1959) 1–63.
[25] A. Ben-Naim, Open J. Biophys. 1 (2011) 1–7.
[26] H.J. Bakker, Nature 491 (2012) 533–535.
[27] J.G. Davis, K.P. Gierszal, P. Wang, D. Ben-Amotz, Nature 491 (2012) 582–585.
[28] P. Wernet, D. Nordlund, U. Bergmann, M. Cavalleri, M. Odelius, H. Ogasawara, L.Å. Näslund, T.K. Hirsch, L. Ojamäe, P. Glatzel, L.G.M. Pettersson, A. Nilsson, Science 304 (2004) 80.
[29] J.D. Bernal, R.H. Fowler, J. Chem. Phys. 1 (1933) 515.
[30] D.T. Bowron, M.H. Krisch, A.C. Barnes, J.L. Finney, A. Kaprolat, M. Lorenzen, Phys. Rev. B. 62 (2000) R9223–R9227.
[31] L. Näslund, D.C. Edwards, P. Wernet, U. Bergmann, H. Ogasawara, L.G.M. Pettersson, S. Myneni, A. Nilsson, J. Phys. Chem. A 109 (2005) 5995–6002.
[32] I. Waluyo, D. Nordlund, U. Bergmann, D. Schlesinger, L.G.M. Pettersson, A. Nilsson, J. Chem. Phys. 140 (2014) 244506.
[33] I. Waluyo, C. Huang, D. Nordlund, U. Bergmann, T.M. Weiss, L.G.M. Pettersson, A. Nilsson, J. Chem. Phys. 134 (2011) 64513.
[34] A.H. Narten, H.A. Levy, Science 165 (1969) 447–454.
[35] I. Waldner, A. Bassen, H. Bertagnolli, K. Tödheide, G. Strauss, A.K. Soper, J. Chem. Phys. 107 (1997) 10667.
[36] L. Kong, X. Wu, R. Car, Phys. Rev. B 86 (2012) 134203.
[37] T.D. Kühne, R.Z. Khaliullin, Nat. Commun. 4 (2013) 1450.
[38] F. Fernandez-Alonso, D.L. Price (Eds.), Neutron Scattering – Fundamentals. Experimental Methods in the Physical Sciences, vol. 44, Academic Press Inc./Elsevier Science, San Diego, CA, 2013.
[39] C.G. Windsor, Pulsed Neutron Scattering, Taylor & Francis, London, 1981.
[40] A.K. Soper, Mol. Phys. 107 (2009) 1667–1684.
[41] G. Placzek, Phys. Rev. 86 (1952) 377–388.
[42] J.-P. Hansen, I.R. McDonald, Theory of Simple Liquids, 2nd Edition, Academic Press, London, 1986.
[43] T.E. Faber, J.M. Ziman, Philos. Mag. 11 (1965) 153–173.
[44] A.K. Soper, G.W. Neilson, J.E. Enderby, R.A. Howe, J. Phys. C Solid State Phys. 10 (1977) 1793–1801.
[45] G.J. Herdman, G.W. Neilson, J. Mol. Liq. 46 (1990) 165–179.
[46] G.W. Neilson, P.E. Mason, S. Ramos, D. Sullivan, Philos. Trans. R. Soc. A Math. Phys. Eng. Sci. 359 (2001) 1575–1591.
[47] P.E. Mason, G.W. Neilson, C.E. Dempsey, J.W. Brady, J. Am. Chem. Soc. 128 (2006) 15136–15144.
[48] P.E. Mason, J.M. Cruickshank, G.W. Neilson, P. Buchanan, Phys. Chem. Chem. Phys. 5 (2003) 4686.

[49] P.E. Mason, G.W. Neilson, C.E. Dempsey, A.C. Barnes, J.M. Cruickshank, Proc. Natl. Acad. Sci. U.S.A. 100 (2003) 4557–4561.
[50] A. Zeidler, P.S. Salmon, H.E. Fischer, J.C. Neuefeind, J.M. Simonson, H. Lemmel, H. Rauch, T.E. Markland, Phys. Rev. Lett. 107 (2011) 145501.
[51] A.K. Soper, C.J. Benmore, Phys. Rev. Lett. 101 (2008) 65502.
[52] J.E. Enderby, D.M. North, P.A. Egelstaff, Philos. Mag. 14 (1966) 961–970.
[53] F.G. Edwards, J.E. Enderby, R.A. Howe, D.I. Page, J. Phys. C Solid State Phys. 8 (1975) 3483.
[54] I. Petri, P. Salmon, H. Fischer, Phys. Rev. Lett. 84 (2000) 2413–2416.
[55] G.W. Neilson, J.E. Enderby, Proc. R. Soc. Lond. A Math. Phys. Eng. Sci. 390 (1983) 353.
[56] A.K. Soper, R. Silver, Phys. Rev. Lett. 49 (1982) 471–474.
[57] A.K. Soper, ISRN Phys. Chem. 2013 (2013) 1–67.
[58] T. Springer, Quasielastic neutron scattering for the investigation of diffusive motions in solids and liquids, Springer Tracts in Modern Physics, vol. 64, Springer, Berlin Heidelberg, 1972, pp. 1–100.
[59] G.J. Safford, P.S. Leung, A.W. Naumann, P.C. Schaffer, J. Chem. Phys. 50 (1969) 4444–4467.
[60] P.S. Leung, G.J. Safford, J. Phys. Chem. 74 (1970) 3696–3709.
[61] J. Qvist, H. Schober, B. Halle, J. Chem. Phys. 134 (2011) 144508.
[62] A. Arbe, P. Malo De Molina, F. Alvarez, B. Frick, J. Colmenero, Phys. Rev. Lett. 117 (2016) 185501.
[63] L. Liu, S.H. Chen, A. Faraone, C.W. Yen, C.Y. Mou, Phys. Rev. Lett. 95 (2005) 117802.
[64] S.-H. Chen, L. Liu, E. Fratini, P. Baglioni, A. Faraone, E. Mamontov, Proc. Natl. Acad. Sci. U.S.A. 103 (2006) 9012–9016.
[65] W. Doster, S. Busch, A.M. Gaspar, M.S. Appavou, J. Wuttke, H. Scheer, Phys. Rev. Lett. 104 (2010) 98101.
[66] F. Franks (Ed.), The Physics and Physical Chemistry of Water, Springer New York, Boston, MA, 1972.
[67] L. Van Hove, Phys. Rev. 95 (1954) 249–262.
[68] P.A. Egelstaff, An Introduction to the Liquid State, Clarendon Press, Oxford, 1992.
[69] M.P. Allen, D.J. Tildesley, Computer Simulation of Liquids, Academic Press, Amsterdam, 1987.
[70] C.G. Gray, K.E. Gubbins, The Theory of Molecular Fluids—(The International Series of Monographs on Chemistry, 9) Vol. I: Fundamentals, Oxford University Press, New York, 1984.
[71] P. Egelstaff, Neutron scattering, electronic excitations and chemical reactions, in: R. Silver (Ed.), Proc. 1984 Work. High Energy Excit. Condens. Matter, Los Alamos National Laboratory, Los Alamos, NM, 1984, pp. 626–645.
[72] A.K. Soper, M.G. Phillips, Chem. Phys. 107 (1986) 47–60.
[73] A.K. Soper, GudrunN and GudrunX: Programs for Correcting Raw Neutron and X-ray Diffraction Data to Differential Scattering Cross Section, RAL Tech. Rep. RAL-TR-2011-013, 2011. http://purl.org/net/epubs/work/56240, Accessed 19 February 2015.
[74] E.N.D. Library, http://www.nndc.bnl.gov/endf/b7.1/index.html, Accessed 20 January 2017.
[75] Y.S. Badyal, M.L. Saboungi, D.L. Price, S.D. Shastri, D.R. Haeffner, A.K. Soper, J. Chem. Phys. 112 (2000) 9206–9208.
[76] J.M. Sorenson, G. Hura, R.M. Glaeser, T. Head-Gordon, J. Chem. Phys. 113 (2000) 9149–9161.
[77] R.L. McGreevy, L. Pusztai, Mol. Simul. 1 (1988) 359–367.
[78] H. Rietveld, Acta Crystallogr. 22 (1967) 151–152.
[79] A.K. Soper, Chem. Phys. 202 (1996) 295–306.

[80] A.K. Soper, Mol. Phys. 99 (2001) 1503–1516.
[81] A.K. Soper, Phys. Rev. B. 72 (2005) 104–204.
[82] L.B. Skinner, C. Huang, D. Schlesinger, L.G.M. Pettersson, A. Nilsson, C.J. Benmore, J. Chem. Phys. 138 (2013) 74506.
[83] W.E. Thiessen, A.H. Narten, J. Chem. Phys. 77 (1982) 2656.
[84] A.K. Soper, Chem. Phys. 258 (2000) 121–137.
[85] L. Pusztai, Phys. Rev. B. 60 (1999) 11851–11854.
[86] Z. Steinczinger, P. Jóvári, L. Pusztai, J. Mol. Liq. 228 (2016) 6–11.
[87] R.A. Buckingham, Proc. R. Soc. A Math. Phys. Eng. Sci. 168 (1938) 264–283.
[88] A.K. Soper, J. Phys. Condens. Matter. 22 (2010) 404210.
[89] A.K. Soper, Empirical Potential Structure Refinement—EPSRshell: A User's Guide, Version 18, RAL Tech. Rep. RAL-TR-2011-012, May 2011. http://purl.org/net/epubs/work/56239, Accessed 19 February 2015.
[90] J.L. Finney, A. Hallbrucker, I. Kohl, A.K. Soper, D.T. Bowron, Phys. Rev. Lett. 88 (2002) 225503.
[90a] N.H. Rhys, A.K. Soper, L. Dougan, J. Phys. Chem. B 119 (2015) 15644–15651.
[91] A.H. Narten, M.D. Danford, H.A. Levy, Discuss. Faraday Soc. 43 (1967) 97–107.
[92] L.B. Skinner, C.J. Benmore, J.C. Neuefeind, J.B. Parise, J. Chem. Phys. 141 (2014) 214507.
[93] M.-C. Bellissent-Funel, J. Teixeira, L. Bosio, J.C. Dore, J. Phys. Condens. Matter. 1 (1989) 7123–7129.
[94] A.K. Soper, Water and ice structure in the range 220–365 K from radiation total scattering experiments, in: P.G. Debenedetti, M.A. Ricci, F. Bruni (Eds.), Proc. Int. Sch. Phys. "Enrico Fermi" Course 187 Water Fundam. as Basis Underst. Environ. Promot. Technol, IOS/SIF, Amsterdam/Bologna, Varenna, 2015, pp. 151–171.
[95] A. Botti, F. Bruni, M.A. Ricci, A.K. Soper, J. Chem. Phys. 109 (1998) 3180–3184.
[96] G.W. Neilson, D.I. Page, W.S. Howells, J. Phys. D: Appl. Phys. 12 (1979) 901–907.
[97] M.-C. Bellissent-Funel, L. Bosio, J. Chem. Phys. 102 (1995) 3727.
[98] M.-C. Bellissent-Funel, Europhys. Lett. 42 (1998) 161–166.
[99] A.V. Okhulkov, Y.N. Demianets, Y.E. Gorbaty, J. Chem. Phys. 100 (1994) 1578–1588.
[100] A.K. Soper, M.A. Ricci, Phys. Rev. Lett. 84 (2000) 2881.
[101] A.K. Soper, J. Phys. Chem. B 119 (2014) 9244.
[102] J.L. Finney, D.T. Bowron, Phys. Rev. Lett. 89 (2002) 205503.
[103] S. Klotz, G. Hamel, J.S. Loveday, R.J. Nelmes, Phys. Rev. Lett. 89 (2002) 285502.
[104] H. Ohtaki, T. Radnai, Chem. Rev. 93 (1993) 1157–1204.
[105] R. Mancinelli, A. Botti, F. Bruni, M.A. Ricci, A.K. Soper, Phys. Chem. Chem. Phys. 9 (2007) 2959–2967.
[106] A.K. Soper, A. Luzar, J. Chem. Phys. 97 (1992) 1320–1331.
[107] H.J.C. Berendsen, J.R. Grigera, T.P. Straatsma, J. Phys. Chem. 91 (1987) 6269–6271.
[108] R. Leberman, A.K. Soper, Nature. 378 (1995) 364–366.
[109] E.C. Hulme, A.K. Soper, S.E. McLain, J.L. Finney, Biophys. J. 91 (2006) 2371–2380.
[110] S.E. McLain, A.K. Soper, A. Watts, J. Phys. Chem. B 110 (2006) 21251–21258.
[111] S.E. McLain, S. Imberti, A.K. Soper, A. Botti, F. Bruni, M.A. Ricci, Phys. Rev. B 74 (2006) 94201.
[112] S. Díaz-Moreno, S. Ramos, D.T. Bowron, J. Phys. Chem. A 115 (2011) 6575–6581.
[113] D.T. Bowron, S. Díaz-Moreno, J. Phys. Chem. B 113 (2009) 11858–11864.
[114] D.T. Bowron, E.C. Beret, E. Martin-Zamora, A.K. Soper, E. Sánchez Marcos, J. Am. Chem. Soc. 134 (2012) 962–967.

[115] G. Pálinkás, E. Hawlicka, K. Heinzinger, Chem. Phys. 158 (1991) 65–76.
[116] E. Hawlicka, R. Grabowski, J. Phys. Chem. 96 (1992) 1554–1557.
[117] S. Dixit, W.C.K. Poon, J. Crain, S. Dixit, W.C.K. Poon, J. Phys. Condens. Matter. 12 (2000) L323–L328.
[118] S. Dixit, J. Crain, W.C.K. Poon, J.L. Finney, A.K. Soper, Nature 416 (2002) 829–832.
[119] L. Dougan, R. Hargreaves, S.P. Bates, J.L. Finney, V. Réat, A.K. Soper, J. Crain, J. Chem. Phys. 122 (2005) 174514.
[120] J.L. Finney, D.T. Bowron, A.K. Soper, J. Phys. Condens. Matter. 12 (2000) A123–A128.
[121] T. Yamaguchi, T. Takamuku, A.K. Soper, J. Neutron Res. 13 (2005) 129–133.
[122] J.J. Towey, A.K. Soper, L. Dougan, J. Phys. Chem. B 116 (2012) 13898–13904.
[123] S.G. Chou, A.K. Soper, S. Khodadadi, J.E. Curtis, S. Krueger, M.T. Cicerone, A.N. Fitch, E.Y. Shalaev, J. Phys. Chem. B 116 (2012) 4439–4447.
[124] O. Kirichek, A.K. Soper, B. Dzyuba, W.V. Holt, R. Soc. Open Sci. 3 (2016) 150655.
[125] E. Shalaev, A.K. Soper, J. Phys. Chem. B 120 (2016) 7289–7296.
[126] D. Stauffer, Introduction to Percolation Theory, Taylor & Francis, London, 1985.
[127] O. Glatter, O. Kratky, Small Angle Scattering, Academic Press, London, 1982.
[128] A.K. Soper, Chem. Phys. Lett. 590 (2013) 1–15.
[129] L. Zoranić, R. Mazighi, F. Sokolić, A. Perera, J. Chem. Phys. 130 (2009) 124315.
[130] A. Perera, R. Mazighi, B. Kežić, J. Chem. Phys. 136 (2012) 174516.
[131] J.G. Kirkwood, F.P. Buff, P.F. Buff, J. Chem. Phys. 19 (1951) 774–777.
[132] E. Owczarek, M. Rybicki, E. Hawlicka, Chem. Phys. 363 (2009) 78–87.
[133] T.H. Elmer, Porous and reconstructed glasses, in: Engineering Materials Handbook, Vol. 4: Ceramics and Glasses, ASM International, Materials Park, OH, 1992, pp. 427–432.
[134] N.T. Skipper, A.K. Soper, J.D.C. McConnell, K. Refson, Chem. Phys. Lett. 166 (1990) 141.
[135] M.A. Ricci, V. Tudisca, F. Bruni, R. Mancinelli, E. Scoppola, R. Angelini, B. Ruzicka, A.K. Soper, J. Non-Cryst. Solids. 407 (2015) 418–422.
[136] C.T. Kresge, M.E. Leonowicz, W.J. Roth, J.C. Vartuli, J.S. Beck, Nature 359 (1992) 710–712.
[137] K. Yoshida, T. Yamaguchi, S. Kittaka, M.-C. Bellissent-Funel, P. Fouquet, J. Chem. Phys. 129 (2008) 54702.
[138] D. Liu, Y. Zhang, C.-C. Chen, C. Mou, P.H. Poole, S. Chen, Proc. Natl. Acad. Sci. U.S.A. 104 (2007) 9570–9574.
[139] Z. Wang, K. Ito, J.B. Leão, L. Harriger, Y. Liu, S.-H. Chen, J. Phys. Chem. Lett. 6 (2015) 2009–2014.
[139a] A.K. Soper, D.T. Bowron, Chem. Phys. Lett. (in press, 2017).
[140] R. Mancinelli, F. Bruni, M.A. Ricci, J. Phys. Chem. Lett. 1 (2010) 1277–1282.
[141] R. Mancinelli, J. Phys. Condens. Matter. 22 (2010) 404213.
[142] A.K. Soper, Proc. Natl. Acad. Sci. 108 (2011) E1192.
[143] A.K. Soper, J. Phys. Condens. Matter 24 (2012) 64107.
[144] P.I. Ravikovitch, S.C.O. Domhnaill, A.V. Neimark, F. Schueth, K.K. Unger, Langmuir 11 (1995) 4765–4772.
[145] A.K. Soper, J. Phys. Chem. B 119 (2015) 9244–9253.
[146] H.J. Bakker, J.L. Skinner, Chem. Rev. 110 (2010) 1498–1517.
[147] R. Hargreaves, D.T. Bowron, K. Edler, J. Am. Chem. Soc. 133 (2011) 16524–16536.

Chapter 4

Ionic Liquids and Neutron Scattering

Olga Russina*,[1] and Alessandro Triolo†,[1]
Department of Chemistry, University of Rome "Sapienza", Rome, Italy; †Istituto Struttura della Materia, Consiglio Nazionale delle Ricerche, Rome, Italy
[1]*Corresponding authors: e-mail: olga.russina@uniroma1.it; Alessandro.Triolo@artov.ism.cnr.it*

Chapter Outline

4.1 INTRODUCTION

Room-temperature ionic liquids (ILs) represent an exciting class of materials that is attracting a great deal of attention in the last couple of decades. Being composed solely of ionic species and being characterized by low melting points (by definition below 100°C), they represent a very peculiar class of solvent media. Together with supercritical fluids, they were referred to as neoteric solvents [1], promising to be active players in an environmentally responsible approach to modern chemistry and chemical processing. Since Paul Walden's discovery of ethylammonium nitrate (hereinafter indicated as EAN) in 1914 [2], the interest toward this class of compounds has been raising: they have a number of smart properties that fully justify this attention. Being ionic in nature, they show negligible vapor pressure (although cases when they can be distilled without appreciable decomposition have been reported [3]); also, their elevated thermal and electrochemical stability, the nonflammability and remarkable solvation capability toward a large variety of compounds suggest to consider them as green solvent media. Another

Experimental Methods in the Physical Sciences, Vol. 49. http://dx.doi.org/10.1016/B978-0-12-805324-9.00004-2

important feature is the striking fine-tuning of their chemical-physical properties (e.g., density, viscosity, hydrophobicity, and melting point) that can be achieved by minor changes to the constituent cations and anions. An outstanding aspect of the phenomenology related to ILs is the possibility to custom prepare so-called task-specific ILs that have explicitly formulated to address specific technological requirement, by introduction of targeted functionalities in their chemical structure [4–7]. Such a high flexibility in performance that can satisfy different requirements in a variety of applicative fields led to the indication of ILs as *designer solvents* [8]. Nowadays, ILs find application in a wealth of technological fields ranging from extraction and separation to synthesis, (bio-)catalysis, electrochemistry, pharmaceutics, biomass treatment, etc. In parallel with the ever-growing interest in this class of compounds, research has been flourishing in the last couple of decades aiming at providing a robust and comprehensive understanding of IL properties in terms of micro- and mesoscopic models, trying to correlate bulk performances with microscopic properties [9–25]. These strong efforts in exploring microscopic properties in ILs are witnessed by a large number of dedicated reviews [19,20,22,26–32] and special issues [33–39].

In this blooming scenario, the role played by neutron science is large: the quest for establishing firm correlations between macroscopic performances and micro- to mesoscopic structural and dynamic features cannot help but exploit the large potentialities offered by the wide range of techniques that make use of neutrons (see, e.g., contributions in the first volume of this series: "Neutron Scattering—Fundamentals," edited by F. Fernandez-Alonso and David L. Price [40]).

In the last decade, it emerged very clearly that ILs are characterized by an enhanced level of structural and consequently dynamic complexity. An intricate hierarchy of spatial scales is found to ubiquitously characterize the structural correlations in ILs. Driven by molecular dynamic simulations [19,20,41–55], research soon shifted toward the exploration of microscopic and later mesoscopic structural and dynamic properties in ILs. In this context, then neutron and X-ray scattering techniques turned out to be the best fitting experimental techniques that could provide a microscopic insight into ILs from the point of view of both spatial and temporal scales. While in this contribution we will mainly focus on the contribution that neutron techniques succeeded in providing to the exploration of ILs, we need to mention that, at least from the structural point of view, X-ray scattering has been and keeps on being an extremely powerful tool to access spatial scales from sub-Å up to microns. When possible, we tried to provide the flavor for the strong complementarity between these neutrons and X-ray studies, and we mention that recently two important contributions [31,32] provided similar information from a more general point of view, that is, less focused on neutron scattering than our present one.

Overall, this contribution aims at providing a general overview of the role of neutron scattering in addressing general and highly specific structural and dynamic properties in ILs. While some studies are still limited by the available facilities brilliance, on the other hand (a), the steadily increasing flux that becomes more and more available (in perspective, the operation of new sources, such as the European Spallation Source (ESS [56]) will provide unprecedented levels of brilliance, thus making experiments so far unimaginable, become feasible) and (b) the unique role played by isotopic substitution in accessing very specific and detailed features in the structural and dynamic properties of ILs make the use of neutron scattering an essential research tool for scientists involved in IL-focused research.

Hereinafter, we will indicate with ionic liquids (ILs) the whole class of ionic compounds that show melting point below 100°C. A representative selection of cation and anion species that when paired lead to ILs is shown in Figure 1. The specific choice of one cation and one anion from this selection will lead to an ionic compound that in most of the cases will show melting point below 100°C.

However, it is noteworthy that each specific pair of ions will be characterized by its own set of chemical-physical parameters. Accordingly, one way to modulate these parameters, maintaining the compound into the ILs realm, is just to change either the cation or the anion. One of the specific features of ILs is the fact that most of them are characterized by a short/medium length alkyl tail. Many studies focused on the role of side alkyl tails in affecting the bulk properties of ILs, and as a matter of fact, this parameter (i.e., the number of methylene units along the chain) is one of those used to fine-tune the chemical-physical properties of ILs [21,57]. For example, Figure 2 shows the role of the side-chain length/anion [58–61] nature in affecting the melting point of the resulting ionic compound for the class of 1-alkyl-3-methylimidazolium-based (see Figure 1) ILs: Therein one can appreciate how by simply modulating the length of the alkyl tail connected to the imidazolium ring; one can widely change the melting point and, for example, the phase diagram (e.g., in the case of [$C_n$mim][BF_4], 1-alkyl-3-methylimidazolium tetrafluoroborate salts, when the alkyl tail is long enough, one can observe the existence of a high-temperature smectic phase as a precursor of the isotropic liquid one that is not observed for the case of short chains [21]).

The role played by this side chain on a wide range of ILs' properties is complex. In many cases, the high level of asymmetry introduced by the presence of the alkyl tail in the cation leads to a destabilization of the crystalline phase, thus favoring a drop in the melting point. On the other hand, when long enough alkyl tails are connected to the cation (or occasionally to the anion), they can organize in a polyethylene-like way, thus favoring the crystallization of the compound. Overall, in the intermediate length range between 2–4 and

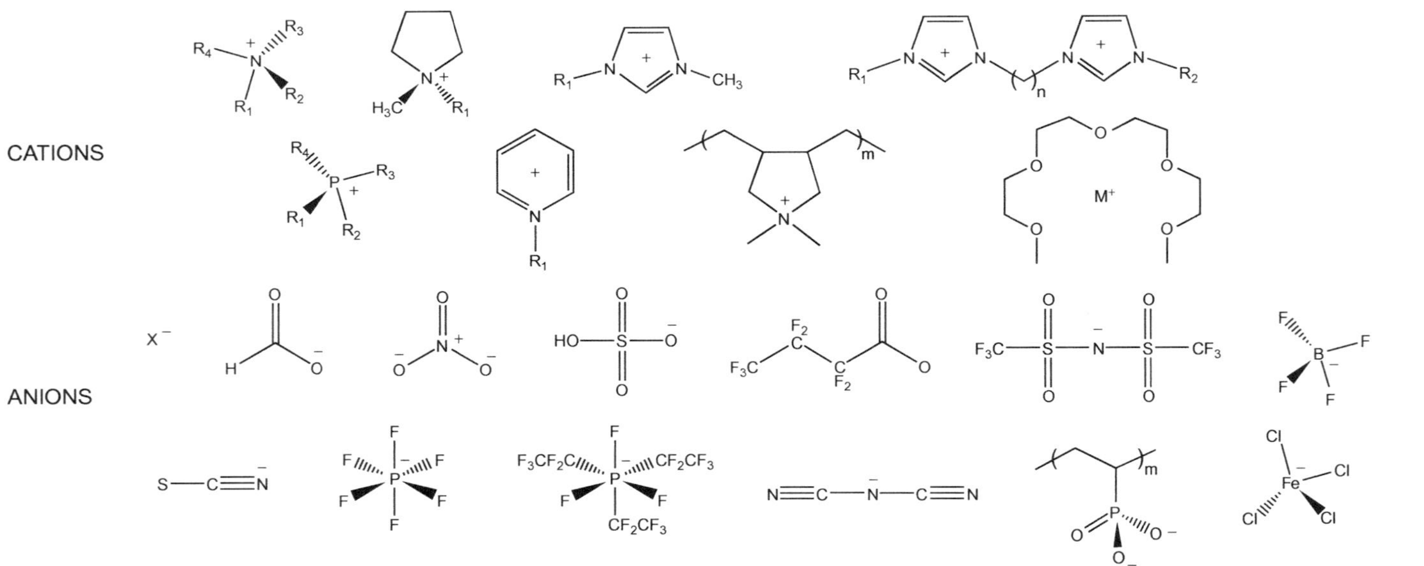

FIGURE 1 Selected chemical structures of representative cations and anions commonly encountered in protic, aprotic, dicationic, polymeric, magnetic, and solvate ionic liquids. From left to right, the cations (top row) include: ammonium, pyrrolidinium, 1-methyl-3-alkylimidazolium ([$C_n$mim]), 1,3-bis[3-methylimidazolium-1-yl]alkane; (second row) phosphonium, pyridinium, poly(diallyldimethylammonium), metal (M+) tetraglyme. The anions include (third row) halides, formate, nitrate, hydrogen sulfate, heptafluorobutyrate, bis(perfluoromethylsulfonyl)imide, tetrafluoroborate, (bottom row) thiocyanate, hexafluorophosphate, tris(pentafluoroethyl)trifluorophosphate, dicyanamide, poly(phosphonic acid), and tetrachloroferrate. *Reprinted from Ref. [31] Open Access ACS AuthorChoice.*

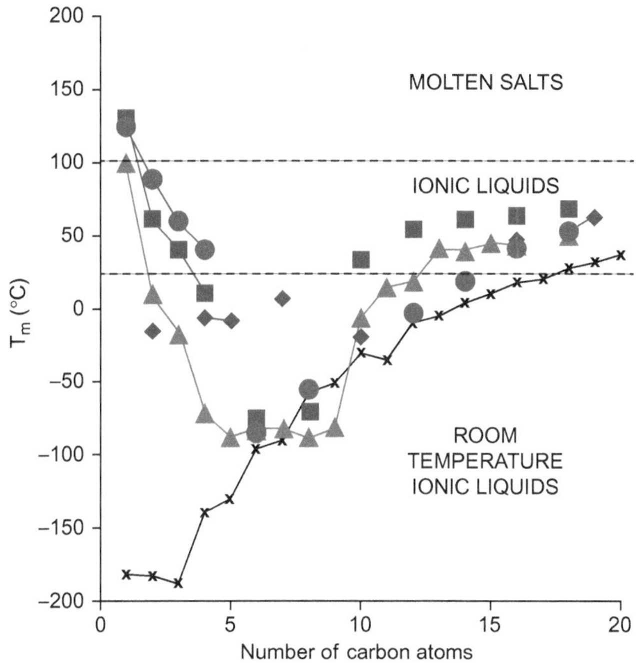

FIGURE 2 Melting point (T_m) as a function of cation alkyl chain length (n) for salts systems with 1-alkyl-3-methylimidazolium ([$C_n$mim]) cations. n-Alkane data (n = 1−20, crosses) are also shown for comparison. Data referto the following anions: hexafluorophosphate (PF_6, red ■), tetrafluoroborate (BF_4, green ▲), bis(perfluoromethyl-sulfonyl)imide (Tf2N, purple ◆), and chloride (Cl, blue ●). *Adapted from Ref. [31] Open Access ACS AuthorChoice.*

10–12 methylene groups, alkyl tails tend to favor the liquid state of the chosen IL. Moreover, it is clear, ILs being ionic compounds, that coulomb interactions are one of the strongest driving forces in affecting structure and dynamics in ILs; on the other hand, by progressively increasing the alkyl chain length, the amount of interaction energy related to dispersive interactions between the tails increases, while the overall coulomb term related to the interaction of the ionic moieties remains almost unaffected. This was shown in the case of a large set of ionic liquids, for example, by Shimizu et al. who reported the different contributions (either dispersive or electrostatic) to total cohesive energy [62]; by increasing the chain length, the dispersive term grows proportionally without appreciable change in the coulomb term. Similar trends but with some important exceptions were more recently detected by Izgorodina et al. [63].

Another important aspect that can affect structural and dynamic properties in ILs is the capability of developing hydrogen-bonding correlations. These correlations can play a role in determining the position, orientation, and spatial distribution of the anion around an aprotic cation such as dialkylimidazolium

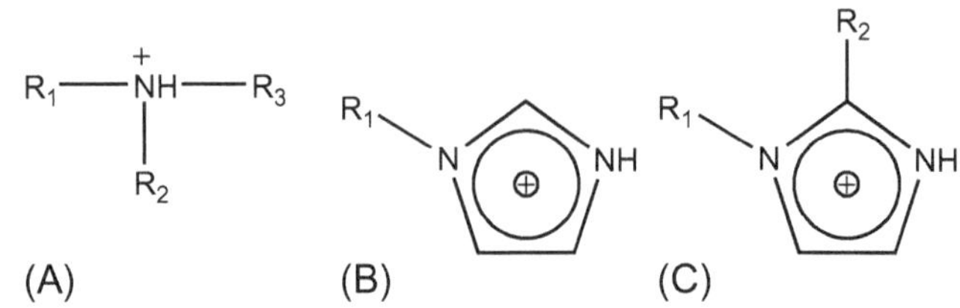

FIGURE 3 Representative cations often encountered in Protic Ionic Liquids, including (A) primary, secondary, or tertiary ammonium cations (where R1, R2, and/or R3 can be hydrogen atoms), (B) 1-alkylimidazolium cations, (C) 1-alkyl-2-alkylimidazolium cations. *Adapted with permission from Ref. [80]. Copyright 2008 American Chemical Society.*

[$C_n$mim] [64–79]. In particular, there is a class of ionic liquids where these interactions are expected to be relevant and this is the protic ILs (see Figure 3 for a selection of PILs' cations) [29,80–83].

An example of these compounds is the ethylammonium nitrate that was one of the first ILs discovered. These ILs are a subset of the IL category and are characterized by being formed as a combination of a Bronsted acid and a Bronsted base; due to this specific feature, PILs possess a proton available for hydrogen bonding. As such, one can expect that hydrogen bonding will play a large role in affecting a variety of their properties, including the existence of an extended ionic network mediated by HB [84] and dynamic [85] and structural [70,86–92] features. These interactions are expected not only to affect bulk chemical-physical properties but also to determine the structural and dynamic features in ILs at micro- and mesoscopic level.

In this contribution, we will describe how neutron techniques can help in exploring the correlation between micro- and mesoscopic scales and bulk macroscopic phenomenology.

4.2 STRUCTURE IN IONIC LIQUIDS SYSTEMS: MICROSCOPIC AND MESOSCOPIC CORRELATIONS

4.2.1 Structure of Neat Compounds

To our knowledge, Takahashi et al. were the first who accounted for structural properties of an ionic liquid using neutron diffraction techniques [93]. Diffraction data from $AlCl_3$/1-ethyl-3-methylimidazolium chloride ([C2mim] [Cl]) mixtures at different $AlCl_3$ concentrations were collected and rationalized in terms of ab initio calculations accounting for the structural properties of the isolated ions and the different complexes. This investigation provides a first hint of the high level of structural information that can be extracted out of neutron diffraction data on this class of materials. On the other hand, the joint use of different selectively deuterated samples with the same chemical composition can further enhance this information, thus providing a robust description of structural features in the liquid state. Other studies aimed at exploiting the advantage of using selectively deuterated samples that might provide complementary data set to be simultaneously analyzed to extract structural information. For example, in 1994, [C2mim][Cl]-HCl mixtures were investigated at the GLAD diffractometer at IPNS (the United States), using different H/D substitutions at the imidazolium ring and HCl, providing information on the interactions involving the H atoms and a first indication of π-stacking of the imidazolium rings ([94] and references therein).

In the last 15 years, there has been a flowering of structural studies focused on ionic liquids and their mixtures using neutron diffraction studies. In 2001, a careful study based on the simultaneous analysis of differently isotopically labeled samples was focused on 1,3-dimethylimidazolium chloride, [C1C1im][Cl] [95]. This investigation provided the first example of how long tradition experimental and analysis techniques used for more conventional liquids, such as water ([96,97]; please also refer to another chapter in this volume by A. Soper dealing with the use of neutron scattering to assess structural properties in liquid water and other aqueous systems) could be efficiently applied to the more complex realm of ionic liquids to explore their structural complexity. Hardacre and coworkers used four different isotopically labeled systems to extract complementary neutron diffraction data at SANDALS (ISIS, the United Kingdom) [98] and simultaneously modeled these data sets using the empirical potential structure refinement approach (EPSR). The latter [96,97,99] is based on the reverse Monte Carlo technique, aiming at providing a series of simulation snapshots that are consistent with the provided experimental neutron and/or X-ray diffraction data sets. EPSR accomplishes this task, imposing intramolecular structural constraints, a likely choice of interaction potentials and the bulk density and then using the diffraction data to derive a perturbation to the reference interatomic potential aiming at maximizing the agreement between computed and experimental scattering data. It is straightforward to realize that the more independent experimental data sets are available for a given system (e.g., different neutron scattering data sets obtained from differently selectively deuterated samples or joint use of X-ray diffraction pattern), the more robust will be the estimation of the perturbation potential that will determine the final series of simulation snapshots representing the bulk structure of the system. Out of these simulation snapshots, a series of radial, angular, dihedral, orientational, spatial correlation functions can then be extracted that, we stress again, are compatible with the experimental diffraction data [100,101].

The original work focused on [C1C1im][Cl] was soon extended by Hardacre and coworkers to describe also the microscopic structure in [C1C1im][PF_6] [102] and [C1C1im][Tf2N] [103]. This series of works [19] focused on the structure of ILs with the same cation (i.e., 1,3-dimethylimidazolium) and different anions; accordingly, this sequence provided useful insights on how structure is affected by the anion nature. This class of ionic liquids is characterized by a distinct out-of-phase alternation of ionic shells as a direct consequence of charge ordering: a central ion will be surrounded by a shell of counterions and, at larger distances, by another shell of similarly charged ions. This onion-like structure is characterized by a corresponding alternation spatial scale that, in the case of the mentioned series of ILs, depends on the anions nature. Hardacre et al. found that a central cation is surrounded by an anion shell, containing ca. 7 ions, at a distance of ~4.2, 4.5, and 5.2 Å for [Cl] [95], [PF_6] [102], and [Tf2N] [103], respectively. An outer cation shell has been found at a distance of 5.5, 6.3, and 7.0 Å for [Cl], [PF6], and [Tf2N], respectively [19]. Such an expansion of the liquid structure is the direct consequence of the bulkier and less accommodating size and shape of the anion, and in the limit of [Tf2N], this leads to a substantial disappearing of the characteristic long-range correlations that feature in the other ILs. While the spherical shape of the [Cl] and [PF_6] anions allows them to strongly interact with several cations through hydrogen bonds, this is more difficult for the [Tf2N] anion. Moreover, it was observed that, upon increasing the anion size, opposite ions are more and more forced to occupy mutually excluding volumes.

Hydrogen bonding between the halogen anion and the ring hydrogens has been detected in the crystalline phases [65,66,104,105] and in the liquid state for the case of [Cl] [95]. However, when considering [PF_6] [102] and [Tf2N] [103] anions, lower and lower correlation is found between the anion and the ring hydrogens. Accordingly, it was reported that the anion density axial to the ring decreases following the order [Cl] > [PF_6] > [Tf2N] while the opposite trend is found for anion density above or below the imidazolium ring.

Cation-cation correlations depend also on the anion's nature. While in the case of [Cl] [95] indication of π-π stacking is found, together with methyl-methyl correlations, the former is not found in the case of neither [PF_6] [102] nor [Tf2N] [103].

Overall, while a strong resemblance has been found between the structural correlations in the liquid state with those in the crystal phase for the cases of [Cl] [95] and [PF_6] [102], no such a similarity is found when considering the [Tf2N] [103] anion.

Further studies on neat aprotic ILs appeared subsequently, thus extending our understanding on this class of materials [106–108]. The Hardacre's team explored [C2mim][AcO], an IL that has a great relevance in biomass treatment and as such deserves a peculiar attention to rationalize its performances in terms of microscopic structural details. The anion is peculiar as it is built up by an apolar hydrophobic methyl group and a hydrophilic carboxylate moiety that is also bidentate. Furthermore, the study considers a cation with an ethyl chain, thus exploring an interesting phenomenology in ILs, as we will discuss. In this case, the joint use of neutron diffraction data from different isotopically labeled samples with the EPSR approach revealed that a near-to-planar correlation between neighbor cations occurs with the side tails in a configuration "top to tail" rather than "tail to tail," due to steric reasons. Moreover, the cation-anion correlation is characterized by a thick anion shell around a central cation, with the anions forming a bilayer-like structure with the charged groups toward the cations and the apolar hydrophobic methyl groups clustered in a central domain [106].

A recent study by Norman and coworkers was focused on a series of samples based on the [C2mim] cation and three different anions, namely, alaninate, serinate, and glycinate; making use of different isotopic substitutions, they provide several neutron diffraction data sets that were rationalized using the EPSR approach [108]. These systems turn out to be characterized by a weak cation-anion correlation, as witnessed by the absence of a second coordination shell, at odd with the behavior found in the vast majority of the investigated ILs. Moreover, while a weak cation-cation correlation has been observed, a distinctly nonnegligible correlation, even at distances as large as 10 Å, between anions has been detected, due to the formation of anion-anion interactions through hydrogen bonding. One interesting aspect explored by Norman et al. in their paper is the role played by chirality on structure. They looked for specific effects that might have arisen between the R and S form of alaninate and serinate, but only small effects could be detected in the anion-anion association.

A recent interesting study that focuses on a new class of materials that is attracting interest in the framework of green chemistry and is related to ionic liquids refers to the structural exploration of a deep eutectic solvent (DES) [109–114], namely, reline, a binary mixture of choline chloride (CC, an ionic compound) and urea (at ratio CC-urea = 1:2) [115]. In this study, Hammond and coworkers make use of isotopic substitution to extract neutron scattering data from a liquid reline sample at 303 K and model these data using the EPSR approach. Despite the existence of some purely computational studies on this class of materials [111,116–118], the information that can be extracted from EPSR studies validated through experimental diffraction data is extremely valuable to provide a robust structural picture of these interesting biocompatible new solvents. The authors could detect the major role played by hydrogen-bonding network in determining the overall structural properties. Hydrogen-bonding correlations are found not only between the chloride anion and urea but also between the cation (choline) and chloride; thus, the existence of a complex species formed by one cation, one anion, and the neutral urea was proposed; overall, it seems that an essentially stoichiometric cage develops around a central chloride anion. Such a concerted organization of the coexisting species turns out to be functional for the developing of deep eutectic mixtures whose tendency to crystallize gets highly frustrated: this argument is alternative to previous proposal of frustration being due to enhanced charge delocalization [117]. While these observations will need further confirmation on other members of the DES family, nevertheless, this study provides a clear example of how a neutron-diffraction-validated computational study can provide additional information as compared with pure computational investigations.

Another emerging class of unconventional ionic liquids is solvate ionic liquids (SILs). These are a subclass of ILs consisting of a metal cation (e.g., Li^+) that is bound to a stoichiometric number of coordinating ligands (e.g., glymes) through strong Lewis acid-base interactions [119–121]. The most studied SILs consist of mixtures of Li-based salts and glymes, leading to compounds such as [Li(G4)][Tf2N], where the LiTf2N inorganic salt is combined with tetraglyme, tetraethylene glycol dimethyl ether (indicated as G4), in a 1:1 molar ratio. Many of these compounds are liquid at ambient conditions. The bulk organization of these systems is characterized by the complexation of the Li cation by the ether groups belonging to the glyme, leading to a positively charged entity that is neutralized by the anion. The environment surrounding the Li cation depends strongly on the anion basicity, as the latter will enter in competition with glyme's ether groups when it comes to coordinate the cation. High anion basicity will lead to few Li-glyme correlations and strong Li-anion ones, thus maintaining a large fraction of glyme uncoordinating the cation. These features are very important in determining the macroscopic performances of these compounds. A compound with a small fraction of free glymes will consist mostly of ionic species, and no molecular ones (glymes) will be present in the bulk state; this situation leads to so-called *good* SILs. On the other hand, when the anion efficiently removes glymes from the Li environment, the bulk state of the mixture will be a mixture of ionic and molecular species leading to *poor* SILs. In their recent study, Murphy and coworkers use neutron scattering on isotopically substituted samples to explore the details of structural differences between good and poor SILs ([Li(G4)][Tf2N] and [Li(G4)][NO_3], respectively); namely, how the change in the anion species, such as [Tf2N] and [NO_3] (that leads to the difference between good or bad SIL), affects coordination numbers and geometry around the Li cation [121]. Using the EPSR to interrogate the experimental diffraction data, the authors observe a large difference in the G4 (tetraglyme) coordination numbers toward lithium cations, due to the mentioned different basicity of the two anions that compete against tetraglyme in complexing lithium cations. As a consequence, a strongly inhomogeneous distribution of lithium species is found in the bulk state of [Li(G4)][NO_3], while [Li(G4)] [Tf2N] is characterized by stable complex cations with Li^+ well solvated and uniformly distributed in the bulk (see Figure 4).

Some of the mentioned studies investigate imidazolium-based ILs with alkyl chain longer than methyl, such as ethyl. The effects of such a change are not so large; however, driven by other results obtained on the basis of molecular dynamic simulations and X-ray scattering experiments, several studies focused on neutron scattering exploration of ILs bearing long alkyl tails that, as we will discuss, induce a large degree of complexity on the structure. While the typical onion-like alternation of ionic layers around a central

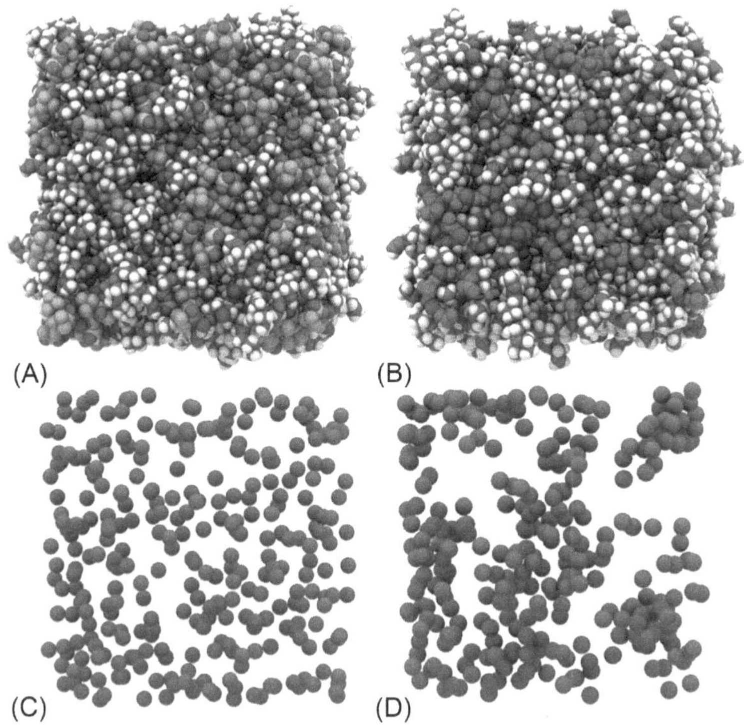

FIGURE 4 Snapshots generated from EPSR description of neutron diffraction data from [Li(G4)][Tf2N] (Left) and [Li(G4)][NO_3] (Right) at 298 K. (A) Shows all [Li(G4)][Tf2N] atoms; 300 G4, 300 Li cations and 300 Tf2N anions. (B) Shows all [Li(G4)][NO_3] atoms; 300 G4, 300 Li cations and 300 NO_3 anions. (C) Shows the 300 Li cations in [Li(G4)][Tf2N] and (D) shows the 300 Li cations in Li(G4)][NO_3]. H atoms are shown in white, C atoms in gray, N atoms in blue, O atoms in red, F atoms in brown, S atoms in yellow and Li atoms in green. *Reproduced from Ref. [121] with permission from the PCCP Owner Societies.*

given ion is a reasonable organization in a situation where one can assume a relatively homogeneous charge distribution in the ions, this is not the case when dealing with ILs bearing long alkyl tails. These moieties are built up by CH_2 units that interact mostly via dispersive interaction, rather than by coulomb ones. This leads to a chemical architecture that is typical for amphiphilic compounds where a hydrophilic moiety is covalently bound to a hydrophobic one and the two moieties show different affinities between each other and toward a selective solvent. A large fraction of ILs is characterized by such an amphiphilic composition, and this leads to an almost ubiquitous structural complexity that is the direct consequence of the poor affinity between polar and apolar moieties in ILs. This behavior is experimentally fingerprinted by the occurrence of well-defined diffraction features at low momentum-transfer values. The existence of such a kind of mesoscopic organization in ILs has been proposed in the past on the basis of molecular dynamic simulations [41–43,45,46,54]. Some of these studies even focused on the calculation of X-ray-weighted static structure factor for the structurally segregated systems that proposed the existence of a low-Q peak in the X-ray or neutron diffraction patterns. Shortly after, a series of experimental investigations focused mostly on X-ray scattering studies of these structural features, highlighting the existence of low-Q diffraction features directly related to the polar-apolar alternation occurring in ILs with medium-long alkyl tails and providing useful information on how chemical architecture, mixture composition, and external parameters such as temperature and pressure affect this mesoscopic organization.

In this scenario that is mostly occupied by X-ray scattering experiments, we wish to stress the role played by neutron diffraction in providing complementary and sometimes unique information as compared with X-ray diffraction.

The mesoscopic organization in ionic liquids has attracted great attention in the last decade. This is one of the most peculiar features that has been detected in this class of materials and directly stems from the amphiphilic nature of ionic liquids. Consistently, with the spatial scale of the mesoscopic order associated to the polar versus apolar alternation in ILs, the corresponding diffraction feature falls in a Q-range that is somehow intermediate between small- and wide-angle X-ray/neutron scattering. The first study in this context dates back to 2007, when a series of imidazolium-based ILs where explored using small/wide-angle X-ray scattering (SWAXS) techniques [122]. In Figure 5, those first results are shown: they are representative of the overall phenomenology that was soon to be discovered in this field.

A low-Q peak can be observed in the momentum-transfer range where conventional, unstructured liquids do not show diffraction features, and the corresponding experimental data tend toward the thermodynamic compressibility limit. On the other hand, in ILs bearing long enough alkyl tails, the onset and progressive development of a low-Q peak can be appreciated when increasing the chain length. Not only the peak amplitude increases with increasing chain length, but also the peak position shifts toward smaller Q-values. Systematic studies have been conducted exploring the role of anion species at variable alkyl chain length on the imidazolium cation [123,124], showing that, apart from the case of very bulky anions, there is no large effect played by their nature. Several studies explored the role of the cation nature, investigating imidazolium, pyrrolidinium [125], piperidinium [126], phosphonium [127], ammonium [128–130], and morpholinium [131] and highlighting the specific features deriving from cations' peculiarities. Use of neutron diffraction focused mostly on imidazolium- and ammonium-based ILs. In 2010, Hardacre and coworkers developed a careful work on a series of 1-alkyl-3-methylimidazolium hexafluorophosphate $[C_n mim][PF_6]$ (with alkyl = butyl, hexyl, and octyl; n = 4, 6, 8) that were selectively deuterated

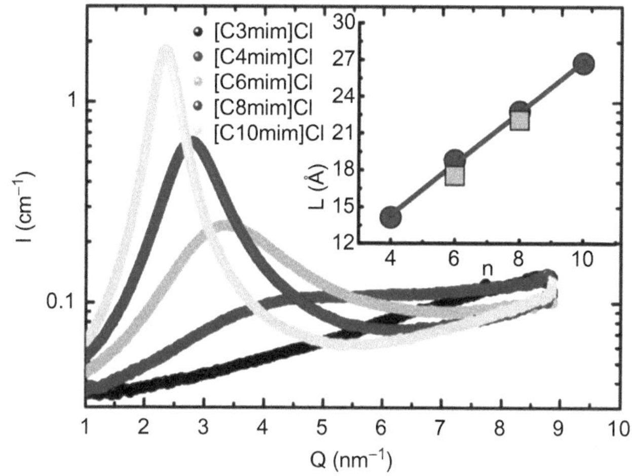

FIGURE 5 X-ray diffraction patterns from the series of ionic liquids: [C$_n$mim][Cl], for n = 3, 4, 6, 8, 10, at 25°C. In the inset the spatial correlation $L = 2\pi/Q_M$, Q_M being the interference peak position, is plotted (circles) as a function of n, the alkyl chain length, for n ≥4. The corresponding data are also reported for the cases n = 6 and 8 for [C$_n$mim][BF$_4$] (squares). *Reprinted with permission from Ref. [122]. Copyright 2007 American Chemical Society.*

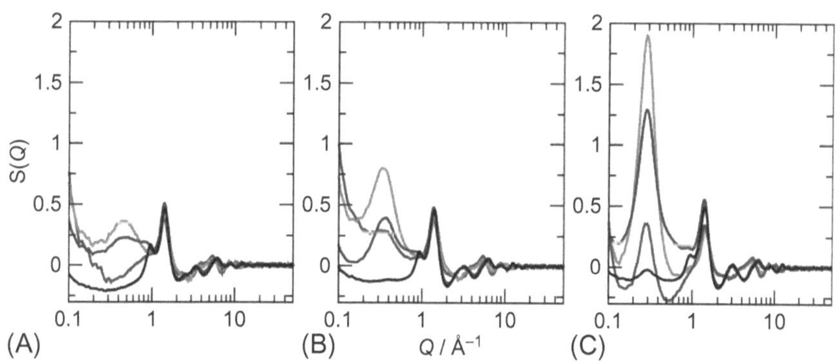

FIGURE 6 Experimental differential cross-sections as a function of Q (S(Q)) for fully protiated (red), perdeuteriated (black), methyl deuteriated (green), and 1-alkyl deuteriated (blue)1-butyl-3-methylimidazolium hexafluorophosphate ([C4mim][PF$_6$]) (A), 1-hexyl-3-methylimidazolium hexafluorophosphate ([C6mim][PF$_6$]) (B), and 1-octyl-3-methylimidazolium hexafluorophosphate ([C8mim][PF$_6$]) (C). *Reprinted from Ref. [132], with the permission of AIP Publishing.*

in different portions of the cations, and by means of neutron diffraction, they obtained a useful data set to explore the issue of the mesoscopic order in imidazolium-based ILs (see Figure 6) [132].

This data set shows how the low-Q portion is affected by selective deuteration of cation's moieties, and although some analysis of MD data accounting for the different deuteration levels was presented (see, e.g., [133]), so

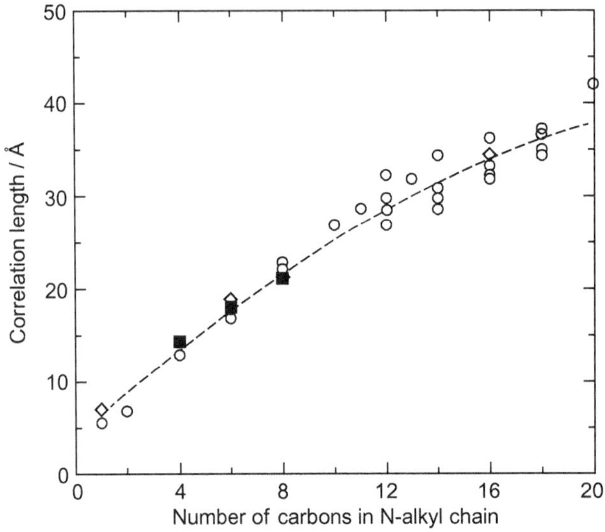

FIGURE 7 Correlation of the real space correlation distances associated with the position of the low Q peak in the neutron diffraction S(Q) data for [$C_n$mim][PF_6], with n = 4, 6, and 8 (closed squares) with the corresponding correlation distances obtained from X-ray scattering data on 1-alkyl-3-methylimidazolium ionic liquids (open symbols) (see the original paper for more details and the corresponding literature [129]). *Reprinted from Ref. [132], with the permission of AIP Publishing.*

far, these highly instructive data sets have not been fully exploited to describe the mesoscopic order in ILs, on the basis of robust validation from diffraction data. The real space distance corresponding to the low-Q peaks detected in this series of samples was plotted, together with data from other experimental sources in Ref. [132] (see Figure 7), and the authors use the observed trend to support their case for the low-Q diffraction feature to be related to the expansion of the second shell cation-cation coordination lattice, due to the interspacing of alkyl tails. While this work provides very valuable neutron diffraction data sets, the model proposed to rationalize the low-Q diffraction features is a minimalistic one that does not take into consideration privileged (or mutually exclusive, hence structuring) correlations between similar polar-polar or apolar-apolar (or different polar-apolar, respectively) moieties.

This kind of structural correlations in the meantime has been clearly identified by means of wise interpretation of molecular dynamic-based studies [49,50,52], and the role of *polar versus apolar* moieties mutual exclusion in determining the structural organization that leads to the low-Q diffraction feature has been strongly validated also experimentally, using X-ray diffraction [134–142]. Neutron scattering has been used to deal with this issue in the case of protic ILs. Shortly after the experimental discovery of mesoscopic order in ILs on the basis of X-ray scattering [122,123], neutron diffraction was successfully used to detect this feature also in a series of alkylammonium nitrates (viz, ethyl- and propylammonium nitrate, EAN and PAN, respectively) [130]; the authors attempted to describe the neutron diffraction data from selectively deuterated samples using a local smectic order model. One remarkable aspect of this study is that therein the existence of mesoscopic organization is detected for ILs bearing an alkyl tail as short as ethyl: this is at odd with the observation that in aprotic ILs, such as imidazolium-based ones, it is possible to detect experimental evidence of the low-Q diffraction feature only for chains longer than butyl. Since this work, protic ILs, in particular those based on alkylammonium cations, have been the subject of a series of neutron diffraction studies that complemented the information obtained from X-ray diffraction [129,143]. The subsequent works on PAN [144] and EAN/EtAN [91] (where EtAN corresponds to ethanolammonium nitrate) have been fundamental, together with related works based on X-ray diffraction techniques and focused on aprotic ILs [134–136], to clarify matters on the major role played by amphiphilicity or polar versus apolar dualism in ILs architecture to induce mesoscopic segregation. These studies allowed detecting that when the side chain is no longer an alkyl one, but contains either a hydroxyl group [91] or an ether one [134–136], the driving force toward alternation between polar and apolar moieties vanishes, making the structure at mesoscopic scale more homogeneous. On the basis of these studies, it was confirmed that due to steric considerations related to the volume ratios of alkyl to ionic components in the cations, the bicontinuous structure that develops cannot be characterized by a high degree of curvature, thus leading to a locally lamellar organization; on the other hand, when the hydroxyl group is connected to the tail (in the case of EtAN), the solvophobic correlations between neighbor tails are disrupted. Overall, the structure of simple PILs such as EAN or PAN has been found to be well described in terms of a bicontinuous morphology or a sponge-like one (see Figure 8), where a rather regular alternating arrangement of polar and apolar domains propagates over mesoscopic scales.

Further neutron diffraction studies on neat PILs were focused on addressing the nature of hydrogen-bonding interactions in these materials. This is an interesting issue, as this kind of interaction plays a role in driving structural correlations similarly to coulomb interactions. By careful inspection of neutron diffraction data from a series of selectively isotopically labeled alkylammonium-based PILs (viz, EAN, PAN, EASCN, EAHS, EAF, and BASCN, where EA, PA, and BA refer to ethyl, propyl, and butyl ammonium cations, while N, SCN, HS, and F refer to the nitrate, thiocyanate, hydrogen sulfate, and formate anions, respectively), Hayes and coworkers monitor H-bond lengths and angles and highlighted the role of ion packing in determining the geometric features of H-bonds in PILs (see Figure 9) [90]; the

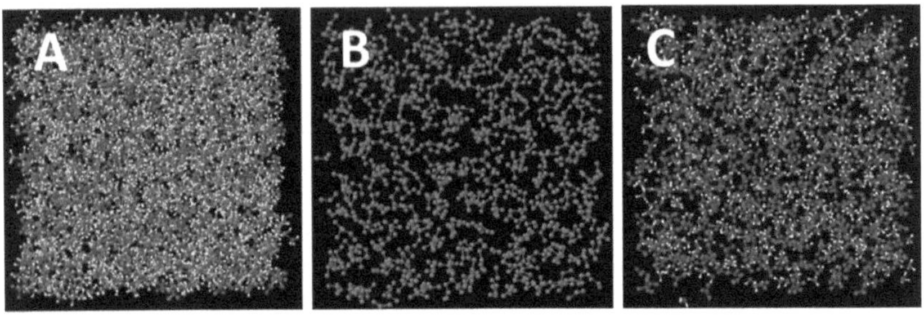

FIGURE 8 Snapshots generated from EPSR description of neutron diffraction data from propylammonium nitrate (PAN) at thermal equilibrium (298 K). From left to right, the front face of the 3D simulation box of (A) 500 PA cations+500 NO_3 anions (B) Apolar –C–C– domains only (ammonium and nitrate moieties omitted) and (C) Nitrate anions only (500 PA cations omitted) is presented. Atoms coloring are: C (gray), H (white), N (blue), O (red). *Reproduced from Ref. [144] with permission from the PCCP Owner Societies.*

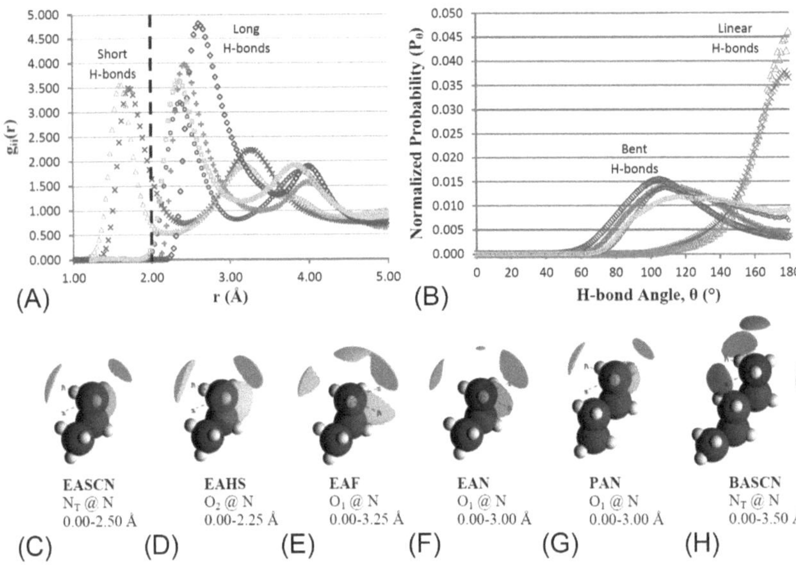

FIGURE 9 Description of structural organization in Protic Ionic Liquids, obtained after interrogation of Neutron Scattering data using the EPSR approach [88]. (A) and (B) describe the pair distribution and angular distribution functions for hydrogen bonding correlations in a series of PILS, respectively (EAHS (green, △), EASCN (red, ×), EAF (blue, +), EAN (○), PAN (pink, □), and BASCN (purple, ◇)). (C)–(H) Spatial Distribution functions of H-bond acceptor location as a function of distance and angular position from the H-bond donor (ammonium nitrogen) for different PILS. *Reproduced with permission from Ref. [90] (Copyright 2013 Wiley-VCH Verlag GmbH & Co. KGaA).*

authors detect a wide spectrum of H-bonding scenarios as a consequence of changes in anion nature and cation alkyl chain length, that is, those features that affect the polar versus apolar alternation and the ion packing in these materials. The latter induces a strong modulation in the nature of H-bonding, whose strength and geometric properties eventually turn out to be determined by the steric constraints induced by the bicontinuous nanostructure found in PILs. Overall, it is found that rather than driving nanostructure in PILs, H-bonds are instead affected by the polar versus apolar alternation in these materials.

Another recent systematic study focused on the role of cation side-chain length and anion nature in affecting micro- and mesoscopic structure in a set of PILs using neutron diffraction [92]. This characterization confirms that the low-Q peak related to the mesoscopic alternation between polar and apolar domains is rather anion-insensitive but strongly depends on the alkyl side-chain length, similarly to what had been observed in aprotic ILs. The nature of isotopic substitution can also affect the detection of such a diffraction feature: due to contrast reasons, fully deuterated samples do not show such a peak that is instead observable when dealing with partially deuterated samples. Overall, the mesoscopic scenario is the one of a polar/apolar segregated morphology for all the studied systems: this information is extracted from optimization of structure through the EPSR approach and is evidenced in a series of snapshots where the spatially heterogeneous organization is easily detected (see Figure 10) [92].

On a more local scale, the hydrogen-bonding three-dimensional network is spatially localized throughout the polar domain; the interaction of the anion with the alkyl chain hydrogens is only limited. The strength and geometry of the H-bond network depends on the anion nature and side-chain length. As an extreme example EASCN and BASCN that share the same anion but bear different (ethyl and butyl, respectively) side chains are characterized by a substantially different organization of the H-bonding network: EASCN has relatively short (1.71 Å) and linear H-bonds, while BASCN has much longer (2.61 Å) and bent (average angle $\sim$110°) H-bonds. On the other hand, the hydrophobic clustering of the side chains is witnessed by the strong methyl C–C pair distribution function peak centered at ca. 4 Å and corresponding to an average number of neighbor methyl groups of ca. 3 Å (see Figure 11).

This behavior is common to all the other ILs that show the low-Q peak: a distinct clustering of the methyl groups is observed, with a distinct interdigitation of neighbor tails, and indicates the occurrence of apolar domains. Especially in the case of longer side chain, this turns out to determine a bilayer-like organization on a local scale. In their contribution, Hayes et al. further developed their analogy [130] of the mesostructured in ILs with the one observed in surfactant self-assembly phases: they define a surfactant packing parameter as a_{alkyl}/a_{polar}, where $a_{alkyl} = v/l$ (v = hydrophobic chain volume and l = chain length) and apolar = a_0, the area occupied at the polar-apolar interface and obtain that the explored PILS are characterized by packing parameters near 1, consistently with bicontinuous structures with near-zero curvature and distinctly far from the condition that characterizes direct or reverse micelles.

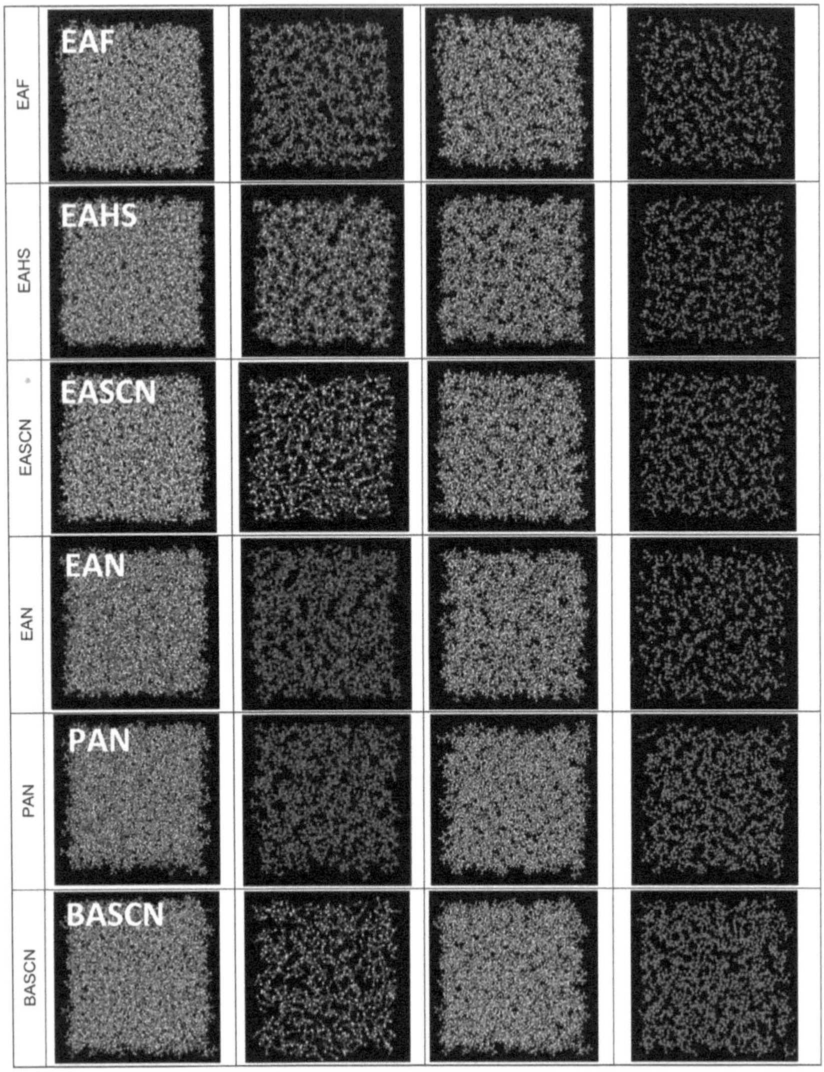

FIGURE 10 Selected EPSR snapshots from the analysis of neutron diffraction data on EAF, ethylammonium formate; EAHS, ethylammonium hydrogen sulfate; EASCN, ethylammonium thiocyanate; EAN, ethylammonium nitrate; PAN, propylammonium nitrate; BASCN, butylammonium thiocyanate, respectively (rows from top to bottom). The columns show different representations of the bulk structure: all atoms (column 1), anions only (column 2), cations only (column 3), and cation alkyl carbons (column 4). C is gray, His white, N is blue, O is red, and S is yellow. Carbon atoms on anions are orange in column 1 only. *Reprinted with permission from Ref. [92]. Copyright (2014) American Chemical Society.*

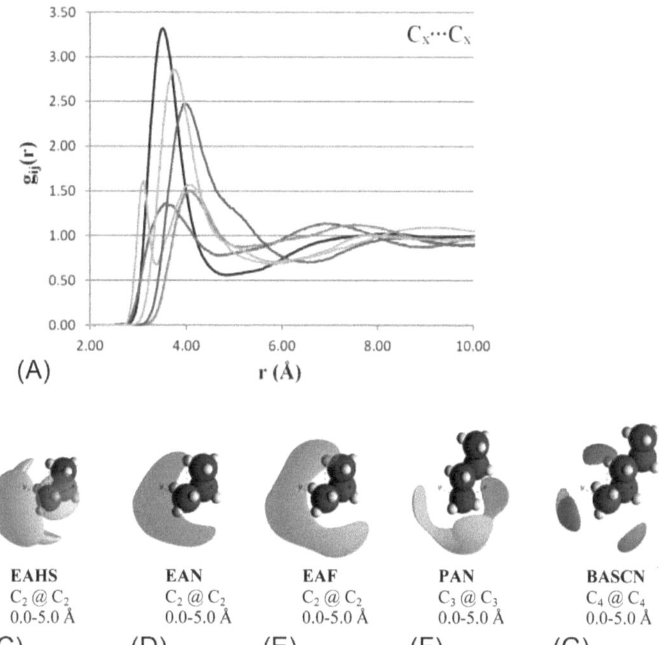

FIGURE 11 (A) Partial methyl–methyl pair distribution functions, g(r), between cation alkyl chains for PILs EAHS (green), EASCN (red), EAF (blue), EAN (black), PAN (pink), and BASCN (purple) and (B–G) spatial distribution functions of methyl carbon location around a reference methyl carbon. *Reprinted with permission from Ref. [92]. Copyright (2014) American Chemical Society.*

4.2.2 Structure of Binary Mixtures Containing Ionic Liquids

Another important area where neutron diffraction provided very useful information is binary mixtures of ILs with other organic, inorganic, or macromolecular compounds. Studies on binary mixtures were focused on micro- and mesoscopic spatial scales, and accordingly, different techniques were used to explore these features: not only wide-angle neutron scattering but also small-angle neutron scattering that allowed accessing the large-scale organization of as complex systems as ILs mixed with molecular liquids close to phase separation and with polymers and block copolymers (that will be discussed in Section 4.2.3).

Contextually to the first explorations of the neat ionic liquids, in the decade 2000–2010, important contributions to the understanding of structural properties of mixtures of aprotic ILs with organic compounds appeared. Liquid clathrate phases were detected using neutron diffraction in binary mixtures of 1-alkyl-3-methylimidazolium containing ionic liquids with hexafluorophosphate, bis(trifluoromethylsulfonyl)imide, tetrafluoroborate, and chloride anions with aromatic compounds such as benzene or toluene. These phases occur when mixing the IL with an excess of the aromatic compound; the IL-rich phase has a reproducible, although nonstoichiometric, composition. Occasionally, these phases can crystallize, and the structure of [C1mim][PF_6]:0.5 benzene has been resolved, leading to a scenario where benzene is stacked between two neighbor imidazolium rings through π-π correlations (see Figure 12) [145].

The structures of liquid mixtures of [C1mim][PF_6] and benzene were studied using neutron diffraction at 2:1 and 1:2 mole ratios [146]. Clearly defined shells of anionic and aromatic species are defined around a reference cation at 5–10 Å distance; on the other hand, an expansion of the cation-cation coordination was detected, fingerprinting the occurrence of intrusion of the benzene ring between neighbor rings, leading to a sandwich structure similar to the one detected by crystallography. This leads to a homogeneous distribution of benzene in the IL environment, without the formation of benzene droplets. A related study focusing on a 1:1 mixture of an electron-rich aromatic compound, such as 1-methylnaphthalene and [1-methyl-4-cyanopyridinium][Tf2N], that bears an electron-withdrawing cyano group used neutron diffraction to reveal the establishment of coplanar correlation between the aromatic compound and the cations, leading to stacking of alternate cyanopyridinium and methylnaphthalene groups, without relevant influence on the anion-anion organization, and the anion progressively shifted from its coordination toward

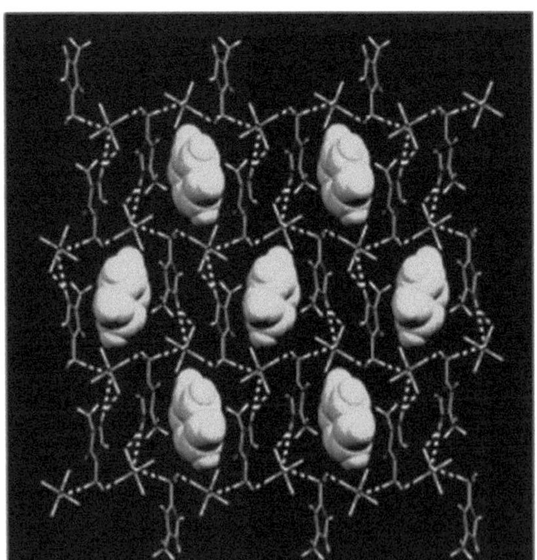

FIGURE 12 Packing in the crystal structure of [C1mim][PF_6]·0.5 benzene showing channels where benzene (shown in green) is included. *Reproduced from Ref. [145] with permission from The Royal Society of Chemistry.*

the cation by the aromatic moiety [147]. This strong effect seems to be the direct consequence of the presence of the electron-withdrawing moiety, as the similar system [C1mim][PF6]/benzene was not characterized by the same extent of anion substitution in the cation solvation shell by the aromatic compound. In the former case, the formation of charge-transfer complex (as a consequence of the π-cation facial stacking, competing with the cation-anion interactions) has been proposed [147]. A similar phenomenology could be detected by neutron diffraction in 1:2 binary mixtures of phenol and [N-methylpyridinium][Tf2N], where at 333 K a homogeneous liquid phase is obtained with the aromatic phenol ring interacting mostly with the cation and the anion playing a role of simply charge neutralization surrounding this complex [148]. Overall, these studies converge in indicating a structural scenario where privileged interactions between aromatic compounds and ionic liquids possessing a large aromatic character are defined: this could pave the way to smart approaches for separation and extraction processes.

The high ability of ILs in dissolving biomass is among the most intriguing properties of this class of materials [11,149], and much effort is paid in understanding and optimizing this capability. Two reports appeared dealing with neutron diffraction studies of mixtures of 1-alkyl-3-methylimidazolium containing ionic liquids and glucose. Chloride- and acetate-based ILs are the most efficient ILs in dissolving biomass [11,150,151]. A report from Young and coworkers interprets neutron diffraction data from a protiated mixture of [C1mim][Cl]/β-glucose at 5:1 mole ratio: Privileged interactions between the hydroxyl groups of the sugar and the chloride anions are detected; furthermore, limited H-bond cation-sugar interaction is found (although dispersive interaction between these species were highlighted), prompting for the possibility of optimizing this correlation and enhance the dissolution capability of ILs toward sugars [152]. Another candidate for efficient biomass treatment is [C2mim][CH_3COO], and this salt was the chosen IL for a careful neutron diffraction study of IL/β-glucose at 6:1 molar ratio, using different isotopic substitutions (see Figure 13) [153]. While on one hand no major differences are found in the ionic liquid structure as compared with the pure IL [106], close contacts between the acetate anions and the hydroxyl groups of the sugar through strong, short, and almost linear H-bonds are found. This interaction leads, at least indirectly, to the disruption of the anion-cation networking.

A careful inspection of the hydrogen-bonding architecture across the anion-sugar network allowed the rationalization of the improved performance of acetate-based salts in dissolving biomass as compared with chloride-based ILs. If on one hand the spherical, nondirectional nature of the H-bonding associated to the chloride anion leads to the establishment of interactions with the sugar that disrupt sugar-sugar correlations, on the other hand, the chloride anion can also act as a binder between different sugar moieties. Acetate anions instead tend to coordinate two hydroxyl groups belonging to the same

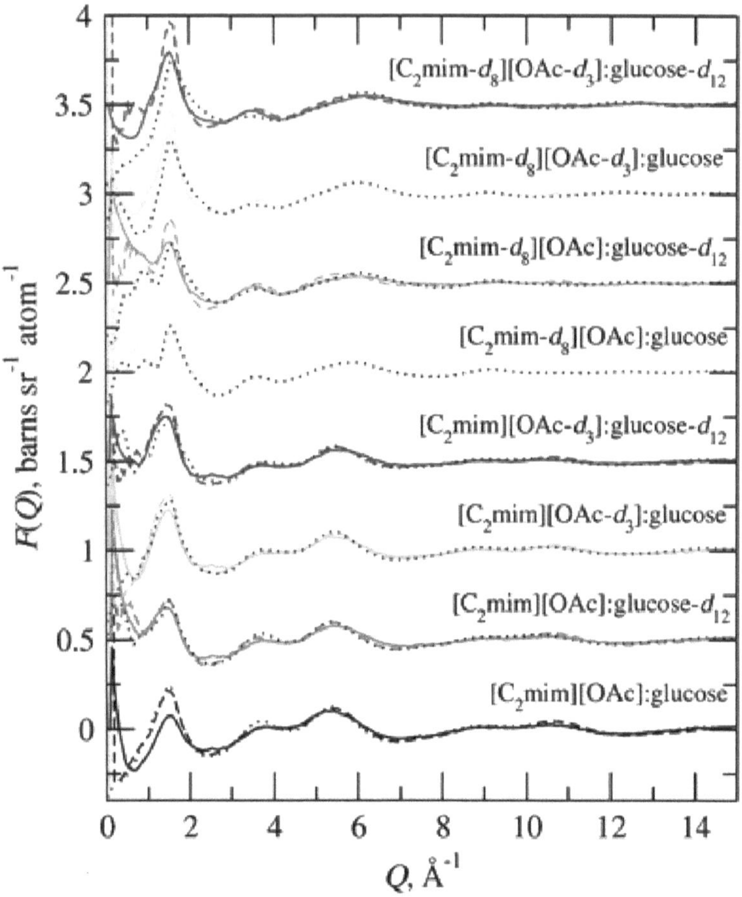

FIGURE 13 Experimental neutron scattering data (continuous lines), EPSR refined (dotted lines) and MD simulated (dashed lines) distinct scattering profiles of different isotopically substituted 6:1 [C2mim][CH_3COO]:glucose mixtures. *Reproduced from Ref. [153] with permission from The Royal Society of Chemistry.*

sugar, thus decreasing the probability to form sugar clusters mediated by interactions with the anion [153].

The complementary use of X-ray and neutron scattering provided great opportunities in the last few years to probe the structure of protic ionic liquids mixed with ionic/molecular compounds. As outlined in the previous section dealing with the studies of neat PILs, their structure is influenced by several factors: coulomb interactions and dispersive correlations dictate much of the structural organization in these interesting materials. However, hydrogen bonding, with all the multifacets peculiarities that were mentioned above, also plays a role in influencing the interactions between the polar portions in PILs. In this complex scenario, mixtures of PILs with compounds that can interact with the polar moieties are recently attracting a great attention. Among those important studies that took advantage of selective deuteration to explore the morphology of these mixtures using neutron scattering, we remind the investigation of ethylammonium nitrate (EAN) and water, a very important mixture both from applicative and academic points of view. In 2012, Hayes and coworkers explored how water gets dissolved in the protic ionic liquid par excellence, EAN [154]. This study focused on the mixture EAN-water = 1:6, and by modeling different data sets corresponding to various isotopic substitutions (see Figure 14), Hayes and coworkers highlighted the structural organization of water dispersed in EAN. Notably, the amount of water dispersed in EAN is not enough to disrupt the mesoscopic sponge-like organization found in neat EAN, and a distinct low-Q peak centered at ca. 0.6 $Å^{-1}$ is still present in the mixture (see Figure 14).

Among the relevant results extracted from the analysis of these data, it emerges that despite the large amount of water present in the mixture, the latter does not behave as a homogeneous water solution. EAN maintains its mesoscopic organization, and water can only marginally affect it; the polar versus apolar dualism present in the EAN architecture drives the mesoscopic organization toward a differentiation between the coulomb and H-bond interacting moieties and the alkyl tails that interacts through dispersive forces. On the other hand, the local structure of water is strikingly similar to the one observed in bulk water; both first and second shell of water oxygen-oxygen correlations are preserved at unaltered positions, witnessing a substantial maintaining of the tetrahedral water environment. Accordingly, Hayes et al.

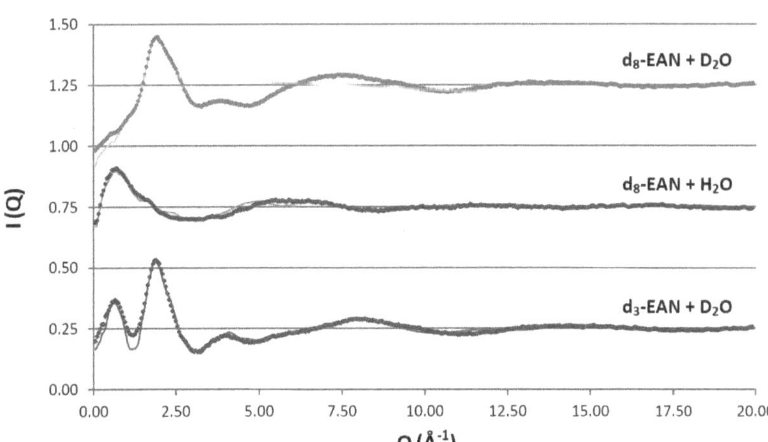

FIGURE 14 Experimental (dots) and EPSR-fitted (solid line) neutron scattering patterns for EAN/water mixtures in a molar ratio of 1:6 at 298 K; d3-EAN + D_2O (bottom), d8-EAN + H_2O (middle), d8-EAN + D_2O (top). Data for different isotopic substitutions are offset for clarity. *Reproduced with permission from Ref. [154] (Copyright 2012 Wiley-VCH Verlag GmbH & Co. KGaA).*

arrived to the conclusion that two different kinds of microenvironment, nearly unaltered EAN and nearly unaltered water, simultaneously keep on coexisting. Water then plays a complex role in interacting with the polar EAN's moieties, leading to a change of the polar/apolar interface curvature, transforming the morphology from sponge-like (as in neat EAN) into a branched, cylinder-like mesh one. This study provides a first example of the complex structural behavior of PILs when mixed with polar or amphiphilic compounds.

Recently, a joint X-ray/neutron diffraction study focused on the exploration of the structure of a binary mixture of EAN and dimethyl sulfoxide (DMSO), namely, for EAN-DMSO = 1:1.5 (see Figure 15) [155].

In this study, in-house X-ray diffraction and (isotopically substituted) neutron diffraction data from SANDALS (ISIS) were jointly interpreted using the EPSR approach to provide information on the structural organization of this mixture. DMSO is well known to interact in a complex way with water [156–159], and the strict analogy between the H-bonding network between water and EAN [84,85] prompts for similar behavior for water-DMSO and EAN-DMSO mixtures. In the mentioned work, this similarity is confirmed as it is shown that the proposed increase in DMSO-DMSO correlations due to hydrophilic association mediated by hydrogen-bonding correlations that is found in water-DMSO [158] takes place also in EAN-DMSO mixtures [155]. Concomitantly, a loss in the DMSO methyl-methyl correlations is found, confirming that the enhanced DMSO association is not related to hydrophobic correlations. DMSO as a polar aprotic compound is able to affect EAN structure; cation-anion correlations get enhanced, while apolar correlations are weakened, due to an increasing correlation between apolar groups belonging to the cation and DMSO. Such an effect has been interpreted as a steric consequence of the hydrogen-bonding-mediated interaction

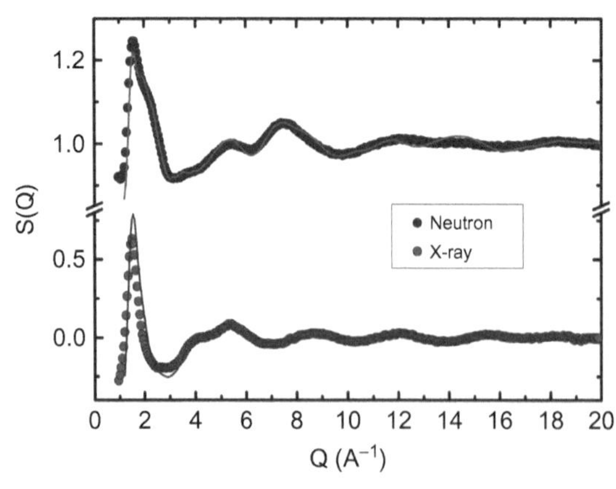

FIGURE 15 X-ray (bottom) and neutron (top) diffraction patterns from EAN/DMSO and d3-EAN/d9-DMSO, respectively, with EAN molar fraction = 0.4 at ambient conditions. The continuous lines correspond to the EPSR modeling of the experimental data. *Reprinted from Ref. [155] Copyright (2015), with permission from Elsevier.*

between cation and DMSO. The H-bonding correlation between DMSO and the ammonium hydrogens coexists with the one between the same donor and the nitrate anion. Both H-bonds are found to be moderate: Their length is ca. 2.4 Å, and they are not linear (see Figure 16). Further extending the analogies

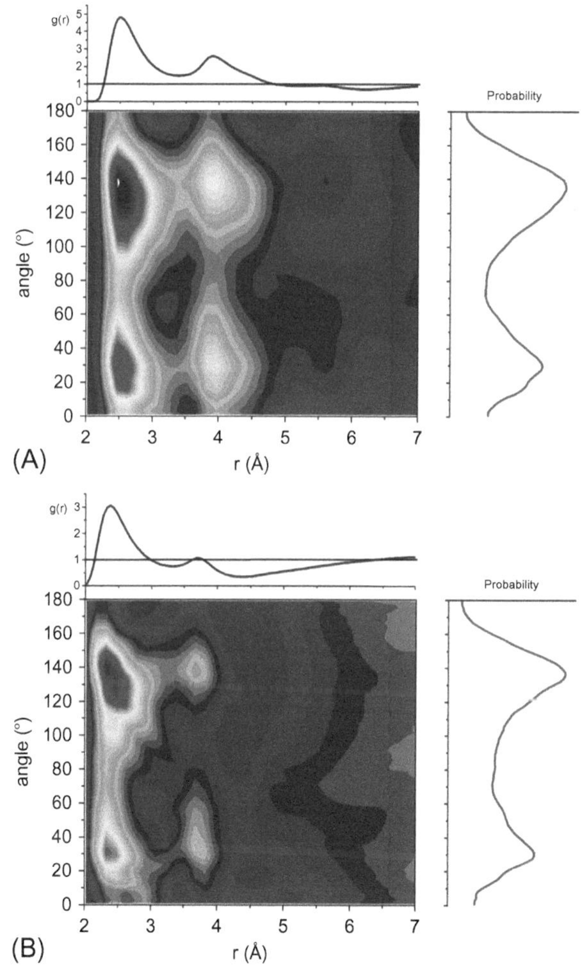

FIGURE 16 (A) Combined angular (N1–H1···OO) and radial (H1···OO) distribution functions in the EAN–DMSO mixture, illustrating the hydrogen bonding correlation between ammonium and nitrate species. In the upper inset the g(r) N1–H1···OO is plotted; in the right inset, the angular distribution function for the angle N1–H1···OO is plotted under the condition $2.3 < r_{H1 \cdots OO}$ (Å) < 2.7. (B) Combined angular (N1–H1···OD) and radial (H1···OD) distribution functions in the EAN–DMSO mixture, illustrating the hydrogen bonding correlation between ammonium and DMSO's oxygen species. In the upper inset the g(r) N1–H1···OD is plotted; in the right inset, the angular distribution function for the angle N1–H1···OD is plotted under the condition $2.1 < r_{H1 \cdots OD}$ (Å) < 2.7. *Reprinted from Ref. [155] Copyright (2015), with permission from Elsevier.*

with water-DMSO (but at odd with methanol-DMSO mixtures), the study describes the existence of a recurrent structural *leit-motif* of a central ammonium head that is coordinated, through H-bonding, by two DMSO molecules. This evidence supports the interpretation that a well-defined stoichiometric complex might be responsible for the structural organization in EAN-DMSO; the latter would be held together by the extensive H-bonding network in these mixtures.

Other studies used neutron and X-ray to explore the structure in PILs mixed with ionic compounds. Hayes et al. studied binary mixtures of EAN and EtAN (ethanolammonium nitrate) mixed with a series of inorganic nitrate salts (Li, Mg, Ca, and Al nitrates were studied) [160]. This study was followed by X-ray-scattering-based explorations on mixtures of EAN (and related PILs) with $LiNO_3$ [161–163]. In agreement with MD simulations [164], the addition of the mentioned inorganic salts does not affect much the structure of EAN over the whole explored concentration range (for the case of X-ray studies focused on $LiNO_3$, the saturation limit was reached [161]). The PIL sponge-like morphology is maintained, and a careful X-ray study highlighted a progressive change of the characteristic size for polar-apolar alternation in mixtures of $LiNO_3$ and X-AN (where X=E, P, and B; ethyl, propyl, and butyl, respectively) [162] that is partially confirmed by the neutron study focused on one EAN mixture [160]. On the other hand, EtAN turns out to be more affected by the $LiNO_3$ addition, as it is shown that its morphology changes from clustered into sponge-like: the inorganic salt addition triggered the amphiphilic behavior that this EtAN did not show in its neat state [160]. The expulsion of EtAN's hydroxyl groups from the polar region to accommodate therein the lithium cations led those –OH moieties to cluster outside of the polar domains, thus facilitating the differentiation between polar and apolar moieties that was not stable in EtAN's neat state. In all the probed mixtures, the lithium cation is found to be dissolved in the polar portion of the PIL mesostructure and heavily solvated by nitrate anions. The structure of the polar domains is left rather unaffected by the introduction of the lithium salt. On the other hand, although neutron diffraction data [160] at a concentration EAN-$LiNO_3$=10:1 do not fully confirm this picture, analysis of X-ray diffraction data [161–163] at a concentration EAN-$LiNO_3$ =5.5:1 (close to saturation) indicates that $LiNO_3$ tends to be organized in a way that resembles its crystalline state; moreover, the X-ray study highlighted a further level of segregation of lithium ions: Not only they are distributed inside the polar domain, but also they tend to cluster in microregions where they are organized in a crystalline-like way. This peculiar organization [161] of inorganic salts dispersed in an amphiphilic compound such as EAN is to be fully confirmed yet and deserves further exploration.

Another interesting neutron-scattering-based study of mixtures of PILs with ionic compounds is the one by Murphy et al., where PAN (propylammonium nitrate) and EtAN (ethanolammonium nitrate) are mixed with their chloride counterparts (PACl and EtACl, respectively) [165]. This study can be considered an extension of the investigation of alkylammonium nitrate salts mixed with $LiNO_3$. While in the previously discussed series the mixture is constituted by a common anion and two different cations (one of them being the small lithium one), in the presently discussed study, the mixture is made of one common cation and two different anions (nitrate and chloride, respectively). Chloride-poor mixtures were studied (molar fraction = 0.15) aiming at understanding the role of small additions of this small anion. As a result of the fitting of several neutron scattering data sets on different isotopically substituted mixtures using the EPSR approach, a series of simulation snapshots were obtained providing detailed insight on the micro- and mesoscopic organization of these mixtures. The small size of the chloride anion makes it easier for this anion to approach the ammonium polar head, shifting the nitrate from the coordination of this moiety. Overall, in PAN-PACl mixtures, the mesoscopically segregated morphology is maintained, although the introduction of chloride affects the alkyl tail conformation, so that the local environment of the terminal methyl group is changed. In EtAN/EtACl, where, as discussed, no sponge-like structure exists, but only clusters determine the mesoscopic organization, the introduction of chloride strongly affects the morphology. Chloride can efficiently interact not only with the ammonium head but also with the terminal hydroxyl group, and indications are obtained that a single cation can bind to a single chloride anion in a bidentate way, through the adoption of a gauche or eclipsed conformation of the chain. Such a dramatic change in the chain conformation induces a strong reduction of the cluster organization in the mixture. Although the use of the EPSR approach to describe effects that depend on the torsional barriers along alkyl chains might not be the most precise tool to explore this phenomenology, this study provides very interesting hints on how even a small amount of chloride ions might dramatically affect the morphology of PILs.

A research field that benefitted from neutron scattering experiments is the one related to binary mixtures of protic ionic liquids with alcohols. These investigations provide very stimulating information on the complex structural and thermodynamic scenario of amphiphilic compounds (such as alcohols) dissolved into amphiphilic solvents (such as PILs). This is a very recent activity that is involving both neutron and X-ray scattering experiments and is focused over not only medium/wide-angle neutron/X-ray scattering but also small- (a potentially ultra-small) angle scattering techniques, as the relevant spatial scales are spread over several nanometers and even larger than that, as we will describe. The first of such studies was addressing the nanostructure of ethylammonium formate (EAF), a representative protic IL, mixed with glycerol, a polyol that, when mixed with water, is well known as a cryoprotective agent. Taking advantage of isotopic substitution, Murphy et al. collected neutron diffraction data on different equivolume mixtures that were simultaneously modeled using the EPSR approach [166]. In these mixtures, the sponge-like structure found in PILs was detected as well; the low-Q peak fingerprinting the mesoscopic organization is however shifted toward smaller Q-values when adding glycerol to neat EAF, thus indicating the occurrence of a swelling of the mesoscopic structure. There are not major changes in the structure of EAF, after adding glycerol. Only it can be appreciated that glycerol tends to insert between the cation polar moieties, separating them and thus leading to a growth in the size of the characteristic distances of the sponge-like structure. Glycerol, however, does not penetrate into the apolar portion of the structure and remains confined in the polar one. Overall, the structure is then dictated by the polar versus apolar differentiation and the packing restraints that determine where glycerol molecules have to be located. The authors find that glycerol interacts through hydrogen bond mostly with the EAF (either the cation or the anion, acting as an acceptor or donor, respectively) and negligible hydrogen-bonding correlations are detected between glycerol molecules. Another peculiar aspect of these mixtures is that glycerol if found to be more conformationally disordered than in its neat state, a wider distribution of conformers is found, as a consequence of the more complex distribution of hydrogen-bonding sites and lower strengths of these correlations in the mixtures. This looks to be an interesting effect of the complex PIL structural environment on amphiphilic molecules that can interact in diverse ways with both cations and anions.

An even more complex structural scenario has been later on detected using a combination of X-ray and neutron diffraction studies. After the preliminary X-ray-scattering-based studies by Greaves and coworkers [167], it emerged that, similarly to related water-alcohol mixtures [168–170], PILs mixed with alcohols are characterized by the occurrence of an excess low-Q scattering. This behavior is strictly related to the existence of an upper consolation point in EAN-octanol mixtures that was detected by various techniques [171–174], including small-angle neutron scattering [175]. In their study, Bonetti and Calmettes collected SANS data from a homogeneous mixture of EAN and deuterated octanol in the vicinity of its critical point (EAN's critical molar fraction = 0.760 and critical temperature = 29.160°C). As expected for close to criticality, these systems are characterized by growing scattering amplitude at low Q, as temperature gets closer and closer to the critical one. In the paper, the authors describe the SANS patterns at temperature approaching the critical point with a model accounting for concentration and density fluctuation contributions to the critical scattering. While the latter are relatively small at the still far from critical temperature conditions explored in the work, the concentration fluctuations were not negligible and accounted for using the Fisher-Langer expansion [176]. The observed critical component to the concentration-fluctuation correlation function allows detecting a critical correlation length of approximately 5 Å, in agreement with light-scattering determinations [172,177] and supporting the conclusion that this mixture displays an Ising-like behavior [171,174,177,178]. More recently, small-angle X-ray scattering data on a wider range of systems (viz, mixtures of PILs (EAN, PAN, and EAF)) with linear alcohols (C2-, C4-, C6-, C8-, C10-, and C12-OH, where Cn indicates the number of carbon atoms in the linear alcohols) have been presented at ambient conditions [179]. This study extends the previous work from Greaves et al. [167] and provides a detailed collection of SAXS data that clarify the situation concerning mesoscopic organization in these important mixtures. The SAXS patterns reported in Ref. [179] (see Figure 17) show clear indications of the development of a distinct low-Q excess scattering contribution upon changes in concentration for the mixtures.

While below ca. 0.3 $Å^{-1}$ the pure components (neat PILs and alcohols) do not show any significant feature, upon mixing these components, gradually, a diffraction feature develops, whose amplitude reaches a maximum at a nominal concentration depending on the alcohol chain length (for a given PIL) and then progressively drops; this behavior is ubiquitous in PIL/alcohol mixtures, despite the fact that all the mentioned systems remain macroscopically homogeneous, without intervening phase separation, at the probed experimental conditions. The development of such a new, larger length scale in these mixtures is presently attracting a great attention. Greaves et al. [167] and Jiang et al. [179] interpreted these evidences in terms of the formation of large aggregates [167] or even postulating the existence of small micelles or microemulsions [179], related to alcohol aggregates dispersed in the PIL matrix. While quantitative differences between EAN and EAF mixtures with alcohols have been related [179] to the different nature of hydrogen-bonding network in these PILs, the common qualitative scenario is relatively independent on the specific nature of the PIL and alcohol. In the interpretation from Jiang et al., alcohol molecules tend to form clusters that might also contain alkylammonium tails oriented in such a way to play the role solubilizing the clusters into the PIL-rich matrix; the PIL cation would then simultaneously play the double role of solvent (the matrix where the clusters are dispersed) and cosurfactant. This scenario reminds the surfactant-free microemulsion that is detected in the pre-ouzo regime in water-ethanol-octanol ternary systems [180–182] or in more complex systems [183].

The complex structural scenario emerging from mixing two amphiphilic compounds such as ethylammonium nitrate and methanol (the shortest liquid PIL and linear alcohol, respectively) was the topic of a joint neutron/X-ray scattering and computational study [184]. Although the amplitude of the SAXS low-Q scattering excess in these mixtures is of smaller extent than in the case of longer-chain alcohols, nevertheless, methanol also shows SAXS patterns that resemble the ones observed in Figure 17. The concentration value where the excess scattering amplitude reaches a maximum is significantly different from other alcohols and corresponds to ca. EAN's molar fraction = 0.15. The joint description of neutron diffraction patterns from differently isotopically labeled mixtures of EAN-methanol (at EAN's molar fraction = 0.15 and XEAN = 0.15) and the X-ray diffraction pattern from an

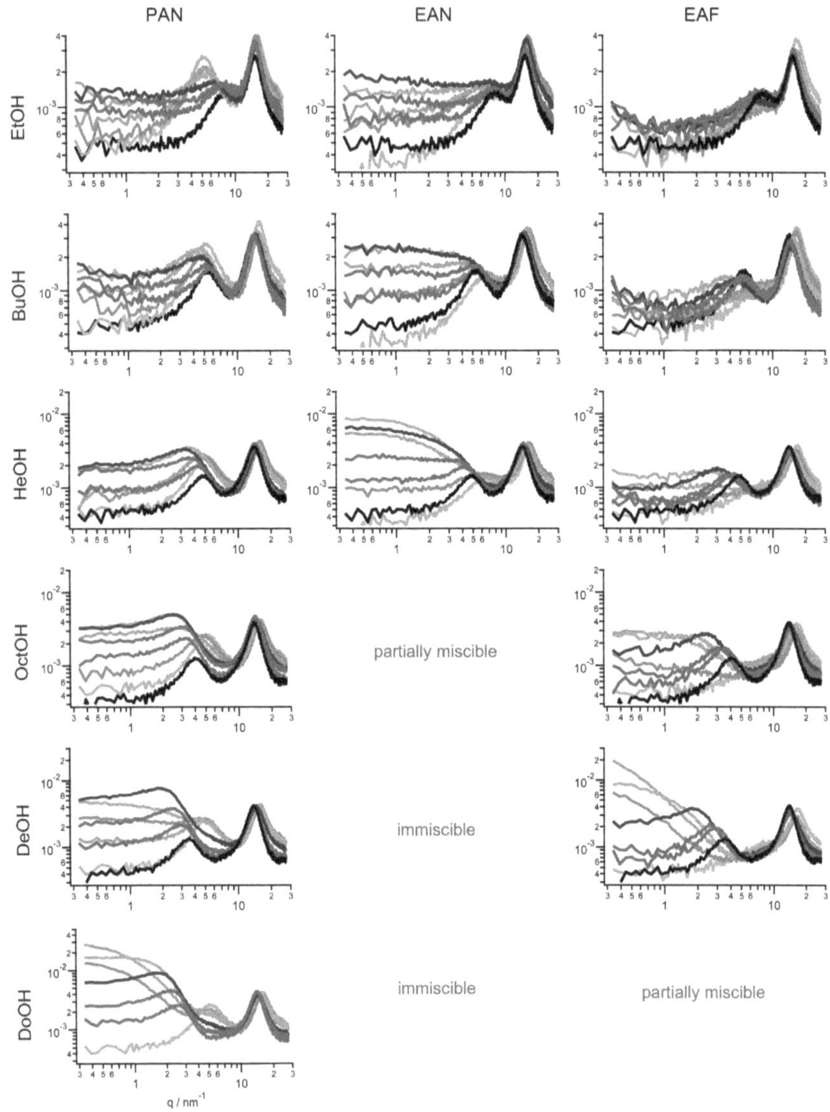

FIGURE 17 Small-wide-angle X-ray scattering patterns of amphiphilic Ionic Liquid (PAN, EAN and EAF, left, center and right panels, respectively) + aliphatic n-alcohol mixtures of various compositions at 25°C: yellow, IL; brown, 10 wt% alkanol; green, 20 wt%; light blue, 40 wt%; blue, 60 wt%; purple, 80 wt%; red, 90 wt%; black, alkanol. *Reprinted from Ref. [179] Open Access ACS AuthorChoice.*

unlabeled mixture using the EPSR approach provided a detailed insight into this mixture (see Figure 18).

EAN-methanol mixtures are characterized by cations and anions that are still strongly correlated, despite the large excess of methanol that does not

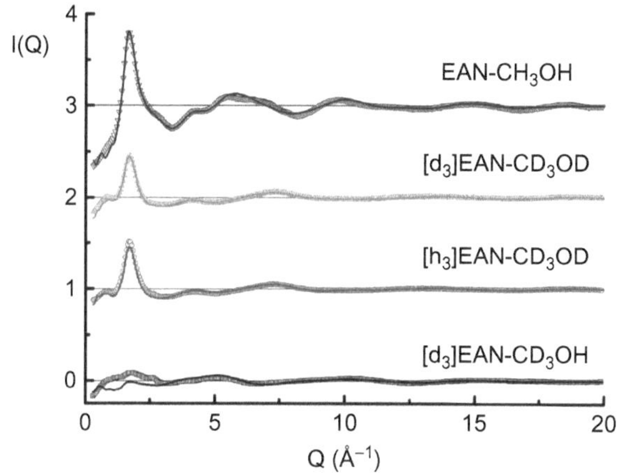

FIGURE 18 Experimental (symbols) and EPSR-fitted (lines) neutron/X-ray scattered intensity for EAN/methanol mixtures with EAN molar fraction = 0.14 at 298 K. From bottom to top: neutron data from [d3]EAN–CH_3OH, [h3]EAN–CD_3OD, [d3]EAN–CD_3OD, and (top) X-ray data from EAN–CH_3OH. Data are offset for clarity. *Reprinted with permission from Ref. [184]. Copyright (2014) American Chemical Society.*

succeed in efficiently separating the counterions. Nevertheless, strong hydrogen bonds develop that capture methanol in such a way that its hydroxyl group is engaged in H-bonding (where methanol acts both as acceptor and donor) in the polar part of EAN's mesoscopic organization and its methyl group is embedded into the apolar domain formed by EAN's ethyl groups. At the chosen concentration, the low-Q peak that characterizes neat EAN's X-ray and neutron diffraction patterns either vanishes or is submerged by the low-Q excess that originates from the scattering from large clusters whose shape and lifetime fluctuates. Russina et al. then proceed evaluating the distribution of clusters sizes due to the fragmentation of EAN's sponge-like morphology: it turns out that while in neat EAN ionic species clusters (i.e., those clusters formed by those cations and anions whose mutual distance is below a given cutoff) simply create a percolating network extending over the whole simulation box, in the case of mixtures with XEAN = 0.15, ionic species tend to form a highly heterogeneous distribution of clusters. It is noteworthy that, in agreement with the progressive decrease of the low-Q excess amplitude, the corresponding cluster-size distribution gets narrower and narrower when changing the mixture concentration (e.g., for the case of XEAN = 0.3) (see Figure 19). Overall, then despite the similar amphiphilic nature of EAN and methanol, when mixed, the two compounds tend to organize in a way that is far from being the usual polar versus apolar differentiation found in the case of other IL/molecular liquid systems. The presence of an excess of methanol seems to be incapable to disrupt the strong correlation between counter ions but on the other hand succeeds in breaking apart the sponge-like organization into a highly heterogeneous distribution of ionic clusters [184].

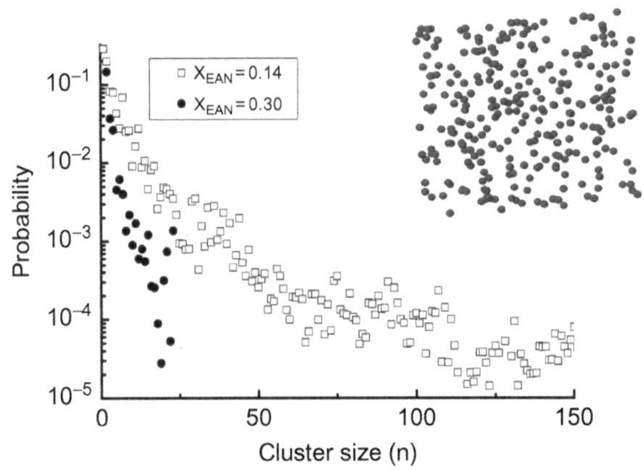

FIGURE 19 Cluster size distributions for ammonium nitrate clusters in EAN–CH_3OH mixtures with EAN molar fraction = 0.14 and 0.30 (yellow and black symbols, respectively). In the inset, a snapshot of a slice of the simulation box shows the clustering of the ammonium and nitrate ions for the case EAN molar fraction = 0.14 (only nitrogen atoms are shown). *Reprinted with permission from Ref. [184]. Copyright (2014) American Chemical Society.*

A very recent paper from Murphy et al. [185] deals with a highly related system, namely, PAN (propylammonium nitrate)/octanol, a system that was previously shown to behave in line with the trends described so far [179]. A careful inspection by means of neutron scattering of several differently isotopically labeled binary mixtures at various concentrations has been fully described using the EPSR approach. The authors report that their study allows detecting two different structural length scales: a large one associated to aggregated octanol interpenetrated by PAN and a smaller one related to the polar/apolar alternation that characterizes the bulk structure of neat PAN. The complex structure of octanol clusters is described in Ref. [185]; it is found that PAN cations tend to organize coating the octanol drops, forming a layer where the octanol emerging tails are embedded into the propyl tail environment of the PAN cation. A distinct presence of alcohol hydroxyl groups is also found inside PAN's apolar region, and accordingly, these hydroxyl groups not only are involved into hydrogen-bonding correlations, as one might expect, but also are found in this unexpected environment. Over a mesoscopic scale, octanol clusters are found to evolve from small, isolated ones (for octanol content lower than 20% v/v) to an extended network, upon increase in octanol concentration. While the study provides support to the claim that PILs can act as cosurfactant and solvents when mixed with other compounds, indications are also provided that a structural readaptation of the mesoscopic organization in PILs might occur in order to accommodate the guest aggregate. Despite the accurate description of robust experimental data sets, we believe however that the exploration of this class of systems (not only PILs—alcohols, in particular, but also ILs—amphiphilic molecular liquids, in general) will still need further deep investigations. Many of the neutron scattering data sets that were successfully described using the EPSR approach covered only limited Q-windows; in particular, the low-Q region where the peculiar scattering excess feature is located in such a kind of mixtures is not accessed by the reported experiments, mostly due to the technical reasons in dealing with inelasticity effects. Also, several simulation studies (e.g., Refs. [163,186]), although describing accurately the short-range organization in these mixtures, missed to detect the existence of long-range complex morphology, indicating that interatomic potentials might not be fully adequate yet to account for the complex phenomenology observed experimentally. Clearly, more extended experimental data sets (e.g., SANS or even USANS and SAXS/USAXS) would be required to provide a robust experimental basis to further elaborate on the complex behavior of these amphiphile/amphiphile mixtures.

The use of the SANS technique to access density/concentration fluctuation related to close-to-separation processes was also reported in the case of aprotic ionic liquids/water mixtures, by Perera and coworkers [187]. This is an interesting study where SANS is used, together with vapor-pressure data, to determine concentration fluctuations and Kirkwood-Buff integrals (KBIs) [188] (see Figure 20), aiming at rationalizing the behavior of mixtures close to demixing that were previously described in terms of a structural model invoking the existence of micellar aggregates to account for the low-Q scattering features [189].

As a matter of fact, the amphiphilic nature of most ILs led to a large number of studies focusing on their behavior in water, leading to the formation of a wealth of structural organizations, including micelles and microemulsions [190]. In this context, SANS and SAXS played a large role in accessing the structural properties of these aggregates.

One of the first such studies focuses on mixtures of heavy water and selected imidazolium-based ILs, namely, [C4mim][BF_4], [C8mim][Cl], and [C8mim][I] at salt concentrations below 80 vol%. Difficulties on obtaining a robust structural model based on micellar-like aggregates were reported for the first IL (in agreement with the results reported by Perera et al. [187]), while mixtures with the two other salts could be satisfactorily described in terms of core-shell ellipsoids, whose interaction is described with an Hayter-Penfold hard-sphere structure factor. The possibility that at higher concentrations coexisting phases might be found was proposed to rationalize the fact that the model does not account for the experimental data sets. Short-chained ILs are less prone to aggregate into micellar-like clusters, as observed by Kusano and coworkers [191], who used SANS to show that water mixtures of [C4mim][X], with [X] = Cl-, Br-, and trifluoromethanesulfonate, do not show any low-Q feature at odd with the behavior shown by [C4mim] [BF_4], thus confirming the observation that the low-Q scattering from this salt's aqueous solution is most likely related to critical phenomena [187]

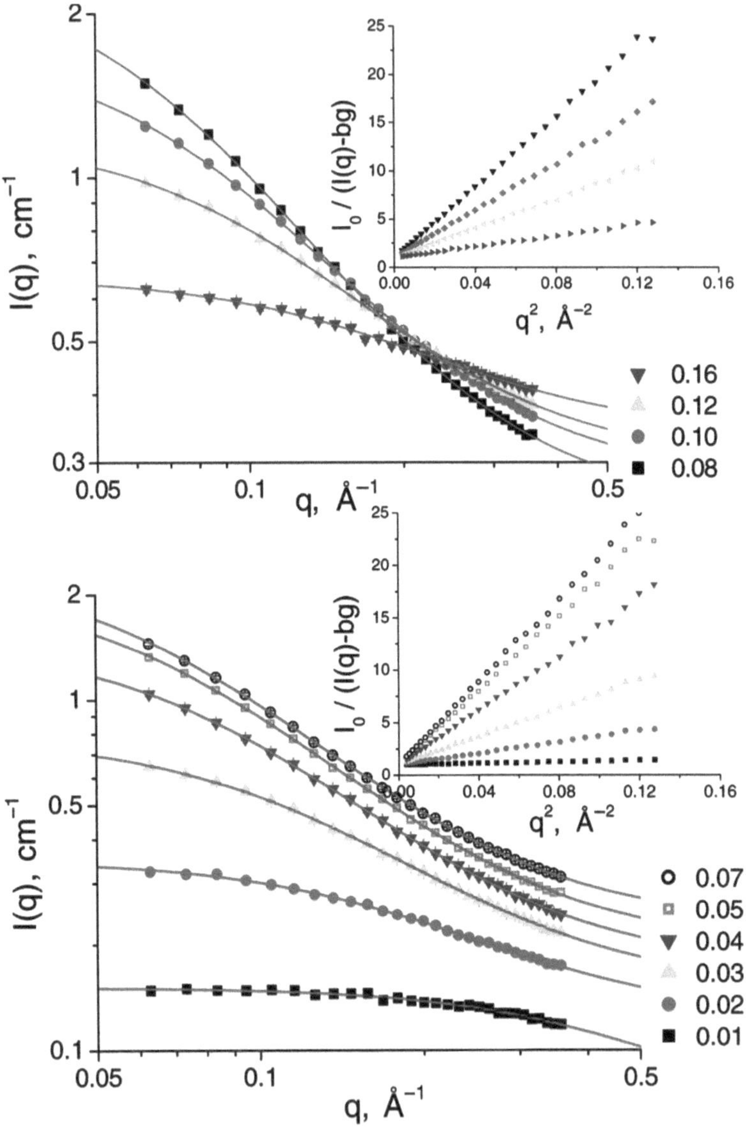

FIGURE 20 Small-angle neutron scattering patterns of aqueous solutions of ionic salt [C4mim][BF_4] measured at 25°C. Mole fraction of salt is indicated. The lines are fits accounting for statistical concentration fluctuations. Insets show the same data represented in 1/I(q) vs q^2 plots. *Adapted with permission from Ref. [187] Copyright (2008) American Chemical Society.*

rather than to the formation of structured entities like micelles, despite recent SANS works keep on mentioning the existence of micelles in these mixtures [192]. Similar observations were obtained using SANS experiments on binary mixtures of [$C_n$mim][bis(trifluoromethanesulfonyl)amide (Tf2N)]/methanol

[193] and of mixtures of [C12mim][NO_3]/benzene [194], where a highly heterogeneous structural organization is found in the methanol-rich/benzene-rich concentration window.

4.2.3 Macromolecules and Other Large Scale Aggregates Dissolved in ILs

While the previously described studies refer to ILs as surfactants in different molecular liquids, other studies exist where ILs behave instead as solvents and accordingly more IL-specific effects can be observed. For example, several studies focused on mixtures of ILs and homopolymers [195–200] or block copolymers. These studies aimed at understanding the solvating performances of ILs toward macromolecular compounds and rationalizing the complex behavior that occurs when IL-philic and IL-phobic moieties are covalently bound to form block copolymers. Neutron scattering techniques provided a great contribution in this context.

The first of such studies on homopolymers dates to 2006, when Triolo and coworkers explored the morphology of polyethylene oxide (PEO) [201] (an important biomimetic, synthetic polymer that finds a range of important applications in medicine, biology, electrochemistry, etc.) when dissolved in an IL. This polymer (in its deuterated form, d-PEO, to achieve bulk contrast conditions) was dissolved in a representative hydrogenous aprotic IL, namely, [C_4mim][BF_4], to deliver homogeneous solutions. In this solvent, PEO fully dissolves organizing itself in a random-coil conformation, when the solution is dilute enough. SANS data were collected as a function of PEO content in the dilute concentration regime at ambient temperature, and data were modeled in terms of Debye's model for the chain-like architecture of flexible polymers [202], delivering fundamental information such as R_g, the radius of gyration of the polymer random coil dispersed in an homogeneous solvent matrix. A similar model [203] (that includes excluded volume effects) has been later used by Werzer and coworkers to model SANS data from dilute and semidilute solutions of d-PEO dissolved in EAN [204], a protic ionic liquid. In both papers, the concentration dependence of the radius of gyration was reported, and similar dependences have been observed in the different ILs. These two studies allowed detecting some differences between the solvent nature of the IL toward PEO: While [C_4mim][BF_4] behaves as good solvent toward this polymer, EAN is instead θ-solvent, implying a worse solvent quality; in other words, monomer-monomer interactions are more energetically favorable than monomer-solvent ones, leading to a change in the overall polymer conformation. In both studies, it was highlighted that in the dilute regime, the dependence of polymer random-coil size on the concentration follows the same trend ($Rg \sim c^{-0.25}$). In the semidilute regime, Werzer et al. found a stronger concentration dependence ($Rg \sim c^{-1}$), fingerprinting a rapid change of PEO's network mesh size, as a consequence of the transition from the sphere-like coil organization in the dilute regime into a more globular and elongated morphology in the semidilute range.

Polymers dissolved in ILs show a complex phase diagram; several of them are characterized by the existence of a lower critical solution temperature (LCST), above which phase separation occurs; a rationalization for this phenomenon considers the occurrence of IL-specific structure-forming solvation features and negative changes in entropy of mixing [197–199]. PEO and other ether-like homopolymers show such a kind of behavior that attracts much attention because of its interesting applicative potentialities in extraction, separation, pharmacology, etc. The LCST behavior of PEO dissolved in ILs has been studied in detail by Lodge and coworkers [200]. They used the SANS technique to explore the temperature dependence of the structural conformation of PEO in the solution of deuterated [C2mim][BF_4]; they observed a progressive increase of the low-Q scattering intensity as a consequence of the raising composition fluctuations upon approaching the phase boundary. An Ornstein-Zernike [205] analysis allowed determining the spinodal temperature corresponding to infinity amplitude of the low-Q excess scattering. In this way, the phase diagram was obtained together with cloud-point determinations, thus reporting both bimodal and spinodal curves for the mentioned systems. The distinct asymmetry of these LCST phase diagram provides a stimulus to develop theories related to polymer phase behavior in ionic media (see Figure 21).

Cloud-point determination and SANS were used to determine the phase diagram in the case of another homopolymer showing LCST behavior when dissolved in an IL. Lodge et al. studied poly(n-butyl methacrylate) (PnBMA) in the ionic liquids [C2mim][Tf2N] and [C4mim][Tf2N] and in their binary mixtures [206]. The study has the goal to relate the performance of a mixed solvent to its neat components, a practice that is common when aiming at fine-tuning the chemical-physical properties of a medium. Similarly to the previously described study, SANS data were used to extract information on the spinodal temperature of each mixture and construct the corresponding phase diagram containing information on bimodal and spinodal points. These parameters are found to linearly scale with the content in [C4mim][Tf2N], thus providing a robust tool to modulate the demixing behavior. Estimation of the effective interaction parameter, χ_{eff}, and the polymer random-coil size, R_g, was done by modeling the SANS data; χ_{eff}'s temperature dependence was then described in terms of an enthalpic and an entropic contribution, showing that the solvent composition can affect the entropy of mixing (that decreases upon increasing [C4mim][Tf2N] content). This observation is consistent with the enhanced structural order that [C4mim][Tf2N] introduces in the mixed solvent, due to the apolar tail self-assembly that determines a more ordered polymer solvation, because of stronger alkyl-alkyl interactions (cation-cation or cation-polymer).

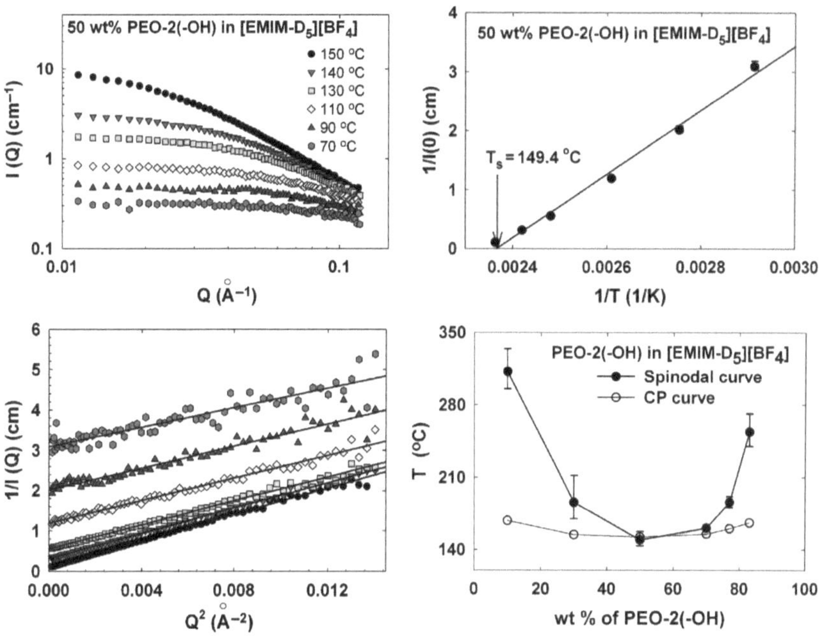

FIGURE 21 (Left, top) Small-angle neutron scattering patterns for 50 wt% of PEO-2(−OH) in [d5-C2MIM][BF_4] at different temperatures. (Left, bottom) Ornstein–Zernike plots for the same data sets. (Right, top) 1/I(0) vs 1/T plot for 50 wt% of PEO-2(−OH) in [d5-C2MIM][BF_4]. The intercept of the 1/I(0) vs 1/T plot represents the spinodal temperature (T_s). (Right, bottom) Temperature–composition phase diagram for PEO-2(−OH) in [d5-C2MIM][BF_4]. The Cloud Points (CP) were determined from the transmittance measurements. *Adapted with permission from Ref. [200]. Copyright (2012) American Chemical Society.*

Another polymer that shows the LCST behavior is poly(benzyl)methacrylate (PBnMA); it was studied by SANS in its binary mixtures with ILs, such as [C2mim][Tf_2N] [207]. The SANS study was conducted as a function of temperature at a given composition where the LCST behavior had been detected and the data nicely indicate that upon crossing the LCST the conventionally observed random-coil behavior gets overwhelmed by a further low-Q scattering contribution that corresponds to the development of structural heterogeneities over a spatial scale of several hundreds of nanometers (estimated on the basis of light-scattering experiments) that coexist with the remaining isolated polymer random coils. These studies lead to the interesting observation of differences in the molecular mechanism behind the occurrence of an LCST in aqueous or in IL media: while in former medium the entropic change is due to the disruption of structured water clusters around the polymer, in ILs, it originates from a desolvation of the polymer [207]. A related thermoresponsive polymer has been very recently investigated by SANS when dissolved in a representative IL, namely, [C2mim][Tf_2N] [208]. Similarly to the case of PBnMA, poly(2-phenylethyl methacrylate (PPhEtMA)) has been found to show an LCST behavior when dissolved in the mentioned IL: across that threshold a distinct change in the SANS pattern is observed fingerprinting the progressive development of large aggregates of polymer separating from IL solution. By fitting the SANS data, the authors could extract information on the random-coil size and the effective interaction parameter, χ_{eff}, that delivers information on the extent of single-contact intermolecular interactions [209]. While the random-coil size is not affected upon approaching the LCST threshold, χ_{eff} increases with increasing temperature, and in Figure 22, a comparison between the trends observed for PBnMA and PPhEtMA dissolved in [C2mim][Tf_2N] is shown. The comparison between the two trends indicates a decrease in the absolute values of enthalpy of mixing, resulting in a decrease of the LCST threshold, when considering PPhEtMA compared with PBnMA, in agreement with the experimental observations [208].

Differently from the behavior shown in the above described cases, poly (N-isopropylacrylamide) (pNIPAm) when dissolved in [C2mim][Tf2N] is well known to show an upper critical solution temperature (UCST) behavior (it is noteworthy that the same polymer shows an LCST when dissolved in water [210]); the polymer is soluble at high temperature, but below a given threshold, it demixes from the solvent to deliver a two-liquid systems. Similarly to the case of LCST, UCST situations are very interesting, as they allow switching from a thermodynamic state to another by a change, in this case, of temperature. Accordingly, the mentioned systems has been largely explored to

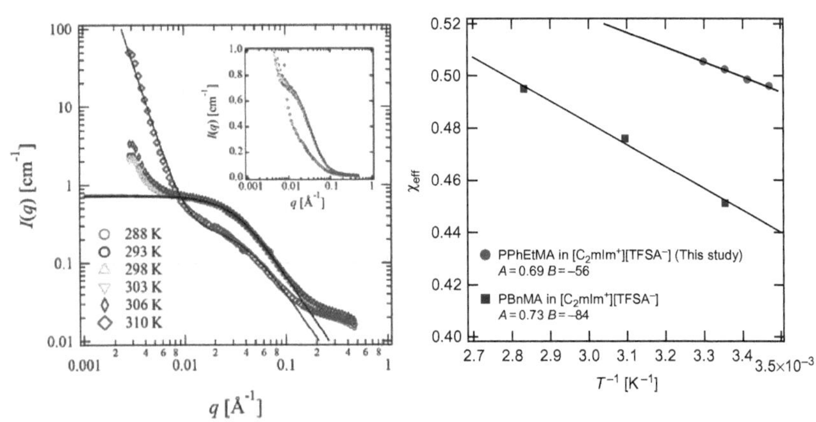

FIGURE 22 (Left) Small-angle neutron scattering patterns for PPhEtMA in d8-[C2mim][Tf2N] solution (c = 16 mg/mL) at various temperatures shown in the form of the log–log plot. The solid lines account for fits with a model of interacting Random coils. Data at high temperature required an additional low Q contribution accounting for large scale inhomogeneities. Data are also shown in the form of the semi-log plot in the inset. (Right) Temperature dependence of the effective interaction parameter, χ_{eff}, of PPhEtMA in d8-[C2mim][Tf2N] obtained from the curve fitting shown in the left figure, together with the corresponding values for PBnMA in d8-[C2mim][Tf2N]. The solid lines are linear fits to $\chi_{eff} = A + B/T$; the fitted values for A and B are shown in the graph. *Adapted from Ref. [208] with permission from the PCCP Owner Societies.*

rationalize this behavior, and recently, Asai et al. applied SANS to explore these mixtures, as a function of both concentration and temperature [211]. In line with expected trend, SANS data from a given pNIPAm/[C2mim][Tf2N] mixture show a distinct low-Q scattering upturn, when decreasing the temperature from above the UCST (see Figure 23). By modeling their data with a Debye function accounting for interaction effects, the authors observe that, upon approaching the UCST, the radius of gyration progressively decreases, indicating coil shrinkage upon approaching the cloud point while the effective interaction parameter becomes larger when decreasing temperature, thus reaching the theta condition at 45°C [211].

Very recently, synthetic homopolymers were also dissolved in other IL-related media and investigated using the SANS technique. In a series of papers [212,213], Chen and coworkers studied PEO dissolved in a solvate IL (SIL) [119]. SILs were discussed above, and they can be considered as a subclass of ILs, consisting of a metal cation bound to a stoichiometric number of coordinating ligands via strong Lewis acid-base interactions [121]. In the first paper, Chen et al. describe how PEO is dissolved in [Li(G4)][Tf2N], where [Li(G4)] is the positively charged complex of lithium and tetraglyme. Being this cation paired to a weakly basic anion (Tf2N), the anion does not

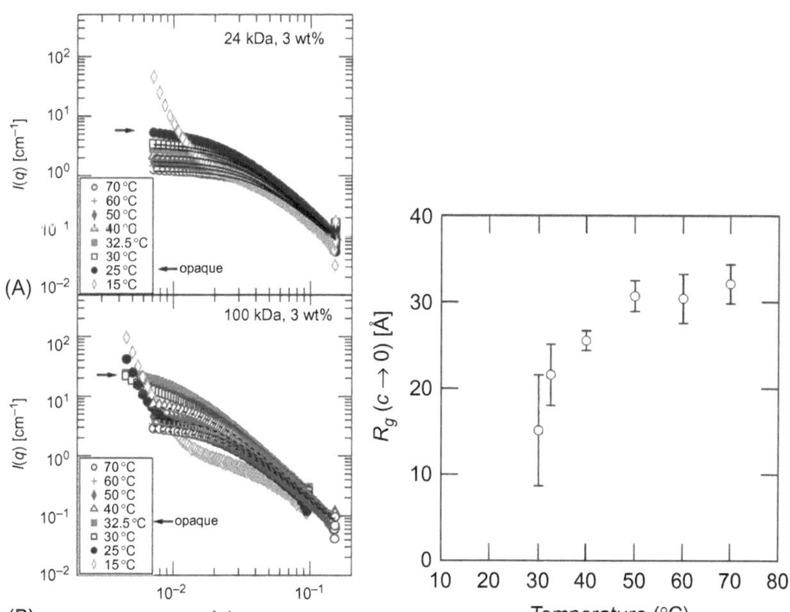

FIGURE 23 (Left) Temperature dependence of small-angle neutron scattering patterns for 3 wt% pNIPAm/IL solution: (A) 24 kDa and (B) 100 kDa. (Right) Temperature dependence of the radius of gyration (Rg) at c → 0 obtained from the analysis for 24 kDa samples shown in figure. *Adapted with permission from Ref. [211]. Copyright (2013) American Chemical Society.*

efficiently compete with tetraglyme for the Li cation, and accordingly, stable Li(G4) complexes exist. This study applied the same approach previously mentioned for studying PEO in EAN or $[C_4mim][BF_4]$ in order to probe the SIL solvent quality toward the homopolymer; estimations of the second virial coefficient indicate that the mentioned SIL is a solvent of intermediate quality between EAN and $[C_4mim][BF_4]$ toward PEO. It is also worth mentioning that the Rg versus concentration power-law trend for the dilute regime in this system is identical to the one observed for EAN and $[C_4mim][BF_4]$ ($Rg \sim c^{-0.25}$), thus supporting the hypothesis that this trend might be a common feature of polymers (or PEO) dissolved in an ionic medium. Using zero-average contrast (ZAC) conditions (i.e., a proper mixture of hydrogenated and deuterated PEO, such as to exclude from the measured scattering pattern the contribution arising from intermolecular correlations) [214], Chen et al. were able to characterize the polymer conformational features in the concentrated regime, excluding intermolecular effects; using the Guinier-Porod model [215], they extracted values for the radius of gyration over a concentration range that exceeds the semidilute one and observed that the polymer size maintains unaltered under these conditions [204]. This important work then provided an interesting R_g versus c-map that describes the three different concentration regimes and how R_g changes across them; notably, these transitions correspond to the related changes detected in the viscosity of these solutions (see Figure 24).

Recently, Chen et al. further extended this study, investigating by SANS a series of PEO mixtures with three different SILs, namely, [Li(G3)][Tf2N], [Li(G4)][BETI], and [Li(G4)][ClO4], where Li(G3) is the positively charged complex of lithium and triglyme [213]. Using both bulk contrast and ZAC experimental conditions, the radius of gyration's dependence from concentration was determined, allowing the determination of the border between the different concentration regimes in these mixtures. The locations of c* and c**, the transition concentrations for dilute/semidilute and semidilute/concentrated regimes, respectively, were found to depend on the SIL nature, thus allowing to determine the solvent quality toward PEO. Overall, a validation of the structural model that proposes the privileged correlation between PEO and lithium cation as responsible for the solubility of PEO in SILs is provided.

Another report accounted for the conformational features of a synthetic polymer, namely, poly(vinylpyrrolidone) (PVP), dissolved in a deep eutectic solvent (DES) composed by choline chloride and urea in a 1:2 molar ratio, using the SANS technique [216]. Using a deuterated solvent, the authors achieved enough bulk contrast with the hydrogenated polymer to observe the scattering related to the random-coil organization of the latter in the dilute/semidilute regime. By comparing with the behavior of the same polymer dissolved in D_2O, the authors observe that, in DES, PVP tends to be more swollen (in the semidilute regime) and is better solvated by the solvent,

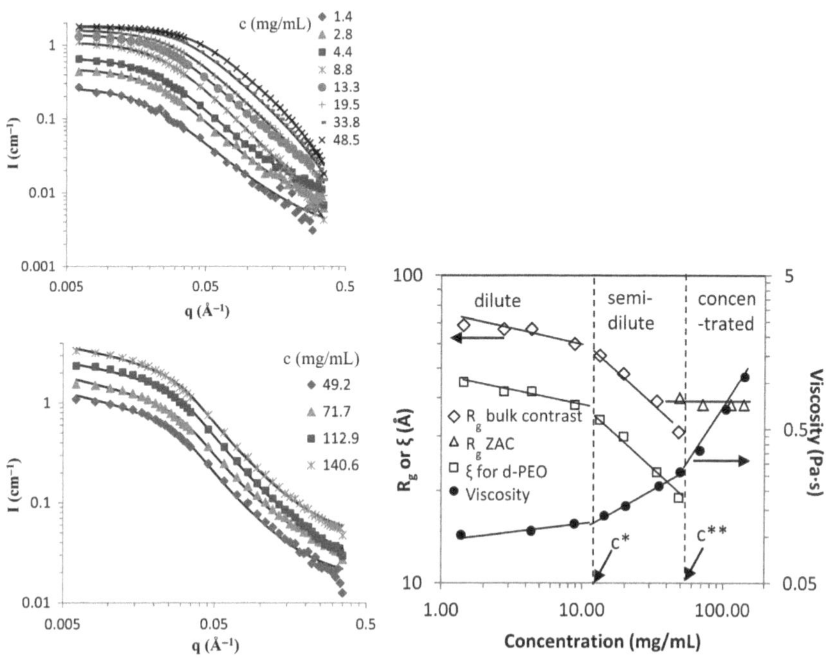

FIGURE 24 (Left, top) Small-angle neutron scattering patterns for d-PEO-[Li(G4)][Tf2N] solutions (dilute, semi-dilute conditions). (Left, bottom) Small-angle neutron scattering patterns for d-PEO-[Li(G4)][Tf2N] solutions at ZAC (concentrated conditions). (Right) PEO's radius of gyration, Rg, and correlation length, ξ, as function of PEO concentration (open symbols) and viscosity of PEO-[Li(G4)][Tf2N] solutions as function of PEO concentration (closed circles) in PEO-[Li(G4)][Tf2N] solutions. *Adapted from Ref. [212] with permission from the PCCP Owner Societies.*

maintaining its flexible coil conformation, than in water. Accordingly, they arrive to the conclusion that the DES is a better solvent for PVP.

The aggregation process in ILs of common surfactants or larger block copolymers leading to a wealth of organized entities such as micelles is another technologically appealing field where SANS has been successfully applied. Since the work of Evans in the 1980s [217], it has emerged the capability of ILs (in particular EAN) to support the formation and stabilization of micelles of alkyltrimethylammonium bromides and alkylpyridinium bromides at 50°C and for Triton X-100 at 20°C. Nowadays, this capability is well understood, and several examples of ILs acting as solvent media supporting the aggregation of amphiphilic (macro-) molecular compounds have been presented [80,218–223]. One of the first SANS studies of such a phenomenology dates to 2006, when Patrascu and coworkers investigated alkyl polyethyleneglycol ethers (often indicated as CiEj, with Ci and Ej being the number of methylene and ether groups, respectively) dissolved in [C4mim][PF_6]; although the authors did not make use of deuteration, contrast was sufficient to deliver a SANS pattern that contained relevant information on the existence of globular aggregates with size of approximately 15 Å [223]. A much more careful characterization of these aggregates was later on presented by Araos and Warr, who applied SANS to explore solutions in selectively deuterated d5-EAN of CiEj's as a function of alkyl and ethoxy chain length, concentration, and temperature [224]. This study confirms that CiEj surfactants in ILs tend to aggregate into conventionally structured micelles with a solvophobic, hydrocarbon core surrounded by a corona of ethoxy groups. Several analogies were deduced with the behavior of the same surfactants in water, and the existence of a concentration offset was rationalized in terms of a weaker solvophobic driving force in EAN, as compared with water. The authors observed as well that nonionic surfactants form more highly curved aggregates in EAN than in water, due to the different molar volume of the two solvents that causes enhanced steric repulsions between the micelles' shells [224].

Using deuterated CiEj's and selectively deuterated propylammonium nitrate (PAN), Atkin et al. further extended this exploration [221] and investigated the nature of CiEj-PAN-oil ternary systems, leading to microemulsion organized entities. In this study, they also detect additional evidences of the existence of a cosurfactant role of the propylammonium cation on the assembled entities.

Recently, a careful investigation described the formation of aggregates of primary quaternary dodecylammonium-based salts in EAN, EASCN, EtAN, and EtASCN PILs [225]. This study aims at comparing the aggregation behavior to the one in water. Using the SANS technique, the authors explore several effects, including the role of the anion's nature and of hydrogen-bonding capability and the amphiphilic capability of the solvent. Micelles from primary ternary dodecylammonium-based salts are not affected by the inherent IL's nanostructure, and it is just the average solvent polarity that affects their formation and geometric features. More specific effects emerge when dealing with quaternary dodecylammonium-based salts whose chemical structure does not allow them to interact through hydrogen bonding with the solvent. In this case, the solvent nanostructure does play a major role in determining thermodynamic and geometric properties of the micellar aggregates (see Figure 25) [225].

Using SANS, Lopez-Baron et al. investigated mixtures of hexadecyltrimethylammonium bromide (CTAB) and EAN, detecting the existence of three concentration regimes [226]. Upon increasing concentration, the noninteracting spherical micellar aggregates begin to appreciate each other's presence, and eventually, a transition from the isotropic (normal micellar) to a hexatic [227] phase is observed. The latter observation is quite interesting and related to the specific properties of IL medium that imparts an enhanced surface curvature to the micelle-solvent interface so to increase the surfactant concentration at this interface.

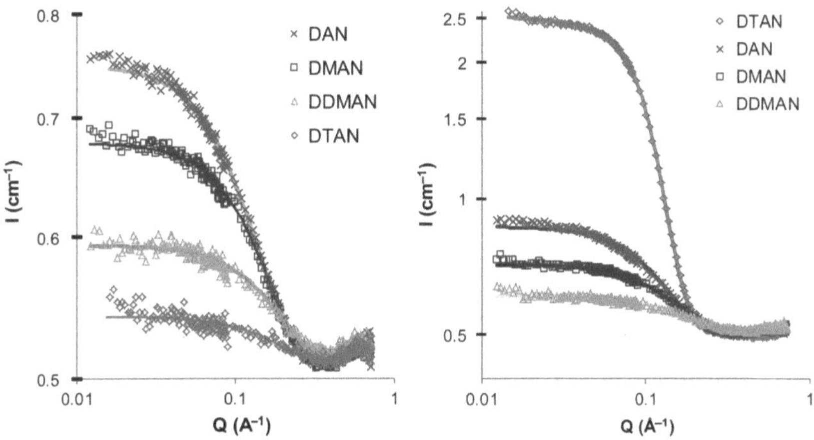

FIGURE 25 (Left) Small-angle neutron scattering patterns from 5 wt% dodecylammonium nitrate surfactants in EAN at 60°C. (Right) Small-angle neutron scattering patterns from 5 wt% dodecylammonium nitrate surfactants in EtAN at 60°C. While the patterns from DAN, DMAN and DDMAN in the two solvents are very similar, a large difference is detected for the behavior of the quaternary ammonium salt DTAN. (DAN = dodecylammonium nitrate, DMAN = dodecylmethylammonium nitrate, DDMAN = dodecyldimethylammonium nitrate; DTAN = dodecyltrimethylammonium nitrate). *Reproduced from Ref. [225]. Published by The Royal Society of Chemistry.*

In another SANS-USANS study, Lopez-Baron et al. investigated didodecyldimethylammonium bromide (DDAB) dissolved into EAN, reporting that this system can spontaneously organize into vesicles [228]. This study is very interesting in showing the potentialities of joint SANS-USANS in determining the structural properties of large-size aggregates that form spontaneously and remain stable in a PIL, such as EAN. The authors managed to describe the vesicles in terms of core-shell model, fitting data over a wide Q-range. The high background related to the hydrogen content is limited; however, the amount of information could have been extracted by a more detailed modeling at high Q (see Figure 26).

Larger size macromolecules that display amphiphilic behavior in ILs have also been studied by SANS techniques. Among the most representative studies those referring to Pluronic (PEOx-PPOy-PEOx, where PEO and PPO represent polyethylene oxide and polypropylene oxide moieties, respectively) can provide an idea of the success of this technique in providing details on thermodynamic and structural properties of the corresponding aggregates. In 2009, Atkin and coworker explored the phase diagram and structural features of a series of three different Pluronic block copolymers (viz, P65, L81, and L121) dissolved in EAN at different temperatures and compositions [220]. Using partially deuterated EAN, they studied the SANS patterns for 1 and 10 wt% mixtures at two temperatures (25 and 63°C) to detect the random-coil conformation in the dilute regime and room temperature and,

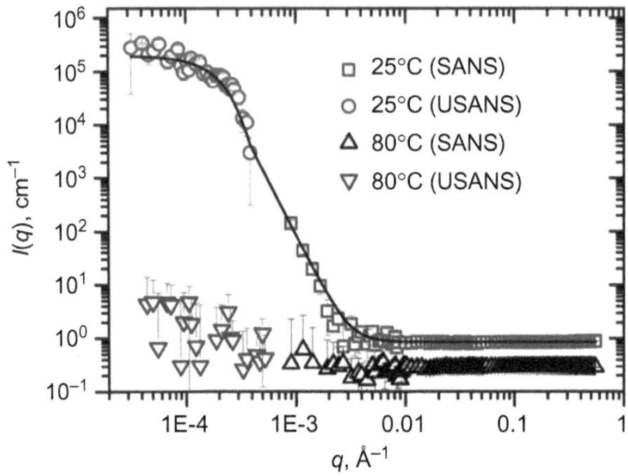

FIGURE 26 Ultra- (USANS) and small-angle neutron scattering (SANS) patterns from didodecyldimethylammonium bromide dissolved in EAN at 25°C and 80°C. At the lower temperature, the low Q excess scattering is modeled in terms of a model accounting for the existence of large vesicles. *Reprinted with permission from Ref. [228]. Copyright (2012) American Chemical Society.*

when in the high concentration regime, the nature of the aggregates at high temperature. At high temperature, when the solubility of PPO in EAN decreases, a wealth of structural organizations is detected, when changing the copolymer: P65 forms micelles; L81 forms lamellar stacks at 1 wt% and unilamellar vesicles at 10 wt%, while L121 organizes into bilayer. In a more recent paper, Lopez-Baron and coworkers notably use SANS and USANS to carefully investigate related systems composed by Pluronic block copolymer (viz, F127, P123, and L121 that have similar solvophobic moiety) dissolved in deuterated EAN [229]. The concentration/temperature-dependent SANS studies highlight the transitions from spherical micelles to face-centered and body-centered cubic supramolecular crystals, for the first two copolymers. L121 instead, due to its lower affinity to the solvent, performs a more complex phase diagram, forming vesicles, elongated micelles, and a nematic phase upon increasing concentration and temperature (see Figure 27).

The use of USANS technique in the context of these structured systems is highly indicated; the authors succeed to model USANS-SANS data sets over more than three decades in Q, using a core+multiple shells form factor or cylindrical form factor model to account for the morphology of L121 at low or high concentration and room temperature, respectively. Moreover, at high temperature, the USANS-SANS patterns indicate the existence of a multilamellar organization; as the copolymer concentration increases, a transition from a multilamellar vesicle toward a macrophase separation is detected by USANS.

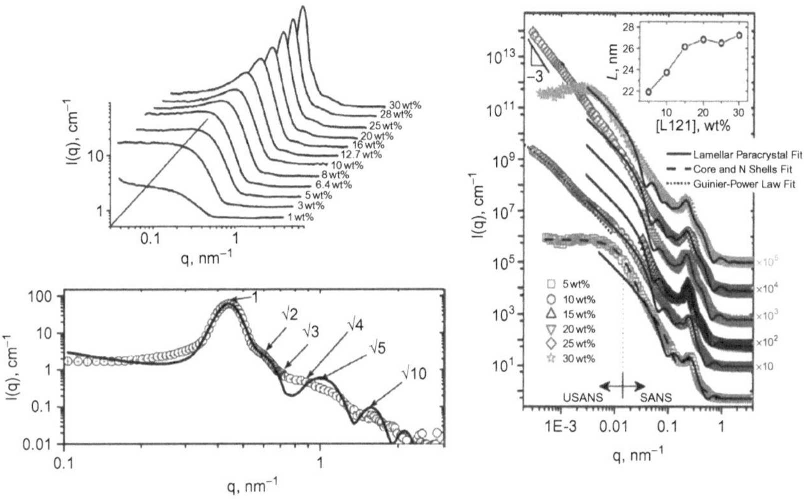

FIGURE 27 (Left top) Small-angle neutron scattering patterns from Pluronic F127 dissolved in d3-EAN at various concentrations and 25°C. (Left, bottom) Small-angle neutron scattering patterns from Pluronic P123 dissolved in d3-EAN at 30 wt%. concentration and 25°C. Solid line is the best fit to body centered cubic (BCC) lattice models with paracrystalline distortion. (Right) Combined ultra- and small-angle neutron scattering from Pluronic L121 dissolved in d3-EAN at various concentrations and 60°C. Solid, dashed, and dotted lines are best fits to the lamellar paracrystal, the core with N shells, and the Guinier–Porod models, respectively. Data are shifted by multiplying by the indicated factors for clarity. *Adapted with permission from Ref. [229]. Copyright (2014) American Chemical Society.*

4.3 RELAXATION PROCESSES IN IONIC LIQUID SYSTEMS

Quasi-elastic neutron scattering (QENS) is a powerful technique that provides access to relaxation phenomena occurring in matter at spatial and temporal scales of high relevance when aiming at describing correlation between bulk properties and microscopic features.

As outlined elsewhere (see, e.g., contributions in Chapters 1 and 6 in Ref. [40]), QENS techniques probe a spatial range between angstrom and nanometer and can access a timescale spanning from fraction of picosecond to hundreds of nanoseconds. Furthermore, by taking advantage of selective deuteration, soft matter systems, in general, and ionic liquids in particular can get explored at a very accurate level, distinguishing different moieties' relaxations or accessing collective dynamic processes.

Typically, a QENS characterization of a system as complex as an ionic liquid, which is characterized by an intricate hierarchy of structural and dynamic features, requires the use of different QENS spectrometers in order to access complementary momentum transfer and energy windows. Furthermore, the synergic use of other experimental techniques, such as rheology, dielectric spectroscopy, dynamic light scattering, and NMR, and molecular dynamic computational ones can be of great impact in addressing complex issues in the dynamic behavior of ionic liquids. Examples of this approach exist, and the high impact of ILs on applicative themes is expected to further boost the use of multitechnique approaches to investigate them.

Among the interesting properties of ILs, the fact that many of them can be easily supercooled from their liquid into the amorphous solid (glass) state is very attractive as it provides nice opportunities to probe the existence and nature of activated processes over a wide temperature window that is not always accessible in the case of molecular or ionic compounds. Accordingly, several dynamic studies based on QENS were focused on exploring the temperature dependence of the relaxation processes, detecting a variety of dynamic events.

The first report of incoherent QENS data sets on imidazolium-based ionic liquids appeared in 2003 [230], where, using a time-of-flight spectrometer with intermediate resolution performance, a paradigmatic IL, namely, hydrogenated [C4mim][PF_6], was explored over the temperature window of 250–320 K, between 1.4 and 2.2 Å^{-1} with a resolution of 90 μeV at FOCUS at SINQ [231] and NEAT at HZBerlin [232]. This study highlighted several features later to be discussed in other reports. In particular, similarly to other glass formers, the incoherent dynamic structure factor, $S_i(Q,E)$, from this IL is characterized by the occurrence of at least two well-separated dynamic processes, one occurring at picosecond, with limited temperature dependence and another, much slower, temperature-dependent process (see Figure 28). The main feature of the slower process is its highly non-Arrhenius and non-Debye nature, consistently with its interpretation as the fingerprint of the main structural relaxation due to the diffusion process.

The nice complementarity between different QENS spectrometers and the powerful opportunity offered by selective deuteration of IL's moieties has been shown in another paper on [C4mim][PF_6] [233]. There either the side methyl or the side butyl groups were selectively deuterated with the goal to distinguish the incoherent dynamic structure-factor contribution arising from one moiety or the other, considering the large incoherent-scattering cross section of hydrogen compared with the one of deuterium. In that paper, also backscattering techniques were applied to approach the slow dynamics in ILs. In particular, elastic and inelastic fixed-energy scans (EFEs and IFEs, respectively) were applied for the first time to the study of dynamic processes in ILs. While the first approach is quite common in the exploiting of backscattering spectrometers to study soft matter systems (see, e.g., [234]), the latter is much less common, due to the low flux available at these instruments; only recently, the improved backscattering IN16B at ILL [235,236] or BASIS at SNS-ORNL [237] are providing the possibility to collect this highly informative kind of experiments on a routine basis. EFEs and IFEs

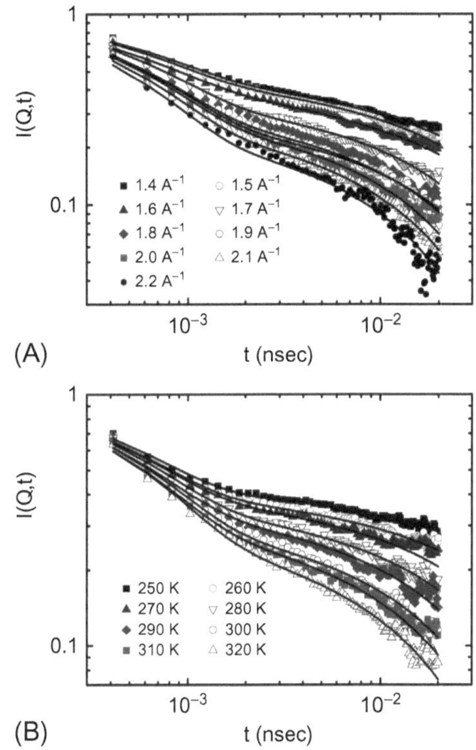

FIGURE 28 (A) Temperature dependence of I(Q,t) for $[C4mim][PF_6]$ at $Q=2\,Å^{-1}$. (B) Momentum transfer dependence of I(Q,t) for $[C4mim][PF_6]$ at T = 320 K. The symbols refer to experimentally derived data, while the lines represent the fit in terms of a bimodal relaxation. *Reprinted from Ref. [230], with the permission of AIP Publishing.*

are specific kinds of QENS experiments where one probes the relaxation process occurring in a sample, by monitoring either the elastic or the inelastic signal, respectively, as a function of an external parameter, such as temperature. In typical EFEs or IFEs experiments, one monitors the number of neutrons that are scattered with no energy change (considering the instrumental resolution) or with a given energy exchange (typically in the range of a few μeVs), respectively, upon changing the sample temperature at a given rate. It turns out that a given relaxation process manifests itself as a sigmoid-like step or a peak in the case of an EFEs or IFEs, respectively (see Figures 29 and 30).

Appropriate modeling of these curves can deliver a very accurate insight into the relaxation dynamics in ILs. Since the original proposals of EFEs [234,238] that were applied to amorphous polymers, it has emerged the great power of this experimental tool, to explore the relaxation map of complex

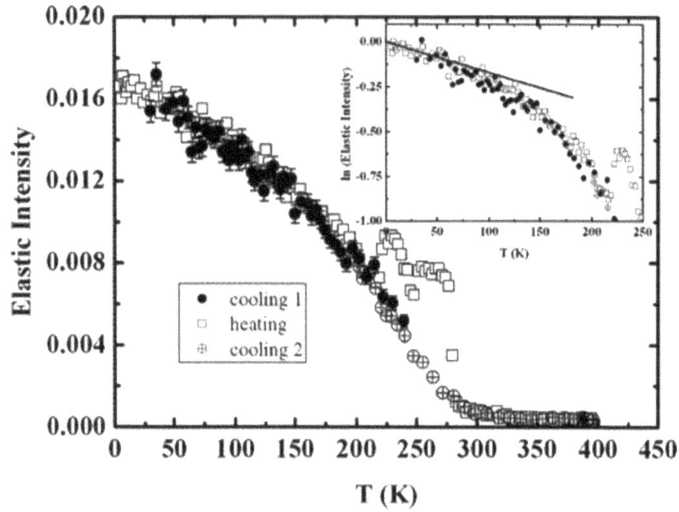

FIGURE 29 Elastic fixed-energy (EFE) scans collected on [C4mim][PF_6] at fixed Q = 1.89 $Å^{-1}$. In the inset, the natural logarithm of the EFE intensity is plotted versus temperature, to highlight the initial linear behavior for T > 60 K. Above that temperature, relaxation processes become active at the time scale probed by the spectrometer. *Reprinted with permission from Ref. [233]. Copyright 2005 American Chemical Society.*

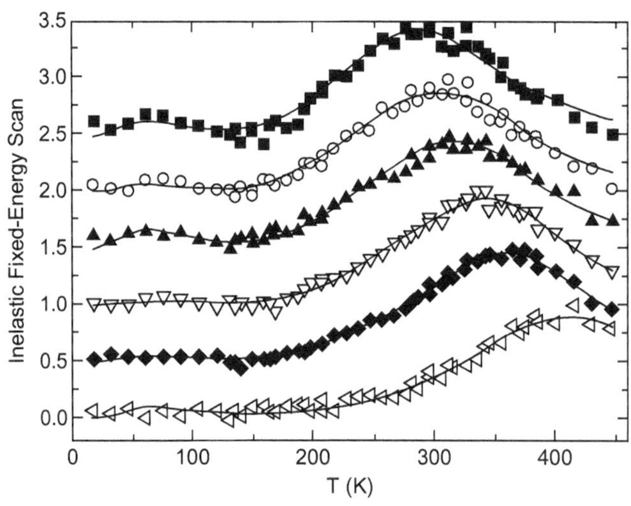

FIGURE 30 Inelastic fixed-energy (IFE) scans collected on [C4mim][PF_6], at different Q values (Q = 0.56, 1.02, 1.29, 1.52, 1.72, and 1.87 $Å^{-1}$, respectively, from top to bottom). The lines correspond to the fits in terms of three different relaxation processes: namely two, low temperature, localized processes and the high temperature diffusive one. *Reprinted with permission from Ref. [233]. Copyright 2005 American Chemical Society.*

amorphous systems. Assuming that below the glass transition the local relaxation processes can be described in terms of simple Debye ones with an Arrhenius temperature dependence or as a combination of Debye processes with a distribution of activation energies, the description of such local process in terms of their activation energy is pretty straightforward, as exemplified in some examples [233,234,239].

In the previously mentioned paper [233], the joint use of selectively deuterated samples and EFEs/IFEs kind of experiments over large Q/T ranges allowed distinguishing between the relaxation processes involving either the methyl group of the side butyl chain, in a clear-cut and undebatable way (Figure 31).

Furthermore, the Q-dependence of the IFE scan allows the separation between Q-independent and Q-dependent peaks that are associated to localized and diffusive processes, respectively. A simple comparison between the data sets reported in the first QENS experiment and the ones stemming from the EFEs and IFEs allows understanding the great power of these latter techniques, especially when coupled to selective deuteration of specific moieties in the sample material: A wealth of features becomes accessible and modeling provides useful details on their nature. Furthermore, the slow process (occurring at high temperature) whose features do not depend on deuteration

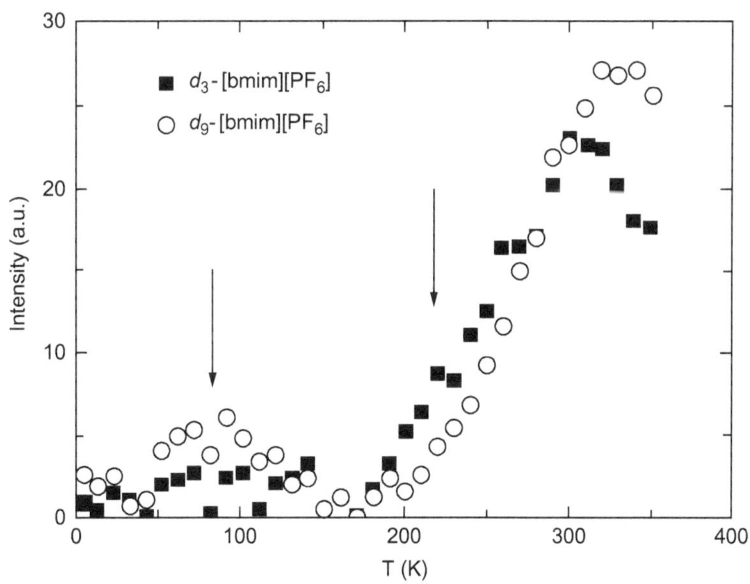

FIGURE 31 Comparison between inelastic fixed-energy (IFE) scans collected on two different selectively deuterated samples, d9-[C4mim][PF_6] and d3-[C4mim][PF_6], at a selected Q value (Q = 1.52 Å^{-1}). *Reprinted with permission from Ref. [233]. Copyright 2005 American Chemical Society.*

of different moieties and whose peak position depends on Q is associated to the diffusive relaxation that can be modeled in terms of a non-Arrhenius and non-Debye process.

The integration of QENS experiments with molecular dynamic simulations is a useful approach to explore relaxation processes at atomistic detail. Currently, probably due to the difficulty in determining interatomic potentials that might accurately describe not only structure but also dynamics, there are not many of such joint studies. We are aware of just one paper dealing with [C2mim][Br] that attempts at this complex task and succeeds in identifying the localized motion of the side ethyl chain that is responsible for the low-temperature QENS feature [240] (Figure 32).

Recently, EFEs and IFEs were reported for selectively deuterated pyridinium-based ILs, where a local process was detected and modeled with two different geometric models related to the methyl group relaxation, while only initial portion of the slower, diffusive process was observed in the IFEs, thus preventing an accurate determination of the diffusive process [241].

The use of EFEs, joint with the analysis of $S_i(Q, E)$ data sets, led to a study of proton dynamics in a different kind of ionic liquid (viz, a protic one, N,N,N′,N′-tetramethylguanidinium bis(perfluoroethylsulfonyl)imide ([H2NC(dma)2][BETI])) [242], where the authors describe the liquid-state data sets from a backscattering instrument in terms of two distinct diffusion processes. They describe these diffusive processes in terms of two Lorentzian contributions; despite the accurate geometric description of the data that is shown in terms of such a model, the fact that diffusive processes are described in terms of Debye terms represents, in our view, a limitation of the model. The description of diffusive processes in terms of a simple Debye process (a simple exponential in time domain or, alternatively, a Lorentzian function in energy domain) is quite common in the study of ionic liquids [242–249] [240]. We remain of the opinion that the complex nature of the main relaxation process in these systems and in other glass formers would require a more elaborate description, such as the so-called stretched exponential or Kohlrausch-Williams-Watts approach [250]. As a matter of fact, Fourier transform of QENS data into the time domain [230,251] or neutron spin-echo (NSE) data [252–254] converge in indicating that diffusive processes in these compounds and in several other glass formers are distinctly non-Debye processes.

Overall, the exploitation of time-of-flight and backscattering techniques to probe the incoherent dynamic structure factor from a large variety of imidazolium-based ILs indicates the existence of (a) a fast, librational process that is largely unaffected by chemical details and shows a limited temperature dependence; (b) a (combination of) localized process(es), whose nature is still a matter of debate, as it may result from the combination of different motions that depending on the energy and Q-range probed is described in different

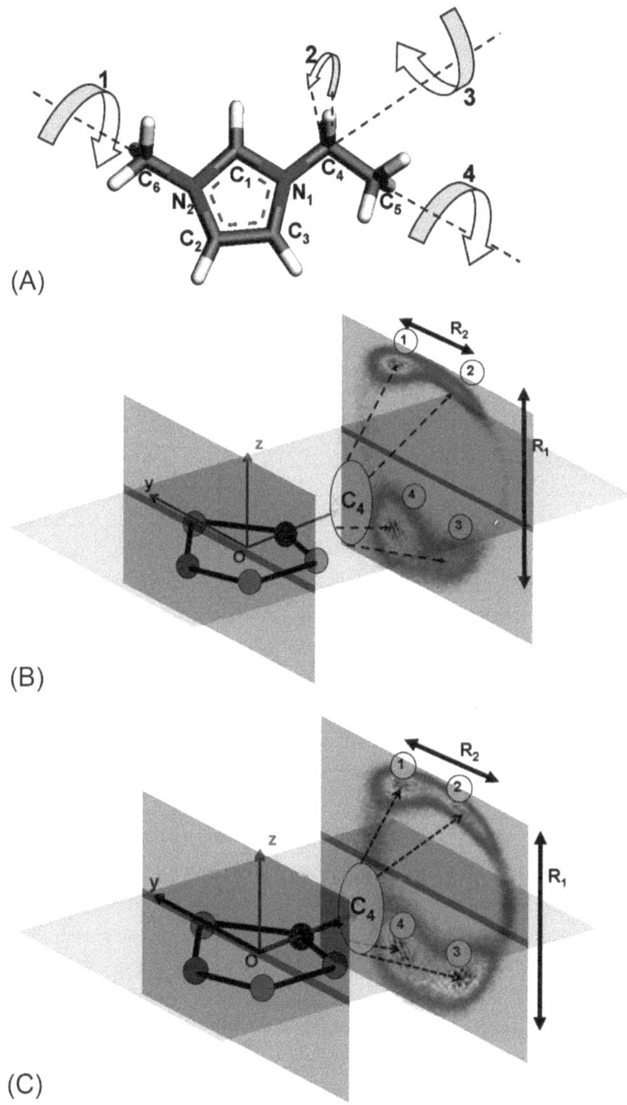

FIGURE 32 (A) Possible reorientational motions of the [C2mim] cation; (B, C) contour plots of the average spatial distribution of the terminal methyl carbon atom, C5, relative to the N1 nitrogen, representing the ethyl group reorientations denoted as 3 in (A), derived from the simulations in the solid and liquid phases, respectively. *Reprinted with permission from Ref. [240]. Copyright 2010 American Chemical Society.*

ways; and (c) a diffusive process that is strongly related to the rheological properties.

Recently, QENS techniques have also been applied to the investigation of ILs under less common conditions, such as under nanoconfinement.

In recent works based on QENS data collected at the BASIS backscattering instrument at SNS, interesting new features are detected on the existence of a fast diffusive process, modeled through a Lorentzian term, experienced by ions under confinement conditions, and provided one that distinguishes them from the immobilized ones at the confining walls [246,247]. The long-range diffusive behavior in glyceline, a deep eutectic mixture composed of choline chloride and glycerol (1:2 molar ratio) has been recently investigated using backscattering techniques that access subnanosecond dynamics [249]. By measuring QENS patterns from two different samples where either choline or glycerol is selectively deuterated, the authors could distinguish the diffusive dynamics of the two hydrogenated components, thus highlighting that choline diffuses over a faster timescale than glycerol, despite its larger size. A rationalization for this counter-intuitive observation could be provided, considering the complex hydrogen-bonding network, which glycerol develops with chloride anions that strongly inhibits the diffusion of the former component.

A more specific discussion is required to account for those studies that involve fully deuterated IL samples. Replacing hydrogen atoms with deuterium isotope is a very efficient way to enhance to coherent neutron scattering cross section at the expense of the incoherent one and thus access information on collective dynamics rather than self-particle one (see, e.g., contributions in Chapters 1 and 6 in Ref. [40]). While there are a few studies conducted using TOF instruments on fully deuterated ionic liquids [245,255], the technique that best fits the exploration of complex, collective relaxations in soft mater and in ILs, in particular, is NSE [256,257]. This technique directly probes the coherent dynamic structure factor, $S_c(Q,E)$, in temporal domain (thus directly delivering the intermediate scattering function, I(Q, t)), making data exploration, modeling, and interpretation more direct and straightforward than other techniques that deliver data in the energy domain. Studies based on NSE measurements so far conducted on fully deuterated ILs are still a few. These studies have been focused on imidazolium-based ILs, and they accessed the collective dynamics at momentum-transfer values corresponding to the three main diffraction peaks for $Q < 2.0$ Å^{-1}. As described in the previous section on structural properties, these peaks fingerprint the existence of a hierarchy of structural correlations, whose time decay can be monitored by means of the NSE technique directly in the temporal domain. The first of such studies on ILs dates to 2009, when the relaxation map of [C6mim][Tf2N] was studied with different techniques, including NSE [252]. In Figure 33, the static structure factor S(Q) obtained with neutron scattering on a fully deuterated sample of [C6mim][Tf2N] is shown. There three different peaks can be observed that, as mentioned above, correspond to the correlations between adjacent moieties ($Q = 1.4$ Å^{-1}), charged parts ($Q = 0.8$ Å^{-1}), and polar-apolar portions ($Q = 0.4$ Å^{-1}), respectively. In that paper, the first two structural correlation

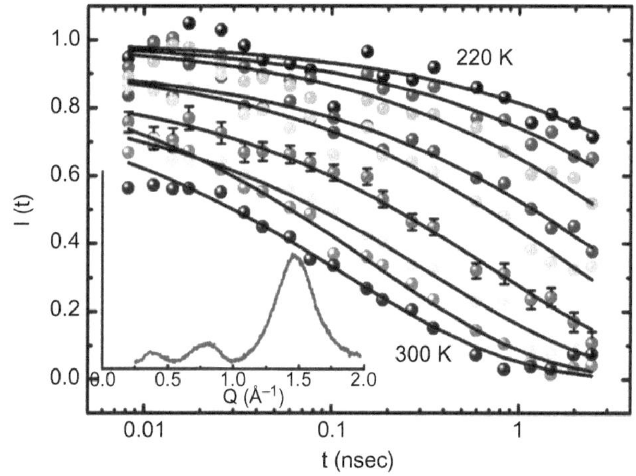

FIGURE 33 Dynamic structure factor I(Q,t) at the main peak position ($Q = 1.4\ \text{Å}^{-1}$) for fully deuterated [C6mim][Tf2N], at selected temperatures between 220 and 300 K (every 10 K). The continuous lines represent the fit of the experimental data in terms of the stretched exponential model. In the inset the neutron diffraction pattern of fully deuterated [C6mim][Tf2N] at 4 K is presented. *Reprinted with permission from Ref. [252] Copyright {2009} American Chemical Society.*

decays were studied as a function of temperature, and in Figure 33, these trends for $220 \leq T\ (K) \leq 300$ are shown for $Q = 1.4\ \text{Å}^{-1}$.

In the accessed temporal window, it appears that only a relaxation process can be detected. It is to be stressed that a temperature-dependent prefactor A determines the fitting model: $I(Q, t) = \cdot\exp(-t/\tau)^{\beta}$, where A accounts for faster dynamics than ca. 1 psec (such as a faster relaxation process or a Debye-Waller term), τ is the characteristic relaxation time, and β is the so-called stretched exponent that accounts for deviations from simple exponential decay. In the present case, one finds that $\beta \sim 0.5$–0.6 that is a typical range of values for the stretched exponential in glass formers and indicates the complexity of the main relaxation process that can be considered as either build up by a combination of simple, Debye processes or, alternatively, characterized by an intrinsic heterogeneity of the relaxation process that cannot be described in terms of Debye-like processes. This problem is very well known in the field of polymers and other glass formers, and the considerations developed in those fields fully apply to the case of ionic liquids. The temperature dependence of the characteristic time for this relaxation can be directly extracted from the experimental data sets, using the a.m. model, and the average relaxation time is given by $<\tau> = \tau/\beta \cdot \Gamma(1/\beta)$, where Γ is the gamma function. The average relaxation time for this process nicely follows the master curve obtained for the main structural relaxation process that is dictated by diffusion. In Figure 34, the relaxation map for this IL, containing a series of

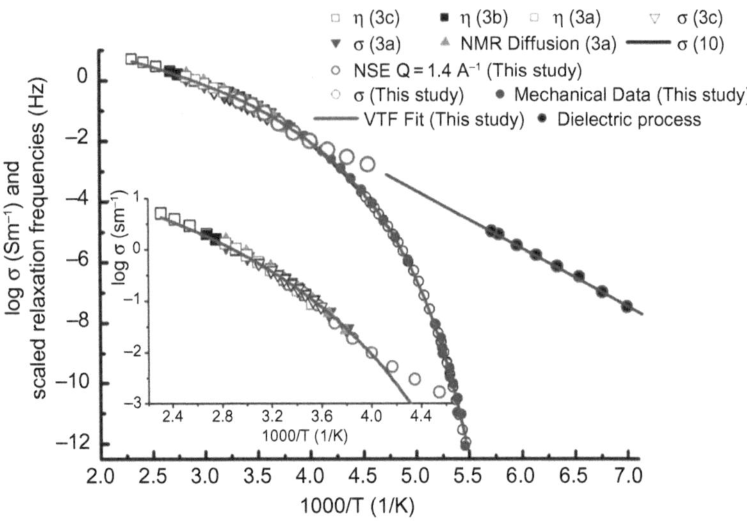

FIGURE 34 Primary process relaxation map for [C6mim][Tf2N], as obtained integrating data from different sources (see the original paper for the different sources) including Neutron Spin Echo data at Q = 1.4 Å^{-1} (red open circles). In the inset a magnification of the high temperature portion of the whole data sets is presented. *Reprinted with permission from Ref. [252] Copyright {2009} American Chemical Society.*

relaxation data sets referring to the main relaxation process and including <τ>'s from the fit of Figure 33, is shown.

This relaxation process is highly non-Arrhenius, and, as well established, this behavior can be described in terms of the Vogel-Fulcher-Tammann trend: $\tau = \tau_o \cdot \exp[B/(T - T_o)]$. It can be seen that the high-temperature regime (above ca. 250 K) follows the trend of diffusive process; on the other hand, a deviation from this trend is observed at lower temperatures, where the relaxation times seem to follow the Arrhenius-like trend of a dielectrically active process. Similarly, to the case for Q = 1.4 Å^{-1}, the structural relaxation process detected at 0.76 Å^{-1} for the same deuterated IL follows a non-Debye, non-Arrhenius trend. In Figure 35, data from this relaxation are reported, and they are described using the same model outlined for the previous case.

The main relaxations at these Q-values follow the same diffusive-like trend with a stretched exponential $\beta \sim 0.5$–0.6. This observation is quite remarkable if we consider that these structural correlations refer to charged moieties correlations (Q = 0.7 Å^{-1}) and adjacency correlations (Q = 1.4 Å^{-1}). NSE data for other ILs have been collected at the Q-values corresponding to charged moieties correlations [258,259], namely, for [C8mim][Cl], [C8mim][Tf2N], and [C8mim][PF_6]. All these ILs show similar NSE data sets as the ones previously discussed. The larger dynamic window accessible in the experiments reported in Refs. [258,259] allows

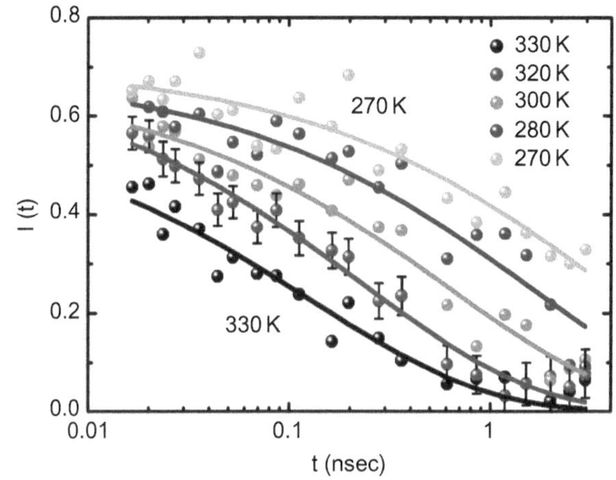

FIGURE 35 Dynamic structure factor I(Q,t) at the low Q peak position (Q = 0.76 Å^{-1}) for fully deuterated [C6mim][Tf2N], at selected temperatures between 270 and 330 K. The continuous lines represent the fit of the experimental data in terms of the stretched exponential model. *Reprinted with permission from Ref. [252] Copyright {2009} American Chemical Society.*

better determination of the faster dynamics acting as a precursor for the slow relaxation process, and this faster process has been described, similarly to the case of incoherent-scattering experiments, in terms of a simple Debye-like, exponential process. This process shows a weak, if at all, temperature dependence and corresponds to a fast process in the timescale of picoseconds. This process has been associated to fast relaxations of the side alkyl tail [258,259], or alternatively, it can be interpreted as an in-cage, complex, collective relaxation of the cation as a whole [260]. At longer temporal scales, the caging network relaxes, and the diffusive behavior activates. The relaxation for the slow process, related to the ionic motions, also in these cases follows the stretched exponential trend with $\beta \sim 0.5$. The temperature dependence for the average time follows the viscosity trend for a limited temperature window and then deviates from it to merge with an Arrhenius-activated process.

The situation is more complicated in the case of the structural correlation decay for the polar-apolar alternation that is fingerprinted by the peak at ca. 0.3–0.5 Å^{-1} (depending on the alkyl tail length). As mentioned above, this structural correlation reflects the existence of a complex separation between polar and apolar moieties that are mutually excluding each other. The structural properties of these heterogeneities have been only recently been understood from a structural point of view, but there is essentially no information on their dynamic nature. The Yamamuro group published the only NSE data existing on the structural correlation decay at Q = 0.28–0.29 Å^{-1} for the case of [C8mim][Cl], [C8mim][Tf2N], and [C8mim][PF_6] [258,259]. While [C8mim][Cl] [258] shows unclear results, presumably due to its high

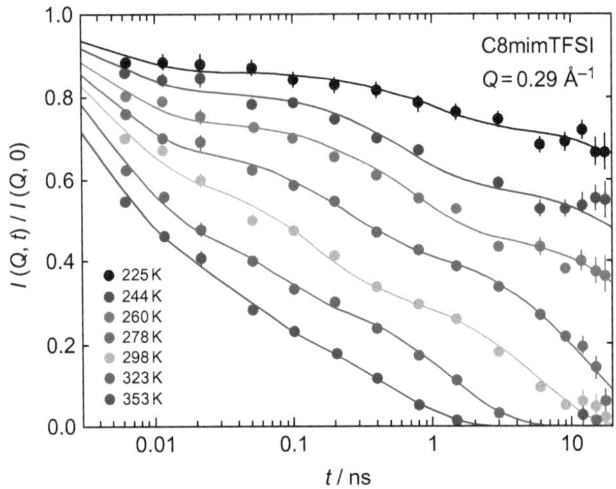

FIGURE 36 Dynamic structure factor I(Q,t) at the low Q peak position (Q=0.29 $Å^{-1}$) for fully deuterated [C8mim][Tf2N], at selected temperatures between 225 and 353 K. The continuous lines represent the fit of the experimental data in terms of three different simple exponential decays. *Reprinted with permission from Ref. [259] Copyright {2013} American Chemical Society.*

viscosity, the two other ILs could be explored in a more reliable way [259]. In Figure 36, NSE data collected on [C8mim][Tf2N] at Q=0.29 $Å^{-1}$ as a function of temperature over more than three decades in time are reported, together with the corresponding modeling in terms of a combination of three different exponential terms.

The experimental (I(Q,t)) data and corresponding fit in terms of the same model are comparable with the case of [C8mim][PF_6] [259]. The modeling of these data delivers information on the existence of three different contributions to the overall correlation decay. While the faster process turns out to be essentially temperature-independent, the two slower processes show an Arrhenius-like trend. More importantly, Yamamuro and coworkers extracted from their fits the fractions of the different relaxations and highlighted that the fraction of the slowest process strongly increases when decreasing temperature, at the expenses of the fraction associated to the faster process. They interpreted this evidence as a fingerprint of the existence of two classes of structural correlations: an ordered and highly correlated portion that becomes more and more dominant at low temperature and a more disordered portion that instead dominates at higher temperature. Such a scenario is consistent with the polar versus apolar segregation model that is commonly used to account for the morphology in these systems, as the authors propose that "the slower relaxation arises from the (ordered) nanoscale structure, while the faster ones are from the broken (disordered) part of the structure" [259].

4.4 CONCLUSION AND PERSPECTIVE

Ionic liquids represent one of the hottest research fields in the last couple of decades. The potential role that these compounds might play in the fields of green chemistry and environmentally sustainable technologies is such that great efforts are being paid in understanding and rationalizing IL properties, aiming at exploiting them for conventional and still to be developed applications. In this scenario, neutron scattering techniques are playing a great role in driving researchers' attention toward a comprehensive correlation between bulk performances and micro- to mesoscopic structural and dynamic properties.

Being techniques that access spatial scales between angstrom and several nanometers and temporal scales from fraction of picosecond to several nanoseconds, neutron-based techniques provide a complete suite of investigation tools to probe relevant relaxation processes and structural correlations that affect and determine macroscopic features.

Of course, the chemical complexity and variety of ILs represent a challenge for such techniques: the necessity to systematically probe different samples with slightly different chemical compositions (e.g., changing alkyl chain length or anion), the laborious (and expensive) opportunity to probe different isotopically substituted compounds, the rather limited brilliance of conventional neutron source, and the limited access to large-scale facilities that can only provide neutron beams, all this are limiting factors for the advance of neutron-scattering-based knowledge on ILs. On the other hand, the objectively unique opportunity to exploit isotopic substitutions to access smarter levels of understanding of both structural and dynamic properties and the continuous advance in neutron facilities technology (either as improvement of existing facilities/beamlines or as development of new, brilliant neutron sources, such as the ESS) counterbalance the a.m. difficulties and make the future of neutron scattering applied to ILs brilliant.

Nowadays, we can say that much of the knowledge on structural and dynamic properties of neat ILs has been achieved and a great deal of understanding of the intricate microscopic properties in these compounds has been obtained. On the other hand, mixtures of two or three (or even more) ILs will remain composed of ionic species and possibly have low melting point, thus leading to new ILs [261,262] with an estimated number of binary ILs quantified in 10^6 compounds and that of ternary ILs quantified in 10^{18} compounds (to be compared with <1000 molecular solvents in use today) [13]; it is clear that much will need to be explored even in the field of ILs!

The exploration of the intricate interplay between coulomb, dispersive, hydrogen-bonding (just to name the most relevant) interactions occurring in mixtures of ILs with other ionic compounds or molecular liquids will also require large efforts and the synergic use of different techniques such as

X-ray scattering, NMR, and light spectroscopies together with computational tools (such as molecular dynamics) to support neutron scattering techniques. These multidisciplinary and multitechnique approaches are becoming more and more common nowadays, and it can be foreseen that ILs exploration will greatly benefit from this approach.

Another field where neutrons are envisaged to greatly advance knowledge will be that of mixtures of ILs with macromolecules and biomacromolecules, given the wealth of environments that ILs can offer for these complex systems that will greatly expand their range of applications and exploitation. These systems are characterized by a very spatial extent and very slow relaxation dynamics, thus pushing for the limits of structural and dynamic characterizations using neutron techniques, pushing for routinely use of, for example, ultra-small-angle neutron scattering and NSE techniques.

We can imagine as well a development in the use of neutron techniques focused to the exploration of more unconventional IL-based systems such as mixtures of ILs with gaseous compounds (e.g., IL-CO_2 mixtures have been proposed as a green compound with appealing solvating properties [18]), or mixtures of ILs with rigid/soft (functionalized) nanoparticles or nanoorganized systems (e.g., graphene or similar technologically appealing systems) or recently developed compounds, such as deep eutectic solvents (that are often fully biocompatible, thus significantly dropping their environmental impact) or solvate ionic liquids.

REFERENCES

[1] K.R. Seddon, J. Chem. Technol. Biotechnol. 68 (1997) 351.
[2] P. Walden, Bull. Acad. Imper. Sci. St. Petersburg 8 (1914) 405.
[3] M.J. Earle, J.M.S.S. Esperança, M.A. Gilea, J.N. Canongia Lopes, L.P.N. Rebelo, J.W. Magee, K.R. Seddon, J.A. Widegren, Nature 439 (2006) 831.
[4] J.H. Davis Jr., Chem. Lett. 33 (2004) 1072.
[5] A.E. Visser, R.P. Swatloski, W.M. Reichert, J.H. Davis, R.D. Rogers, R. Mayton, S. Sheff, A. Wierzbicki, Chem. Commun. (2001) 135.
[6] A.E. Visser, R.P. Swatloski, W.M. Reichert, R.D. Mayton, S. Sheff, A. Wierzbicki, J.H. Davis, R.D. Rogers, Environ. Sci. Technol. 36 (2002) 2523.
[7] R. Giernoth, Angew. Chem. Int. Ed. Engl. 49 (2010) 2834.
[8] M. Freemantle, Chem. Eng. NEWS 76 (1998) 32.
[9] T. Welton, Chem. Rev. 99 (1999) 2071.
[10] R. Sheldon, Chem. Commun. 2001 (2001) 2399.
[11] R.P. Swatloski, S.K. Spear, J.D. Holbrey, R.D. Rogers, J. Am. Chem. Soc. 124 (2002) 4974.
[12] N.V. Plechkova, K.R. Seddon, Chem. Soc. Rev. 37 (2008) 123.
[13] R.D. Rogers, K.R. Seddon, Science 302 (2003) 792.
[14] M.J. Earle, K.R. Seddon, Pure Appl. Chem. 72 (2000) 1391.
[15] M. Armand, F. Endres, D.R. MacFarlane, H. Ohno, B. Scrosati, Nat. Mater. 8 (2009) 621.
[16] V.I. Pârvulescu, C. Hardacre, Chem. Rev. 107 (2007) 2615.
[17] J.P. Hallett, T. Welton, Chem. Rev. 111 (2011) 3508.
[18] L.A. Blanchard, D. Hancu, E.J. Beckman, J.F. Brennecke, Nature 399 (1999) 28.

[19] C. Hardacre, J.D. Holbrey, M. Nieuwenhuyzen, T.G. Youngs, Acc. Chem. Res. 40 (2007) 1146.

[20] L.P.N. Rebelo, J.N.C. Lopes, J.M.S.S. Esperanca, H.J.R. Guedes, J. Lachwa, V. Najdanovic-Visak, Z.P. Visak, Acc. Chem. Res. 40 (2007) 1114.

[21] J.D. Holbrey, J. Chem. Soc. Dalton Trans. (1999) 2133.

[22] H. Weingaertner, Angew. Chem. Int. Ed. 47 (2008) 654.

[23] C. Chiappe, D. Pieraccini, D.B. Zhao, Z.F. Fei, P.J. Dyson, Adv. Synth. Catal. 348 (2006) 68.

[24] H. Olivier-Bourbigou, L. Magna, D. Morvan, Appl. Catal. A Gen. 373 (2010) 1.

[25] W. Xu, L.-M. Wang, R.A. Nieman, C.A. Angell, J. Phys. Chem. B 107 (2003) 11749.

[26] E.W. Castner, J.F. Wishart, J. Chem. Phys. 132 (2010) 120901.

[27] E.W. Castner, C.J. Margulis, M. Maroncelli, J.F. Wishart, E.W.C. Jr., Annu. Rev. Phys. Chem. 62 (2011) 85.

[28] T.L. Greaves, C.J. Drummond, Chem. Soc. Rev. 42 (2013) 1096.

[29] T.L. Greaves, C.J. Drummond, Chem. Rev. 115 (2015) 11379.

[30] T.L. Greaves, C.J. Drummond, Chem. Soc. Rev. 37 (2008) 1709.

[31] R. Hayes, G.G. Warr, R. Atkin, Chem. Rev. 115 (2015) 6357.

[32] T. Murphy, R. Atkin, G.G. Warr, Curr. Opin. Colloid Interface Sci. 20 (2015) 282.

[33] J.F. Wishart, E.W. Castner, J. Phys. Chem. B 111 (2007) 4639.

[34] F. Endress, Phys. Chem. Chem. Phys. 12 (2010) 1648.

[35] F. Endres, Phys. Chem. Chem. Phys. 14 (2012) 5008.

[36] R. Ludwig, E. Maginn, S. Balasubramanian, ChemPhysChem 13 (2012) 1603.

[37] P. Wasserscheid, W. Schröer, J. Mol. Liq. 192 (2014) 1.

[38] J. Jacquemin, M. Bendová, J. Solution Chem. 44 (2015) 379.

[39] O. Russina, W. Schroer, A. Triolo, J. Mol. Liq. 210 (2015) 161.

[40] F. Fernandez-Alonso, D.L. Price (Eds.), In: Neutron Scattering – Fundamentals, Experimental Methods in the Physical Sciences, vol. 44, Academic Press, London, 2013.

[41] Y. Wang, G.A. Voth, J. Am. Chem. Soc. 127 (2005) 12192.

[42] Y. Wang, G.A. Voth, J. Phys. Chem. B 110 (2006) 18601.

[43] Y. Wang, W.E.I. Jiang, T. Yan, G.A. Voth, Acc. Chem. Res. 40 (2007) 1193.

[44] Y. Wang, W. Jiang, G.A. Voth, in: J.F. Brennecke, R.D. Rogers, K.R. Seddon (Eds.), Ion. Liq. IV, Not Just Solvents Anymore, ACS Symposium Series, vol. 975, American Chemical Society, Washington, DC, 2007, p. 272.

[45] S.M. Urahata, M.C.C. Ribeiro, J. Chem. Phys. 120 (2004) 1855.

[46] S.M. Urahata, M.C.C. Ribeiro, J. Chem. Phys. 122 (2005).

[47] S.M. Urahata, M.C.C. Ribeiro, J. Chem. Phys. 124 (2006) 74513.

[48] Z. Hu, C.J. Margulis, Acc. Chem. Res. 40 (2007) 1097.

[49] H.V.R. Annapureddy, H.K. Kashyap, P.M. De Biase, C.J. Margulis, J. Phys. Chem. B 114 (2010) 16838.

[50] H.K. Kashyap, J.J. Hettige, H.V.R. Annapureddy, C.J. Margulis, Chem. Commun. (Camb) 48 (2012) 5103.

[51] Z. Hu, C.J. Margulis, Proc. Natl. Acad. Sci. U. S. A. 103 (2006) 831.

[52] J.C. Araque, J.J. Hettige, C.J. Margulis, J. Phys. Chem. B 119 (2015) 12727.

[53] M.F.C. Gomes, J.N.C. Lopes, A.A.H. Padua, Thermodynamics and Micro Heterogeneity of Ionic Liquids, in: Kirchner B. (Ed.), Ionic Liquids, Springer, Berlin, Heidelberg, 2010, pp. 161–183.

[54] J.N. Canongia Lopes, A.A.H. Padua, J. Phys. Chem. B 110 (2006) 3330.

[55] J.N. Canongia Lopes, M.F. Costa Gomes, A.H. Pádua, J. Phys. Chem. B 110 (2006) 16816.

[56] www.europeanspallationsource.se (last accessed 09.01.2017).

[57] C. Červinka, A.A.H. Pádua, M. Fulem, J. Phys. Chem. B 120 (2016) 2362.

[58] I. López-Martin, E. Burello, P.N. Davey, K.R. Seddon, G. Rothenberg, ChemPhysChem 8 (2007) 690.
[59] A.E. Bradley, C. Hardacre, J.D. Holbrey, S. Johnston, S.E.J. McMath, M. Nieuwenhuyzen, Chem. Mater. 14 (2002) 629.
[60] S.V. Dzyuba, R.A. Bartsch, Chem. Phys. Chem. 3 (2002) 161.
[61] C.M. Gordon, J.D. Holbrey, A.R. Kennedy, K.R. Seddon, J. Mater. Chem. 8 (1998) 2627.
[62] K. Shimizu, M. Tariq, M.F.C. Gomes, L.P.N. Rebelo, J.N.C. Lopes, J. Phys. Chem. B 114 (2010) 5831.
[63] E.I. Izgorodina, D. Golze, R. Maganti, V. Armel, M. Taige, T.J.S. Schubert, D.R. MacFarlane, Phys. Chem. Chem. Phys. 16 (2014) 7209.
[64] K. Dong, S. Zhang, D. Wang, X. Yao, J. Phys. Chem. A 110 (2006) 9775.
[65] A. Elaiwi, P.B. Hitchcock, K.R. Seddon, N. Srinivasan, Y.-M. Tan, T. Welton, J.A. Zora, J. Chem. Soc. Dalton Trans. (1995) 3467.
[66] A.G. Avent, P.A. Chaloner, M.P. Day, K.R. Seddon, T. Welton, J. Chem. Soc. Dalton Trans. (1994) 3405.
[67] K. Fumino, A. Wulf, R. Ludwig, Angew. Chem. Int. Ed. Engl. 47 (2008) 8731.
[68] A. Wulf, K. Fumino, R. Ludwig, Angew. Chem. Int. Ed. Engl. 49 (2010) 449.
[69] K. Fumino, T. Peppel, M. Geppert-Rybczyńska, D.H. Zaitsau, J.K. Lehmann, S.P. Verevkin, M. Köckerling, R. Ludwig, Phys. Chem. Chem. Phys. 13 (2011) 14064.
[70] K. Fumino, A. Wulf, R. Ludwig, Phys. Chem. Chem. Phys. 11 (2009) 8790.
[71] K. Dong, S. Zhang, Chem. A Eur. J. 18 (2012) 2748.
[72] J. Thar, M. Brehm, A.P. Seitsonen, B. Kirchner, J. Phys. Chem. B 113 (2009) 15129.
[73] K. Dong, Y. Song, X. Liu, W. Cheng, X. Yao, S. Zhang, J. Phys. Chem. B 116 (2012) 1007.
[74] C. Roth, T. Peppel, K. Fumino, M. Köckerling, R. Ludwig, Angew. Chem. Int. Ed. Engl. 49 (2010) 10221.
[75] M. Kohagen, M. Brehm, Y. Lingscheid, R. Giernoth, J. Sangoro, F. Kremer, S. Naumov, C. Iacob, J. Kärger, R. Valiullin, B. Kirchner, J. Phys. Chem. B 115 (2011) 15280.
[76] E.I. Izgorodina, D.R. MacFarlane, J. Phys. Chem. B 115 (2011) 14659.
[77] P.A. Hunt, C.R. Ashworth, R.P. Matthews, Chem. Soc. Rev. 44 (2015) 1257.
[78] R.P. Matthews, T. Welton, P.A. Hunt, Phys. Chem. Chem. Phys. 16 (2014) 3238.
[79] R.P. Matthews, T. Welton, P.A. Hunt, Phys. Chem. Chem. Phys. 17 (2015) 14437.
[80] T.L. Greaves, C.J. Drummond, Chem. Rev. 108 (2008) 206.
[81] H. Markusson, J.-P. Belières, P. Johansson, C.A. Angell, P. Jacobsson, J. Phys. Chem. A 111 (2007) 8717.
[82] C.A. Angell, N. Byrne, J.-P. Belieres, Acc. Chem. Res. 40 (2007) 1228.
[83] J. Belieres, C. Angell, J. Phys. Chem. B (2007) 4926.
[84] K. Fumino, A. Wulf, R. Ludwig, Angew. Chem. Int. Ed. Engl. 48 (2009) 3184.
[85] J. Hunger, T. Sonnleitner, L. Liu, R. Buchner, M. Bonn, H.J. Bakker, J. Phys. Chem. Lett. 3 (2012) 3034.
[86] K. Fumino, V. Fossog, K. Wittler, R. Hempelmann, R. Ludwig, Angew. Chem. Int. Ed. 52 (2013) 2368.
[87] K. Fumino, P. Stange, V. Fossog, R. Hempelmann, R. Ludwig, Angew. Chem. Int. Ed. Engl. 52 (2013) 12439.
[88] P. Stange, K. Fumino, R. Ludwig, Angew. Chem. Int. Ed. Engl. 52 (2013) 2990.
[89] K. Fumino, V. Fossog, P. Stange, K. Wittler, W. Polet, R. Hempelmann, R. Ludwig, ChemPhysChem 15 (2014) 2604.
[90] R. Hayes, S. Imberti, G.G. Warr, R. Atkin, Angew. Chem. 125 (2013) 4721.

[91] R. Hayes, S. Imberti, G.G. Warr, R. Atkin, Phys. Chem. Chem. Phys. 13 (2011) 3237.
[92] R. Hayes, S. Imberti, G.G. Warr, R. Atkin, J. Phys. Chem. C 118 (2014) 13998.
[93] S. Takahashi, K. Suzuya, S. Kohara, N. Koura, L.A. Curtiss, M.-L. Saboungi, Z. Phys. Chem. 209 (1999) 209.
[94] F.R. Trouw, D.L. Price, Annu. Rev. Phys. Chem. 50 (1999) 571.
[95] C. Hardacre, J.D. Holbrey, S.E.J. McMath, D.T. Bowron, A.K. Soper, J. Chem. Phys. 118 (2003) 273.
[96] A.K. Soper, Chem. Phys. 202 (1996) 295.
[97] A.K. Soper, Chem. Phys. 258 (2000) 121.
[98] http://www.isis.stfc.ac.uk/instruments/sandals/ (last accessed 09.01.2017).
[99] A.K. Soper, Mol. Simul. 38 (2012) 1171.
[100] M. Brehm, B. Kirchner, J. Chem. Inf. Model. 51 (2011) 2007.
[101] O. Hollóczki, M. Macchiagodena, H. Weber, M. Thomas, M. Brehm, A. Stark, O. Russina, A. Triolo, B. Kirchner, ChemPhysChem 16 (2015) 3325.
[102] C. Hardacre, S.E.J. McMath, M. Nieuwenhuyzen, D.T. Bowron, A.K. Soper, J. Phys. Condens. Matter 15 (2003) S159.
[103] M. Deetlefs, C. Hardacre, M. Nieuwenhuyzen, A.H. Padua, O. Sheppard, A.K. Soper, J. Phys. Chem. B 110 (2006) 12055.
[104] A.K. Abdul-Sada, A.M. Greenway, P.B. Hitchcock, T.J. Mohammed, K.R. Seddon, J.A. Zora, J. Chem. Soc. Chem. Commun. (1986) 1753.
[105] A.J. Arduengo, H.V.R. Dias, R.L. Harlow, M. Kline, J. Am. Chem. Soc. 114 (1992) 5530.
[106] D.T. Bowron, C. D'Agostino, L.F. Gladden, C. Hardacre, J.D. Holbrey, M.C. Lagunas, J. McGregor, M.D. Mantle, C.L. Mullan, T.G.A. Youngs, J. Phys. Chem. B 114 (2010) 7760.
[107] C. Hardacre, J.D. Holbrey, C.L. Mullan, M. Nieuwenhuyzen, T.G.A. Youngs, D.T. Bowron, J. Phys. Chem. B 112 (2008) 8049.
[108] S.E. Norman, A.H. Turner, T.G.A. Youngs, RSC Adv. 5 (2015) 67220.
[109] A.P. Abbott, G. Capper, D.L. Davies, H.L. Munro, R.K. Rasheed, V. Tambyrajah, Chem. Commun. (2001) 2010.
[110] A.P. Abbott, G. Capper, D.L. Davies, R.K. Rasheed, V. Tambyrajah, Chem. Commun. (2003) 70.
[111] D. Shah, F.S. Mjalli, Phys. Chem. Chem. Phys. 16 (2014) 23900.
[112] A.P. Abbott, D. Boothby, G. Capper, D.L. Davies, R.K. Rasheed, J. Am. Chem. Soc. 126 (2004) 9142.
[113] E.L. Smith, A.P. Abbott, K.S. Ryder, Chem. Rev. 114 (2014) 11060.
[114] C. D'Agostino, R.C. Harris, A.P. Abbott, L.F. Gladden, M.D. Mantle, Phys. Chem. Chem. Phys. 13 (2011) 21383.
[115] O.S. Hammond, D.T. Bowron, K.J. Edler, Green Chem. 18 (2016) 2736.
[116] S.L. Perkins, P. Painter, C.M. Colina, J. Phys. Chem. B 117 (2013) 10250.
[117] G. García, M. Atilhan, S. Aparicio, Chem. Phys. Lett. 634 (2015) 151.
[118] H. Sun, Y. Li, X. Wu, G. Li, J. Mol. Model. 19 (2013) 2433.
[119] C.A. Angell, Y. Ansari, Z. Zhao, Faraday Discuss. 154 (2012) 9.
[120] C. Zhang, A. Yamazaki, J. Murai, J. Park, T. Mandai, K. Ueno, K. Dokko, M. Watanabe, J. Phys. Chem. C 118 (2014) 17362.
[121] T. Murphy, S.K. Callear, N. Yepuri, K. Shimizu, M. Watanabe, J.N. Canongia Lopes, T. Darwish, G.G. Warr, R. Atkin, Phys. Chem. Chem. Phys. 18 (2016) 17224.
[122] A. Triolo, O. Russina, H.-J. Bleif, E. Di Cola, J. Phys. Chem. B 111 (2007) 4641.
[123] A. Triolo, O. Russina, B. Fazio, R. Triolo, E. Di Cola, Chem. Phys. Lett. 457 (2008) 362.

[124] O. Russina, A. Triolo, L. Gontrani, R. Caminiti, D. Xiao, L.G. Hines Jr., R.A. Bartsch, E.L. Quitevis, N.V. Plechkova, K.R. Seddon, J. Phys. Condens. Matter 21 (2009) 424121.
[125] C.S. Santos, N.S. Murthy, G.A. Baker, E.W. Castner, J. Chem. Phys. 134 (2011) 1211011.
[126] A. Triolo, O. Russina, B. Fazio, G.B. Appetecchi, M. Carewska, S. Passerini, J. Chem. Phys. 130 (2009) 1645211.
[127] L. Gontrani, O. Russina, F. Lo Celso, R. Caminiti, G. Annat, A. Triolo, J. Phys. Chem. B 113 (2009) 9235.
[128] T.L. Greaves, D.F. Kennedy, S.T. Mudie, C.J. Drummond, J. Phys. Chem. B 114 (2010) 10022.
[129] X. Song, H. Hamano, B. Minofar, R. Kanzaki, K. Fujii, Y. Kameda, S. Kohara, M. Watanabe, S.-I. Ishiguro, Y. Umebayashi, J. Phys. Chem. B 116 (2012) 2801.
[130] R. Atkin, G.G. Warr, J. Phys. Chem. B 112 (2008) 4164.
[131] O. Russina, R. Caminiti, A. Triolo, S. Rajamani, B. Melai, A. Bertoli, C. Chiappe, J. Mol. Liq. 187 (2013) 252.
[132] C. Hardacre, J.D. Holbrey, C.L. Mullan, T.G.A. Youngs, D.T. Bowron, J. Chem. Phys. 133 (2010) 74510.
[133] M. Macchiagodena, L. Gontrani, F. Ramondo, A. Triolo, R. Caminiti, J. Chem. Phys. 134 (2011) 114521.
[134] O. Russina, A. Triolo, Faraday Discuss. 154 (2012) 97.
[135] O. Russina, A. Triolo, L. Gontrani, R. Caminiti, J. Phys. Chem. Lett. 3 (2012) 27.
[136] A. Triolo, O. Russina, R. Caminiti, H. Shirota, H.Y. Lee, C.S. Santos, N.S. Murthy, E.W. Castner, Chem. Commun. (Camb) 48 (2012) 4959.
[137] K. Shimizu, C.E.S. Bernardes, A. Triolo, J.N. Canongia Lopes, Phys. Chem. Chem. Phys. 15 (2013) 16256.
[138] H.K. Kashyap, C.S. Santos, R.P. Daly, J.J. Hettige, N.S. Murthy, H. Shirota, E.W. Castner, C.J. Margulis, J. Phys. Chem. B 117 (2013) 1130.
[139] O. Russina, F. Lo Celso, M. Di Michiel, S. Passerini, G.B. Appetecchi, F. Castiglione, A. Mele, R. Caminiti, A. Triolo, Faraday Discuss. 167 (2013) 499.
[140] A.B. Pereiro, M.J. Pastoriza-Gallego, K. Shimizu, I.M. Marrucho, J.N.C. Lopes, M.M. Piñeiro, L.P.N. Rebelo, J. Phys. Chem. B 117 (2013) 10826.
[141] Y. Shen, D.F. Kennedy, T.L. Greaves, A. Weerawardena, R.J. Mulder, N. Kirby, G. Song, C.J. Drummond, Phys. Chem. Chem. Phys. 14 (2012) 7981.
[142] J.J. Hettige, J.C. Araque, C.J. Margulis, J. Phys. Chem. B 118 (2014) 12706.
[143] L. Gontrani, E. Bodo, A. Triolo, F. Leonelli, P. D'Angelo, V. Migliorati, R. Caminiti, J. Phys. Chem. B 116 (2012) 13024.
[144] R. Hayes, S. Imberti, G.G. Warr, R. Atkin, Phys. Chem. Chem. Phys. 13 (2011) 13544.
[145] J.D. Holbrey, W.M. Reichert, M. Nieuwenhuyzen, O. Sheppard, C. Hardacre, R.D. Rogers, Chem. Commun. (Camb) (2003) 476.
[146] M. Deetlefs, C. Hardacre, M. Nieuwenhuyzen, O. Sheppard, A.K. Soper, J. Phys. Chem. B 109 (2005) 1593.
[147] C. Hardacre, J.D. Holbrey, C.L. Mullan, M. Nieuwenhuyzen, T.G.A. Youngs, D.T. Bowron, S.J. Teat, Phys. Chem. Chem. Phys. 12 (2010) 1842.
[148] A.H. Turner, J.D. Holbrey, J. Solution Chem. 44 (2015) 621.
[149] H. Wang, G. Gurau, R.D. Rogers, Chem. Soc. Rev. 41 (2012) 1519.
[150] R.C. Remsing, R.P. Swatloski, R.D. Rogers, G. Moyna, Chem. Commun. (2006) 1271.
[151] S. Zhu, Y. Wu, Q. Chen, Z. Yu, C. Wang, S. Jin, Y. Ding, G. Wu, Green Chem. 8 (2006) 325.
[152] T.G.A. Youngs, C. Hardacre, J.D. Holbrey, J. Phys. Chem. B 111 (2007) 13765.

[153] T.G.A. Youngs, J.D. Holbrey, C.L. Mullan, S.E. Norman, M.C. Lagunas, C. D'Agostino, M.D. Mantle, L.F. Gladden, D.T. Bowron, C. Hardacre, Chem. Sci. 2 (2011) 1594.
[154] R. Hayes, S. Imberti, G.G. Warr, R. Atkin, Angew. Chem. Int. Ed. Engl. 51 (2012) 7468.
[155] O. Russina, M. Macchiagodena, B. Kirchner, A. Mariani, B. Aoun, M. Russina, R. Caminiti, A. Triolo, J. Non-Cryst. Solids 407 (2015) 333.
[156] D.H. Rasmussen, A.P. MacKenzie, Nature 220 (1968) 1315.
[157] A.K. Soper, A. Luzar, J. Chem. Phys. 97 (1992) 1320.
[158] S.E. McLain, A.K. Soper, A. Luzar, J. Chem. Phys. 127 (2007) 174515.
[159] P.P. Wiewiór, H. Shirota, E.W. Castner, J. Chem. Phys. 116 (2002) 4643.
[160] R. Hayes, S.A. Bernard, S. Imberti, G.G. Warr, R. Atkin, J. Phys. Chem. C 118 (2014) 21215.
[161] O. Russina, R. Caminiti, T. Méndez-Morales, J. Carrete, O. Cabeza, L.J. Gallego, L.M. Varela, A. Triolo, J. Mol. Liq. 205 (2015) 16.
[162] T. Méndez-Morales, J. Carrete, J.R. Rodríguez, Ó. Cabeza, L.J. Gallego, O. Russina, L.M. Varela, Phys. Chem. Chem. Phys. 17 (2015) 5298.
[163] L.M. Varela, T. Méndez-Morales, J. Carrete, V. Gómez-González, B. Docampo-Álvarez, L.J. Gallego, O. Cabeza, O. Russina, J. Mol. Liq. 210 (2015) 178.
[164] T. Méndez-Morales, J. Carrete, Ó. Cabeza, O. Russina, A. Triolo, L.J. Gallego, L.M. Varela, J. Phys. Chem. B 118 (2014) 761.
[165] T. Murphy, S.K. Callear, G.G. Warr, R. Atkin, Phys. Chem. Chem. Phys. 18 (2016) 12797.
[166] T. Murphy, R. Hayes, S. Imberti, G.G. Warr, R. Atkin, Phys. Chem. Chem. Phys. 16 (2014) 13182.
[167] T.L. Greaves, D.F. Kennedy, N. Kirby, C.J. Drummond, Phys. Chem. Chem. Phys. 13 (2011) 13501.
[168] S. Dixit, J. Crain, W.C.K. Poon, J.L. Finney, A.K. Soper, Nature 416 (2002) 829.
[169] T. Morita, K. Nishikawa, Chem. Phys. Lett. 389 (2004) 29.
[170] K. Nishikawa, T. Iijima, J. Phys. Chem. 97 (1993) 10824.
[171] H. Weingaertner, T. Merkel, S. Kashammer, W. Schroer, S. Wiegand, Berichte Der Bunsen-Gesellschaft - Phys, Chem. Chem. Phys. 97 (1993) 970.
[172] W. Schröer, S. Wiegand, H. Weingaertner, Ber. Bunsen. Ges. Phys. Chem. 97 (1993) 975.
[173] T. Heimburg, S.Z. Mirzaev, U. Kaatze, Phys. Rev. E 62 (2000) 4963.
[174] A. Oleinikova, M. Bonetti, J. Chem. Phys. 104 (1996) 3111.
[175] M. Bonetti, P. Calmettes, Int. J. Thermophys. 19 (1998) 1555.
[176] M.E. Fisher, J.S. Langer, Phys. Rev. Lett. 20 (1968) 665.
[177] M. Bonetti, C. Bagnuls, C. Bervillier, J. Chem. Phys. 107 (1997) 550.
[178] M. Bonetti, A. Oleinikova, C. Bervillier, J. Phys. Chem. B 101 (1997) 2164.
[179] H.J. Jiang, P.A. FitzGerald, A. Dolan, R. Atkin, G.G. Warr, J. Phys. Chem. B 118 (2014) 9983.
[180] M.L. Klossek, D. Touraud, T. Zemb, W. Kunz, ChemPhysChem 13 (2012) 4116.
[181] O. Diat, M.L. Klossek, D. Touraud, B. Deme, I. Grillo, W. Kunz, T. Zemb, J. Appl. Cryst. 46 (2013) 1665.
[182] S. Schöttl, D. Touraud, W. Kunz, T. Zemb, D. Horinek, Colloids Surf. A Physicochem. Eng. Asp. 480 (2015) 222.
[183] R.F. Hankel, P.E. Rojas, M. Cano-Sarabia, S. Sala, J. Veciana, A. Braeuer, N. Ventosa, Chem. Commun. (Camb) 50 (2014) 8215.
[184] O. Russina, A. Sferrazza, R. Caminiti, A. Triolo, J. Phys. Chem. Lett. 5 (2014) 1738.
[185] T. Murphy, R. Hayes, S. Imberti, G.G. Warr, R. Atkin, Phys. Chem. Chem. Phys. 18 (2016) 12797.

[186] B. Docampo-Álvarez, V. Gómez-González, T. Méndez-Morales, J. Carrete, J.R. Rodríguez, Ó. Cabeza, L.J. Gallego, L.M. Varela, J. Chem. Phys. 140 (2014) 2145021.
[187] L. Almásy, M. Turmine, A. Perera, J. Phys. Chem. B 112 (2008) 2382.
[188] J.G. Kirkwood, F.P. Buff, J. Chem. Phys. 19 (1951) 774.
[189] J. Bowers, C.P. Butts, P.J. Martin, M.C. Vergara-Gutierrez, R.K. Heenan, Langmuir 20 (2004) 2191.
[190] R.L.M. Blesic, M.H. Marques, N.V. Plechkova, K.R. Seddon, A. Lopes, Green Chem. 9 (2007) 481.
[191] T. Kusano, K. Fujii, M. Tabata, M. Shibayama, J. Solution Chem. 42 (2013) 1888.
[192] J. Gao, N.J. Wagner, Langmuir 32 (2016) 5078.
[193] T. Shimomura, K. Fujii, T. Takamuku, Phys. Chem. Chem. Phys. 12 (2010) 12316.
[194] T. Takamuku, T. Shimomura, K. Sadakane, M. Koga, H. Seto, Phys. Chem. Chem. Phys. 14 (2012) 11070.
[195] N. Winterton, J. Mater. Chem. 16 (2006) 4281.
[196] T. Ueki, M. Watanabe, Bull. Chem. Soc. Jpn. 85 (2012) 33.
[197] R. Tsuda, K. Kodama, T. Ueki, H. Kokubo, S. Imabayashi, M. Watanabe, Chem. Commun. (2008) 4939.
[198] H.-N. Lee, T.P. Lodge, J. Phys. Chem. Lett. 1 (2010) 1962.
[199] K. Kodama, R. Tsuda, K. Niitsuma, T. Tamura, T. Ueki, H. Kokubo, M. Watanabe, Polym. J. 43 (2011) 242.
[200] H.-N. Lee, N. Newell, Z. Bai, T.P. Lodge, Macromolecules 45 (2012) 3627.
[201] A. Triolo, O. Russina, U. Keiderling, J. Kohlbrecher, J. Phys. Chem. B 110 (2006) 1513.
[202] P. Debye, J. Phys. Colloid Chem. 51 (1947) 18.
[203] B. Hammouda, Adv. Polym. Sci. 106 (1993) 87.
[204] O. Werzer, G.G. Warr, R. Atkin, J. Phys. Chem. B 115 (2011) 648.
[205] H.E. Stanley, Introduction to Phase Transitions and Critical Phenomena, Oxford University Press, New York, 1971.
[206] M.L. Hoarfrost, Y. He, T.P. Lodge, Macromolecules 46 (2013) 9464.
[207] K. Fujii, T. Ueki, K. Niitsuma, T. Matsunaga, M. Watanabe, M. Shibayama, Polymer (Guildf) 52 (2011) 1589.
[208] K. Hirosawa, K. Fujii, T. Ueki, Y. Kitazawa, K.C. Littrell, M. Watanabe, M. Shibayama, Phys. Chem. Chem. Phys. 18 (2016) 17881.
[209] H. Benoit, M. Benmouna, Polymer (Guildf) 25 (1984) 1059.
[210] H.G. Schild, Prog. Polym. Sci. 17 (1992) 163.
[211] H. Asai, K. Fujii, T. Ueki, S. Sawamura, Y. Nakamura, Y. Kitazawa, M. Watanabe, Y.-S. Han, T.-H. Kim, M. Shibayama, Macromolecules 46 (2013) 1101.
[212] Z. Chen, P.A. FitzGerald, G.G. Warr, R. Atkin, Phys. Chem. Chem. Phys. 17 (2015) 14872.
[213] Z. Chen, S. McDonald, P.A. Fitzgerald, G.G. Warr, R. Atkin, Phys. Chem. Chem. Phys. 18 (2016) 14894.
[214] M. Benmouna, B. Hammouda, Prog. Polym. Sci. 22 (1997) 49.
[215] B. Hammouda, J. Appl. Cryst. 43 (2010) 716.
[216] L. Sapir, C.B. Stanley, D. Harries, J. Phys. Chem. A 120 (2016) 3253.
[217] D.F. Evans, A. Yamauchi, R. Roman, E.Z. Casassa, J. Colloid Interface Sci. 88 (1982) 89.
[218] T.L. Greaves, A. Weerawardena, C. Fong, C.J. Drummond, Langmuir 23 (2007) 402.
[219] T.L. Greaves, A. Weerawardena, C. Fong, C.J. Drummond, J. Phys. Chem. B 111 (2007) 4082.
[220] R. Atkin, L.-M. De Fina, U. Kiederling, G.G. Warr, J. Phys. Chem. B 113 (2009) 12201.
[221] R. Atkin, S.M.C. Bobillier, G.G. Warr, J. Phys. Chem. B 114 (2010) 1350.

[222] Z. Chen, P.A. Fitzgerald, Y. Kobayashi, K. Ueno, M. Watanabe, G.G. Warr, R. Atkin, Macromolecules 48 (2015) 1843.
[223] C. Patrascu, F. Gauffre, F. Nallet, R. Bordes, J. Oberdisse, N. de Lauth-Viguerie, C. Mingotaud, ChemPhysChem 7 (2006) 99.
[224] M.U. Araos, G.G. Warr, Langmuir 24 (2008) 9354.
[225] A. Dolan, R. Atkin, G.G. Warr, Chem. Sci. 6 (2015) 6189.
[226] C.R. López-Barrón, N.J. Wagner, Langmuir 28 (2012) 12722.
[227] R. Pindak, D.E. Moncton, S.C. Davey, J.W. Goodby, Phys. Rev. Lett. 46 (1981) 1135.
[228] C.R. López-Barrón, D. Li, L. DeRita, M.G. Basavaraj, N.J. Wagner, J. Am. Chem. Soc. 134 (2012) 20728.
[229] C.R. López-Barrón, D. Li, N.J. Wagner, J.L. Caplan, Macromolecules 47 (2014) 7484.
[230] A. Triolo, O. Russina, V. Arrighi, F. Juranyi, S. Janssen, C.M. Gordon, J. Chem. Phys. 119 (2003) 8549.
[231] https://www.psi.ch/sinq/focus/ (last accessed 09.01.2017).
[232] https://www.helmholtz-berlin.de/zentrum/zukunft/neat_en.html (last accessed 09.01.2017).
[233] A. Triolo, O. Russina, C. Hardacre, M. Nieuwenhuyzen, M.A. Gonzalez, H. Grimm, J. Phys. Chem. B 109 (2005) 22061.
[234] B. Frick, L.J. Fetters, Macromolecules 27 (1994) 974.
[235] https://www.ill.eu/instruments-support/instruments-groups/instruments/in16b/description/instrument-layout/ (last accessed 09.01.2017).
[236] B. Frick, J. Combet, L. van Eijck, Nucl. Instrum. Methods Phys. Res. Sect. A Accel. Spectrom. Detect. Assoc. Equip, 669 2012, p. 7.
[237] https://neutrons.ornl.gov/basis (last accessed 29.12.2016).
[238] H.H. Grapengeter, B. Alefeld, R. Kosfeld, Colloid Polym. Sci. 265 (1987) 226.
[239] O. Russina, A. Triolo, Y. Aihara, M.T.F. Telling, H. Grimm, Macromolecules 37 (2004) 8653.
[240] B. Aoun, M.A. González, J. Ollivier, M. Russina, Z. Izaola, D.L. Price, M.-L. Saboungi, J. Phys. Chem. Lett. 1 (2010) 2503.
[241] T. Burankova, E. Reichert, V. Fossog, R. Hempelmann, J. Peter, J. Mol. Liq. 192 (2014) 199.
[242] E. Mamontov, H. Luo, S. Dai, J. Phys. Chem. B 113 (2009) 159.
[243] Y. Inamura, O. Yamamuro, S. Hayashi, H.-O. Hamaguchi, Phys. B 386 (2006) 385.
[244] E. Mamontov, G.A. Baker, H. Luo, S. Dai, ChemPhysChem 12 (2011) 944.
[245] J.P. Embs, T. Burankova, E. Reichert, R. Hempelmann, J. Phys. Chem. B 116 (2012) 13265.
[246] S.M. Chathoth, E. Mamontov, S. Dai, X. Wang, P.F. Fulvio, D.J. Wesolowski, Europhys. Lett. 97 (2012) 66004.
[247] S.M. Chathoth, E. Mamontov, P.F. Fulvio, X. Wang, G.A. Baker, S. Dai, D.J. Wesolowski, Europhys. Lett. 102 (2013) 16004.
[248] T. Burankova, R. Hempelmann, V. Fossog, J. Ollivier, T. Seydel, J.P. Embs, J. Phys. Chem. B 119 (2015) 10643.
[249] D.V. Wagle, G.A. Baker, E. Mamontov, J. Phys. Chem. Lett. 6 (2015) 2924.
[250] G. Williams, D.C. Watts, Trans. Faraday Soc. 66 (1970) 80.
[251] M. Kofu, M. Tyagi, Y. Inamura, K. Miyazaki, O. Yamamuro, J. Chem. Phys. 143 (2015) 234502.
[252] O. Russina, M. Beiner, C. Pappas, M. Russina, V. Arrighi, T. Unruh, C.L. Mullan, C. Hardacre, A. Triolo, J. Phys. Chem. B 113 (2009) 8469.
[253] J.L. Bañuelos, G. Feng, P.F. Fulvio, S. Li, G. Rother, N. Arend, A. Faraone, S. Dai, P.T. Cummings, D.J. Wesolowski, Carbon 78 (2014) 415.

[254] K.L. Van Aken, J.K. McDonough, S. Li, G. Feng, S.M. Chathoth, E. Mamontov, P.F. Fulvio, P.T. Cummings, S. Dai, Y. Gogotsi, J. Phys. Condens. Matter 26 (2014) 284104.
[255] T. Burankova, R. Hempelmann, A. Wildes, J.P. Embs, J. Phys. Chem. B 118 (2014) 14452.
[256] D. Richter, J.B. Hayter, F. Mezei, B. Ewen, Phys. Rev. Lett. 41 (1978) 1484.
[257] F. Mezei, W. Knaak, B. Farago, Phys. Rev. Lett. 58 (1987) 571.
[258] O. Yamamuro, T. Yamada, M. Kofu, M. Nakakoshi, M. Nagao, J. Chem. Phys. 135 (2011).
[259] M. Kofu, M. Nagao, T. Ueki, Y. Kitazawa, Y. Nakamura, S. Sawamura, M. Watanabe, O. Yamamuro, J. Phys. Chem. B 117 (2013) 2773.
[260] B. Frick, D. Richter, Science 267 (1995) 1939.
[261] G. Chatel, J.F.B. Pereira, V. Debbeti, H. Wang, R.D. Rogers, Green Chem. 16 (2014) 2051.
[262] H. Niedermeyer, J.P. Hallett, I.J. Villar-Garcia, P.A. Hunt, T. Welton, Chem. Soc. Rev. 41 (2012) 7780.

Chapter 5

Catalysis

Peter W. Albers*[,1], David Lennon† and Stewart F. Parker‡
*Evonik Technology & Infrastructure GmbH, Hanau (Wolfgang), Germany; †School of Chemistry, University of Glasgow, Glasgow, United Kingdom; ‡ISIS Facility, Rutherford Appleton Laboratory, Didcot, United Kingdom
[1]Corresponding author: e-mail: peter.albers@evonik.com

Chapter Outline

Experimental Methods in the Physical Sciences, Vol. 49. http://dx.doi.org/10.1016/B978-0-12-805324-9.00005-4

5.1 INTRODUCTION—WHY NEUTRONS FOR CATALYSIS?

Other chapters in this book series have described the basic theory of neutron scattering [1] and detailed the reasons why neutron scattering is an invaluable tool for the investigation of condensed phase systems [2,3]. This chapter will describe some of the applications of neutrons in catalytic science. The focus will be on heterogeneous catalysis because this is the area where almost all of the work has been carried out. Homogeneous catalysis studies involving neutrons are very rare but offer opportunities that will be described in Section 5.6.3. Applications of QENS to catalysts are largely the study of diffusion of molecules in microporous materials such as zeolites, metal-organic frameworks (MOF), and related structures. This is the subject of a dedicated chapter elsewhere in this book. Thus, this chapter will focus mainly on applications of oxides and metals, and in particular, supported metal catalysts.

Perhaps the first question to be asked is why neutrons are useful to studies of catalysts. The answers are not immediately obvious; after all neutrons generally probe the bulk of the material, in contrast, heterogeneous catalysis invariably occurs at a surface; catalytic processes are generally run at or above room temperature and some as high as 1200 K, whereas neutron vibrational spectroscopy (inelastic neutron scattering-INS), which is the most commonly used technique, is generally carried out below 30 K; neutron scattering is an inherently insensitive technique, and surface species are often present in submonolayer concentrations.

All of these difficulties are very real. The triumph of neutron scattering is that it has been able to either overcome them or turn them to advantage. The need for surface sensitivity means that most experiments are difference measurements; a reference of the clean surface is recorded; the surface is then modified by addition of a material and a second dataset acquired; the difference between the two shows how the surface has been modified. This places stringent requirements on instrument stability and reproducibility of sample positioning; fortunately, modern INS instruments have sufficient stability and reproducible sample positioning can be designed into the experimental procedure. The need to measure at low temperature for INS studies means that a "react and quench" approach (e.g., [4,5]) is generally used to provide snapshots along the reaction coordinate. Current work suggests that with a careful choice of system and spectrometer, measurements at up to ~500 K may be feasible. We note that low temperature is only a requirement for INS studies; neutron diffraction and quasielastic neutron scattering (QENS) are routinely carried out well above room temperature. The problem with insensitivity is overcome by using large samples. For new catalysts, providing the 10–50 g needed can be challenging; for industrial catalysts, this is not a problem. The use of large samples means that concerns about reproducible sampling are greatly ameliorated, and this is regarded by industry as a positive factor in neutron scattering. One major advantage of neutron scattering is that there are a wide range of materials (including aluminum, steel, the null scattering alloy TiZr, quartz, and sapphire) that can be used for cell construction.

This area has been reviewed several times over the years [6–12]; consequently, we will largely restrict the examples to work in the last 10 years or so. The major change that has occurred in this period for INS studies of catalysts is the use of direct geometry spectrometers. Over the last 20 years, their use at spallation sources has revolutionized the study of magnetism [3]. However, their use in molecular spectroscopy and catalysis has been much slower. Partly, this was due to the scarcity of instruments, but it was also partly due to a lack of understanding by the chemistry community as to how the instruments could benefit their science. As described in more detail elsewhere [13–15], there are three major advantages of direct geometry instruments:

(1) Sensitivity
(2) Access to small momentum transfer (i.e., low Q)
(3) Momentum transfer resolved spectra

The sensitivity arises from the very large detector areas that are typical of modern direct geometry instruments; thus, MAPS at ISIS has $\sim 16\,m^2$ of detector area; in contrast, the indirect geometry spectrometer TOSCA has $\sim 0.4\,m^2$. The improved sensitivity is crucial; the investigation of methane reforming catalysts or of hydrogenation catalysts at realistic precious metal concentrations and gas pressures would not have been possible without the use of MAPS. This factor is becoming less important with the current generation of indirect geometry instruments; VISION at SNS is on a more powerful source, has a high m-number guide and a larger detector area, Lagrange at ILL has a much increased collection area over its predecessor IN1BeF and a guide (similar to that of VISION) will be installed on TOSCA at ISIS in 2016, resulting in an average tenfold increase of incident flux.

The second factor is the key difference between direct and indirect geometry spectrometers; the former can access a much larger range in Q and crucially can always access smaller Q. This is shown in Figure 1A that shows the (Q,ω) plot for iodomethane (methyl iodide-CH_3I) recorded with incident energies (E_i) of 4840 and 2017 cm^{-1} [14]. ($1\,cm^{-1} = 0.124$ meV). The solid black lines show the trajectories of the forward and backscattering banks on TOSCA (note that all low final energy indirect spectrometers, for example, VISION and Lagrange, follow trajectories that lie either on or between those shown). It can be seen that large energy transfer involves large momentum transfer. This is critical because the intensity, S, of an INS transition is given by [1]:

$$S(Q,n\omega_i) \propto \frac{(QU_i)^{2n}}{n!} \exp\left(-(QU_{Tot})^2\right)\sigma$$

where ω_i is the ith mode at transition energy ω, $n=1$ for a fundamental, 2 for a first overtone or binary combination, 3 for a second overtone or ternary

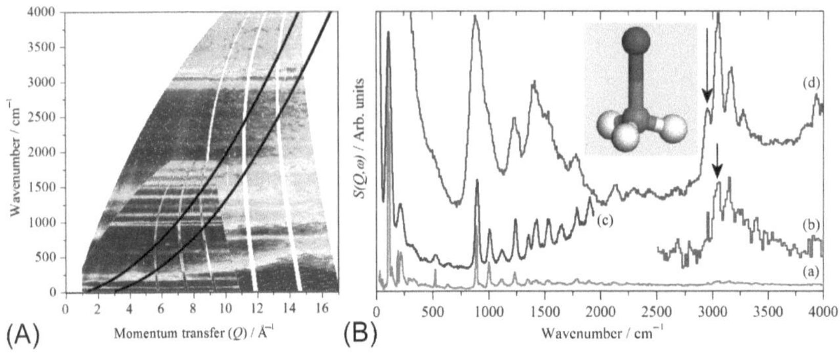

FIGURE 1 (A) (Q,ω) plot of iodomethane recorded on MAPS (ISIS) at 20 K with incident energies of 4840 and 2017 cm^{-1}. The solid lines are the trajectories of an indirect geometry instrument with a final energy of ~40 cm^{-1} for scattering angles of 45° (upper trace) and 135° (lower trace). (B) INS spectra of iodomethane recorded at 20 K on (a) TOSCA (ISIS), (b) TOSCA × 10 ordinate expansion of the 2500–4000 cm^{-1} region, and (c) and (d) MAPS with incident energies of 4840 and 2017 cm^{-1}, respectively. (c) and (d) are the data shown in (A) summed for the detectors at $4 \leq Q \leq 10$ Å^{-1}. *Reprinted with permission from Ref. [14].*

combination *etc*., and Q is the momentum transfer. Higher-order transitions, that is, $n \geq 2$, depend on Q^{2n}, and these become significant at large Q and, in particular, phonon wings (combinations between internal modes and the external density of states) become very strong. In parallel, there is an overall dampening of intensity caused by the Debye-Waller factor. The combination of these two processes means that the fundamental transitions are largely swamped by the nearby higher-order processes, rendering them unobservable. With a direct geometry spectrometer, by only using the detectors in forward scattering, that is, at low Q, high-energy transitions are readily observable. This is illustrated in Figure 1B, where the data for $4 \leq Q \leq 10$ Å^{-1} obtained from Figure 1A is compared with the spectrum of iodomethane recorded on TOSCA. From the TOSCA spectrum (Figure 1Ba), it would be assumed that the arrowed peak at 3043 cm^{-1} was the unresolved C–H stretch modes. Inspection of the MAPS spectrum (Figure 1Bd) has considerably better signal-to-noise ratio and shows a peak at 2948 cm^{-1} (arrowed) that is not apparent in the TOSCA data. In fact, this peak is the unresolved C–H stretch modes; the peaks at 3043, 3161, and 3272 cm^{-1} are combinations of the C–H stretch and the intense librational mode at 112 cm^{-1} (CH+lib, CH+2 × lib, and CH+3 × lib, respectively).

The third factor is less useful, because as can be seen from Figure 1, the Q information is lost as the data are usually coadded across a range in Q to improve the signal-to-noise ratio. Nonetheless, there are occasions when the information is useful: to distinguish fundamentals from higher-order transitions [16,17] and to distinguish phonon and magnetic process (see Section 5.5.2).

5.2 EXPERIMENTAL

5.2.1 Design of Experiments

For academic and industrial research on catalysts and catalytic material, neutron scattering offers a unique way to probe the proton dynamics over a broad energy range. To obtain the optimum informational output, sufficient sample quantities of adequate proton content are needed together with the choice of the most suitable direct or indirect geometry spectrometers for the individual project [13,18]. Relevant spectral regions and experimental conditions, equipment, and operational procedures available on-site should be identified and discussed in advance together with the instrument scientists and regarding the beam schedule of the facility.

Complementary analytic data about the individual catalyst's composition, morphology, and structure at the macro-, micro- and nanoscale; purity, porosity, oxidation states of the elements; and surface area [19,20]—as far as possible and available—are essential support for a successful experiment on complex technical catalysts and to supply or ease the data evaluation in advance.

Spectroscopic studies of real catalysts from a technical reaction system are complicated by the requirements that have to be fulfilled by a neutron experiment. For example, there is the need for high sensitivity and high surface specificity of the method so as to analyze monolayer or submonolayer coverage of reactants, intermediates, products, and other adsorbed species such as solvent molecules, contaminants, oligomers or polymers, cokes, and poisons, adherent to the catalyst surface. Furthermore, it is highly desirable to monitor physico-/chemical properties in the topmost atomic layers under varying catalyst conditions including hydrogenation/dehydrogenation, oxidation, activation, aging, and poisoning and to measure the catalyst under realistic gas pressures and at realistic element compositions. In certain cases, it is possible to take advantage of the benefits, and the synergies, with existing surface science work on single-crystal surfaces and well-defined reference materials and to utilize and translate this information to the characterization of complex commercial catalysts.

To obtain surface sensitivity, a high surface-to-bulk ratio is needed together with a hydrogenous adsorbate to provide contrast. A catalyst sample has to be removed from its source whether this is a production laboratory, an industrial plant, a medium-scale reactor, a pilot plant, or a bypass pipe of an on-stream catalyst testing device. A literally *in situ* study in a chemical process during synthesis is not possible. Material has to be sealed and transported to a high-neutron flux facility and has to be measured in the cold state, after quenching or shock freezing. Therefore, the properties that are monitored in an INS experiment cannot be regarded as rapid online research in a plant under in-process conditions but—depending on the kind of catalyst and its handling, on safety measures, and, correspondingly, on the dedicated layout of the sample can or reaction cell (Section 5.2.3), the sample environment requirements [21,22], and the quenching technique—the observations can directly reflect the operating conditions in the real catalyst application. Furthermore, *in situ* equipment can be designed to be used at the large-scale neutron facility to allow for a simulation of catalyst operation conditions followed by the immediate neutron experiment afterward or in alternating subsequent catalytic cycles.

As a first step of catalyst characterization with neutrons, usually a feasibility check and not a full "operando" program is needed. A "snapshot" of the catalyst's properties frozen in a certain condition or reaction step can be sufficient to supply essential information of the proton dynamics, which often already provides completely new and unexpected information and insights.

The quantity of catalyst needed to provide optimum conditions has to be worked out in advance with respect to the individual hydrogen concentration in order to avoid distinct multiple scattering contributions to the spectrum on the one hand as well as keeping far below the detection limit for the relevant molecular or atomic species on the other. For catalysts such as Raney® nickel or cobalt or for fuel cell catalysts of enhanced precious metal loading (20–60 wt%), smaller sample volumes are needed as for fully impregnated or core/shell grades of, for example, supported precious metal catalysts at concentrations of 10, 5, 2, or $<1\%$, which are of common use in chemical technology.

An essential prerequisite for an experiment is the knowledge about the amount of water or strongly adherent traces of residual humidity or solvent in the usually porous, high-surface-area catalytic materials with respect to potentially excessive background contributions.

5.2.2 Sampling

The advantage of being able to probe bulk quantities of catalysts in a single-neutron experiment is highly attractive and convincing from the chemical engineering point of view. For powder-type supported catalysts with a narrow precious metal size distribution and often homogeneous loading, it is usually not a problem to obtain representative samples [23]. The efforts in sample selection and separation may be enhanced in the case of used catalyst pellets from large-scale chemical reaction systems with catalyst inventory up to the several tens of tonnes scale and additional contaminants, corrosion products, or coke deposits. This requires dedicated strategies of sampling and close cooperation with the plant crew and sampling experts to generate laboratory quantities of about 0.5–50 g from primary samples of 10^2–10^4 kg including sieving, fractioning, selecting, or homogenizing—if needed—[24,25].

However, neutron scattering often allows obtaining results much more straightforwardly and representative than is possible with the more usual analytic techniques.

5.2.3 Cell Design

The high penetration of neutrons in matter allows the direct, nondestructive characterization of macroscopic amounts of pelletized fixed-bed catalysts and supported nanoparticles and related materials such as carbonaceous and oxide supports, cokes and corrosion products, and organic molecules and solvent mixtures for referencing and background subtraction. These can be supported, loaded, or conditioned in dedicated *in situ* cells or flow-through reactors. Sample cells can be manufactured from aluminum, Zircaloy, chemical reactor steels, quartz, or other materials. Sample cans or reaction vessels with a wall thickness and sealing that are sufficient to allow for safe handling of an activated, air-sensitive or pyrophoric, hydrogenated catalyst and to pressurize and heat the catalyst to realistic operational level can be penetrated by the neutron beams. Thus, unperturbed analysis of pyrophoric materials such as Raney® nickel, finely divided precious metal blacks, hydrogenated fuel cell catalysts containing nanometer-sized alloyed particles, or catalysts in contact with flammable residual solvents is possible. Catalysts and materials of catalytic relevance may be taken directly from the technical application, sealed, for example, under argon and measured in the original state or after further treatment [11].

For working out the optimum cell design, the inner diameter of the available cryostats, the size and shape of the neutron beam at a given spectrometer, and the geometry between incident beam, catalyst sample, and neutron detector banks have to be considered in advance.

Some typical sample cans are depicted in Figure 2. The choice of sample size, cuvette layout, wall thickness, and material often has to be a compromise between matching the detection limits for a given experimental task and catalyst material, beam time available, safety issues in reducing wall thickness, and potential loss of spectral quality due to multiple scattering and corresponding undesirable background contributions to the spectra.

In a typical INS experiment, a catalyst is measured in a sealed reaction vessel or bypass can under vacuum, ambient conditions (as taken from the original source or from the chemical reaction), and, if possible, equilibrium partial pressure of reactants or enhanced gas pressures. It is quenched to liquid-nitrogen temperature as fast as possible to retain the conditions of a frozen sorption equilibrium as well as possible. Finally, the precooled sample is loaded into a cryostat, cooled further to below 30 K and the spectrum measured.

For samples containing $>6\times 10^{21}$ protons, adequate spectra can be obtained on a timescale of several hours to about one day. For quantitative and

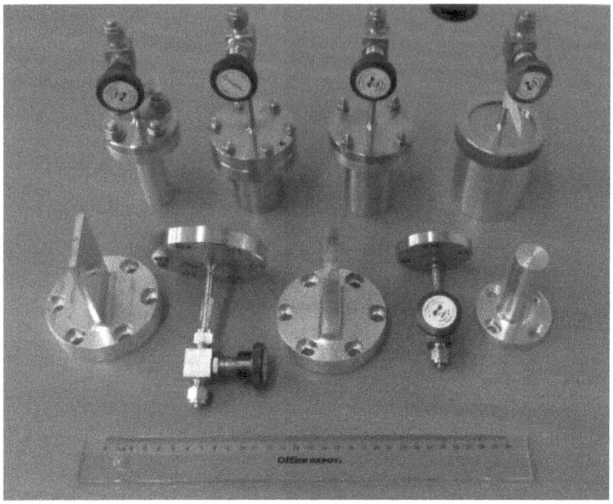

FIGURE 2 Size and shape of typical sample cuvettes and some applications. Top, left to right—small Al cylinder for measuring, for example, precious metal blacks; medium-sized stainless steel cylinder for *in situ* hydrogenation/dehydrogenation experiments on supported catalysts at pressures up to *ca.* 1 bar; medium-sized Al cylinder for measurements on catalyst powders of low active metal loading, pellets, extrudates, or other granular supports; large volume Al can for measuring fluffy, low-density oxide powders of low proton concentration level. Bottom, left to right—thin slit-shaped rectangular Al cuvette for metal hydride and liquid samples to reduce the risk of multiple scattering; slit-shaped can with enhanced path length for hydrides at lower hydrogen level; Al can, for example, high-density mixed-metal-oxide catalysts.

semiquantitative comparison, the neutron spectra can be normalized to the amount of sample in the beam. Most catalyst experiments are difference measurements—the spectrum of the catalyst with and without an adsorbate present is recorded and the difference taken.

In catalysis, the concentration of the relevant species often is low, and subtraction of a proper background can be a major task.

After the neutron experiment, the samples have to be stored on-site due to neutron activation. The radioactivity will decrease to background levels within a period of a few hours up to several months depending on the atomic number and isotopes of the catalyst components.

5.2.4 Choice of Complementary Methods

Neutron scattering has proved to be a very powerful technique to extend and to supplement the characterization of catalytic materials by, for example, sorption analyses and porosimetry, electron microscopies such as transmission electron microscopy (TEM), scanning electron microscopy (SEM), energy dispersive X-ray microanalyses (EDX) in the TEM and SEM, and surface analysis and surface science techniques such as X-ray photoelectron spectroscopy (XPS), secondary ion mass spectrometry (SIMS), or high-resolution electron energy loss spectroscopy (HREELS), to name a few [19,20]. INS also strongly complements infrared and Raman spectroscopies [26]. Additionally, because INS experiments are so dependent on sufficient hydrogen being present for detection, data from quantitative hydrogen analysis by gas evolution/hot extraction and thermal desorption spectroscopies or gravimetric gas absorption analyzer techniques [27] are also very valuable.

Neutron diffraction (ND) and X-ray diffraction (XRD) are clearly a complementary pair [28]. All of the advantages of ND (ability to detect light atoms (especially hydrogen) in the presence of heavy atoms, no form factor, and distinguishing adjacent elements in the periodic table) are valuable for catalyst studies (see Sections 5.5.4 and 5.5.5). XRD offers much better sensitivity (particularly synchrotron XRD) and so facilitates time-resolved studies; it is also an indispensable catalyst characterization method.

Extended X-ray absorption fine structure (EXAFS) and X-ray absorption near-edge structure (XANES) spectroscopies using synchrotron radiation provide information on the degree of alloying in nanosized particles that can be aggravated or not be possible by conventional X-ray diffraction in the *ca.* <3 nm region as the Scherrer equation [29] reaches its limit of applicability. Changes in local coordination and bond lengths can be monitored complementary to INS data on hydrogen in surface sites and site occupation inside of supported nanoparticles [30]. The potential of EXAFS in studying industrial catalysts has been recognized and utilized for decades [31–33]. Pioneering work was carried out in the 1980s, for example, by researchers from Exxon, Boeing, or BP. Today, an increasing number of spallation neutron sources or nuclear reactors are colocated with synchrotron radiation sources (ILL/ESRF Grenoble, ISIS/Diamond Harwell, and 2020 ESS/MAX IV in Lund). In principle, this should enable and foster cooperative research projects between the neutron and synchrotron communities. Combined studies of catalyst characterization by INS/EXAFS and joint studies by small-angle X-ray scattering and contrast match techniques in small-angle neutron scattering together with TEM are attractive possibilities.

In order to fully exploit the capabilities of neutrons and synchrotrons for studies of catalysis, it is essential to have a cadre that has been trained to use, and has experience of, both techniques. This was one of the motivations for establishing the UK Catalysis Hub [34] at the STFC Rutherford Appleton Laboratory that hosts both ISIS and Diamond.

In Section 5.4.3, the potential of the complementary use of different neutron spectrometers, which allows a coherent and complete picture of catalyst's properties, is outlined under the focus on fuel cell application.

5.2.5 Industrial Aspects

For industrial users of neutron beam facilities, there are a variety of access mechanisms. Peer-reviewed access for fundamental studies in precompetitive research of common interest and strategic feasibility studies at long-term access, which is to be published in the public domain, is frequently used. Commercial work and troubleshooting under nondisclosure agreements and fast time access are possible.

From the technological point of view, neutron scattering on catalytically relevant materials in fundamental and in proprietary research offers unique capabilities. Since in industrial catalysis the reaction volumes usually are large, integrated information on representative amounts of material is essential. Specimen sizes range typically around 10–150 g per catalyst sample. Under certain conditions, and with good sampling, one single experiment may give information on the catalyst equivalent of some cubic meters of reaction volume. In chemical engineering and process technology, this is highly attractive [35] since it allows easy access to relevant data that are sufficiently averaged on a macroscopic level. This is of relevance in characterizing fresh and used powder-type precious metal catalysts as well as fixed-bed catalyst pellets or highly porous systems such as reforming catalysts from petrochemical plants.

Commercial catalysts have morphological, physical, and chemical peculiarities such as high porosity, a chemically complex and geometrically rugged surface, and a high dispersion of supported nanoparticles. These may result in a strong absorption of electromagnetic radiation, or the electrical conductivity can limit the use of state-of-the-art equipment in catalyst characterization in the industrial laboratory at the production site or in central R&D.

Under such circumstances, a catalyst may be an ideal candidate for INS: the complementary use of neutron scattering may uniquely provide the solution. This was illustrated by investigating a phenomenon in catalyst deactivation in a large-scale industrial plant on the trail of catalyst poisons. In this case, neutrons were utilized as a remedy of last resort in an explorative—but highly successful—approach. Results of direct practical relevance were obtained (see Section 5.4.2). The results demonstrated that neutron scattering can directly bridge the gap between the operating catalyst and related surface science work on well-defined crystal or reference surfaces [36].

A catalyst's activity in selective hydrogenation was significantly hampered. Although the palladium-based catalyst had been thoroughly analyzed using the complete analytic spectrum of industrial laboratories, no clear reason was apparent. The catalytic activity could not be regenerated sustainably to an economic useful level. While the catalyst continued to be active to the uptake of hydrogen, it was no longer able to provide sufficient adsorption capability for the reactants—in this case, large aromatic molecules—that should have accumulated on the surface to be selectively hydrogenated.

Over a period of about 10 months of varying performance of a catalyst running in a continuous hydrogenation process, catalyst samples were monitored using INS. The "snapshot" or "react and quench" procedure was chosen: discharging catalyst samples from the plant at different times of operation and residual activity level, sampling of a pyrophoric catalyst, handling under argon, and INS measurements of large quantities of catalyst each. Measurements of finely divided, pyrophoric matter with adherent residual organics, before and after solvent extraction, after hydrogenation/dehydrogenation in the INS can to monitor the surface chemistry and hydrogen storage capability of the palladium catalyst, after liquid-phase oxidation or after competitive gas exposure (carbon monoxide).

As a consequence of the surprising INS data, the operating conditions of the reactor were changed, and the catalyst was again able to be used as intended for the synthesis.

In conclusion, INS opened completely new perspectives at the molecular scale by comparing the proton dynamics of catalyst samples from industry at realistic conditions just by adequate timing and sample transfer from the running chemical plant to the running neutron plant under fast access conditions.

5.3 PRODUCTION, FORMATION AND USE OF CATALYSTS

5.3.1 Catalyst Supports: Carbons

Due to their high surface area (up to about 1500 $m^2\ g^{-1}$ or more) and high porosity (up to about 1.5 ml g^{-1} volume in total of macro-, meso-, and micropores), activated carbons [37] derived from various natural sources are routinely used, for example, for the production of supported chemical catalysts for catalytic hydrogenations and other applications. Pure synthetic grades of carbon black [38] derived from natural oil and gas are used, for example, as very clean supports for the production of fuel cell catalysts such as polymer electrolyte membrane catalysts (PEM) and also catalysts for chemical syntheses in the production of, for example, intermediates or pharmaceuticals. The production of such carbons can be described as the removal of hydrogen from feedstocks.

The surface chemistry regarding C/O, C/N, and other heterofunctional groups is accessible by X-ray photoelectron spectroscopy (XPS) and chemical titration techniques (Ref. [38], pp. 175). The hydrogen content of such carbons shows strong variations, albeit at a comparatively low concentration. The usual range is about 20.000 $\mu g\ g^{-1}$ H in activated carbons of moderate graphiticity, about 2.000–4.000 $\mu g\ g^{-1}$ for typical furnace blacks or purified activated carbons and as low as to 200 $\mu g\ g^{-1}$ or less in acetylene blacks or graphitized carbon blacks. Revealing the corresponding proton dynamics down to the trace concentration level of hydrogenous species is a straightforward target for INS—if a suitable neutron flux and high-resolution spectrometer such as TOSCA is available; carbon black is not black—for inelastic incoherent neutron scattering.

Figure 3 illustrates how dramatically the controlled removal of only about 3900 $\mu g\ g^{-1}$ of hydrogen by thermal treatment under argon changes the morphology and spectral signature of a typical grade of carbon black synthesized

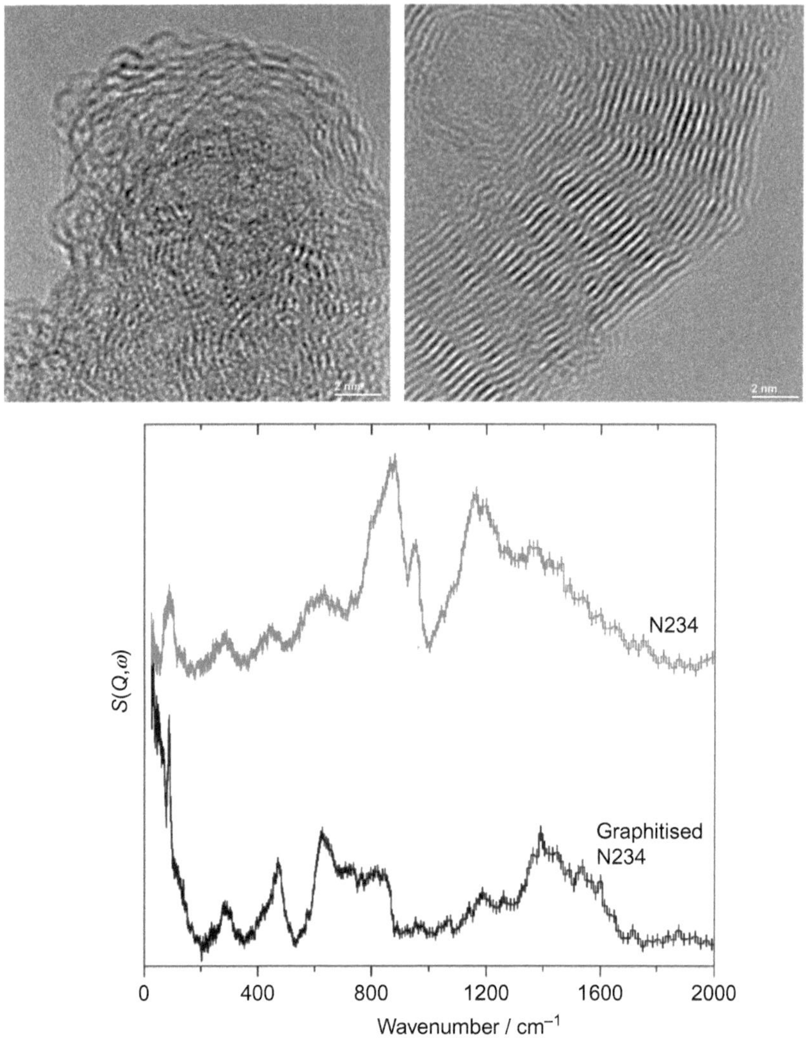

FIGURE 3 Top: TEM images of furnace black N234 original (left) and graphitized (right); bottom: the corresponding INS spectra of N234; upper trace: as produced, 4200 μg/g H; lower trace: after graphitization treatment, 290 μg/g H. Recorded on TOSCA.

by furnace reactor technology from paracrystalline disorder to enhanced graphitic surface ordering of graphitic sheets of increased size.

According to the model derived from transmission electron microscopy (TEM), the basic structural units (BSU's) of carbon blacks and other finely divided carbons consist of small graphite-like platelets that are arranged in turbostratic disorder to form spherical structures [37]. These entities must be terminated by (few) oxygen-, nitrogen-, or sulfur-containing functional groups and, predominantly, by protons [39]—otherwise, these flat open sp^2 structures would not be stable and fullerenes or nanotubes would be formed. Curvature is induced by nonhexagonal rings, for example, corannulene-type entities [40]. Due to the absorption of electromagnetic radiation and the electrical conductivity of such finely divided carbons, these protons in edge termination can be studied exclusively by inelastic neutron scattering [41].

Neutron studies of carbons have demonstrated that the fine structure of the out-of-plane aryl C–H bending mode region at 600–1000 cm^{-1} shows distinct variations, which are not simply correlated with the hydrogen content but with the grade of carbon and the structure as revealed by TEM [42]. It was found to be possible to model the fine structure of the BSU's by approximating the spectra by DFT simulations of the vibrational spectra of molecular structures of isolated, vicinal, or ternary C–H substitutional patterns [43] and, recently, by periodic DFT of graphene sheets with isolated (H1) and vicinal (H2) protons (Figures 4 and 5) [42]. For activated carbon, a H1/H2 ratio of

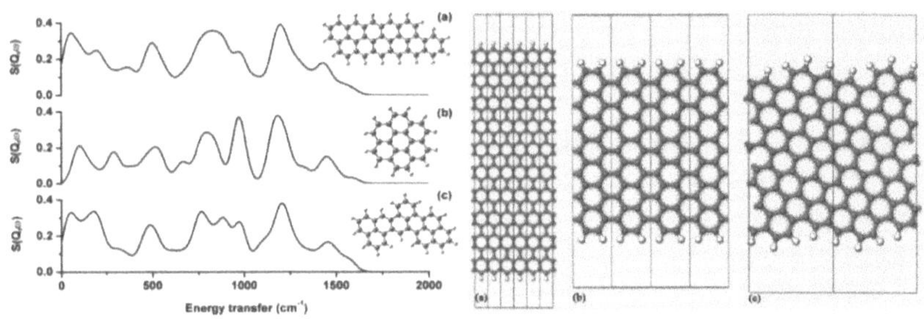

FIGURE 4 Left: calculated INS spectra of model graphite fragments. Right: graphene plane (a) [100], (b) [110], and (c) [310]. *Reprinted with permission from Refs. [42,43].*

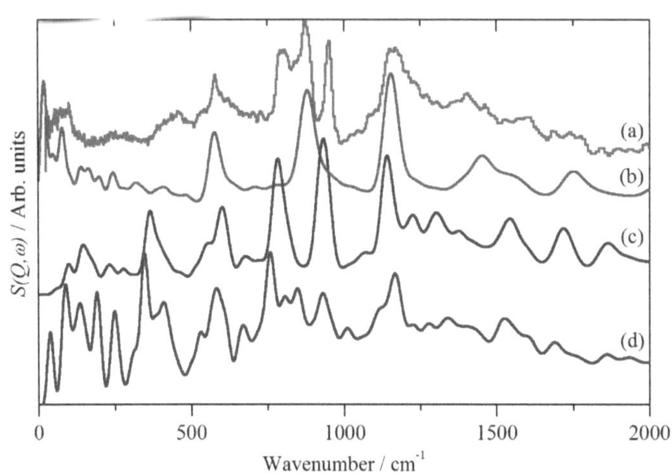

FIGURE 5 (a) INS spectrum of glassy carbon, (b) H1 termination, (c) H2 termination, and (d) mixed H1 and H2. *Reprinted with permission from Ref. [42].*

2:1 was determined, 4:1 for a furnace black, 1.5:1 for anthracite coal, and 2:1 for type 1 glassy carbon that was used in an INS study on graphitizing and nongraphitizing carbons.

5.3.2 Catalyst Supports: Oxides

The transparency of oxides such as silica to neutrons means that bulk material properties become accessible, whereas infrared spectroscopy is restricted to the surface regions. The detection of the well-known O–H stretch vibrations is difficult with indirect geometry INS spectrometers (although readily observable with direct geometry instruments at a pulsed spallation source), whereas low-frequency torsional and deformational modes appear with high intensity.

Above 1400 cm^{-1}, infrared spectroscopy provides detailed information about the surface properties of silica, particularly regarding hydroxyl groups. However, silica is largely opaque to infrared radiation at energies <1400 cm^{-1} because of the intense lattice absorption. The low-energy vibrational bands of SiO_2 can be probed by INS revealing the vibrational density of states of both surface species such as hydroxyls and also of the interior of the silica (Figures 6 and 7). Neutron scattering on vitreous silica led to the conclusion that bands at about 800–1200 cm^{-1} are due to Si-O stretching modes and those at 300–400 cm^{-1} to Si-O bending modes [44].

The influence of process steps such as wetting, agglomeration, pelletization, and calcination at $T < 1250$ K, for example, of amorphous silica to generate oxidic catalysts supports can be studied. For finely divided amorphous silica, the calcination temperature is chosen below the transition temperature to the crystalline phases of quartz: tridymite or cristobalite. With increasing posttreatment temperature and compression treatment, the X-ray amorphous material exhibits somewhat enhanced short-range structural ordering as indicated by slightly increasing contributions of the broad bands around 200–450, 700–900, and 1000–1300 cm^{-1} (Figure 7) [45].

The low-energy (*ca.* <150 cm^{-1}) phonon density of states of fumed and of precipitated silica is typical of completely amorphous [46] silica materials (Figure 6). This observation is in excellent agreement with data from X-ray diffraction [47] and high-resolution transmission electron microscopy (Figure 6). The amorphicity is retained during granulation posttreatments (Figure 7).

Details on the complementarity between INS and infrared spectroscopy for the case of alumina supports will be presented in Section 5.5.1.1.

5.3.3 Fresh Catalysts: Water and OH-Groups

Supported precious metal catalysts are mostly prepared by wet impregnation techniques. In the course of catalyst manufacture and depending on the final

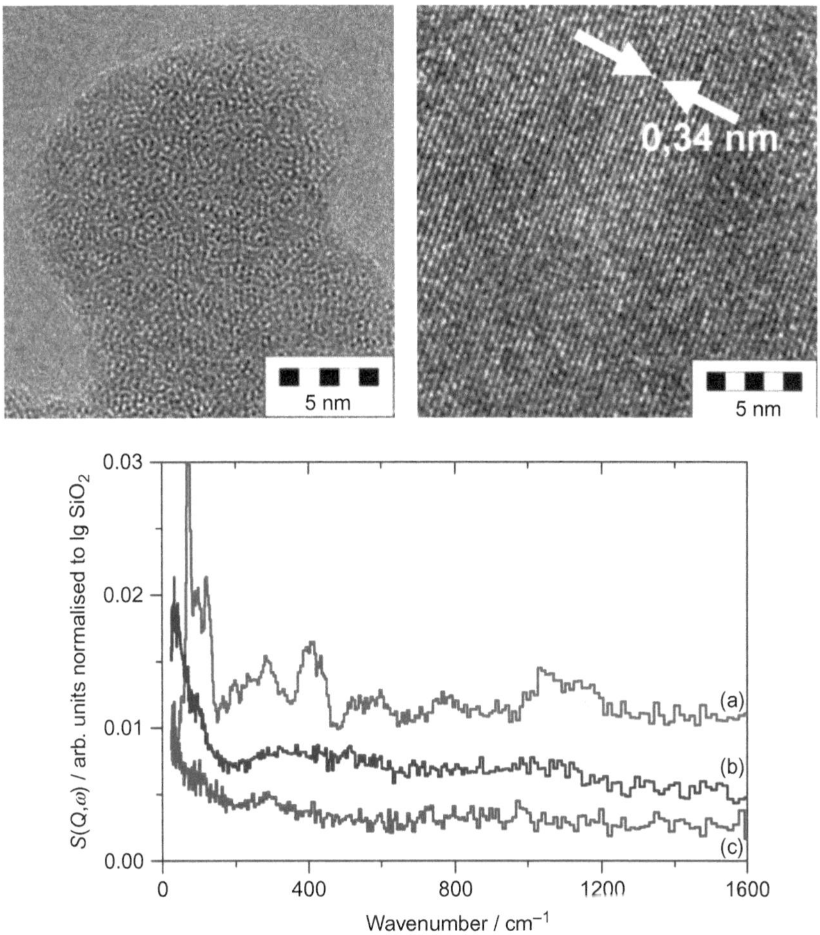

FIGURE 6 Top: TEM images of amorphous (left) and crystalline (right, quartz) silica. Bottom: INS spectra of (a) quartz, (b) precipitated silica, and (c) fumed (pyrogenic) silica (TOSCA).

application, they are dried under varying conditions and are reduced or modified to adjust the desired dispersion, morphology, and surface chemistry in terms of surface polarity and reduction state of the precious metal. The chemical valence of the supported particles can be routinely determined by means of XPS. The chemical shift and fine structure of the photoelectron signals are usually broad and of complex structure for fresh, unreduced, or reduced catalyst's surfaces due to the presence of different chemical states of the supported precious metal, and uncertainties may occur in discriminating the contributions of stoichiometric and substoichiometric oxides on the one hand and hydroxyl groups and water on the other. Since hydrogen cannot be

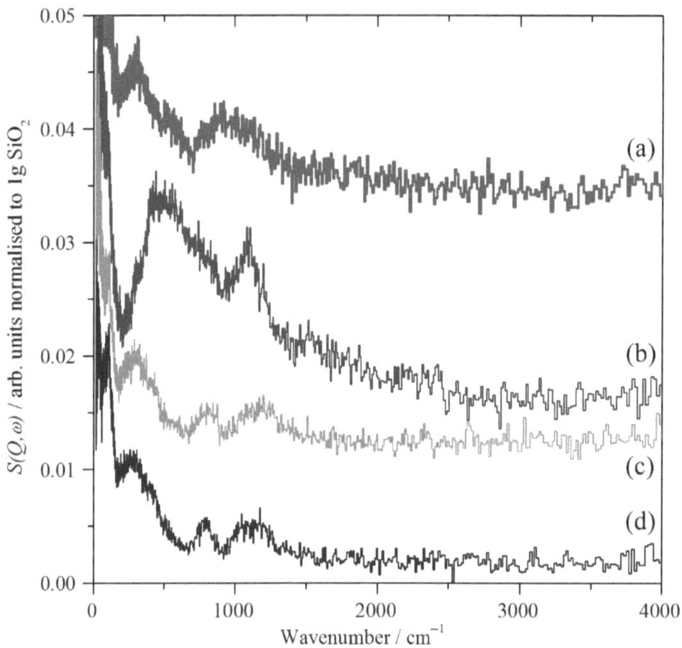

FIGURE 7 INS spectra of pyrogenic silica (TOSCA). (a) Original, powder, fluffy. (b): Agglomerated, wetted. (c): Calcined at $T < 1250$ K, and (d) calcined and pelletized catalyst support.

detected by XPS, complementary information from INS is needed to clarify the situation.

For a freshly prepared supported Pd/C catalyst, the characteristic Pd-O–H bending mode and its first overtone are observed at 940 and 1880 cm^{-1} (Figure 8) [30] and a broad O–H stretching mode (surface hydroxyls) at 3520 cm^{-1} with a broad shoulder at *ca.* 3400 cm^{-1} (water) [48].

Surface hydroxyls on a freshly prepared supported Pt/C catalysts were also observed (Figure 9) at 1050 cm^{-1} using the beryllium filter option of the hot-neutron triple-axis spectrometer IN1-Lagrange (ILL). It was observed that both surface hydroxyls and adsorbed water could be displaced by dosing with carbon monoxide.

For the case of a completely alloyed supported platinum/ruthenium, fuel cell catalyst CO dosing (Figure 10) caused the formation of water (indicated by the librational modes at 570 cm^{-1} and the H–O–H scissor modes at 1670 cm^{-1}) and the loss of hydroxyls (bending mode at 1085 cm^{-1}), suggesting that CO reacts with the hydroxyls, as observed for hydrous palladium oxide [49]. Such findings are of relevance in understanding and further improving the CO tolerance of polymer electrolyte membrane (PEM) fuel cell catalysts, for example, by controlled alloying.

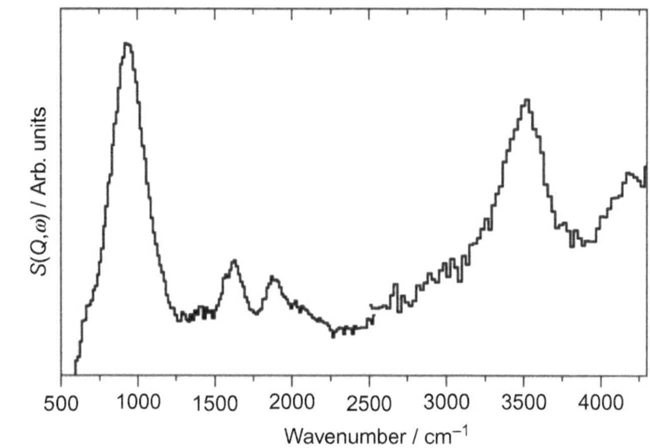

FIGURE 8 INS spectra of a freshly prepared Pd (20%)/C catalyst recorded on MERLIN (ISIS) (direct geometry chopper spectrometer) with incident energies of 2825 (500–2500) and 4840 (2500–4300) cm^{-1}. *Reprinted with permission from Ref. [30].*

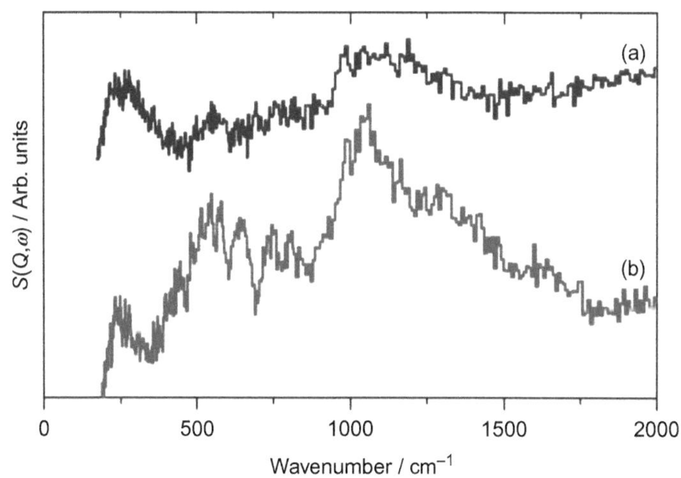

FIGURE 9 INS spectrum recorded on IN1-Lagrange (ILL) of freshly prepared Pt (40%)/C; (a) after first dose and (b) after second CO dose. *Reprinted with permission from Ref. [30].*

In this context, there is an unexploited potential, and strong synergies, between neutron experiments on the proton dynamics (Section 5.4.3) and synchrotron radiation investigations experiments by X-ray absorption spectroscopy (extended X-ray absorption fine structure EXAFS, Section 5.2.4) on the degree of alloying in supported bi- or multimetallic nanoparticles.

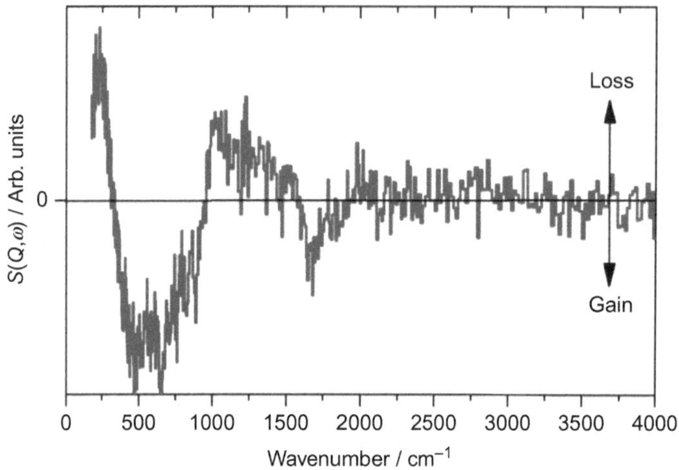

FIGURE 10 INS spectra recorded on IN1-Lagrange of a freshly prepared Pt (40%) Ru (20%)/C fuel cell catalyst after reaction with CO. *Reprinted with permission from Ref. [30].*

5.3.4 Pearlman's Catalyst

Pearlman's catalyst [50,51], nominally written as Pd $(OH)_2/C$, is widely used in organic syntheses for hydrogenation and C–C coupling reactions [52,53]. A prominent practical feature is that in spite of the high precious metal loading, this versatile material is nonpyrophoric, and handling is easy.

Contrary to the accepted view, the combination of INS, TEM, and XPS revealed that Pearlman's catalyst as prepared and after drying consists of carbon-supported (mostly) nanoparticulate hydrous palladium oxide capped with a monolayer of hydroxyls hydrogen-bonded to a few layers of water: a core-shell structure of $C/PdO/OH/H_2O$. The conventional formulation Pd $(OH)_2/C$ from the macroscopic point of view is ruled out by the different spectral signatures of surface hydroxyls and stoichiometric hydroxides (Figure 11). Furthermore, a minor fraction of the palladium is present as a reduced species [54].

5.3.5 Catalyst Reduction Step

In further stages of catalyst preparation and activation, forming gas or other agents are applied to remove residual surface hydroxyls and traces of water and to generate the precious metal in a useful oxidation state. The formation of certain finely divided hydride phases can be investigated by INS (Section 5.4) that also supplies information on changes of the carbon support.

In Figure 12, the shape of a section of a piece of a palladium/activated carbon powder-type catalyst is shown in an SEM micrograph. The primary

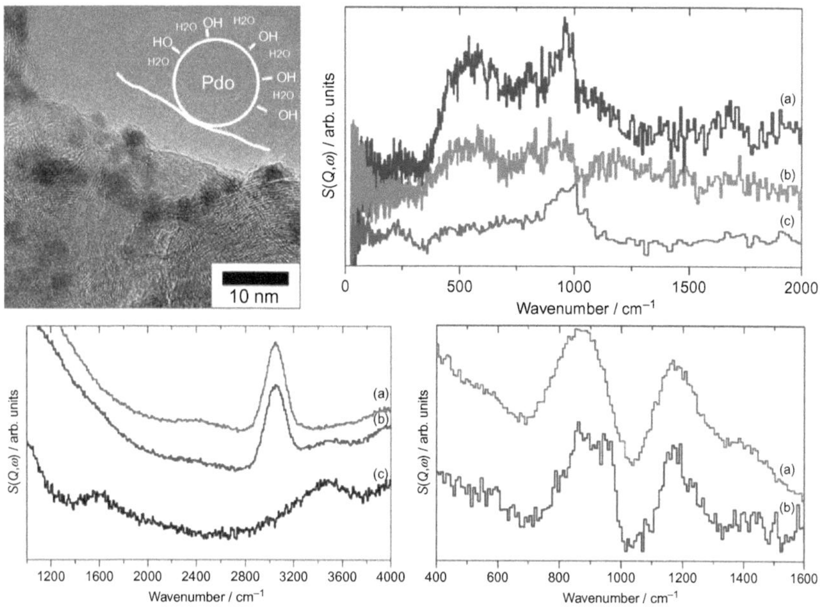

FIGURE 11 Top, left: TEM micrograph and schematic sketch of the catalyst's structure; top, right: comparison of difference spectra of dried samples: (a) 20 wt% Pd/C, (b) 5 wt% Pd/C, and (c) hydrous palladium oxide dried at 100°C (TOSCA). Bottom, left: (a) carbon support, (b) dried 20 wt% Pd/C, and (c) difference spectrum (MAPS); bottom, right: (a) carbon support and (b) dried 20 wt% Pd/C, deformation region (MAPS). *Reprinted with permission from Ref. [54].*

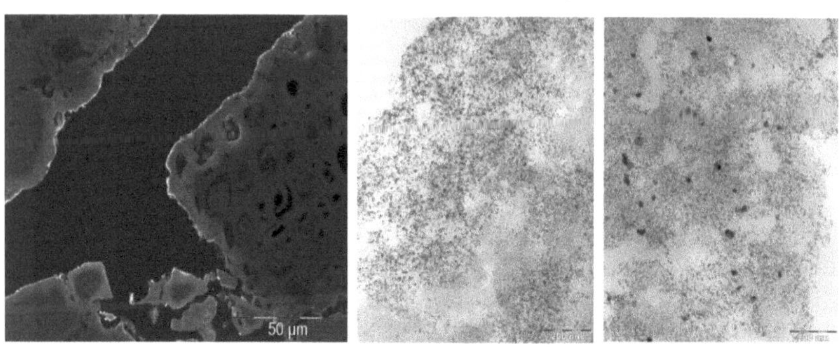

FIGURE 12 left: backscattered electron image of a palladium (enriched in the bright surface area) on activated carbon (darker area with vascular structure of the pristine wood retained) hydrogenation catalyst; middle and right: TEM images of finely divided supported palladium catalysts with good (middle) and—partly—poorer (right) precious metal dispersion on different activated carbon support qualities. *Reprinted with permission from Ref. [55].*

particle size of the precious metal particles has to be identified by means of TEM. INS allows the comparison of the properties of activated carbons [43,55] in the search of parameters to improve and stabilize precious metal dispersion and, therefore, catalyst performance.

It has been observed that the impregnation of an activated carbon support with platinum was accompanied by an increase in hydrogen content of about 900 μg g^{-1}, and, correspondingly, the relative intensity of the 880 cm^{-1} part of the band of the out-of-plane C–H bending mode region decreased and the relative intensity of the 820 and 970 cm^{-1} bands increased [43]. This indicates that during precious metal impregnation and subsequent reduction, not only deposition and reduction of platinum species may occur (to be investigated by XPS and EXAFS) but also hydrogen-related chemical changes of the support material.

Evidence for carbothermal reactions during catalyst formation treatment was also observed in PtRu/C fuel cell catalysts of different degree of alloying [56].

Such studies allow monitoring and optimization of the reactions between the surface of the catalyst support (derived from natural resources or synthetic), the precious metal entities, and the reduction agents in catalyst manufacture in more detail, from the deposition of molecular species of subnanometer dimension to nucleation and particle growth at the nanoscale and above. The combination of INS, SEM, TEM, XPS, and EXAFS data provides valuable information for improving catalyst preparation strategies and procedures.

5.3.6 Used Catalysts: Preferential Adsorption

Monitoring, for example, of liquid-phase catalytic processes by means of chromatographic or mass spectroscopic investigations and by online spectroscopy, such as near-infrared process control probes or by discontinuous sampling usually, provides knowledge about the composition of the organic or aqueous phase and the quantity of products and by-products and, therefore, the selectivity of a catalyst in a certain reaction. INS is able to supply the "mirror image" information: to reveal what is usually inaccessible to infrared or Raman spectroscopy and what is invisible in transmission electron microscopy, to supply answer to the question "what is going on at the catalyst's surface itself?". One example is the detection of a preferentially adsorbed isomer at a used catalyst's surface [57].

The Heck reaction is one of the most important C–C coupling reactions in organic synthesis. The olefination of aryl halides is usually carried out in homogeneous solution, catalyzed by palladium complexes. Immobilization of the palladium catalyst on a carbon support has been used as a method to overcome difficulties in catalyst separation and recovery. A quasihomogeneous reaction mechanism involving a Pd dissolution/reprecipitation process was observed: palladium complexes or colloidal particles of palladium in the solution are the catalytically active species. Control of the palladium leaching and redeposition at the end of the reaction could be achieved to give an excellent recovery of the precious metal from the reaction mixture [58].

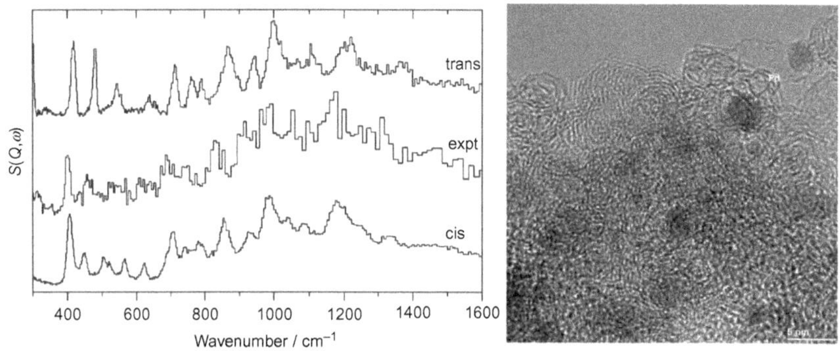

FIGURE 13 Left: INS spectra (TOSCA, ISIS). Top: *trans*-stilbene; middle: 5% Pd/C catalyst after use in the Heck reaction; bottom: *cis*-stilbene. Right: TEM image of the used catalyst; activated carbon support (*ca.* 1500 m^2/g surface area and paracrystalline) and finely divided isolated primary particles of palladium (*ca.* 3–5 nm, crystalline). *Reprinted with permission from Ref. [57].*

Figure 13 (left) shows the INS spectrum of a 5%Pd/C catalyst after use in the C–C coupling reaction of bromobenzene and styrene to *trans*-stilbene. Even after solvent extraction to remove the complex organic reaction mixture, (products, base, and salts), the INS spectrum of the used catalyst shows the sharp bands typical of molecular species. These are superimposed on the C–H out-of-plane bending bands of the activated carbon catalyst support. Comparison with the spectra of *trans*- Figure 13 (top spectrum) and *cis*-stilbene (bottom spectrum) showed that residual amounts of, predominantly, *cis*-stilbene are left on the surface of the 5%Pd/C catalyst, whereas all other organic material was successfully removed. The *trans*-stilbene was released/desorbed already into the liquid reaction mixture and/or removed by the washing procedures prior to the INS experiment. Since a yield of ~90% of *trans*-stilbene was measured in the product mixture, it follows that *cis*-stilbene is preferentially adsorbed on the surface of the pine-derived activated carbon support.

Usually, it is assumed that educts and products are adsorbed to a similar degree. This is not the case in the present example. Preferential adsorption is also of relevance for catalytic hydrogenation reactions.

INS of the catalyst's surface itself is, therefore, highly complementary to the large set of methods in process control which supervise the liquid or gaseous reaction media and increases the knowledge about the reasons for product selectivity of a given process.

5.4 HYDROGENATION CATALYSTS

5.4.1 Hydrogen in Supported Nano-Particles

Neutron studies on the adsorption, dissociative absorption, diffusion and site occupation of hydrogen in metals, alloys [59,60], and intermetallic compounds [61], for example, under the aspect of suitability as hydrogen storage materials [62] have been performed for decades [63]. A major focus is the understanding and improvement of absorption activity and capacity, long-term stability in hydrogenation/dehydrogenation cycles and adjusting plateau pressure, and monitoring hysteresis effects and probing mechanical resistance versus enhanced hydrogen embrittlement.

Most of the corresponding work has been performed on compact matter, finely divided metal blacks and granular-shaped intermetallic compounds. However, there are unexploited synergies in studying the size dependence of hydrogen-related properties such as hydrogen storage capacity or the relative proportions of different hydride phases in catalytic materials, with primary particle sizes from the micro down to the nanometer scale. It was shown that β-phase palladium hydride has a higher activity and lower selectivity in catalytic hydrogenation reactions such as the selective hydrogenation of ethyne than the α-phase hydride [64].

From gas volumetric work on the hydrogen absorption of nanoparticulate palladium, it is known that the hydrogen solubility in single-sized palladium clusters does change as a function of primary particle size (5, 3, and 2 nm) and the absorption isotherms differ compared with larger particles or to bulk palladium (Figure 14). The solubility in the α-phase region is enhanced by a factor of five- to tenfold compared with bulk palladium, and the miscibility gap is narrowed compared to bulk palladium from 0.58 H/Pd for polycrystalline Pd at 20°C down to 0.28 H/Pd for small Pd clusters. Hydrogen isotherms of nanosized palladium clusters resemble the bulk palladium-hydrogen system above the critical point with enhanced low-concentration solubility and reduced high concentration solubility and do not plateau but a sloped isotherm for 3.0 nm Pd-H clusters compared with 5.2 nm Pd-H clusters with a small plateau. Also, the hydrogen absorption behavior of nanocrystalline palladium (grain size < 10 nm, XRD; 8 nm) is significantly higher in the α-phase region than in the case of coarser palladium grains, but the maximum solubility is 50% lower. This particular behavior of lower hydrogen absorption at lower

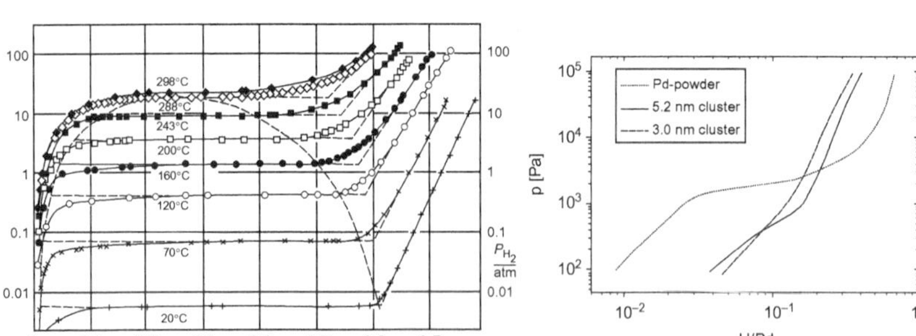

FIGURE 14 Left: hydrogen absorption isotherms of bulk palladium; p(n) isotherms, including the critical region; n:H/Pd [65,66]. Right: hydrogen isotherms of palladium powder and clusters of different size. *Reprinted with permission from Ref. [65,67].*

particle size [67–71] compared with the usual hydrogen in metal systems [59,60,65,72,73] may mean that INS may come to its detection limit. Higher hydrogen solubilities at the surface and in subsurface sites compared with the bulk are also observed [74].

Depending on the location of the hydrogenous entities (organic species, molecular adsorbed, atomically chemisorbed or interstitially dissolved hydrogen) INS can be utilized as a bulk technique or as a surface science technique in the case of supported palladium nano-particles (dissolved protons vibrating and diffusing in the interstitial sites of the palladium entities) [59,75]) due to the huge hydrogen storage capacity of this element on the one hand [66,72,73] and supported platinum nano-particles on PEM fuel cell catalysts on the other (atomic surface-hydrogen, Section 5.4.3).

5.4.1.1 The Lindlar Catalyst: Controlled Poisoning

Controlled moderation on hydrogenation activity and target-product selectivity of supported palladium is known and utilized in fine-tuning chemical syntheses. The Lindlar catalyst is applied in the stereo selective catalytic hydrogenation of triple bonds to double bonds, especially the enantioselective hydrogenation of alkynes to *cis*-alkenes. The catalyst is named after Herbert Lindlar (Hoffmann-La Roche) [76–79]; it consists, for example, of 5% palladium on a calcium carbonate support. The activity of the palladium is moderated by adding lead and quinolines, the effect being that the hydrogenation is stopped at the alkene stage and no overhydrogenation to alkanes occurs. An example of industrial application is the synthesis of vitamin A, which includes an alkyne reduction step after coupling of β-ionone with ethynylmagnesium bromide.

In addition to the poisoning/moderation effect of the lead on active sites [80] to the organic reactant, it is also of interest to what extent the interactions between lead and the precious metal may influence hydrogen-related properties of supported nanodispersed palladium entities, since the α- and β-phase palladium hydride have different activities and selectivities in catalytic hydrogenation reactions (Section 5.4.1 [64]). It is assumed that in hydrogenation reactions, β-phase hydride provides activity and α-phase hydride selectivity. For alloy catalysts, this may be further complicated by decreasing primary particle size and other differences between core/shell systems and homogeneous alloyed nanoparticles.

Due to the comparatively low precious metal loading (5%Pd+3.5%Pb/$CaCO_3$) and to the observation that the hydrogen storage capability decreases with decreasing particle size, high-sensitivity neutron spectrometers are needed for nondestructive investigation of the given catalyst system under the access of hydrogen *in situ*. Therefore, the work has been carried out using MAPS at ISIS to focus on intensity rather than high resolution as provided by TOSCA. The influence of particle shape, size, and controlled poisoning on hydrogen storage properties was studied (Figure 15) [81].

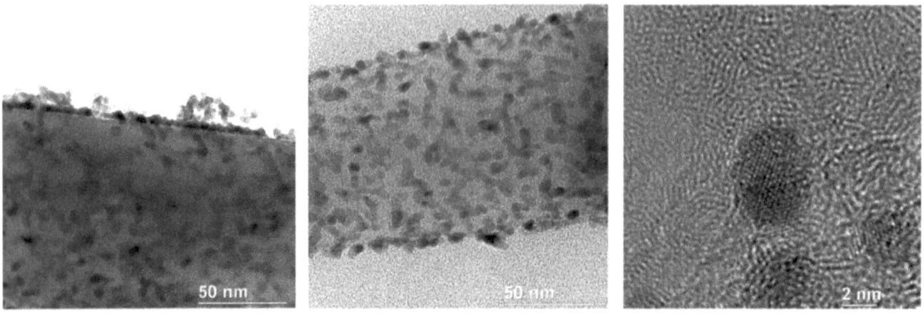

FIGURE 15 Left, middle: Lindlar catalyst, palladium/lead aggregates of intergrown primary structures; right: 5% palladium on activated carbon, isolated primary particles. *Reprinted with permission from Ref. [81].*

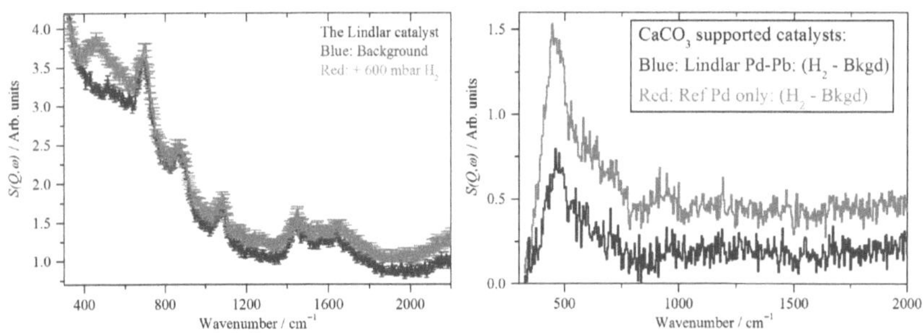

FIGURE 16 Left: INS spectrum of aggregates of supported palladium (Figure 15) consisting of *ca.* 3.7 nm primary particles, measured under 600 mbar hydrogen equilibrium sorption pressure (upper trace) and after evacuation of the sample can (lower trace). Right: background subtracted spectra of a palladium only (upper trace) and the Lindlar catalyst (lower trace) under 600 mbar both with the same primary particle and aggregate sizes. *Reprinted with permission from Ref. [81].*

Both the α- and β-hydride phases in supported Pd only and Pd/Pb-alloy catalysts were detected (Figure 16). The addition of lead to the palladium on calcium carbonate in catalysts of identical primary particle size (3.7 nm) and shape and of identical size of the particle aggregates (10–15 nm) caused a decrease of the β-phase hydride formation by a factor of two. Reference measurements on a catalyst containing exclusively isolated primary particles (Figure 15, right) of palladium of about 3.2 nm size at comparable hydrogen loading revealed an even lower degree of formation of β-phase palladium hydride than the aggregated palladium nanoparticles in Pd/$CaCO_3$ and in the Lindlar catalyst (factor about 3.5). For the isolated primary particles, the long-range phase coherence that is necessary for the phase transition is lower than for the aggregates of *ca.* 10–15 nm size.

This INS result suggests that for catalyst selectivity in the hydrogenation of triple to double bonds, not only the presence of modifiers at the surface but also a reduced, moderated hydrogen storage function is needed to achieve product selectivity in the technical synthesis. With less hydrogen available, the reaction stops at an earlier step and alkane production is prevented.

INS complements high-precision hydrogen isotherm measurements on nanoscale stabilized Pd-H clusters [67–71]. Quantitative information on the impact of changing size and shape in particle design on hydrogen storage capabilities in nanoparticles of industrial catalysts is available (Figures 15 and 16).

5.4.1.2 Effects of Alloying on Hydrogen Storage and Catalytic Activity

The catalytic hydrogenation of nitroarenes, for example, of nitrobenzene to aniline or of dinitrotoluenes (DNT) to toluene diamines (TDA) is a major industrial activity. TDA is a key component in the production of toluene diisocyanate (TDI) that is used for the manufacture of polyurethanes (PU). These reactions are prominent examples for the need and economic impact of steering and optimizing catalyst selectivity at—simultaneously—adequate catalyst activity. Economics favor lower reaction pressure and temperature that can be achieved by improving the catalyst's properties. The reaction rate and selectivity in the hydrogenation can be enhanced by changing from nickel- to palladium-based catalysts. Usage of other noble metals such as platinum and additional promoters can further improve reaction efficiency and lower the amount of catalyst needed. The addition of platinum to palladium increases activity, and the addition of iron increases selectivity: the enhancement of activity by the addition of platinum needs to be moderated by iron in order to avoid aromatic ring hydrogenation [82].

The addition of Pt and Fe results in better selectivity and long-term stability for large-scale production. As was already reported by DuPont in the late 1960s, the Pd-Pt-Fe/C trimetallic catalyst has become a "workhorse for the hydrogenation of aromatic nitro compounds" [83].

Neutron scattering on MAPS gave a quantitative assessment of the relative impact of different influences on the hydrogen storage capability in supported palladium-based nanoparticles (Figure 17). The addition of Fe to form a ternary alloy has an even stronger impact on decreasing β-phase hydride formation than the addition of lead in the Lindlar catalyst (Figures 15 and 16, and Table 1; factor 3.2 *vs* 2.2). Together with the deaggregation of Pd by alloying with Pt (Figure 17), this leads to a factor of 4.7, whereas the enhancement of the proportion of isolated primary particles of palladium in the case of a porous activated carbon support leads to a factor of 3.5. Knowledge of these effects in finely divided supported particles offers clear paths for further catalyst optimization: alloying appears to allow fine-tuning catalytic activity by changing morphological and chemical properties of the supported particles at the nanoscale.

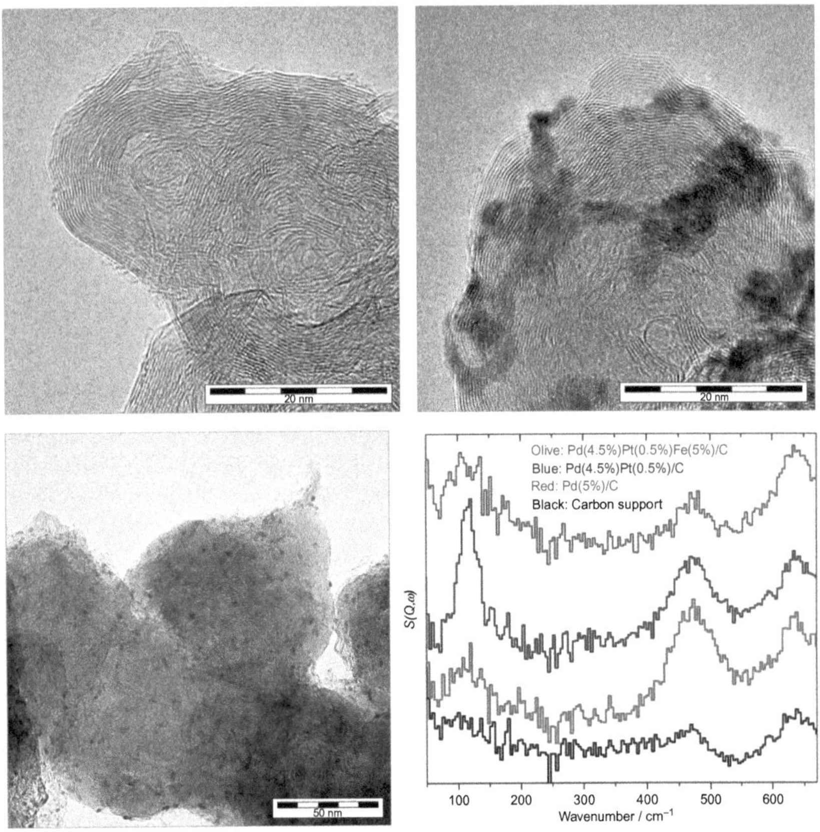

FIGURE 17 TEM images of a carbon black catalyst support of enhanced graphiticity (top, left), Pd(5%)/C edge decoration by chain-like deposited palladium aggregates on the support (top, right), and Pd(4.5%)Pt(0.5%)/C bimetallic alloy catalyst with isolated primary particles (bottom, left); INS spectra of the catalysts under 750 mbar hydrogen and of the carbon black support (bottom, right). At 480 cm^{-1} supported β-palladium hydride, vibrations of the interstitial protons on the octahedral sites; at 120 cm^{-1} rotational transition of physisorbed molecular hydrogen at the catalyst's surface. *Reprinted with permission from Ref. [84].*

INS enables the direct study of protons and their vibrational dynamics inside of supported 2 nm sized particles by probing bulk quantities of catalyst under realistic hydrogen pressure in one single spectrum. Adsorbed molecular hydrogen can be detected as well (Figure 17).

5.4.2 Methyls on Pd: Catalyst Deactivation

Uncontrolled and accidental poisoning or deactivation of catalysts by various mechanisms can be a considerable financial burden in the chemical industry and is still a focus of academic and technological research. A better understanding of deactivation processes is essential for improving and optimizing

TABLE 1 Ratios of Normalized INS Peak Integrals: Quantification of the Effect of Alloying, Disaggregation, Moderation, and Controlled Poisoning in Supported Palladium Catalysts on the Hydrogen Storage Capacity of Supported Nanosized Catalyst Particles

Composition of catalyst	Effect on hydrogen storage capability	Factor of decrease in formation of β-PdH_x
5%Pd/(4.5%Pd+0.5%Pt) on C	Deaggregation by alloying with Pt	1.5
(4.5%Pd+0.5%Pt)/(4.5%Pd+0.5%Pt+5%Fe) on C	Moderating Pd/Pt with Fe	3.2
5%Pd/(4.5%Pd+0.5%Pt+5%Fe) on C	Deaggregation, alloying, and moderation with Fe	4.7
5%Pd/(5%Pd+3.5%Pb) on $CaCO_3$	Poisoning Pd with Pb	2.2
5%Pd on $CaCO_3$ / 5%Pd on C	Aggregates versus isolated primary particles	3.5

Reprinted with permission from Ref. [84].

process conditions and the catalysts themselves and for circumventing premature catalyst degradation in order to minimize additional costs and shutdowns. To reveal the various reasons for loss of activity or selectivity under unusual or poorly chosen operating conditions and to identify unknown phenomena is a major task in applied catalyst research.

In the catalytic hydrogenation of C=O to C-OH groups in aromatic/polyaromatic species, a premature loss of activity and unexpected activity variation over time of a well-established standard-type palladium catalyst were observed, which led to uneconomic operation conditions of a large-scale plant. Employment of a broad variety of catalyst characterization methods by online and on-site analytic methods, as well as in central research and external institutes and application laboratories, failed to reveal the cause(s) of the problem. No typical catalyst poisons, loss of surface area, corrosion phenomena, and catalyst coking were observed. Due to the properties of the catalyst, methods such as infrared, Raman, and nuclear magnetic resonance spectroscopies were inapplicable [11,36,85].

It was concluded that something had blocked the surface of the catalyst and was strongly adsorbed or chemically bound and so could not be removed by the usual reactivation procedures or by solvent extraction. Surface analytic measurements were performed on extracted catalyst samples taken from the reactor during active and periods of lowered activity and from active reference plants. Time-of-flight secondary ion mass spectrometry (ToF-SIMS) did not give evidence for critical surface contamination by inorganic elements and XPS did not show systematic variation in the concentration level of non-extractable carbon deposits or significant differences in the chemical valence state of the palladium. It was concluded that organic surface species of changing molecular structure might be the cause of the problem, but they could not be unambiguously identified by evaluating the XPS C1s signal's fine structure and the ToF-SIMS fragmentation patterns measured at varying primary ion beam current densities.

Therefore, it was decided to try out the surface science potential of neutron scattering to verify the presence of H/C-containing molecules at the nonmicroporous catalyst's surface using TOSCA *via* rapid access scheme.

In a first step, a solvent extracted catalyst from an early stage of a deactivation event was measured after *in situ* hydrogenation in the neutron sample can and after evacuation (Figure 18). Formation and decomposition of palladium hydride was observed in spite of the presence of molecular surface species that appear as sharp signals and that are resistant to hydrogenation and dehydrogenation at room temperature [11,85].

In the second step, measurements on changes in the topmost atomic layers such as the formation, transformation, or degradation of adsorbed species over time at different stages of a deactivation process were performed. A pyrophoric catalyst as taken directly from the reaction system with adherent mixtures of solvent and product molecules has been measured (Figure 19, left, (a)). Sharp bands are observed superimposed to the broad background

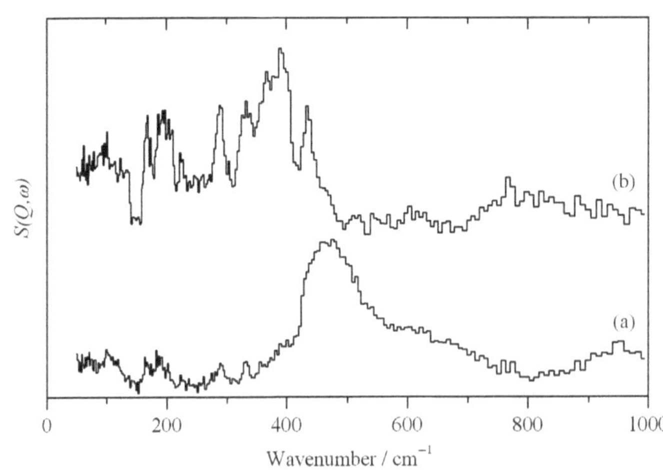

FIGURE 18 INS spectrum (TOSCA) of a Pd catalyst at an early stage of deactivation. (a) Under 1.5 bar H_2. The spectrum is largely that of β-PdH (broad signal at *ca.* 480 cm^{-1}). (b) After pumping at room temperature to decompose the β-PdH (x3 ordinate expansion relative to (a), sharp signals at *ca.* 291, 333, 371, 394, and 437 cm^{-1}, enhanced multiplet splitting). *Reprinted with permission from Ref. [11].*

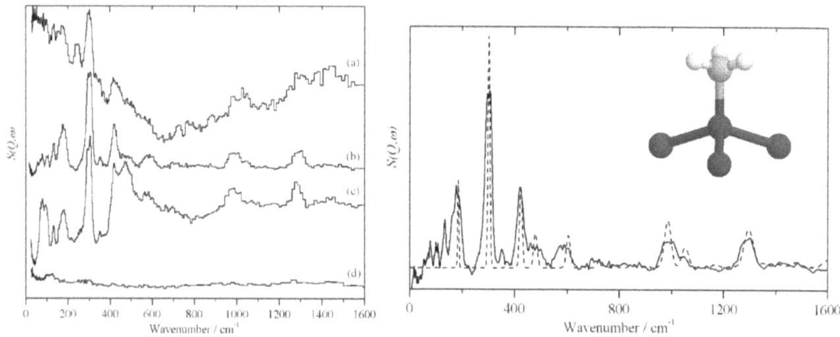

FIGURE 19 Left: Comparison of the INS spectra (TOSCA) of industrial palladium catalysts. (a) Spectrum of the deactivated catalyst as taken directly from the hydrogenation process and sealed under argon, (b) same sample as in (a) after solvent extraction, (c) same sample as in (a) and (b) after *in situ* hydrogenation up to 1.5 bar of hydrogen equilibrium pressure, and (d) a long-term used, but still active catalyst as measured after cleaning by solvent extraction under the same conditions as the deactivated sample. The spectra are normalized to the same total mass. Right: Comparison of the INS spectrum (TOSCA and ISIS) of the deactivated catalyst (solid line) and the results from the Wilson GF matrix method analysis of the spectrum (dashed line). The model used is shown in the top right-hand corner, Pd-CH_3 in C_{3v}. *Reprinted with permission from Ref. [11,36].*

that is attributed to the presence of residual organic reaction mixture. After extraction, the sharp bands of nonextractable, simple molecules remained (Figure 19b). The corresponding chemical species are largely resistant to extraction treatment and also to hydrogenation/dehydrogenation cycles (Figure 19c). This spectral signature is completely missing from a reference sample from a running sister plant, without deactivation phenomena (Figure 19d). Computer simulation of the spectral features of methylidyne, methylene, ethyl, and ethylidyne groups revealed that the only fit with reasonable quality and chemically reasonable frequencies was obtained for a surface methyl group.

Measurements on catalyst samples, which were taken from the running plant over several months, revealed that the normalized INS band intensities at 301 cm^{-1} perfectly correlated with the decreasing or increasing macroscopic hydrogenation activity of the catalyst running in the plant at changing efficiency level.

One conclusion was that the species detected at the beginning of a deactivation event might have been precursor molecules (of geminal substitution [86]) (Figure 18) and the formation of simple molecular entities to the methyl group level was observed over time.

The observations on the stability of the methyl species are in line with conclusions on the frustration of methane formation in Fischer-Tropsch catalysis [87] (see also Section 5.5.2) and also on the slow step in CO methanation being the reaction between methyl and hydrogen [88]. Here, CH_x $(x=0\text{–}3)$ species play a primary role.

5.4.3 Fuel Cell Catalysts

Supported platinum catalysts are used, for example, in catalytic hydrogenations of aromatic molecules such as nitroaromatic compounds to anilines (Section 5.4.1.2), the conversion of imines to amines, in reductive amination or alkylations, and many other processes [89]. Furthermore, supported and unsupported platinum catalysts are used in the synthesis of HCN from CH_4 and NH_3 [90], the oxidation of ammonia [91] in the production of nitric acid, automotive exhaust gas purification systems [92], and in fuel cell catalysts [93–95].

Improving the efficiency of fuel cell catalysts at substantially lower precious metal concentration level is a major task in catalyst optimization. This includes the fine-tuning of particle size, shape, and degree of alloying in bi- or multimetallic catalyst qualities.

With varying size, there is a correlation between the fraction of Pt surface atoms on the (100) and (111) crystal faces and the specific electrocatalytic activity. For cuboctahedral particles, an optimum mass activity at 3.5 nm particle size diameter was reported [96]. Details on the statistics of surface atoms and surface sites on metal catalysts are given in Refs. [97,98]. It is reported that (111) and (110) faces are less effective for oxygen reduction than the (100) face [99,100].

INS provides direct spectroscopic access to the dissociative chemisorbed, atomic species of hydrogen at the platinum particle's surface [101,102] even when the Pt is supported on high-surface-area carbon blacks [38,103]. These are conductive powders and, therefore, offer ideal synergies to electrochemical testing, cyclo-voltammetry, and high-resolution transmission electron microscopy together with statistical evaluations of the TEM images. Measurements under *in situ* hydrogen loading in the INS sample can at realistic pressure conditions allow evaluation of the changes in surface site occupation. This is for cubic and cuboctahedral environment in technical fuel cell catalysts of narrow primary particle distribution functions and how they depend on the average particle size and catalyst preparation technique.

Differing size and morphologies of the precious metal particles, as determined by TEM, can be correlated with different relative contributions of H on low and high-coordination sites as studied by proton dynamics *via* "site titration with atomic H." Introduction of disorder/defects in real nanoparticles supported on high surface area supports and changes of particle morphology under the influence of hydrogen can be studied to give input for improving future catalyst syntheses and long-term stability.

A central task is the identification and promotion of active centers and discrimination from "spectator" sites in search for particle properties that are really relevant for performance.

A range of different spectrometers is needed to obtain a complete picture of the catalyst's properties. Combining the information allows to study technical fuel cell catalysts under different aspects:

(1) Site occupation Pt_x-H, on top, twofold, threefold, and fourfold bridge, and on PtRu/C of different degree of alloying: INS on TOSCA [56,102]
(2) Detection of on-top hydrogen $\nu_{Pt\text{-}H}$ at 2080 cm^{-1}: INS on MAPS [104] (vibrational spectroscopy on TOSCA is ideal below 1600 cm^{-1} and on MAPS above 1600 cm^{-1})
(3) Pt-H distance (0.188 nm): neutron diffraction on SANDALS [105]
(4) Detection of Pt-OH and Pt_xRu_y-OH groups: INS on TOSCA [106]
(5) Studying catalytic dissociation of dihydrogen: quasielastic neutron scattering on OSIRIS [104])

5.4.4 Hydrogenation of Nitriles with Raney® Metals

Raney®-type catalysts are fine-grained metals with high (10–50 $m^2\ g^{-1}$) surface areas. They are produced by dissolution of aluminum with a concentrated aqueous alkaline solution from binary Al-metal alloys [107]. The most well known is Raney® nickel, but there are also Raney® cobalt, copper, and ruthenium. Raney® nickel is an excellent hydrogenation catalyst and will readily reduce benzene to cyclohexane. It is used commercially for the reduction of nitriles (–C≡N) to primary amines (–CH_2–NH_2), although secondary and tertiary amines occur as undesirable by-products. Raney® cobalt has better selectivity to primary amines but is less active [108].

Raney® metals are ideal materials for INS spectroscopy; they are available in large quantities, there is no support to complicate the spectra and they have good surface area; consequently, they have been extensively investigated. The INS spectrum of hydrogen on Raney® Ni is shown in Figure 20 [105]; this is similar to previous work [109,110] but has better resolution. The major peaks at 900 and 1050 cm^{-1} are assigned to the asymmetrical and symmetrical Ni–H stretch modes respectively of hydrogen in a threefold site on the (111) facet (which is the lowest energy surface for an fcc metal such as Ni). The assignment is by analogy to hydrogen on Ni(111) studied by surface science methods including high-resolution electron energy loss spectroscopy (HREELS) [111] and is confirmed by a periodic density functional (DFT) calculation of a monolayer of hydrogen on Ni(111). There is significant vibrational dispersion in this system, thus the calculation is for the complete Brillouin zone. The calculated spectrum does not account for all the spectral intensity, particularly in the 500–800 cm^{-1} regions, and there are clearly other facets present. These are most likely to be (100) and (110) [110].

It was also shown that it is possible to directly measure the Ni–H distance using neutron total scattering. This is also a difference method; diffractograms are recorded without and with adsorbed hydrogen; the Fourier transform of the difference diffractogram yields the radial distribution function of the adsorbate from which the interatomic distances can be extracted by inspection. Figure 21 shows the result; the negative-going peaks are Ni–H distances, the first peak at 1.68 Å is the Ni–H bond distance, and the peaks at 3.00 and 4.00 Å are the

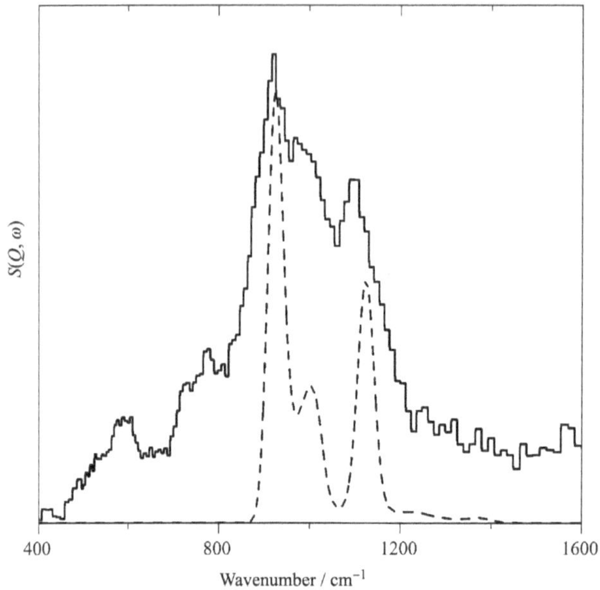

FIGURE 20 INS spectrum recorded at 20 K on TOSCA at ISIS of adsorbed hydrogen on a Raney® nickel catalyst (solid line). Also shown (dashed line) is a DFT simulation for one monolayer of hydrogen on Ni(111). The calculation is for the complete Brillouin zone; the transition energies have been scaled by 0.93. *Reprinted with permission from Ref. [105].*

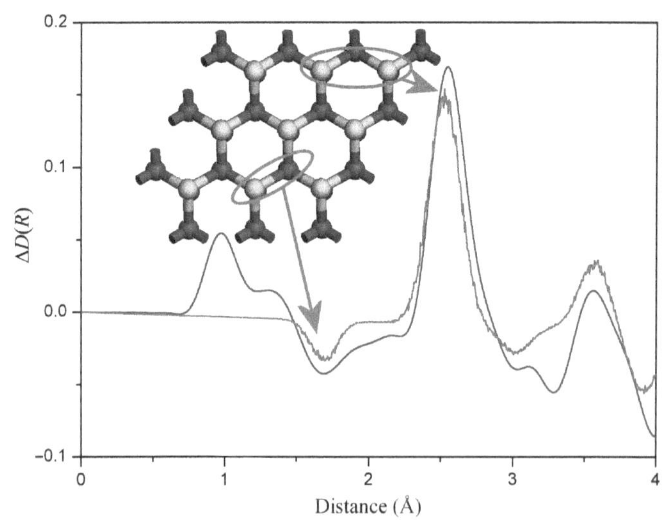

FIGURE 21 Difference pair distribution function, $\Delta D(R)$, for hydrogen on Raney® nickel (black) at room temperature. (The peak at ~1.00 Å is nonphysical and results from a truncation error in the Fourier transform.) Also shown is the pair distribution $D(R)$ function (gray) generated from the optimized structure (also used for the phonon calculation) shown in the inset. The spectrum is abscissa scaled such that the large peaks at 2.54 Å have the same amplitude. No ordinate scaling has been applied. *Reprinted with permission from Ref. [105].*

distance to the nickel atom below the hydrogen and to the corresponding atom in the adjacent cells. (Nickel and hydrogen have positive and negative scattering lengths, respectively. A negative scattering length means that the incident wave undergoes a 180° phase change on scattering [1].)

The adsorption of hydrogen on Raney® cobalt has also been studied [112,113]. This results in a much more complex spectrum with a broad band of intensity from ~300 to 1100 cm^{-1}. In this case, there appears to be approximately equal populations of all the low-energy facets: (100), (110), and (111).

A major use of Raney® metals is the hydrogenation of nitriles to amines, for example, adiponitrile to hexamethylenediamine that is used to produce Nylon:

$$NC(CH_2)_4CN + 4H_2 \rightarrow H_2N(CH_2)_6NH_2$$

The adsorption of acetonitrile ($CH_3C{\equiv}N$) on Raney® nickel results in weak chemisorption as the INS spectrum of acetonitrile is essentially unperturbed (the modes shift down by only 10 cm^{-1} or so) [114]. Gas-phase catalytic hydrogenation of acetonitrile over Raney® nickel [115] showed that while both weakly and strongly bonded atomic hydrogen were present, INS spectroscopy demonstrated that only the weakly bonded (presumably on-top sites) was active for hydrogenation. Acetonitrile and reactive hydrogen compete for the same sites, while amines and acetonitrile are adsorbed on different sites; thus, the product does not poison the catalyst.

In contrast, adsorption of acetonitrile-D3 ($CD_3C{\equiv}N$) [112] (Figure 22) or butyronitrile ($CH_3CH_2CH_2C{\equiv}N$) [113] on Raney® cobalt results in strongly

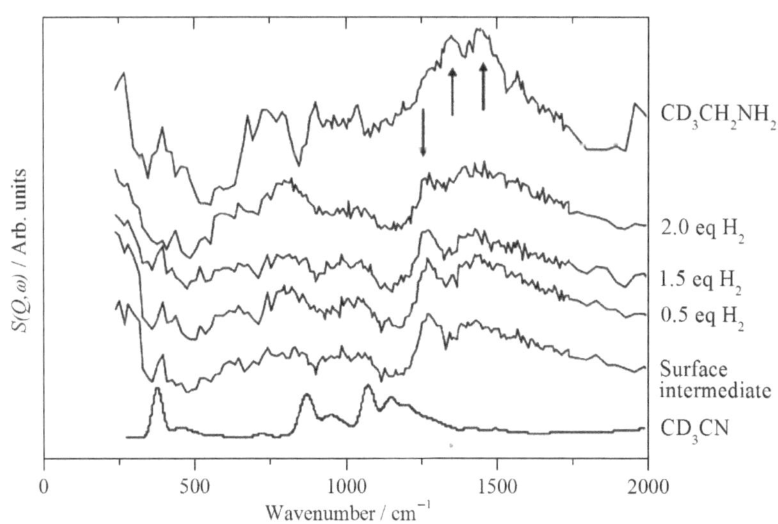

FIGURE 22 INS spectra of hydrogen coadsorbed with CD_3CN on activated Raney® Co recorded with IN1BeF at the ILL. The amount of hydrogen was varied in the range H_2/CD_3CN 0, 0.5, 1.5, and 2.0. The CD_3CN loading was 0.18 molecules per surface metal atom. For comparison, the spectrum of $CD_3CH_2NH_2$ adsorbed on activated Raney® Co is also given (top) and the DFT calculated spectrum of CD_3CN (bottom). The spectrum of activated Raney® Co has been subtracted from all the experimental spectra shown. *Reprinted with permission from Ref. [113].*

perturbed species. The use of acetonitrile-D3 is interesting because the methyl group is a spectator in this process (NMR spectroscopy shows that less than 1% of the methyl protons exchange); thus, the changes that occur are directly associated with the hydrogenation of the nitrile. The authors concluded that the nitrile reacted preferentially with hydrogen bound to threefold sites on the Co(001) plane. On the surface, a mixture of a nitrene-like ($CD_3CH_2\ddot{N}=Co$) intermediate and coadsorbed $CD_3CH_2NH_2$ was formed, as indicated by the strong CH_2 twisting mode at 1278 cm^{-1}, which decreased in intensity relative to the other CH_2 vibrational modes, when the amount of hydrogen was increased.

5.5 CATALYSTS FOR COMMODITY CHEMICALS

5.5.1 Methyl Chloride Synthesis

Methyl chloride is an important industrial product with a global capacity of *ca.* 900 ktonne. It is a valuable starting material in the production of higher chlorinated products, silicones, and methyl cellulose. It also has applications as a methylating agent [116]. One method of the large-scale manufacture of methyl chloride is performed *via* the hydrochlorination of methanol over a solid acid catalyst at elevated temperatures (Eq. 1). Selectivity to methyl chloride is reduced by the reaction shown in Eq. (2), where the alcohol is converted to dimethyl ether. An improved methyl chloride synthesis catalyst will catalyze reaction (1) more effectively than reaction (2). A catalyst that has received considerable application is γ-alumina [117].

$$CH_3OH + HCl \rightarrow CH_3Cl + H_2O \tag{1}$$

$$2CH_3OH \rightarrow CH_3OCH_3 + H_2O \tag{2}$$

As part of a program to develop a new generation methyl chloride synthesis catalyst based on η-alumina rather than γ-alumina, INS has been used to probe aspects of the surface chemistry of the candidate replacement alumina. A review of this topic by Parker and Lennon appeared in 2014 [118], and subsequently, Lennon and Winfield reviewed aspects of the surface chemistry of relevant reactions over η-alumina by a variety of techniques [119]. Further, reaction profiles presented at a recent Faraday Discussions meeting demonstrate how modified η-alumina catalysts can lead to improved product yields [120]. This section is limited to reviewing how INS has been able to provide valuable information on (i) the surface acidity of η-alumina, (ii) the interaction of methanol on η-alumina, and (iii) the interaction of hydrogen chloride on η-alumina.

5.5.1.1 The Surface Acidity of η-Alumina

Pyridine chemisorption and inspection by vibrational spectroscopy, typically infrared spectroscopy, is a well-established procedure for investigating the

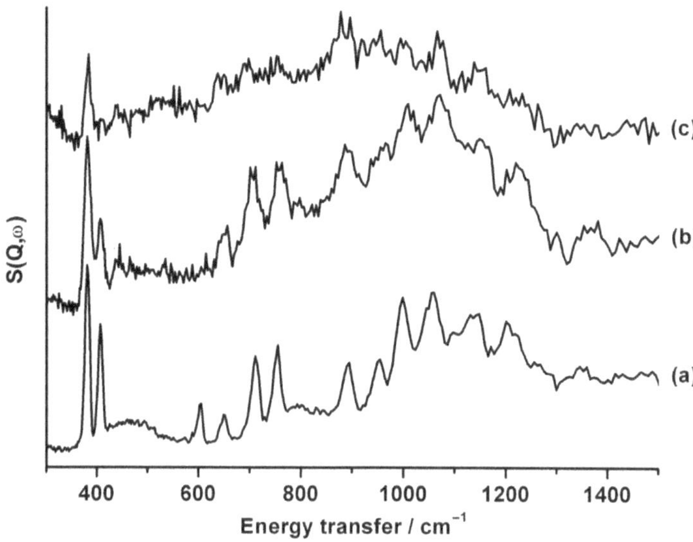

FIGURE 23 INS spectra recorded at 20 K of (a) solid pyridine, (b) background subtracted η-Al_2O_3 + 896 μmol pyridine g^{-1}, and (c) the chemisorbed pyridine spectrum after heating the adsorbed overlayer to 473 K in flowing helium for 30 min. *Reprinted with permission from Ref. [122].*

surface acidity of solid acid catalysts [121]. Temperature-programed infrared measurements classify the acid site distribution of η-alumina as follows: weak, medium, and strong Lewis acid sites [122]. Comparable measurements using INS for pyridine adsorption on η-alumina recorded using the TOSCA spectrometer are shown in Figure 23 for the region 300–1500 cm^{-1}. These measurements complement infrared measurements that, due to strong Al-O phonon modes, are restricted to energies above 1100 cm^{-1}.

Figure 23a shows the INS spectrum of pyridine, where distinct bands below *ca.* 1400 cm^{-1} are evident. The spectrum of a pyridine overlayer (896 μmol pyridine g^{-1}) on activated η-alumina is shown in Figure 23b. This amount of pyridine retention by the catalyst involves a substantial physisorption component, so not unsurprisingly, the overlayer spectrum is similar to the reference spectrum. However, it is noted that the ν_{6a} A_1 mode about 605 cm^{-1} is absent in the spectrum of the adsorbed overlayer. The reason why this particular mode uniquely should be attenuated on adsorption is unknown. Additionally, a weak but distinct feature at 440 cm^{-1} is also evident in Figure 23b. This could arise as a result of shifts in the pyridine 16a and/or 16b modes at 380 and 406 cm^{-1}, respectively.

Figure 23c presents the spectrum after the overlayer has been warmed to 473 K in a continuous stream of helium. This thermal treatment desorbs physisorbed pyridine and chemisorbed pyridine residing in all of the weak Lewis acid sites and *ca.* 35% of medium Lewis acid sites [122]. Thus, Figure 23c corresponds to the INS spectrum of pyridine residing in predominantly medium and strong Lewis acid sites. Overall, the spectrum is characterized by a general reduction in scattering intensity, with four notable exceptions. Firstly, the loss of the 16a out-of-plane skeletal bending mode at 408 cm^{-1} is observed, while the 16b feature at 382 cm^{-1} remains present. Figure 23a and (b) shows that the 16a mode is inherently weaker than the 16b mode, and its loss in Figure 23c is attributed to a reduction in intensity to background levels, rather than any site-specific desorption process. Secondly, the relatively weak signal at 437 cm^{-1} remains intact after the heating process, suggesting that this feature arises from pyridine in strong Lewis acid sites. The 16a requires a relatively large shift to high frequency of 57 cm^{-1}, whereas a more modest 31 cm^{-1} is required for the 16b mode. Figure 23c shows the 16a mode to exhibit the greatest amplitude of motion, and therefore, tentatively, the weak feature remaining at 437 cm^{-1} is assigned to a perturbation of this mode. Thirdly, with respect to the pyridine reference spectrum (Figure 23a) modest shifts of 10, −15, and +9 cm^{-1} are seen, respectively, for the ν_{11}, ν_{10a}, and ν_{18a} modes. Finally, the absence in Figure 23b and c of any obvious relative intensity enhancement of the 18a in-plane C–H deformation about 1060 cm^{-1} confirms the absence of substantial Brønsted acidity for this catalyst, as this feature dominates the INS spectrum of the pyridinium ion [123]. Collectively, these results indicate that the INS spectrum of an adsorbed probe molecule offers some potential in evaluating surface acidity. It is complementary to the infrared spectrum, and the two very different methodologies provide the full vibrational spectrum of the chemisorbed base. Extending the dynamic range in this way provides additional boundary conditions against which theoretical efforts to define further the alumina/pyridine interaction can be evaluated and refined. It is acknowledged that supplementary infrared spectroscopy and temperature-programed desorption measurements have enabled the medium-strength Lewis acid site to be subdivided into two components: medium-weak and medium-strong Lewis acid sites [122].

5.5.1.2 The Interaction of Methanol on η-Alumina

The interaction of methanol on η-alumina is central to the hydrochlorination reaction under consideration (Eq. 1). Figure 24a shows the INS spectrum obtained for a sample of solid methanol recorded using the TOSCA indirect geometry spectrometer. The intense structured feature below 400 cm^{-1} is due largely to the methyl torsion. The complex band shape at 690–780 cm^{-1} is assigned to the out-of-plane O–H bend. The two other major features, at 1150 and 1450 cm^{-1}, are assigned, respectively, to the CH_3 rock and a mode resulting from the overlap of CH_3 deformation modes and the in-plane O–H bend [124]. Figure 24b shows the INS spectrum for a saturated chemisorbed overlayer of methanol on η-alumina prepared at

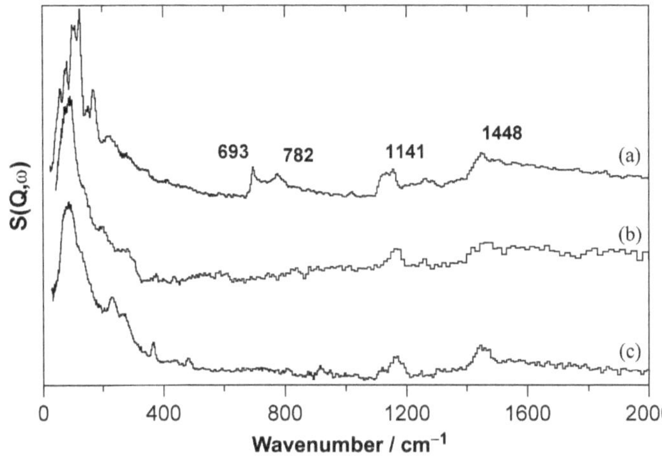

FIGURE 24 INS spectra recorded using TOSCA of (a) methanol, (b) background subtracted spectrum of a saturated chemisorbed overlayer of methoxy on η-alumina (1.12 mmol CH_3OH g^{-1}) prepared at 300 K, and (c) $Al(OCH_3)_3$. *Reproduced with permission from Ref. [124].*

300 K. The relatively weak spectrum is indicative of the low density of active sites present on the alumina surface. The spectrum is characterized by the intense CH_3 torsion at 84 cm^{-1}, the CH_3 rock at 1170 cm^{-1}, and the CH_3 deformation modes at 1460 cm^{-1}. Further refinement of the assignments is hindered by the low signal-to-noise ratio of Figure 24b and the matter of resolution broadening for bands at higher energies, which becomes noticeable for features above *ca.* 1000 cm^{-1}. The spectrum is clearly different from that of methanol, as indicated by the absence of the out-of-plane O–H bend at 690–780 cm^{-1}. Figure 24b is assigned to chemisorbed methoxy species, formed from the dissociative adsorption of methanol.

Confirmation of this assessment of the alumina/methanol adsorption system was achieved by recording the INS spectrum of a model compound, $Al(OCH_3)_3$. This spectrum is shown in Figure 24c. There is reasonable agreement with the spectrum obtained for adsorbed methanol, in particular the coincidence of bands at 87, 1170, and 1460 cm^{-1}. The difference between Figure 24b and c is the result of the three interacting methoxy groups in $Al(OCH_3)_3$, whereas the low density of active sites on the η-alumina surface means that the surface methoxy groups do not interact with one another.

Methanol when heated in the presence of an activated alumina catalyst produces dimethyl ether (DME) (Eq. 2). Figure 25a shows the INS spectrum obtained after adsorption of excess methanol on η-alumina followed by subsequent heating to 473 K in a closed cell. A reference spectrum of solid dimethyl ether is presented in Figure 25b. In the region 1100–1600 cm^{-1}, Figure 25a is broadly similar to that reported for adsorbed methanol (Figure 24b). These similarities occur as methanol and DME share similar

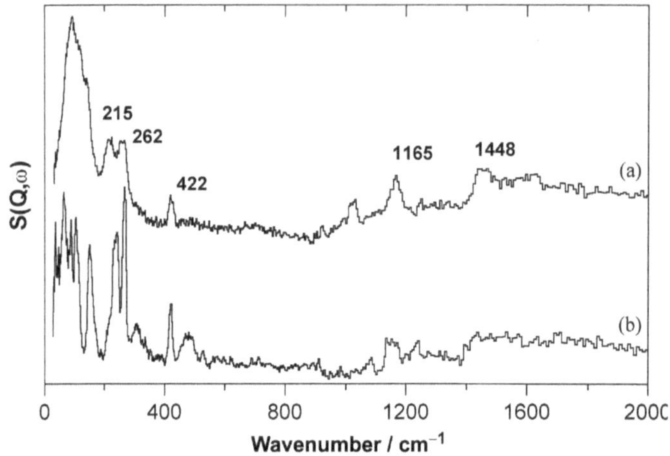

FIGURE 25 INS spectra recorded using TOSCA of (a) background subtracted spectrum of η-alumina dosed with excess methanol (1.57 mmol CH_3OH g^{-1}) at 300 K, isolated and warmed to 473 K for 15 min, and (b) dimethyl ether. *Reproduced with permission from Ref. [124].*

structural motifs, that is, the CH_3 rock and CH_3 deformation peaks. However, distinct bands at 218, 260, and 420 cm^{-1} are present in Figure 25a but absent in Figure 25b. By comparison to the reference spectrum (Figure 25b), these bands are respectively assigned as the ν_{11} A_2 out-of-phase methyl torsion, the ν_{21} B_2 in-phase methyl torsion, and the ν_7 A_1 C–O–C bend of dimethyl ether [124]. The band at 480 cm^{-1} in Figure 25b is assigned to a combination mode of the C–O–C bend and the lattice mode at 62 cm^{-1} (480 $cm^{-1} \approx 417 + 62$ cm^{-1}). Its absence in Figure 25a indicates that the dimethyl ether is disordered on the surface.

Thus, Figure 25a can be explained in terms of formation of dimethyl ether from the dimerization of surface methoxy species. However, while the INS spectrum can distinguish the formation of DME, for instance, *via* the growth of the O-C-O bend of dimethyl ether at 420 cm^{-1}, it cannot profile the loss of surface methoxy units. This is because the comparable molecular structure of these moieties means that they share methyl torsions, rocks, and deformations that are essentially indistinguishable from each other. Separate temperature-programed desorption measurements using a continuously flushed tubular reactor and a mass spectrometer indicate that the methoxy → DME conversion is incomplete at 473 K [125]. Therefore, Figure 25a represents a contribution from dimethyl ether formed from a combination of surface methoxy units, plus a quantity of unreacted methoxy species. The changes noted as a chemisorbed overlayer prepared at 300 K (Figure 25b) is warmed to 473 K (Figure 25a) indicate that the isolated INS cell can act as a batch reactor, where further warming would presumably convert all of the methoxy species to dimethyl ether.

A further set of INS experiments were undertaken on this adsorption system using the direct geometry MARI spectrometer and exploiting the capability of such instruments in being able to measure neutron scattering intensity simultaneously as a function of energy transfer and momentum transfer [13]. Those studies were able to explain an anomalous feature observed in diffuse reflectance infrared spectroscopy measurements at 2600 cm^{-1}, thereby unambiguously defining the stable surface species that result from methanol exposure to an activated η-alumina catalyst.

5.5.1.3 *The Interaction of Hydrogen Chloride on η-Alumina*

The adsorption of HCl on alumina is thought to occur *via* two distinct pathways [126]. The first involves the dissociation of HCl onto aluminum oxygen pairs forming new hydroxyl groups and Al-Cl species **(I)**. The second pathway also proceeds via dissociative adsorption of the HCl but via a hydroxyl exchange mechanism, where chlorine replaces a surface hydroxyl group coordinated to an aluminum ion **(II)**.

Figure 26 presents the INS spectrum for activated η-alumina catalyst recorded using the MARI spectrometer operating at incident energies of (a) 4840 and (b) 1290 cm^{-1}. A broad feature about 3700 cm^{-1} that ranges from *ca.* 3600 to 3800 cm^{-1} with a maximum at 3676 cm^{-1} is assigned to the OH stretch of hydroxyl groups present at the alumina surface. The peak at 905 cm^{-1} is assigned to the deformation mode of surface hydroxyl groups [127]. The intense feature at *ca.* 200 cm^{-1} is assigned to the out-of-plane deformation of surface hydroxyl groups, γ-(OH). This mode approximates to a torsional motion of the hydroxyl group [127].

How HCl exposure modifies the INS spectrum of the η-alumina catalyst is indicated in Figure 27. Spectra are recorded at a primary energy of 1280 cm^{-1} in order to optimize resolution of the hydroxyl deformation modes. The spectrum for the activated catalyst is shown in Figure 27a, whereas Figure 27b shows the spectrum corresponding to a saturation exposure of HCl. Figure 27c shows the difference spectrum (i.e., adsorbed spectrum (b) and background spectrum (a)).

The HCl dosed spectrum (Figure 27b) shows some attenuation of the γ-(OH) feature at *ca.* 210 cm^{-1} and changes are apparent in the δ-(OH) region, particularly about 1000 cm^{-1}, indicating a shift to higher wave number for this feature. The background subtracted spectrum (Figure 27c) demonstrates these perturbations to the hydroxyl signals more clearly. The negative inflection about 180–210 cm^{-1} indicates the loss of hydroxyl groups contributing to the γ-(OH) mode. A broad feature at *ca.* 550 cm^{-1} is discernible; this is assigned to the librational mode of water [41], which is formed as a consequence of process **(II)**. In addition, the hydroxyl deformation mode shows an increase in intensity at *ca.* 1000 cm^{-1}, which is resolvable into two peaks at 990 and 1050 cm^{-1}. Formation of hydroxyl groups, via process **(I)**, will lead to a positive feature about 1000 cm^{-1}, which is attributed to the (OH) deformation mode [ref].

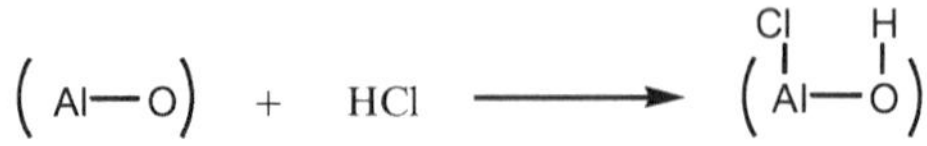

(I)

(II)

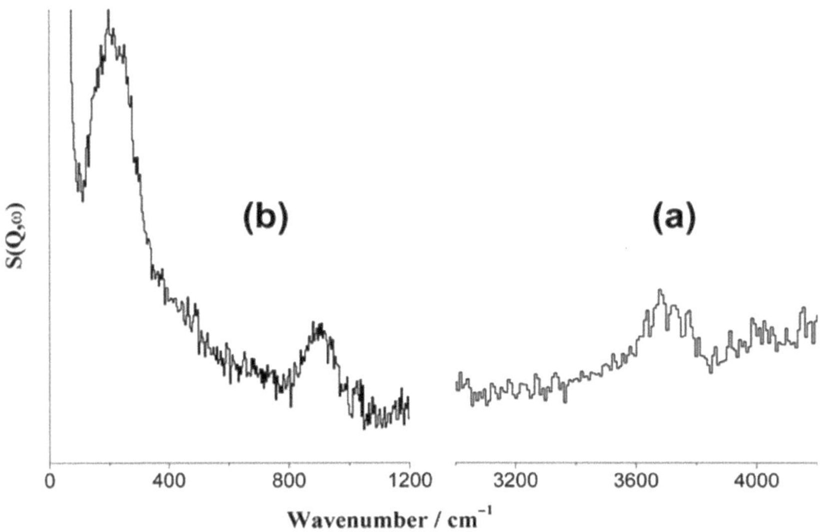

FIGURE 26 INS spectra of the η-alumina catalyst after activation at 623 K recorded using the MARI spectrometer with an incident energy of (a) 4840 cm^{-1} (3000–4200 cm^{-1}) and (b) 1290 cm^{-1} (50–1200 cm^{-1}). *Reproduced with permission from Ref. [126].*

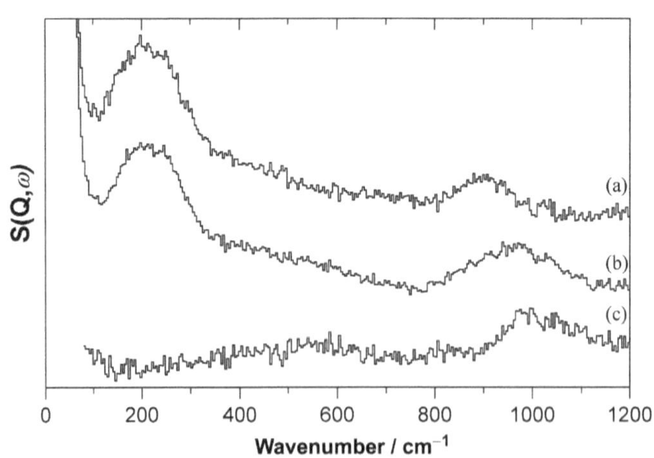

FIGURE 27 INS spectra recorded using the MARI spectrometer with an incident energy of 1280 cm^{-1} for (a) the η-alumina background and (b) a saturated chemisorbed overlayer of HCl on η-alumina and (c) the difference spectrum (b)–(a). *Reprinted with permission from Ref. [127].*

In a similar manner to that mentioned above when considering the η-alumina/methanol adsorption system (Section 5.5.1.2), INS measurements supplemented infrared spectroscopic measurements, with the former helping to explain anomalous features seen in the optical spectra.

5.5.1.4 Site-Selective Chemistry

Pyridine adsorption/desorption INS measurements on the activated η-alumina catalyst helped a comprehensive description for the acid site distribution of the candidate catalyst to be developed (Section 5.5.1.1). Thereafter, INS investigations on how the individual reagents react with the η-alumina catalyst were undertaken. Firstly, methanol adsorption on the η-alumina catalyst confirmed methoxy species to be the only adsorbed species and, moreover, eliminated doubts about a band seen in IR measurements at $2600\ cm^{-1}$ (Section 5.5.1.2). Secondly, HCl chemisorption was seen to perturb the surface hydroxyl population in a manner consistent with previously postulated chemical pathways (Section 5.5.1.3). Collectively, this awareness of substrate/adsorbate interactions enabled researchers to describe conventional microreactor studies in terms of defined surface species occupying specific active sites [127]. Just recently, that analysis has been extended to define a role for chemical modifiers; specifically, how certain modifiers can induce site-selective chemistry to ultimately improve the atom economy of this particular industrial chemical manufacturing process [120].

5.5.2 Fischer-Tropsch Catalysis

Fischer-Tropsch synthesis (FTS) uses synthesis gas (CO and H_2) obtained from resources such as coal, natural gas, and biomass to produce a variety of hydrocarbon products [128]. Several variants of the process operate around the world, with the catalysts predominantly based on Fe- and Co-containing materials [129]. Whereas recent large-scale unit operations have tended to feature Co catalysts [130], Fe-based catalysts are used extensively, and, furthermore, they uniquely provide the possibility of accessing olefins and alkanes, which otherwise tend to dominate the FTS product slate [131,132].

In 2003, Hans Schulz, citing an article by Albers and coworkers [36], highlighted the potential of neutron scattering spectroscopy to determine the presence of stable surface species and the relevance of that work to FTS [87]. Indeed, whereas conventional analytic techniques such as Mössbauer spectroscopy, X-ray diffraction, and temperature-programed oxidation applied to Fe-based systems are able to characterize the catalysts in terms of solid-state chemistry (hematite, magnetite, and iron carbides) and the presence of carbonaceous materials, importantly, INS is one of the few probes that can provide information as to how hydrogen is partitioned within the catalyst matrix [13].

A significant advancement in this position was realized in 2012 when Hamilton and coworkers obtained the INS spectrum [133] of an iron-based

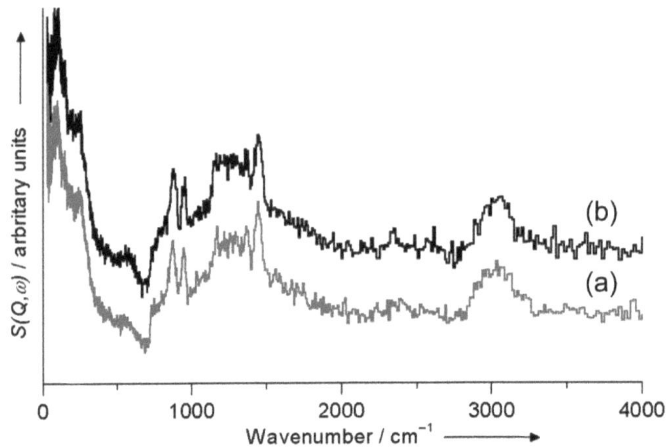

FIGURE 28 INS spectra of F-T catalyst samples obtained with the TOSCA spectrometer in the 20–4000 cm^{-1} region. (a) The toluene extracted sample and (b) toluene/CH_2Cl_2 extracted sample. *Reprinted with permission from Ref. [133].*

TABLE 2 Main Features and Associated Vibrational Assignments for the INS Spectrum of the Iron-Based Catalysts After Extraction in Toluene

Energy/cm^{-1}	Assignment
3050	Aromatic/olefinic C–H stretch
2940	Aliphatic C–H stretch
1441	CH_2 scissors of a cycloalkane
1368	Aromatic C–H wag
1168	In-plane deformation of aromatic C–H
938	*Cis/trans*-wag of olefinic CH_2/out-of-plane deformation of aromatic C–H
865	Out-of-plane C–H deformation
231	Methyl torsion
89	Fe-O phonon

Reproduced with permission from Ref. [133].

FTS catalyst that had been extracted from a coal-to-liquids FTS plant in Secunda (South Africa). Prior to spectral acquisition, product was removed from the sample by solvent extraction. Figure 28 represents a rather distinct spectrum; band assignments are presented in Table 2.

The spectrum is attributed to a hydrocarbonaceous overlayer with approximately equal numbers of H atoms bonded to sp^2 and sp^3 hybridized C atoms.

Moreover, it was postulated that the hydrocarbonaceous species could possibly play a role in the process chemistry; namely, moderating the supply of reagents within the catalyst matrix during the CO dissociative adsorption and CH_x chain propagation processes [133].

Two further studies examined the characteristics of an iron-based FTS catalyst using the CO hydrogenation reaction at ambient pressure as a test reaction [134,135]. This reaction was selected because it involves $Fe/CO/H_2$ surface chemistry and produces gaseous products (e.g., methane) that will not compromise postreaction INS spectra. Specifically, in the presence of long-chain saturated hydrocarbon product, INS would not be able to discern stable surface species that may contribute to the molecular transformation process. The two studies explored how different catalyst activation procedures influence catalyst performance and led to a suggestion that the hydrocarbonaceous overlayer may be an integral component of the working catalyst matrix [135].

In a further illustration as to how INS can provide new insight in to well-established heterogeneously catalyzed reaction systems, Warringham and coworkers used the MARI spectrometer to analyze a high-specification iron-based FTS catalyst. This precatalyst exhibited superior surface area and a more homogeneous crystallite size than the laboratory sample examined in the preceding studies that was prepared using a manual procedure [136]. The new catalyst sample was prepared using an automated preparative procedure that ensured better control of the precipitation stage of the catalyst preparative process.

Figure 29a shows the INS spectrum in the 1800–4000 cm^{-1} region of the clean catalysts prior to reaction recorded using the MAPS spectrometer.

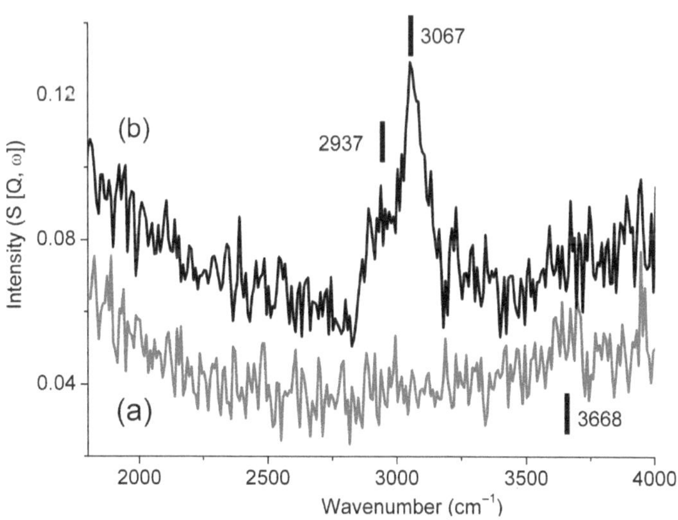

FIGURE 29 The INS spectrum recorded at an incident energy of 4840 cm^{-1} of the iron catalyst after (a) dehydrating conditions at 373 K and (b) CO hydrogenation conditions at 623 K for 6 h. *Reprinted with permission from Ref. [136].*

Retaining ambient pressure CO hydrogenation as a test reaction, Figure 29b shows the INS spectrum after 6 h time-on-stream at 623 K.

Figure 29b shows the spectrum of the reacted sample, which is characterized by a sharp feature at 3067 cm^{-1} and a low-energy shoulder at 2937 cm^{-1}. These features can be assigned to ν(C–H) modes: sp^2 hybridized carbon, that is, aromatic/olefinic ν(C–H), at 3067 cm^{-1}, and sp^3 hybridized carbon, that is, aliphatic ν(C–H), at 2937 cm^{-1}.

The dehydrated catalyst spectrum (Figure 29a) is effectively featureless, which is in dramatic contrast to the spectrum reported for the hematite precatalyst prepared using manual methods that displayed a strong ν(O–H) band corresponding to surface hydroxyl groups [134]. This contrast established that INS spectroscopy can distinguish how different preparative methods can affect hydroxyl populations at the catalyst surface.

INS spectra of the prereaction and postreaction catalyst samples in the region 700–1500 cm^{-1} are presented in Figure 30. The spectrum of the dehydrated sample (Figure 30a) is characterized by two features located at 649 cm^{-1} and 810 cm^{-1}. The feature at 649 cm^{-1} is tentatively ascribed to the Fe-O phonon mode of the α-Fe_2O_3 starting material [136]. Assignment of the 810 cm^{-1} peak caused some difficulty. Tentatively, it was assigned to a magnetic transition of the hematite lattice, a magnon, or a spinon, but further evidence was sought to bolster this assignment (see below).

The reacted catalyst spectrum (Figure 30b) is characterized by bands centered at 1447, 1362, 1165, 947, 874, and 584 cm^{-1}. The 1447 cm^{-1} band is assigned to semicircle ring deformation modes that are possibly linked with

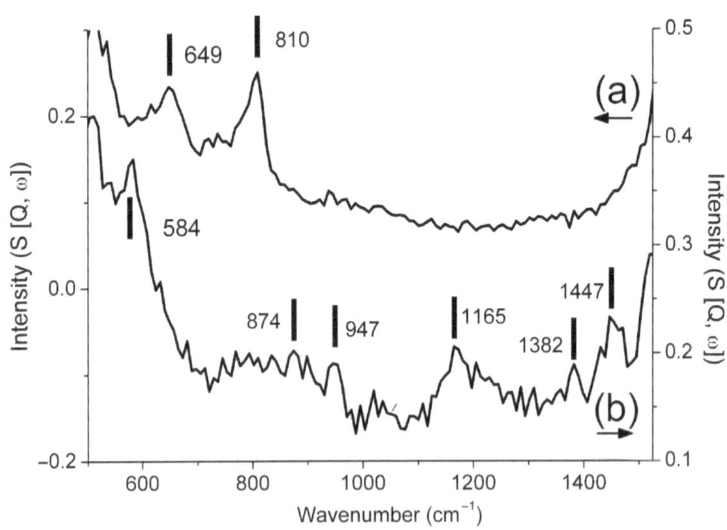

FIGURE 30 The INS spectra recorded at an incident energy of 1613 cm^{-1} of the scaled-up iron catalyst after (a) dehydrating conditions at 373 K and (b) CO hydrogenation conditions at 623 K for 6 h. *Reprinted with permission from Ref. [136].*

a δ-(CH) mode associated with perimeter carbons of an extended polycyclic aromatic network. The weak 1362 cm^{-1} feature is assigned to an in-plane ring deformation of a naphthalene-type molecule, and the 1165 cm^{-1} band is assigned to a CC–H in-plane deformation mode of a polyaromatic hydrocarbon. The feature at 947 cm^{-1} is assigned to alkenic δ-(C–H). The band just discernible at 874 cm^{-1} is assigned to an out-of-plane CH deformation of either an olefinic or aromatic group. The feature at 584 cm^{-1} is tentatively assigned to the A_{1g} Fe-O phonon mode of Fe_3O_4. Overall, these assignments show the presence of a predominantly olefinic/aromatic hydrocarbonaceous overlayer, which additionally possesses some aliphatic character, as demonstrated by the ν(C–H) mode observed at 2937 cm^{-1}.

As mentioned above, the relatively intense band at 810 cm^{-1} in Figure 30a was tentatively attributed to a spinon feature of α-Fe_2O_3. A spinon is a collective excitation of the spins that comprise the magnetic sublattice [137], which in the case of α-Fe_2O_3 consists of antiferromagnetically aligned iron ions. As a result of the magnetic moment of the neutron, INS has the capability to probe the magnetic properties of materials [3]. Direct geometry instruments such as MARI can simultaneously determine both energy and momentum exchange [13]; this capability was used to further investigate the origins of the 810 cm^{-1} feature. Correlated energy transfer (ω, cm^{-1}) and momentum transfer (Q, $Å^{-1}$) plots are presented in Figure 31 for the dehydrated iron oxide sample and the reacted sample. Figure 31(left) presents the two-dimensional INS spectrum of the dehydrated precatalyst recorded at an incident energy of 1613 cm^{-1}. In fact, it is the integrated neutron scattering intensity over the full range of Q correlated with energy transfer that yields the INS spectrum presented in Figure 30a. The 810 cm^{-1} band of Figure 30a is evident in Figure 31(left) at low Q but its intensity is observed to rapidly diminish with increasing Q.

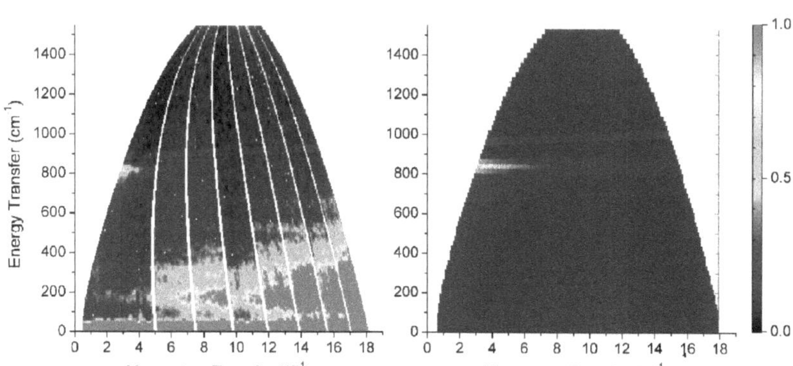

FIGURE 31 Comparison of neutron scattering intensity as a function of energy and momentum transfer (mitre plots) for the iron oxide catalyst after (left) dehydrating conditions at 373 K, recorded at an incident energy of 1613 cm^{-1}, and (right) the simulated powder-averaged spin wave dispersion of α-Fe_2O_3. *Reprinted with permission from Ref. [136].*

Figure 31(right) shows the calculated powder-averaged spin wave dispersion for hematite, the energy and profile of which is in excellent agreement with the experimental observation (Figure 31(left)). The peak of intensity near $810\ cm^{-1}$ arises from a peak in the density of states of spin waves at this energy, unambiguously confirming this band to be a spinon. This heightened confidence in the origins of the strong band seen in the INS spectrum for the iron oxide precatalyst (Figure 30a) and the observation that this mode, a spinon of α-Fe_2O_3, is absent from the spectrum of the reacted catalyst (Figure 30b) indicate that the spinon transition plays no part in the CO hydrogenation chemistry studied here. However, given that the presence and intensity of this feature is expected to reflect, in part, physical attributes such as particle size and iron oxide crystallite homogeneity [138,139]—properties thought to influence catalytic performance—a more detailed analysis of this feature with respect to different catalyst formulations may be productive in future catalyst developments. Clearly, INS is providing useful insight in to the physicochemical properties of technical grade FTS catalysts.

5.5.3 Synthesis of Acrylics—Methyl Methacrylate

Methyl methacrylate ($CH_2{=}C(CH_3)COOCH_3$, MMA) is the monomer used in the production of poly(methyl methacrylate), a major product of the acrylics manufacturing sector that is marketed under trade names such as Plexiglas, Lucite, and Perspex. The polymer has wide application in areas such as aviation and automobile components as well as various display devices. The conventional route to MMA is the acetocyanohydrin process. HCN is produced, for example, in the Andrussow oxidation process from CH_4, NH_3, and O_2 or in the Degussa process (BMA) from CH_4 and NH_3 [90] using platinum-based catalysts. HCN is then reacted with acetone to form ultimately MMA [140,141]. However, throughout the world, new routes using a variety of reagents are currently being developed [142]. One such process is the Alpha process that has been developed and commercialized by Lucite International [141]. This is a two-stage process involving a homogeneously catalyzed reaction that combines ethene, methanol, and carbon monoxide to produce methyl propanoate ($CH_3COOCH_2CH_3$). In the second stage, a heterogeneous catalyst is used to react the ester with formaldehyde (H_2CO) to yield MMA (Figure 32). A representative catalyst for the second stage of this particular MMA manufacturing process is cesium nitrate supported on a high-surface-area silica [144].

A survey of the literature indicated a paucity of studies examining the interaction of oxygenates and esters in particular on oxide surfaces; therefore, work was undertaken to explore the interaction of methyl propanoate on a representative silica, namely, Fuji Q-10 silica spheres [145]. As methyl propanoate has been neither structurally nor spectroscopically characterized, initial efforts concentrated on a comprehensive description of the solid state structure and vibrational spectroscopy of methyl propanoate in the gas, liquid,

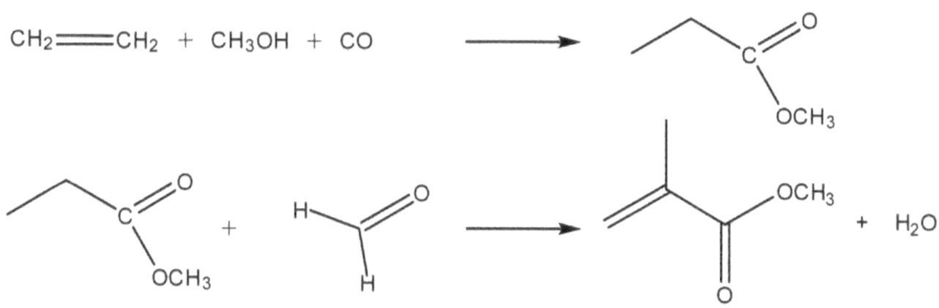

FIGURE 32 The Alpha process: Stage 1 (top)—the manufacture of methyl propanoate from ethene, methanol, and CO (homogeneous catalysis). Stage 2 (bottom)—the production of methyl methacrylate from the reaction of methyl propanoate and formaldehyde over a heterogeneous catalyst. *Reprinted with permission from Ref. [143].*

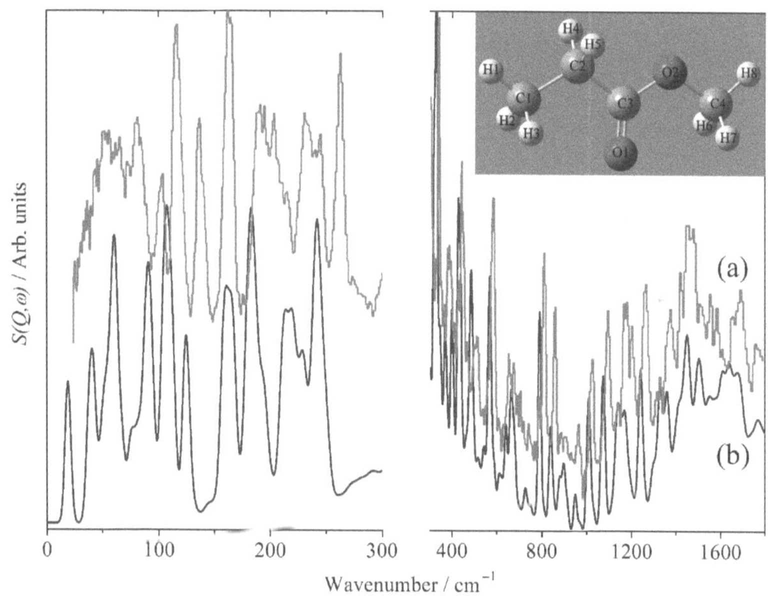

FIGURE 33 INS spectra of (a) solid methyl propanoate at 20 K and (b) CASTEP calculated solid state spectrum. The range 300–1800 cm^{-1} is ordinate expanded × 4 relative to the 0–300 cm^{-1} region. The inset shows the structure and numbering scheme. *Reprinted with permission from Ref. [143].*

and solid phases, with assignments supported by periodic density functional theory (DFT) [143]. The resulting analysis then enabled the adsorption geometry of methyl propanoate on silica to be evaluated. INS spectra were recorded using both indirect geometry (TOSCA) and direct geometry (MAPS) spectrometers [13]. The work was not extended to consider Cs-doped silica samples.

Figure 33a shows the INS spectrum of solid methyl propanoate at 20 K (TOSCA spectrometer), while Figure 33b shows the simulated spectrum based on periodic DFT calculations performed using the CASTEP code.

TABLE 3 Vibrational Transition Energies of Methyl Propanoate in the Solid State and Adsorbed On Silica

Mode[a]	Solid state/cm^{-1}	Adsorbed on SiO_2/cm^{-1}
C=O stretch	1728	1726
C4 methyl asym def	1469, 1456	1468
C4 methyl sym deformation	1445, 1432	1442
Methylene scissors	1417	1427
C1 methyl sym deformation	1384	1384
Methylene wag	1356	1364
C1 methyl rock	1092	1090
C1 methyl rock	1024	1029
C1–C2+O1–C4 out-of-phase stretch	962	953
C2–C3+C3–O1 in-phase stretch	858	859
Methylene rock	809	802
C=O out-of-plane bend	581	577
C2–C3–O1 in-plane bend	443	445
C1–C2–C3+C3–C1–C4 out-of-phase and in-plane bend	339	343
C1 methyl torsion	263, 244, and 231	225
C1 methyl torsion and C3–O1 torsion	204, 192, and 163	170

[a]*See the inset of Figure 33 for the numbering scheme.*

In Figure 33a and b, the pattern in the range 100–300 cm^{-1} seen in the experimental spectrum is repeated by the calculated spectrum. Below 100 cm^{-1}, the acoustic modes occur; these are seen in the INS spectrum but are not present in the calculation because it is a Γ-point only calculation where the acoustic modes have zero energy. Table 3 lists modes that are of relevance to the investigation of the adsorbed species.

Methyl propanoate was adsorbed on activated silica and the INS spectrum recorded. Figure 34 shows the C–H and O–H stretch region of activated silica (Figure 34a), while Figure 34b shows the spectrum after adsorption of methyl propanoate. Figure 34 clearly shows the O–H stretch downshift from

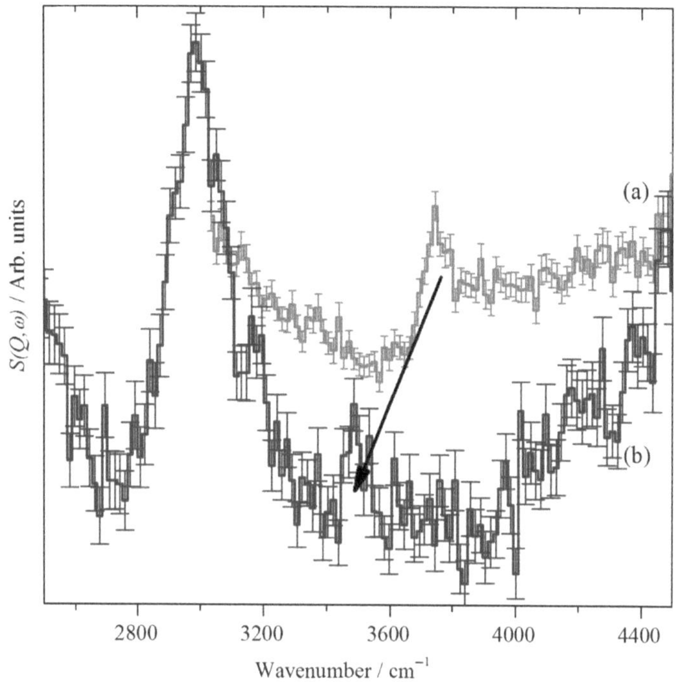

FIGURE 34 INS spectra of (a) silica after activation at 623 K and (b) methyl propanoate on silica. *Reprinted with permission from Ref. [143].*

3750 to 3485 cm^{-1} as a result of methyl propanoate hydrogen bonding to the hydroxyls.

As stated in Sections 5.3.1 and 5.3.2, one of the advantages of INS spectroscopy is that common catalyst supports such as carbon or silica are essentially transparent to neutrons, which is also of relevance for alumina and mixed metal oxides used, for example, in automotive exhaust gas catalysis and many other applications. The benefits of this attribute are shown in Figure 35 that compares the solid-state spectrum of methyl propanoate (Figure 35a) with the difference spectrum, generated by subtraction of the activated silica from that of the adsorbed material (Figure 35b). Over the range 400–1600 cm^{-1}, the spectra are comparable, confirming molecular adsorption. This is not the case in the 0–400 cm^{-1} region; in the solid-state spectrum, there are many more bands. This arises from four molecules with two distinct conformations in the unit cell [143].

Methyl propanoate has two possible hydrogen-bond acceptor sites: the carbonyl oxygen (Figure 36A) or the ester oxygen (Figure 36B).

In order to discriminate between the two possibilities, the two systems were modeled with the Gaussian 03 code and the results are shown in Table 4.

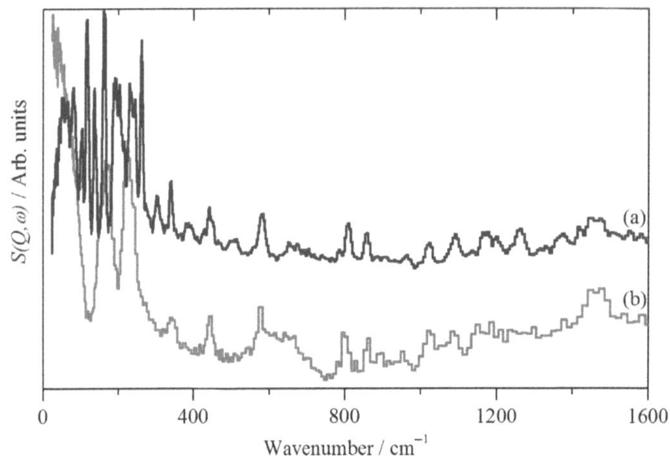

FIGURE 35 INS spectra of (a) methyl propanoate and (b) methyl propanoate on silica, after subtraction of the activated silica. *Reprinted with permission from Ref. [143].*

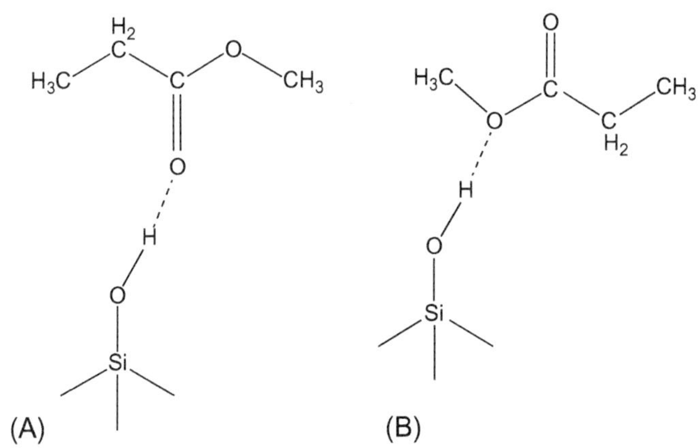

FIGURE 36 Possible adsorption modes of methyl propanoate on silica: (A) *via* the carbonyl oxygen or (B) *via* the ester oxygen. *Reprinted with permission from Ref. [143].*

For the species adsorbed *via* the carbonyl oxygen (Figure 36A), the calculated vibrational transition energy for the O–H stretch undergoes a large downshift, the carbonyl stretch downshifts, and the skeletal stretch mode slightly upshifts. For adsorption *via* the ester oxygen (Figure 36B), there is a somewhat smaller downshift in the O–H stretch mode, the carbonyl upshifts, and the skeletal mode downshifts. The experimentally observed pattern is in almost quantitative agreement with the predictions of adsorption *via* the carbonyl oxygen (Figure 36A). Therefore, Figure 36A defines the adsorption geometry for methyl propanoate on the silica, with the silica providing Brønsted acid sites that bind the ester *via* the carbonyl group. Further work

TABLE 4 Selected Calculated Vibrational Transition Energies (cm^{-1}) for Silica, Free and Adsorbed Methyl Propanoate

	SiO_2	Methyl propanoate	Adsorbed *via* carbonyl O		Adsorbed *via* ester O		Experimental
O–H stretch/cm^{-1}	3765		3508	(−257)	3576	(−189)	(−265)
C=O stretch/cm^{-1}		1730	1699	(−31)	1746	(+16)	(−36)
C2–C3 + C3–O1 in-phase stretch/cm^{-1}		836	842	(+6)	808	(−28)	(+12)

Values in brackets are the difference between the free anc the adsorbed species.

examining the adsorption and reaction of reagents on a Cs modified silica would be helpful in understanding the specific interactions associated with the industrial process.

5.5.4 Methanol Synthesis

Methanol is one of the world's largest commodity chemicals; in 2013 global methanol demand was estimated to be 6.5×10^{10} kg [146]. Methanol is manufactured by the reaction:

$$CO + CO_2 + 5H_2 \rightarrow 2CH_3OH + H_2O$$

The process employs a $Cu/ZnO/Al_2O_3$ catalyst and is typically carried out at 50 bar and 523 K. The generally [147] (but not universally [148]) accepted mechanism is that CO_2 is adsorbed as formate, which is then sequentially hydrogenated to adsorbed methoxy, which then desorbs as methanol. However, there are still unresolved questions as to the nature of the active catalyst and the details of the mechanism.

To attempt to understand the structure of the active catalyst, small-angle neutron scattering (SANS) [149] and neutron powder diffraction (NPD) [149–153] have been employed. SANS is rarely used for catalyst studies which is surprising because supported metal particles are typically in the 2–20 nm range that is ideal for SANS. In this instance, SANS and NPD of a model $Zn_{0.92}Cu_{0.08}O$ catalyst showed that on annealing in hydrogen, metallic copper precipitates on the surface of distorted ZnO crystallites. At room temperature, hydrogen is associated with 10 nm Cu clusters [149].

In order to better understand a working catalyst, Schlögl, Kandemir, and Behrens have developed a flow cell that enables NPD studies at up to 60 bar at 603 K with online gas analysis by mass spectrometry [151], which spans the operating range of the commercial catalyst. The advantages of NPD are shown in Figure 37 that compare NPD and data from a laboratory X-ray source [152]; NPD provides a much larger Q range allowing better characterization of nanomaterials.

The NPD studies of a working $Cu/Zn/Al_2O_3$ catalyst [152,153] showed that the metallic Cu component is nanostructured and has many and different kinds of defects; in particular, there is a high concentration of stacking faults. The bulk defects were stable upon changing the gas atmosphere to partial pressures as applied in industrial methanol synthesis. The results also showed that there was no inclusion of reactive species into the bulk, for example, oxide or carbide formation, and that the formation of α-brass (Cu(65%)Zn(35%)) during methanol synthesis can be excluded. Alloy formation and annealing of the bulk defects was observed only above 603 K, well beyond the 523 K used industrially. Under industrial reaction conditions, the nanoparticles and the nonideal nature of Cu were found to be relatively stable. The results indicate that the interaction between Cu and

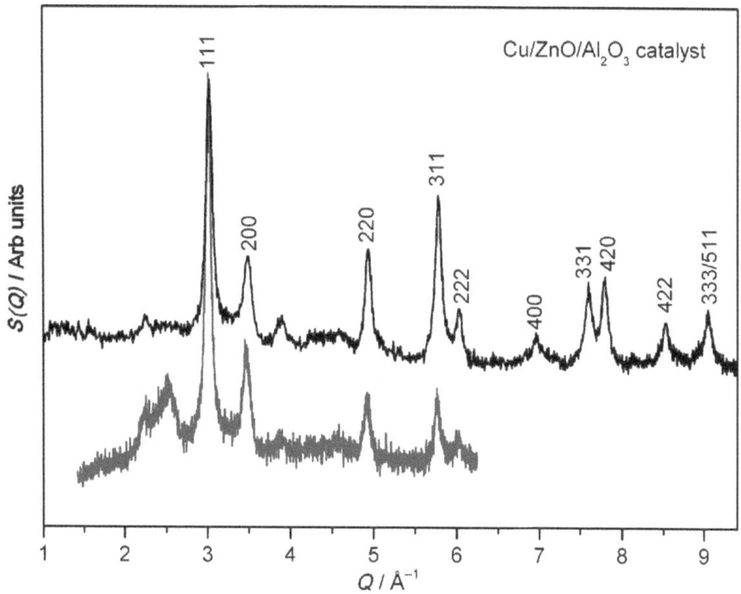

FIGURE 37 Diffraction pattern of a freshly reduced Cu/Zn/Al_2O_3 catalyst acquired at room temperature with NPD (black pattern, upper trace) and laboratory X-ray diffraction ($Cu_{K\alpha}$, gray pattern, lower trace). Peaks that are due to Cu are indexed. *Reprinted with permission from Ref. [152].*

ZnO during methanol synthesis involves dynamics of the poorly crystalline ZnO component rather than the Cu particles, and this affects only the Cu surface.

The surface species present are of significant interest, both during the activation process and on a working catalyst [154,155]. Figure 38 shows INS spectra of a newly prepared $Zn_{0.92}Cu_{0.08}O$ catalyst and the changes resulting from various treatments [154]. The bands at 162, 452, and 545 cm^{-1} present in all the spectra are bulk phonons of ZnO [156] and are most clearly seen after high-temperature treatment (Traces 4 and 5). The broadband centered at ~900 cm^{-1} in the fresh sample (Trace 1) and after mild calcination (Trace 3) is assigned to the bending modes of surface hydroxyls on the ZnO.

Reduction in hydrogen (Trace 2) results in a major change in the spectrum. The bands at 810 and 880 cm^{-1} are assigned to the bending modes of ZnH and ZnOH, respectively, formed by heterolytic dissociation of H_2 on the surface [156]. The weak feature at 1710 cm^{-1} may be the Zn-H stretch mode. By comparison to extensive INS work on formate-containing model compounds [157] and formate on model surfaces [158], the strong band at 262 cm^{-1} and the weak shoulders at 1100 and 1370 cm^{-1} are assigned to adsorbed bidentate formate on copper, which is confirmed by a DFT investigation of formate on the low-index surfaces of Cu [159]. The strong peaks at

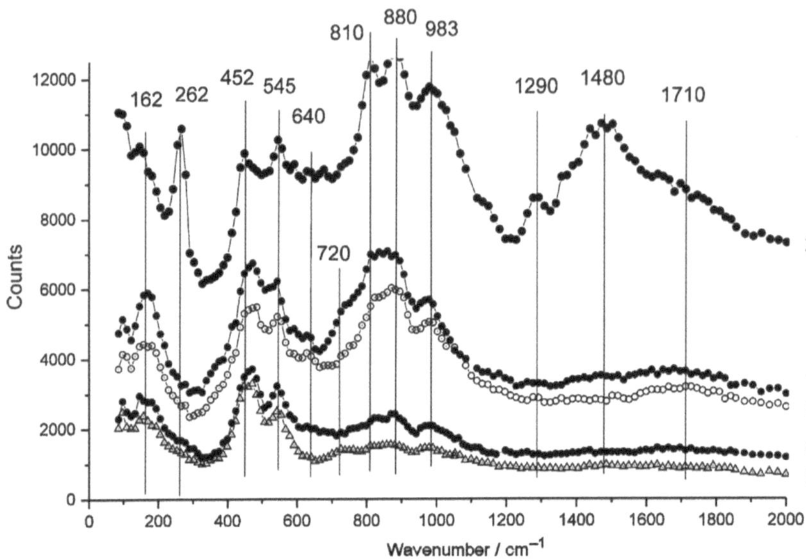

FIGURE 38 Experimental INS spectra recorded on IN1BeF at the ILL of $Cu_{0.08}Zn_{0.92}O$. The sample was treated: (1) in Ar at 623 K and sealed under Ar; (2) in Ar at 623 K, in H_2 at 473 K and sealed under H_2; (3) in Ar at 623 K, in H_2 at 473 K, in 0.5–1% O_2 in Ar at 523 K and sealed under Ar; (4) in Ar at 623 K, in H_2 at 753 K and sealed under H_2; and (5) in Ar at 753 K and sealed under Ar. *Reprinted with permission from Ref. [154].*

980, 1290, and 1480 cm^{-1} were assigned to hydroxyl bending modes with different hydrogen bonding strengths. While this assignment is plausible for the 980 cm^{-1} feature, 1290 and 1480 cm^{-1} are too high to be M–O–H deformation modes. A parallel infrared study in [154] shows that significant quantities of carbonates are generated by heating in hydrogen. Carbonates would be essentially invisible to neutrons (due to the small scattering cross sections), but hydroxycarbonates would have significant intensity, and this would be a reasonable alternative assignment.

Figure 39 shows difference (reacted and clean) INS spectra of a Cu/ZnO after methanol synthesis at 6.31 bar and 523 K in CO/H_2, 3a, and CO_2/H_2, 3b [155]. The peaks labeled M in both spectra at 2940, 1450, 1160, and 95 cm^{-1} are assigned as the C–H stretch, the C–H deformations, and methyl rock and methyl torsion, respectively, of chemisorbed methoxy. These can be unambiguously assigned to methoxy by comparison to a model compound, $Al(OCH_3)_3$ [124]. Physisorbed methanol can be ruled out by the absence of the features relating to the O–H group of methanol: the O–H stretch at 3200 cm^{-1} and the C–O–H bend at 750 cm^{-1}. In contrast to the reaction in CO/H_2, formate is also present on the surface: the characteristic in-plane C–H deformation mode is clearly seen at 1375 cm^{-1}, the weaker out-of-plane deformation is just visible at 1055 cm^{-1} [157,158]. Unfortunately, the mode

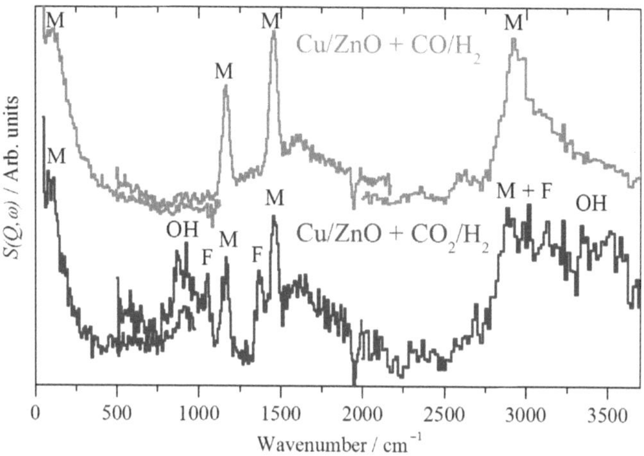

FIGURE 39 Difference INS spectra recorded on MAPS at ISIS of Cu/ZnO after methanol formation with (a) CO/H_2 and (b) CO_2/H_2 feed. Peaks labeled M are methoxy, F are formate, and OH are hydroxyls on Cu. The region 4000–2000 cm^{-1} was recorded with an incident energy $(E_i) = 4840\ cm^{-1}$, 2000–500 cm^{-1} with $E_i = 2420\ cm^{-1}$, and 500–0 cm^{-1} with $E_i = 1210\ cm^{-1}$. *Reprinted with permission from Ref. [155].*

of formate coordination (mono- or bidentate) cannot be distinguished here as the diagnostic modes are the symmetrical and asymmetrical C-O stretches that are invisible to INS. However, a computational study [159] shows that the bidentate form bridging two copper atoms is the lowest energy structure on all the low-index copper surfaces. A second difference between the reactions is the presence of hydroxyls, bands at 3430 and 915 cm^{-1} for the O–H stretch and bend, respectively. The transition energy of the hydroxyl bending mode suggests that the hydroxyls are on the Cu component rather than the ZnO. Thus, there are distinct differences between the behavior of the catalyst depending on the feedstock.

Methanol synthesis has been the subject of extensive theoretical work. The Gibbs free energy calculated [160] by DFT for the CO/H_2 reaction shows that the only stable hydrogenous intermediate is methoxy. In contrast, for the CO_2/H_2 reaction, both formate and methoxy are stable intermediates and that hydroxyls should be present on the metal surface. These results are completely consistent with this model.

5.5.5 Ammonia Synthesis

Ammonia synthesis is among the most important processes in the chemical industry [161]. It was commercialized at BASF over a century ago by Carl Bosch based on the fundamental work of Fritz Haber. This success led to the large-scale production of artificial fertilizers, and today, approximately

80% of the worldwide ammonia output of 1.77×10^{11} kg (2015) [162] is used for this purpose. Without this chemical process, the current global population of seven billion could not be sustained.

The process is conducted at typical temperatures of 773 K and pressures around 200 bar, resulting in ammonia concentrations in the exhaust gas of up to 18 vol%. To achieve a highly active iron catalyst, magnetite (Fe_3O_4) is promoted by fusing it with irreducible oxides (K_2O, Al_2O_3, and CaO) in an oxide melt. The fused material is mechanically granulated and carefully reduced to give the α-Fe catalyst, which is commonly termed "ammonia iron" and is still used industrially.

Ammonia synthesis is deceptively simple:

$$N_2 + 3H_2 \rightarrow 2NH_3$$

However, understanding the mechanism has occupied much of the last 50 years and earned a Nobel prize (to Gerhard Ertl in 2007). The rate-limiting step is the dissociative chemisorption of dinitrogen on the iron surface, and it has been proposed that on the working catalyst, there are surface layers of Fe_2N and γ-Fe_4N.

NPD was used [163,164] to investigate the phases present *in situ*. The reaction was carried out using an N_2/D_2 (1:3) feed (D_2 was used to reduce the incoherent background) on a real catalyst (supplied by BASF) in a tubular fixed-bed reactor made of a nickel alloy at 698 K and 75 bar, somewhat below the industrial conditions. This was dictated by safety and materials considerations. Figure 40 shows the results from the catalyst after initial reduction, at the start of ammonia synthesis, after 88 h time-on-stream (TOS), and at room temperature after the reaction (top to bottom). It is clear that the α-Fe phase is essentially unchanged throughout the reaction and that bulk iron nitrides are not formed. This is surprising, because at the ammonia partial pressure presented in the reactor, nitriding should readily occur. Close inspection shows the presence of a small amount of magnetite that has disappeared by 26 h TOS. The α-Fe peaks are anisotropically broadened compared with those of a reference iron sample. This can be interpreted as "structural anisotropy" of the ammonia iron caused by defects and size and strain effects. It seems likely that the unexpected stability of the α-Fe phase in the catalyst is related to these microstructural effects.

5.5.6 Hydrogen Storage and Production *via* Oxyhydrides

$CeNi_XO_Y$ oxides were originally investigated as hydrogenation catalysts, but it was noted that they could absorb as much hydrogen as the intermetallic compound $CeNi_5$ [165]. XRD shows that $CeNi_XO_Y$ oxides can be described as a mixture of nickel oxide with ceria modified by the insertion of a part of nickel in its lattice (i.e., a solid solution) [166]. (Ternary compounds $CeM_{0.5}Ni_XO_Y$ ($M = Zr$ or Al, $0 \leq X \leq 3$) are also possible.) The size of the nickel oxide varies

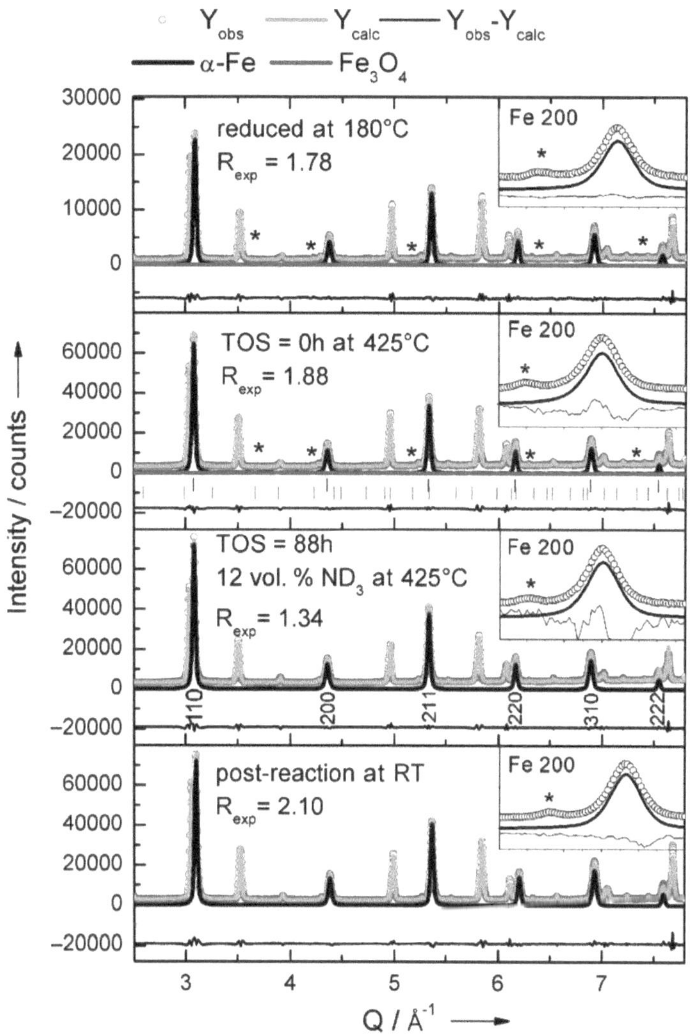

FIGURE 40 Neutron diffraction patterns recorded on SPODI at FRM II of the ammonia synthesis catalyst under different conditions. The black line indicates the fitted contribution of the α-Fe phase to the patterns. From top to bottom: Reduced, initial catalyst in 4.4 bar D_2 at 453 K (the dark gray line is the profile of magnetite, peaks marked by asterisks are from the Ni reactor tube.). Prereaction catalyst at 698 K under 75 bar syngas at TOS = 0 h. *In situ* reaction state at 698 K under 75 bar syngas, which is converted to yield 12 vol% ND_3 at TOS = 88 h. Postreaction catalyst in 75 bar Ar at room temperature. The insets show the magnification of the 200 peak of α-Fe. *Reprinted with permission from Ref. [164].*

considerably from clusters to a crystallized material, depending on the X value and the experimental conditions. The materials have been investigated by a range of characterization techniques; however, only INS is able to provide information on the state of the hydrogen [166,167].

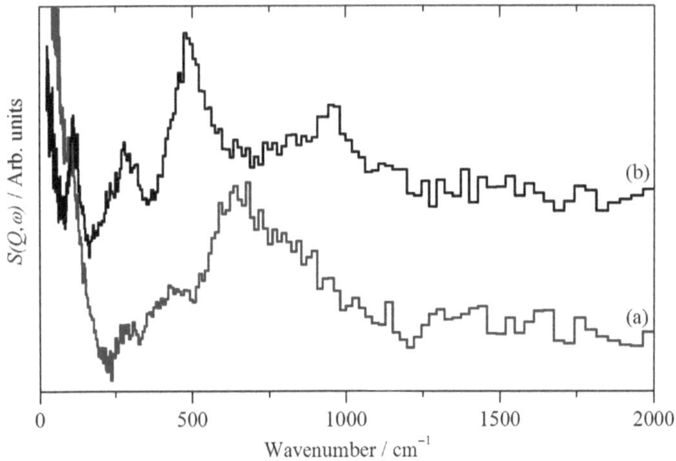

FIGURE 41 INS spectra of $CeNi_{0.5}O_Y$ [167] recorded on TOSCA at ISIS. Catalyst after (a) calcination at 573 K in air and (b) reduction at 473 K in H_2.

More recently, it has been found that $CeNi_XO_Y$ functions as an efficient catalyst for the low-temperature steam reforming of ethanol [168–171]:

$$C_2H_5OH + 3H_2O \rightarrow 6H_2 + 2CO_2$$

The INS spectra of $CeNi_XO_Y$ and $CeM_{0.5}Ni_XO_Y$ are very similar [166–171]. Figure 41 shows data for $CeNi_{0.5}O_Y$. The catalyst after calcination shows a strong broadband at 650 cm^{-1}, and weaker features at 425 and 280 cm^{-1} (Figure 41a). These are assigned as deformation modes of surface hydroxyls. The same features are seen on 5 nm CeO_2 [172]; however, on commercial CeO_2, which has a much larger particle size, the 650 cm^{-1} peak is much weaker. This is consistent with the 650 cm^{-1} peak being a hydroxyl bending mode (the larger particles mean the surface area is less, so there are fewer hydroxyls). The lower energy peaks occur within the vibrational density of states of ceria [173] and may be bulk phonons of ceria.

On reduction in hydrogen, the 650 cm^{-1} peak is removed, and new modes at 480 and 950 cm^{-1} appear (Figure 41b). The 480 cm^{-1} peak is assigned [166–171] to a hydride (H^-) inserted into the Ni-substituted ceria lattice. (An alternative possibility is that it is the bending mode of a hydroxyl on a reduced metal site). The 950 cm^{-1} peak is assigned to surface hydrogen on nickel metal. Close inspection of Figure 41b shows a shoulder at ~1150 cm^{-1}; together, these are consistent with surface hydrogen on Ni (see Figure 20 in Section 5.4.4). Exposure to oxygen removes most of the bands produced by the reduction step.

The closely related materials $Mg_2AlNi_XH_ZO_Y$ (X = 1, 3, and 12) that are nanooxyhydrides derived from hydrotalcites have been shown to be highly

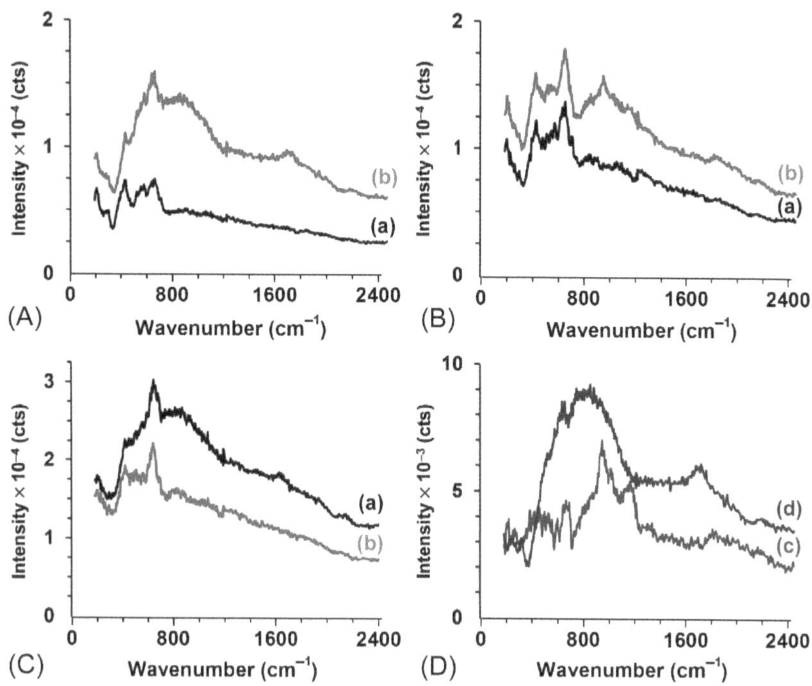

FIGURE 42 INS spectra of $Mg_2AlNi_XO_Y$ recorded on Lagrange at the ILL. INS spectra of $Mg_2AlNi_XO_Y$ nanocomposites (a) treated in vacuum at 473 K for 2 h and (b) treated in H_2 at 723 K for 10 h. x = (A) 12, (B) 3, and (C) 1. (D) INS spectra of $Mg_2AlNi_XO_YH_Z$ after subtraction of spectra of $Mg_2AlNi_XO_Y$ (c) (Bb-Ba) X = 3 and (d) (Ab-Aa) X = 12. *Reprinted with permission from Ref. [174].*

active, stable, and cheap catalysts for H_2 production at only 323 K (after initiation at 473 K) from a concentrated ethanol (14% in mol) in water mixture as a model of bioethanol [174]. Figure 42A–C shows the effect of reduction in H_2 at 723 K for the X = 12, 3, and 1 materials and in Figure 42D the difference spectra for the X = 12 and 3 samples. The spectra were assigned as a mixture of metal hydrides: MgH_2, AlH_3, and NiH, although it was recognized that the experimental conditions were far from those reported [175] for the formation of NiH (24 h exposure to a H_2 pressure of 3 GPa at 573 K, sample recovered below 200 K). Inspection of Figure 42A and 42 C shows this assignment is not correct, as seen in Figure 42D, reaction with hydrogen has formed water [47] for the X = 1 and 12 samples. The reduction conditions (100% H_2 at 723 K for 10 h) will easily reduce NiO to metallic Ni; thus, for X = 3, the difference spectrum (Figure 42D), bears a striking resemblance to hydrogen on Raney Ni (see Figure 20 in Section 5.4.4), and the spectrum is assigned as surface hydrogen on Ni metal, not as a bulk hydrogen-in-metal phase.

5.6 OUTLOOK: EXPERIMENTAL LIMITS AND NEW PERSPECTIVES

5.6.1 Non-Hydrogenous Adsorbates

Neutron scattering is a flux-limited technique, a situation that is an immutable part of the technique. The situation is worse still for inelastic scattering because the energy analysis requires that most of the neutrons are rejected either from the incident beam (direct geometry) or from the scattered beam (indirect geometry) [176]. Over the last 60 years, the brightness (neutrons $cm^{-2}\ s^{-1}$) has increased roughly a hundredfold; by comparison, synchrotron brightness has increased by a factor of $\sim 10^{12}$. This has required that neutron facilities use the available neutrons more effectively. Consequently, there have been improvements in neutron guide technology (guides up to $m = 7$ are now available) to improve the efficiency of neutron transport. Installation of a guide on TOSCA at ISIS will result in an average tenfold increase in incident flux across the entire energy transfer range [177]. Also, instruments are designed with, or upgraded with, collection areas as large as possible. Made possible by the rapid increase in computer power, the other key innovation is to get as much out of the data as possible and *ab initio*, and molecular dynamics simulations are becoming increasingly essential components of data analysis.

All of these factors are particularly acute for catalytic studies, because often only a small fraction of the sample is responsible for the scattering of interest. The need for "sufficient sample to see something" has meant that only hydrogenous adsorbates have been studied to date (2016). INS studies of catalysts have only been possible by virtue of the large incoherent cross section of ^1H; as much of catalysis is concerned with the transfer of hydrogen between reagents and products, this is especially fortuitous. Nonetheless, there are major areas of catalysis that involve nonhydrogenous reactants. These are particularly important in pollution control and include CO and NO_x oxidation in automotive catalysts as well as SO_2 and CO_2 capture and utilization. The small cross sections of C, N, O, and S mean that these have been invisible to INS, although the perturbations they cause to the local framework may be detectable, as was shown for SO_2 and CO_2 in a MOF [178].

This situation is changing; the current generation of INS spectrometers led by VISION at SNS has sufficient sensitivity to directly observe these materials. Figure 43 shows that only 2.5 mmol of CO_2 are easily seen adsorbed on an oxidized activated carbon [179]. The added benefit that INS provides is that for chemisorbed species, it will be possible to observe the substrate-adsorbate stretch and deformation modes that provide direct information on the strength of the interaction. The upgrades to TOSCA [177] and the evolution of IN1BeF into Lagrange [180] mean that these instruments should also be able to detect these modes. For the next-generation instrument VESPA [181] proposed at the ESS, these studies should be standard.

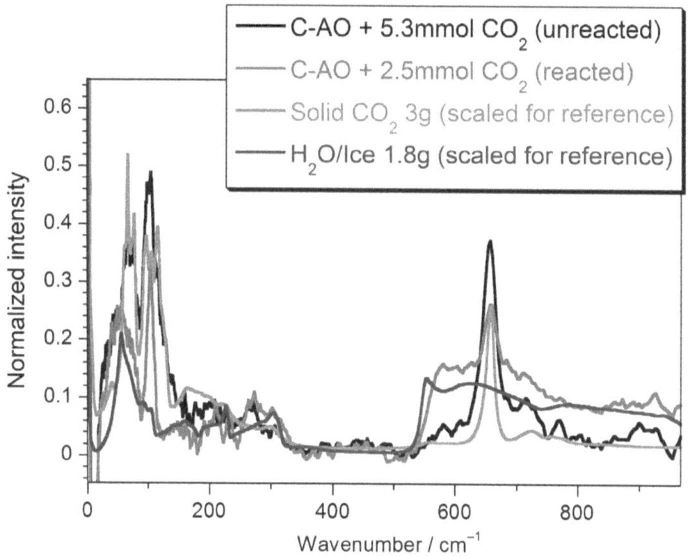

FIGURE 43 The difference INS spectra before and after CO_2 dosing in oxidized activated carbon, in comparison with the reference spectra for bulk solid CO_2 and H_2O. Signal from the background and the oxidized activated carbon has been subtracted. *Reprinted by permission from Ref. [179].*

5.6.2 *Operando* Spectroscopy

The current trend in catalysis research is to study a working catalyst under conditions that are as realistic as possible regarding temperature and pressure and simultaneously to measure as many spectroscopic and physical parameters as possible. As a minimum, this usually means a probe of the catalyst with real-time product analysis. This approach has been enthusiastically adopted at synchrotrons with multiple probes (e.g., EXAFS/XANES, XRD, infrared and/or Raman, and UV/vis) possible together with online mass spectrometry for gas analysis [182].

Neutron scattering has been slower to adopt the methodology, probably because the longer timescales of neutron experiments have meant that they have tended to be static measurements and *operando* techniques are most useful for dynamic processes. This is changing; the NPD studies of methanol [151–153] and ammonia [164,165] synthesis catalysts had online gas analysis during the diffraction measurements.

INS is much less favorable for *operando* because most measurements are made at 20 K. The only example [183] so far (2016) where a reaction (CO oxidation) was studied by INS under *operando* conditions used a particularly favorable combination of conditions (the catalyst was a pure material, the reaction occurred at room temperature at a convenient (i.e., slow) rate). The work was carried out on a direct geometry spectrometer where, by only using the small Q data, it was possible to ameliorate the detrimental effect of the large

Debye-Waller factor. This achieved a very modest time resolution of 30 min or so per spectrum. However, an increase in detected flux leads directly to a corresponding increase in time resolution. Thus, *operando* INS experiments are conceivable with 1 min time resolution as instrumentation improves.

Operando QENS measurements are potentially feasible, although there are none to date. Backscattering spectrometers at spallation sources are ideally suited to measure NPD patterns simultaneously, as exemplified by OSIRIS at ISIS, thus increasing the range of probes. Measurement of QENS and Raman spectra has been demonstrated [184]. The same system has been used to measure INS and Raman spectra simultaneously as a function of temperature to study polymorphism in *cis*-platin [185]. An upgrade to allow the study of adsorbed gases for catalytic studies is underway.

5.6.3 Homogeneous Catalysis

As stated in the Introduction, neutron scattering studies of homogeneous catalysis are very rare. The principal reason is that the catalyst is usually only present in low concentration in solution and because neutrons "see everything," the scattering is dominated by the solvent. There is an extensive literature [186] on solute and ion solvation in solution by total neutron scattering, but this is largely studying the organization of the solvent around the species of interest. Systems relevant to homogeneous catalysis include OsO_4 in *t*-butanol [187] (used for the oxidation of alkenes to *cis*-diols), Pd^{2+} [188] (see Section 5.3.4), and Cu^{2+} [189] (both Pd^{2+} and Cu^{2+} are used in the Wacker process for the production of aldehydes from alkenes). The challenge is to observe an intermediate, for example, a palladium-ethene complex in the Wacker process, and this becomes a question of sensitivity. Total scattering is very demanding on signal-to-noise ratio, and a significant increase in detected flux is probably needed before such experiments become feasible.

INS measurements do not yield useful data from liquids at room temperature (the enormous Debye-Waller factor obliterates the spectral information). Thus, any measurement will be of a frozen solution. INS studies of ions in frozen aqueous solution have been made [190], but as with the total scattering measurements, it is the organization of the water that is observed. For homogeneous catalysis, INS suffers from the same problem of a low concentration of the species of interest. The use of a deuterated solvent (provided it does not exchange with the catalyst) can reduce the solvent background. The increased sensitivity of the latest instruments means that such studies are feasible but are likely to remain rare because there are many techniques that are more easily able to study solutions: NMR, infrared, and Raman spectroscopies, to name a few.

5.6.4 Conclusions

Neutron scattering studies of catalysts are a flourishing field. In the last 10 years or so, the recognition of the advantages of direct geometry instruments has enabled the study of systems with realistic metal loadings and expanded the spectral range to include the highly diagnostic C–H/N-H/O–H stretch region at >2800 cm^{-1}. Over the same period, the use of NPD under *operando* conditions to study catalysts in conditions that are close to those used industrially is a very exciting development.

The use of *ab initio* and molecular dynamics simulations for data analysis is a trend that will only increase. Neutrons and calculations are a symbiotic pairing; each informs and challenges the other. Neutron observables are readily calculable providing a stringent test of the calculation. Having validated the calculation, it can then be "data mined" to give information that is not available in the neutron data such as atomic charges and electronic structure.

The continuous upgrading of instruments at ILL (France) and ISIS (UK) is already yielding results. The potential of the new generation of instruments at the high-power sources SNS (USA) and J-PARC (Japan) is beginning to be explored and exploited. The next-generation source, the ESS (Sweden), will offer unprecedented capabilities for neutron scattering studies of catalysts.

REFERENCES

[1] D.L. Price, F. Fernandez-Alonso, An introduction to neutron scattering, in: F. Fernandez-Alonso, D.L. Price (Eds.), Experimental Methods in the Physical Sciences, Neutron Scattering, vol. 44, Academic Press, 2013, pp. 1–136.

[2] M.R. Johnson, G.J. Kearley, Dynamics of atoms and molecules, in: F. Fernandez-Alonso, D.L. Price (Eds.), Experimental Methods in the Physical Sciences, Neutron Scattering, vol. 44, Academic Press, 2013, pp. 415–469.

[3] F. Fernandez-Alonso, D.L. Price (Eds.), In: Experimental Methods in the Physical Sciences, Neutron Scattering, vol. 48, Academic Press, 2015.

[4] P.N. Jones, E. Knözinger, W. Langel, R.B. Moyes, J. Tomkinson, Surf. Sci. 207 (1988) 159–176.

[5] X. Liao, R. Gerdts, S.F. Parker, L. Chi, Y. Zhao, M. Hill, M.O. Jones, Z. Jiang, Phys. Chem. Chem. Phys. 18 (2016) 17311–17319.

[6] T.J. Udovic, R.D. Kelley, Neutron scattering of hydrogen in catalysts, in: Z. Paal, P.G. Menon (Eds.), Hydrogen Effects in Catalysis, Marcel Dekker, New York, 1988, pp. 167–182.

[7] J.M. Nicol, Spectrochim. Acta A 48 (1992) 313–327.

[8] H. Jobic, Applications of neutron scattering to catalysis, in: B. Imelik, J.C. Vedrine (Eds.), Catalyst Characterization, Plenum Press, New York, 1994, pp. 347–375.

[9] G. Sankar, R. Raja, J.M. Thomas, D. Gleeson, Advances in the determination of the architecture of active sites in solid catalysts, in: G. Centi, B. Wichterlova, A.T. Bell (Eds.), NATO Science Series, Series II: Mathematics, Physics and Chemistry, vol. 13, Springer, AA Dordrecht, 2001, pp. 95–114.

[10] P.C.H. Mitchell, S.F. Parker, A.J. Ramirez-Cuesta, J. Tomkinson, Vibrational Spectroscopy with Neutrons, with Applications in Chemistry, Biology, Materials Science and Catalysis, World Scientific, Singapore, 2005 (Chapters 7 and 11).

[11] P.W. Albers, S.F. Parker, Inelastic incoherent neutron scattering in catalysis research, in: B.C. Gates, H. Knözinger (Eds.), Advances in Catalysis, vol. 51, Academic Press, London, 2007, pp. 99–132.

[12] H. Jobic, Neutron scattering, in: M. Che, J. Vedrine (Eds.), Characterization of Solid Materials and Heterogeneous Catalysts, vol. 1, Wiley-VCH, Weinheim, 2012, pp. 185–209.
[13] S.F. Parker, D. Lennon, P.W. Albers, Appl. Spectrosc. 65 (2011) 1325–1341.
[14] S.F. Parker, F. Fernandez-Alonso, A.J. Ramirez-Cuesta, J. Tomkinson, S. Rudic, R.S. Pinna, G. Gorini, J. Fernández Castañon, J. Phys. Conf. Ser. 554 (2014) 012003.
[15] S.F. Parker, D. Lennon, J. Phys. Conf. Ser. 746 (2016) 012066.
[16] S.F. Parker, S.M. Bennington, A.J. Ramirez-Cuesta, G. Auffermann, W. Bronger, H. Herman, K.P.J. Williams, T. Smith, J. Am. Chem. Soc. 125 (2003) 11656–11661.
[17] J. Tomkinson, S.F. Parker, D.J. Lennon, Chem. Phys. 133 (2010) 034508.
[18] C.G. Windsor, Pulsed Neutron Scattering, Taylor and Francis Ltd., London, 1981.
[19] G. Ertl, H. Knözinger, F. Schüth, J. Weitkamp (Eds.), 2nd ed., In: Handbook of Heterogeneous Catalysis, vol. 2, Wiley-VCH, Weinheim, 2008.
[20] M. Che, J. Vedrine (Eds.), In: Characterization of Solid Materials and Heterogeneous Catalysts, vol. 1, Wiley-VCH, Weinheim, 2012, pp. 185–209.
[21] R. Warringham, D. Bellaire, S.F. Parker, J. Taylor, R.A. Ewings, C.M. Goddway, M. Kibble, S.R. Wakefield, M. Jura, M.P. Dusman, R.P. Tooze, P.B. Webb, D. Lennon, J. Phys. Conf. Ser. 554 (2014) 012005.
[22] M.G. Kibble, A.J. Ramirez-Cuesta, C.M. Goodway, B.E. Evans, O. Kirichek, J. Phys. Conf. Ser. 554 (2014) 012006.
[23] E.A. Smith, The Sampling and Assay of the Precious Metals, 2nd ed., Charles Griffin & Co. Ltd., London, 1947.
[24] P. Gy, Sampling of Particulate Matter, Theory and Practise, Elsevier, Amsterdam, 1979.
[25] M. Lüschow, Sampling of Precious Metals, in: G. Kraft (Ed.), Sampling in the Non-Ferrous Metals Industry, Trans. Tech Publications, Clausthal-Zellerfeld, 1993.
[26] S.F. Parker, Inelastic neutron scattering spectroscopy, in: J.M. Chalmers, P.R. Griffiths (Eds.), Handbook of Vibrational Spectroscopy, vol. 1, John Wiley & Sons, Chichester, 2002, pp. 838–852.
[27] P.R. Stonage, M.J. Benham, D.K. Ross, C. Manwaring, I.R. Harris, Z. Phys. Chem. 181 (1993) 125.
[28] C.J. Howard, H. Kisi, Applications of Neutron Powder Diffraction, OUP, Oxford, 2012.
[29] P. Scherrer, Göttinger Nachrichten Math. Phys. 2 (1918) 98–100.
[30] S.F. Parker, D. Adroja, M. Jimenez-Ruiz, M. Tischer, K. Möbus, S.D. Wieland, P. Albers, Phys. Chem. Chem. Phys. 18 (2016) 17196–17201.
[31] J.H. Sinfelt, G.H. Via, F.W. Lytle, Catal. Rev. Sci. Eng. 26 (1984) 81–140.
[32] D.C. Koningsberger, R. Prins (Eds.), X-Ray Absorption: Principles, Applications, Techniques of EXAFS, SEXAFS and XANES, Wiley, New York, 1987.
[33] R.W. Joyner, Applications of synchrotron radiation in catalysis, NATO ASI Series, vol. 398, Springer, Netherlands, Dordrecht, 1993, pp. 249–265.
[34] http://www.rc-harwell.ac.uk/UKCatalysisHub.
[35] P. Albers, G. Prescher, K. Seibold, S.F. Parker, Chem. Eng. Technol. 22 (1999) 135–137.
[36] P. Albers, H. Angert, G. Prescher, K. Seibold, S.F. Parker, Chem. Commun. 17 (1999) 1619–1620.
[37] R.C. Bansal, J.-B. Donnet, F. Stoeckli (Eds.), Active Carbon, Marcel Dekker, New York, 1988.
[38] J.-B. Donnet, R.C. Bansal, M.-J. Wang (Eds.), Carbon Black Science and Technology, Marcel Dekker, New York, 1993, p. 95. p. 101.
[39] E. Koberstein, E. Lakatos, M. Voll, Ber. Bunsen. Phys. Chem. 75 (10) (1971) 1104–1114.
[40] Y. Hu, O.A. Shenderova, Z. Hu, C.W. Padgett, D.W. Brenner, Rep. Prog. Phys. 69 (6) (2006) 1847–1895.

[41] P.C.H. Mitchell, S.F. Parker, A.J. Ramirez-Cuesta, J. Tomkinson, Vibrational Spectroscopy with Neutrons: with applications in Chemistry, Materials Science and Catalysis, World Scientific, Singapore, 2005. p. 487.
[42] W. Albers, W. Weber, K. Möbus, S.D. Wieland, S.F. Parker, Carbon 109 (2016) 239–245.
[43] P.W. Albers, J. Pietsch, J. Krauter, S.F. Parker, Phys. Chem. Chem. Phys. 5 (2003) 1941.
[44] M. Arai, A.C. Hannon, A.D. Taylor, T. Otomo, A.C. Wright, R.N. Sinclair, D.L. Price, Trans. Am. Crystallogr. Assoc. 27 (1991) 113–131.
[45] P.W. Albers, G. Michael, H. Lansink-Rotgerink, M. Reisinger, S.F. Parker, Z. Naturforsch. 67b (2012) 1016–1020.
[46] P. Albers, M. Maier, M. Reisinger, B. Hannebauer, R. Weinand, Cryst. Res. Technol. 50 (2015) 846–865.
[47] B.E. Warren, Z. Krist. 86 (1933) 349–358.
[48] J.-C. Li, J.D. Londono, D.K. Ross, J.L. Finney, S.M. Bennington, A.D. Taylor, J. Phys. Condens. Matter 4 (1992) 2109.
[49] S.F. Parker, Chem. Commun. 47 (2011) 1988.
[50] W.M. Pearlman, Tetrahedron Lett. 8 (1967) 1663–1664.
[51] Pearlman's catalyst: Comprehensive Organic Name Reactions and Reagents, John Wiley & Sons, Inc. 2010, 2143–2146.
[52] S.R. Angle, R.M. Henry, J. Org. Chem. 62 (1997) 8549–8552.
[53] L. Yin, J. Liebscher, Chem. Rev. 107 (2007) 133–173.
[54] P.W. Albers, K. Möbus, S.D. Wieland, S.F. Parker, Phys. Chem. Chem. Phys. 17 (2015) 5274.
[55] P. Albers, R. Burmeister, K. Seibold, G. Prescher, S.F. Parker, D.K. Ross, J. Catal. 181 (1999) 145–154.
[56] P.W. Albers, W. Weber, K. Kunzmann, M. Lopez, S.F. Parker, Surf. Sci. 602 (2008) 3611.
[57] P. Albers, J.G. Krauter, D.K. Ross, R.G. Heidenreich, K. Köhler, S.F. Parker, Langmuir 20 (2004) 8254.
[58] R.G. Heidenreich, K. Köhler, J. Pietsch, J.G.E. Krauter, J. Mol. Catal. A Chem. 182–183 (2002) 499.
[59] D.K. Ross, P.F. Martin, W.A. Oates, R. Khoda Baksh, Z. Phys. Chem. 144 (1979) 221.
[60] D.K. Ross, Neutron scattering studies of metal hydrogen systems, in: H. Wipf (Ed.), Hydrogen in Metals III, Topics in Applied Physics, vol. 73, Springer, Heidelberg, 1997, pp. 153–214.
[61] D. Richter, R. Hempelmann, R.C. Bowman, Dynamics of hydrogen in intermetallic hydrides, in: L. Schlapbach (Ed.), Hydrogen in Intermetallic Compounds II, Topics in Applied Physics, vol. 67, Springer, Heidelberg, 1992, pp. 97–163.
[62] L. Schlapbach, A. Züttel, Nature 414 (2001) 353–358.
[63] D.K. Ross, D.L. Roach, Inelastic and quasi-elastic neutron scattering, in: Neutron Scattering and Other Nuclear Techniques for Hydrogen in Materials, Springer for Research and Development, 2016, pp. 245–276. Part of the series Neutron Scattering Applications and Techniques.
[64] A. Borodziński, G.C. Bond, Catal. Rev. 48 (2006) 91–144.
[65] H. Frieske, E. Wicke, Ber. Bunsen. Phys. Chem. 79 (1973) 48–52.
[66] E. Wicke, H. Brodowski, Hydrogen in palladium and palladium alloys, in: G. Alefeld, J. Völkl (Eds.), Hydrogen in Metals II, Topics in Applied Physics, vol. 29, Springer, Berlin, Germany, 1978, pp. 73–155.
[67] A. Pundt, M. Suleiman, C. Bähtz, M.T. Reetz, R. Kirchheim, N.M. Jisrawi, Mater. Sci. Eng. B 108 (2004) 19–23.

[68] A. Pundt, C. Sachs, M. Winter, M.T. Reetz, D. Frisch, R. Kirchheim, J. Alloys Compd. 293–295 (1999) 480–483.
[69] C. Sachs, A. Pundt, R. Kirchheim, M. Winter, M.T. Reetz, D. Frisch, Phys. Rev. B 64 (2001) 075408-1–075408-10.
[70] M. Suleiman, N.M. Jisrawi, O. Dankert, M.T. Reetz, C. Bähtz, R. Kirchheim, J. Alloys Compd. 356–357 (2003) 644.
[71] A. Pundt, M. Dornheim, M. Guerdane, H. Teichler, H. Ehrenberg, M.T. Reetz, N.M. Jisrawi, Eur. Phys. J. D 19 (2002) 333–337.
[72] F.A. Lewis, The Palladium Hydrogen System, Academic Press, London, UK, 1967.
[73] F.A. Lewis, Platinum Met. Rev. 26 (1) (1982) 20–27.
[74] A.M. Doyle, Sh.K. Shaikhutdinov, H.-J. Freund, J. Catal. 223 (2004) 444.
[75] G. Sicking, Ber. Bunsen. Phys. Chem. 76 (1972) 790.
[76] H. Lindlar, Helv. Chim. Acta 446 (1952) 35.
[77] H. Lindlar, R. Dubuis, E.J. Corey, Organic Synth. 5 (1973) 880.
[78] H. Lindlar, R. Dubuis, Palladium catalyst for partial reduction of acetylene, Org. Synth. Coll. 5 (1973) 880–886.
[79] H. Lindlar, R. Dubuis, Palladium catalyst for partial reduction of acetylene, Org. Synth. Coll. 46 (1966) 89–92.
[80] R. Schlögl, K. Noack, H. Zbinden, A. Reller, Helv. Chim. Acta 70 (1987) 627–679.
[81] P.W. Albers, K. Möbus, C.D. Frost, S.F. Parker, J. Phys. Chem. 115 (2011) 24485.
[82] K. Möbus, D. Wolf, H. Benischke, U. Dittmeier, K. Simon, U. Packruhn, R. Jantke, S. Weidlich, C. Weber, B. Chen, Top. Catal. 53 (2010) 1126.
[83] J.R. Kosak, Chemical Industries, Catalysis of Organic Reactions, vol. 68, Dekker, New York, 1996. pp. 31–41.
[84] K. Möbus, E. Grünewald, S.D. Wieland, S.F. Parker, P.W. Albers, J. Catal. 311 (2014) 153.
[85] P.W. Albers, S. Bösing, H. Lansink Rotgerink, D.K. Ross, S.F. Parker, Carbon 40 (2002) 1549–1558.
[86] J. Tomkinson, Personal Communication.
[87] H. Schulz, Top. Catal. 26 (2003) 73.
[88] C.S. Kellner, A.T. Bell, J. Catal. 70 (1981) 418.
[89] P. Rylander, Catalytic Hydrogenation in Organic Synthesis, Academic Press, New York, 1979.
[90] H. Klenk, A. Griffiths, K. Huthmacher, H. Itzel, H. Knorre, C. Voigt, O. Weiberg, Hydrogen cyanide, in: W. Gerhartz, Y. Yamamoto, F.T. Campbell, R. Pfefferkorn, J.F. Rounsaville (Eds.), fifth ed., Ullmann's Encyclopedia of Industrial Chemistry, vol. A8, VCH, Weinheim, 1987, pp. 159–163.
[91] M. Thiemann, E. Scheibler, K.W. Wiegand, Nitric acid, nitrous acid and nitrogen oxides, in: W. Gerhartz, Y. Yamamoto, F.T. Campbell, R. Pfefferkorn, J.F. Rounsaville (Eds.), fifth ed., Ullmann's Encyclopedia of Industrial Chemistry, vol. A17, VCH, Weinheim, 1987, pp. 293–339.
[92] K.C. Taylor, Automotive catalytic converters, in: J.R. Anderson, M. Boudart (Eds.), Catalysis-Science and Technology, Springer, Berlin, 1984, p. 140.
[93] W.R. Grove, Philos. Mag. 14 (86) (1839) 127.
[94] G.J. Acres, J.C. Frost, G.A. Hards, R.J. Potter, T.R. Ralph, D. Thompsett, G.T. Burstein, G.J. Hutchings, Catal. Today 38 (1997) 393.
[95] E. Auer, A. Freund, J. Pietsch, T. Tacke, Appl. Catal. A 173 (1998) 259.
[96] K. Kinoshita, J. Electrochem. Soc. 137 (1990) 845.
[97] R. Van Hardeveld, F. Hartog, Surf. Sci. 15 (1969) 189.

[98] R. Van Hardeveld, F. Hartog, Influence of metal particle size in nickel-on-aerosil catalysts on surface site distribution, catalytic activity and selectivity, in: D.D. Eley, H. Pines, P.B. Weisz (Eds.), Advances in Catalysis, Academic Press, New York, 1972, p. 75.
[99] M.L. Sattler, P.N. Ross, Ultramicroscopy 20 (1986) 21.
[100] P.N. Ross, J. Electrochem. Soc. 126 (1979) 78.
[101] A.J. Renouprez, H. Jobic, J. Catal. 113 (1988) 509.
[102] P. Albers, E. Auer, K. Ruth, S.F. Parker, J. Catal. 196 (2000) 174.
[103] P. Albers, K. Seibold, G. Prescher, B. Freund, S.F. Parker, J. Tomkinson, D.K. Ross, F. Fillaux, Carbon 37 (1999) 437–444.
[104] S.F. Parker, C.D. Frost, M. Telling, P. Albers, M. Lopez, K. Seitz, Catal. Today 114 (2006) 418.
[105] S.F. Parker, D.T. Bowron, S. Imberti, A.K. Soper, K. Refson, E.S. Lox, M. Lopez, P. Albers, Chem. Commun. 46 (2010) 2959–2961.
[106] P.W. Albers, M. Lopez, G. Sextl, G. Jeske, S.F. Parker, J. Catal. 223 (2004) 44.
[107] B. Zeifert, J.S. Blasquez, J.G.C. Moreno, H.A. Calderón, Rev. Adv. Mater. Sci. 18 (2008) 632–638.
[108] M.G. Banwell, M.T. Jones, T.A. Reekie, B.D. Schwartz, Org. Biomol. Chem. 12 (2014) 7433–7444.
[109] R.R. Cavanagh, R.D. Kelley, J.J. Rush, J. Chem. Phys. 77 (1982) 1540–1547.
[110] H. Jobic, A. Renouprez, J. Chem. Soc. Faraday Trans. 1 80 (1984) 1991–1997.
[111] H. Yanagita, J. Sakai, T. Aruga, N. Takagi, M. Nishijima, Phys. Rev. B 56 (1997) 14952–14955.
[112] A. Chojecki, H. Jobic, A. Jentys, T.E. Müller, J.A. Lercher, Catal. Lett. 97 (2004) 155–162.
[113] P. Schärringer, T.E. Müller, A. Jentys, J.A. Lercher, J. Catal. 263 (2009) 34–42.
[114] F. Hochard, H. Jobic, G. Clugnet, A.J. Renouprez, Catal. Lett. 21 (1993) 381–389.
[115] F. Hochard, H. Jobic, J. Massardier, A.J. Renouprez, J. Tomkinson, J. Mol. Catal. A: Chem. 95 (1995) 165.
[116] K. Weissermel, H.-J. Arpe, Industrial Organic Chemistry, third ed., VCH Wiley, Weinheim, 1997. p. 72.
[117] T.P. Beebe Jr., J.E. Crowell, J.T. Yates Jr., J. Phys. Chem. 92 (1988) 1296.
[118] S.F. Parker, D. Lennon, Acc. Chem. Res. 47 (2014) 1220–1227.
[119] J.M. Winfield, D. Lennon, A case study: surface chemistry and surface structure of catalytic aluminas, as studied by vibrational spectroscopy of adsorbed species, in: E. Kemnitz, V. Parvulescu (Eds.), New Materials for Catalytic Applications, Elsevier, Amsterdam, ISBN: 9780444635877 2016. (Chapter 7).
[120] A.R. McInroy, J.M. Winfield, C.C. Dudman, P. Jones, D. Lennon, Faraday Discuss. 188 (2016) 467–479.
[121] C. Morterra, G. Magnacca, Catal. Today 27 (1996) 497.
[122] D.T. Lundie, A.R. McInroy, R. Marshall, J.M. Winfield, P. Jones, C.C. Dudman, S.F. Parker, C. Mitchell, D. Lennon, J. Phys. Chem. B 109 (2005) 11592–11601.
[123] D. Lennon, D.T. Lundie, S.D. Jackson, G.J. Kelly, S.F. Parker, Langmuir 18 (2002) 4667.
[124] A.R. McInroy, D.T. Lundie, J.M. Winfield, C. Dudman, P. Jones, S.F. Parker, J.W. Taylor, D. Lennon, Phys. Chem. Chem. Phys. 7 (2005) 3093–3101.
[125] A.R. McInroy, D.T. Lundie, J.M. Winfield, C. Dudman, P. Jones, D. Lennon, Langmuir 21 (2005) 11092–11098.
[126] A.R. McInroy, D.T. Lundie, J.M. Winfield, C.C. Dudman, P. Jones, S.F. Parker, D. Lennon, Catal. Today 114 (2006) 403–411.

[127] A.R. McInroy, D.T. Lundie, J.M. Winfield, P. Jones, C.C. Dudman, D. Lennon, J. Appl. Catal. B: Environ. 70 (2007) 606–610.

[128] J.v.d. Loosdrecht, F.G. Botes, I.M. Ciobica, A. Ferreira, P. Gibson, D.J. Moodley, A.M. Saib, J.L. Visagie, C.J. Westrate, J.W. Niemantsverdriet, Comprehensive Inorganic Chemistry II, Elsevier, Oxford, 2013. pp. 525–557.

[129] M.E. Dry, Appl. Catal. A 138 (1996) 319.

[130] M.E. Dry, Stud. Surf. Sci. Catal. 152 (2004) 533.

[131] H.M.T. Galvis, J.H. Bitter, C.B. Khare, M. Ruitenbeek, A.I. Dugulan, K.P. de Jong, Science 335 (2012) 835.

[132] H.M.T. Galvis, K.P. de Jong, ACS Catal. 3 (2013) 2130–2149.

[133] N.G. Hamilton, I.P. Silverwood, R. Warringham, J. Kapitán, L. Hecht, P.B. Webb, R.P. Tooze, S.F. Parker, D. Lennon, Angew. Chem. Int. Ed. 52 (2013) 5608–5611.

[134] N.G. Hamilton, R. Warringham, I.P. Silverwood, J. Kapitán, L. Hecht, P.B. Webb, R.P. Tooze, W. Zhou, C.D. Frost, S.F. Parker, D. Lennon, J. Catal. 312 (2014) 221–231.

[135] R. Warringham, N.G. Hamilton, I.P. Silverwood, C. How, P.B. Webb, R.P. Tooze, W. Zhou, C.D. Frost, S.F. Parker, D. Lennon, Appl. Catal. A Gen. 489 (2015) 209–217.

[136] R. Warringham, A.R. McFarlane, D.A. MacLaren, P.B. Webb, R.P. Tooze, J. Taylor, R.A. Ewings, S.F. Parker, D. Lennon, J. Chem. Phys. 143 (2015) 174703.

[137] S.J. Blundell, Magnetism in Condensed Matter, Oxford University Press, Oxford, 2001.

[138] A.H. Morrish, Canted Antiferromagnetism, Hematite, World Scientific, Singapore, 1994.

[139] S.N. Klausen, K. Lefmann, P.A. Lindgård, K.N. Clausen, M.F. Hansen, F. Bødker, S. Mørup, M. Telling, J. Magn. Magn. Mater. 266 (2003) 68.

[140] K. Nagai, Appl. Catal. A Gen. 221 (2001) 367.

[141] A. Tullo, Chem. Eng. News 87 (42) (2009) 22–23.

[142] J. Tai, R.J. Davis, Catal. Today 123 (2007) 42.

[143] A.R. McFarlane, H. Geller, I.P. Silverwood, R.I. Cooper, D.J. Watkin, S.F. Parker, J.M. Winfield, D. Lennon, Phys. Chem. Chem. Phys. 18 (2016) 17210–17216.

[144] K. AlGhamdi, J.S.J. Hargreaves, S.D. Jackson, Base catalysis with metal oxides, in: S.D. Jackson, J.S.J. Hargreaves (Eds.), Metal Oxide Catalysis, Wiley-VCH, Weinheim, 2009, pp. 819–843.

[145] S.D. Jackson, G.J. Kelly, D. Lennon, React. Kinet. Catal. Lett. 70 (2000) 207.

[146] http://www.methanol.org/Methanol-Basics/The-Methanol-Industry.aspx (last accessed on June 10, 2016).

[147] K.C. Waugh, Catal. Lett. 142 (2012) 1153–1166.

[148] Y. Yang, D. Mei, C.H.F. Peden, C.T. Campbell, C.A. Mims, ACS Catal. 5 (2015) 7328–7337.

[149] V.A. Trounov, A.E. Sokolov, V.T. Lebedev, O.P. Smirnov, A.I. Kurbakov, J. van den Heuvel, E. Batyrev, T.M. Yurieva, L.M. Plyasova, G. Török, Phys. Solid State 48 (2006) 1291–1297.

[150] V.A. Trounov, V.T. Lebedev, A.E. Sokolov, Y.S. Grushko, G. Török, J. van den Heuvel, E. Batyrev, T.M. Yurieva, L.M. Plyasova, Crystallogr. Rep. 52 (2007) 474–478.

[151] T. Kandemir, D. Wallacher, T.C. Hansen, K.-D. Liss, R.N. d'Alnoncourt, R. Schlögl, M. Behrens, Nucl. Instrum. Methods Phys. Res. Sect. A 673 (2012) 51–55.

[152] T. Kandemir, F. Girgsdies, T.C. Hansen, K.-D. Liss, I. Kasatkin, E.L. Kunkes, G. Wowsnick, N. Jacobsen, R. Schlögl, M. Behrens, Angew. Chem. Int. Ed. 52 (2013) 5166–5170.

[153] T. Kandemir, I. Kasatkin, F. Girgsdies, S. Zander, S. Kühl, M. Tovar, R. Schlögl, M. Behrens, Top. Catal. 57 (2014) 188–206.

[154] A.A. Khassin, H. Jobic, G.A. Filonenko, E.V. Dokuchits, A.V. Khasin, T.P. Minyukova, N.V. Shtertser, L.M. Plyasova, T.M. Yurieva, J. Mol. Catal. A 373 (2013) 151–160.

[155] T. Kandemir, M. Friedrich, S.F. Parker, F. Studt, D. Lennon, R. Schlögl, M. Behrens, Phys. Chem. Chem. Phys. 18 (2016) 17253–17258.

[156] J. Howard, I.J. Braid, J. Tomkinson, J. Chem. Soc. Faraday Trans I 80 (1984) 225–235.

[157] R.P. Holroyd, S. Poulston, M. Bowker, P.C.H. Mitchell, S.F. Parker, J. Chem. Soc. Faraday Trans. 93 (1997) 2569–2577.

[158] S. Poulston, R.P. Holroyd, M. Bowker, S.F. Parker, P.C.H. Mitchell, Surf. Sci. 402–404 (1998) 599–603.

[159] A. Chutia, I.P. Silverwood, M.R. Farrow, D.O. Scanlon, P.P. Wells, M. Bowker, S.F. Parker, C.R.A. Catlow, Surf. Sci. 653 (2016) 45–54.

[160] F. Studt, M. Behrens, E.L. Kunkes, N. Thomas, S. Zander, A. Tarasov, J. Schumann, E. Frei, J.B. Varley, F. Abild-Pedersen, J.K. Nørskov, R. Schlögl, ChemCatChem 7 (2015) 1105–1111.

[161] R. Schlögl, Ammonia synthesis, in: G. Ertl, H. Knözinger, F. Schüth, J. Weitkamp (Eds.), Handbook of Heterogeneous Catalysis, Wiley, Weinheim, 2008.

[162] L.E. Apodaca, US Geological Survey, 2011. http://minerals.usgs.gov/minerals/pubs/commodity/nitrogen/mcs-2016-nitro.pdf (last accessed on June 10, 2016).

[163] T. Kandemir, F. Girgsdies, I. Kasatkin, E. Kunkes, K.-D. Liss, V.K. Peterson, R. Schlögl, M. Behrens, J. Phys. Conf. Ser. 340 (2012) 012053.

[164] T. Kandemir, M.E. Schuster, A. Senyshyn, M. Behrens, R. Schlögl, Angew. Chem. Int. Ed. 52 (2013) 12723–12726.

[165] C. Lamonier, A. Ponchel, A. D'Huysser, L. Jalowiecki-Duhamel, Catal. Today 50 (1999) 247–259.

[166] C. Lamonier, E. Payen, P.C.H. Mitchell, S.F. Parker, J. Mayers, J. Tomkinson, Hydrogen species in cerium-nickel oxide catalysts: Inelastic and compton neutron scattering studies, in: A. Corma, F.V. Melo, S. Mendioroz, J.J.G. Fierro (Eds.), Studies in Surface Science and Catalysis, vol. 130, Elsevier, Amsterdam, 2000, pp. 3161–3166.

[167] L. Jalowiecki-Duhamel, S. Debeusscher, H. Zarrou, A. D'Huysser, H. Jobic, E. Payen, Catal. Today 138 (2008) 266–271.

[168] L. Jalowiecki-Duhamel, H. Jobic, E. Payen, Int. J. Nucl. Hydrogen Prod. Appl. 2 (2009) 148–158.

[169] C. Pirez, M. Capron, H. Jobic, F. Dumeignil, L. Jalowiecki-Duhamel, Angew. Chem. Int. Ed. 50 (2011) 10193–10197.

[170] W. Fang, C. Pirez, S. Paul, M. Capron, H. Jobic, F. Dumeignil, L. Jalowiecki-Duhamel, ChemCatChem 5 (2013) 2207–2216.

[171] C. Pirez, W. Fang, M. Capron, S. Paul, H. Jobic, F. Dumeignil, L. Jalowiecki-Duhamel, Appl. Catal. A 518 (2016) 78–86.

[172] R. Juarez, S.F. Parker, P. Concepción, A. Corma, H. Garcia, Chem. Sci. 1 (2010) 731–738.

[173] K. Clausen, W. Hayes, J.E. Macdonald, R. Osborn, M.T. Hutchings, A. Magerl, J. Chem. Soc. Faraday Trans. II 83 (1987) 1109–1112.

[174] W. Fang, C. Pirez, S. Paul, M. Jiménez-Ruiz, H. Jobic, F. Dumeignil, L. Jalowiecki-Duhamel, Int. J. Hydrogen Energy 41 (2016) 15443–15452.

[175] V.E. Antonov, V.K. Fedotov, B.A. Gnesin, G. Grosse, A.S. Ivanov, A.I. Kolesnikov, F.E. Wagner, Europhys. Lett. 51 (2000) 140–146.

[176] F. Fernandez-Alonso, D.L. Price (Eds.), Neutron Scattering – Fundamentals, Academic Press, Oxford, 2014 (Chapter 3).

[177] R.S. Pinna, S. Rudic, S.F. Parker, G. Gorini, F. Fernandez-Alonso, EPJ Web Conf. 83 (2015) 03013.

[178] S. Yang, J. Sun, A.J. Ramirez-Cuesta, S.K. Callear, W.I.F. David, D.P. Anderson, R. Newby, A.J. Blake, J.E. Parker, C.C. Tang, M. Schroeder, Nat. Chem. 4 (2012) 887–894.

[179] T.J. Bandosz, M. Seredych, E. Rodríguez-Castellón, Y. Cheng, L.L. Daemen, A.J. Ramírez-Cuesta, Carbon 96 (2015) 856–863.

[180] A. Ivanov, M. Jiménez-Ruiz, J. Kulda, J. Phys. Conf. Ser. 554 (2014) 012001.

[181] https://europeanspallationsource.se/sites/default/files/vespa_proposal.pdf (last accessed on June 10, 2016).

[182] A.M. Beale, S.D.M. Jacques, B.M. Weckhuysen, Chem. Soc. Rev. 39 (2010) 4656–4672.

[183] S.F. Parker, Chem. Commun. 47 (2011) 1988–1990.

[184] M.A. Adams, S.F. Parker, F. Fernandez-Alonso, D.J. Cutler, C. Hodges, A. King, Appl. Spectrosc. 63 (2009) 727–732.

[185] M.P. Marques, R. Valero, S.F. Parker, J. Tomkinson, L. Batista de Carvalho, J. Phys. Chem. B 117 (2013) 6421–6429.

[186] D.T. Bowron, S.D. Moreno, J. Phys. Condens. Matter 25 (2013) 454213.

[187] S.D. Moreno, D.T. Bowron, J. Phys. Chem. B 106 (2002) 11876–11886.

[188] D.T. Bowron, E.C. Beret, E. Martin-Zamora, A.K. Soper, E.S. Marcos, J. Am. Chem. Soc. 134 (2012) 962–967.

[189] D.T. Bowron, M. Amboage, R. Boada, A. Freeman, S. Hayama, S.D. Moreno, RSC Adv. 3 (2013) 17803–17812.

[190] A.M. Gaspar, M. AlvesMarques, M.I. Cabaço, M.I. de Barros Marques, A.I. Kolesnikov, J. Tomkinson, J.-C. Li, J. Phys. Condens. Matter 16 (2004) 6343–6364.

Chapter 6

Sorbate Dynamics in Zeolite Catalysts

Alexander J. O'Malley*,†,1 and C. Richard A. Catlow*,†,‡
*UK Catalysis Hub, Rutherford Appleton Laboratory, Didcot, United Kingdom; †Cardiff Catalysis Institute, Department of Chemistry, Cardiff University, Cardiff, United Kingdom; ‡Department of Chemistry, University College London, London, United Kingdom
1Corresponding author: e-mail: OMalleyA@cardiff.ac.uk

Chapter Outline

6.1 INTRODUCING ZEOLITE CATALYSTS

Zeolites are crystalline aluminosilicate materials containing pores, channels, and cavities of molecular dimensions, typically 3–15 Å [1]. The chemical

Experimental Methods in the Physical Sciences, Vol. 49. http://dx.doi.org/10.1016/B978-0-12-805324-9.00006-6

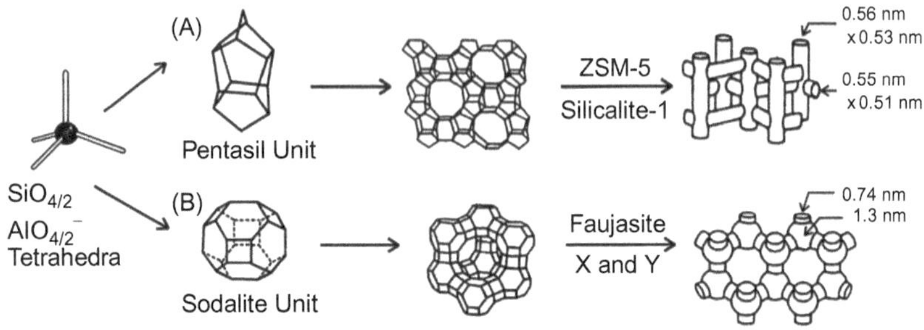

FIGURE 1 Schematics of the framework structures of (A) ZSM-5 (MFI) and (B) zeolite Y (FAU) and their composite building units formed from the TO_4 tetrahedron. *Figure adapted and reprinted with permission from Ref. [2].*

composition of zeolites is generally limited to the elements Si, Al, and O; however, these elements combine in a vast number of ways giving way to the observed wide range of topologies. The primary structural unit of a zeolite is the tetrahedron (TO_4), where the T-site is a silicon or aluminum atom. The $[SiO_4]^{4-}$ and $[AlO_4]^{5-}$ tetrahedra (shown in Figure 1) are linked at the corners to create the observed structures. Other tetrahedrally based classes of microporous solids include aluminophosphates (ALPOs) and silicoaluminophosphates (SAPOs).

Zeolite framework structures are classified by a three-letter code; there are currently 229 framework structures approved by the International Zeolite Association Structural Commission [3]. The two structures relevant to this chapter are the MFI and FAU (faujasite) frameworks shown in Figure 1 (a) and (b), respectively. The MFI framework has channels of ~5.5 Å width, which are straight in the 010 direction. However, in the 100 direction, the channels have a sinusoidal (or "zig-zag") structure. The channels meet at intersections as shown schematically in Figure 1(A). The faujasite structure is composed of sodalite cages connected by their hexagonal faces to create the tetrahedrally connected supercages shown in Figure 1(B). These supercages are 13 Å in diameter, connected by windows with a diameter of 7.5 Å.

When the siliceous framework has a silicon T-site replaced with aluminum, a negative charge is induced on the adjacent oxygen atom. In order to maintain a neutral framework structure, a cation is necessary to compensate for the induced charge. These cations may be alkali metals, alkali earth metals, transition metals, or lanthanides located within the channel/cage region of the zeolite. The charge may also be compensated by a proton that is bonded to the framework oxygen as shown in Figure 2. These protons are weakly bound and thus very acidic; indeed, the Brønsted acid sites form the basis of zeolitic acidic catalysis, which is integral to the petrochemical industry.

In addition to the higher activity of zeolite catalysts, the shape-selective nature of the catalysis is another defining feature of zeolite-catalyzed conversions. The selectivity arises from the zeolite pores having molecular

FIGURE 2 Alumination of the siliceous framework and resultant protonation of the adjacent O atom forming the Brønsted acid hydroxyl site.

dimensions, so the pore size or shape has a direct influence on the observed products of the reaction. Shape-selectivity may manifest itself in a number of ways, in particular reactant/product shape-selectivity and product molecular transport shape-selectivity. Reactant shape-selectivity involves the diffusion of one reactant being hindered, perhaps due to its size or shape preventing it from reaching the active site. Product selectivity involves the hindered diffusion of a bulkier product molecule, so smaller products are formed and recovered preferentially. Detailed discussions on understanding and influencing this shape selectivity can be found in Refs. [4,5].

6.2 STUDYING DYNAMICS IN ZEOLITES

Studying the dynamic processes governing the industrial applications of zeolites is difficult, as the majority take place inside the zeolite channels. The importance of understanding such dynamics can however be illustrated by considering the heterogeneous catalytic process. Rate-limiting steps to the reaction progress can be found in either the diffusion of the reactants toward the active site, adsorption onto the active site, conversion at the active site, or the diffusion of the products away. The diffusion of sorbates inside the zeolite channels is not only crucial to reaction efficiency, but also selectivity, as separation processes depend on a significant difference in the diffusivity of the components involved. Aside from molecular transport, understanding the localized dynamical behavior at adsorption sites such as rotational freedom and residence times can give a crucial insight into guest-host interactions in such processes.

We consider first the study of molecular diffusion of sorbates in zeolites using different methods. In particular, we discuss how translational diffusion characteristics may be probed by quasielastic neutron scattering (QENS), and how the interpretation of the resulting data may be assisted by parallel molecular dynamics simulations. We then discuss how QENS may be used to characterize and quantify localized motions of sorbed molecules in zeolite catalysts.

6.2.1 Studying Molecular Diffusion in Zeolites

There are two main approaches to studying the diffusion of sorbates in microporous solids, based on either macroscopic or microscopic methods. Macroscopic methods usually involve measuring mass transfer in the presence of a concentration gradient, or concentration differences as a function of length.

Such techniques include gravimetry, zero-length column (ZLC) chromatography, and membrane permeation methods. For an extensive review of macroscopic studies of diffusion in zeolites and the principles governing such methods, we refer the reader to that of Kärger [6]. Microscopic methods differ in that they are concerned with the stochastic motions of molecules and can probe molecular diffusion without the limit of a concentration gradient. These methods include pulsed field gradient (PFG) NMR and quasielastic neutron scattering (QENS). The use of the latter and its complementarity with molecular dynamics simulations will form the basis of this chapter. For context, an attempt to summarize some key characteristics of the more prominent macro- and microscopic methods is given in Table 1 [7–12,14].

An important advantage of the microscopic methods is that they are directly comparable with computational simulations of diffusivity based on the molecular dynamics (MD) method. The majority of the studies discussed in this chapter exploit this comparability. As a large number of microscopic studies into hydrocarbon diffusion have been undertaken using these techniques, we have selected studies that allow us to explain clearly the concepts involved in using QENS to decipher dynamical behavior of sorbates in zeolites, together with some recent developments where the complementarity of the QENS and MD has been used to gain detailed insight into such dynamics.

6.2.2 Quantifying Translational Diffusion

In this section, we summarize the theoretical models used to quantify translational diffusion. Our account has been adapted from Refs. [12,13].

The incoherent scattering function (outlined in chapter 1.2 of volume 44) is used in this case as the much larger scattering cross section of ^{1}H hydrogen allows for a much larger scattering of the hydrocarbon, giving it more visibility over the inorganic material of the zeolite. As a result, only the motions of the sorbate are detected. The incoherent scattering function obtained is the four-dimensional Fourier transform of the self-part of the van Hove correlation function (outlined in chapter 1.4.1 of volume 44), in essence, a probability density in three-dimensional space that a particle will be displaced by a vector r within time t at equilibrium. In the scattering function, $\mathbf{Q}$ and ω are the Fourier transformed variables of $\mathbf{r}$ and t. Small Q involves average values over long distances, and slow motions taking place over long times in $G_s(\mathbf{r},t)$ will correspond to small ω. When the characteristic time of the motion is shorter than the inverse of the instrumental resolution, a broadening of the scattering function will be observed (shown in figure 1.18, chapter 1.45 of volume 44). However, when the timescale of movement is longer than the inverse of the instrumental resolution (slower than the machine can measure), the shape of the peak is given by the instrumental resolution function, approximated by a Gaussian, and an instrument of higher energy resolution must be used. The broadening of the scattering function must be convoluted with the instrumental resolution function in order to be quantified.

TABLE 1 Summary of Prominent Experimental Methods for Diffusion Measurement Both Macroscopic (Gravimetry [7], Frequency Response [8], Zero-Length Column [9], and Single-Crystal Membrane Permeation [10]) and Microscopic (Pulsed Field Gradient NMR [11], Quasielastic Neutron Scattering [12,13], and Neutron Spin-Echo [14])

Technique	Description	Advantages/limitations (+/−)
Gravimetric methods	The sample is subjected to a step change in the pressure of the surrounding atmosphere; the time dependence of changes in the amount adsorbed is monitored	+ Accessible experimentally + Can study large systems accessing both macro-, meso, and micropore diffusion − Affected by external heat and mass transfer resistances − Unreliable for faster diffusion processes due to the response time of the measurement
Frequency response methods	The pressure of the sorbate is varied sinusoidally; the phase angle and amplitude of the weight changes are measured. Alternatively, the pressure response of a sorbate system to sinusoidal volume variation may be measured	+ Accessible experimentally + Can study large systems accessing both macro-, meso, and micropore diffusion + Sensitive to a wide range of diffusivities − Also affected by heat and mass transfer resistances
Zero-length column	The zeolite is equilibrated with the sorbate under controlled conditions. The desorption curve is measured when the carrier is switched to a nonadsorbing stream, and the effluent concentration is monitored	+ Not affected by heat and mass transfer phenomena − Limited by the system response time − The resistance of an external sorbate film around crystals may also affect the accuracy requiring a high flow rate to remove it. Sample size is also very small
Single-crystal membrane permeation	A single crystal is embedded in a substance such as epoxy resin; the "inflow" side is exposed to the diffusant gas, and the pressure increase is measured in constant volume on the "outflow" side of the membrane	+ Not affected by external heat transfer and mass transfer phenomena − Difficult experimentally as large crystals are needed (>100 μm) − Technique is limited to certain samples that can produce larger crystals, and heterogeneities will cause inconsistencies in measurement − Diffusant gas may bypass the embedded crystal

Continued

TABLE 1 Summary of Prominent Experimental Methods for Diffusion Measurement Both Macroscopic (Gravimetry [7], Frequency Response [8], Zero-Length Column [9], and Single-Crystal Membrane Permeation [10]) and Microscopic (Pulsed Field Gradient NMR [11], Quasielastic Neutron Scattering [12,13], and Neutron Spin-Echo [14])—cont'd

Technique	Description	Advantages/limitations (+/−)
Pulsed field gradient NMR	Radiofrequencies are applied to the sample to generate a spin-echo NMR signal. Pairs of magnetic field gradients from the sample are superimposed to label spin positions. The spin-echo amplitude is recorded as a function of pulsed gradients, and spin-echo attenuation is analyzed to deduce the diffusivity	+ Can measure self-diffusivity on the micro-milliscale + Nondestructive method with fast data collection − Range of diffusivity measurable is limited by spin relaxation time − Larger zeolite crystals are needed to prevent the sorbate from reaching the crystallite boundary on the time scale of the experiment − Presence of paramagnetic species, for example, Cu^{2+} restricts the experiment to shorter observation times
Quasielastic neutron scattering	Small differences in the energy of incident and scattered neutrons (μeV-meV) are evaluated as a function of neutron momentum transfer. The dependence is fit to models describing diffusivity and translational diffusion mechanisms	+ Allows self-diffusivity (10^{-9}–10^{-10} m^2s^{-1}) measurements and diffusion mechanism elucidation on the nanoscale − Instrumental access is severely limited, and time length of a single measurement is of the order of hours
Neutron spin-echo	Energy differences between incident and scattered polarized neutrons are determined from spin vector differences; similar to quasielastic neutron scattering, the neutron energy transfer is evaluated as a function of momentum transfer to obtain diffusivity	+ Allows self- and transport diffusivity (10^{-11}–10^{-12} $m^2\ s^{-1}$) measurements and diffusion mechanism elucidation on the nanoscale − Instrumental access is even more limited than QENS, and time length of a single measurement is longer than a QENS experiment

6.2.2.1 Self-Diffusion – Isotropic and Jump Models

In the case of unhindered, isotropic diffusion as defined by Fick's law,

$$\frac{\partial p(\mathbf{r},t)}{\partial t} = D_s \nabla^2 p(\mathbf{r},t) \tag{1}$$

where $p(\boldsymbol{r},t)$ is the probability of finding an atom at position $\boldsymbol{r}$ at time t; the self-part of the van Hove correlation function $G_s(\mathbf{r},t)$ (described in chapter 1.4.1 of volume 44) is the solution. The scattering function obtained is a Lorentzian function:

$$S(\mathbf{Q},\omega) = \frac{1}{\pi}\frac{D_s Q^2}{\omega^2 + (D_s Q^2)^2} \tag{2}$$

At sufficiently low Q values (larger distances), the half width at half maximum (HWHM) of this Lorentzian function has a DQ^2 dependence (D is the diffusion coefficient); thus a plot of the HWHM as a function of Q^2 would give a straight line (shown in Figure 3), though this analysis is only valid for larger distances (smaller Q values).

When considering sorbate diffusion in zeolites, jump diffusion models tend to be used. These consider a molecule that is stationary on a site for a defined period which then jumps a given distance to another site. At larger momentum transfers, we observe deviations from the linearity of the DQ^2 dependence as the system is not continuous at the shorter length scales probed at these larger wave vector values.

A model commonly used to describe such behavior was formulated by Chudley and Elliot (discussed in chapter 1.4.5, volume 44). Their model of

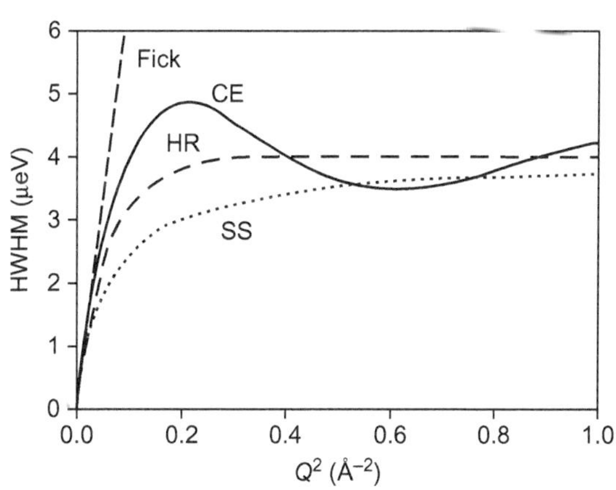

FIGURE 3 Q dependences of the HWHM observed when molecules exhibit Fickian behavior and jump diffusion consistent with the Chudley and Elliot (CE), Hall and Ross (HR), and Singwi and Sjölander (SS) models. *Figure reprinted with permission from Ref. [12].*

Chudley and Elliot [15] was originally developed to describe liquid systems with local order, but is widely used for systems such as ion diffusion in lattices through to diffusion of sorbates in microporous solids. It proposes that for an average time interval τ, an atom remains on a given site vibrating about a center of equilibrium. After this time, the atom moves rapidly to another site in a negligible jump time compared with the time spent at the residence site. The Chudley and Elliott model assumes a constant jump length d, with the HWHM dependence on Q being

$$\Delta\omega(Q) = \frac{1}{\tau}\left(1 - \frac{\sin(Qd)}{(Qd)}\right) \tag{3}$$

One may then obtain the Fickian diffusion coefficient as at low Q; expanding up to the third order in terms of Qd gives us

$$\Delta\omega(Q) = \frac{(Q^2 d^2)}{(6\tau)} \tag{4}$$

And we can again obtain a DQ^2 dependence with a diffusion coefficient of

$$D_s = \frac{d^2}{6\tau} \tag{5}$$

Other models deal with jump distances that are not constant but consider a jump length distribution. These include the Singwi-Sjölander model [16] originally derived for a model of water where the jump length distribution is given as

$$\rho(r) = \frac{r}{r_0^2}\exp\left(-\frac{r}{r_0}\right) \tag{6}$$

And the HWHM broadening with Q is

$$\Delta\omega(Q) = \frac{1}{6\tau}\,\frac{Q^2\langle r^2\rangle}{1 + Q^2\langle r^2\rangle/6} \tag{7}$$

where $\langle r^2\rangle$ is the mean square jump length, given as $6r_0^2$.

A Gaussian jump distance distribution is given by the Hall and Ross model [17]:

$$\rho(r) = \frac{2r^2}{r_0^3(2\pi)^{1/2}}\exp\left(-\frac{r^2}{2r_0^2}\right) \tag{8}$$

where the broadening is

$$\Delta\omega(Q) = \frac{1}{\tau}\left(1 - \exp\left(-\frac{Q^2\langle r^2\rangle}{6}\right)\right) \tag{9}$$

The diffusion coefficient for both these jump models can be calculated from $D_s = \frac{\langle r^2 \rangle}{6\tau}$. Figure 3 shows the plotted Q dependencies for Fickian and the above jump diffusion models, for $D_s = 1 \times 10^{-9}\ m^2s^{-1}$. We note that at low wave vector transfers, all jump models show the D_sQ^2 dependence.

6.2.3 Localized Motions in Zeolites

The dynamics associated with localized sorbate motions in zeolites is of interest from a number of perspectives, which can include probing rotational/translational coupling, the changes in physicochemical behavior of confined liquids, or the motion of molecules in restricted rotational/orientational potentials provided by the zeolite framework. Preferred siting in the pore system may be understood from the observed accessible rotational geometries, and interactions with the catalyst active sites may be considered in detail through probing such motions. In this section, we outline some procedures for both characterizing and quantifying rotational and localized translational motions of sorbed molecules.

6.2.3.1 Rotational Dynamics

The experimental elastic incoherent structure factor (EISF) is the fraction of the total intensity that is elastic and is introduced in section 6.8 of volume 44. It is used to characterize rotational motions of molecules, while the width of the Lorentzian functions describing the quasielastic component may be used to quantify the dynamics of the protons involved. Specifically, we are interested in how the EISF varies with Q. This variation may be fit to various models describing, for example, isotropic diffusion around a sphere (or diffusion within a sphere), uniaxial jump rotation around an n-fold axis (e.g., threefold in the case of a methyl group or sixfold in the case of a rotating benzene molecule). We will discuss examples of the above in the context of zeolite-sorbed molecules later and refer the reader to figure 6.2.2 in volume 44 for examples of plotted EISFs for diatomic molecules of differing bond lengths and molecules with increased n-fold rotations. Numerous examples of experimental and theoretical EISFs are shown in Sections 6.4 and 6.5. We will first introduce some of these models in more detail. For a detailed derivation of these models, we refer the reader to chapter 6 of Ref. [13].

Isotropic rotation is characterized by a molecule whose reorientation takes place through a series of small-angle, random rotations so that no most probable orientation exists on a time average. The scattering law as derived by Sears [18] for this form of rotation is given by

$$S(Q,\omega) = A_0(Q)\delta(\omega) + \sum_{l=1}^{\infty} A_l(Q)\frac{1}{\pi}\frac{l(l+1)D_R}{[l(l+1)D_R]^2 + \omega^2} \quad (10)$$

With the EISF, $A_0(Q)$ is given as

$$A_0(Q) = j_0^2(Qr) \tag{11}$$

And any quasielastic structure factors (QISFs) are given as

$$A_l(Q) = (2l+1)j_l^2(Qr) \tag{12}$$

where r is the radius of rotation, j_l is the lth spherical Bessel function, an D_R is the rotational diffusion coefficient. Plots of both the EISF and QISFs for a methane molecule undergoing isotropic rotation ($r = 1.1$ Å) are shown in figure 9 of Ref. [12].

If there is a potential barrier hindering rotation in a particular axis of a molecule, one may measure the rotational diffusion, where the trajectory of each proton is a circle. The scattering function is still a Gaussian resolution function plus a series of Lorentzian functions, but it is now dependent on the angle θ between the axis of rotation and the direction of Q.

We consider first the case of continuous rotation on a circle; here, the scattering function is given by

$$S(\mathbf{Q},\omega) = J_0^2(Qr\sin\theta)\delta(\omega) + 2\sum_{l=1}^{\infty} J_l^2(Qr\sin\theta)\frac{1}{\pi}\frac{D_R l^2}{(D_R l^2)^2 + \omega^2} \tag{13}$$

where the first and second terms are the EISF and QISFs, respectively. Equation (13) cannot be used for powder samples typical for zeolite studies, because no expression exists for the average angle θ between the axis of rotation and the direction of $\mathbf{Q}$. However, it has been shown that for models of jump rotation over N equally spaced sites, with a sufficiently large N (>6), the scattering function does not change as N increases. The approximation of jump rotation over N sites (outlined later) may then be used.

For a rotation with jumps between two equivalent sites, such as a framework H-bonded water molecule rotating around its C_2 axis, the scattering law (averaged for a powder sample) may be given as

$$S(Q,\omega) = A_0(Q)\delta(\omega) + A_1(Q)\frac{1}{\pi}\frac{2\tau}{4+\omega^2\tau^2} \tag{14}$$

where τ is the mean residence time on each site.

The elastic and quasielastic structure factors are then given as

$$A_0(Q) = \frac{1}{2}[1 + j_0(Qd)] \tag{15}$$

and

$$A_1(Q) = \frac{1}{2}[1 - j_0(Qd)] \tag{16}$$

where d is the distance between the two sites.

A jump rotation model between three equidistant sites on a circle (such as a rotating methyl group on a framework-bound methoxy group) gives the scattering law below:

$$S(Q,\omega) = A_0(Q)\delta(\omega) + A_1(Q)\frac{1}{\pi}\frac{3\tau}{9+\omega^2\tau^2} \quad (17)$$

The elastic and quasielastic structure factors are given as

$$A_0(Q) = \frac{1}{3}\left[1 + 2j_0\left(Qr\sqrt{3}\right)\right] \quad (18)$$

and

$$A_1(Q) = \frac{2}{3}\left[1 - 2j_0\left(Qr\sqrt{3}\right)\right] \quad (19)$$

where r is the radius of the circle on which the jumps take place, for example, 0.99 Å for protons on a methyl group.

Finally, we consider jump rotation around N equivalent sites on a circle. Here, the scattering law is given by

$$S(Q,\omega) = A_0(Q)\delta(\omega) + \sum_{l=1}^{N-1} A_l(Q)\frac{1}{\pi}\frac{\tau_l}{1+\omega^2\tau_l^2} \quad (20)$$

with elastic and quasielastic structure factors given as

$$A_l(Q) = \frac{1}{N}\sum_{n=1}^{N} j_0\left[2Qr\sin\left(\frac{n\pi}{N}\right)\right]\cos\left(\frac{2l\,n\pi}{N}\right) \quad (21)$$

Here, r is the radius of gyration, which in the case of benzene is 2.5 Å (where $N=6$), and j_0 is the spherical Bessel function of zero order. The quasielastic terms are expressed as a sum of Lorentzians whose widths τ_l^{-1} are related to the mean time between successive jumps.

6.2.3.2 *Diffusion in a Confined Volume*

There are variety of physical systems where the diffusion of molecules confined in media or restricted to movements in certain dimensions are of interest. These include lamellar structures, polymers, clay minerals, and of course diffusion of sorbates in zeolites. Volino and Dianoux [19] developed a model to describe a scattering molecule undergoing translational motions in a confined spherical volume of radius r_{conf}. This scattering model is based on the general problem of a particle diffusing in a potential field of spherical symmetry, where the potential is low inside the sphere's volume but infinite outside of it. Detailed derivation of this model may be found in Ref. [19] which has been used in systems of confined water in Nafion membranes [20] and Vycor glass [21].

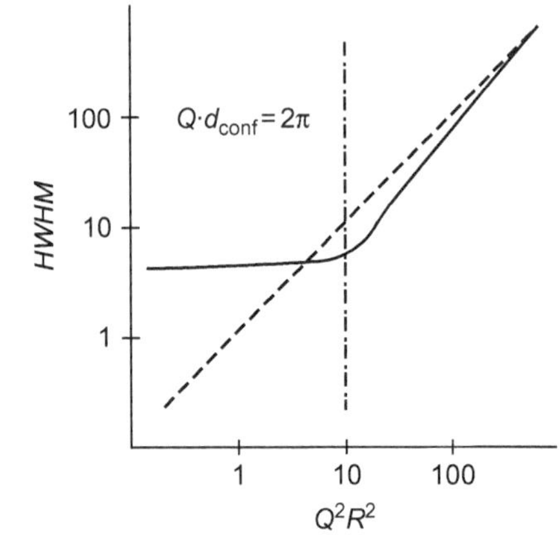

FIGURE 4 Q dependence of the HWHM for a molecule diffusing inside a sphere. Where the dashed line is the Q dependence of the corresponding unconfined Fickian behavior. *Figure adapted and reprinted with permission from Ref. [22].*

The EISF in this model is given as

$$A_0(Q) = \left[\frac{3j_1\left(Qr_{conf}\right)}{Qr_{conf}} \right]^2 \tag{22}$$

where j_1 is the spherical Bessel function of the first kind.

In the scattering function, one observes an elastic peak and a combined quasielastic Lorentzian line with a width depending on the values of diffusion constant D and radius of the confining sphere r_{conf}. The Q^2 versus HWHM dependence of this Lorentzian will follow the trend shown in Figure 4, which shows that the half width at half maximum of the quasielastic line is practically constant and equal to $4.33D/r_{conf}$ at low Q up until $Q = \pi/r_{conf}$. At this point, the EISF becomes very small, and the dependence of the HWHM approaches the DQ^2 law [22].

If the Q dependence after $Q = \pi/r_{conf}$ deviates from linearity, one may be observing jump diffusion within the confined volume and fit an appropriate model [21,22].

6.3 COMPLEMENTARITY BETWEEN QENS AND MD SIMULATIONS

Computational methods such as molecular dynamics (MD) simulations may also be used to measure diffusion coefficients by directly tracking the time resolved movement of molecules through space. The complementarity stems from MD and QENS measuring over the same nanometer and nanosecond scales. In terms of self-diffusion coefficient measurements, it is common for simulations to yield higher values of the diffusion coefficient than experimental methods such as PFG-NMR and QENS [12]. This observation is attributed first to the usual practice of modeling the zeolite lattice as a perfect siliceous crystal, whereas defects such as extraframework ions, Brønsted acid sites, and terminal hydroxyls are inevitable in real zeolite samples and are known to inhibit diffusion. Also, many MD simulations are conducted in the limit of infinite dilution, removing all the sorbate-sorbate interactions (rarely possible experimentally); these interactions often slow the diffusion of sorbates through the channels, thus reducing the diffusion coefficient. It is also observed that the diffusion coefficients of *n*-alkanes in zeolites decrease monotonically with chain length, attributed to the fact that the longer the alkane, the higher the diffusion barrier because more atoms have to be displaced from their position, necessitating more activation energy. There have also in many earlier simulations been limitations on the potential models used, as discussed in the next section.

6.3.1 Recent Improvements in MD Simulations

The large length and timescales associated with diffusion in zeolites require force-field-based (or so-called "classical") simulations. Though these are of a lower level of theory than quantum mechanical simulations, detailed qualitative and quantitative information may be obtained about the diffusing sorbate. We note that extensive reviews of classical (and quantum mechanical) simulations of adsorption, diffusion, and reaction processes in zeolites are given by Smit and Maesen [5] and more recently by Van Speybroeck et al. [23]. We will center the discussion in this section on the approximations necessary historically for modeling these systems and their potential shortcomings for quantitative (and indeed qualitative) study of hydrocarbon diffusion in zeolites. We will also describe how recent studies using more detailed models have overcome these difficulties.

Though computational methods have long been established in zeolite science as highlighted by the aforementioned reviews, growth in computational power has allowed for increasingly realistic simulations. In earlier work, a common approximation made to reduce computational expense was the use of a rigid framework to represent the zeolite structure [5]. Given that in any zeolite simulation, the vast majority of the atoms belong to the zeolite framework itself, fixing the zeolite atoms in their crystallographic positions will cut the cost of these simulations significantly. This approximation, however, may have been at least partially responsible for the sizeable discrepancies between QENS measurements [24,25] and MD simulations of *n*-alkane diffusion in silicalite [26].

The role of framework flexibility on the diffusivity of shorter alkanes in silicalite was systematically studied by Leroy and coworkers [27]. It was found that diffusion was enhanced by framework flexibility for the shorter alkanes (methane and butane) at the lowest loadings, as sorbate-sorbate interactions dominated at higher loadings rendering the effect of framework flexibility negligible. The dependence of self-diffusivity on chain length however remained unchanged with the framework model. Other examples include higher diffusion coefficients for *n*- and isobutane in MFI using a flexible framework [28], along with studies showing faster diffusion for methane in a flexible MFI lattice (at low loadings) but little influence in LTA frameworks [29].

It is often argued that for molecules that do not fit tightly in the zeolite pores, this approximation is reasonable; as with tighter fitting molecules, pore-breathing is more likely to facilitate potential barrier crossing. It is also important to remember that it is crucial to model accurately the interactions between the framework species; only with reliable parameters for these interactions will the required accuracy in framework flexibility be achieved. We should also note that the aforementioned simulations were all carried out using a united-atom hydrocarbon model as opposed to an all-atom hydrocarbon model. The details of these hydrocarbon models will now be outlined.

When simulating hydrocarbons, there are a number of models available. The most realistic of course is the all-atom (AA) model, where all carbon and hydrogen atoms are explicitly represented with their associated potentials. The united-atom (UA) approach is, however, a very popular choice. CH_3, CH_2, and CH groups are represented by single atoms, reducing considerably the computational cost. However, in the present context, it has severe limitations, reducing barriers to diffusion as the potential is too "smooth," and it may not represent properly the larger surface area for intermolecular interactions between the chain and the channel walls. Indeed, the united-atom model is known to give higher local mobilities of *n*-tridecane melts [30], perhaps due to these factors combining to decrease the energy barriers to mobility and allowing the molecule to move more freely.

Another representation used in diffusion studies is the anisotropic united-atom model (AUA) proposed by Toxvaerd for *n*-alkanes [31]. While each force center is located on the carbon in the conventional united-atom potentials, the force center is shifted in the AUA potential and is placed between the carbon and hydrogen atoms of the related group. A further issue is that the majority of force fields for hydrocarbons were originally developed for biological applications, which give very accurate descriptions of behavior at ambient conditions. However, many simulations of diffusion within zeolites are carried out at much higher temperatures to mimic industrial conditions. Thus, it is critical that the force field gives an accurate description across a range of temperatures and pressures. Examples are force fields that have been shown to replicate the liquid-vapor coexistence curve over a wide range of conditions, such as the TraPPE-UA force field [32].

O'Malley and Catlow [33] aimed to assess systematically the role of framework flexibility and hydrocarbon models over a range of longer *n*-alkanes in the silicalite framework. The study performed simulations using a united-atom model with a rigid zeolite cage, an all-atom hydrocarbon model with a rigid zeolite cage, and an all-atom model with a flexible zeolite cage. It was found that the use of a rigid cage resulted in the so-called "window effect" with a periodic dependence of diffusivity on chain length

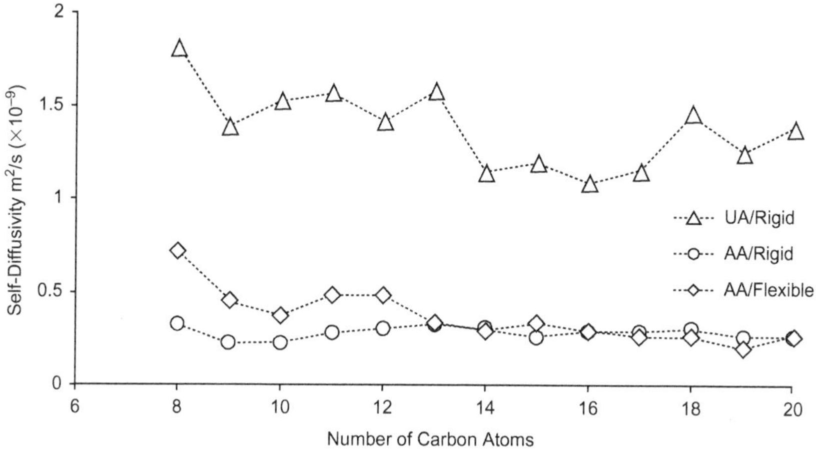

FIGURE 5 The dependence of diffusivity on chain length for *n*-alkanes in silicalite using three different model systems, comparing the united-atom/rigid framework simulation, all-atom/rigid framework simulation, and all-atom/flexible framework simulation. *Figure reprinted with permission from Ref. [33].*

as shown in Figure 5. The use of a united-atom hydrocarbon model gives diffusion coefficients up to an order of magnitude higher than those obtained using an all-atom hydrocarbon model. All diffusion coefficients obtained using a flexible framework and all-atom hydrocarbon model not only gave the same trend as QENS studies of the same systems [24] (a monotonic decrease in diffusivity with chain length) but also the absolute values of the self-diffusion coefficient were all within an order of magnitude of these QENS studies. The conclusion was that the potential of the united-atom hydrocarbon model is indeed too smooth, allowing more free mobility and increased diffusivity. Also, the study suggests that the structural energy trapping phenomena observed in previous simulations are merely a product of the rigid framework zeolite model.

Upon use of a flexible zeolite model, these energy traps dissipate too quickly during the framework breathing process to affect significantly sorbate diffusion. The remaining sources of discrepancy when using the contemporary models were attributed to the use of a perfect silicalite crystal in the simulations and the condition of infinite dilution eliminating sorbate-sorbate interactions. If the systems were modeled at experimental alkane loadings, the sorbate-sorbate interactions would lower the diffusion coefficients, bringing them closer to values obtained using QENS. We address this discrepancy directly in Section 6.4. In the following sections, we will review a number of examples where the above principles are applied to understand the dynamics of sorbed molecules in zeolites. Notably, the majority of examples involve the concerted use of QENS with MD simulations, illustrating the capabilities of the concerted use of the techniques.

6.4 HYDROCARBON BEHAVIOUR IN MFI ZEOLITES

A significant amount of work both theoretical and experimental has been carried out studying the diffusion of hydrocarbons in the MFI zeolite structure [27,34–45], the framework type of industrially important ZSM-5, and its siliceous analogue, silicalite. The diffusion of *n*-alkanes longer than C_8 in this framework has been of more recent interest and has been studied using QENS in both Na-ZSM-5 [25] and silicalite [24]. We will first discuss in detail how the dynamics of these *n*-alkanes can be studied using QENS, with emphasis on the use of contemporary molecular dynamics simulations to obtain better quantitative agreement with measured diffusion coefficients. We then discuss studies of branched alkanes in this structure and, finally, aromatic species.

6.4.1 *n*-Alkane Diffusion in Silicalite and Na-ZSM-5

We begin with a selection of studies where the dynamics of *n*-alkanes in silicalite and Na-ZSM-5 are measured. The effect of counterion presence on *n*-octane diffusion was studied by Leroy and Jobic [46] using the backscattering spectrometer IN16 [47] at the Institut Laue-Langevin, who measured the D_s in silicalite and Na-ZSM-5. A single Lorentzian was needed to fit the data (some of the QENS spectra for the silicalite sample are shown in Figure 6, left). The Singwi and Sjölander jump diffusion model from Eqs. (6) and (7) was then used to fit the broadening of the spectra, giving the D_s at 300 K of $2.0 \times 10^{-10}\,m^2\,s^{-1}$.

This value of D_s was four times larger than that in Na-ZSM-5, and the influence of the cation presence on mobility is clear from the QENS spectra at

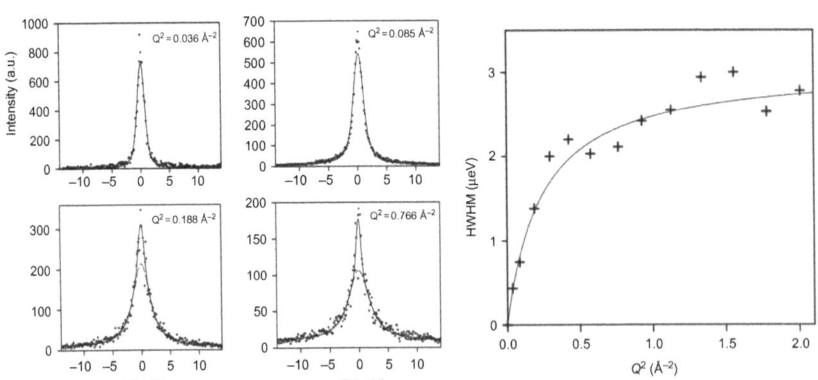

FIGURE 6 (left) QENS spectra obtained and fitted for *n*-octane in silicalite at 300 K (1.56 mol/uc loading). (right) Broadenings derived for the spectra, fitted to the Singwi and Sjölander jump diffusion model. *Figure adapted and reprinted with permission from Ref. [46].*

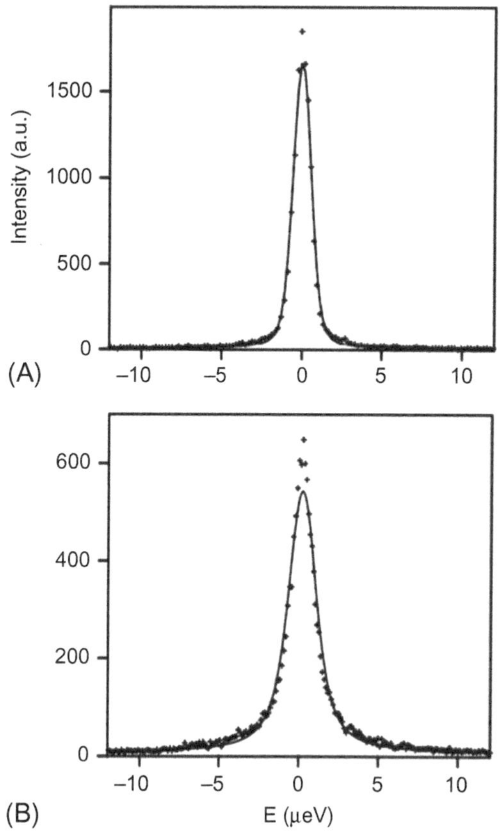

FIGURE 7 (A) QENS spectra obtained and fitted for *n*-octane (B) Na-ZSM-5, 0.9 mol/uc and (bottom) silicalite (1.56 mol/ucl) at 300 K, $Q=0.29\ \text{Å}^{-1}$. *Figure adapted and reprinted with permission from Ref. [46].*

$Q=0.29\ \text{Å}^{-1}$ in Figure 7. The spectra were recorded with the same spectrometer, where the broadening is significantly more pronounced for the silicalite structure.

The diffusion of longer *n*-alkanes was also studied in both Na-ZSM-5 and silicalite. Again, it was found that a single Lorentzian function could fit the data accurately, with the broadening described by the Singwi and Sjölander jump diffusion model. The diffusion coefficients obtained as a function of chain length are plotted with other methods including PFG-NMR, MD and hierarchical simulations; the loadings of the QENS experiments were approximately 28 H atoms per unit cell (Figure 8).

It was noted that, at 300 K, the measured D_s values of the QENS experiments and the hierarchical simulations were in close agreement. For example, for decane, the difference is less than a factor of two, considering the experimental errors of the QENS values (30% from C8 to C12 and 50% for C14 and

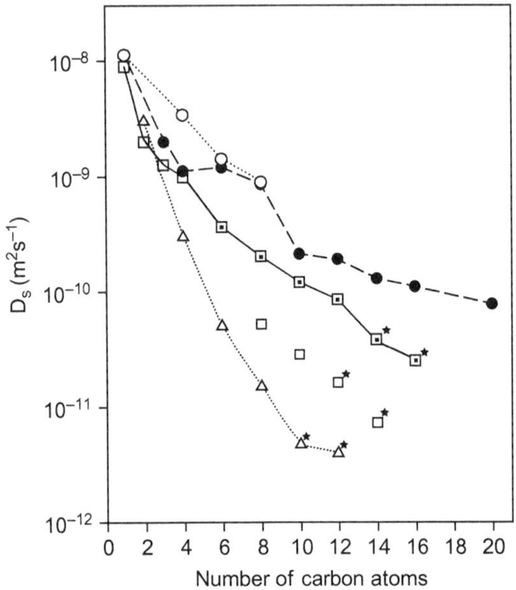

FIGURE 8 Self-diffusivities obtained at 300 K from MD simulations (○), hierarchical simulations (●), QENS in Na-ZSM-5 (□), QENS in silicalite-1(■), and PFG-NMR (Δ). * = extrapolation to 300 K. *Figure adapted and reprinted with permission from Ref. [24].*

C16) and the error associated with the coarse-graining procedures of the simulations; this can be considered as not significant. It was also noted that the D_s values from the simulations were consistently higher than the QENS measured values, which was attributed to the use of infinitely dilute conditions for the simulation. The activation energies, however, did not quite replicate such a close agreement, particularly for the longest alkanes as discussed later. However, it was noted that the NSE technique would give more accurate values due to its ability to probe slower motions.

Closer agreement was found in more recent atomistic molecular dynamics simulations for these systems [48]. The correct loadings were used (therefore accounting for sorbate-sorbate interactions), employing a flexible zeolite framework and explicit atom model of the hydrocarbon, displayed in Figure 9.

The diffusion coefficients for all chain lengths, at all temperatures, are listed in Table 2, with those calculated at 300 K plotted in Figure 10. As expected, the experimentally observed monotonic decrease in diffusion coefficient with chain length was observed. Importantly, all the calculated diffusion coefficients were within a factor of five of the QENS measurements, with the lowest correlation being with hexadecane at 300 K, differing from QENS studies by a factor of 3.68. From the plot in Figure 10, comparing all values between the current study and those obtained at infinite dilution in

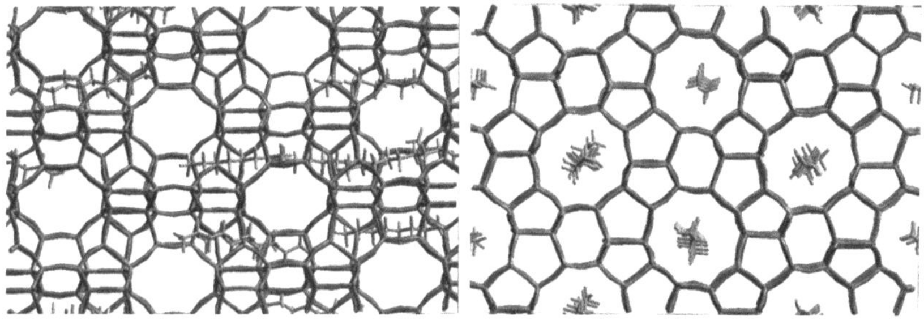

FIGURE 9 A view of the system for octane preproduction run at 300 K viewed down the (left) 010 direction and (right) 100 direction. *Figure reprinted with permission from Ref. [48].*

TABLE 2 A Comparison of D_s Values in $m^2\ s^{-1}$ Obtained by the Discussed MD Simulations [48], Hierarchical Simulations (HS) [49], and QENS Studies [24], for all Alkanes at all the Temperatures Studied

N	T (K)	D_s (MD)	D_s (QENS)	D_s (HS)	D_s (MD)/D_s (QENS)
8		$6.09\ 10^{-10}$	$2.0\ 10^{-10}$	$8.5\ 10^{-10}$	3.05
10		$3.31\ 10^{-10}$	$1.2\ 10^{-10}$	$2.1\ 10^{-10}$	2.76
12	300	$2.10\ 10^{-10}$	$8.5\ 10^{-11}$	$1.9\ 10^{-10}$	2.47
14		$1.23\ 10^{-10}$	$3.8\ 10^{-11}$	$1.3\ 10^{-10}$	3.23
16		$9.20\ 10^{-11}$	$2.5\ 10^{-11}$	$1.1\ 10^{-10}$	3.68
8		$8.83\ 10^{-10}$	$2.8\ 10^{-10}$		3.15
10		$5.90\ 10^{-10}$	$1.8\ 10^{-10}$		3.27
12	350	$3.91\ 10^{-10}$	$1.4\ 10^{-10}$		2.79
14		$2.38\ 10^{-10}$	$6.7\ 10^{-11}$		3.55
16		$1.81\ 10^{-10}$	$5.1\ 10^{-11}$		3.55
8		$1.09\ 10^{-9}$	$3.9\ 10^{-10}$		2.79
10		$7.20\ 10^{-10}$	$2.7\ 10^{-10}$		2.66
12	400	$5.48\ 10^{-10}$	$2.3\ 10^{-10}$		2.38
14		$3.70\ 10^{-10}$	$1.2\ 10^{-10}$		3.08
16		$3.28\ 10^{-10}$	$1.05\ 10^{-10}$		3.12

Ref. [33], one can see a significant reduction in the diffusion coefficient for dodecane, tetradecane, and hexadecane that are lower by a factor of 3, 3.8, and 5, respectively (which may be due to improved statistics, rather than sorbate-sorbate interactions, as the molecules stay in their separate channel

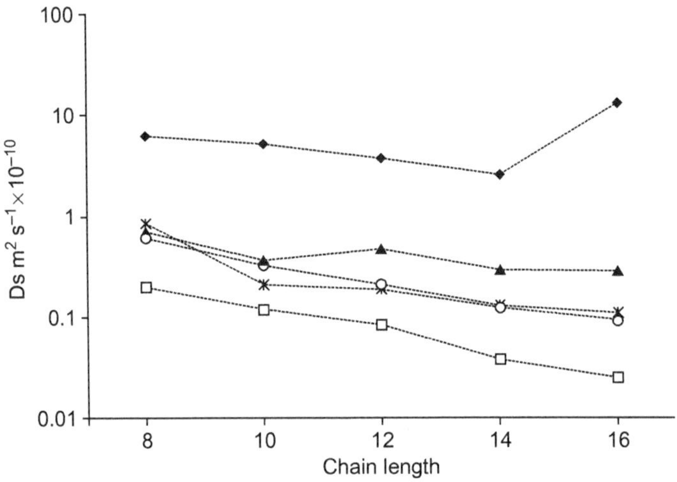

FIGURE 10 A plot of the calculated diffusion coefficients from the MD simulations using experimental loadings (O)[48], MD simulations at infinite dilution[33] (▲), MD simulations using more primitive models[26] (◆), Hierarchical simulations[49] (*), and QENS studies[24] (□). *Figure reprinted with permission from Ref. [48].*

systems). However, for octane and decane, the diffusion coefficient values are still very similar to those obtained at infinite dilution.

In Figure 10, the diffusion coefficients obtained at 300 K are also plotted in comparison with the previous MD simulations using a united-atom model and rigid zeolite cage [26] and QENS measurements [24]. This plot clearly shows the great improvement in accuracy of MD simulations when a more sophisticated and realistic model is used. When the diffusion coefficients obtained from the most sophisticated MD study are compared with those obtained from hierarchical simulations [49], a close agreement is shown. Almost identical values for dodecane and tetradecane were obtained, and the most significant difference occurs for decane, with a difference of a factor of 1.5. Despite the sophisticated model and use of experimental loadings, the diffusion coefficients obtained by the present study are higher than those obtained experimentally, which can probably, as noted, be attributed to the use of a perfect silicalite crystal in the simulations, allowing for faster, less hindered diffusion on the nanometer scale. An experimental sample would have defects such as silanol nests and grain boundaries on the nanometer scale and also surface effects from each crystallite, which are not accounted for in the periodic perfect lattice employed by the simulation.

The activation energy values obtained from this study are tabulated and plotted in comparison with the values obtained by QENS and hierarchical simulations in Table 3 and Figure 11, respectively. One can see that the trend (a roughly monotonic increase) of the calculated activation energies follows that observed by QENS. The absolute values in activation energy show close

TABLE 3 The Calculated Activation Energies in kJ mol^{-1} from the Discussed MD Simulations, QENS Studies [24], and Hierarchical Simulations [49]

N	E_a (MD)	E_a (QENS)	E_a (HS)
8	5.83	7.1	4.3
10	7.77	8.6	11.2
12	9.56	10.3	11.6
14	10.40	12.3	13.1
16	12.60	15.2	11.3

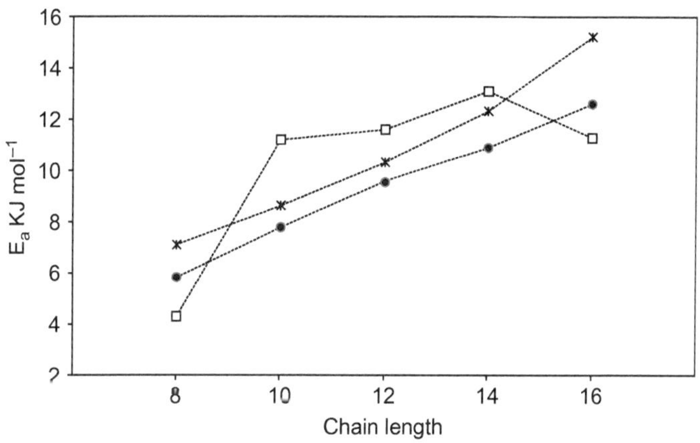

FIGURE 11 A plot of the activation energy of diffusion with chain length, for the MD studies at experimental loadings (●), QENS studies [24] (*), and hierarchical simulations[49] (□). *Figure reprinted with permission from Ref. [48].*

agreement with experiment for octane, decane, and dodecane (within 1.5 kJ mol^{-1}); the agreement is slightly less so for tetradecane and hexadecane with a difference of 1.9 and 2.6 kJ mol^{-1}, respectively. It was noted in neutron spin-echo studies in Ref. [50] that for longer alkanes, redistribution of the molecules throughout the crystal was observed at higher temperatures, allowing the molecules to cross internal barriers and access other parts of the crystal. The activation energy is then observed to be higher as the energy necessary for crossing the barriers is larger than that for diffusing in the regular channel structure. It then appeared that the longer the chain, the more significant the influence of the internal barriers on the redistribution. As the simulations were carried out using a periodic perfect lattice, this redistribution behavior is therefore not possible due to the even distribution of alkanes in the supercell being repeated infinitely. All calculated values of E_a are slightly but consistently below those obtained experimentally, which can be attributed (as with the diffusion coefficients) to the use of a perfect silicalite crystal structure. A perfect structure would have both fewer and lower energy barriers to diffusion than a real system containing the aforementioned defects so would cause the variation of diffusion coefficient with temperature to be lower in our case.

The trend and agreement in activation energies between the most sophisticated MD results and the QENS experiments are a marked improvement compared with that of the hierarchical simulations, suggesting that although the coarse-grained method is accurate at 300 K, this accuracy is not maintained at higher temperatures.

6.4.2 Branched Alkane Diffusion in Silicalite and Na-ZSM-5

Though the majority of work studying molecular diffusion in the MFI framework focuses on *n*-alkanes, branched alkanes are just as prevalent in fluid catalytic cracking and molecular sieving systems under industrial conditions. An understanding of the branched alkane diffusion behavior is crucial to gauge the effect of molecular shape on both catalytic cracking and molecular sieving characteristics. Experimental studies of isobutane in silicalite and ZSM-5 have employed a number of techniques including the zero-length column method [51,52] showing a strong, inversely proportional dependence on loading. Supported membrane techniques have also been used comparing straight and branched alkanes [53], showing that self-diffusivity of isobutane is lower by a factor of 8 compared with *n*-butane. Microscopic techniques have rarely been used to study these systems, mainly due to the low resolution of instrumentation that leads previously to significant errors in measured diffusion coefficients of the slower moving branched sorbates. However, a comparative study of supported membrane and QENS experiments of isobutane in ZSM-5 (performed on IN16 [47], ILL Grenoble) gave diffusivities an order of magnitude lower for the QENS measurements, consistent with literature differences between macroscopic and microscopic techniques [54]. The HWHM broadenings of the elastic peak measured for isobutane in ZSM-5 are shown in Figure 12 (left) along with a QENS spectrum at $Q = 0.8\ \text{Å}^{-1}$. We note that the broadenings are very small; in fact, the spectra show very small broadenings with respect to the resolution function (15%) indicating very low mobility (far lower than for longer *n*-alkanes).

The broadenings at low Q were too small to be determined. However, at higher Q values, the broadenings could be fitted with the Singwi and Sjölander jump diffusion model, with a mean jump distance of 4.5 Å and residence time of 6 ns. The self-diffusion coefficient of isobutane in ZSM-5 zeolite was found to be about 3 orders of magnitude lower than that of *n*-butane in the same

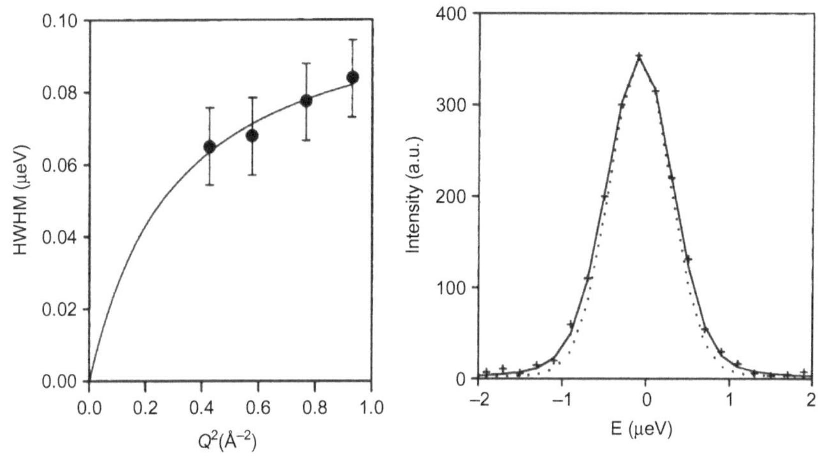

FIGURE 12 (left) Broadenings obtained for isobutane in ZSM-5 at 570 K and (right) QENS spectra obtained at 570 K (Q = 0.87 Å). The dotted line is the resolution function. *Figure adapted and reprinted with permission from Ref. [54].*

zeolite [56]. However, the inability to measure at low Q lead to errors in D_s of a factor of 4, much larger than the usual error, ~50% [56,57] [58] (Figure 12).

Recent work has used the neutron spin-echo method and fully atomistic MD simulations to measure isobutane diffusion in silicalite, in the temperature range of 450–550 K [55]. The simulations were carried out using contemporary flexible zeolite potential and explicit atom hydrocarbon model. Experimentally, the NSE method is especially appropriate because of its ability to measure diffusion coefficients of very slow moving sorbates in smaller pore zeolites with very high precision, due to the high resolution of the instruments. The NSE experiments were performed using the NSE spectrometer [59] at the ELLA neutron guide hall at the FRJ-2 reactor in Jülich (which at the time had been transferred to the MLZ in Garching).

Normalized intermediate scattering functions ($I(Q,t)$) obtained experimentally for isobutane in silicalite at different temperatures are shown in Figure 13 at two Q values. The faster decay of the scattering functions observed with increasing temperature at a given Q value reflects the increase of the diffusivity. The experimental data are well fitted by an isotropic diffusion model [12]. Figure 14 shows a comparison with experiment for both 3-D and 1-D diffusion models; the 3-D diffusion model fits better with the experimental points than 1-D diffusion, which implies that isobutane molecules are able to explore both straight and sinusoidal channels, while previous QENS and NSE measurements showed that n-alkanes longer or equal to C_8 diffuse mainly along the straight channels [25,50].

Temporal decays extracted from the curves in Figure 13 can be expressed as energy broadenings as depicted in Figure 15, when plotted as a function

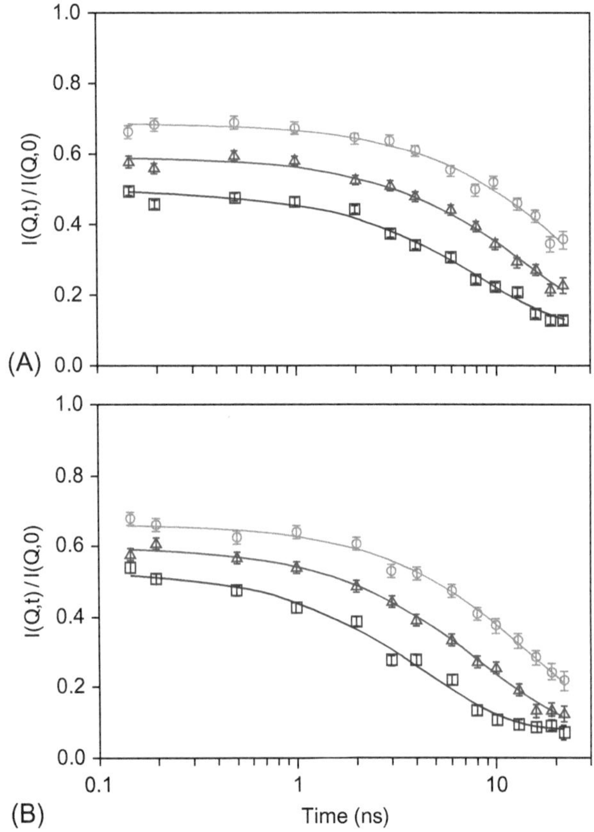

FIGURE 13 Comparison between experimental and calculated normalized intermediate scattering functions obtained for isobutane in silicalite for $Q = 0.2$ Å^{-1} (A) and 0.3 Å^{-1} (B), at 444 K (○), 491 K (△), and 550 K (□). The solid lines correspond to a 3-D diffusion model. *Figure reprinted with permission from Ref. [55].*

of Q^2. In Figure 15, the trend is a characteristic of jump diffusion. The jump model fitted incorporates a fixed jump length of 10 Å so agrees with the Chudley-Eliott model [15] previously used in the case of benzene in ZSM-5 [60]. The diffusivities were then extracted from the low-Q range [12].

The experimental diffusion coefficients are reported in Table 4 and plotted in Figure 16. The activation energy obtained from an Arrhenius representation is 22.6 kJ mol^{-1}. Previous QENS measurements were performed on a backscattering spectrometer on the same system, using ZSM-5 instead of silicalite [54]. The diffusion coefficients were about one order of magnitude lower in ZSM-5, with an estimated activation energy of 17 kJ mol^{-1}. We note the activation is counterintuitively lower than that in silicalite. However, since the broadenings were small compared with the instrumental resolution in that study, the errors were much larger than in this study.

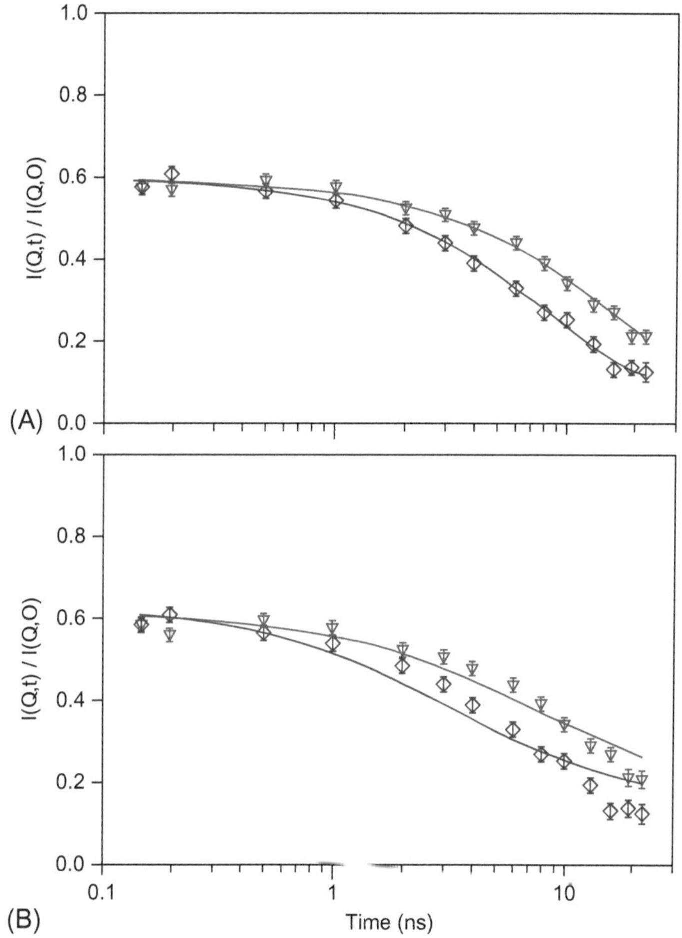

FIGURE 14 Comparison between experimental and calculated normalized intermediate scattering functions obtained for isobutane in silicalite at 491 K, for $Q=0.2$ (▽) and 0.3 Å^{-1} (◇). The solid lines are calculated with (A) a 3-D diffusion model and (B) a 1-D diffusion model. *Figure reprinted with permission from Ref. [55].*

Concerning the MD simulations, the calculated diffusion coefficients are all within a factor of 7.5 of experimental results. At 550 K, where the temperatures match exactly agreement to a factor of 5 was obtained. All calculated diffusion coefficients were in the upper limit of the experimental results, which as in previous sections, can be attributed to the use of a perfect periodic silicalite crystal structure in these simulations. The activation energy for diffusion was calculated from the Arrhenius plots shown in Figure 16, with a value of 19.2 kJ mol^{-1}:3.4 kJ mol^{-1} (~15%) lower than that observed by the NSE experiments.

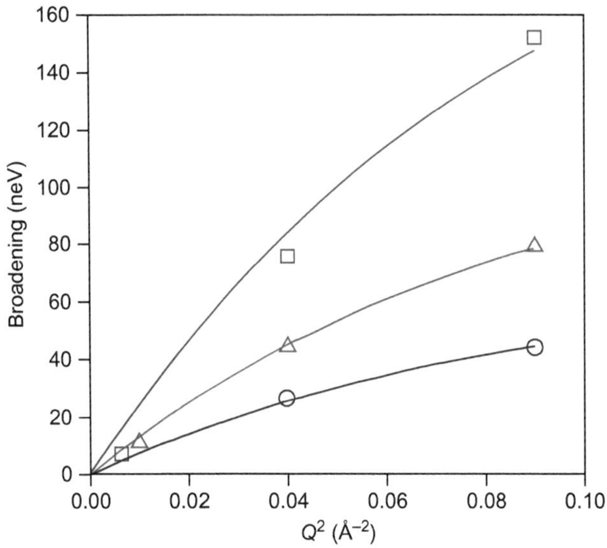

FIGURE 15 Broadenings obtained for isobutane in silicalite at 444 K (○), 491 K (△), and 550 K (□). The solid lines correspond to a Chudley-Elliot jump diffusion model. *Figure reprinted with permission from Ref. [55].*

TABLE 4 Calculated D_s Values in $m^2\ s^{-1}$ for Isobutane at all Temperatures

T (K)	D_s (NSE)	D_s (MD)
444	1.2×10^{-11}	
450		9.1×10^{-11}
491	2.1×10^{-11}	
500		1.5×10^{-10}
550	3.9×10^{-11}	2.2×10^{-10}
E_a (kJ mol^{-1})	22.6	19.2

Table reprinted with permission from Ref. [55].

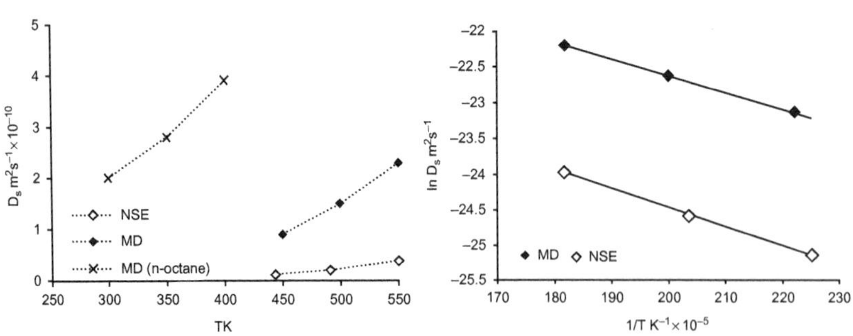

FIGURE 16 (left) Comparison of diffusion coefficients of isobutane from NSE (◇) and MD (◆) compared with previous simulation of n-octane diffusion in Ref. [48] (×). (right) Arrhenius plots from the MD simulations (◆) and NSE experiments (◇). *Figure reprinted with permission from Ref. [55].*

It is notable that the diffusion coefficients obtained for isobutane are significantly lower than those obtained for *n*-alkanes in Ref. [48]. The simulated D_s of *n*-octane (also plotted in Figure 16) at 400 K is higher by a factor of 12 than that of isobutane at 450 K, illustrating the effect of alkane branching on the diffusion in the 5.5 Å silicalite pore, even when framework flexibility is taken into account. This hindrance is also illustrated by the much higher activation energy, compared with that of *n*-octane (5.8 kJ mol^{-1}).

To gain a greater understanding of the diffusion of isobutane in the silicalite framework, the mean square displacement plot in the possible directions was examined to observe preference for a specific channel system, which can also be illustrated by the use of a trajectory plot to show specific locations of molecules. The average directional MSD plot of all isobutane molecules at 450 K is shown in Figure 17, which indicates that the molecules are diffusing in both channel systems but show preference for the 100 direction (the sinusoidal channels).

Though this result contrasts with previous studies into *n*-alkane diffusion, where the 010 channel is preferred, the ability to explore both the straight and sinusoidal channels would agree with the 3-D diffusion observed by NSE. The trajectory plot of this molecule between 3000 and 5000 ps is shown in Figure 18, illustrating molecular trapping in the very small section of the sinusoidal channel between intersections (ca. 3 Å in length), then movement between two adjacent sinusoidal channel segments via the straight channel segment connecting them. The movement in the 010 direction is rapid, taking

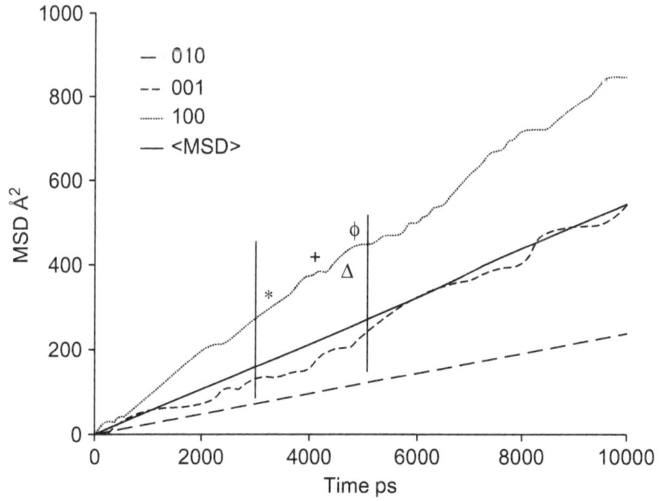

FIGURE 17 MSD plot in all directions of an individual isobutane molecule in silicalite at 450 K. Each symbol signifies the assigned portion of the trajectory plot shown in Figure 18. *Figure reprinted with permission from Ref. [55].*

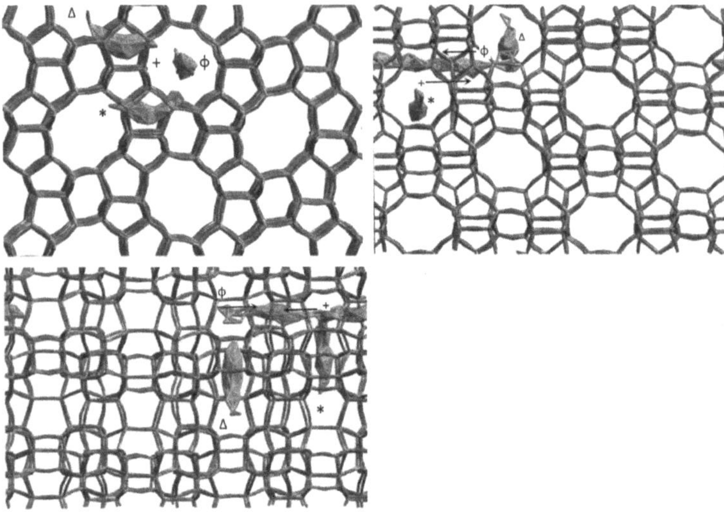

FIGURE 18 Trajectory plot of isobutane in silicalite at 450 K between 3000 and 5000 ps of the MD simulation run. The symbols represent the appropriate portion of the associated MSD plot in Figure 17. *Figure reprinted with permission from Ref. [55].*

place back and forth between the two trapping sinusoidal segments, indicative of the jump diffusion mechanism. The 10 Å jump length between sinusoidal segments matches that observed in the NSE experiments. However, it is not possible from experiment to determine whether the jumps occur between the intersections or between sinusoidal channels. Previous simulations all showed that the preferred siting is in the intersections between channels, where it is assumed the molecule would spend the residence time between jumps. The present simulation suggests that the molecule spends the residence time in a 3 Å sinusoidal channel section. This preference can be attributed to the favorable interactions between the molecule and channel walls, enabled by the increased access and geometric freedom offered by the use of a flexible silicalite framework. However, in previous simulations, the rigid framework, used due to computational restrictions, would have prevented sustained access to this region due to steric hindrance and associated instability. Thus, the lowest potential energy region in the rigid framework would be the intersections. The breathing of the framework in this simulation allows for both sustained access and framework relaxation around the isobutane accommodating this preferred siting. The isobutane then makes the 10 Å jumps between adjacent sections of sinusoidal channel. Oscillation across the 3 Å channel section and passage via the straight channel segments are sufficient to make the diffusion appear isotropic in the NSE experiment. This study in particular illustrates how the use of more sophisticated MD models not only can give accurate quantitative agreement with QENS/NSE but also may reveal different qualitative behaviors of the sorbate.

The studies discussed so far have shown how recent improvements in MD models have lead to closer agreement between diffusion coefficients and also new qualitative insights into alkane behavior MFI zeolites. We will now discuss some studies of localized molecular motions in this class of zeolite catalysts.

6.4.3 Localised Hydrocarbon Motions in ZSM-5: Benzene

As mentioned previously, ZSM-5 is used to convert methanol into hydrocarbons such as alkanes and olefins. Another important set of products in particular is aromatics such as benzene, which are of particular interest for mobility studies due to the similarity in size between the molecule and the MFI channels (~5.5 Å). This similarity in size leads to translational diffusion being too slow, and beyond the resolution limit of time-of-flight and backscattering spectrometers. The study of localized motions such as molecular rotation has formed the majority of studies of aromatic dynamics in zeolites (though NSE experiments have quantified benzene diffusion on the order of $1 \times 10^{-14}\ m^2\ s^{-1}$ at 300 K [60]). We now discuss studies of benzene rotation in ZSM-5 zeolites through analysis of the experimental EISF.

Mitra et al. studied benzene dynamics in H-ZSM-5 and Ca-ZSM-5 at room temperature [61], using the medium resolution quasielastic spectrometer at the Dhruva reactor, Trombay [62] at a loading of approximately 3 molecules per unit cell (mol/uc). After the experimental EISF was obtained (Figure 19), three models were fitted to the data: the twofold jump rotation model where the

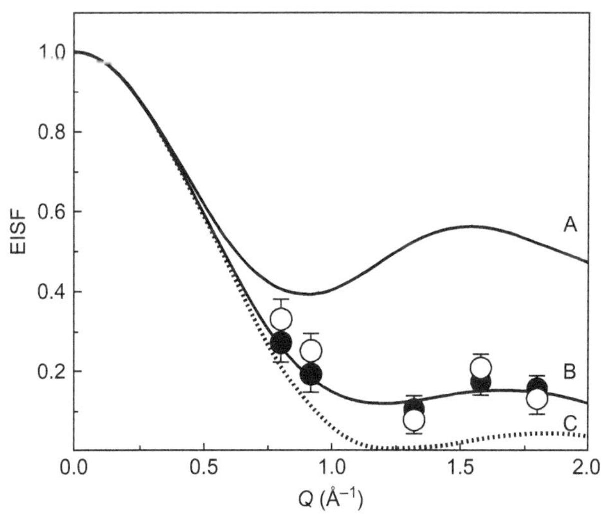

FIGURE 19 EISF variation as a function of Q for benzene in H-ZSM-5 (●) and Ca-ZSM-5 (○), where line A is the twofold jump model, line B is the sixfold jump model, and line C is the isotropic rotational diffusion model. *Figure reprinted with permission from Ref. [61].*

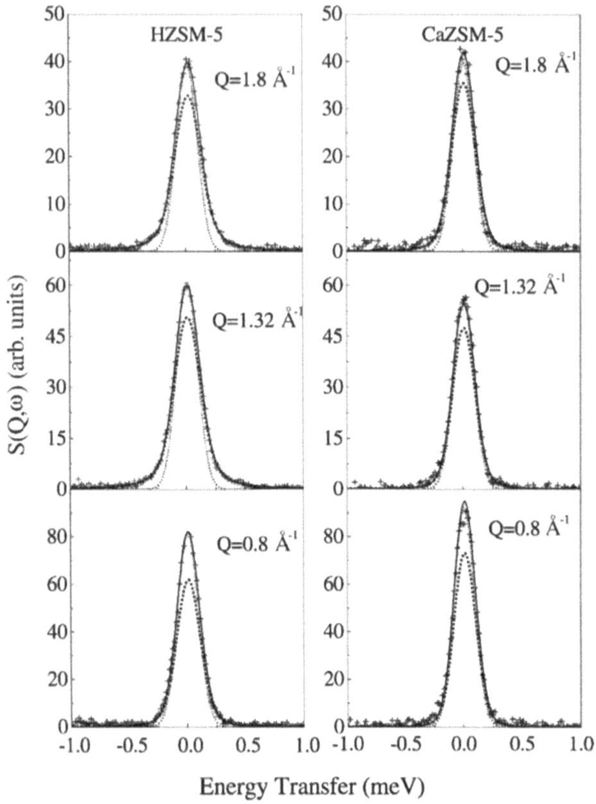

FIGURE 20 QENS spectra for benzene in H-ZSM-5 and Ca-ZSM-5, solid line is the total fit, dotted line is the instrumental resolution, and the dashed line is the quasielastic component. *Figure reprinted with permission from Ref. [61].*

EISF is given by Eq. (16) (line A in Figure 19) that falls above the experimental data in Figure 20. The sixfold jump rotation model was also considered, where the EISF is given in Eq. (21) (line B in Figure 19), which matches the experimental data well, and the isotropic rotational diffusion model, EISF in Eq. (11) (line C in Figure 19), which fell below the experimental points.

The scattering law in Eq. (21) was then used to estimate the residence time τ between jumps on the sixfold axis $\left(\text{where } \tau_l^{-1} = 2\tau^{-1}\sin^2\left(\frac{\pi l}{N}\right)\right)$. Good fits were obtained upon using $\tau = 16.5$ ps in H-ZSM-5 and 32 ps in Ca-ZSM-5. The fits at different Q values are shown in Figure 20 (the dotted line is the instrumental resolution).

The observed residence times are similar to those of solid benzene (~19 ps). Upon observing the FTIR band shifting of the out-of-plane bending modes to higher frequencies than that of liquid phase benzene, they concluded that confined benzene molecules are subject to a significant pressure effect, leading to behavior similar to that of solid benzene.

The residence times in H-ZSM-5 disagree with previous QENS studies [63], which we will discuss now, using the IN6 spectrometer [64] (ILL Grenoble) where the effect of loading was considered by measuring at 3 mol/uc and 8 mol/uc.

First, the shape of the EISF for different benzene rotations was considered as before: That of rotational diffusion on a circle ($N = \infty$), sixfold uniaxial rotation around a circle, and two different twofold uniaxial rotations. The different twofold uniaxial rotation models are based on the two different axes illustrated in Figure 21 (noted as C_2 and C_2'). The isotropic rotation models, the C_2' axis-based twofold rotation, and models of rotational diffusion on a circle were discarded. However, the sixfold uniaxial rotation around a circle and the twofold rotations around the C_2 axis show the same shape as the experimental data (Figure 22). Significantly, the EISF has a temperature dependence that would later be explained.

The EISF plots in Figure 22 show that isotropic rotational diffusion is never reached even for the lowest loading at the highest temperature, suggesting significant interaction between benzene and the zeolite. Convergence toward C_6' axis uniaxial rotational diffusion is found with increasing temperature at 3 mol/uc. However, the motion is more hindered at lower temperature as shown in the EISF plots. The hindrance is more pronounced at 8 mol/uc with a lower amplitude of motion observed suggesting the presence of significant benzene-benzene interactions. However at 90 K, the EISFs looked very

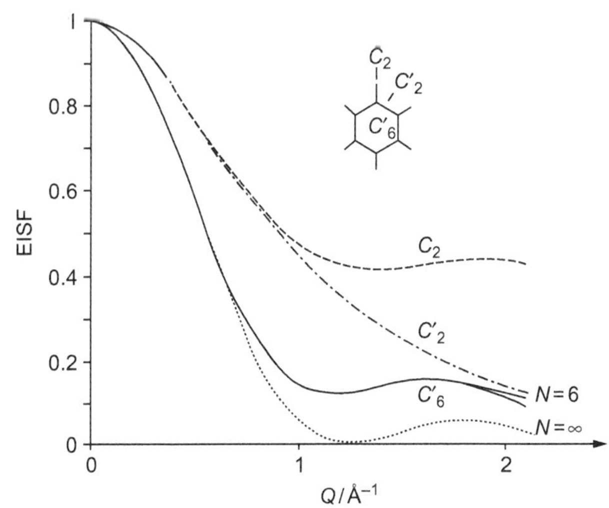

FIGURE 21 Theoretical EISFs obtainable for benzene rotation corresponding to different rotational models described in the text. *Figure reprinted with permission from Ref. [63].*

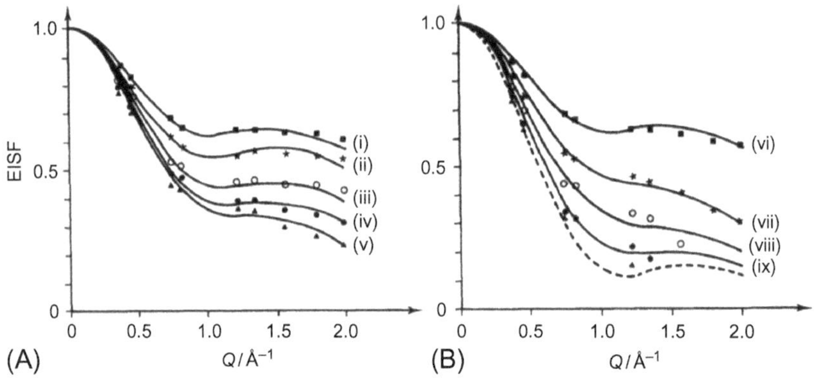

FIGURE 22 Experimental EISFs obtained for benzene in ZSM-5 at loadings of (A) 8 mol/uc and (B) 3 mol/uc at 90 K (■), 150 K (★), 220 K (○), 300 K (●), and 400 K (△) The dashed line is that of the theoretical EISF of isotropic rotation ($R = 2.5$ Å). *Figure and reprinted with permission from Ref. [63].*

similar at both the higher and lower loading—suggesting that at low temperature, the static potential is larger than that of benzene-benzene interactions. Jobic et al. then considered an EISF model of uniaxial rotation in an N-fold cosine potential [65] that considers both the well-defined orientations and the probability of a molecule to be in an intermediate state.

$$A_0(Q) = \frac{1}{\pi J_0^2(\gamma'_N)} \int_0^{\pi} j_0(2Qr\sin x)\, J_0(2\gamma'_N \cos Nx)\, dx \tag{23}$$

with J_0 being a Bessel function of the first kind, N being the number of wells in the potential and γ'_N related to the potential barrier, as

$$\gamma'_N = \frac{V_N}{2k_B T} \tag{24}$$

and

$$V(\phi) = -\frac{V_N}{2}\cos(N\phi) \tag{25}$$

The simulated EISFs using different values of $\Delta\phi$ (the average angle of rotation) are shown in Figure 22 (solid lines), giving satisfactory agreement with experimental data. The mean correlation times were then deduced from the width of a single Lorentzian fitted to the data and plotted against $1/T$ in Figure 23, not appearing to follow Arrhenius behavior. The correlation times at least for intermediate temperatures do not differ particularly suggesting that the loading does not appear to change the dynamics of rotation even if the geometries are affected due to packing. As mentioned, the correlation time obtained was the same as liquid benzene at 300 K at 3.2 ps.

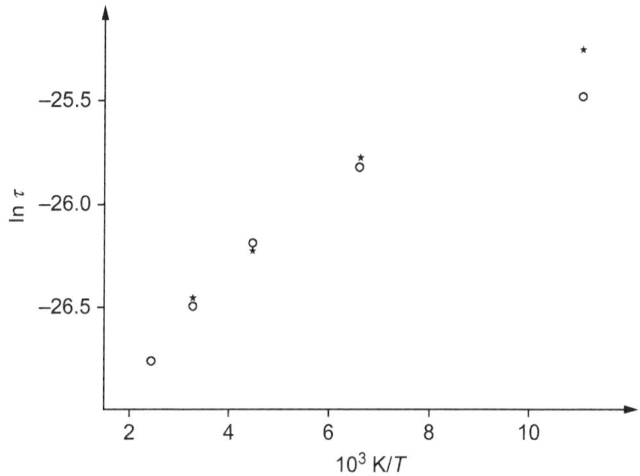

FIGURE 23 Temperature dependence of mean correlation times for C_6 uniaxial rotation in a cosine potential of benzene in ZSM-5 at (●) 3 mol/uc and (○) 8 mol/uc. *Figure reprinted with permission from Ref. [63].*

6.5 HYDROCARBON BEHAVIOUR IN THE FAUJASITE ZEOLITES

The study of hydrocarbon dynamics in faujasite zeolites is of great importance due to its relevance to processes associated with cracking of hydrocarbons in the petrochemical industry by zeolite X and zeolite Y [1]. Zeolite X is characterized by having a Si/Al ratio between 1–1.5, whereas zeolite Y has higher Si/Al ratios. The structure is composed of tetrahedrally connected supercages 12 Å in diameter, which are connected by windows with a diameter of roughly 7.5 Å [2]. The open three-dimensional pore system results in diffusion being faster and general motions being far less hindered than in other zeolites such as MFI [4,5]. It is therefore useful in separation and reaction processes involving large molecules leading to its many applications in fluid catalytic cracking (FCC) catalysts. With these applications in mind, we will discuss studies of hydrocarbon dynamics in faujasite structures including the diffusion and rotation of propane, pentane, and its isomers.

6.5.1 Propane Dynamics in NaY

6.5.1.1 Propane Diffusion

We discuss first a combined study of classical MD and QENS of translational diffusion of propane in zeolite NaY. Sayeed and coworkers used the quasielastic spectrometer [62] at the Dhruva reactor, Trombay, with a loading of 4 molecules per supercage [66]. They found a single Lorentzian could fit the quasielastic part (Figure 24, left) suggesting that only translational motion is contributing

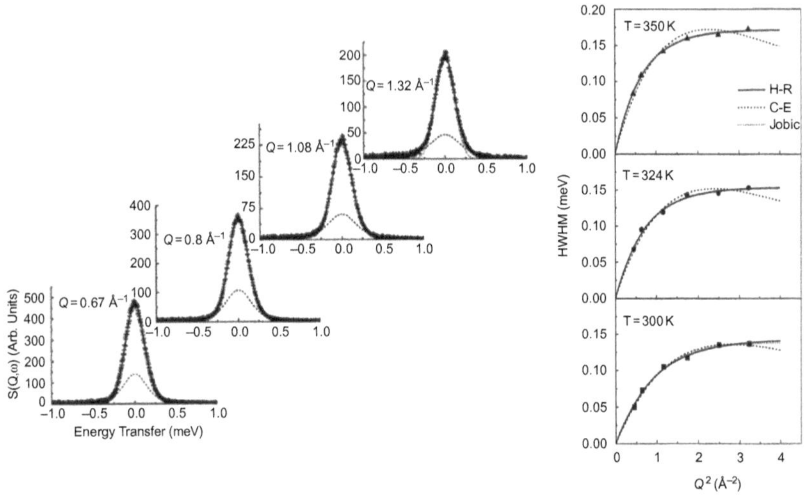

FIGURE 24 (left) QENS spectra at different Q values for propane in zeolite NaY at 324 K. (right) Q-dependence of Lorentzian HWHM of the QENS spectra fitted with the three mentioned jump diffusion models. *Figure and reprinted with permission from Ref. [66].*

to the data, as rotational dynamics are at a rate beyond the range observable with the present energy window of the spectrometer.

The HWHM versus Q^2 plots for the three temperatures are shown in Figure 25, right. The Chudley-Elliot (CE), Hall-Ross (HR), and Jobic models were all fitted to the data. The oscillatory behavior associated with the Chudley-Elliot model was not observed, but the Hall-Ross and Jobic models fitted the data equally well. In terms of dynamical parameters, at 300 K, the HR model gave a jump distance and residence time of 2.5 Å and 4.6 ps, respectively, compared with 3 Å and 3.8 ps at 350 K, giving diffusion coefficients of 2.3–4 × 10^{-9} m^2 s^{-1}.

Using the MD simulation trajectory, the theoretical self-intermediate scattering function $I(\mathbf{Q},t)$ can be calculated using

$$\langle I(\mathbf{Q},t)\rangle = \frac{1}{N}\sum_{i=1}^{N}\left\langle \frac{\sin(|\mathbf{Q}||\boldsymbol{R}(t+t_0)-\boldsymbol{R}(t_0)|)}{|\mathbf{Q}||\boldsymbol{R}(t+t_0)-\boldsymbol{R}(t_0)|}\right\rangle \tag{26}$$

where $\mathbf{R}(t)$ is the position vector of the ith propane molecule at time t and N is the number of molecules used in the MD simulation. The calculated $I(\mathbf{Q},t)$ plots are shown in Figure 26 at 6 different wave vector transfers at 300 K.

The diffusion models all yield a Lorentzian form for $S(\mathbf{Q},\omega)$ versus ω with a HWHM of $\Gamma(Q)$. The equivalent in the time domain is an exponential form, where $I(Q,t) \propto \exp(-\Gamma(Q)t)$. They found that a single exponential does not fit the $I(Q,t)$ obtained from the MD data. A linear sum of three-weighted

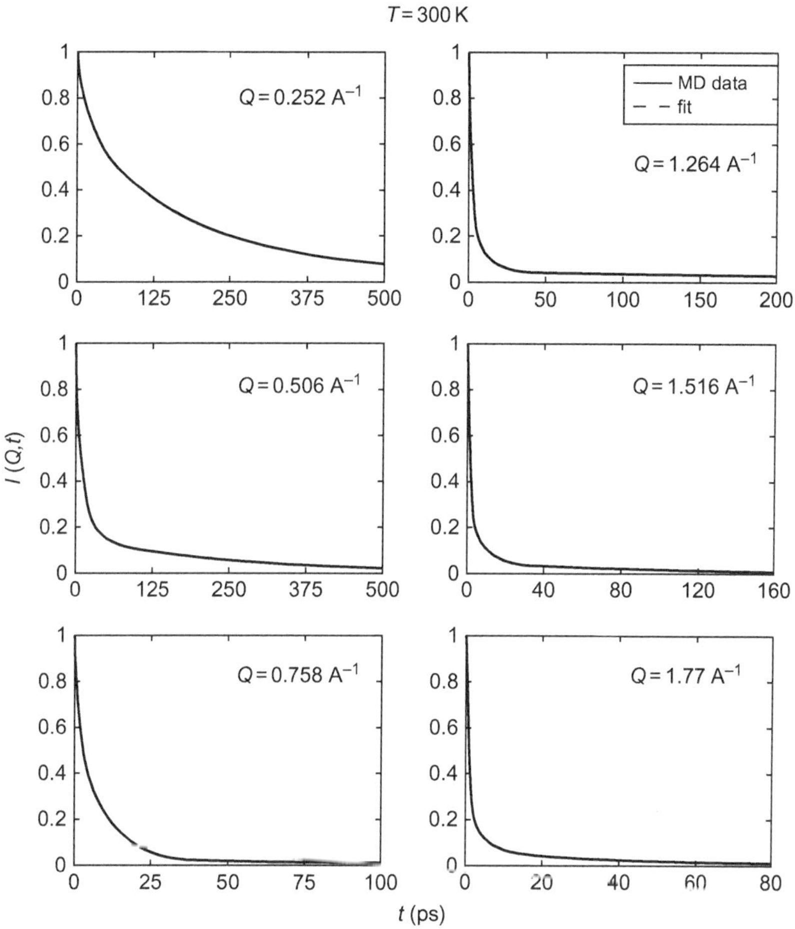

FIGURE 25 *I*(*Q*,*t*) versus t plots from MD simulations of propane in zeolite NaY. *Figure adapted and reprinted with permission from Ref. [66].*

exponentials was then used to fit the total $I(Q,t)$: one each for fast, intermediate, and slow diffusion.

An example of this combination is shown in Figure 26. The intermediate component has a decay constant over the measured Q values of ~7–70 ps, similar to that observed by the QENS experiment. The variation of this decay constant $\Gamma_2(Q)$ with Q^2 is shown in Figure 27 (left), which exhibits the Q dependence of jump diffusion. Fits were attempted using the CE, HR, and Jobic models. The quality of the fit to the simulated data is reasonable for all three models; however, the Jobic model gave the largest errors. At 300 K, the jump distance attributed to the HR model is 4.2 Å, decreasing to 3.2 Å at 350 K, with residence times of 8.3 ps and 6.1 ps at 300 and 350 K,

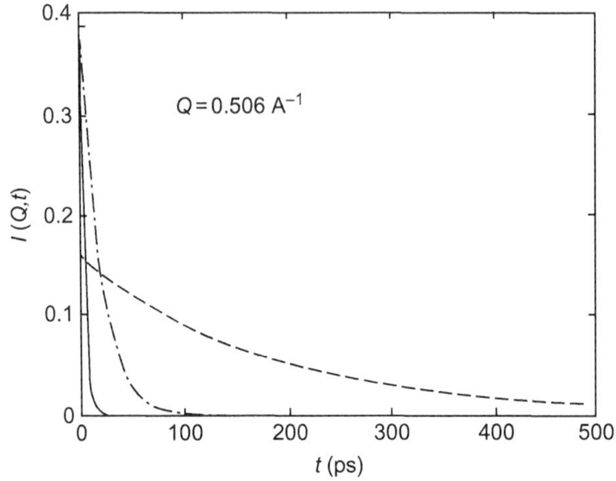

FIGURE 26 Plots showing the three different exponential decays associated with the simulated $I(Q,t)$ at 324 K. *Figure adapted and reprinted with permission from Ref. [66].*

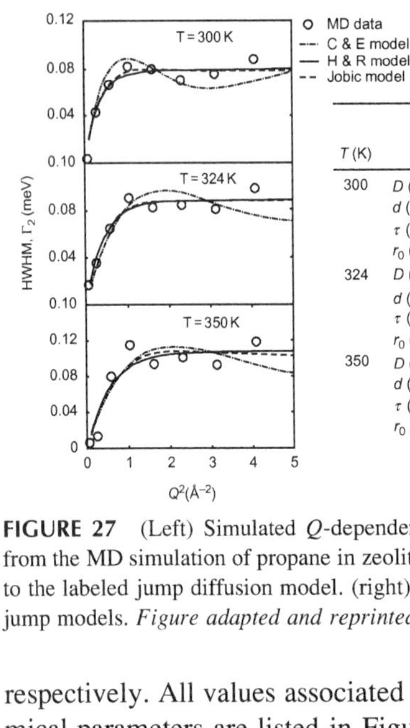

		MD simulation study				QENS experiment		
T (K)	parameter	H–R	C–E	Jobic	msd	H–R	C–E	Jobic
300	D ($\times10^{-9}$ m^2/s)	3.6	3.7	2.0	1.32	2.3	2.0	1.3
	d (Å)	4.2	4.5	3.2		2.5	2.7	2.0
	τ (ps)	8.3	9.0	8.5		4.6	5.9	4.9
	r_0 (Å)			1.5				0.9
324	D ($\times10^{-9}$ m^2/s)	3.2	2.2	1.2	1.9	3.2	2.7	0.5
	d (Å)	3.8	3.3	2.3		2.9	2.9	1.2
	τ (ps)	7.4	8.2	7.4		4.3	5.3	4.3
	r_0 (Å)			1.7				1.5
350	D ($\times10^{-9}$ m^2/s)	3.0	2.3	1.8	2.0	4.0	3.2	0.7
	d (Å)	3.3	3.1	2.6		3.0	3.0	1.3
	τ (ps)	6.1	7.0	6.4		3.8	4.7	3.8
	r_0 (Å)			1.2				1.6

FIGURE 27 (Left) Simulated Q-dependence plots of Γ_2 with single points obtained directly from the MD simulation of propane in zeolite NaY. The different lines fit to the points correspond to the labeled jump diffusion model. (right) Table of dynamical parameters comprising the fitted jump models. *Figure adapted and reprinted with permission from Ref. [66].*

respectively. All values associated with the experimental and simulated dynamical parameters are listed in Figure 27 (right).

The MD simulations also supported the use of a single Lorentzian to fit the QENS spectra, which represented only translational motion. MD snapshots in Figure 28 show the difference in magnitude between the time it takes for a proton to rotate through π radians compared with the movement of the molecular center of mass. From the MD run, an estimated width of a spectrum from

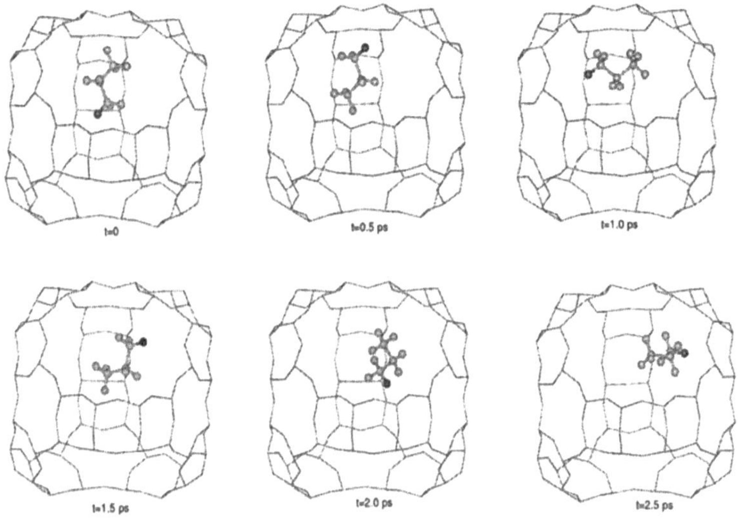

FIGURE 28 Snapshots from the MD simulation showing the translational and rotational motion. A single hydrogen of a CH_3 group in propane is shown darker to distinguish it from others. Rotation through an angle π can be seen within 0.5 ps. During this period, little or no translation is seen. *Figure reprinted with permission from Ref. [66].*

this rate of rotation would be 30 times wider than that from translational motion (too wide for the energy window used in their experimental setup).

The rotation was studied specifically using a different spectrometer and compared with MD simulations, which will be addressed in the next section.

6.5.1.2 Propane Rotation

Mukhopadhyay and coworkers used the triple-axis spectrometer (TAS) [67] at Dhruva, Trombay, to study specifically propane rotation in NaY [68], which happens on a much faster timescale than translational motion with no coupling between the two sets of dynamics. TAS has an energy resolution in inverted geometry of ~3.3 meV (FWHM), and measurements were taken in the Q range of 0.8–2.5 Å^{-1} at 300 K. The extracted EISF is shown in Figure 29, with the model of isotropic rotational diffusion and uniaxial rotational diffusion around a circle; the isotropic rotational model appeared to fit the data.

The isotropic model was then used to fit the experimental $S(\mathbf{Q},\omega)$ and determine the rotational diffusion constant. When a radius of gyration of 1.89 ± 0.02 Å was employed (the average distance of a hydrogen atom from the center of mass of propane), very good fits at all Q values were obtained as shown in Figure 30 (left), with a $D_R = 1.05\ (\pm 0.09) \times 10^{12}\ \text{s}^{-1}$ obtained from the fitting (variation with Q shown in Figure 30 (right)). The D_R values in Figure 30 are given in meV, where 1 meV = 1.519 ps^{-1}.

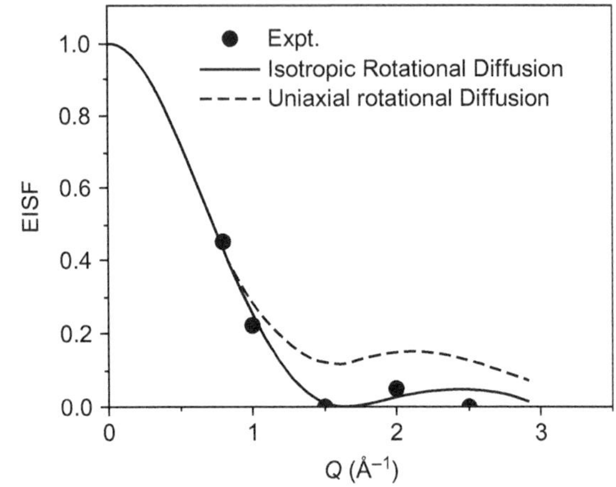

FIGURE 29 Experimental EISF for propane in zeolite NaY at 300 K, fitted to the labeled models. *Figure reprinted with permission from Ref. [68].*

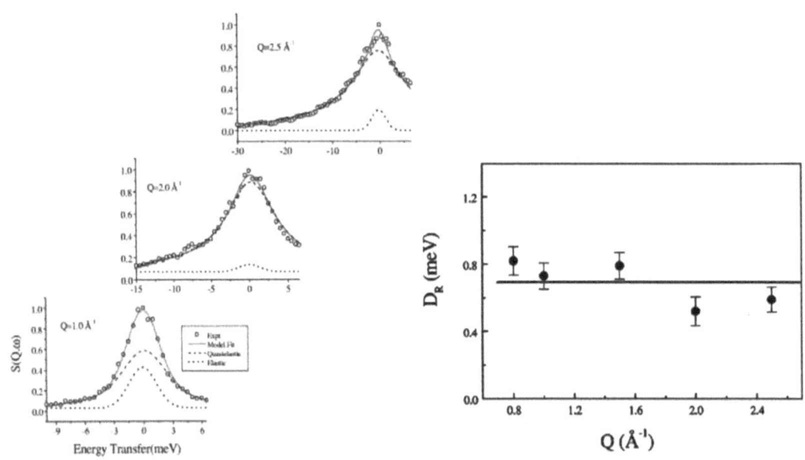

FIGURE 30 (left) Experimental $S(\mathbf{Q},\omega)$ fitted with that calculated from the isotropic diffusion model for propane in zeolite NaY. (right) Q-dependence of diffusion constant D_R obtained from the fitted model. *Figure adapted and reprinted with permission from Ref. [68].*

The MD simulations directly calculated the theoretical rotational intermediate scattering function using

$$\langle I^{rot}(\mathbf{Q},t)\rangle = \frac{1}{N}\sum_{i=1}^{N}\left\langle \frac{\sin\left(|\mathbf{Q}||r_i(t+t_0)-r_i(t_0)|\right)}{|\mathbf{Q}||r_i(t+t_0)-r_i(t_0)|}\right\rangle \tag{27}$$

where $r_i(t)$ is the radius vector of the ith hydrogen atom at time t and N is the number of hydrogens in the propane molecule. The intermediate scattering functions for six different Q values are plotted in Figure 31. To fit the

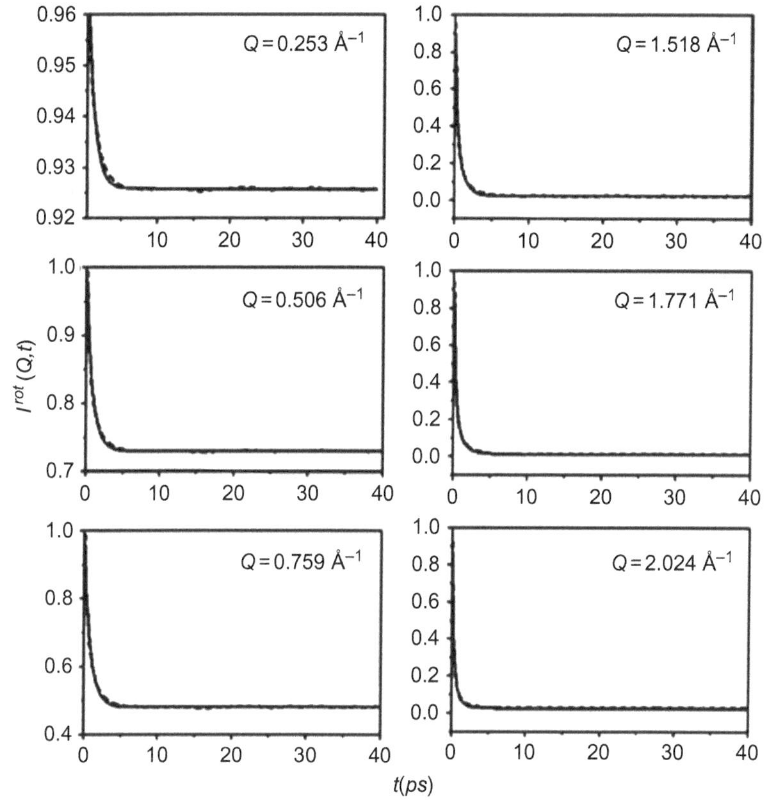

FIGURE 31 Simulated plot of $I(\mathbf{Q},t)$ versus time at 300 K for propane in zeolite NaY. The solid line is obtained directly from the MD simulation, while the dashed line is the fit with the isotropic diffusion model. *Figure adapted reprinted with permission from Ref. [68].*

intermediate scattering function obtained from the MD simulation directly to an isotropic rotational diffusion model, the time Fourier transform of Eq. (10) was used, giving

$$\langle I^{rot}(\mathbf{Q},t)\rangle = \sum_{l=0}^{\infty}(2l+1)j_l^2(QR)e^{-l(l+1)D_R t} \tag{28}$$

where R, the radius of gyration, was obtained at 1.88 Å. D_R was calculated at $0.82\times10^{12}\text{s}^{-1}$. The values of both R and D_R fitted to the MD data agree well with those obtained from the QENS experiments (Figure 31).

6.5.2 Pentane Isomer Dynamics in NaY – The Levitation Effect

We now consider an interesting phenomenon in the dynamical behavior of molecules in microporous materials, mainly probed by simulation, that of the so-called levitation effect [70,71]. In Section 6.4.2, we reported a study comparing *n*- and isobutane in MFI zeolites, finding significantly lower self-diffusivities of branched analogues due to steric hindrance of translational motion in constricted MFI channels. The levitation effect considers that self-diffusivity exhibits a maximum when the diameter of the diffusant is comparable with the pore diameter; the comparability is represented by the levitation parameter

$$\gamma = \frac{2\sigma_{opt}}{\sigma_w} \tag{29}$$

where σ_{opt} is the distance where the interactions between the zeolite wall and the sorbate are most favorable and σ_w is the channel width. A maximum in the diffusivity is observed when this parameter is unity. The effect is considered to have its roots in the dispersion interaction between guest and host, where a mutual cancellation of forces at an optimum small distance between the sorbate and the walls allows for increased mobility. The reduced force is a consequence of the forces along diagonally opposite directions being equal in magnitude, leading to their effects on the sorbate being canceled resulting in the effective "floating" of the molecule.

This hypothesis was tested by Borah et al. through the study of linear and branched pentanes in NaY [69], which all have the same mass but a differing cross-sectional diameter. NaY was chosen due to its pore dimensions relative to pentane isomers, in order to ensure that the levitation parameter for the isomers may lie in the "anomalous regime" where the maximum in diffusivity is seen for a Lennard-Jones fluid in zeolite Y (as opposed to the linear regime, where self-diffusivity varies as a function of $1/\sigma_s^2$ (σ_s=diffusant diameter)).

QENS experiments were carried out using IN6 [64] at the ILL, Grenoble at 300–360 K, with spectra selected in a Q range of 0.25–1.2 Å^{-1}. The experimental and calculated spectra at four Q values at 300 K are shown in Figure 32 (left) for neopentane, along with the Q-dependence corresponding

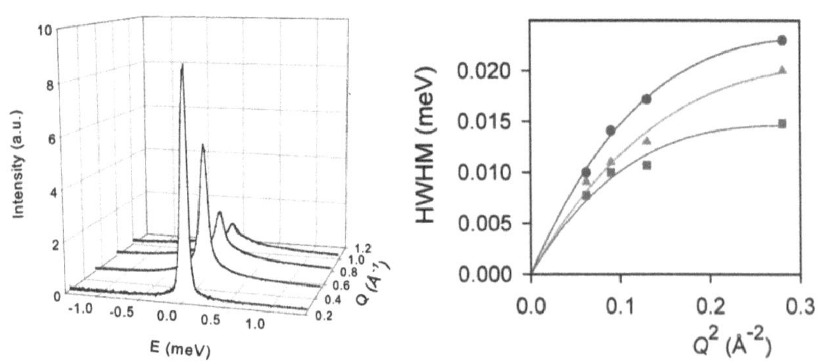

FIGURE 32 (left) QENS spectra obtained for neopentane at 300 K and (right) HWHM versus Q^2 plots for neopentane (●), *n*-pentane (▲), and isopentane (■). *Figure adapted and reprinted with permission from Ref. [69].*

TABLE 5 **D_s Values at 300 K and Activation Energies for the Three Pentane Isomers Obtained by QENS and MD. The Levitation Parameter γ is Also Listed**

		D ($\times 10^{19}$ m^2/s)		E_a (kJ/mol)	
Pentane isomer	γ	*QENS*	*MD*	*QENS*	*MD*
n-Pentane	0.71	2.27	3.99	7.1	9.4
Isopentane	0.86	2.46	5.40	5.9	5.9
Neopentane	0.96	3.17	6.13	5.8	5.0

Table reprinted with permission from Ref. [69].

to translational motion of all three pentane isomers at 300 K. The self-diffusivities were then determined from low Q values where the contributions from rotational motions were negligible. The values obtained for D_s are listed in Table 5. The trend at 300 K is D_s(neo) > D_s(isopentane) > D_s(n-pentane), the reverse of what one would expect given the lower cross-sectional diameter of n-pentane. The same trend is observed by the MD simulations, with both techniques giving a lower activation energy of diffusion for neopentane, also listed in Table 5.

Generally, there is a strong correlation between the activation energy of diffusion and the potential energy landscape of the system. The authors postulated that the energy landscape associated with the anomalous regime would be shallower with only small amplitude energetic undulations in a given direction, compared with a system associated with the linear regime. This effect would be particularly exhibited at window regions in the NaY structure, which would act as bottlenecks to the diffusion. Through averaging the hydrocarbon-zeolite potential energy interactions over all window crossovers, the potential energy profile as a function of d (distance from the plane of the 7.5 Å window regions) is plotted in Figure 33 (left), where d = 0 corresponds to the 12-ring window in NaY. For n-pentane, the energy maximum is at the d = 0 coordinate; however, for neopentane, there is a minimum.

The potential energy landscape around this bottleneck was then correlated with the distribution of cage residence times obtained from the MD simulation plotted in Figure 33 (right) at 200 K, where the average cage residence times are 13.6, 8.0, and 6.3 ps for n-, iso, and neopentane, respectively. The trend in cage residence times is also correlated with those obtained from QENS, where the Q-dependence plots in Figure 33 suggest that the shortest residence times between jumps are those of neopentane. A particularly obvious illustration of the difference in mobility between n-pentane and neopentane is shown by the trajectory plots in Figure 34, where the movement of a typical sorbate is

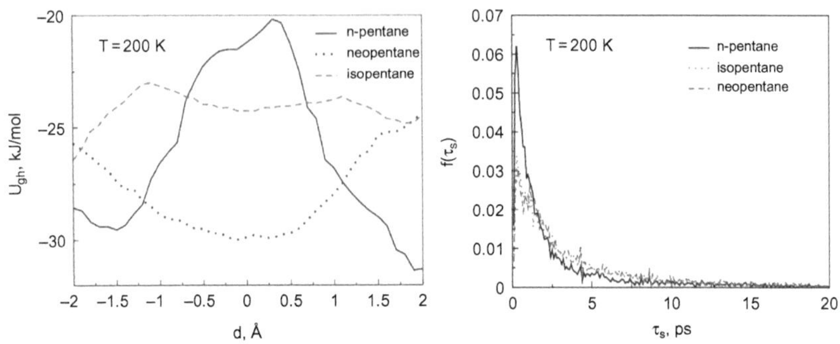

FIGURE 33 (left) Potential energy landscape of different isomers in NaY at 200 K as a function of d, the distance of molecular center of mass from the plane of the 12-ring window. (right) Distribution of cage residence times for the three isomers at 200 K. *Figure adapted and reprinted with permission from Ref. [69].*

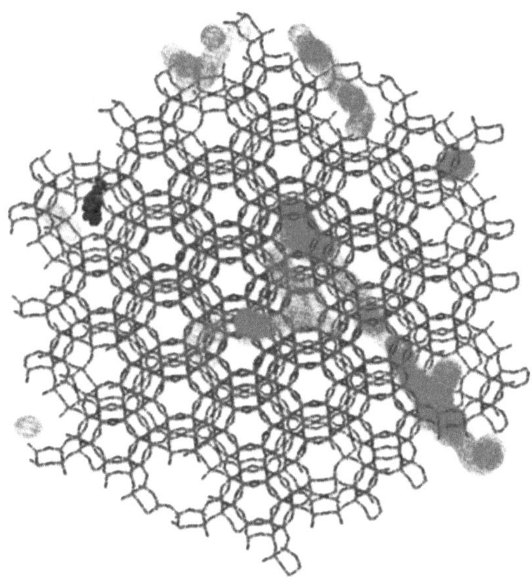

FIGURE 34 Trajectory plots of the center of mass of *n*-pentane (blue) and neopentane (red) in the NaY framewok from the MD simulations at 200 K taken over 200 ps. *Figure reprinted with permission from Ref. [69].*

tracked over 200 ps. It is clear that the neopentane is far more free to explore a larger amount of the void space in the zeolite.

The study concluded that the self-diffusivities are correlated with the levitation parameter (0.96 for neopentane, 0.86 for isopentane, and 0.76 for *n*-pentane) and that the activation energy of diffusion is related to the observed minimum in potential energy at the diffusion bottleneck at the 12-ring window between zeolite cages, providing direct verification for the levitation effect.

6.6 DYNAMICS OF METHANOL IN FAUJASITE AND MFI ZEOLITES

Since the seminal publication of Chang and Silvestri [72], the zeolite-catalyzed conversion of methanol to hydrocarbons (MTH) in ZSM-5 has been intensively studied, with the first commercialization of the MTH process in New Zealand in 1985 [73]. Recent demand for light olefins has also made conversions such as the methanol-to-olefins (MTO) process highly desirable [74,75]. Faujasite zeolites, though primarily used for the fluid catalytic cracking of heavy oil fractions to lighter hydrocarbons [76,77], have been studied for the side chain alkylation of aromatics with alkali metal substituted Y zeolites using methanol [78,79] and the conversion of methanol to dimethylether [80,81]. In this section, we discuss QENS studies of methanol adsorbed in ZSM-5, where rotational/localized diffusive motions are probed. We also discuss studies of methanol diffusion in HY where the diffusivities obtained are compared with other faujasite samples.

6.6.1 Localised Motions in ZSM-5

Using the instrument IN6 [64] at ILL, Grenoble, Jobic and coworkers studied the dynamics of methanol adsorbed on H-ZSM-5 in a temperature range of 273–425 K at a loading of 10 mol/uc [82]. The experimental EISF was first evaluated using different models such as rotations of the methyl group alone through reorientations of 120° (the OH group is considered fixed). The theoretical EISF is given by Eq. (16), where d in this case is the H-H near-neighbor distance (1.79 Å). The calculated EISF is given as the dotted line in Figure 35 (left), where the experimental curves fall well below the predicted curve at all temperatures. The isotropic rotational diffusion model was also considered as in Eq. (11). To deduce the radius of total molecule rotation, r, the bond lengths and bond angles from microwave spectroscopy were implemented (O-H = 0.95 Å, C-O = 1.425 Å, and C-H = 1.094 Å while $\theta_{H\text{-}C\text{-}H} = 108.3°$ and $\theta_{C\text{-}O\text{-}H} = 108.4°$) leading to a mean radius of gyration of 1.48 Å. The EISF plot is given by the dot-dashed line in Figure 35 (left), where one can see that the experimental motions are less isotropic than the model predicts even at the highest temperatures. An intermediate was then considered, where two models of jump rotation around a circle were combined: that of 2π/3 jumps for the methyl group, convoluted with 2π/6 jumps of the molecule around the O-zeolite axis in the geometry shown in Figure 35 (right).

The theoretical EISF for this combination is given by

$$A_0(Q) = \frac{3}{4}\left\{1/6\left[1+2j_0(QR)+2j_0\left(3^{\frac{1}{2}}QR\right)+j_0(2QR)\right]1/3[1+2j_0(QR')]\right\} + \frac{1}{4}\left\{1/6\left[1+2j_0(QR'')+2j_0\left(3^{\frac{1}{2}}QR''\right)+j_0(2QR'')\right]\right\} \tag{30}$$

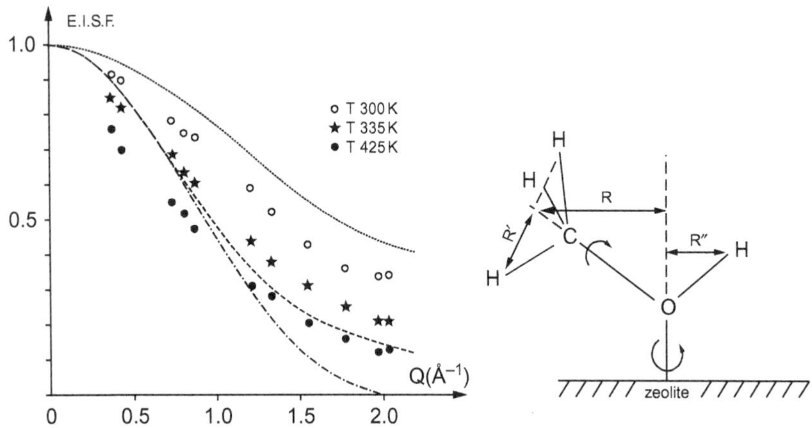

FIGURE 35 (left) Experimental EISF of methanol in H-ZSM-5 at 300–425 K. Theoretical EISFs also plotted for models as described in the text. (right) Adsorption geometry considered when the first three theoretical EISF models are calculated. *Figure adapted and reprinted with permission from Ref. [82].*

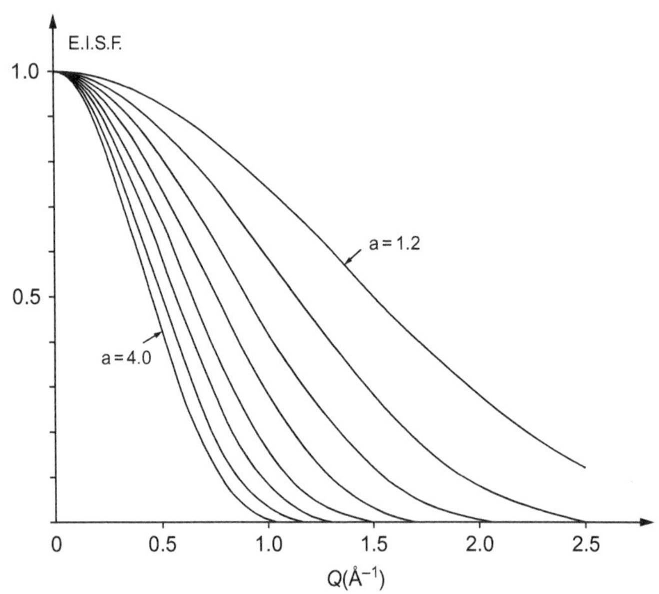

FIGURE 36 Calculated EISFs for particle diffusing in a sphere of radius r as in Eq. (20), a increases in intervals of 0.4 Å for each plot. *Figure adapted reprinted with permission from Ref. [82].*

where $R = 1.45$ Å (total molecule rotation around the O-zeo axis), $R' = 1.79$ Å (CH_3 rotation), and $R'' = 0.77$ Å (OH rotation around the O-zeo axis). This combination is plotted in Figure 35 (dashed line), and while falling within the experimental points, the predicted elastic intensity is too low for lower temperatures.

The possibility of diffusion in a confined spherical volume (of radius a) was then considered, using the model of Volino and Dianoux where the EISF is given in Eq. (23) and plotted in Figure 37 between $a = 1.2–4.0$ in intervals of 0.4 Å.

This EISF model alone is not sufficient to fit the data. However, a total EISF model was then considered by combining two models: one that describes mobile molecules (that of confined diffusion as in Figure 36) and one that describes immobile molecules (that of uniaxial rotation of the methyl group, shown as the dotted line in Figure 35). The percentage of mobile molecules as a function of temperature is shown in Figure 37 (left), calculated from high Q values of the experimental EISF. This percentage was then implemented in weighting the total EISF model between the model for mobile and immobile molecules. Significantly, improved agreement was found between the model and experimental EISF shown in Figure 37 (right). The temperature dependence of the EISF was also accounted for through weighting the model for the number of mobile molecules.

It is important to note that other adsorption geometries were considered, two of which are shown in Figure 38. In these models, the molecule

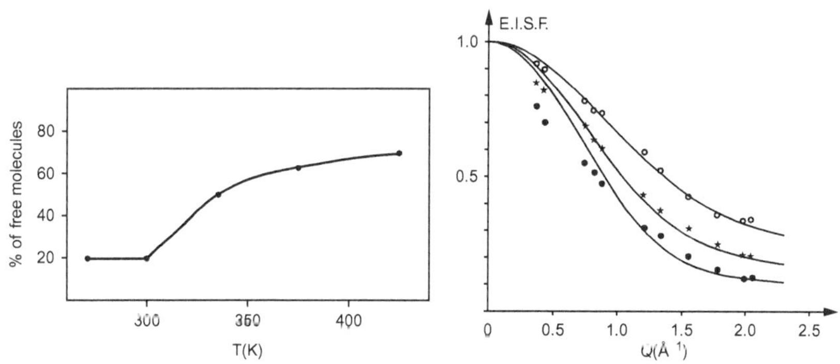

FIGURE 37 (left) Percentage of free methanol molecules as a function of temperature calculated from elastic intensities at high Q. (right) Experimental EISFS fitted to the combined theoretical model described in the text. *Figure adapted reprinted with permission from Ref. [82].*

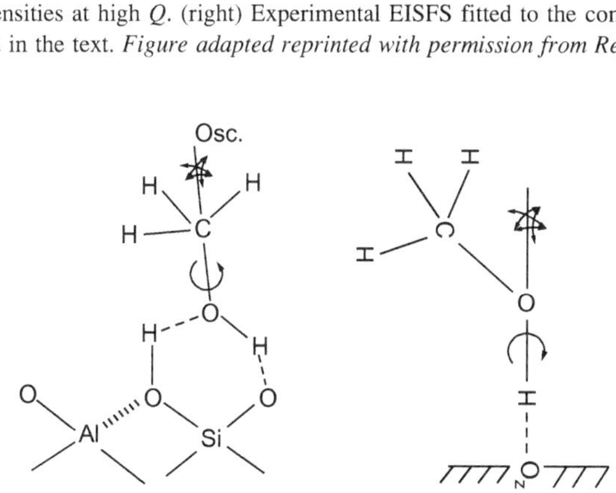

FIGURE 38 Adsorption geometries considered for the EISF models described in Eqs. (30) and (31) where the axis is allowed to fluctuate. *Figure adapted reprinted with permission from Ref. [82].*

is H-bonded to the zeolite either through the Brønsted acid sites or through H-bonding to framework oxygen through the methanol hydroxyl H. The models for these systems allow for rapid fluctuations of the O-CH_3 axis about its equilibrium position. This fluctuation is described by the function

$$f(\alpha) = \frac{\delta}{2sh\delta} \exp(\delta \cos\alpha) \tag{31}$$

α being the angle between the instantaneous and mean axis position and δ characterizes the width of the angular distribution. For the geometry in Figure 38 (left), the theoretical EISF is given as

$$A_0(Q) = \sum_{l=0}^{\infty} (2l+1) j_l^2(QR) S_l^2(\delta) P_l^2(\cos\mu) \tag{32}$$

where R is the distance between the methanol oxygen and the methyl protons (2.063 Å) and μ is the angle between the OH bond and the O-CH_3 axis (30.1°). S_l are orientational order parameters calculated by recurrence, and P_l are Legendre polynomials of order l. The geometry in Figure 38 (right) was also considered where the methyl group rotates around the O-H-O_{zeo} axis, and this axis may also fluctuate. The EISF in this case is given by

$$A_0(\mathbf{Q}) = \left[3\sum_{l=1}^{\infty} (2l+1) j_l^2(QR) S_l^2(\delta) P_l^2(\cos\mu) + \sum_{l=1}^{\infty} (2l+1) j_l^2(QR_{\text{OH-O}_{\text{Zeo}}}) S_l^2(\delta) \right] / 4 \tag{33}$$

where $R = 3.633$ Å (the radius of rotation of methyl protons around the OH-O_{zeo} axis), $\mu = 27.68°$, and $R_{OH-OZeo} = 1.684$ Å.

These models alone did not successfully reproduce the experimental EISF. However, if either of the models are combined and weighted with the confined diffusion model, the data could be fitted. Allowing the O-CH_3 axis to oscillate as in these models did lead to an estimated precision of ±15% in the proportion of mobile species; however, the general model of combining confined diffusion with methyl group rotation appears to be correct.

6.6.2 Methanol Diffusion in HY

The translational diffusion of methanol was measured in zeolite HY (Si/Al = 30) [83] using the OSIRIS spectrometer [84] at the ISIS neutron source, and diffusion coefficients were compared with molecular dynamics simulations at a loading of 6 mol/uc. The QENS spectra were fitted with a single Lorentzian to describe the quasielastic broadening. The Q^2 versus HWHM plots of the fitted Lorentzian are shown in Figure 39.

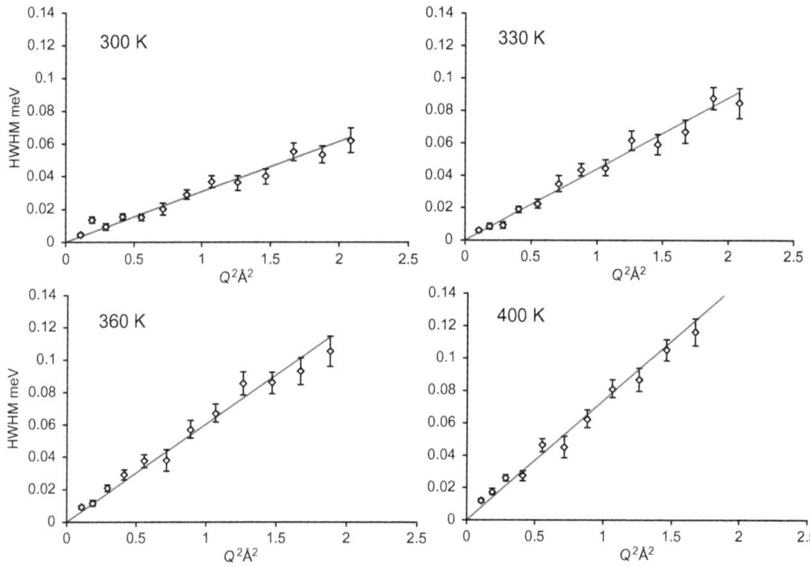

FIGURE 39 *Q* dependencies of the HWHM of the Lorentzian component of each QENS spectrum for methanol in HY at all 4 temperatures. The linear dependence with Q^2 suggests unhindered diffusion. *Figure reprinted with permission from Ref. [83].*

The DQ^2 dependence suggests that the diffusion of methanol is unhindered on the timescale of the spectrometer, which has a resolution of 24 μev at FWHM. These results contrast with the measurements of propane in zeolite NaY discussed in 5.5.1.1, where the propane was found to undergo jump diffusion. Diffusion coefficients were extracted from the low *Q* measurements, where rotational contributions were minimized. The measured diffusion coefficients are in a range of $2–5 \times 10^{-10}$ m^2 s^{-1}. The Arrhenius dependence, plotted in Figure 40, gives an activation energy of 8.8 kJ mol^{-1}.

Possible explanations for the difference in observed diffusion behavior between methanol in HY and propane in zeolite NaY could be found in the smaller size of the methanol molecule, or the use of a HY sample instead of NaY. The difference in counterion can cause significant differences in mobility. For example, previous MD simulations drew comparison between NaY (Si/Al = 2.4) and siliceous Y [85], showing that the presence of extraframework counterions can significantly inhibit the diffusion (with D_s values lower by a factor of 22 at higher temperatures) with an activation energy higher by a factor of 4 with the counterions present. The increase was attributed to a strong interaction between the methanol oxygen and the Na^+ counterion. Longer MD simulations [86] of this system found that activation energies for long-range (intercage) motion decrease with loading and that at lower loadings surface-mediated (intracage) diffusion dominates. Cation behavior upon methanol adsorption was also studied [87], showing that the extraframework

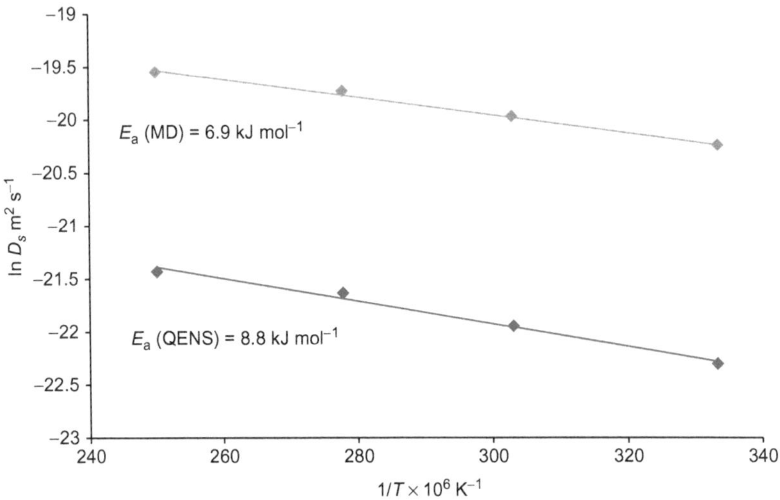

FIGURE 40 Arrhenius plots and calculated E_a for methanol in HY from diffusion coefficients measured by QENS and MD. *Figure reprinted with permission from Ref. [83].*

Na^+ in certain crystallographic sites can move to the center of the supercage, hindering methanol motion owing to its interaction with the molecules. These observations are notable in the context of this comparative study, as the protons associated with the Brønsted acid sites in these HY samples are not able to perform such migration to cage centers and windows so should not hinder methanol motion to this extent.

The MD simulations performed in tandem with the QENS studies of methanol in HY included the presence of Brønsted acid sites in the same ratio as the experimental sample. Simulated diffusion coefficients were calculated in the range of $1.6–3.2 \times 10^{-9}$ m^2 s^{-1}, which are all within a factor of eight of the experimental D_s, but consistently higher. The calculated values were expected to be at the upper limit of experiment due to the use of a perfect crystal with evenly distributed Brønsted acid sites free of defects as mentioned in Section 6.3. The defects may be particularly significant with this silica HY sample, which as noted previously [88] was synthesized through steam dealumination. This synthesis method would leave significant defects such as silanol nests that may act as strong methanol adsorption sites, hindering diffusion. The hindrance may also be illustrated by the higher activation energy measured by QENS. The measured activation energy from MD, calculated from the Arrhenius plot in Figure 40, is 6.9 kJ mol^{-1}, which is lower by $\sim$2 kJ mol^{-1} than that measured by QENS, possibly due to the additional energy necessary to overcome the binding energy to silanol nests or other defects such as extraframework aluminum, or grain boundaries on the nanoscale.

With respect to previous simulations, the calculated diffusion coefficients in HY were lower than those obtained in the siliceous structure [74], by roughly a factor of 3 at 400 K, with an activation energy higher by 1.1 kJ mol^{-1}. This observation is reasonable due to the presence of Brønsted acid sites in our simulated zeolite structure, allowing for favorable H-bonding interactions between the acid sites and the methanol molecules slowing diffusion. Examples of these simulated H-bonds are shown in Figure 41, showing both end-on and side-on geometries, observed to be stable configurations in previous studies [89,90]. However, the diffusion coefficients calculated in NaY are lower by a factor of 68 at 400 K than the simulated HY system. The Si/Al ratio in Ref. [86] was 2.4, compared with 30 in this HY study. This far higher population of Na^+ coordination sites would explain the significant reduction in methanol mobility, further illustrated by the activation energy in NaY of 24 kJ mol^{-1}.

From Figure 42, we observe a rational trend in measured diffusivities given the systems under study, and the methods used when studying methanol in faujasite systems. The highest diffusivities were calculated when using MD simulations in the siliceous Y structure, as the coordination sites acting as

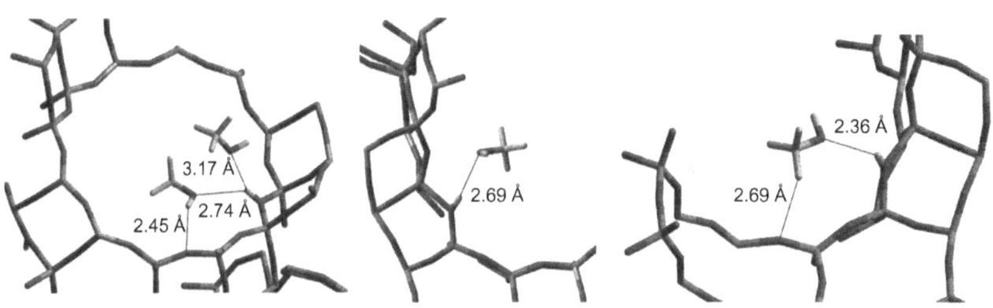

FIGURE 41 Observed H-bonded configurations of methanol in HY. *Figure adapted reprinted with permission from Ref. [83].*

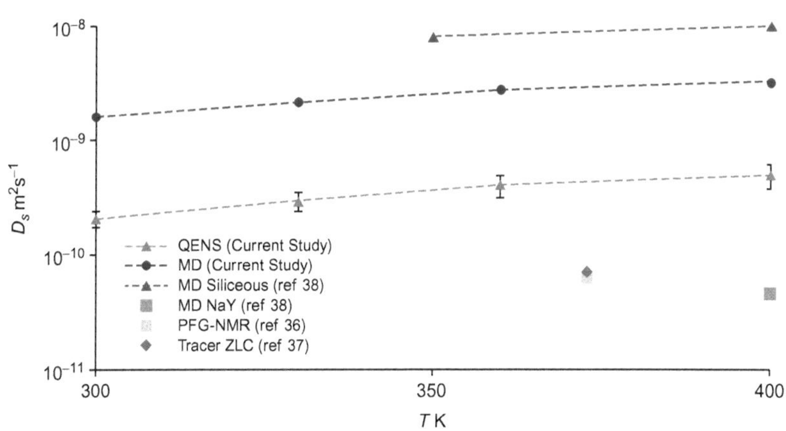

FIGURE 42 Plot of diffusion coefficients of methanol in different faujasite systems between 300 and 400 K measured using different methods. *Figure reprinted with permission from Ref. [83].*

potential barriers for diffusion are not present as in the other systems. The second highest diffusivities are observed by the MD simulations in the HY structure, slower than the siliceous simulations due to H-bonding observed between the methanol and the Brønsted acid sites. The third highest diffusivities are measured in the HY structure using QENS, which we expect as MD simulations usually yield higher diffusivities than experiment, due to the use of a perfect crystal as previously mentioned. The slowest diffusivities are measured in NaY, with similar values obtained using PFG-NMR and tracer ZLC in methods [91,92] and MD simulations in Ref. [81], where the high concentration of counterions (Si/Al = 1.2 in the experiments and 2.4 for the MD simulations) slow diffusion through strong interactions of the methanol oxygen with the counterion.

6.7 CONCLUSION

The case studies summarized above clearly show that QENS techniques can provide unique information on both quantitative and qualitative aspects of molecular motions of sorbates in microporous catalysts and that the power of the method is greatly enhanced when used in conjunction with complementary molecular dynamics simulation techniques. With an increased focus on the need to obtain such dynamical information in order to optimize catalytic processes, the use of the techniques is expected to grow and potentially play a significant role in the design of new microporous catalysts. Moreover, their scope is expected to expand with developments in both neutron instrumentation and computational resources, and with growing computational power, other modeling approaches based on electronic structure calculations will contribute increasingly to the field. Indeed, the concerted QENS/MD simulation approach can be anticipated as playing a major future role in zeolite and catalytic science.

ACKNOWLEDGMENTS

We are very grateful to the authors for the case studies used in this chapter, particularly Prof. Hervé Jobic and Prof. Ramaprosad Mukhopadhyay, and also for helpful discussion before writing this chapter. Prof. Nora De Leeuw is thanked for funding under the EPSRC grant EP/K009567/2. The authors are also grateful to Dr. Stewart Parker and Prof. Felix Fernandez-Alonso for their consultation. The STFC and the ISIS Neutron and Muon source are thanked for funding and beamline facilities access. The UK Catalysis Hub is kindly thanked for resources and support provided via our membership of the UK Catalysis Hub Consortium and funded by EPSRC (grants EP/K014706/2, EP/K014668/1, EP/K014854/1, EP/K014714/1 and EP/M013219/1).

REFERENCES

[1] S.M. Auerbach, K.A. Carrado, P.K. Dutta, Handbook of Zeolite Science and Technology, CRC Press, Boca Raton, 2003.
[2] J. Weitkamp, Solid State Ion. 131 (2000) 175–188.

[3] http://www.iza-structure.org/databases/ (last accessed January 12, 2017).
[4] J. Weitkamp, L. Puppe, Catalysis and Zeolites: Fundamentals and Applications, Springer-Verlag Science & Business Media, Berlin Heidelberg, 2013.
[5] B. Smit, T.L. Maesen, Chem. Rev. 108 (2008) 4125–4184.
[6] J. Kärger, D. Ruthven, Diffusion in zeolites, Handbook of Zeolite Science and Technology, CRC Press, Boca Raton, 2003, p. 341.
[7] J. Karger, D.M. Ruthven, Zeolites 9 (1989) 267–281.
[8] Y. Yasuda, J. Phys. Chem. 86 (1982) 1913–1917.
[9] M. Eic, D.M. Ruthven, Zeolites 8 (1988) 40–45.
[10] D.T. Hayhurst, A.R. Paravar, Zeolites 8 (1988) 27–29.
[11] J. Kärger, H. Pfeifer, W. Heink, Principles and applications of self-diffusion measurements by nuclear magnetic resonance, in: J. Waugh (Ed.), Advances in Magnetic Resonance, Academic Press Inc., San Diego, 1988, pp. 1–89.
[12] H. Jobic, D.N. Theodorou, Microporous Mesoporous Mater. 102 (2007) 21–50.
[13] M. Bée, Quasielastic Neutron Scattering: Principles and Applications in Solid State Chemistry, Biology, and Materials Science, Adam Hilger, Bristol, 1988.
[14] F. Mezei, The Principles of Neutron Spin Echo, Springer-Verlag, Berlin Heidelberg, 1980. pp. 1–26.
[15] C. Chudley, R. Elliott, Proc. Phys. Soc. 77 (1961) 353.
[16] K. Singwi, A. Sjölander, Phys. Rev. 119 (1960) 863.
[17] P.L. Hall, D. Ross, Mol. Phys. 42 (1981) 673–682.
[18] V. Sears, Can. J. Phys. 44 (1966) 1299–1311.
[19] F. Volino, A. Dianoux, Mol. Phys. 41 (1980) 271–279.
[20] F. Volino, M. Pineri, A. Dianoux, A. De Geyer, J. Polym. Sci. Polym. Phys. Ed. 20 (1982) 481–496.
[21] M.-C. Bellissent-Funel, S. Chen, J.-M. Zanotti, Phys. Rev. E 51 (1995) 4558.
[22] R. Zorn, Eur. Phys. J. Spec. Top. 189 (2010) 65–81.
[23] V. Van Speybroeck, K. Hemelsoet, L. Joos, M. Waroquier, R.G. Bell, C.R.A. Catlow, Chem. Soc. Rev. 44 (2015) 7044–7111.
[24] H. Jobic, D.N. Theodorou, J. Phys. Chem. B 110 (2006) 1964–1967.
[25] H. Jobic, J. Mol. Catal. A: Chem. 158 (2000) 135–142.
[26] R.C. Runnebaum, E.J. Maginn, J. Phys. Chem. B 101 (1997) 6394–6408.
[27] F. Leroy, B. Rousseau, A. Fuchs, Phys. Chem. Chem. Phys. 6 (2004) 775–783.
[28] A. Bouyermaouen, A. Bellemans, J. Chem. Phys. 108 (1998) 2170–2172.
[29] S. Kar, C. Chakravarty, Chem. Phys. Lett. 352 (2002) 294–300.
[30] W. Paul, D.Y. Yoon, G.D. Smith, J. Chem. Phys. 103 (1995) 1702–1709.
[31] S. Toxvaerd, J. Chem. Phys. 107 (1997) 5197–5204.
[32] M.G. Martin, J.I. Siepmann, J. Phys. Chem. B 102 (1998) 2569–2577.
[33] A.J. O'Malley, C.R.A. Catlow, Phys. Chem. Chem. Phys. 15 (2013) 19024–19030.
[34] M. Jiang, M. Eic, S. Miachon, J.-A. Dalmon, M. Kocirik, Sep. Purif. Technol. 25 (2001) 287–295.
[35] R. Krishna, J. Van Baten, Chem. Eng. J. 140 (2008) 614–620.
[36] A.O. Koriabkina, A.M. de Jong, D. Schuring, J. van Grondelle, R.A. van Santen, J. Phys. Chem. B 106 (2002) 9559–9566.
[37] D. Lin, V. Ducarme, G. Coudurier, J. Vedrine, Stud. Surf. Sci. Catal. 46 (1989) 615–623.
[38] M. Yu, J.C. Wyss, R.D. Noble, J.L. Falconer, Microporous Mesoporous Mater. 111 (2008) 24–31.
[39] J. Fried, S. Weaver, Comput. Mater. Sci. 11 (1998) 277–293.
[40] F. Jianfen, B. Van de Graaf, H. Xiao, S. Njo, J. Mol. Struct. THEOCHEM 492 (1999) 133–142.

[41] P. Wu, A. Debebe, Y.H. Ma, Zeolites 3 (1983) 118–122.
[42] L. Song, Z.-L. Sun, L.V. Rees, Microporous Mesoporous Mater. 55 (2002) 31–49.
[43] C.L.J. Cavalcante, D.M. Ruthven, Ind. Eng. Chem. Res. 34 (1995) 177–184.
[44] D.M. Ruthven, M. Eic, E. Richard, Diffusion of C8 aromatic hydrocarbons in silicalite. Zeolites 11 (7) (1991) 647–653.
[45] A. Zikanova, M. Bülow, H. Schlodder, Zeolites 7 (1987) 115–118.
[46] F. Leroy, H. Jobic, Chem. Phys. Lett. 406 (2005) 375–380.
[47] B. Frick, A. Magerl, Y. Blanc, R. Rebesco, Phys. B Condens. Matter 234 (1997) 1177–1179.
[48] A. O'Malley, C. Catlow, Phys. Chem. Chem. Phys. 17 (2015) 1943–1948.
[49] E.J. Maginn, A.T. Bell, D.N. Theodorou, J. Phys. Chem. 100 (1996) 7155–7173.
[50] H. Jobic, B. Farago, J. Chem. Phys. 129 (2008) 171102.
[51] J.R. Hufton, D.M. Ruthven, Ind. Eng. Chem. Res. 32 (1993) 2379–2386.
[52] W. Zhu, A. Malekian, M. Eić, F. Kapteijn, J. Moulijn, Chem. Eng. Sci. 59 (2004) 3827–3835.
[53] B. Millot, A. Methivier, H. Jobic, H. Moueddeb, J. Dalmon, Microporous Mesoporous Mater. 38 (2000) 85–95.
[54] B. Millot, A. Méthivier, H. Jobic, H. Moueddeb, M. Bée, J. Phys. Chem. B 103 (1999) 1096–1101.
[55] A.J. O'Malley, C.R.A. Catlow, M. Monkenbusch, H. Jobic, J. Phys. Chem. C 119 (2015) 26999–27006.
[56] H. Jobic, M. Bee, J. Caro, Translational mobility of n-butane and n-hexane in ZSM-5 measured by quasi-elastic neutron scattering, in: Proceedings of the 9th International Zeolite Conference, 1993, p. 121.
[57] H. Jobic, M. Bée, G.J. Kearley, Zeolites 12 (1992) 146–151.
[58] H. Jobic, M. Bée, G.J. Kearley, Zeolites 9 (1989) 312–317.
[59] M. Monkenbusch, R. Schätzler, D. Richter, Nucl. Instrum. Methods Phys. Res. Sect. A Accel. Spectrom. Detect. Assoc. Equip. 399 (1997) 301–323.
[60] H. Jobic, M. Bée, S. Pouget, J. Phys. Chem. B 104 (2000) 7130–7133.
[61] S. Mitra, A. Tripathy, N. Gupta, R. Mukhopadhyay, Appl. Phys. A 74 (2002) s1308–s1310.
[62] R. Mukhopadhyay, S. Mitra, S. Paranjpe, B. Dasannacharya, Nucl. Instrum. Methods Phys. Res. Sect. A Accel. Spectrom. Detect. Assoc. Equip. 474 (2001) 55–66.
[63] H. Jobic, M. Bée, A.J. Dianoux, J. Chem. Soc. Faraday Trans. 1 Phys. Chem. Condens. Phases 85 (1989) 2525–2534.
[64] Y. Blanc, Le Spectrometre a temps de vol IN6, ILL-Report 83BL21G, 1983.
[65] A. Dianoux, F. Volino, Mol. Phys. 34 (1977) 1263–1277.
[66] A. Sayeed, S. Mitra, A. Anil Kumar, R. Mukhopadhyay, S. Yashonath, S. Chaplot, J. Phys. Chem. B 107 (2003) 527–533.
[67] S. Chaplot, R. Mukhopadhyay, P. Vijayaraghavan, A. Deshpande, K. Rao, Pramana 33 (1989) 595–602.
[68] R. Mukhopadhyay, A. Sayeed, S. Mitra, A.A. Kumar, M.N. Rao, S. Yashonath, S. Chaplot, Phys. Rev. E 66 (2002) 061201.
[69] B.J. Borah, H. Jobic, S. Yashonath, J. Chem. Phys. 132 (2010) 144507.
[70] P.K. Ghorai, S. Yashonath, P. Demontis, G.B. Suffritti, J. Am. Chem. Soc. 125 (2003) 7116–7123.
[71] S.Y. Bhide, S. Yashonath, Mol. Phys. 102 (2004) 1057–1066.
[72] C.D. Chang, A.J. Silvestri, J. Catal. 47 (1977) 249–259.
[73] C. Maiden, Stud. Surf. Sci. Catal. 36 (1988) 1–16.

[74] U. Olsbye, S. Svelle, M. Bjørgen, P. Beato, T.V. Janssens, F. Joensen, S. Bordiga, K.P. Lillerud, Angew. Chem. Int. Ed. 51 (2012) 5810–5831.
[75] M. Bjørgen, S. Svelle, F. Joensen, J. Nerlov, S. Kolboe, F. Bonino, L. Palumbo, S. Bordiga, U. Olsbye, J. Catal. 249 (2007) 195–207.
[76] P.B. Venuto, E.T. Habib Jr., Fluid catalytic cracking with zeolite catalysts, Marcel Dekker, Inc., New York, 1979.
[77] E. Vogt, B. Weckhuysen, Chem. Soc. Rev. 44 (2015) 7342–7370.
[78] W.S. Wieland, R.J. Davis, J.M. Garces, J. Catal. 173 (1998) 490–500.
[79] A. Palomares, G. Eder-Mirth, M. Rep, J. Lercher, J. Catal. 180 (1998) 56–65.
[80] J. Fei, Z. Hou, B. Zhu, H. Lou, X. Zheng, Appl. Catal. A Gen. 304 (2006) 49–54.
[81] D. Jin, B. Zhu, Z. Hou, J. Fei, H. Lou, X. Zheng, Fuel 86 (2007) 2707–2713.
[82] H. Jobic, A. Renouprez, M. Bee, C. Poinsignon, J. Phys. Chem. 90 (1986) 1059–1065.
[83] A.J. O'Malley, V.G. Sakai, I.P. Silverwood, N. Dimitratos, S.F. Parker, C.R.A. Catlow, Phys. Chem. Chem. Phys. 18 (2016) 17294–17302.
[84] M.T. Telling, K.H. Andersen, Phys. Chem. Chem. Phys. 7 (2005) 1255–1261.
[85] D. Plant, G. Maurin, R. Bell, J. Phys. Chem. B 110 (2006) 15926–15931.
[86] D.F. Plant, G. Maurin, R.G. Bell, J. Phys. Chem. B 111 (2007) 2836–2844.
[87] G. Maurin, D. Plant, F. Henn, R.G. Bell, J. Phys. Chem. B 110 (2006) 18447–18454.
[88] A.J. O'Malley, S.F. Parker, A. Chutia, M.R. Farrow, I.P. Silverwood, V. García-Sakai, C.R.A. Catlow, Chem. Commun. 52 (2016) 2897–2900.
[89] C. Zicovich-Wilson, P. Viruela, A. Corma, J. Phys. Chem. 99 (1995) 13224–13231.
[90] A.J. O'Malley, A. Logsdail, A. Sokol, C. Catlow, Faraday Discuss. 188 (2016) 235–255.
[91] F. Meunier, P. Gray, J. Kärger, Z. Xu, D. Ruthven, Zeolites 14 (1994) 242–249.
[92] S. Brandani, D.M. Ruthven, Zeolites 15 (1995) 494–495.

Chapter 7

Atomic Quantum Dynamics in Materials Research

Carla Andreani[*,†], Roberto Senesi[1,*,†,‡], Matthew Krzystyniak[§,¶], Giovanni Romanelli[§] and Felix Fernandez-Alonso[§,|]

*Università degli Studi di Roma "Tor Vergata", Rome, Italy; †Museo Storico della Fisica e Centro Studi e Ricerche Enrico Fermi, Rome, Italy; ‡Consiglio Nazionale delle Ricerche, CNR-IPCF, Messina, Italy; §ISIS Facility, Rutherford Appleton Laboratory, Didcot, United Kingdom; ¶Nottingham Trent University, Nottingham, United Kingdom; |Department of Physics & Astronomy, University College London, London, United Kingdom

1Corresponding author: e-mail: roberto.senesi@uniroma2.it

Chapter Outline

Experimental Methods in the Physical Sciences, Vol. 49. http://dx.doi.org/10.1016/B978-0-12-805324-9.00007-8

Nothing tends so much to the advancement of knowledge as the application of a new instrument.

Sir Humphrey David

7.1 AIM AND SCOPE

Quantum Mechanics (QM) in text books as well as in chemistry and materials science publications is commonly strictly applied to electrons. These are the lightest particles among the stable bricks composing the elements of the periodic table and all the way up in size and complexity up to proteins. Historically, QM was first applied to electrons only and a sharp dynamic separation with respect to the much heavier nuclei was applied known as the Born-Oppenheimer Approximation (BOA). Nowadays, a large part of computer simulations available for condensed matter systems describes the motion of nuclei through classical Molecular Dynamics (MD) based on Newton's laws of motion. This can be a rough simplification of a system under study, and the reader should not be surprised that particles ca. 2000 times heavier than the electron, as the proton and neutron are, can in some cases only be treated as quantum objects, *e.g.*, in neutron scattering [1–4] or in the case of nuclear fusion.

The need to treat atoms across the periodic table as truly quantum objects, in particular at *low temperatures*, is presented in what follows. This is an experimental fact, as it is shown later by Deep Inelastic Neutron Scattering (DINS) experiments, where Nuclear Momentum Distributions (NMDs) are measured and values of the mean kinetic energy of atoms ($\langle E_K \rangle$) are obtained. Neutrons are a powerful probe of structure and dynamics of condensed systems as they directly collide on nuclei without electrically interact with surrounding electrons. Experimental measurements of NMDs require scattering at energy and momentum transfers, which are much larger than the characteristic energy and inverse length scales of the system under study, as being roughly defined by the atomic binding energy and the inverse interatomic distance [5]. When applied to molecular systems, DINS shares complementary insights to X-ray Compton scattering since, by measuring nuclear and electron momentum distributions, respectively, the two techniques pinpoint the nuclear and electronic quantum effects on intramolecular scale energetics and structure [5].

By making use of high energy and momentum transfer, DINS is placed at the extreme momentum and energy ranges of the typical neutron spectroscopic techniques applied to the study of materials. This mainly explains why so far DINS has been exploited less than INS and QENS spectroscopies.

Recent studies capitalize on the DINS evolution and assessment in the last decades [6–8], at spallation neutron sources such as ISIS pulsed neutron and muon source (UK) [9] or earlier at the Intense Pulsed Neutron Source

IPNS [10–18]. As a Compton-like scattering process [19–21], DINS is able to measure the nuclear momentum on a very short time frame [22]. As a large number of scattering events are collected during a measurement, it is possible to reconstruct the statistical distribution of momenta as the result of the ensemble averages yielding the NMD.

NMDs can be easily defined in the case noninteracting atoms, as in the kinetic theory of gases by Maxwell and Boltzmann (MB) with the result that the mean kinetic energy of atoms in a many-particle system is proportional to the macroscopic thermodynamic temperature and to the number of degrees of freedom via the Boltzmann constant k_B. In this chapter, we quantitatively replace this result by introducing (a) interactions between atoms in a condensed system in the form of phonon frequencies ω and (b) the additional constant at the base of QM, the Plank constant $\hbar$. The first result will be that at low temperatures, the nuclear kinetic energy is not dependent on the thermodynamic temperature any longer and does not vanish in the limit $T \to 0$ but stays constant at the Zero-Point Energy (ZPE) level. Figure 1 anticipates some results of Section 7.2 showing a comparison between values of the atom mean kinetic energy obtained from an MB model at $T = 10$ K and 300 K and those obtained in QM for a particle in a confining box potential. As in the case of H bond lengths are of the order of ~1 Å, one can think of typical sizes for the confining potential on an atom in the order of fractions of Å, therefore being convinced on what an underestimate for such energies would be to restrict oneself to the MB approximation.

The need for QM corrections to the classical dynamics of nuclei is testified by the increasing number of approaches for computer simulations and

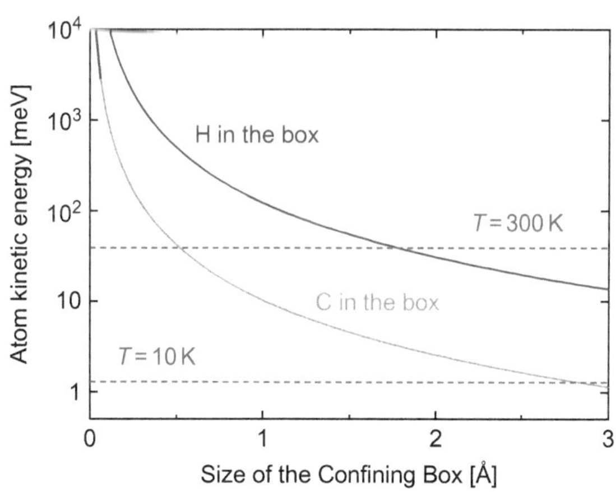

FIGURE 1 Comparison of atom kinetic energies from MB model at $T = 10$ K and 300 K (horizontal-dashed lines) and the prediction from QM for hydrogen and carbon atoms as a function of the three-dimensional square-box size.

theoretical models that introduce nuclear ZPEs as well as other Nuclear Quantum Effects (NQEs). Among these, Path Integral Molecular Dynamics (PIMD) [23] is the state of the art among computer simulations treating atoms as quantum objects. However, within some limits as the harmonic approximation for the local potential, simulated and experimental Vibrational Densities of States (VDoS) can be used to quantitatively predict the value of $\langle E_K \rangle$ and of ZPEs.

Section 7.2 explains with simplified models why ZPEs constitute an essential ingredient in an accurate description of the properties of a given material, the goal being achieved by recalling a standard exercise in QM textbooks, that is the particle in a box. In particular, it will be shown how, for a size of the potential affecting a single atom of the order of the Å or less, ZPE greatly exceeds the thermal contribution from an MB approach. This phenomenon is related to an important (and easily measurable) broadening of the NMD that is therefore a thermometer for the effective and local temperature of an atom rather than for the thermodynamic and macroscopic temperature. Section 7.3 explains how it is possible to experimentally access NMDs in the DINS regime. The special case of the VESUVIO spectrometer is treated in some detail for this is at present the only spectrometer where a user-oriented scientific program is conduced. Some of the most recent results of the technique are then presented in Section 7.4 through three examples spanning the range of capabilities of DINS investigations. Finally, Section 7.5 summarizes the main points of the chapter and suggests how to make the best use of data collected in a DINS experiment as well as commenting on the many and promising ways forward for this relatively recent and unique technique.

7.2 CONCEPTUAL FRAMEWORK

7.2.1 Atoms Are Quantum Objects

Confinement is widely believed to induce phenomena that can only be described using the language of QM [8]. On the one hand, particles of atomic size confined at the nanometer scale exhibit discrete energy levels observed spectroscopically. On the other hand, it is confinement that is the origin of the subtle interplay between the amount of different sources of information an observer can infer about different microscopic properties, referred to as observables in the QM language. The most celebrated model used as example to illustrate the confinement-induced differences between classical and quantum systems is the particle in a box model (also known as the infinite potential well or the infinite square well). This section will show how the particle in a box model can be used to intuitively explain the basic atomic property directly accessible in DINS, the NMD.

The simplest form of the particle in a box model considers a one-dimensional system [24–26]. The walls of a one-dimensional box may be visualized as regions of space with an infinitely large potential energy. Conversely, the interior of the box has a constant, zero potential energy. Mathematically, this behavior can be modeled using a one-dimensional potential function, $V(x)$, defined in such a way that it vanishes in a region of space, say in the range $-L < x < L$. This is the only region where a particle can be found, for elsewhere, *i.e.*, $x \leq -L$ and $x \geq L$, the potential is extremely large compared with the energy of the particle, say $V(x) = \infty$ (see Fig. 2)

In classical systems, according to Newton's Laws, the particle can move at any speed within the box and it is no more likely to be found at one position than another. In QM, the wave function gives the most fundamental description of the behavior of a particle. The wave function, $\psi(x, t)$, can be found by solving the Schrödinger equation for the system

$$i\hbar\frac{\partial}{\partial t}\psi(x,t) = -\frac{\hbar^2}{2M}\nabla^2\psi(x,t) + V(x)\psi(x,t). \tag{1}$$

The wave function inside the box oscillates through space and time. Mathematically, $\psi(x, t)$ can be cast into a product of a time-dependent part, $\exp(-i\omega t)$ with ω introduced in Eq. (3), and a time-independent (spatial) part, $\psi(x)$. The spatial part of the wave function, $\psi(x)$, can be further cast in specific terms depending on the regions of space. This is the first effect induced by the confinement with profound consequences for the amount of information that one can infer about the particle in terms of the measurement of its position, momentum, and energy. Firstly, in the region of space "behind the walls," $x \leq -L$ and $x \geq L$, $\psi(x, t) = 0$. This means that, for the limiting case of infinitely steep and high box walls, the particle cannot be found and its properties measured both classically and

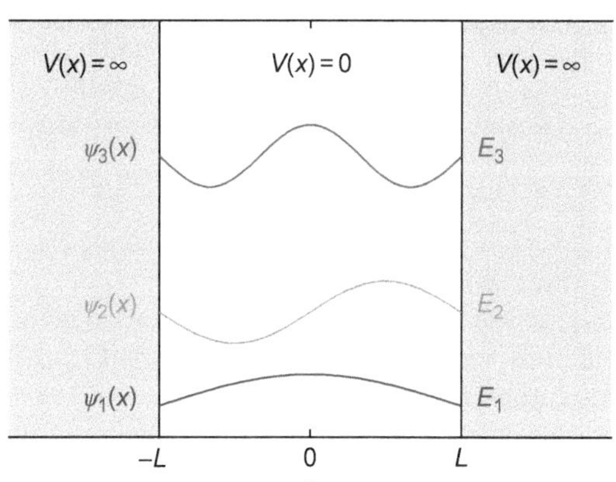

FIGURE 2 The barriers outside a one-dimensional box have infinitely large potential, while the interior of the box has a constant, zero potential.

quantum-mechanically. For the remaining region of space, $-L < x < L$, the opposite is the case and the particle is characterized by a wave function, $\psi(x)$ of the form:

$$\psi(x) = A\exp(ikx) + B\exp(-ikx), \tag{2}$$

with A and B two constants to be fixed. The dispersion relation for a nonrelativistic quantum particle is defined as

$$\mathcal{E} = \hbar\omega = \frac{\hbar^2 k^2}{2M}, \tag{3}$$

where $\mathcal{E}$ in this case is both the total energy and the kinetic energy, for there is no contribution from $V(x) = 0$ within the box. The condition of vanishing wave function on the barrier, $\psi(x = 0, t) = \psi(x = L, t) = 0$, calls for the quantization of the wave vector, $k_n = n\pi/2L$, with n integer $n \geq 1$, as well as a quantization on energies through Eq. (3),

$$\mathcal{E}_n = \hbar\omega_n = \frac{n^2\pi^2\hbar^2}{2ML^2}. \tag{4}$$

It can be shown that the Hamiltonian of a particle in a box commutes with the parity operator thus enabling to classify the solutions of the Schrödinger equation, given by Eq. (2), according to their parity [27]. Thus the full solution of the one-dimensional particle-in-a-box problem displays the following parity-adopted form:

$$\psi_n(x) = \begin{cases} \sqrt{\frac{1}{L}}\cos\left(\frac{\pi n x}{2L}\right), & n\text{-odd}, -L < x < L, \\ \sqrt{\frac{1}{L}}\sin\left(\frac{\pi n x}{2L}\right), & n\text{-even}, -L < x < L, \\ 0, & x \leq -L \text{ or } x \geq L. \end{cases} \tag{5}$$

The first few solutions of the problem are shown in Fig. 2.

As opposed to the classical case when the lowest energy of a particle in the box would be $\mathcal{E}_{cl} = 0$, *i.e.*, the particle at rest in its position, in the QM case there exist a nonvanishing ZPE $\mathcal{E}_1 = \frac{\pi^2\hbar^2}{8ML^2}$ that defines the ground state of the system. Moreover, the particle behaves as a stationary wave inside the box, with spatial wave function $\psi_1(x)$.

As a wave, the particle is delocalized both in space and momentum space, the two representations being linked by a Fourier transform between the two canonical variables x and $p = \hbar k$, namely

$$\phi(p) = \frac{1}{\sqrt{2\pi}}\int dx e^{ikx}\psi(x). \tag{6}$$

The expectation value for the momentum $\langle p\rangle$, calculated using the momentum representation defined in Eq. (6),

$$\langle p\rangle = \int dpn(p)p = \int dp|\phi(p)|^2 p = 0, \tag{7}$$

yields a zero value regardless of the state (n) of the particle for stationary solutions.

The NMD can be characterized by the variance, $\sigma_p^2 = \langle p^2\rangle$ [27]:

$$\sigma_p^2 = \langle p^2\rangle = \int dp\, n(p)(p-\langle p\rangle)^2 = \int dp|\phi(p)|^2 p^2 = \frac{\hbar^2 n^2 \pi^2}{4L^2} = 2M\mathcal{E}_n. \tag{8}$$

Also the variance of the NMD calculated in the simple particle-in-a-box model is a manifestation of a more general rule applicable in QM. Using the results obtained in Eq. (8) one can write:

$$\begin{aligned}\sigma_p^2 &= \int d\mathcal{E}\psi^\dagger(\mathcal{E})(p-\langle p\rangle)^2\psi(\mathcal{E})\\ &= \int d\mathcal{E}\psi^\dagger(\mathcal{E})p^2\psi(\mathcal{E}) = 2M\int d\mathcal{E}\psi^\dagger(\mathcal{E})E_k\psi(\mathcal{E}) = 2M\langle E_k\rangle,\end{aligned} \tag{9}$$

where E_k is the kinetic energy operator.

As the square of the wave function, $\rho(x) = |\psi(x)|^2$ represents the probability density to find the particle in the neighborhood of x, the square of the Fourier transform of the wave function $n(p) = |\phi(p)|^2$ is the NMD and represents the probability for a particle to have a momentum p and can be written, in the case of the particle in a box, as [27]:

$$n(p) = \begin{cases} \dfrac{8n^2\pi}{\hbar}\dfrac{\cos^2\left(\dfrac{pL}{\hbar}\right)}{\left(\dfrac{4p^2L^2}{\hbar^2} - n^2\pi^2\right)}, & n\text{-odd},\\[2ex] \dfrac{8n^2\pi}{\hbar}\dfrac{\sin^2\left(\dfrac{pL}{\hbar}\right)}{\left(\dfrac{4p^2L^2}{\hbar^2} - n^2\pi^2\right)}, & n\text{-even}.\end{cases} \tag{10}$$

The product of the squares of uncertainties on the position, σ_x^2, and momentum, σ_p^2, of a particle, according to the Heisenberg uncertainty principle, has to be greater than or equal to $\frac{\hbar^2}{4}$. In the case of a particle in the box, one has (see Appendix A.1):

$$\sigma_x^2\sigma_p^2 = \frac{\hbar^2}{4}\left[\frac{n^2\pi^2}{3} - 2\right]. \tag{11}$$

This product increases with increasing n, and its minimum value in the ground state is $\frac{\hbar^2}{4}\left[\frac{\pi^2}{3}-2\right]$. The uncertainty product is minimized by Gaussian wave functions of the form [27]:

$$\psi_0(x)=\left(\frac{M\omega}{\pi\hbar}\right)^{\frac{1}{4}}\exp\left(-\frac{M\omega}{2\hbar}x^2\right). \tag{12}$$

Moreover, the Gaussian wave function is the ground-state (corresponding to the energy $\mathcal{E}_0=\hbar\omega/2$) solution of the Schrödinger equation for the harmonic oscillator oscillating with frequency ω

$$\hat{H}=\frac{\hat{p}^2}{2M}+\frac{1}{2}M\omega^2\hat{x}^2, \tag{13}$$

where $\hat{x}$ is the position operator and $\hat{p}$ is the momentum operator, given by $\hat{p}=-i\hbar\frac{\partial}{\partial x}$, with general solutions for the nth quantum state, corresponding to the energy $\mathcal{E}_n=\hbar\omega(n+\frac{1}{2})$, given by [27]:

$$\psi_n(x)=\frac{1}{\sqrt{2^n\,n!}}\cdot\left(\frac{M\omega}{\pi\hbar}\right)^{1/4}\exp\left(-\frac{M\omega x^2}{2\hbar}\right)\cdot H_n\left(\sqrt{\frac{M\omega}{\hbar}}x\right), \tag{14}$$

where $n=0,1,2,\ldots$ and H_n are the Hermite polynomials defined in Appendix A.2.

It is worth noting that the ground-state eigenfunction of the single-dimensional harmonic oscillator in the momentum representation is also a Gaussian function:

$$\phi(p)=\frac{1}{\sqrt{2\pi}}\int dx e^{ikx}\psi_0(x)=\left(\frac{1}{\pi M\omega\hbar}\right)^{\frac{1}{4}}\exp\left(-\frac{p^2}{2\hbar M\omega}\right). \tag{15}$$

In the single-dimensional harmonic oscillator case the uncertainty principle, applied to wave functions for the nth state corresponding to the energy value $\mathcal{E}_n$, reads $\sigma_{x,n}\sigma_{p,n}=\frac{\hbar}{2}(2n+1)$. In the specific case of the ground state, $\sigma_{x,0}\sigma_{p,0}=\frac{\hbar}{2}$. Thus the absolute lower bound of the uncertainty product of position and momentum, attained by the quantum harmonic oscillator, is by only 29% lower than the value of $\sigma_x\sigma_p$ obtained for the single-dimensional particle in a box model.

7.2.2 Epithermal Neutrons Probe the Quantum Character of Atoms

Following the basic aspects introduced in Sec. 7.2.1, the formalism of the quantum harmonic oscillator can be used to define a typical scale for the momentum of a quantum particle. If we measure energy in units of ω and position x in units of $dL = \sqrt{\frac{\hbar}{M\omega}}$, then the Hamiltonian of the harmonic oscillator simplifies to [27]

$$H = -\frac{1}{2}\frac{d^2}{dx^2} + \frac{1}{2}x^2. \tag{16}$$

In DINS, a system of units is widely used for the characterization of position, momentum, and energy scales, where the momentum transfer is given in units of $Å^{-1}$, energy transfer in meV, and atomic mass in amu. With this choice of units, Planck's constant is given by $\hbar = 2.04458$ (meV amu)$^{1/2}$/Å. To illustrate the above, we can consider the hydrogen dynamics in a condensed system. The wave function in position space can be sampled along a spatial grid whose extent can be evaluated based on the range of frequencies of typical projected VDoS (pVDoS). The highest vibration frequency exhibited by molecules containing hydrogen is due to the stretching mode of the hydrogen molecule H_2 of energy of ca. 500 meV. Consequently, the characteristic length scale, $dL = \sqrt{\frac{\hbar}{M\omega}}$, extends to ca. 0.1 Å. It is worth mentioning that the typical amplitude of a nuclear oscillation around the equilibrium position of a local harmonic potential is of the same order of 0.1 Å. Using the momentum operator one can come up with a natural extent of the wave function in momentum space being $dp = \hbar/dL = \sqrt{\hbar M\omega}$. Thus the hydrogen wave function in momentum space has a natural extent of ca. 20$Å^{-1}$ in case of a stretching mode of the energy of ca. 500 meV. For a high-resolution measurement in momentum space, the extent of the momentum wave function should be sampled with a dense enough sampling interval of the order of $dp = 0.1$ $Å^{-1}$, which corresponds to the de Broglie wavelength of epithermal neutrons.

In the case of a statistical ensemble of harmonic quantum particles at a given temperature T, one finds that the probability densities that specific values of position and momentum variables occur are both Gaussian functions (see Appendix A.3). In the case of the NMD, the variance associated with such distribution is

$$\sigma^2 = \hbar M\omega\left(\langle N\rangle + \frac{1}{2}\right), \tag{17}$$

where $\langle N\rangle = 1/\left[\exp\left(\frac{\hbar\omega}{kT}\right) - 1\right]$ and we simplified the notation $\sigma_p \equiv \sigma$, as variances in the following sections are only referred to NMDs. In case of the ensemble at absolute zero temperature, $T = 0$, or alternatively, in case of a harmonic system spectrum dominated by a high-frequency vibration such as $\hbar\omega/kT \gg 1$, $\langle N\rangle \to 0$, the system will be in the ground state with $\sigma_0^2 = \hbar M\omega/2$. The variance of momentum then corresponds to the same variance of a pure state of a single-dimensional quantum mechanical oscillator with $\mathcal{E}_0 = \hbar\omega/2$. This is an important result in many respects. In case of light-weight atomic species like hydrogen in molecular systems, the intramolecular high-frequency vibrations can be, to a good degree of approximation, described within the three-dimensional harmonic oscillator model. As a consequence, for the calculation of the variance of the NMD of an ensemble of hydrogen atoms inside molecules or a solid-state system at temperature T, the high-frequency approximation can be applied with impunity for each frequency of the relevant vibrational spectrum. On the whole, the total variance of the NMD becomes then a sum of contributions $\sigma^2 = \sum_\lambda \hbar M\omega_\lambda/2$. This behavior governs the NMD of hydrogenous and other lightweight masses in condensed matter systems and molecules in a wide range of temperatures from $T = 0$ K up to the room temperature.

For an isotropic solid, σ^2 may be obtained by means of the pVDoS, $G_M(\omega)$, that is determined as the contribution from a given atom to the total VDoS, $G(\omega) = \sum_M G_M(\omega)$. $G_M(\omega)$ is defined by the following sum over all q points in the first Brillouin zone (1BZ) and over all phonon bands

$$G_M(\omega) = \frac{1}{3N_q}\sum_{\mathbf{q}}^{N_q}\sum_{\lambda}^{N_\lambda}\mathbf{e}^\lambda(\mathbf{q})^2\delta(\omega - \omega_\lambda(\mathbf{q})), \tag{18}$$

where $\mathbf{e}^\lambda(\mathbf{q})$ are the polarization vectors associated with the vibrational modes λ of energy $\omega_\lambda(\mathbf{q})$ and N_q is the number of points in the 1BZ. For a mode λ, the polarization vectors $\mathbf{e}^\lambda(\mathbf{q})$ are N three-dimensional vectors associated with N atoms, showing the amplitude and direction along which the atom oscillates in each mode.

Equation (18) reflects the fact that, in case of a real solid, not all vibrational frequencies ω_λ contribute with equal weighting to the definition of σ^2, but are subject to weighting over all phonon dispersion relations for all phonon branches in a crystal.

If we now leave the restricting case of $\hbar\omega/k_BT \gg 1$ and reintroduce a population factor $\langle N\rangle$, we obtain [28, 29]

$$\sigma^2 = \frac{\hbar M}{2}\int_0^\infty \omega G_M(\omega)\coth\left(\frac{\hbar\omega}{2k_BT}\right)d\omega. \tag{19}$$

As NMDs are accessed via spectroscopic methods, the immediate consequence of this is that in the sum, $\sum_\lambda \hbar M\omega_\lambda/2$, weights $|c_\lambda|^2$ must appear describing the effective coupling of given microscopic probes (radiation or projectiles) with every single mode of the relevant vibrational spectrum ω_λ. In case of DINS with epithermal neutrons, $|c_\lambda|^2$ arise from the projections of neutron momentum transfer, $\mathbf{Q}$, onto the normal modes of vibrations projected onto a given nucleus, $|c_\lambda|^2 = \left[\hat{\mathbf{Q}}\cdot\mathbf{e}^\lambda(\mathbf{Q})\right]^2$ [22, 30, 31]:

$$\sigma\left(\hat{\mathbf{Q}}\right)^2 = \frac{\hbar M}{2}\frac{1}{N_Q}\sum_{\mathbf{Q}}^{N_Q}\sum_{\lambda}^{N_\lambda}\left[\hat{\mathbf{Q}}\cdot\mathbf{e}^{\lambda}(\mathbf{Q})\right]^2\omega_\lambda(\mathbf{Q})\coth\left(\frac{\hbar\omega_\lambda(\mathbf{Q})}{2k_BT}\right). \tag{20}$$

Equation (20), in the limit of $\hbar\omega_\lambda(\mathbf{Q})/2k_BT \gg 1$, becomes:

$$\sigma\left(\hat{\mathbf{Q}}\right)^2 = \frac{\hbar M}{2}\frac{1}{N_Q}\sum_{\mathbf{Q}}^{N_Q}\sum_{\lambda}^{N_\lambda}\left[\hat{\mathbf{Q}}\cdot\mathbf{e}^{\lambda}(\mathbf{Q})\right]^2\omega_\lambda(\mathbf{Q}). \tag{21}$$

Equation (20) is then a single-crystal version of the low-temperature/high-frequency limit of the variance of NMD in the harmonic limit, as seen by the DINS technique.

Introducing a set of orthogonal directions x, y, z, spherically averaged variances of NMDs are obtained from Eq. (20) using [22, 30, 31]

$$\sigma^2 = \frac{1}{3}\left[\sigma_x^2 + \sigma_y^2 + \sigma_z^2\right]. \tag{22}$$

In a DINS experiment on a liquid or a powder harmonic system, representing a simple extension of the single-dimensional quantum harmonic oscillator case to three dimensions, a spatially averaged NMD σ, may be compared against the classical limit of an MB distribution at the same temperature T for a three-dimensional system obeying $3k_BT/2 = 3\hbar^2\sigma_{cl}^2/2M$. Using this classical relation between T and σ, one can write [32]

$$\sigma_{cl} = \frac{\sqrt{Mk_BT}}{\hbar}. \tag{23}$$

The classical values, σ_{cl}, provide a *lower* bound to the actual values of this observable, as chemical binding leads to an effective confinement of the atom under consideration and a concomitant increase in kinetic energy. Within the harmonic approximation, this excess kinetic energy strictly arises from the cumulative contribution of the ZPEs of *all* normal modes projected onto a given nucleus as shown mathematically by Eq. (18). It is important to stress that DINS provides direct experimental access to these kinetic energies irrespective of the specific nature (harmonic or not) of the underlying nuclear dynamics [32]. From these considerations, we can also define a quantum-energy excess (QE) as the ratio of the squares of actual NMD widths relative to classical predictions, namely,

$$\mathrm{QE} = \frac{\sigma^2}{\sigma_{cl}^2} = \frac{\hbar^2}{Mk_BT}\sigma^2. \tag{24}$$

Mathematically, both σ^2 and T play an identical role in terms of approaching the classical limit (QE = 1) as either of the two (or both) increases.

7.2.3 Atoms and Their Local Chemical Environment

As a result of the nuclear spatial confinement, atoms experience NQEs such as the ZPEs introduced in the case of the particle in a box discussed in Sec. 7.2.1. If we model the interaction between two atoms within a bond with a harmonic spring, the nuclear total energy $\mathcal{E}(x)$ is given as a function of the interatomic distance x, $\mathcal{E}(x)=\frac{1}{2}f(x-x_0)^2$, then there are two parameters characterizing the bond, the force constant f, and the average position x_0. The values of these parameters may be derived from conventional spectroscopic measurements, relying on measurements of energy differences between discrete vibrational levels, in terms of bond length and vibrational frequency.

Contrary to other spectroscopic techniques, DINS allows to perform absolute energy measurements, thus being independent of any modeling to access the ZPEs due to the confining potential. In the limit of the harmonic approximation, the virial theorem relates the mean total energy to the mean kinetic energy as

$$2\langle\mathcal{E}\rangle=\langle E_K\rangle. \tag{25}$$

Moreover, the knowledge of the shape of NMD inferred from a DINS measurement allows the reconstruction of the shape of the effective potential curve, *e.g.*, the value of f in the previous case, in a mass-resolved manner, *i.e.*, for individual nuclei present in a sample of interest. In general, the local curvature of the potential will be the function of the sum of contributions from different vibrational modes a given nucleus is involved in, weighted by their frequencies. Therefore, when a given bond has got more than one vibrational mode associated with it (*e.g.*, stretching and bending mode in case of atoms involved in hydrogen bonds), DINS characterizes the net effect of the interplay between bond strengthening and weakening action of different vibrational modes for different nuclei and thus provides direct and model-independent measures of binding environment for a specific nuclide by means NMDs.

The relation between bond characteristics (*e.g.*, its strength and geometry) and DINS observables ($\langle E_K\rangle$ and NMD) can be best understood for an Einstein solid of characteristic frequency ω_E, defined via a VDoS of the form $G_E(\omega) = \delta(\omega - \omega_E)$. In the low-temperature limit, each atom is an independent harmonic oscillator. Thus the wavevector-energy relationships of phonons, referred to as phonon dispersion relations, cannot be accounted for. Nevertheless, the model is very useful to describe cases when little energy is transferred between the atoms in the solid. Important practical cases include: (a) hydrogen atoms in interstitial sites between metal atoms with hydrogen atoms showing "local modes" effectively vibrating as if a cage and metal atoms displacements due to hydrogen motion being very small, and (b) heavy impurity atoms in a host of lighter atoms with heavier atoms producing "resonance modes" with pVDoS dominated by a peaked spectrum at low frequency [33]. In these cases, increasing the bond strength, *f*, increases the bond vibrational frequency ω_E, which leads to an increase of the width of NMD $\sigma^2 = \hbar M \omega_E/2$.

Having in mind some simple models linking DINS observables to VDoS, we can now try to tackle some more realistic cases. The strength of the bond between two atoms of the same element is strongly influenced by the size of the interacting orbitals. When two different atoms combine, an additional degree of freedom starts playing a role in bond strength, namely the degree of energy match between the valence electron orbital energies of the interacting orbitals [34]. As a consequence of this, there is a nonmonotonic relationship between the bond length of different diatomic molecules and their bond vibrational frequencies. The situation is even more complicated in molecules containing more than two atoms where one atom may be involved in bonds with more than one neighboring atom. Thus, the bond strength is usually measured in units of bond dissociation energy or bond energy for simple molecules. Bond-dissociation energy can be defined as the standard enthalpy change when a bond is cleaved by homolysis, usually defined at standard conditions of 298 K at atmospheric pressure [34]. Except for diatomic molecules, the bond-dissociation energy differs from the bond energy. While the bond-dissociation energy is the energy of a single chemical bond, bond energy is the average of all the bond-dissociation energies of the bonds in a molecule. A more general quantity, applicable to condensed systems no matter what their physical state is (liquids, glasses, amorphous systems, and crystals) as well as to gasses, is the standard enthalpy of formation [34], defined as the change of enthalpy during the formation of 1 mole of the compound from its constituent elements, with all substances in their standard states at 1 atmosphere. A plot of the hydrogen kinetic energy measured with DINS at standard conditions versus standard enthalpy of formation for a number of hydrogen-containing compounds is shown in Fig. 3. The compounds include LiH, an ionic system measured in polycrystalline form [35]; ZrH_2, a standard DINS calibration sample [36]; lithium imide, Li_2NH, a material of interest for hydrogen storage [23]; hydrogen fluoride, HF, [30]; water under ambient conditions [37]; the metastable phase of cesium hydrogen sulfate, $CsHSO_4$ [32]; sodium bifluoride, $NaHF_2$, a water-insoluble sodium source for use in oxygen-sensitive applications, such as metal production [38]; and the squaric acid, $C_4H_2O_4$, a carbon-based antiferroelectric material [39]. The values of the standard enthalpy of formation have been taken from a collection of standardized thermodynamic measurements [40].

In principle, the electron density of crystals contains all of the information required to complete a classification of bonding types in condensed matter systems. In this spirit, about 60 years ago, van Arkel [41] and Katelaar [42]

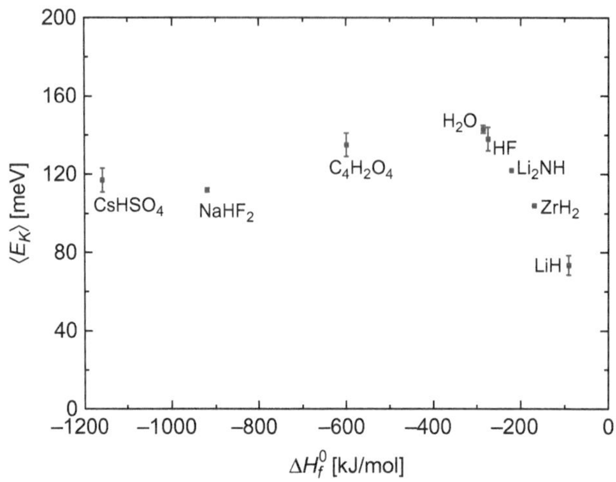

FIGURE 3 Hydrogen kinetic energy measured with DINS at standard condition versus standard enthalpy of formation for a number of hydrogenous compounds. See text for details.

proposed an outstanding classification of binary crystals along an ionic-covalent-metallic triangular diagram. The diagram was motivated by the realization that while there really is no such thing as a pure electrostatic bond, there is nonetheless a meaningful distinction between compounds labeled "ionic" and "covalent," based on how closely matched in energy the valence orbitals of the compound constituent atoms are [34]. Large difference results in more ionic compounds and a small difference results in more covalent bonding [34]. In the van Arkel-Katelaar diagram (also referred to as van Arkel-Katelaar triangle), for species composed of just two elements, X and Y, the difference between their orbital energies is plotted against the average of these energies, with the baseline indicating the average valence orbital energies of pure elements. Pure elements are arranged in order to decreasing orbital energies. Cesium is found at the extreme left, and fluorine at the extreme right. The compound with the greatest difference in the orbital energies of its constituent atoms is cesium fluoride, and this forms the uppermost point of the triangle. Metals in general tend to be found in the lower left-hand corner of the triangle, and compounds with more covalent bonding tend to be found in the lower right-hand corner. The top of the triangle contains the compounds with the greatest differences in orbital energies—the ionic compounds. The transitions between these different regions are seamless [34] (see Fig. 4 adopted from [34]).

In a modern description, the triangle is based on the relative electronegativity of their atomic components: the covalency-ionicity of the XY compound is provided by the difference in electronegativity, whereas the average electronegativity measures the metallicity-covalency character [43].

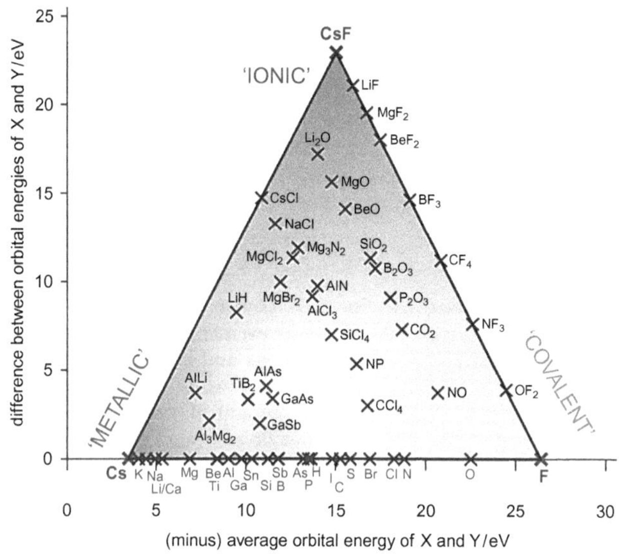

FIGURE 4 van Arkel-Katelaar diagram, see text for details. *Reprinted with permission from Ref. [34].*

The classical van Arkel-Katelaar diagram has been revisited many times [44–47], analyzing the role of different electronegativity scales, including non-binary compounds, or even adding a fourth vertex to the triangle to represent van der Waals bonding and other forms of molecular interactions. Analyzing the results shown in Fig. 3 in the spirit of the van Arkel-Katelaar diagram, one can see a trend showing up that the compounds exhibiting metallic-ionic character (LiH, ZrH_2, Li_2NH) lie on the steepest part of the plot, whereas the compounds with more covalent-ionic character (HF, H_2O, $CsHSO_4$, $NaHF_2$, $C_4H_2O_4$) form a plateau with weak dependence of the kinetic energy of the hydrogen on the standard enthalpy of formation of the compound.

7.2.4 Link to Materials Modeling and Theory

As it was introduced in Eq. (20), widths of NMDs can be calculated based on the knowledge of phonon dispersion curves. This approach can be referred to as the first step toward QM rather than a proper treatment of nuclei as quantum particles. In fact, the existence of a phonon quantum of energy $\hbar\omega_\lambda$ does not naturally imply the existence of a ZPE of the atom. However, in a harmonic potential the energy difference between the ground state and the one where all phonons have been excited, $\mathcal{E}_1 - \mathcal{E}_0 = \sum_\lambda \hbar\omega_\lambda$, can be though as twice the energy of the ground state, $\mathcal{E}_0 = \sum_\lambda \hbar\omega_\lambda/2 = (\mathcal{E}_1 - \mathcal{E}_0)/2$. On the other hand, the determination of NMDs in DINS experiments does not rely on such an assumption and interpretation and a value of the mean kinetic energy can be obtained whatever the complexity of the system under study or the possibility to calculate phonon dispersion curves. Consequently, DINS is the method of choice to either (1) benchmark new methodological developments in the theoretical description of condensed matter, including subtle quantum-mechanical effects; or (2) interpret and add value to experimental studies, particularly on complex systems.

NQEs have historically first been considered when generating free energy models [48]. In this approach, plane-wave expansion is employed to compute approximate canonical partition functions [49]. Using this approach, an approximate nuclear quantum-corrected formula for the NMD can also be obtained which leads to the definition of an effective temperature. That is, the approximate distribution $n(p)$ is still Gaussian in the momenta with the standard deviation $\sigma = \sqrt{Mk_B T_{M,eff}}$, but with an increased effective temperature $T_{M,\ eff}$ for each particle [48]:

$$T_{M,\ eff} \approx T + \frac{\beta^2 \hbar^2}{12 k_B M} \langle (\nabla_M U)^2 \rangle_{cl}, \tag{26}$$

where $\beta = 1/k_B T$, $\nabla_M U$ specifies differentiation with respect to the Cartesian coordinates of one of the particles of mass M, U is the total system potential energy, and the averaging utilizes classical mechanics. This effective temperature suggests that quantum effects can possibly be modeled by simply increasing the temperature in a way related to each particle's mass and the mean square forces on it during its classical motion [48].

If we insert the harmonic oscillator potential into the temperature correction formula (Eq. (26)), we get [48]

$$T_{M,\ eff} \approx T + \frac{\beta^2 \hbar^2 \omega^2}{12} T. \tag{27}$$

Using the vibrational physics tools described in previous sections one can say that the neglect of NQEs provides the lover conservative bound on the effective temperature and thus variance of NMD as being exclusively determined by the pVDoS calculated within harmonic approximation. Any excess of variance of momentum will thus come from the NQEs. At room temperature, $\beta^{-1} \approx 208\ \text{cm}^{-1}$ in typical units for vibrational frequencies, so the correction is $25(\tilde{\nu}/208)^2$ [48]. If we assume a frequency of 294 cm^{-1}, we get a temperature correction of 50 K. Thus the quantum effects on the structural properties of liquid water are roughly equivalent to a 50 K temperature rise in the classical liquid [48].

Apart from water, there exist a number of physical systems, in which quantum effects are important [48]. The prototype quantum liquid is liquid helium with its well-known exotic properties. This liquid requires a full quantum treatment and has been examined using Path Integral (PI) simulation [50]. A second highly quantum example is the solvated electron with its interaction with atomic cores in the solution modeled with a pseudopotential [51–53]. Liquid hydrogen is another example with significant quantum effects [54–56]. Liquids which exhibit quantum effects of smaller magnitude are methane [57, 58], neon [55, 57, 58], and even nitrogen [59].

The problem with the approach using the effective temperature (see Eqs. (26) and (27)) is that $T_{M,\, eff}$ depends on the oscillator frequency as well as on the physical target temperature. In the case of a multidimensional oscillator (that can be taken to be a decent model for a solid at low temperature) one would need to associate a different temperature to each normal mode [60]. This difficulty is, *inter alia*, responsible for the use of PI formulation of QM as the approach to simulate the NQEs. The most widely used PI method is PIMD. Other PI methods include *Ring-Polymer* and *Centroid MD*, all of which assume the validity of the BOA to separate nuclear and electronic motions. The BOA is generally very satisfactory, except when the system evolves in an electronic excited state or for a few ultra-fast chemical reactions. In essence, in the PI formulation the description of the quantum problem for nuclear motions is achieved via the use of a number of replicas (or beads) which are connected to each other via harmonic springs in a sequential fashion, and these are propagated in *imaginary time*, *i.e.,* a time-independent quantum system is mapped onto a time-dependent classical one in the limit of an infinite number of beads. Two main reasons are responsible for the success of the PI approach to simulate the NQEs [48]. Firstly, this method gives an intuitive physical picture that adapts naturally to the classical limit: variational approaches and resulting approximate theories emerge quite easily. Secondly, the PI formulation of equilibrium statistical mechanics is well-suited for simulation. The analogy to classical polymer simulations is direct, meaning that the quantum simulation merely requires extending the number of degrees of freedom and employing either Monte Carlo or MD procedures to generate the equilibrium distributions. For this reason, the majority of finite-temperature condensed-phase quantum simulations of large systems have utilized PI methods due to the favorable scaling compared with other quantum methods.

Importantly for the discussion earlier, DINS is capable of measuring the excess effective temperature due to NQEs and thus proves to be a unique method for benchmarking *ab initio* methods going beyond popular MD approaches (such as PI) assuming classical atoms and using Newton's equations of motion to calculate their temporal behavior. In case of the PI approach, the existence of well-defined normal modes of vibration cannot be taken generally as a good starting point, and NMDs need to be calculated explicitly in order to obtain DINS observables for comparison with experimental data. PI allows for a numerically exact evaluation of NMD from the density matrix $\rho(\mathbf{r},\mathbf{r}') = \langle \mathbf{r}|\exp(-H/k_BT)|\mathbf{r}'\rangle$. NMD is then given by the following orientationally averaged Fourier relationship [61]

$$n(p) = \left\langle \frac{1}{2\pi\hbar^3} \int d\mathbf{r} d\mathbf{r}' \exp\left[\frac{i}{\hbar}\mathbf{p}\cdot(\mathbf{r}-\mathbf{r}')\right], \frac{\rho(\mathbf{r},\mathbf{r}')}{Z} \right\rangle_{\Omega}, \tag{28}$$

where Z is the partition function

$$Z = \int d\mathbf{r} \rho(\mathbf{r},\mathbf{r}). \tag{29}$$

Importantly, Z is an integral over diagonal terms, whereas $n(p)$ includes off-diagonal elements and, therefore, its evaluation is computationally costly. However, increasingly efficient methods continue to be developed to speed up these calculations [62–64] thereby facilitating the interpretation of NQEs such as tunneling, particle delocalization, and isotope effects [64, 65].

7.3 PRACTICE OF SPECTROSCOPY WITH EPITHERMAL NEUTRONS

7.3.1 Anatomy of a Measurement

Investigation of atomic quantum dynamics using epithermal neutrons aims at the measurement of observables directly related to the QM character of the atomic momenta and kinetic energies. As such, it complements measurements accessing properties related to average atomic positions and average structure. From the experimental viewpoint, it is necessary to know either the wavelength or energy of the neutrons being used. Neutron beams from pulsed neutron sources are rich in epithermal neutrons, and an easy and efficient way to analyze neutron energies is the Time of Flight (ToF) method [66]. Neutron velocities and energies are determined through a measurement of the neutron flight time over paths from the source to the detector. The spectra will then consist of neutron counts as a function of the ToF, t:

$$t = \frac{L_i}{v_i} + \frac{L_f}{v_f} = A\left(\frac{L_0}{E_i^{1/2}} + \frac{L_1}{E_f^{1/2}}\right), \tag{30}$$

where L_0 is the distance from source to sample, L_1 is the distance from sample to detector, v_i and v_f are the incident and scattered neutron velocities, E_i and E_f are the corresponding energies, respectively, and A is a constant equal to $72.3\ \text{eV}^{1/2}\mu\text{s m}^{-1}$. The energy transfer is defined as:

$$E = E_i - E_f, \tag{31}$$

and the magnitude of the momentum transfer by:

$$\hbar Q = B\left(E_i + E_f - 2\sqrt{E_i E_f}\cos\phi\right)^{1/2}, \tag{32}$$

with $B = \sqrt{2m_n}/\hbar = 21.97\ \text{Å}^{-1}\ \text{eV}^{-1/2}$, ϕ being the scattering angle, and m_n being the neutron mass.

In a scattering experiment with epithermal neutrons, the Impulse Approximation (IA) [66] is used to describe the energy transfer distribution of the scattered neutrons:

$$E = \frac{\hbar^2 Q^2}{2M} + \frac{\hbar \mathbf{Q} \cdot \mathbf{p}}{M}, \tag{33}$$

where M is the mass of the struck atom and $\mathbf{p}$ its momentum (in the laboratory frame) before the scattering. The scattered intensity is thus centered at an energy transfer corresponding to the recoil energy:

$$E_r = \frac{\hbar^2 Q^2}{2M}, \tag{34}$$

where E_r corresponds to the recoil energy of an initially stationary atom of mass M. For a collection of masses in a sample, the scattering will become distributed in recoil peaks, each one centered at the recoil energy of a specific mass. The broadening of a recoil peak is directly related to the NMD and mean kinetic energy, as described in the first volume of this series [66]: within the framework of the IA, the scattering function (dynamic structure factor) is given by

$$S(\mathbf{Q}, E) = \int d\mathbf{p}\; n(\mathbf{p})\; \delta\left(E - E_r - \frac{\hbar \mathbf{Q} \cdot \mathbf{p}}{M}\right), \tag{35}$$

where $n(\mathbf{p})$ is the NMD. The equation above shows that in the IA the dynamic structure factor depends on the mass of the scattering atom, and its shape is determined by the NMD. The second moment of the scattering function is given by $E_r^2 + 2E_r\langle p^2 \rangle / 3M$, where $\langle p^2 \rangle$ is the variance of the NMD [66]. The measurement therefore aims at probing isolated single-atom recoil excitations of the form Eq. (33) and to determine their line shape properties [22]. The experimental set up and measurement conditions must therefore be optimized to meet the following requirements, hereafter referred to as *DINS requirements*:

- The scattering intensity is described within the IA and does not contain contributions from collective and/or single-particle excitations other than the recoil excitation. This implies the necessary condition that energy and momentum transfers are much larger than the characteristic energy and inverse length scales of the system under study, as being roughly defined by the atomic binding energy and the inverse interatomic distance [5].
- The wave vector and energy transfers allow to achieve *Mass Selectivity*, that is recoil peaks from individual masses in the sample under study are well separated in the scattering spectra and can be individually analyzed.

The scattering spectrum will then correspond to an *Intensity* vs *mass* histogram, where the Intensity is usually measured as neutron count rate and the ordinate is measured as ToF, which is related, through Eqs. (30) and (31) to energy transfer, the latter being inversely proportional to the mass in the IA.

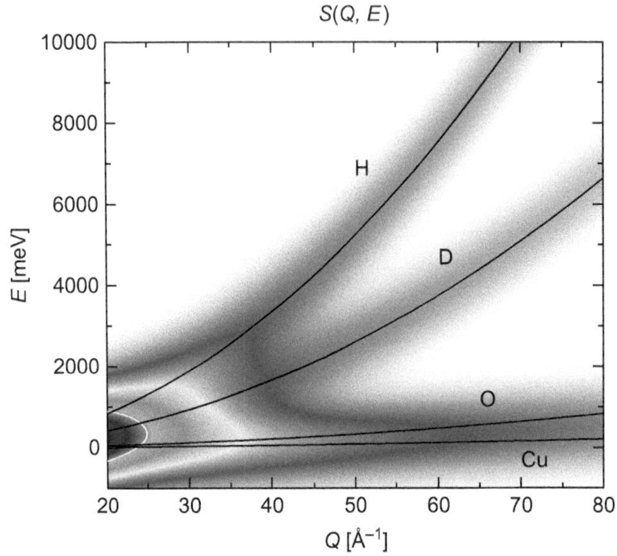

FIGURE 5 Calculated color map of the scattering intensity associated with atomic recoil excitations for hydrogen, deuterium, oxygen, and copper, as a function of energy and wave vector transfers using epithermal neutrons. The stationary atom recoil excitations are reported as continuous black lines. The broadenings due to the NMD are superimposed to these lines and are reported as color intensities.

7.3.1.1 Momentum Transfer, Atomic Recoil, and Mass Selectivity

A practical visual inspection of the expected scattering intensity of a measurement with epithermal neutrons can be obtained by considering a system composed of a collection of atomic masses, and calculating their recoil excitations and the expected broadening. Figure 5 reports a calculated color map of the scattering intensity associated with atomic recoil excitations for hydrogen, deuterium, oxygen, and copper, as a function of energy and wave vector transfers. The recoil excitations define parabolic trajectories in the Q, E space, whose positions depend on the atomic mass. The calculated broadening (one standard deviation), due to the NMD, of the stationary atomic recoil lines determines the color scale. The figure shows that assuming the DINS regime can be reached for wave vector transfers larger than approximately 30 Å^{-1}, as confirmed experimentally [7, 22], effective *Mass Selectivity* can be obtained for wave vector transfers in excess of 50 Å^{-1}. The selectivity decreases for adjacent intermediate and large masses, such as O and Cu. The DINS requirements outlined in the previous section therefore can be optimized by the extension to higher values of energy and wave vector transfers [67]. In the design of a DINS experiment with optimized requirements the neutron spectrometer configuration plays the main role in determining the accessed momentum transfer, accessed recoil trajectories, and mass selectivity.

We briefly recall here the two configurations which have been used for spectroscopy with epithermal neutrons:

1. Inverse geometry: In this configuration v_f is fixed by energy selectors or analyzers, making use of the relation $E_f = m_n v_f^2/2$. In this case the Q, E trajectories crossed by detectors placed at fixed angle positions intersect the recoil trajectories almost tangentially for hydrogen and light masses, with increasing intersection angles for intermediate and heavier masses, as shown in Fig. 6.
2. Direct geometry: In this configuration v_i is fixed by velocity or energy selectors, by making use of the relation $E_i = m_n v_i^2/2$. In this case the Q, E trajectories crossed by detectors placed at fixed angle positions intersect the recoil trajectories almost at right angles for intermediate and heavier masses, as shown in Fig. 6. Mass selectivity is less effective with respect to the indirect geometry case, due to detector trajectories restrictions.

It has to be noticed that, due to kinematic restrictions imposed by Eqs. (30), (31), (32), (34), the hydrogen recoil trajectory does not intersect detector trajectories for detectors placed at scattering angles above 90 degrees. Therefore, DINS signals from hydrogen are recorded at forward scattering angles only. As shown in the figure, the inverse geometry configuration allows an optimization of the DINS requirements, the reference spectrometer of this type being VESUVIO, operating at the ISIS Facility, Rutherford Appleton Laboratory, United Kingdom [67].

7.3.1.2 Basic Observables

The observable quantity in a DINS experiment is a collection of ToF spectra representing a neutron count rate as a function of the time elapsed from the source to the detector. The standard expression for the count rate as a function of t on an inverse-geometry spectrometer is [68]:

$$C(t) = 2\left(\frac{2}{m_n}\right)^{1/2} \frac{E_f^{3/2}}{L_f} I(E_f) D(E_f) N \frac{d^2\sigma}{d\Omega dE_f} \Delta\Omega, \tag{36}$$

where $I(E_i)dE_i$ is the number of incident neutrons per unit time with energies between E_i and $E_i + dE_i$ and $D(E_f)$ is the probability that a neutron of energy E_f is detected. N is the number of atoms illuminated by the incident beam, $\frac{d^2\sigma}{d\Omega dE_f}$ the double differential scattering cross section, and $\Delta\Omega$ the solid angle subtended by a given detector. Information on the atomic quantum dynamics in the system is contained in the double differential scattering cross section, which, for a single atomic species for unpolarized neutrons is related to the dynamic structure factor via [66, 68]

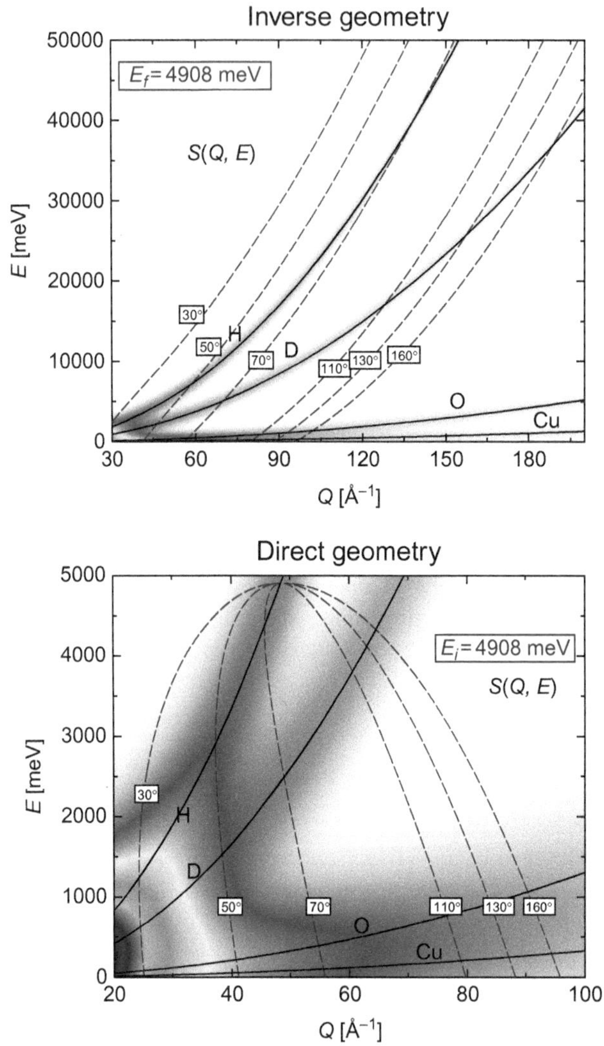

FIGURE 6 Calculated color map of the scattering intensity associated with atomic recoil excitations for hydrogen, deuterium, oxygen, and copper, as a function of energy and wave vector transfers using epithermal neutrons. The stationary atom recoil excitations are reported as continuous black lines. The broadenings due to the NMD are superimposed to these lines and are reported as color intensities. Detector trajectories for scattering angles between 30 degrees and 160 degrees, for inverse and direct geometry configurations, are reported as blue-dashed lines and correspond to $E_f = 4908$ meV and $E_i = 4908$ meV, respectively.

$$\frac{d^2\sigma(E_i, E_f, \phi)}{d\Omega dE_f} = \hbar^{-1}\sqrt{\frac{E_i}{E_f}}\frac{\sigma}{4\pi}S(\mathbf{Q}, E), \tag{37}$$

where σ_s is the total scattering cross section

$$\sigma_s = 4\pi\left[|b|^2 + \left(|b^2| - |b|^2\right)\right], \tag{38}$$

where b is the neutron-scattering length [66, 68]. The dependence of b upon the spin and mass of every isotope in the system brings to the definition of the coherent scattering cross section proportional to $|b|^2$ and the incoherent scattering cross section proportional to $|b^2|-|b|^2$.

For a sample-containing different atomic species of mass M the double differential cross section will be represented by the sum of individual contributions from each mass

$$\frac{d^2\sigma(E_i, E_f, \phi)}{d\Omega dE_f} = \hbar^{-1}\sqrt{\frac{E_i}{E_f}}\sum_M \frac{\sigma_s}{4\pi} S_M(\mathbf{Q}, E), \tag{39}$$

where

$$S_M(\mathbf{Q}, E) = \int d\mathbf{p_M}\ n(\mathbf{p_M})\ \delta\left(E - E_r - \frac{\hbar \mathbf{Q}\cdot \mathbf{p_m}}{M}\right). \tag{40}$$

The observed spectra are then composed of recoil peaks, separated in ToF as a function of the atomic mass, and broadened by the mass-dependent NMD.

A typical example showing scattering spectra from multicomponent samples with consistent mass separation is reported in Fig. 7. The sample was composed of three superimposed polyethylene, niobium, and aluminum slabs. The recoil peaks of hydrogen show a pronounced broadening due to the large quantum effects and the mass-dependent broadening described in Eq. (33). The backscattering spectra, due to the larger wave vector transfer accessed with respect to the forward scattering, allow the spectroscopic mass separation of the heavier Al and Nb signals.

7.3.1.3 From Momentum Space to Chemistry and Materials Science

The connection from the DINS observables to the quantities which directly describe the atomic quantum dynamics in the investigation of the chemical and binding environment in materials is carried out by considering how the wave vector and energy transfers, Q, E, are related to the atomic momentum of the species under study.

In the IA, the scattering variables, Q, E, are coupled by introducing the West variable, which is a measure of the atomic momentum component along the direction of $\mathbf{Q}$ [22]:

$$y = \frac{1}{\hbar}\mathbf{p}\cdot\hat{\mathbf{Q}} = \frac{M}{\hbar^2 Q}(E - E_r), \tag{41}$$

where y is the momentum $\mathbf{p}$ of the nucleus in the initial state projected onto the scattering vector $\mathbf{Q}$, and $\hat{\mathbf{Q}}$ is a unit vector in the direction of the momentum transfer.

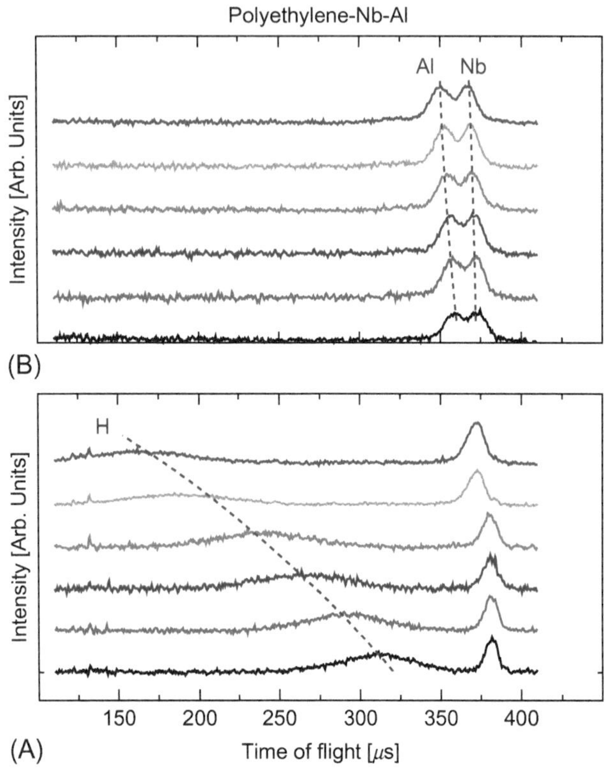

FIGURE 7 ToF spectra from a sample composed of three superimposed polyethylene, niobium, and aluminum slabs. (A) Spectra from a series of detectors at forward scattering angles, shifted upwards for clarity. The blue-dashed line marks the ToF positions of the hydrogen recoil line for each detector spectrum. (B) Spectra from a series of detectors at backscattering angles, shifted upwards for clarity. The blue-dashed lines mark the ToF positions of the aluminum and niobium recoil lines for each detector spectrum.

Therefore, the dynamic structure factor can be rewritten in terms of the new variable, in momentum space, via

$$S_{IA}(\mathbf{Q},E)=\frac{M}{\hbar Q}J_{IA}(y,\hat{\mathbf{Q}}). \tag{42}$$

The function $J_{IA}(y,\hat{\mathbf{Q}})$ is referred to as the *Neutron Compton Profile* (NCP):

$$J_{IA}(y,\hat{\mathbf{Q}})=\hbar\int n(\mathbf{p})\ \ \delta(\hbar y-\mathbf{p}\cdot\hat{\mathbf{Q}})\ d\mathbf{p}. \tag{43}$$

The NCP does not depend on the magnitude of **Q** and is a probability density distribution function. In other words, $J_{IA}(y,\hat{\mathbf{Q}})\,dy$ is the probability that an atom has a momentum of magnitude between $\hbar y$ and $\hbar(y+dy)$ along $\hat{\mathbf{Q}}$. As such, a DINS experiment gives access to observables directly related, through Eq. (42), to the NCP and thus to NMDs.

DINS is particularly suited to the study of disordered materials, as it is not affected by spectral broadenings or suppression of spectral features due to temperature effects or disorder. In the case of isotropic samples, the NCP can be averaged over the directions of $\hat{\mathbf{Q}}$ to yield:

$$J_{IA}(y) = 2\pi\hbar \int_{|\hbar y|}^{\infty} p\, n(p)\, dp. \tag{44}$$

The latter quantity may be formally inverted to derive the NMD via [69]:

$$n(p) = -\frac{1}{2\pi\hbar^3 y}\frac{d}{dy} J_{IA}(y)|_{\hbar y = p}. \tag{45}$$

The mean kinetic energy can be derived from the second moment of the NCP:

$$\langle E_K \rangle = \frac{3\hbar^2}{2M} \int_{-\infty}^{\infty} y^2 J_{IA}(y) dy = \frac{3\hbar^2}{2M} \sigma^2, \tag{46}$$

where σ is the standard deviation of the NCP and of the underlying NMD.

7.3.2 VESUVIO Spectrometer

VESUVIO is a neutron spectrometer specifically designed to measure NMDs in materials using epithermal neutrons. The spectrometer, in its current configuration, has been in operation since 2005 at the ISIS pulsed neutron and muon source, Rutherford Appleton Laboratory, United Kingdom [67, 70]. VESUVIO makes use of a polychromatic incident neutron beam characterized by an energy spectrum $I(E_i)$, peaked at about 0.03 eV and with an $E_i^{-0.9}$ tail in the epithermal region.

Incident neutrons with initial energy E_i travel over a primary flight path L_i from the moderator to the sample. After scattering at an angle ϕ, neutrons with final energy E_f travel over a secondary flight path L_f to the detector and are counted [67, 70]. Energy selection on the instrument is performed using an Au resonance filter placed between the sample and detector.

The energy selection and neutron detection on VESUVIO is operated in two modes. In the first mode, referred to as Resonance-Filter (RF), and used for the backscattering detector array, an analyzer foil, placed between sample and detector, strongly absorbs neutrons within narrow energy intervals around nuclear resonances, thus selecting the energy of the scattered neutrons, *i.e.*, $E_f \sim 4.9 \pm 0.15$ eV for the Au foil currently in use. The analyzer foil is cycled in and out of the scattered neutron beam and two measurements are taken, one with the foil between sample and detector (foil-in) and the other with the foil removed (foil-out). DINS ToF spectra then correspond to the difference between the two measurements, which are collected by ^{6}Li-doped glass neutron scintillators. The second mode, referred to as Resonance-Detector (RD) configuration, corresponds to a simultaneous energy analysis and detection. The latter is used forward scattering geometry, and here the foil is employed to define the energy of the scattered neutrons via radiative resonant neutron absorption; the prompt γ cascade following the (n, γ) reaction is recorded by YAP gamma scintillators, which measure the time of arrival of the γ cascade [71–82]. The time stamp of the gamma count coincides with neutron absorption by the foil within ~ps uncertainty, allowing the determination of the initial neutron energy E_i through the ToF technique. Extended descriptions of the RF and RD, together with refined methods to improve the resolution for the energy analysis are described in the reviews reported in References [7, 22, 83].

A schematic layout of the instrument is reported in Fig. 8, and the main instrument parameters are reported in Table 1.

7.3.2.1 Accelerator-Driven Neutron Sources

The practice of spectroscopy with epithermal neutron poses a series of requirements on the neutron source. On the one hand, the incident spectrum must be rich in epithermal neutrons. On the other hand, the ToF method is necessarily used to allow efficient energy analysis and detection. Therefore, the neutron pulse width for epithermal neutrons must be short in order to guarantee a timing resolution to fulfill mass separation and NMD capabilities in DINS experiments. Bearing in mind the energy and velocity ranges of epithermal neutrons, and with reference to Fig. 7, any epithermal neutrons produced and detected are in the microseconds range [7, 22, 66, 68, 83]. This condition

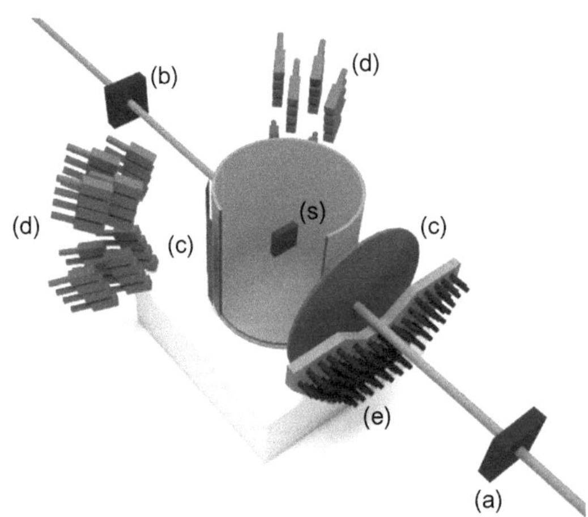

FIGURE 8 Schematic layout of the VESUVIO spectrometer: (a) incident neutron monitor; (b) transmission neutron monitor; (c) Au resonance filters; (d) forward scattering detectors; (e) backward scattering detectors; and (s) sample.

TABLE 1 VESUVIO Instrument Parameters

Moderator	Water at 300 K
L_i	11 m
L_f (forward scattering)	0.60 m (average)
L_f (backscattering)	0.55 m (average)
E_f	Natural Au resonant foils ($E_f \simeq 4.9$ eV)
Forward scattering detectors	64 YAP gamma scintillators
Backscattering detectors	132 ^{6}Li neutron glass scintillators
Neutron monitor	^{6}Li neutron glass scintillator
Incident neutron monitor	8.6 m from moderator
Transmission neutron monitor	13.4 m from moderator

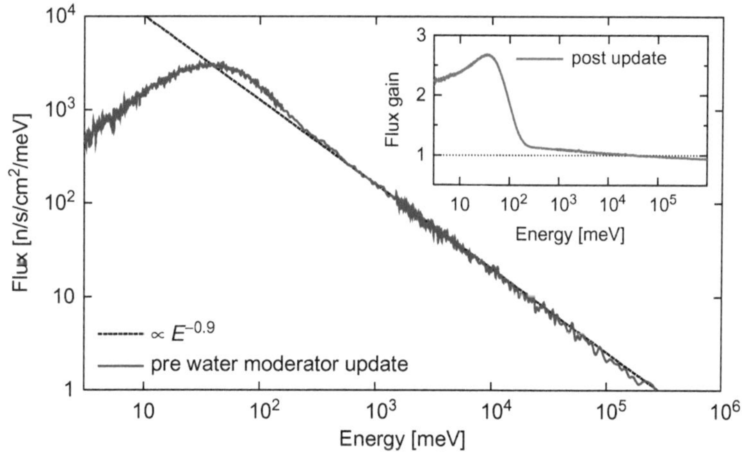

FIGURE 9 Absolute neutron flux on VESUVIO at sample position, as a function of energy. The black-dashed line corresponds to a fit of the experimental data to a power-law of the form $E^{-0.9}$. The inset shows the flux gain as a function of energy following a recent upgrade of the water moderator.

is fulfilled at short-pulse accelerator-driven spallation sources such as ISIS, where, for example, using a water moderator at ambient temperature, the mean emission time for epithermal neutrons is $\sim$1 μs [9, 84]. The use of accelerator-driven neutron sources allows to obtain incident neutron spectra that are rich in epithermal neutrons. Figure 9 reports the absolute flux at the sample position on VESUVIO. A recent upgrade of the water moderator at

TABLE 2 List of the Main Neutron-Converting Nuclear Reactions Used for Epithermal Neutron Detection. Gamma-Ray Producing Reactions are Exemplified by the Resonant Radiative Absorption from Gold

	Reaction	
$n + {}^{3}He$	$\rightarrow$	${}^{3}H + {}^{1}H + 0.764$ MeV
$n + {}^{6}Li$	$\rightarrow$	${}^{4}He + {}^{3}H + 4.80$ MeV
$n + {}^{197}Au$	$\rightarrow$	$Au^{*} \rightarrow \gamma$-rays

ISIS Target Station I has improved the flux at thermal energies, opening up new opportunities for enlarging the scope of the instrument capabilities, through the simultaneous acquisition of DINS, diffraction, resonant capture, and transmission data over several orders of magnitude in neutron energy.

7.3.2.2 Detection and Spectral Analysis of Epithermal Neutrons

The detection of epithermal neutrons is based on neutron-converting nuclear reactions, where a neutron is absorbed and charged particles or gamma rays are produced and then detected by the energy they release in materials.

The main nuclear reactions used for epithermal neutron detection are listed in Table 2.

Once the charged particles or gamma rays have been produced by one of the nuclear reactions listed in Table 2, they can be detected with gaseous, scintillation, or solid-state detectors.

The charged-particle-producing reaction cross sections vary inversely with the neutron velocity, while the gamma-producing cross section, exemplified by the resonant radiative absorption from gold in Table 2, shows intense resonances at eV energies, thus increasing the detection efficiency up to $\simeq$100 eV [7, 71, 83].

The direct spectroscopic measurement of the neutron energy in the epithermal range, based on these reactions, is not possible, because the kinetic information on the neutron cannot be distinguished within the large amount of energy released in the reaction. It is thus necessary to select or analyze the neutron energy before counting. The use of resonant absorption filters showing intense resonance absorptions in the eV range allow to analyze the neutron energy in the spectrometer's inverse geometry configuration. As shown in Section 7.3.2 this is accomplished either by taking difference spectra from ToF measurements with and without the analyzer foil or by measuring the prompt γ cascade following the (n, γ) reaction, as a function of the neutron ToF [7, 22, 66, 68, 83].

7.3.2.3 Strategies for Data Collection and Analysis

Data collection strategies for DINS are driven primarily from the need to optimize counting times, in order to have a desired accuracy in the determination of the NMD and its parameters, such as its variance or non-Gaussian components. The strategies are broadly divided into collection procedures for hydrogen-containing samples, and for samples that do not contain hydrogen. This is due to the kinematic restrictions in the DINS recoil scattering from hydrogen, which is confined to forward scattering and to the large scattering cross section of hydrogen compared with other elements. Thanks to the mass separation capabilities shown in the previous sections, data collection for light, intermediate, and heavy masses is carried out using the backscattering banks. As shown previously the analysis and detection technologies for forward and backward banks are different, therefore the optimization of counting times is broadly split into two strategies. As a reference, within the current configuration, samples containing $\sim$10 mmol of hydrogen require counting times of the order of 12 hours to obtain measurements of the variance of the NMD with $\lesssim$4% uncertainty. For backscattering the counting times depend on the specific sample composition, but as a general rule they are doubled with respect to the case of hydrogen.

The analysis of experimental recoil data aims at accounting for the Doppler broadening of a recoil peak by relating it to the underlying NMD of the target atom. In practice, DINS data analysis requires a quantitative comparison of NMDs obtained from data at different scattering angles. $n(\mathbf{p})$ is eventually calculated using one of the models introduced in the previous sections and model parameters are optimized using data-fitting protocols.

For a sample containing different atoms of mass M, and introducing a mass-dependent resolution function $R_M(t)$, one can write:

$$C(t) = \frac{E_i I(E_i)}{Q} \sum_M A_M M J_M(y_M) \otimes R_M(t), \tag{47}$$

where J_M is the function describing the NCP $J_{IA}(y_M)$ of the atom with mass M and

$$A_M = \frac{2}{L_i} D(E_f) \sqrt{\frac{2E_i}{m_n}} \Delta\Omega N_M b_M^2, \tag{48}$$

with b_M^2 related to the "free" total scattering cross section via the expression $\sigma_M^2 = 4\pi b_M^2 / [1 + \frac{M}{m_n}]^2$ [66]. Corrections accounting for the small deviations from the IA for the finite Q values of the measurements are routinely incorporated into the data treatment using the procedures detailed thoroughly in the first volume of the series and in previous reviews [22, 66]

In order to derive the NMD for a particular mass, it is necessary to subtract all unwanted contributions to the experimental raw data. Such contributions include multiple scattering, the scattering response from species other than the mass of interest, and γ-backgrounds. Multiple scattering is calculated for each scattering angle using a Monte-Carlo procedure described in Ref. [85] and then subtracted from the data. The unwanted scattering contributions from other atoms in the sample are subtracted by fitting the data and subtracting the fits to masses other than that of interest. Once all unwanted components have been removed, the data consist of a series of ToF spectra containing only the recoil peaks of the masses of interest. These spectra are simultaneously fitted to Eq. (47) convolved with the appropriate instrument resolution function. In an isotropic sample, there are typically a total of 64 spectra fitted simultaneously, whereas with a single-crystal sample as many as 300 spectra comprising several crystal orientations relative to the incident beam may be fitted simultaneously [86].

Reference [87] describes recent software developments aimed at providing a unified framework for the analysis of DINS data. In particular, these include the use of Bayesian inference methods [88], to assess the information content of oftentimes sparse data sets, particularly those exhibiting intrinsic mass selectivity. This method is applied Ref. [89] to the analysis of experimental DINS data for an effective model selection from a set of possible NMD "priors."

7.4 WHETTING YOUR APPETITE FOR MORE: CASE STUDIES

Some recent studies using DINS are presented below so to illustrate the diversity of target systems and physical phenomena amenable to investigation using DINS.

7.4.1 From Order to Disorder in Water

The ability of water to create a three-dimensional network of hydrogen bond (HB) makes this molecule extremely versatile. As the structure and intensity of the network can finely change, water helped life in its evolution and still plays an important role in many phenomena in fundamental and applied science. As hydrogen is the central piece of this puzzle, NQEs in water play a critical role up to ambient temperature and are known to be relevant in charge-transfer mechanisms and acid-base chemistry. Theoretically, many models and computational approaches encounter difficulties in reproducing the large librational excursus that liquid water can undergo without breaking the HB network. A key concept related to this feature is that H is not localized in a point, therefore along a single orientation of the HB, but is delocalized as measured in DINS experiments, therefore for each HB, a distribution of orientations depending on the shape of the local potential would best suit the description. A recent example of joint study based on DINS measurements and PIMD simulations [90] is reported as an example below of how fully quantum simulations are needed in order to explain NQEs.

Experimental data on heavy water and ice, D_2O, across the melting point and at room temperature were obtained at VESUVIO spectrometer. As a first qualitative remark, the values of kinetic energy $\langle E_K \rangle$ of both the D and O atoms were found to be well above the MB value of $\frac{3}{2}k_BT \sim 38$ meV at $T = 300$ K. This first result calls for Eq. (24) and the introduction of an effective temperature to explain the broadening in a Gaussian line shape. However, deviations from a Gaussian function could be found in the collected data that were modeled through an anisotropic NMD of the form [90–92]

$$n(p) = \left\langle \frac{1}{\sqrt{8\pi^3}\sigma_z\sigma_x\sigma_y} \exp\left(-\frac{p_z^2}{2\sigma_z^2} - \frac{p_x^2}{2\sigma_x^2} - \frac{p_y^2}{2\sigma_y^2} \right) \right\rangle_\Omega, \tag{49}$$

where the three parameters of the fit σ_α with $\alpha = x, y, z$ describe the anisotropy of the NMD and the confining potential in three orthogonal directions. We note that even if the anisotropic NMD has an explicit dependence on the momentum as a vector, the line shape obtained in a DINS experiment where the sample is disordered needs to be averaged over all possible orientation, therefore the formalism $\langle \ldots \rangle_\Omega$. It can be shown that anisotropy on an NCP appears as an oscillatory deviation from a Gaussian line shape [89] that can then be spotted on good quality data. For each value of σ_α, a contribution to the mean kinetic energy can be defined as

$$\langle E_K \rangle_\alpha = \frac{\hbar^2\sigma^2}{2M}, \qquad \sum \langle E_K \rangle_\alpha = \langle E_K \rangle. \tag{50}$$

Table 3 shows the results from the mentioned study with the values of the mean kinetic energy of D and O as well as the directional contributions obtained through the fit using Eq. (49). Results from both experiment and simulation are reported. In the case of both O and D, a marked anisotropy is found as $\sigma_z > \sigma_y > \sigma_x$ and a good degree of agreement is found between theory and experiment. In the case of the computer simulation, for water molecules changing in orientation with time in the liquid phase, the Transient Anisotropic Gaussian (TAG) approximation was used [63] using a running average of the kinetic energy estimator in a triangular averaging window of 100 fs. Otherwise, in the approximation where the principal axes of the kinetic energy tensor are in average aligned with respect to the geometry of the molecule, one can accordingly rotate the kinetic energy estimator using mean-square displacement (MSD). On the experimental side, the model Eq. (49) is based on the assumption that three axes exist where the correlation among the three motion along such axes are negligible. Therefore, one does not have an *a priori* orientation of the reference of frame with respect to the molecule. Common practice is to define x-axis in the direction perpendicular to the plane of the molecule and the z-axis along the stretching, even if a small rotation of the yz-plane can be expected as coupling between stretch and bend modes increases.

TABLE 3 Comparison Between Theoretical and Experimental Components of the Quantum Kinetic Energy for D and O in Heavy Water, at Different Temperatures. All Values are in meV, and the Theoretical Results have a Statistical Error Bar Smaller Than 0.1 meV. We also Report the Computed Center-of-Mass Mean Kinetic Energy $\langle E_{COM} \rangle$ of the D_2O Molecules. Reported Results are from Ref. [90]

	D [exp]	D [TAG/MSD]	O [exp]	O [TAG/MSD]
D_2O, $T = 300$ K, liquid			$\langle E_{COM} \rangle = 42.1$	
$\langle E_x \rangle$	20.1 ± 1.1	19.5/18.9	15.8 ± 1.7	13.6/13.7
$\langle E_y \rangle$	36.1 ± 2.3	26.1/25.6	19.5 ± 1.3	19.4/20.4
$\langle E_z \rangle$	55.1 ± 2.3	64.6/65.7	26.3 ± 1.5	23.4/22.3
$\langle E_K \rangle$	111.3 ± 3	110.2	61.6 ± 3.1	56.4
D_2O, $T = 280$ K, liquid			$\langle E_{COM} \rangle = 39.5$	
$\langle E_x \rangle$	18.8 ± 1.1	19.4/18.9	16.0 ± 2.3	13.6/13.7
$\langle E_y \rangle$	38.6 ± 2.5	25.7/25.2	21.0 ± 0.6	19.2/20.2
$\langle E_z \rangle$	54.2 ± 2.4	63.6/64.6	24.1 ± 2.1	23.2/22.2
$\langle E_K \rangle$	111.6 ± 2	108.7	61.1 ± 3.1	56.1
D_2O, $T = 274$ K, solid			$\langle E_{COM} \rangle = 39.2$	
$\langle E_x \rangle$	22.5 ± 1.8	20.1/19.8	16.1 ± 2.3	13.7/13.8
$\langle E_y \rangle$	37.4 ± 2.5	26.3/25.9	20.1 ± 1.6	19.0/19.9
$\langle E_z \rangle$	48.1 ± 3.4	61.9/62.4	24.2 ± 1.4	23.0/21.9
$\langle E_K \rangle$	108.0 ± 2	108.3	60.4 ± 4	55.7

The PIMD simulation predicted an opposite change in the x- and z-contributions to $\langle E_K \rangle$ that where qualitatively confirmed from the experimental data. The change of internal contributions leaving the overall value of the mean kinetic energy unchanged is known as Competing Quantum Effects and is a mechanism found beyond the water molecule case and in many systems with HB. This phenomenon was qualitatively explained in the case of H_2O across melting [93] by comparison to INS data. Figure 10 shows analyzed data in the case of liquid water and ice paying particular attention to the opposite shift in librational modes against stretching mode. As the HB gets stronger in the solid phase as opposed to the liquid, departure from free rotations is stronger and the librational frequency becomes larger as the HB networks hinders the orientation of the molecule in a smaller region

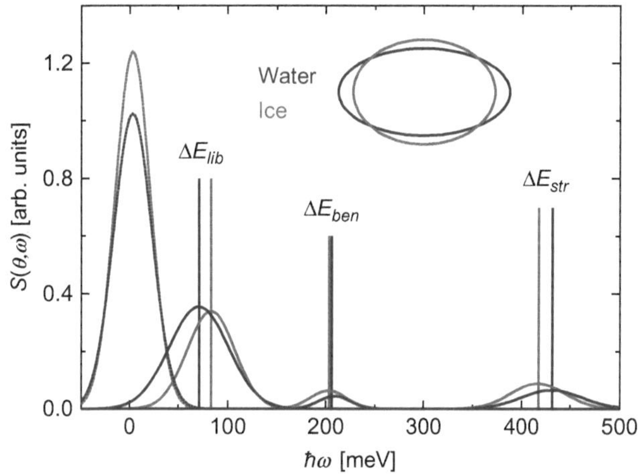

FIGURE 10 Fits on the INS data presented in Ref. [93] in the case of liquid water (blue) and ice (red) at a scattering angle $\theta \sim 20$ degrees. Shifts of the librational, bending, and stretching frequencies across the phase transition are highlighted by vertical lines.

and in a steeper potential. Oppositely, a stronger intermolecular OH interaction competes with the covalent bond within the molecule and lowers the stretch frequency and broadens the potential in the OH...O direction. In this respect it has to be stressed that X-ray Compton scattering applied to condensed phases of water provides complementary sensitivity, pointing to changes in the electronic mean kinetic energy due to intramolecular bond length changes, as the temperature is lowered below the triple point [94] or in confinement [95].

As the value of $\langle E_K \rangle$ can be in principle obtained from the density of states (neglecting anharmonic contributions on each mode), one can express the change in the kinetic energy as

$$\Delta\langle E_K\rangle = \frac{3\hbar}{4}\int_0^{\infty}\frac{\omega}{4}\Delta G_H(\omega)\,d\omega \sim \frac{\Delta E_{lib}+\Delta E_{str}}{4}, \tag{51}$$

where $\Delta E_{lib,str} = \hbar\Delta\omega_{lib,str}$ is the change in the vibrational energy for librations and stretching upon melting. This result only holds in the case of all the modes in the corresponding ground state, as the population factor has been omitted from the integrand, and only in the harmonic approximation.

7.4.2 Molecular Intercalation in Nanostructured Media

Spatial confinement of atoms and molecules can produce a modification of the nuclear wave-function. Changes in NMD and in the value of $\langle E_K \rangle$ occurring as the potential is changed with respect to bulk phases were measured by DINS at the VESUVIO spectrometer in the past. However, the number of experimental investigations on gas- and liquid-uptake using epithermal neutrons is still limited. As an example, the kinetic energy of He atoms adsorbed on activated carbon fibers was measured by Nemirovsky *et al.* [96] finding an increase of about 30% of He mean kinetic energy with respect to the bulk. The adsorption of water in confining media has attracted the largest attention, leading to investigations on water uptake into silica nanopores [97], carbon nanotubes [98], and graphene-based materials [99, 100].

We report here a recent study on molecular hydrogen adsorbed in KC_{24} [101], as a case study where the anisotropy of the hydrogen NMD was unveiled, shedding light upon the orientation of the molecule within the confining material. Moreover, concurrent use of neutron diffraction was crucial to monitor the H_2 uptake as well as the alignment of the ordered layered material invested by the neutron beam. Finally, as a successful comparison with predictions from a state-of-the-art plain wave (PW) Density Functional Theory (DFT).

Among layered media, KC_{24} is an example of nanoporous graphite intercalation compound (GIC) where 2 layers of graphite separate each layer of the intercalated gas. These materials, whose applications include batteries, supercapacitors, and superconductors [102], allow a readily intercalation of alkali metals, forming regular and tunable interlayer spacing structures along the c-axis. H_2 adsorption in GICs allows to highlight the interplay between chemisorption and physisorption as, for example, the breaking of the H–H bond has been observed in KC_8 [103]. KC_{24}, as well as other GICs with similar stoichiometry and Rb or Cs replacing K, reversibly adsorbs H_2 at ambient pressure up to $\sim 2H_2/K$ [104]. As neutron diffraction did not prove a favorable technique to study these systems [102], INS has been utilized for H_2 uptake by RbC_{24} [105] and CsC_{24} [106, 107]. Furthermore, the structure and potential energy landscape of the adsorption site and GIC structure upon gas loading in KC_{24} at low temperatures has been recently investigated using neutron diffraction, INS, and first-principles calculations [108].

In the reported work [101], DINS measurements on VESUVIO were conducted on $(H_2)_xKC_{24}$ for H_2 concentrations in the range $x = 0-2.5$. As VESUVIO allows the concurrent record of diffraction spectra during the DINS measurement, the (003) Bragg reflection at a d-spacing of ~ 2.9 Å was used to monitor the gas within KC_{24}. The increased distance between layers of $\sim 3\%$ [108] was easily detected by the backscattering detectors on the instrument, so that the single peak shown in Fig. 11 for the empty GIC ($x = 0$) is completely shifted to a larger d-spacing when two H_2 molecules have been intercalated per each K atom ($x = 2$). Moreover, in the intermediate case of one gas molecule per alkali metal ($x = 1$), the two Bragg peaks are equally populated, this being an additional check of the uptake levels during the gas-loading process. Diffraction data (not shown) also allowed the determination of the mosaic spread of $\sim 15°$ around the c-axis. The powerful capability of taking diffraction patterns during a spectroscopy experiment has already been used on other indirect geometry instruments at ISIS Facility,

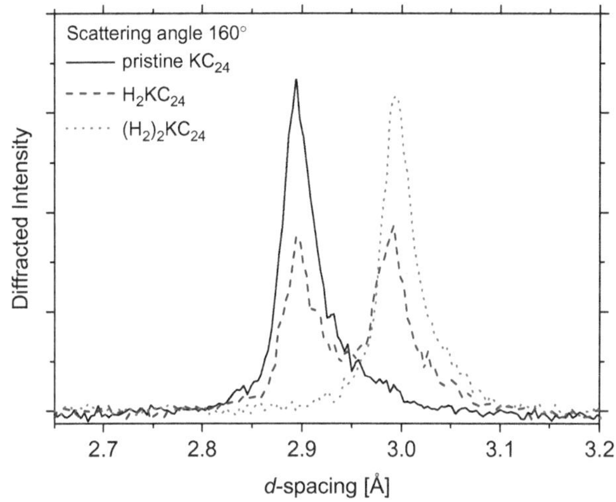

FIGURE 11 Neutron diffraction data at a fixed scattering angle ϕ as a function of H_2 coverage.

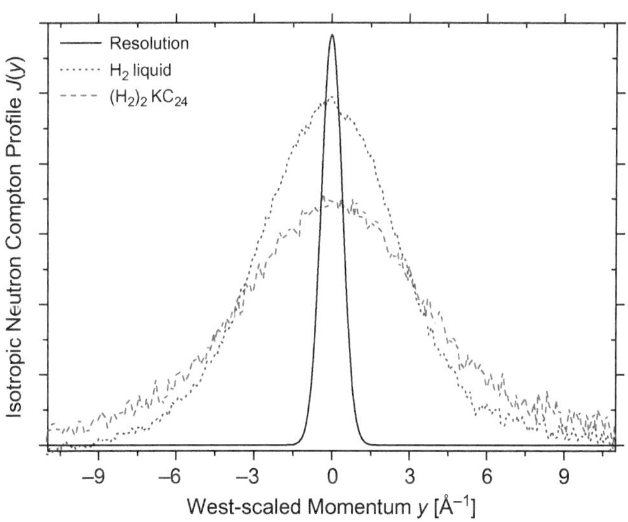

FIGURE 12 Isotropic hydrogen NCPs for $(H_2)_2KC_{24}$ at $T = 1.5$ K and liquid H_2 at $T = 20$ K. For the purposes of comparison, both profiles have been normalized to unity. The blue curve corresponds to the instrumental resolution rescaled by an arbitrary constant. The inset shows the dependence of the total scattered intensity on volumetric dosage.

namely IRIS [109] to study H_2 uptake in the same GIC [108], or OSIRIS [110] and TOSCA [111].

Fig. 12 shows the hydrogen NCP for adsorbed H_2 in the GIC compared with liquid bulk H_2 at 20 K and the experimental resolution of VESUVIO spectrometer. These NCPs correspond to the average over all sample orientations, ϕ, therefore are treated as isotropic NCPs. Qualitatively, the width of the NCP for confined hydrogen is markedly broader than that relative to the bulk, a clear indication of an enhanced interaction of H_2 with the surrounding environment. Indeed, the second moment σ_{iso} of the isotropic NCP, $J_{iso}(y)$, was found to be $\sigma_{iso} = 3.17 \pm 0.02$ and 3.89 ± 0.04 Å^{-1} for bulk and adsorbed hydrogen, respectively. Previously, Mayers [15] had obtained the value 3.34 Å^{-1} using a model for the rotationless (angular moment $J = 0$) bulk molecule, in good agreement with the value aforementioned. The model was based on the convolution of vibrational (*vib*) and translational (*trans*) modes, with the latter intermolecular interactions negligible against former contribution due to the large H–H frequency ~500 meV,

$$\sigma^2_{iso,\ bulk} = \frac{1}{3}\left(\sigma^2_{vib} + 3\sigma^2_{trans}\right), \tag{52}$$

where the contribution from each j-mode is defined at low temperature as $\sigma_j^2 = M\omega_j/\hbar$. In the case of H_2KC_{24}, the observed spectral broadening was related to H_2 librations $\hbar\omega_{lib}$ ~50 meV reported in the literature [108], as well as more energetic translational motions of the molecular adsorbate,

$$\sigma^2_{iso,\ adsorbed} = \frac{1}{3}\left[\sigma^2_{vib} + 2\sigma^2_{lib} + 3\sigma^2_{trans}\right]. \tag{53}$$

Moreover, an estimate of σ_{trans} ~1.6 Å^{-1} was retro-engineered from the measured value of σ_{iso}. This corresponds to a translational energy of ~40 meV, comparable to the simulated value of 27 meV for the vibrational modes of the K–H_2 complex previously obtained [108]. Therefore, translational motions are dominated by interactions between the H_2 molecule and potassium ions in the GIC. Also, the value of σ_{iso} was found constant for $x = 1.0$ and 2.0 within experimental uncertainties, and slightly lower for $x = 2.5$, because of molecules (in excess to $x = 2$) loaded into the sample without being intercalated and confined.

In the same work [101], neutron data were complemented with first-principles simulations using finite-displacement approach [112] on the optimized H_2KC_{14} unit cell. Average results for the atomic species are shown in Table 4, where directional contributions along the c-axis (z-direction) and in the perpendicular plane (x- to y-directions) are reported in addition to the isotropic value. For H atoms, the calculated isotropic value of 3.79 Å^{-1} compares very favorably to the experimental value of 3.89 ± 0.04 Å^{-1}. It can be noticed that how PW-DFT simulation predicts a preferential alignment of the H–H bond along the c-axis, therefore a marked anisotropy of the H NCP. Carbon atoms experience a cylindrical anisotropy reversed with respect to hydrogen, with the larger values of σ_x and σ_y within the graphite plane. The presence of the K atom causes only small differences within the values for different C atoms.

TABLE 4 Calculated NCP Widths Along the Three Crystallographic Directions, Where z is Parallel to the Crystallographic c-axis

Nucleus n	$\sigma_n(\hat{x})/\text{Å}^{-1}$	$\sigma_n(\hat{y})/\text{Å}^{-1}$	$\sigma_n(\hat{z})/\text{Å}^{-1}$	$\sigma_n/\text{Å}^{-1}$
H	2.29	2.13	5.78	3.79
C	13.74	13.74	8.23	12.18
K	1.30	3.24	4.66	3.36

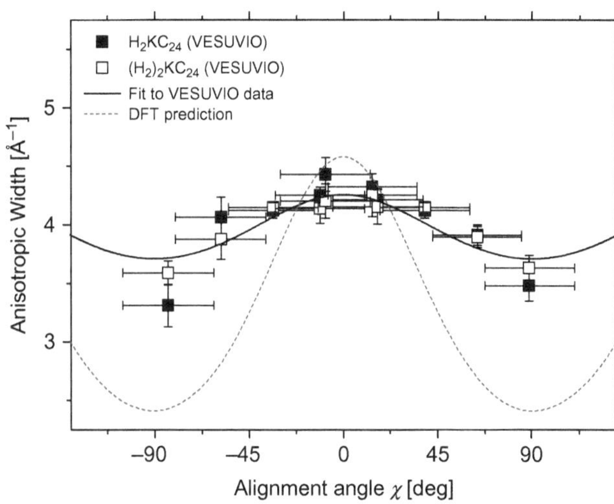

FIGURE 13 Dependence of hydrogen NCP widths on χ for coverages $x = 1$ and 2. For further details, see the text.

To highlight the anisotropy in the NMD, NCPs at several front-scattering angles where grouped so to obtain 10 possible orientations of the sample, spanning the range $\chi = \pm 90°$, where $\chi = 0$ corresponds to complete alignment of the scattering wave vector along the crystallographic c-axis. Figure 13 shows the dependence of the second central moment of the NCP upon the sample orientation, with an expected cylindrical symmetry about $\chi = 0°$. For a quantitative study of the anisotropy shown by the DINS data, the NCP was modeled using two standard deviations: a width along the c-axis, $\sigma_{par} = \sigma_z$, and a width in the perpendicular plane, $\sigma_{per} = \sigma_x = \sigma_y$. The fitting function then reads

$$J(y;\chi) = \frac{1}{\sqrt{8\pi^3 \sigma_{par}^2 \sigma_{per}^4}} \exp\left[-\frac{y^2}{2}\left[\frac{cos^2\chi}{\sigma_{par}^2} + \frac{sin^2\chi}{\sigma_{per}^2}\right]\right]. \tag{54}$$

Once the blurring imposed by the instrument angular resolution and mosaic spread was taken into account, fitted values were found to be $\sigma_{par} = 4.45 \pm 0.07$ Å^{-1} and $\sigma_{per} = 3.55 \pm 0.08$ Å^{-1}, with a corresponding value $\sigma_{iso} \sim 4$ Å^{-1} consistent with the one previously presented and obtained from the fit on $J_{iso}(y)$. The anisotropic prediction from the PW-DFT simulation is reported in Fig. 13 together with the width of the NCPs for the $x = 1, 2$ concentrations. Although a good qualitative agreement between experimental and simulated anisotropies was found, a larger difference between σ_{par} and σ_{per} in the simulation marked a more pronounced dependence on χ.

The anisotropy of the NCP was then related to the most salient features of the underlying H_2 librational wavefunction by expanding it as a series of spherical harmonics in the free-rotor base. Each state of given angular momentum J is associated with the energy $E_J = B_{rot}J(J + 1)$ with $B_{rot} = 7.35$ meV [113]. While a *para*-H_2 has a ground state described by only even-J terms, the interaction with the host matrix introduces a mix with excited states, with the mixture inversely proportional to the energy separation between levels. As for $J > 0$ the energy largely exceeds the one of the ground state, anisotropy of the wavefunction requires a strong orientational potential $V(\chi, \varphi)$ with energy barriers above 100 meV, as was found in the model introduced in Ref. [108]. These results showed how in the physisorption regime is not easy to fix a direction for the H_2 orientation, while binding mechanisms in the chemisorptive regime can explain the strong pinning to the host matrix, with examples of these interactions being those responsible for Kubas-like dihydrogen complexes [114].

In conclusion, it was shown how the effect of an underlying pinning mechanism for the H_2 molecule was measurable in a DINS experiment causing an anisotropic deformation of the NMD of hydrogen and made confirmed that this is a case of chemisorption as opposed to physisorption. Moreover, theoretical approaches and computer simulations need to incorporate NQEs beyond the harmonic and isotropic approximation from the outset.

7.4.3 Beyond Hydrogen: From Lithium to Dental Cements

7.4.3.1 Case of Lithium

NQEs attracted increasing interest in recent years in the field of technological materials. An important example is the role that ZPE effects play in the electron-phonon coupling at the base of superconducting materials [115]. Among the many materials of interest for technology, Lithium-containing materials are of interest in nuclear engineering, detector technology, and hydrogen storage.

Despite the small neutron-scattering cross section of Lithium, several experiments have recently shown how the measurement of its NMD is well within the reach of MAss-selective Neutron SpEctroscopy (MANSE) experiments on VESUVIO. Experiments were performed on pure Lithium [116],

TABLE 5 Values of Li mean Kinetic Energy $\langle E_K \rangle$ Measured by DINS and Reported in the Literature. Experiments were Performed at Ambient Pressure and Several Values of Temperature *T*. More Values from Ref. [116] are shown in Fig. 15

Sample	T [K]	$\langle E_K \rangle_{Li}$ [meV]
Li[a]	100	22.3
	300	42.4
LiF[b]	4	33.7
	100	33.4
	300	48.0
LiH[c]	20	28.1
	300	45.9
LiD[c]	20	28.2
	300	45.9

[a]*From Ref. [116].*
[b]*From Ref. [117].*
[c]*From Ref. [35].*

on Lithium Fluoride [117], and on Lithium Hydride and Deuteride [35]. Also, Lithium in Imide, Li_2NH, was measured and the results compared with quantum thermostat *ab initio* MD [23].

In different materials, the local potential affecting Li nuclei changes as the geometry and chemistry of the lattice change. Therefore, variations in the measured mean kinetic energy are expected and have been measured, as reported in Table 5 for several values of sample temperature. In particular, it is possible to notice that, at fixed temperature, $\langle E_K \rangle_{Li} < \langle E_K \rangle_{LiH} < \langle E_K \rangle_{LiF}$ as the Li nucleus becomes more confined spatially. Interestingly, no isotope effect was observed on the dynamics of Li upon the substitution H→D in Lithium Hydride/Deuteride.

Traditionally, NQEs and in particular ZPEs have been predicted using Debye models. In the Debye theory, the VDoS, $G(E)$, is considered to be quadratic in the phonon energy up to a cut-off E_D related to the Debye temperature Θ, $E_D = k_B\Theta$. Following this approach, an estimate for the mean kinetic energy can be obtained through the integral equation

$$\langle E_K \rangle = 3\int_0^{E_D} \frac{3E^2}{E_D^3}\frac{E}{4}\coth\left(\frac{E}{2k_BT}\right)dE. \tag{55}$$

At very low temperature, the Boltzmann factor equals unity for each phonon energy and one gets $\langle E_K \rangle = \frac{9}{16} E_D$. The Debye model presents several limitations. An example is the dependence of Θ from the thermodynamic temperature, leading to the definition of low-temperature and high-temperature values of Θ. Also, the Debye temperature is a global value of the system, while in order to obtain mass-related quantities such as the mean kinetic energy, a mass-projected density of states, $G_M(E)$, or a mass-related phenomenological quantity should be used.

The second of these limitations can be overcome when a mono atomic system is considered. For example, in a pure Lithium crystal $G(E) \equiv G_{Li}(E)$ by definition. Figure 14 shows the result of a DFT-based first-principles calculation obtained using the CASTEP [112] module in Materials Studio (Accelrys Inc.). Phonon spectra were obtained in Generalized Gradient Approximation (GGA), following the Perdew-Burke-Ernzerhof (PBE) approach, at a temperature $T = 10$ K and using the Finite Displacement method. In the same figure, the VDoS multiplied by the Maxwell-Boltzmann population factor is presented at $T = 100$ K and $T = 300$ K. As previously mentioned, one can estimate the nuclear mean kinetic energy as

$$\langle E_K \rangle = 3 \int_0^\infty G_M(E) \frac{E}{4} \coth\left(\frac{E}{2k_B T}\right) dE. \tag{56}$$

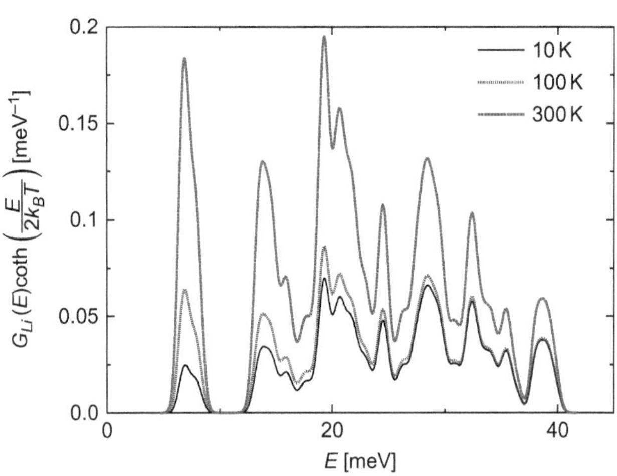

FIGURE 14 VDoS $G(E)$ for elemental Li crystal simulated with the CASTEP module in Materials Studio. The function is reported at the temperature of the simulation $T = 10$ K (black solid line) and multiplied for the Boltzmann population factor $\coth\left(\frac{E}{2k_B T}\right)$ so to reproduce the integrand needed to evaluate $\langle E_K \rangle$. This is reported for $T = 100$ K (blue-dotted line) and $T = 300$ K (red-solid line).

Looking at Fig. 14, one can appreciate how lower-energy modes contribute to the mean kinetic energy at higher temperatures, as opposed to the higher-energy modes equally contributing at all temperatures as the nucleus remains in the ground state and the population factor stays near 1. It is worth noticing that the prediction of the ZPE from both first-principles calculations and Debye model is only accidental as both approaches do not consider nuclei as delocalized quantum particles. However, as far as each phonon in the system is created in a harmonic potential, *i.e.*, as far as a collection of harmonic oscillators is considered, the energy E needed to bring the system from ground state to the first-excited state (therefore creating a phonon of the same energy E) is twice the energy of the ground state (the ZPE $E/2$). Also, this value is proportional to the shape of the local confining potential where the nucleus sits, numerically obtained in the Finite Displacement calculation. There is only one numerical value describing *ontologically* different observables.

Using Eq. (55), values of $\langle E_K \rangle$ from the Debye model in the temperature range $0 < T < 300$ K where obtained and compared with the experimental values from Table 5. As several values of the Debye temperature are found in the literature, shadowed regions are presented as opposed to solid lines. We used: $344 \leq \Theta \leq 448$ K for pure Li [118]; $708 \leq \Theta \leq 734$ K for LiF [119]; and $1030 \leq \Theta \leq 1190$ K for LiH/LiD [120]. Also, using Eq. (56) and the simulated spectrum of Fig. 15, an additional prediction for pure Li is presented as solid line and compared with the Maxwell-Boltzmann behavior for a gas of noninteracting particles, *i.e.*, $\langle E_K \rangle = \frac{3}{2} k_B T$.

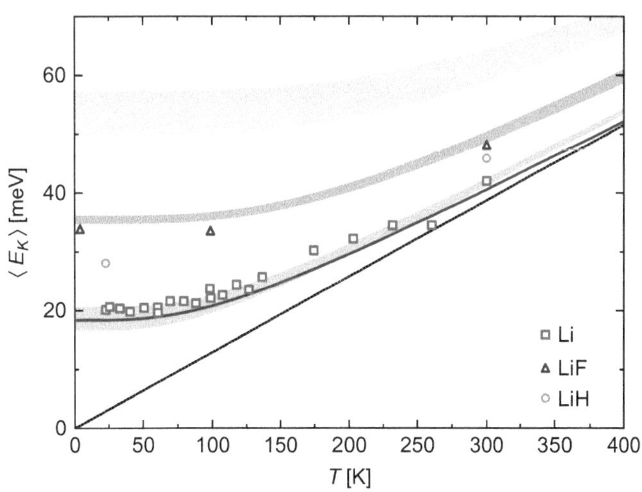

FIGURE 15 Mean kinetic energy of Li in elemental Li (red), LiH (green), and LiF (blue). Markers represent experimental data from DINS measurements at VESUVIO. Shadowed regions represent range for estimates from simple Debye model. Red solid line was obtained from the VDoS reported in Fig. 14. Dashed solid line is the Maxwell-Boltzmann prediction $\frac{3}{2} k_B T$.

In the case of pure Lithium, both predictions from Debye model and MD simulation are consistent with the experimental data or slightly underestimate the measured values. At about 400 K, all curves are confusing with the $\frac{3}{2}k_BT$ prediction and the ZPE is no longer the leading contribution to the mean kinetic energy. At any lower temperature, the Maxwell-Boltzmann prediction is a gross underestimate of the actual $\langle E_K \rangle$ as it does not take into account ZPEs.

Moreover, in the case of LiF the Debye model slightly overestimates the value of the mean kinetic energy. In Ref. [23], a proper Phonon calculation was performed with the partition of the phonon density of states in two contributions projected onto each atom, *i.e.*, $G(E) = G_{Li}(E) + G_F(E)$ and a successful agreement was obtained between simulation and experiment. The only slight overestimate from the Debye model precludes the hypothesis of a partition in atom-projected Debye density of states. The inequality

$$\sum_{a \in \text{atoms}} \langle E_K \rangle_a \neq 3 \int_0^{E_D} \frac{3E^2}{E_D^3} \frac{E}{4} \coth\left(\frac{E}{2k_BT}\right) dE \tag{57}$$

makes the Debye model an inefficient and unpredictable tool in the evaluation of mean kinetic energies.

The inadequacy of the Debye model for poly-atomic systems is much more evident in the case of LiH. First of all, the prediction at low temperatures is about double the experimental value. Second, different Debye temperatures for LiH and LiD do not correspond to isotopic effects on the Lithium $\langle E_K \rangle$. Finally, the larger Debye temperature of LiH as opposed to LiF is an opposite trend as compared to the mean kinetic energy, with a lower value of the former as opposed to the latter.

DFT-based first-principles calculations are nowadays the simplest and more common tool to properly predict NQEs and represents an important improvement with respect to the Debye model. On the other hand, they only work in the harmonic approximation, and curiously just accidentally, as observed above. The proper way is to treat nuclei as quantum particles through the PI formalism originally proposed by Feynmann [121] and more recently applied to nuclei in condensed systems [122].

7.4.3.2 Real-Life Materials

In parallel to the large development for fully quantum simulations of nuclei, recent examples of DINS experiments showed qualitatively interesting results for system far too complex to be reproduced in a computer.

In Ref. [123], DINS was applied to Bioactive glass ionomer cements during setting. Masses heavier than hydrogen showed a temporal evolution in the width of neutron Compton profile that has been related to nonmonotonic advancement in atom cohesion and fracture toughness during the setting of the cement. The system studied in this experiment was essentially different from stoichiometric-controlled standard experiments on VESUVIO, where optimal results in the data analysis are obtained when ratios of relative intensities from each nuclear species are known. Nevertheless, a qualitative analysis of the change in the shape of the partially overlapping peaks can still be analyzed.

In cases where it is not possible to define the stoichiometry of the sample, therefore making it difficult to analyze each mass separately, one can experimentally define the average mass of the system and perform a line shape analysis on the corresponding neutron Compton profile. This can facilitate the qualitative analysis of a time evolution, as in the above-mentioned example.

As a special case of Compton scattering, DINS spectra can be expressed as functions of the probe (neutron) wavelength. In particular, defining the relative change in wavelength

$$\frac{\Delta\lambda_n}{\lambda_n} = \frac{\lambda_f - \lambda_i}{\lambda_f} \tag{58}$$

with $\lambda_n = \lambda_i$, λ_f are the neutron wavelengths before and after the scattering process, respectively, DINS peaks for each mass M are found along the recoil lines

$$\frac{\lambda' - \lambda}{\lambda'} = \frac{\frac{M}{m_n} + 1 - \cos\theta - \frac{M}{m_n}\sqrt{1 - \frac{m_n{}^2}{M^2}\sin^2\theta}}{\frac{M}{m_n} + 1} \tag{59}$$

$$\simeq \frac{1 - \cos\theta - \frac{m_n}{2M}\sin^2\theta}{\frac{M}{m_n} + 1} \tag{60}$$

$$\simeq \frac{m_n}{M}(1 - \cos\theta), \tag{61}$$

with m_n the neutron mass. Eq. (59) is an exact definition of the Compton shift, while Eq. (60) holds whenever $\frac{m_n{}^2}{M^2}\sin^2\theta \ll 1$ and Eq. (61) in the less-general approximation of $M \gg m_n$. En passant, it is worth noticing how the introduced formalism is completely equivalent to what Compton presented last century in dealing with X-rays scattering on electrons [124]:

$$\text{X}-\text{ray}: \frac{\Delta\lambda_\gamma}{\lambda_\gamma} = \frac{h}{m_e c \lambda_\gamma}(1 - \cos\theta) = \frac{E_\gamma}{m_e c^2}(1 - \cos\theta) = \frac{m_\gamma}{m_e}(1 - \cos\theta), \tag{62}$$

where λ_γ is the photon wavelength, $E_\gamma = hc/\lambda_\gamma$ is the photon energy, and $m_\gamma = E_\gamma/c^2$ is the effective photon mass obtained through the Einstein equation linking energy and mass of a particle. For these reasons, DINS is often referred to as Neutron Compton Scattering (NCS).

The mean value of $\Delta\lambda_n/\lambda_n$ weighted by the experimental intensity can be related to an effective mass M^* of the sample whenever mass separation is not completely fulfilled and spectra are composed of only one peak. The experimental intensity must be corrected by incident flux spectrum and other instrument-dependent factors. Moreover, one should keep in mind that contributions from each atom are weighted by specific concentrations and neutron cross sections.

However, the basic principle suggested by Tian *et al.* [123] is to qualitatively relate the change in the width of the peak from each mass to the change in its local potential due to strengthening or weakening of chemical bonds with the surrounding environment.

The difficulty in the analysis of experimental data is matched by the complexity in the simulation of these systems. As the number of atoms increases, the material is amorphous, nonhomogeneous, and disordered, it is not imaginable nowadays to handle such materials through DFT approaches. Yet, it has been suggested how the broadening of the NCP for the effective mass M^* can still be related to macroscopic mechanical properties such as the fracture toughness. This is a promising approach to handle complex systems of relevance for functional materials.

7.5 PERSPECTIVES AND OUTLOOK

In the following remarks, we highlight here the main directions toward ongoing and future developments, strengthening the potential of epithermal neutron instrumentation to expand its capabilities in materials research and directly support the development of new modeling tools, with special emphasis on disordered systems.

7.5.1 Current Capabilities and Beyond

As a general property of inverted-geometry spectrometers, neutrons of all energies reach the sample and can be scattered or adsorbed. Among the inverted-geometry spectrometers, VESUVIO-like instruments oppose to scattered neutrons a *noninvasive filter*, meaning that scattered neutrons of all energies, as well as other particles as γ-rays produced within the sample, can reach the detectors and can be counted. This is in opposition to instruments where the scattered neutrons are selected by intermediate reflections on the filter. In these cases, the secondary path is not a straight line connecting the sample and the detector, with the latter actually shielded against direct radiation from the sample. This unique set-up makes it possible to enrich DINS measurements with a variety of ancillary characterizations such as diffraction, transmission, and Dopplerimetry measurements based on (n, γ) nuclear resonances.

The possibility to collect simultaneous diffraction spectra along with DINS was described in the case study of Sec. 7.4.2. Such a capability allows to detect *in situ* phase transitions in the sample [125], orientation of layered media and adsorption levels [101]. Owing to recent updates at the TS1 water moderator at ISIS, neutron flux in the thermal neutrons region was increased of more than a factor 2, making diffraction measurements faster than ever without compromising the good resolution of the instrument. Such a possibility makes it easier to perform parametric studies of the same sample across phase transitions as the temperature is changed. To this end, it was recently shown that from the difference of two measurements of the same sample at different temperature, it is possible to access values of the mean kinetic energies without any corrections to the data [89].

Transmission measurements are important *per se*, allowing the determination of the total cross section of materials in a broad energy range [126]. Nonetheless, the collection of the transmitted spectrum facilitates the correction procedure in DINS experiments as well as many other measurements based on neutron scattering. In such a large energy-range available, it is possible to look for heavy-metals tracks as the adsorption of epithermal neutrons causes sharp dips in the transmission spectrum at nuclide-specific energies.

Front-scattering YAP detectors on VESUVIO spectrometer have been said to detect γ-rays emitted from the gold foil as a way to fix the neutron final energy. However, photons produced within the sample as a consequence of a nuclear (n, γ) reaction are detected *at the same time* by all front-scattering detectors. When the energy of the resonance is lower than tens of eV, a well-resolved peak appears whose width can be related to NQEs. For some materials as In, Ag, or Ta, the possibility to investigate NQEs on heavy masses with a much higher statistics and much better resolution than in the case of a standard DINS measurement of the related NCP.

In Sections 7.3 and 7.4, we discussed in some length the separation of the spectral responses coming from different masses. In Refs. [127–129], it was noted how Al sample containers and other pieces of equipment produce an NCP at the Al mass position (27 amu $\rightarrow$ about 340 μs on a ToF spectrum) that partially overlaps with sample-related and more interesting peaks, *e.g.*, from C, N, or O. A solution to this problem could be the use of heavier elements instead of Al in order shift the background peak to higher ToF. The second strategy is to lower the backgrounds, for example, reducing the amount of atoms in a sample-container reducing the thickness of its walls or choosing elements with small neutron-scattering cross sections. On the same line, one should note the overlap of signals from atoms with the same mass but located in different chemical environment within a given system. Selective deuteration in the case of DINS experiments is an elegant solution to this problem. Deuterium atoms replacing H of *a structural-chemical type* are measured at the same time as non-replaced hydrogen atoms in a DINS measurement, but at a different position and as a well-separated peak (mass separation between

FIGURE 16 Pictorial representation of the evolution of a VESUVIO-like instrument.

H and D is the largest achievable by this technique). Such a measurement can then be complemented with the one where H/D roles are switched. Moreover, as H, D, and T are all separated within the same spectrum, level uptakes and contamination of technological materials are straightforward measurements using DINS.

An increase of the angular coverage on a VESUVIO-like instrument would decrease drastically the data acquisition times. As it is naively pictured in Fig. 16, an increased angular coverage can be accompanied by the definition of fixed-scattering-angle Debye-Scherrer cones. Along these cones, raw spectra can be directly summed together to improve the data quality as each NCP within the spectra will share the same position for all detectors in the ring. If such geometry is completed by resonant foils moving along the same cones, γ-background is *completely* canceled out during the cycling process, as opposed to the correction that is now needed for front-scattering detectors on VESUVIO.

Although only a small number of DINS studies of single crystals can be found in the literature [130], we should note that layered materials have been used several times to experimentally access cylindrical NMDs [101, 131] providing important information about the way hydrogen-containing molecules are adsorbed therein.

The way data are analyzed and corrected has been largely simplified and made faster thanks to the new routines developed within the MANTID software [87, 132]. These allow an easy visualization of data as function of ToF, y_M, or Compton-shift variables in the case of NCPs, d-spacing in the case of diffraction spectra, incident energy E_i in the case of transmission, as well as the possibility to produce two-dimensional plots when scattering angle or momentum transfer are coupled to ToF and energy transfer for each detector. Dedicated user scripts also allow a straightforward data reduction to obtain values for σ_M^2.

7.5.2 Tackling Disordered Systems

Much of the world around us is not ordered. As DINS observables are related to single-particle dynamics, that is to the local potential on a scale smaller than interatomic distances, disordered systems such as glasses and liquids exhibit NMDs in the very same way as crystalline materials. This can be regarded as a fundamental advantage of DINS that has been only partially exploited so far.

As far as disordered systems are concerned, another neutron-scattering technique comes to mind, namely the total scattering technique introduced in Chapter 1 of Volume 1. Both the total scattering and DINS provide instantaneous snapshots of the systems of atoms, on the scale of the characteristic time scales of excitations. On the whole, DINS appears in the context of this comparison as an ultra-fast structural probe with the structural information encoded in the momentum space, quite similarly to the total scattering technique. However, the description of the consequences of disorder on the neutron observables is provided by DINS in a mass-resolved manner, where each nucleus provides its own mode-averaged picture of both local and "lattice" contribution. On the contrary, in the total scattering technique the disorder-induced shape modification of the scattering function is neutron weighted, *i.e.*, "collapsed" on the axis of the constant momentum transfer Q sum of contributions from all atoms in a given system under consideration weighted by their respective neutron-scattering cross section. In consequence, hydrogeneous disordered systems measured by the total scattering will have their neutronic response always dominated by the hydrogen contribution.

The main difficulty in the progress in the science of the glassy state has been, however, not the lack of the availability of experimental techniques but rather the failures of early modeling attempts to assess the influence of spatial disorder and glassy behavior on spectroscopic observables. The chief problem with early models of glassy dynamics [133, 134] was their mesoscopic character manifesting itself in that they treated short-time dynamics as irrelevant. In consequence, only systems with neutronic response mainly composed of excitations of low (below 1 GHZ) acoustic origin could be modeled with impunity [135]. Later, a much more potent tool entered the scene, namely DFT-based methodology to calculate phonons. An important difference exists, however, between this methodology as applied to crystals or polycrystalline systems on the one hand, and glassy systems on the other hand. Namely, in case of disordered systems an additional preliminary step is required prior to the structure relaxation. A large enough supercell needs to be initially constructed containing a collection of smaller cells realizing all possible atom distributions, from uniform through clustered distribution up to pseudo-random distribution. Then, the usual steps following from structure relaxation are performed for this large supercell [136–138]. This is the size limiting step for the disordered system DFT simulation as large supercells can only be realistically simulated for a small to moderate number of atoms.

Notwithstanding the difficulties associated with computer modeling, progress has been made in DINS work on disordered systems. In this context, it is worth recalling pioneering work on molecular and metallic glasses [139, 140]. To this end, in the first work on hydrogenated and deuterated methanol in both glassy and polycrystalline state [139], classical (not based on the DFT approach) molecular and lattice dynamics calculations were carried out on ensembles composed by rigid molecular units. Then, the DINS response function was modeled for both glass and polycrystal samples as a convolution of that arising from the "lattice" motions and that from single-molecule contributions [141]. Results showed in these references point out that the fingerprints characteristic of glassy dynamics are not confined to low frequencies, as addressed by the "mesoscopic models" described above, but rather, they can extend over at least 14 decades in frequency, from long-time relaxation processes all the way down to microscopic motions. The effects of topological disorder brought forward via glassification are then translated into a measurable modification of atomic NMDs.

In conclusion, crystalline media provide much needed reference for the DINS observables, enabling a dissection and assignment of disorder- and anharmonicity-induced contributions to NMDs and therefore to each atom local dynamics.

FUNDING

The authors gratefully acknowledge the UK Science & Technology Facilities Council (STFC) for financial support, access to beam time at the ISIS Pulsed Neutron and Muon Source, and use of the e-Science SCARF Supercomputing Facility at the Rutherford Appleton Laboratory. This work was partially supported within the CNR-STFC Agreement (2014–2020) concerning collaboration in scientific research at the ISIS Spallation Neutron Source.

A.1 POSITION UNCERTAINTY FOR A PARTICLE IN A BOX

For the steady-state particle in a box, it can be shown that $\langle x \rangle$ corresponds to the center of the box (in this case, $\langle x \rangle = 0$), regardless of the state of the particle. Apart from the average position value, the position distribution of a particle can be characterized by the variance, σ_x^2, as a measure of the uncertainty in position of the particle [27]:

$$\sigma_x^2 = \int dx\rho(x)(x - \langle x \rangle)^2 = \int dx|\psi(x)|^2 x^2 = \frac{L^2}{3}\left[1 - \frac{6}{n^2\pi^2}\right], n = \pm 1, \pm 2, \ldots \tag{A.1}$$

A.2 HERMITE POLYNOMIALS

Hermite polynomials mentioned in Sec. 7.2 are defined as follows:

$$H_n(x) = (-1)^n e^{x^2} \frac{d^n}{dx^n} e^{-x^2}, \tag{A.2}$$

with the first few given by:

$$H_0(x) = 1 \tag{A.3}$$

$$H_1(x) = 2x \tag{A.4}$$

$$H_2(x) = 4x^2 - 2 \tag{A.5}$$

$$H_3(x) = 8x^3 - 12x \tag{A.6}$$

$$H_4(x) = 16x^4 - 48x^2 + 12 \tag{A.7}$$

A.3 POSITION AND MOMENTUM UNCERTAINTIES FOR AN ENSEMBLE OF QUANTUM PARTICLES

In case of a statistical ensemble of quantum particles described by a Hamiltonian, H, characterized by a macroscopic temperature T, the distributions of position and momentum must be obtained using the density matrix operator [25]:

$$\sigma_\rho = \frac{e^{-\frac{H}{k_B T}}}{Tr e^{-\frac{H}{k_B T}}}. \tag{A.8}$$

In case of the single-dimensional quantum harmonic oscillator, the trace of the density operator, calculated in the eigenbasis of H, yields:

$$Tr e^{-\frac{H}{k_B T}} = \sum_{n=0}^{\infty} \int d\mathcal{E} \psi^\dagger(\mathcal{E}) e^{-\frac{H}{k_B T}} \psi(\mathcal{E}) = \sum_{n=0}^{\infty} e^{-\frac{\hbar\omega}{k_B T}\left(n + \frac{1}{2}\right)} = \frac{e^{-\frac{\hbar\omega}{2k_B T}}}{1 - e^{-\frac{\hbar\omega}{k_B T}}}. \tag{A.9}$$

The probability density that a specific value of the particle position, $f(x_0)$, and particle momentum, $g(p_0)$, occurs in a statistical ensemble at temperature T, defined as $f(x_0) = Tr(\rho\delta(x - x_0))$ and $g(p_0) = Tr(\rho\delta(p - p_0))$, can be expressed as [27]:

$$f(x_0) = \frac{1}{\sqrt{\frac{2\hbar\pi}{M\omega}\left(\langle N\rangle + \frac{1}{2}\right)}} \exp\left[-\frac{M\omega x_0^2}{2\hbar\left(\langle N\rangle + \frac{1}{2}\right)}\right] \tag{A.10}$$

and

$$g(p_0) = \frac{1}{\sqrt{\frac{2\hbar\pi}{M\omega}\left(\langle N\rangle + \frac{1}{2}\right)}} \exp\left[-\frac{p_0^2}{2\hbar M\omega\left(\langle N\rangle + \frac{1}{2}\right)}\right] \tag{A.11}$$

where $\langle N\rangle = \frac{1}{e^{\frac{\hbar\omega}{kT}} - 1}$. Both $f(x_0)$ and $g(p_0)$ are Gaussian functions with variances:

$$\sigma_{x_0}^2 = \frac{\hbar}{M\omega}\left(\langle N\rangle + \frac{1}{2}\right) \tag{A.12}$$

and

$$\sigma_{p_0}^2 = \hbar M\omega\left(\langle N\rangle + \frac{1}{2}\right). \tag{A.13}$$

ACKNOWLEDGMENTS

The authors' heartfelt gratitude goes to a growing and vibrant community across the globe driving the field of eV neutron spectroscopy forward into the future, with exciting applications across science and engineering, new and imaginative technical developments, and advances in the theoretical foundations and associated computational tools underpinning a robust description of the quantum behavior of the materials world around us. In particular, we thank the following people for engaging with us in many stimulating discussions and joint ventures using eV neutrons over the years: M.A. Adams, A. Arbe, J. Armstrong, E. Artacho, F. Barroso-Bujans, S.M. Bennington, F.J. Bermejo, L. Bernasconi, J.J. Blostein, D.T. Bowron, J. Boxall, A. Bracco, F. Bresme, F. Bruni, C. Cabrillo, F. Cacialli, R. Car, S.C. Capelli, J.M. Carpenter, C.R.A. Catlow, M. Ceriotti, S. Cerveny, G.A. Chass, P. Collier, J. Colmenero, D. Colognesi, R.A. Cowley, G.J. Cuello, A. D'Angelo, J. Dawidowski, S.O. Diallo, J.C. Dore, M.T. Dove, K. Drużbicki, P.P. Edwards, G. Festa, R. Fernández-Perea, F. Fillaux, A.D. Fortes, V. Garbuio, N.I. Gidopoulos, M. Gigg, H.R. Glyde, M.A. González, G. Gorini, J.R. Granada, G.N. Greaves, J.M.F. Gunn, M.J. Gutmann, A.J. Horsewill, C.A. Howard, S. Imberti, A. Ivanov, S. Jackson, M.R. Johnson, M. Karlsson, G.J. Kearley, A.I. Kolesnikov,

A. Liscio, F. Liscio, C.-K. Loong, S.W. Lovesey, D.T. Major, F. Mallamace, D.E. Manolopoulos, J. Mayers, K.A. McEwen, R.L. McGreevy, D. McPhail, A. Michaelides, S. Moorby, R. Moreh, J.A. Morrone, S. Mukhopadhyay, R.J. Nelmes, R.J. Newport, D. Nixon, J. Norris, C. Offer, E. Oram, E. Pace, V. Palermo, L.C. Pardo, S.F. Parker, M. Parrinello, E. Perelli Cippo, R.J.-M. Pellenq, A. Pietropaolo, D. Pooley, D.L. Price, D. Raspino, K. Refson, G.F. Reiter, N.J. Rhodes, M.A. Ricci, L.A. Rodríguez Palomino, S. Rudić, M.-L. Saboungi, C.G. Salzmann, A. Sartbaeva, E.M. Schooneveld, A.G. Seel, J. Silvestre-Albero, N.T. Skipper, G.P. Škoro, A.K. Soper, U.E. Steigenberger, W.G. Stirling, C. Stock, J.-L. Tamarit, M. Tardocchi, K.V. Tian, J.-C. Tan, A.D. Taylor, J. Tomkinson, P.D. Vaz, K.T. Wikfeldt, C.G. Windsor, I.G. Wood, and M. Zoppi.

REFERENCES

[1] F. Fernandez-Alonso, D.L. Price, Neutron Scattering: Magnetic and Quantum Phenomena, Academic Press, New York, NY, 2015.

[2] F. Fernandez-Alonso, D.L. Price, Neutron Scattering: Fundamentals, Academic Press, New York, NY, 2013.

[3] M.R. Johnson, G.J. Kearley, Dynamics of atoms and molecules, in: F. Fernandez-Alonso, D.L. Price (Eds.), Neutron Scattering: Fundamentals, Academic Press, New York, NY, 2013, pp. 415–469.

[4] D.L. Price, F. Fernandez-Alonso, An introduction to neutron scattering, in: F. Fernandez-Alonso, D.L. Price (Eds.), Neutron Scattering: Fundamentals, Academic Press, New York, NY, 2013, pp. 1–136.

[5] R.N. Silver, P.E. Sokol, Momentum Distributions, Plenum, New York, NY, 1989.

[6] J. Mayers, M.A. Adams, Nucl. Instrum. Methods Phys. Res. A 625 (2011) 47–56.

[7] A. Pietropaolo, R. Senesi, Phys. Rep. 508 (2011) 45–90.

[8] W.H. Miller, Proc. Natl Acad. Sci. USA 102 (19) (2005) 6660–6664.

[9] ISIS, Available from: http://www.isis.stfc.ac.uk (accessed 20.12.16).

[10] P.E. Sokol, K. Sköld, D.L. Price, R. Kleb, Phys. Rev. Lett. 54 (1985) 909–912.

[11] P.E. Sokol, D.A. Peek, R.O. Simmons, D.L. Price, R.O. Hilleke, Phys. Rev. B 33 (1986) 7787–7789.

[12] R. Hempelmann, D. Richter, D.L. Price, Phys. Rev. Lett. 58 (1987) 1016–1019.

[13] W. Langel, D.L. Price, R.O. Simmons, P.E. Sokol, Phys. Rev. B 38 (1988) 11275–11283.

[14] G.J. Kellogg, P.E. Sokol, S.K. Sinha, D.L. Price, Phys. Rev. B 42 (1990) 7725–7733.

[15] J. Mayers, Phys. Rev. Lett. 71 (1993) 1553–1556.

[16] J. Mayers, F. Albergamo, D. Timms, Phys. B Condens. Matter 276 (2000) 811–813.

[17] C. Andreani, C. Pantalei, R. Senesi, J. Phys. Condens. Matter 18 (2006) 5587–5596.

[18] Y. Feng, C. Ancona-Torres, T.F. Rosenbaum, G.F. Reiter, D.L. Price, E. Courtens, Phys. Rev. Lett. 97 (2006) 145501.

[19] G.I. Watson, J. Phys. Condens. Matter 8 (33) (1996) 5955.

[20] V.F. Sears, Phys. Rev. B 30 (1984) 44.

[21] G.B. West, Phys. Rep. C 18 (1975) 263.

[22] C. Andreani, D. Colognesi, J. Mayers, G.F. Reiter, R. Senesi, Adv. Phys. 55 (2005) 377.

[23] M. Ceriotti, G. Miceli, A. Pietropaolo, D. Colognesi, A. Nale, M. Catti, M. Bernasconi, M. Parrinello, Phys. Rev. B 82 (17) (2010). 174306.

[24] P.A.M. Dirac, The Principles of Quantum Mechanics, Oxford University Press, Oxford, 1930.

[25] R.W. Robinett, Quantum Mechanics: Classical Results, Modern Systems, and Visualized Examples, Oxford University Press, Oxford, 1997.

[26] D.J. Griffiths, Introduction to Quantum Mechanics, second ed., Prentice Hall, New York, NY, 2003.
[27] J.B. Brojan, J. Mostowski, K. Wódkiewicz, Quantum Mechanics Workbook, second ed., PWN, Warsaw, 1978.
[28] D. Nemirovsky, R. Moreh, K.H. Andersen, J. Mayers, J. Phys. Condens. Matter 12 (18) 4293.
[29] R. Moreh, D. Nemirovsky, Radiat. Phys. Chem. 61 (36) (2001) 557–558.
[30] M.M. Krzystyniak, T. Abdul-Redah, Phys. Rev. B 82 (2010) 064301.
[31] M. Krzystyniak, F. Fernandez-Alonso, Phys. Rev. B 83 (2011) 134305.
[32] M. Krzystyniak, K. Drużbicki, F. Fernandez-Alonso, Phys. Chem. Chem. Phys. 17 (2015) 31287–31296.
[33] B. Fultz, Prog. Mater. Sci. 55 (4) (2010) 247–352.
[34] J. Keeler, P. Wothers, Chemical Structure and Reactivity: An Integrated Approach, Oxford University Press, New York, NY, 2014.
[35] M. Krzystyniak, S.E. Richards, A.G. Seel, F. Fernandez-Alonso, Phys. Rev. B 88 (18) (2013). 184304.
[36] J. Mayers, G. Reiter, Meas. Sci. Technol. 23 (4) (2012) 045902.
[37] R. Senesi, G. Romanelli, M.A. Adams, C. Andreani, Chem. Phys. 427 (2013) 111–116.
[38] A. Pietropaolo, R. Senesi, Nucl. Instrum. Methods Phys. Res. A 584 (2–3) (2008) 377–382.
[39] M. Krzystyniak, A.G. Seel, S. Richards, M.J. Gutmann, F. Fernandez-Alonso, J. Phys. Conf. Ser. 571 (2014) 012002.
[40] M.W. Chase, C.A. Davies, J.R. Downey, D.J. Frurip, R.A. McDonald, A.N. Syverud, J. Phys. Chem. Ref. Data 14 (11) (1985) 1–926.
[41] A.E. van Arkel, Molecules and Crystals in Inorganic Chemistry, Interscience, New York, NY, 1956.
[42] J.A.A. Katelaar, Chemical Constitution: An Introduction to the Theory of Chemical Bond, Elsevier, New York, NY, 1958.
[43] P. Mori-Sánchez, A.M. Pendás, V. Luana, J. Am. Chem. Soc. 124 (2002) 14721.
[44] L.C. Allen, J. Am. Chem. Soc. 114 (1992) 1510.
[45] L.C. Allen, J.F. Capitani, G.A. Kolks, G.J. Sproul, J. Mol. Struct. 300 (1993) 647.
[46] G.J. Sproul, J. Phys. Chem. 98 (1994) 6699.
[47] L.C. Allen, J. Am. Chem. Soc. 111 (1989) 9003.
[48] T.L. Beck, Quantum contributions to free energy changes in fluids, in: A.W. Castleman Jr., J.P. Toennies, W. Zinth (Eds.), Springer Series in Chemical Physics, Springer, New York, NY, 2007, pp. 387–420.
[49] G. Galli, M. Parrinello, Ab initio molecular dynamics: principles and practical implementation, in: M. Meyer, V. Pontikis (Eds.), Computer Simulation in Material Science, Kluwer, New York, NY, 1991, p. 283.
[50] D.M. Ceperley, Rev. Mod. Phys. 67 (1995) 279–355.
[51] M. Parrinello, A. Rahman, J. Chem. Phys. 80 (1984) 860–867.
[52] D. Laria, D. Chandler, J. Chem. Phys. 87 (1987) 4088–4092.
[53] M. Marchi, M. Sprik, M.L. Klein, J. Phys. Chem. 92 (1988) 3625–3629.
[54] Q. Wang, J.K. Johnson, J.Q. Broughton, Mol. Phys. 89 (1996) 1105–1119.
[55] Q. Wang, J.K. Johnson, J.Q. Broughton, J. Chem. Phys. 107 (1997) 5108–5117.
[56] J.A. Poulsen, G. Nyman, P.J. Rossky, J. Phys. Chem. B 108 (2004) 19799–19808.
[57] L.M. Sese, Mol. Phys. 76 (1992) 1335–1346.
[58] N. Tchouar, F. Ould-Kaddur, D. Levesque, J. Chem. Phys. 121 (2004) 7326–7331.
[59] L.M. Sese, Mol. Phys. 74 (1991) 177–189.

[60] M. Ceriotti, G. Bussi, M. Parrinello, Phys. Rev. Lett. 103 (2009) 30603.
[61] B.J. Berne, D. Thirumalai, Annu. Rev. Phys. Chem. 37 (1986) 401.
[62] L. Lin, J.A. Morrone, R. Car, M. Parrinello, Phys. Rev. Lett. 105 (2010) 110602.
[63] M. Ceriotti, D.E. Manolopoulos, Phys. Rev. Lett. 109 (10) (2012) 100604.
[64] M. Ceriotti, T.E. Markland, J. Chem. Phys. 138 (2013) 014112.
[65] M. Ceriotti, W. Fang, P.G. Kusalik, R.H. McKenzie, A. Michaelides, M.A. Morales, T.E. Markland, Chem. Rev. 116 (13) (2016) 7529–7550.
[66] D.L. Price, F. Fernandez-Alonso, An introduction to neutron scattering, in: F. Fernandez-Alonso, D.L. Price (Eds.), Experimental Methods in the Physical Sciences, Neutron Scattering, vol. 44, 2013, pp. 1–136.
[67] A.G. Seel, R. Senesi, F. Fernandez-Alonso, J. Phys. Conf. Ser. 571 (1) (2014) 011001.
[68] C.G. Windsor, Pulsed Neutron Scattering, Taylor & Francis, London, 1981.
[69] R. Senesi, D. Flammini, G. Romanelli, C. Andreani, Nucl. Instrum. Methods Phys. Res. A 704 (2013) 36–39.
[70] VESUVIO, Available from: http://www.isis.stfc.ac.uk/instruments/VESUVIO (accessed 20.12.16).
[71] C. Andreani, A. Pietropaolo, R. Senesi, G. Gorini, M. Tardocchi, A. Bracco, N. Rhodes, E. Schooneveld, Nucl. Instrum. Methods Phys. Res. Sect. A 481 (1–3) (2002) 509–520.
[72] A. Pietropaolo, C. Andreani, A. D'Angelo, R. Senesi, G. Gorini, S. Imberti, M. Tardocchi, N. Rhodes, E.S. Schooneveld, Appl. Phys. A 74 (2002) 189–190.
[73] M. Tardocchi, A. Pietropaolo, C. Andreani, A. Bracco, A. D'Angelo, G. Gorini, S. Imberti, R. Senesi, N.J. Rhodes, E.M. Schooneveld, Nucl. Instrum. Methods Phys. Res. A 526 (2004) 477–492.
[74] C. Andreani, A. D'Angelo, G. Gorini, S. Imberti, A. Pietropaolo, N.J. Rhodes, E.M. Schooneveld, R. Senesi, M. Tardocchi, Appl. Phys. A 78 (2004) 903–913.
[75] G. Gorini, E. Perelli-Cippo, M. Tardocchi, C. Andreani, A. D'Angelo, A. Pietropaolo, R. Senesi, S. Imberti, A. Bracco, E. Previtali, G. Pessina, N.J. Rhodes, E.M. Schooneveld, Nucl. Instrum. Methods Phys. Res. A 529 (2004) 293–300.
[76] S. Imberti, C. Andreani, V. Garbuio, G. Gorini, A. Pietropaolo, R. Senesi, M. Tardocchi, Nucl. Instrum. Methods Phys. Res. A 522 (2005) 463.
[77] E.M. Schooneveld, J. Mayers, N.J. Rhodes, A. Pietropaolo, C. Andreani, R. Senesi, G. Gorini, E. Perelli-Cippo, M. Tardocchi, Rev. Sci. Instrum. 77 (2006) 5103.
[78] A. Pietropaolo, C. Andreani, A. Filabozzi, E. Pace, R. Senesi, Nucl. Instrum. Methods Phys. Res. A 570 (2007) 498–510.
[79] A. Pietropaolo, C. Andreani, M. Rebai, L. Giacomelli, G. Gorini, E. Perelli Cippo, M. Tardocchi, A. Fazzi, G. Verona Rinati, C. Verona, M. Marinelli, E. Milani, C.D. Frost, E.M. Schooneveld, Europhys. Lett. 92 (2010) 68003.
[80] M. Rebai, C. Andreani, A. Fazzi, C.D. Frost, L. Giacomelli, G. Gorini, E. Milani, E. Perelli Cippo, A. Pietropaolo, G. Prestopino, E. Schooneveld, M. Tardocchi, C. Verona, G. Verona Rinati, Nucl. Phys. B Proc. Suppl. 215 (2011) 313–315.
[81] A. Pietropaolo, C. Andreani, M. Rebai, L. Giacomelli, G. Gorini, E. Perelli Cippo, M. Tardocchi, A. Fazzi, G. Verona Rinati, C. Verona, M. Marinelli, E. Milani, C.D. Frost, E.M. Schooneveld, Europhys. Lett. 94 (2011) 62001.
[82] L. Giacomelli, C. Andreani, A. Fazzi, C.D. Frost, G. Gorini, E. Perelli Cippo, A. Pietropaolo, M. Rebai, H. Schuhmacher, M. Tardocchi, C. Verona, G. Verona Rinati, A. Zimbal, Nucl. Phys. B Proc. Suppl. 215 (2011) 242–246.
[83] E.M. Schooneveld, A. Pietropaolo, C. Andreani, E. Perelli Cippo, N.J. Rhodes, R. Senesi, M. Tardocchi, G. Gorini, Rep. Prog. Phys. 79 (9) (2016) 094301.

[84] R.M. Brugger, A.D. Taylor, C.E. Olsen, J.A. Goldstone, A.K. Soper, Nucl. Instrum. Methods Phys. Res. 221 (1984) 393–407.
[85] J. Mayers, A.L. Fielding, R. Senesi, Nucl. Instrum. Methods Phys. Res. Sect. A 481 (1) (2002) 454–463.
[86] D. Homouz, G. Reiter, J. Eckert, J. Mayers, R. Blinc, Phys. Rev. Lett. 98 (11) (2007) 115502.
[87] S. Jackson, M. Krzystyniak, A.G. Seel, M. Gigg, S.E. Richards, F. Fernandez-Alonso, J. Phys. Conf. Ser. 571 (2014) 012009.
[88] D. Monserrat, A. Vispa, L.C. Pardo, R. Tolchenov, S. Mukhopadhyay, F. Fernandez-Alonso, J. Phys. Conf. Ser. 663 (1) (2015) 012009.
[89] G. Romanelli, M. Krzystyniak, Nucl. Instrum. Methods Phys. Res. A 819 (2016) 84–88.
[90] G. Romanelli, M. Ceriotti, D.E. Manolopoulos, C. Pantalei, R. Senesi, C. Andreani, J. Phys. Chem. Lett. 4 (2013) 3251.
[91] G. Romanelli, F. Fernandez-Alonso, C. Andreani, J. Phys. Conf. Ser. 571 (2014) 012003.
[92] A. Parmentier, J.J. Shephard, G. Romanelli, R. Senesi, C.G. Salzmann, C. Andreani, J. Phys. Chem. Lett. 6 (11) (2015) 2038–2042.
[93] C. Andreani, G. Romanelli, R. Senesi, Chem. Phys. 427 (2013) 106–110.
[94] F. Lehmkuhler, Y. Forov, T. Buning, C.J. Sahle, I. Steinke, K. Julius, T. Buslaps, M. Tolan, M. Hakala, C. Sternemann, Phys. Chem. Chem. Phys. 18 (2016) 6925–6930.
[95] G.F. Reiter, A. Deb, Y. Sakurai, M. Itou, V.G. Krishnan, S.J. Paddison, Phys. Rev. Lett. 111 (2013) 036803.
[96] D. Nemirovski, R. Moreh, K.H. Andersen, J. Mayers, J. Phys. Condens. Matter 11 (1999) 6653.
[97] V. Garbuio, C. Andreani, S. Imberti, A. Pietropaolo, G.F. Reiter, R. Senesi, M.A. Ricci, J. Chem. Phys. 127 (2007) 154501.
[98] G. Reiter, C. Burnham, D. Homouz, P.M. Platzmann, J. Mayers, T. Abdul-Redah, A.P. Moravsky, C.K. Loong, A.I. Kolesnikov, Phys. Rev. Lett. 97 (2006) 247801.
[99] G. Romanelli, R. Senesi, X. Zhang, K.P. Loh, C. Andreani, Phys. Chem. Chem. Phys. 17 (47) (2015) 31680–31684.
[100] G. Romanelli, A. Liscio, R. Senesi, R. Zamboni, E. Treossi, F. Liscio, G. Giambastiani, V. Palermo, F. Fernandez-Alonso, C. Andreani, Carbon 108 (2016) 199–203.
[101] M. Krzystyniak, M.A. Adams, A. Lovell, N.T. Skipper, S.M. Bennington, J. Mayers, F. Fernandez-Alonso, Faraday Discuss. 151 (2011) 171.
[102] T. Enoki, M. Endo, M. Suzuki, Graphite Intercalation Compounds and Applications, Oxford University Press, Oxford, 2003.
[103] A. Lovell, Tuneable graphite intercalates for hydrogen storage, Ph.D. thesis, University College London, 2007.
[104] K. Watanabe, M. Soma, T. Onishi, K. Tamaru, Nat. Phys. Sci. 233 (1971) 160.
[105] A.P. Smith, R. Benedek, F.R. Trouw, M. Minkoff, L.H. Yang, Phys. Rev. B 53 (1996) 10187.
[106] W.J. Stead, P. Meehan, J.W. White, J. Chem. Soc. 84 (1988) 1655.
[107] W.J. Stead, I.P. Jackson, J. McCaffrey, J.W. White, J. Chem. Soc. 84 (1988) 1669.
[108] A. Lovell, F. Fernandez-Alonso, N.T. Skipper, K. Refson, S.M. Bennington, S.F. Parker, Phys. Rev. Lett. 101 (2008) 126101.
[109] ISIS, Available from: http://www.isis.stfc.ac.uk/instruments/IRIS (accessed 20.12.16).
[110] ISIS, Available from: http://www.isis.stfc.ac.uk/instruments/OSIRIS (accessed 20.12.16).
[111] TOSCAweb, Available from: http://www.isis.stfc.ac.uk/instruments/TOSCA (accessed 20.12.16).

[112] S.J. Clark, M.D. Segall, C.J. Pickard, P.J. Hasnip, M.J. Probert, K. Refson, M.C. Payne, Z. Kristallogr. 220 (5–6) (2005) 567–570.
[113] G. Herzberg, Molecular Spectra and Molecular Structure, Krieger Publishing Co., Malabar, 1989.
[114] G.J. Kubas, Chem. Rev. 107 (2007) 4152.
[115] Y. Yang, Y. Kawazoe, Europhys. Lett. 98 (6) (2012) 66007.
[116] A.C. Evans, J. Mayers, D.N. Timms, J. Phys. Condens. Matter 6 (22) (1994) 4197.
[117] A.G. Seel, M. Ceriotti, P.P. Edwards, J. Mayers, J. Phys. Condens. Matter 24 (2012). 365401.
[118] L.M. Roberts, Proc. Phys. Soc. B 70 (8) (1957) 744.
[119] C.V. Briscoe, C.F. Squire, Phys. Rev. 106 (1957) 1175–1177.
[120] B. Yates, G.H. Wostenholm, J.L. Bingham, J. Phys. C 7 (10) (1974) 1769.
[121] R.P. Feynman, Rev. Mod. Phys. 20 (1948) 367–387.
[122] D. Marx, M. Parrinello, J. Chem. Phys. 104 (11) (1996) 4077–4082.
[123] K.V. Tian, B. Yang, Y. Yue, D.T. Bowron, J. Mayers, R.S. Donnan, C. Dobó-Nagy, J.W. Nicholson, D.-C. Fang, A.L. Greer, G.A. Chass, G.N. Greaves, Nat. Commun. 6 (2015). 8631.
[124] H. Compton, Phys. Rev. 21 (1932) 483–502.
[125] C. Andreani, G. Romanelli, R. Senesi, J. Phys. Chem. Lett. 7 (2016) 2216–2220.
[126] J. Dawidowski, J.R. Granada, J.R. Santisteban, F. Cantargi, L.A. Rodríguez Palomino, Appendix: neutron scattering lengths and cross sections, in: F. Fernandez-Alonso, D.L. Price (Eds.), Neutron Scattering: Fundamentals, Academic Press, New York, NY, 2013, pp. 471–528.
[127] A.G. Seel, R. Senesi, F. Fernandez-Alonso, J. Phys. Conf. Ser. 571 (2014) 012001.
[128] A.G. Seel, M. Krzystyniak, F. Fernandez-Alonso, J. Phys. Conf. Ser. 571 (1) (2014) 012006.
[129] A.G. Seel, M. Krzystyniak, R. Senesi, F. Fernandez-Alonso, J. Phys. Conf. Ser. 571 (2014) 012014.
[130] G.F. Reiter, J. Mayers, P. Platzman, Phys. Rev. Lett. 89 (2002) 135505.
[131] A.I. Kolesnikov, G.F. Reiter, N. Choudhury, T.R. Prisk, E. Mamontov, A. Podlesnyak, G. Ehlers, A.G. Seel, D.J. Wesolowski, L.M. Anovitz, Phys. Rev. Lett. 116 (2016) 167802.
[132] O. Arnold, J.C. Bilheux, J.M. Borreguero, A. Buts, S.I. Campbell, L. Chapon, M. Doucet, N. Draper, R. Ferraz Leal, M.A. Gigg, V.E. Lynch, A. Markvardsen, D.J. Mikkelson, R.L. Mikkelson, R. Miller, K. Palmen, P. Parker, G. Passos, T.G. Perring, P.F. Peterson, S. Ren, M.A. Reuter, A.T. Savici, J.W. Taylor, R.J. Taylor, R. Tolchenov, W. Zhou, J. Zikovsky, Nucl. Instrum. Methods Phys. Res. A 764 (2014) 156–166.
[133] U. Buchenau, et al., Low frequency excitations in glasses, in: D. Richter, A.J. Dianoux, W. Petry, J. Teixeira (Eds.), Dynamics of Disordered Materials, vol. 37, Springer, 1989, p. 172.
[134] U. Buchenau, Yu.M. Galperin, V.L. Gurevich, D.A. Parshin, M.A. Ramos, H. Schober, Phys. Rev. B 46 (1992) 27–98.
[135] A. Criado, F.J. Bermejo, M. Garcia-Hernandez, J.L. Martinez, Mol. Phys. 80 (1993) 1263.
[136] S.E. Ashbrook, D.M. Dawson, Acc. Chem. Res. 46 (9) (2013) 1964–1974.
[137] F. Tielens, C. Gervais, J.F. Lambert, F. Mauri, D. Costa, Chem. Mater. 20 (10) (2008) 3336–3344.
[138] A. Benisek, E. Dachs, G. Grodzicki, Phys. Chem. Miner. 42 (3) (2015) 243–249.
[139] F.J. Bermejo, F.J. Mompeán, A. Srinivasan, J. Mayers, A.C. Evans, Phys. Lett. A 189 (1994) 333.
[140] J. Dawidowski, F.J. Bermejo, L.F. Barquín, Phys. Lett. A 214 (1996) 59.
[141] M. Warner, S.W. Lovesey, J. Smith, Z. Phys. B 51 (1983) 109.

Chapter 8

Soft Condensed Matter

Mitsuhiro Shibayama[1]
The University of Tokyo, The Institute for Solid State Physics, Kashiwa, Chiba, Japan
[1]*Corresponding author: e-mail: sibayama@issp.u-tokyo.ac.jp*

Chapter Outline

Experimental Methods in the Physical Sciences, Vol. 49. http://dx.doi.org/10.1016/B978-0-12-805324-9.00008-X

8.1 INTRODUCTION

8.1.1 Soft Condensed Matter

"Soft condensed matter" (or simply called "soft matter") is a generic name for a wide variety of materials including, colloids, polymers, foams, gels, granular materials, liquid crystals, and so on [1]. As schematically illustrated in a size-hierarchy diagram in Figure 1, soft condensed matter is located in the "brackish water area" between "matter" and "life." The components of soft condensed matter are "molecules" of a sub to a few nanometers large and about one order larger than "atoms" constructing "hard condense matter" (condensed

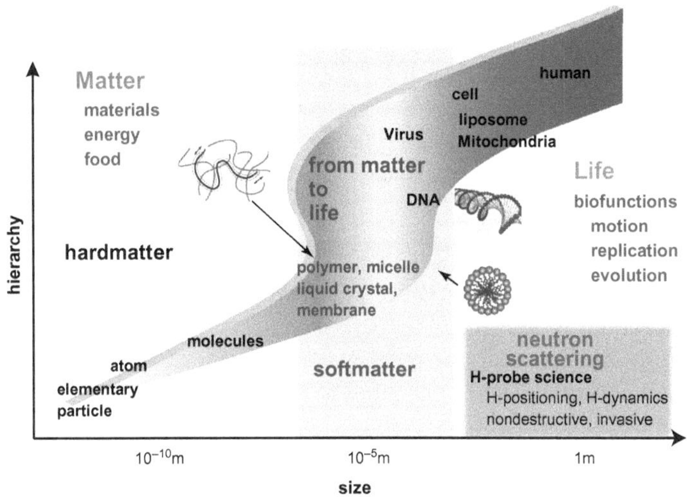

FIGURE 1 Size and hierarchy relationship of matter (hard matter, soft matter) and life. Soft matter is located between matter and life, and is characterized by its complexity, flexibility, and fragility.

matter). As the name literally indicates, soft condensed matter can be easily deformed by weak mechanical force, and self-assembles or undergoes phase transition by thermal agitations at ambient temperature. The "softness" and "flexibility" are associated with large internal degrees of freedom and they provide a rich world of nonlinear/nonequilibrium phenomena in modern condensed matter physics. One of attractive features of soft condensed matter is that the structure and the properties can be tailor-made by chemical modification, physical treatment, hysteresis, and so on. This is one of the reasons why soft condensed matter science has grown rapidly during the last couple of decades as pointed out by de Gennes [1]. Not only from academic points of view but also from industrial applications, soft condensed matter is "a treasure box" that has a variety of applications in livingware, foods, medical/pharmaceuticals, etc.

Soft condensed matter is a very suitable system to be investigated with a probe of "neutron." Typical energy range associated with soft condensed matter is tens of kJ (or a few meV), which is of the same order as those of cold neutrons. In addition, soft condensed matter consists of organic compounds and hence inevitably contains hydrogen. Since neutron is a "hydrogen" probe, neutron scattering has been used for investigations of the structure and the dynamics of soft condensed matter since its first application for determination of the size of polystyrene chains in the early 1970s [2, 3]. This chapter describes the characteristic features of soft condensed matter together with associated scattering theories, applications, and future directions of neutron scattering on soft condensed matter with emphasizing the uniqueness of soft condensed matter.

This chapter consists of basic theory of neutron scattering, instruments, and methodologies, followed by discussions on several soft condensed matter systems mostly from materials science points of view, including polymeric systems, rubbers, and gels, breakthrough works in soft condensed matter, recent topics, and future directions. Furthermore, an appendix, dealing with various scattering functions and contrast variation (CV) method, is attached for the convenience of the reader. The reader may refer a review on "Large-Scale Structures" in the sister book of this series [4], which covers a wide range of soft condensed matter with an emphasis of reflectivity studies on colloidal systems.

8.1.2 Energy Related to Soft Matter

Figure 2 schematically shows the relationship between matter and interaction energy. Neutrons are generated by fission or spallation, and their energy is on the order of 200 MeV. On the other hand, the order of the energy related to various phenomena of soft condensed matter, such as motion, deformation, phase transition, and phase separation, is the thermal energy k_BT at the ambient temperature, i.e., $T \approx 300$ K, and is about 0.6 kcal/mol (20 meV). This energy is very close to that of elastic scattering of cold-to-thermal neutrons, and is much smaller than those of chemical reactions, excitation in hard matter physics (of the order of eV). In order to make use of the neutron for structural characterization and/or studies of dynamics and/or excitation/relaxation, the neutron energy needs to be reduced by moderator to thermal (tens of meV)

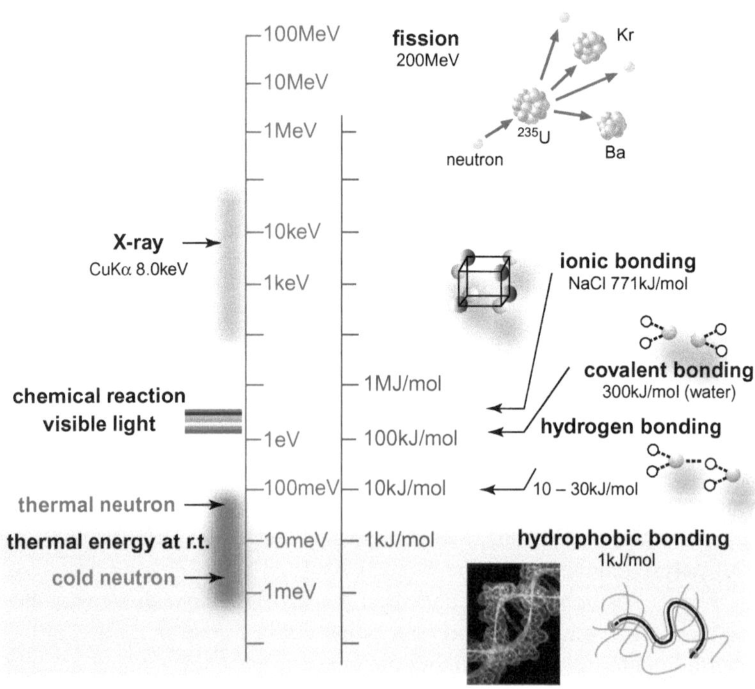

FIGURE 2 Energy landscape of probes, matters, and interactions.

or cold neutrons (a few meV). Note that the energy of X-ray used for crystallography and scattering is around 10 keV. Since this energy is much higher than that of covalent bonds (≈300 kJ/mol; ≈eV), X-ray may cause degradation of soft matter samples. Visible light is also used for structural characterization and studies of dynamics of soft matter as static light scattering (SLS) and dynamic light scattering (DLS). This is because light is complementary to neutron in both space resolution (a few hundred nm to sub μm)(static light scattering) and time resolution (ns to ms)(DLS) (see Figure 4).

As shown in Figure 2, thermal and cold neutrons have energies comparable to those of (≈10-30 kJ/mol) and (≈1 kJ/mol). Because of such low energy of neutrons, neutron can be used an invasive probe for structural characterization of soft condensed matter. Furthermore, neutrons can be used as a probe to study the dynamics of soft condensed matter as will be discussed later, such as reptation of polymer chains in molten state, glass transition or local motion of glassy materials, dynamics of surfactants, emulsions, and colloids, etc.

8.1.3 Energy Dispersion

The energy of electromagnetic waves, such as X-ray and visible light, is proportional to the inverse of the wavelength, λ,

$$E_{\mathrm{n}} = \hbar\omega = \frac{hc}{\lambda} \tag{1}$$

where c is the velocity of light, ω is the angular frequency, and h is the Planck constant ($\hbar \equiv h/2\pi$). On the other hand, neutrons and electrons are particle waves and their energies are proportional to the inverse square of λ, as follows

$$E_{\mathrm{n}} = \frac{h^2}{2m_i\lambda^2} \tag{2}$$

where (m_i) is the mass of neutron $m_{\mathrm{n}} = 1.675 \times 10^{-27}$ kg (i = n) or electron $m_{\mathrm{e}} = 9.109 \times 10^{-31}$ kg (i = e). The dispersion relationship between the energy and wavelength is shown in Figure 3.

In the case of electron with the wavelength $\lambda = 0.1$ nm, its energy is about 100 eV, while that of neutron is about 50 meV because the mass of neutron is about 2000 times larger than that of electron. This energy range is the same as excitation/relaxation of hard condensed matter.

8.1.4 Dynamics

Soft condensed matter has a hierarchy of motion/dynamics due not only to its shape and hierarchical structure but also to intra and intermolecular interactions. As a result, time or energy/frequency window is quite large. In order to cover this large spatiotemporal window, various types of neutron-scattering

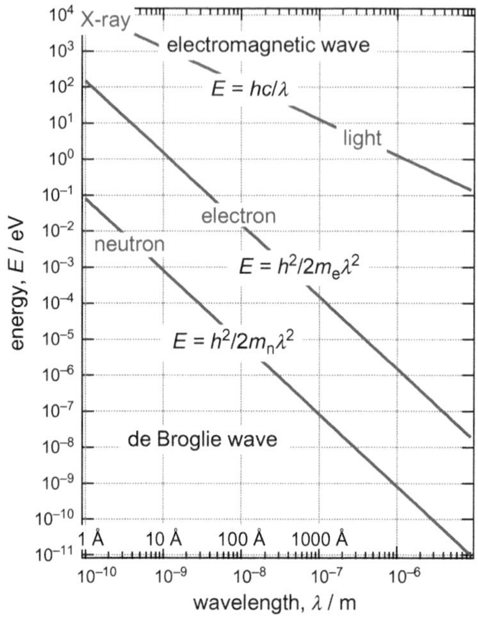

FIGURE 3 Energy dispersion of electromagnetic waves, neutrons, and electrons.

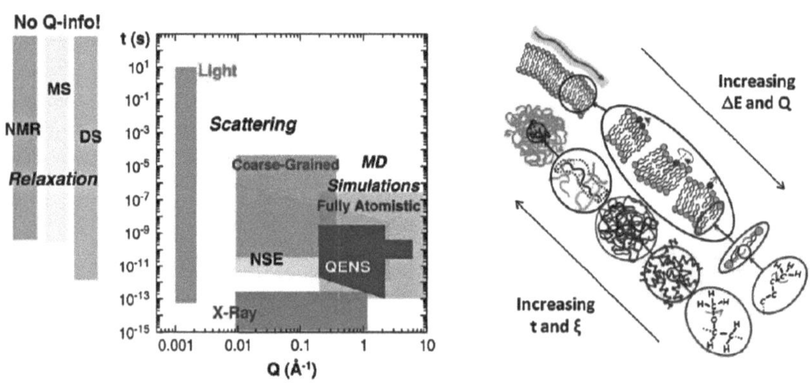

FIGURE 4 Left: Time/space window of experimental and computation techniques. Note that Q is inversely proportional to length. Right: Examples of dynamical modes in polymers and lipid bilayers in the timescales and length scales covered by neutron spectrometers. *Reproduced from Ref. [5].*

instruments are developed. Figure 4 (left) shows time/space window of experimental and computational tools. Here, Q is a reciprocal length measure, called the modulus of the momentum transfer as will be defined in the next section (Equation (3)). The larger Q the smaller space or distance is. There are various scattering techniques to study dynamics; quasielastic neutron scattering (QENS) (Section 8.3.3), neutron spin echo (NSE) (Section 8.3.3.2), X-ray scattering, and light scattering, each of which occupies a limited space of the window. Molecular dynamics (MD) simulations (fully atomistic and coarse-grained MD simulations) are more or less complementary to these scattering methods as shown in the figure. On the other hand, relaxation techniques, such as, NMR, DS (dielectric spectroscopy), MS (mechanical spectroscopy), cover a wide range of time window, but they do not provide any structural information. Figure 4 (right) schematically illustrates the hierarchy of bilayers (upper) and amorphous polymers (lower). By zooming-out (increasing time t or correlation length ξ), the structures become coarse from a molecular structure to higher-order structure (such as, vesicles and aggregated polymer chains). Increase in energy ΔE and Q vector, on the other hand, corresponds to zoom-in. For example, QENS spectrometers cover timescales in the range $\approx 10^{-13}$ to 10^{-9} s and length scales in the range ≈ 1 to 30 Å.

As briefly shown in this section and will be extensively discussed in the following sections, neutron scattering is a powerful tool for investigations of the structure and dynamics of soft condensed matter.

8.2 BASIC THEORY OF NEUTRON SCATTERING FOR SOFT CONDENSED MATTER

In this section, basic technical terms and their physical meanings for neutron scattering, such as scattering vector, scattering intensity, scattering length density, and transmission, are reviewed. For further reading, see textbooks in Refs. [6–8].

8.2.1 Scattering Vector and Scattering Intensity

The most important variable in scattering is the scattering vector. The scattering vector $\mathbf{Q}$, which is also called the momentum transfer, is defined as the difference between the incident vector $\mathbf{k}_i$ and the scattering vector $\mathbf{k}_f$ as discussed in Equation (1.5) of Chap. 1 of Ref. [6].

$$\mathbf{Q} = \mathbf{k}_i - \mathbf{k}_f \tag{3}$$

$$|\mathbf{k}_i| \equiv k = |\mathbf{k}_f| \equiv k' = \frac{2\pi}{\lambda} \tag{4}$$

$$Q = |\mathbf{Q}| = 2k\sin\theta \tag{5}$$

where λ is the wavelength of the neutron, k and k' are the wavenumbers of the incident and scattered neutrons, respectively. Note that Q is a reciprocal length measure with the dimension of [length^{-1}].

The scattering intensity from an atom with the scattering length b is given as the differential cross section ($\mathrm{d}\sigma/\mathrm{d}\Omega$) (see Equation (1.3) of Ref. [6])

$$\frac{\mathrm{d}\sigma}{\mathrm{d}\Omega} = b^2 \tag{6}$$

Here, σ and Ω are the scattering cross section and the solid angle, respectively. Note that $(\mathrm{d}\sigma/\mathrm{d}\Omega)$ is independent of Q. The scattering intensity from an assembly of N particles is obtained by summing up $b_i(\mathbf{R}_i)$ and $b_j(\mathbf{R}_j)$ by taking account of the phase shift $\mathbf{Q} \cdot (\mathbf{R}_i - \mathbf{R}_j)$,

$$\frac{\mathrm{d}\sigma}{\mathrm{d}\Omega} = \sum_{i,j}^{N} b_i b_j e^{-i\mathbf{Q} \cdot (\mathbf{R}_i - \mathbf{R}_j)} \tag{7}$$

Here, $b_i(\mathbf{R}_i)$ is the coherent scattering length at the position vector $\mathbf{R}_i$ of the particle i. The scattering intensity from an assembly of particles per unit volume is given by multiplying the number density of the particle N/V to $\mathrm{d}\sigma/\mathrm{d}\Omega$ as follows,

$$\frac{\mathrm{d}\Sigma}{\mathrm{d}\Omega} = \frac{N}{V}\frac{\mathrm{d}\sigma}{\mathrm{d}\Omega} = n_\mathrm{p}\frac{\mathrm{d}\sigma}{\mathrm{d}\Omega} \tag{8}$$

where V is the scattering volume (the irradiated volume) and n_p is the number density of the scattering element, and $\mathrm{d}\Sigma/\mathrm{d}\Omega$ is called the differential scattering cross section per unit volume, i.e., the absolute intensity. Note that here $\mathrm{d}\Sigma/\mathrm{d}\Omega$ and $\mathrm{d}\sigma/\mathrm{d}\Omega$ have the dimensions of $[\mathrm{cm}^{-1}]$ and $[\mathrm{cm}^2]$, respectively.

It should be noted that the wavelength of cold neutron is about 1 nm, while the size of nuclei and the range of nuclear force are about $1/10^5$ of the wavelength. Hence, the interaction potential acting on the incoming neutron and the nucleus can be regarded as a delta function (Fermi pseudopotential), and the scattering amplitude becomes independent of the scattering vector, Q. On the other hand, the atomic scattering amplitude for X-ray is a decreasing function of Q, and its Q dependence has to be taken into account in the analysis of the scattering intensity functions. This is one of the marked differences of neutron scattering from X-ray scattering (see Section 1.1.4.1 of Ref. [6]).

8.2.2 Coherent Scattering and Incoherent Scattering

The scattering length b is a complex number, where the imaginary part is related to absorption and is given by,

$$b = b' + ib'' \tag{9}$$

where b' and b'' are the real and imaginary parts of b, respectively. Hence, the scattering intensity is given by

$$\frac{\mathrm{d}\sigma}{\mathrm{d}\Omega} = \left|\langle k'|b|k\rangle\right|^2 \tag{10}$$

Note that neutron scattering is dependent also on isotopes of the constituent elements and the relative alignments of spins. Hence, the scattering intensity is rewritten as

$$\frac{d\sigma}{d\Omega} = \sum_{s,s'} P_s \sum_{l,l'} e^{i\mathbf{Q}\cdot(\mathbf{R}_l - \mathbf{R}_{l'})} \left| \langle s' | b_{l'}^{*} b_l | s \rangle \right|^2 \tag{11}$$

This equation means that it is necessary to sum up with respect to isotopes (l, l') as well as to relative alignments of spins (s, s'). Here, P_s is the probability of spin s. As a result, the scattering intensity is given by a sum of two components,

$$\frac{d\sigma}{d\Omega} = \bar{b}^2 \sum_{l,l'} e^{i\mathbf{Q}\cdot(\mathbf{R}_l - \mathbf{R}_{l'})} + N\left(\overline{b^2} - \bar{b}^2\right) \tag{12}$$

The first term of the RHS is the coherent scattering related to the relative positions of the scattering elements (a function of Q), and the second term is the incoherent scattering independent of Q. Here, $\bar{b}$ and $\overline{b^2}$ are given by

$$\bar{b} = p^+ b^+ + p^- b^- \tag{13}$$

$$\overline{b^2} = p^+ (b^+)^2 + p^- (b^-)^2. \tag{14}$$

and p^+ and p^- are the probability of the up and down spins, and b^+ and b^- are their scattering lengths, respectively. For example, in the case of hydrogen (^1H), the spin quantum number is $i = 1/2$. Hence, $p^+ = 2(i+1)/2(2i+1) = 3/4$ and $p^- = 2i/2(2i+1) = 1/4$. Since $b^+ = 10.85$ fm and $b^- = -47.5$ fm, the coherent and incoherent scattering cross sections of a hydrogen atom are $\sigma_{coh} = 4\pi\bar{b}^2 = 1.76$ barn (1 barn $= 10^{-24}$ cm^2) and $\sigma_{inc} = 4\pi(\overline{b^2} - \bar{b}^2) = 80.2$ barn, respectively. The value of the incoherent scattering cross section is abnormally larger than other elements (see A.4 Table of Neutron Scattering Lengths and Cross Sections [9]). Hence, it is very important to reduce this incoherent scattering in neutron scattering experiments of soft condensed matter in order to obtain meaningful results.

Figure 5 shows the scattering intensity profiles for polystyrene (PS) spheres with radius of $R = 560$ or 550 Å in D_2O and in H_2O (open symbols), together with their fitting results (solid lines). The horizontal lines with $I_{inc,\ D2O}$ and $I_{inc,\ H2O}$ indicate the levels of the incoherent scattering intensities, i.e., the background intensities. In the case of PS spheres in D_2O, the scattering peaks can be resolved up to the 4th peak, while the 3rd and 4th peaks are buried in the background for the case of H_2O. This strongly demonstrates that it is very important to minimize the contribution from incoherent scattering. In another word, it is of particular importance to design a sample to label with H-species in a D-substituted matrix.

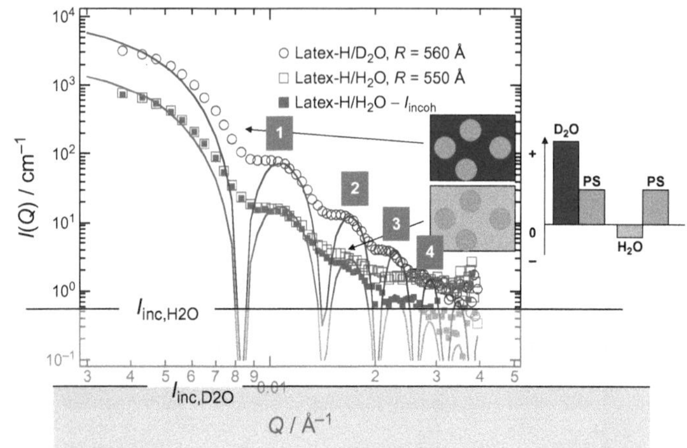

FIGURE 5 Observed small-angle neutron-scattering profiles for polystyrene spheres in D_2O (blue open circles) and in H_2O (red open squares) and calculated scattering curves with spherical form factors with radius $R = 560$ or 550 Å. The yellow and orange bars with $I_{inc,\ D2O}$ and $I_{inc,\ H2O}$ indicate the incoherent scattering intensity for D_2O and H_2O, respectively. The filled red circles indicate the corrected intensity by subtracting I_{H2O}. The bar graph on the right indicates the relative scattering contrast of the components.

8.2.3 Elastic Scattering and Inelastic Scattering

In the case of elastic scattering the differential scattering cross section is given by Eq. (10). On the other hand, the differential scattering cross section for inelastic neutron scattering is given by

$$\left(\frac{\mathrm{d}\sigma}{\mathrm{d}\Omega}\right)_{k,\lambda\to k',\lambda'} = \frac{k'}{k}\left|\langle k'\lambda'|b|k\lambda\rangle\right|^2 \tag{15}$$

By imposing the law of the conservation of energy and momentum transfer, it is rewritten to

$$\left(\frac{\mathrm{d}^2\sigma}{\mathrm{d}\Omega\mathrm{d}\omega}\right)_{k,\lambda\to k',\lambda'} = \frac{k'}{k}\sum_{\lambda} P_\lambda \sum_{\lambda'}\left|\langle k'\lambda'|b|k\lambda\rangle\right|^2 \delta(\hbar\omega + E_\lambda - E_{\lambda'}) \tag{16}$$

Here, P_λ is the wavelength distribution. The total cross section σ_T is a sum of scattering cross section and absorption cross section, i.e., $\sigma_T = \sigma + \sigma_a$.

8.2.4 Scattering Length Density

Since the scattering elements (e.g., nuclei) are much smaller than the wavelength of neutron, they can be regarded to be continuously distributed in the space. Thus it is more convenient to use the scattering length density (distribution), $\rho(\mathbf{r})$ [cm^{-2}], defined by dividing the scattering length b [cm] and by the volume of the element v [cm^3],

$$\rho(\mathbf{r}) = \sum_j b_j \delta(\mathbf{r} - \mathbf{R}_j) \tag{17}$$

If the scattering element is a molecule, the scattering length density is given by

$$\rho_j = \frac{\sum_i b_i}{v_j}. \tag{18}$$

Here, the numerator is a sum of atoms i constituting the molecule $\sum_i$, and the denominator is the volume of the molecule v_j [cm^3] (=the molar volume divided by the Avogadro constant, N_A). By using the scattering length densities, the scattering intensity is rewritten to

$$\frac{d\sigma}{d\Omega}(Q) = \left\langle \left| \sum_j b_j \delta(\mathbf{r} - \mathbf{R}_j) e^{i\mathbf{Q}\cdot\mathbf{R}_j} \right|^2 \right\rangle \rightarrow \left\langle \left| \int d^3 r \rho(r) e^{i\mathbf{Q}\cdot\mathbf{R}_j} \right|^2 \right\rangle \tag{19}$$

As shown in Figure 6, the scattering lengths of common atoms consisting soft condensed matter roughly fall in the range of + (5 ~ 7) fm, while only ^{1}H has a negative value, −3.744 fm. By using this nature, neutron-scattering experiments on soft condensed matter are conducted by H/D substitution on targeting part in the matter. On the other hand, it is very important to note that ^{1}H has an extraordinarily large incoherent scattering cross section. This leads to insignificant incoherent background scattering (e.g., Figure 5) and multiple scattering as will be discussed later.

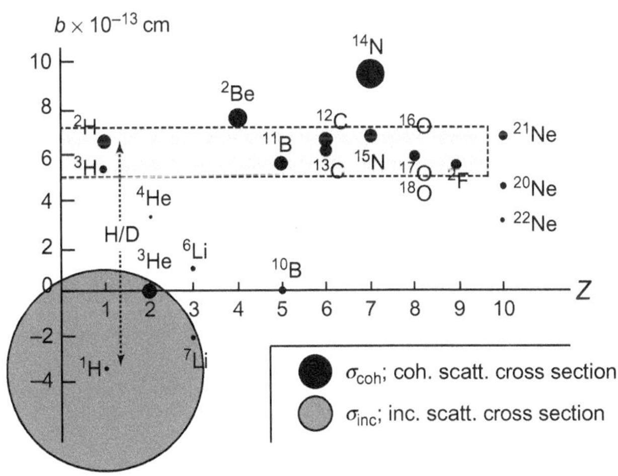

FIGURE 6 Variations of scattering lengths and scattering cross sections of isotopes with the atomic number $Z = 1$ to 10. The shaded area with the dashed line roughly indicates the range of the scattering lengths for elements consisting of soft condensed matter. The black and dark-shaded circles indicate the relative magnitudes of coherent and incoherent scattering cross sections, respectively. Note that the difference in the scattering lengths between H and D is very large as shown with the dotted line with H/D, and the incoherent scattering cross section of ^{1}H is extraordinarily large.

8.2.5 Scattering Length Density Distribution Function and Correlation Function

Let us define the local deviation of the scattering length density $\Delta\rho(\mathbf{r})$ from its average $\overline{\rho}$,

$$\Delta\rho(\mathbf{r}) \equiv \rho(\mathbf{r}) - \overline{\rho} \tag{20}$$

Then, the scattering intensity (the scattering cross section) is given by

$$\frac{d\sigma}{d\Omega} = \int_V d^3r \int_{V'} d^3r' \Delta\rho(\mathbf{r})\Delta\rho(\mathbf{r}')e^{i\mathbf{Q}\cdot(\mathbf{r}-\mathbf{r}')} \tag{21}$$

The Fourier transform of $\Delta\rho(r)$ is the scattering amplitude, $\Delta\hat{\rho}(Q)$, and is given by

$$\Delta\hat{\rho}(Q) \equiv \int_V d^3r \Delta\rho(\mathbf{r})e^{i\mathbf{Q}\cdot\mathbf{r}} \tag{22}$$

By using this, the scattering intensity is written by

$$\frac{d\sigma}{d\Omega} = |\Delta\hat{\rho}(Q)|^2 \tag{23}$$

On the other hand, the autocorrelation function of the scattering length density function $\Delta\rho(\mathbf{r})$ is called "correlation function," and is given by

$$\gamma(\mathbf{r}) \equiv \langle \Delta\rho(\mathbf{r})\Delta\rho(\mathbf{r}') \rangle = \int_{V'} d^3r' \Delta\rho(\mathbf{r})\Delta\rho(\mathbf{r}') \tag{24}$$

Its Fourier transform $S(\mathbf{Q})$ is given by

$$S(\mathbf{Q}) = \int_V d^3r \gamma(\mathbf{r}) \tag{25}$$

This is called the "response function" or the "structure factor," and the quantity observed by scattering experiments. Not only in neutron scattering but also in other scattering experiments, it is an essential process to relate $S(\mathbf{Q})$ to $\Delta\rho(\mathbf{r})$ and to discuss the structure of the matter. In the next sections, various $S(\mathbf{Q})$s related to soft condensed matter physics are introduced and discussed in details. However, before discussing $S(\mathbf{Q})$s, let us make a remark on the transmission of neutron beam for soft condensed matter.

8.2.6 Transmission

Similar to other scattering experiments, it is necessary to measure the transmission of the sample. The transmission, T, is defined by the ratio of the transmitted intensity I_{tr} with respect to the incident intensity I_0 and the Lambert-Beer law holds,

$$T \equiv \frac{I_{tr}}{I_0} = \exp(-\Sigma_{tot} t) \tag{26}$$

Here, Σ_{tot} [cm^{-1}] is the macroscopic total cross section per unit volume and t [cm] is the sample thickness. In the case of neutron scattering, Σ_{tot} is related to the total scattering cross section,

$$\Sigma_{tot} = \sum_i d_{N,i} \sigma_{tot,i}, \tag{27}$$

where $d_{N,\,i}$ [cm^{-3}] and $\sigma_{tot,\,i}$ [cm^2] are the number density and the total cross section of the scattering element i, respectively. For compound i, $d_{N,\,i}$ is given by

$$d_{N,i} = \frac{d_i N_A}{m_i}, \tag{28}$$

Here, d_i [$g \cdot cm^{-3}$] and m_i [$g \cdot mol^{-1}$] are the mass density and molecular weight of compound i, respectively and N_A is Avogadro's number.

In the case of visible light scattering or infrared spectroscopy, absorption by the material is the dominant term that lowers transmission. On the other hand, transmission of neutron scattering on soft condensed matter is dominated by multiple scattering of ^1H because of its large incoherent scattering cross section as discussed above (see Figure 6). Figure 7 shows the experimental results of transmission measurements for mixtures of H_2O and D_2O with different sample thicknesses t and different H_2O compositions ϕ_H [10, 11]. The wavelength of the neutron was $\lambda = 7$ Å. As shown in the figure, the transmission T dramatically decreases by increasing the sample thickness t and the H content. Note that a single master curve is obtained according to Lambert-Beer law (Equation (26)). It is important to note the following. For better results in neutron scattering for soft condensed matter, prepare samples with H as low as possible.

It should be also noted that nominal incoherent scattering is dependent on the sample thickness because of multiple scattering. Shibayama et al. studied the incoherent scattering from a series of mixtures of H_2O and D_2O with different sample thicknesses and confirmed that the observed incoherent scattering is strongly modified by multiple scattering [10, 11]. Let us discuss this problem in more detail.

8.2.7 Multiple Scattering and Background Scattering

The mean-free path of neutron beam Λ [cm] is obtained by $\Lambda = 1/\Sigma_{tot}$. In the case of neutron scattering for soft condensed matter, Σ_{tot} arises mainly from incoherent scattering from H as described in the previous section. The observed mean-free paths of 7 Å-neutron in H_2O and in D_2O were $\Lambda = 0.16$ cm ($\Sigma_{tot,\ H2O} = 6.22\ cm^{-1}$) and 1.70 cm ($\Sigma_{tot,\ D2O} = 0.588\ cm^{-1}$),

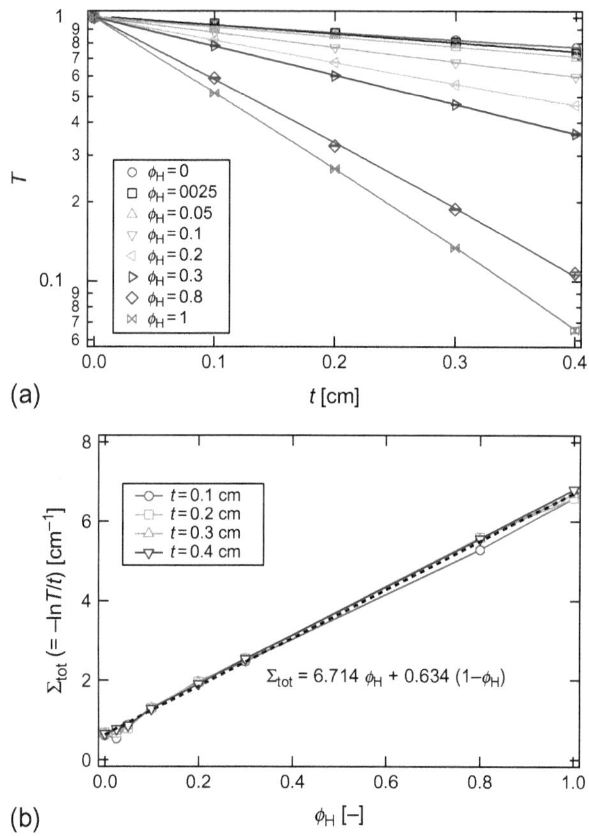

FIGURE 7 Neutron transmission of mixtures of H_2O and D_2O. (a) t-dependence, (b) ϕ_H-dependence. The wavelength of the neutron was $\lambda = 7$ Å. *Reproduced from Ref. [11].*

respectively [10]. This means that strong multiple scattering occurs in an SANS experiment for H-containing material.

Shibayama et al. discussed multiple scattering for a slab-shape sample and proposed the following equation [10, 11],

$$\left(\frac{d\Sigma}{d\Omega}\right)_{inc} = \frac{(e^{\Sigma_{inc} t} - 1)}{4\pi t} = \frac{\Sigma_{inc}}{4\pi} \frac{(e^{\Sigma_{inc} t} - 1)}{\Sigma_{inc} t}. \tag{29}$$

In the third formula of Equation (29), the term $\Sigma_{inc}/(4\pi)$ represents the differential incoherent scattering cross section of the sample ($\equiv d\Sigma_{inc}/d\Omega$) at the zero-thickness limit, and the term $(e^{\Sigma_{inc} t} - 1)/(\Sigma_{inc} t)$ is the correction term for multiple scattering with finite sample thickness of the sample. If the sample is an H-containing sample, Σ_{inc} is very close to Σ_{tot} because attenuation of neutron by transmission is exclusively due to incoherent scattering of H. Hence,

$$\left(\frac{d\Sigma}{d\Omega}\right)_{inc} \cong \frac{\Sigma_{tot}}{4\pi}\frac{(e^{\Sigma_{tot}t}-1)}{\Sigma_{tot}t} = \frac{1}{4\pi}\frac{(e^{\Sigma_{tot}t}-1)}{t} = \frac{1}{4\pi}\frac{1-T}{tT}. \tag{30}$$

Equation (30) means that the absolute incoherent scattering intensity can be determined solely by the sample transmission T and thickness t. We call this T-method which stands for the transmission method. Note that Equation (30) is essentially the same as that obtained by Jacrot [12].

It should also be noted that Equation (30) is independent of absolute intensity calibration. Hence, it is important for incoherent intensity evaluation to measure accurately the scattering intensity of a sample in the absolute intensity scale. In general, the transmission is related not only to multiple scattering but also to absorption. For example, Equation (29) has to be used when absorption is not negligible, e.g., vanadium ($\Sigma_{inc,V} = 0.348\ cm^{-1}$, $\Sigma_{abs,V} = 0.350\ cm^{-1}$). In the following, we examine the applicability and validity of Equation (30).

Figure 8 shows the H-fraction dependence of the zero-thickness differential cross section, $(d\Sigma/d\Omega)_{inc}(t \to 0) \equiv (d\Sigma(t=0)/d\Omega)_{inc}$. As shown in Figure 8, $(d\Sigma(t=0)/d\Omega)_{inc}$ is a linear function of ϕ_H. The dashed line was obtained by fitting the observed data with Equation (31) by adjusting the Jacrot's parameter g_λ, i.e.,

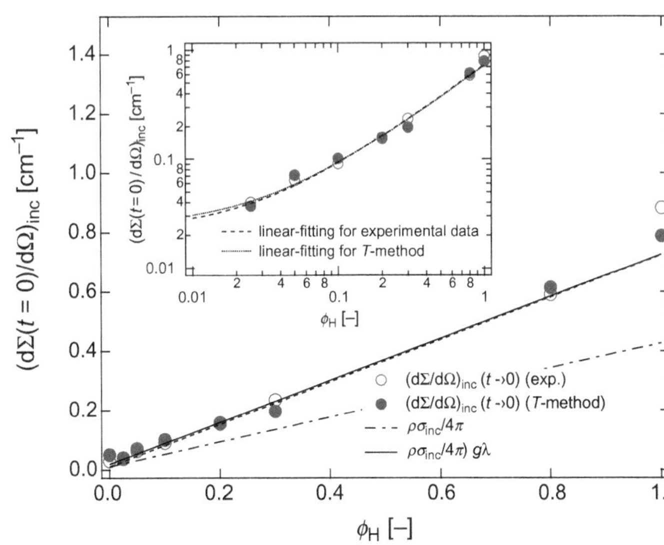

FIGURE 8 Sample-thickness dependence of $(d\Sigma(t=0)/d\Omega)_{inc}$ in the mixtures of D_2O and H_2O. Lines indicate the fit results according to Equation (30). *Reproduced from Ref. [11].*

$$\left(\frac{d\Sigma(t=0)}{d\Omega}\right)_{inc} = \frac{g_\lambda}{4\pi}\left[\phi_H \Sigma^{tbl}_{inc,H2O} + (1-\phi_H)\Sigma^{tbl}_{inc,D2O}\right]. \tag{31}$$

$$g_\lambda \simeq 1/[1-\exp(-0.6\lambda)^{1/2}]. \tag{32}$$

The values of the incoherent cross sections per unit volume for H_2O and D_2O are calculated from the table [9] and are $\Sigma^{tbl}_{inc,\ H2O} = 5.369\ cm^{-1}$ and $\Sigma^{tbl}_{inc,\ D2O} = 0.136\ cm^{-1}$. For $g_\lambda = 1.0$, the calculated values of $(d\Sigma(t=0)/d\Omega)_{inc}$ shown by the chain line is much lower than the observed ones (dashed line). This is partially due to the anisotropic scattering of incoherent scattering for $\lambda \leq 10$Å as discussed by Jacrot. Another possible reason is an error of the absolute intensity calibration for the SANS instrument [10, 11]. Since the reason is unclear at this stage, we simply take into account these unknown error by the g_λ factor in this work. The best fit was obtained with $g_\lambda = 1.7$, as shown by the solid lines in Figure 8 and in the inset. This value is somewhat larger than the value calculated by Equation (32), i.e., $g_\lambda \cong 1.3$ for $\lambda = 7$ Å. The dashed line with filled circles in Figure 8 and in the inset denotes $(d\Sigma(t=0)/d\Omega)_{inc}$s evaluated by the T-method, which shows excellent agreement between the observed ones and those evaluated by the T-method. It partly explains why the total cross section of H_2O is higher than the calculated one.

8.3 NEUTRON SCATTERING INSTRUMENTS AND METHODOLOGIES

As discussed in the previous section, neutron scattering is a very suitable tool for structure/dynamics investigations of soft condensed matter, and neutron scattering has been greatly contributing to soft condensed matter science. Particularly, it is often used to obtain (1) the structural information, e.g., size, shape, and orientation of polymer chains and micelles, (2) information on surface and/or interface, and (3) information about dynamics and motion of polymers, micelles, vesicles, etc. Among various scattering techniques, small-angle neutron scattering (SANS), neutron reflectivity (NR), and inelastic neutron scattering (INS) are suitable for (1), (2), and (3), respectively. Let us review the common features of the scattering theories in soft condensed matter. For further reading on SANS, NR, and INS, the following sections in the book of this series are recommended [6]; Sections 1.4.6.1 [13], 1.4.6.2 [13], and 3.5 [14], respectively.

8.3.1 Small-Angle Neutron Scattering (SANS)

SANS is most commonly used in soft condensed matter [8, 14]. Figure 9 shows the small-angle neutron-scattering instrument, SANS-U [15], owned by the University of Tokyo and installed at Japan Atomic Energy Agency (JAEA). A neutron beam from the research reactor, JRR-3, is monochromatized with a velocity selector, and collimated by a set of pinholes or guided

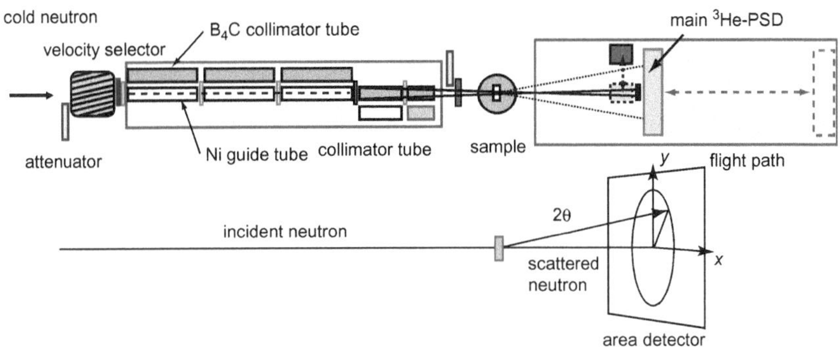

FIGURE 9 Schematic illustration of the optical setup of SANS-U, a small-angle neutron-scattering instrument owned by the University of Tokyo and installed in the research reactor, JRR-3, Japan Atomic Energy Agency. The apparent neutron source can be varied by a set of B_4C collimator and Ni guides. A movable high-resolution area detector with a scintillator and CCD is also mounted in the center. The Q is about 3×10^{-3} Å to 2×10^{-1} Å for the neutron wavelength $\lambda = 7$ Å. *Reproduced from Ref. [15].*

by neutron guides before incidence to samples. A scattered neutron beam is fed by a ^{3}He area detector in the flight path. The detected neutrons are corrected for transmission, background (cell) scattering, detector inhomogeneities, and then scaled to the absolute intensity, $(d\Sigma/d\Omega)(Q)$. In the case of isotropic scattering, it is further circularly averaged with respect to Q and the scattering intensity (the differential cross section), $(d\Sigma/d\Omega)(Q)$ is obtained.

8.3.1.1 Form Factor and Guinier Law

When scattering elements of each volume V_p are scarcely dispersed in a medium, e.g., water, the scattering amplitude $f(Q)$ is given by

$$f(Q) = \int_{V_p} d^3r \Delta\rho(r) e^{i\mathbf{Q}\cdot\mathbf{r}} \tag{33}$$

$$= 4\pi \int_{V_p} d^3r \Delta\rho(r) \frac{\sin(Qr)}{Qr} \tag{34}$$

$$= V_p(\Delta\rho)\left[1 - \frac{Q^2}{6}\int_{V_p} d^3r\rho(r) + \ldots\right]. \tag{35}$$

Here, an orientational average with respect to **Q** and **r** is taken, i.e.,

$$< e^{i\mathbf{Q}\cdot\mathbf{r}} >_{\text{orient}} = \frac{\sin(Qr)}{Qr} = 1 - \frac{1}{3!}(Qr)^2 + \frac{1}{5!}(Qr)^4 - \ldots \tag{36}$$

Hence, the scattering intensity (the differential cross section) is obtained from Equation (35) as follows,

$$\frac{d\Sigma}{d\Omega} \equiv n_p |f(Q)|^2 = n_p (\Delta\rho)^2 F(Q) = n_p V_p^2 (\Delta\rho)^2 \left[1 - \frac{R_g^2 Q^2}{6} + \ldots\right]^2 \tag{37}$$

Here, n_p is the number density of the scattering elements. $F(Q)$ is called the form factor and represents the shape and size of the scattering elements.

When $R_g Q \ll 1$, Equation (37) is approximated to

$$\frac{d\Sigma}{d\Omega} = n_p V_p^2 (\Delta\rho)^2 \exp\left[-\frac{R_g^2 Q^2}{3}\right] \tag{38}$$

This is called *Guinier law*. By plotting observed scattered intensity $\ln[d\Sigma/d\Omega](Q)$ as a function of Q^2, one obtains the radius of gyration of the scattering elements, R_g,

$$R_g^2 \equiv \frac{\int_{V_p} d^3 r \rho(\mathbf{r}) r^2}{\int_{V_p} d^3 r \rho(\mathbf{r})}. \tag{39}$$

8.3.1.2 Porod Law and Scattering Invariant

For large Q, on the other hand, asymptotic behaviors of the scattering intensity are described with the so-called Porod law (a Q^{-4} power law). This power law is applied to any two-phase systems having a sharp interface. According to Porod, the correlation function for isolated particle systems $\gamma_0(r)$ is given by [16]

$$\gamma_0(r) = 1 - \frac{S_p}{4V_p} r + \ldots \tag{40}$$

Here, S_p is the surface area of the particle. As a result, the scattering intensity for an ideal two-phase system is given by

$$\frac{d\Sigma}{d\Omega}(Q \to \infty) = (\Delta\rho)^2 \frac{2\pi S_V}{Q^4}, \tag{41}$$

where S_V is the specific area (the surface area S_p divided by its volume V_p; $S_V \equiv S_p/V_p$). This is so-called *Porod law* and this law is widely applicable to two-phase systems with sharp interface. Furthermore, the *scattering invariant* Q_{inv} is obtained for isotropic two-phase systems by integrating the scattering intensity with respect to Q, which is only dependent on the composition $\phi(0 < \phi < 1)$ and the scattering contrast $\Delta\rho$, as follows,

$$Q_{inv} = 2\pi n_p V_p^2 (\Delta\rho)^2 \phi(1 - \phi) \tag{42}$$

In addition to these fundamental functions commonly used in SANS, a variety of scattering functions are proposed based on various models. By fitting observed scattering functions with these model scattering functions, structural parameters, such as the size, interparticle distance, orientation, the outer and inner radius of micelles/vesicles, the number density of the particles, are evaluated.

8.3.1.3 Ornstein-Zernike and Debye-Bueche Functions

In the case of nonparticulate systems, the *Ornstein-Zernike function* [17] (one-phase systems with density/concentration fluctuations) or *Debye-Bueche function* [18] (two-phase systems) are often used. The Ornstein-Zernike (OZ) function describes a system, which is macroscopically one phase but is microscopically fluctuating in density or concentration, such as density-fluctuating systems near critical point, binary systems, semidilute polymer solutions, and polymer gels.

$$\frac{d\Sigma}{d\Omega}(Q) = \frac{d\Sigma}{d\Omega}(Q=0)\frac{1}{1+\xi^2Q^2}, \tag{43}$$

where ξ is the correlation length characterizing the length scale of the correlation or screening. The OZ function is often used to describe critical phenomena and polymer solutions in semidilute regime. See Appendix A.5 for more details. On the other hand, the Debye-Bueche (DB) function is applied to two-phase systems, such as an "island-and-sea" structure, having a characteristic "island" size of Ξ and is written in the form,

$$\frac{d\Sigma}{d\Omega}(Q) = \frac{d\Sigma}{d\Omega}(Q=0)\frac{1}{\{1+\Xi^2Q^2\}^2}. \tag{44}$$

Note that the asymptotic behaviors of the OZ and DB functions are different, i.e., $\propto Q^{-2}$ and $\propto Q^{-4}$, respectively. The latter corresponds to the Porod law.

8.3.1.4 Interparticle Interference and Structure Factor

In the dilute limit, the scattering intensity is given by

$$\frac{d\Sigma}{d\Omega} = n_p V_p^2 (\Delta\rho)^2 F(Q) \tag{45}$$

However, in most cases, the interparticle interference effect cannot be ignored. The interparticle interference effect is represented by the structure factor $S(Q)$ and the scattering intensity is rewritten to

$$\frac{d\Sigma}{d\Omega} = n_p V_p^2 (\Delta\rho)^2 F(Q) S(Q). \tag{46}$$

Here, $S(Q)$ is related to the radial distributing function $g(r)$ as follows,

$$S(Q) = 1 + 4\pi n_p \int_0^\infty dr[g(r)-1]\frac{\sin(Qr)}{Qr}r^2. \tag{47}$$

The details of interparticle functions are described in Appendix A.

8.3.1.5 Unified Exponential/Power-Law Approach

Scattering functions often comprise a Guinier and associated power-law regime. Beaucage proposed a unified function to cover the two regimes without additional parameter [19].

$$I(Q) = G\exp(-Q^2R_g^2/3) + B(\Delta\rho)^2\left\{\frac{\{\mathrm{erf}(QR_g/6^{1/2})\}^3}{Q}\right\}^P, \tag{48}$$

where G is the exponential prefactor ($G = n_pV_p^2(\Delta\rho)^2$; for neutron scattering) and B is a constant prefactor specific to the type of power-law scattering observed as determined by the regime in which the exponent P falls. For Porod's law scattering, $P = -4$ and $B = 2\pi n_p(\Delta\rho)^2S_p$.

8.3.1.6 Contrast Variation

The scattering function for an n-component system, with component 1 being the reference component, can be written by

$$I(Q) = \sum_{i=2}^{n}(\rho_i - \rho_1)^2S_{ii}(Q) + 2\sum_{1<i<j}^{n}(\rho_i - \rho_1)(\rho_j - \rho_1)S_{ij}(Q). \tag{49}$$

Here, ρ_i is the scattering length density of the ith component and the systems are assumed to be incompressive. By conducting v-SANS experiments with more than n different contrasts (i.e., $v \geq n$) and solve v-simultaneous linear equations with respect to $\{(n - 1) + (n - 1)(n - 2)/2\}$ unknowns, one obtains $\{(n - 1) + (n - 1)(n - 2)/2\}$-partial scattering functions, $S(Q)_{ij}$. For example, for a three-component system ($n = 3$), there are three independent partial scattering functions. This method is called "contrast variation" (CV) method. The CV method is unique to neutron scattering and is one of the most powerful methods of neutron scattering by using the advantage of neutron. The details of the CV method are described in Appendix B and examples are given in Sections 8.7.5.1 and 8.8.5.2.

8.3.2 Neutron Reflectivity (NR)

Since the refractive index of neutron is slightly smaller than unity, it undergoes total reflection when it is incident to a flat and smooth surface. By using this nature, various neutron devices such as neutron guides, monochromators, and neutron polarization devices are developed. Neutron reflectivity (NR) is also used to investigate the surface structure and depth profiling of soft condensed matter since the 1980s [20, 21]. Recently, NR instruments are installed in most neutron facilities [4]. For example, J-PARC (Japan Proton Acceleration Research Complexes) has a horizontal NR, SOPHIA, and a vertical and polarization NR, SHARAKU [14].

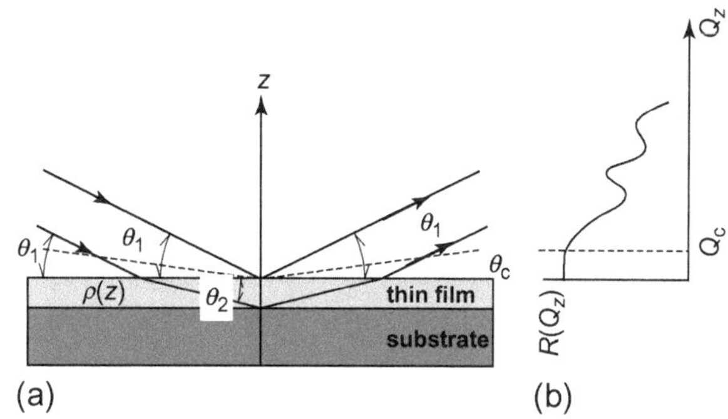

FIGURE 10 Principle of neutron reflectivity. (a) A neutron beam is incident and reflected with the angle θ_1 and is detected by a thin film as a function of Q_z. The neutron-scattering density depth profile $\rho(z)$ can be obtained by NR. Here, θ_c is the critical angle for total reflection. (b) A schematic NR profile $R(Q_z)$.

Figure 10 shows the principle of NR. When a neutron beam is incident to a thin film with a very small incident angle less than the critical angle θ_c, it undergoes total reflection. On the other hand, a neutron beam incident with the incident angle larger than θ_c penetrates into the film and is scattered in the film or reflected at the bottom surface (i.e., the film-substrate interface), and reaches a detector with carrying the information about the scattering length density variation along z-direction, $\rho(z)$. The neutron intensity obtained as a function of Q_z is called neutron reflectivity, $R(Q_z)$. Let us discuss the principle of NR in more detail.

In general, the refractive index of matter can be regarded as a real number except for some elements, such as boron B and cadmium Cd, and is related to the scattering length density ρ as follows,

$$n \cong 1 - \frac{\lambda^2 \rho}{2\pi} \tag{50}$$

As shown in Figure 11, let us define the wave functions and their amplitudes of the incident, reflected, and transmitted neutron beams, φ_{I}, φ_{R}, φ_{T} and a_{I}, a_{R}, a_{T}, respectively, for a neutron beam propagating from medium with refractive index $n = 1$ to a thin film with n. According to the condition of continuity for φ and $\mathrm{d}\varphi/\mathrm{d}z$ at the interface $z = 0$, the following equations have to be satisfied,

$$a_{\mathrm{I}} + a_{\mathrm{R}} = a_{\mathrm{T}} \tag{51}$$

$$a_{\mathrm{I}}\mathbf{k}_{\mathrm{I}} + a_{\mathrm{R}}\mathbf{k}_{\mathrm{R}} = a_{\mathrm{T}}\mathbf{k}_{\mathrm{T}} \tag{52}$$

$$k_{\mathrm{I}} = nk_{\mathrm{T}}. \tag{53}$$

These conditions lead to Snell's equation,

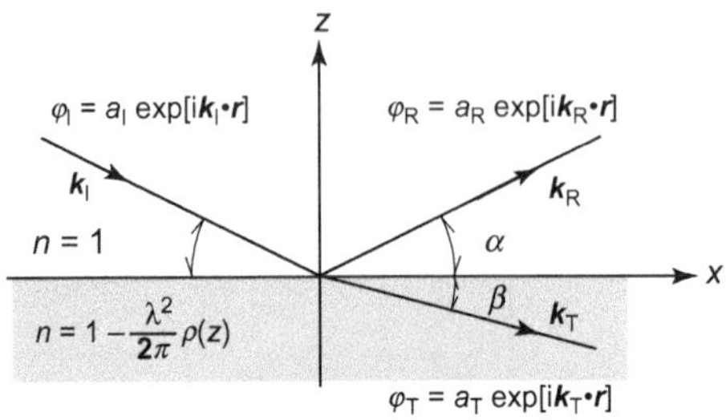

FIGURE 11 Refection of neutron by a smooth surface: Fresnel's law. $\mathbf{k}_\mathrm{I}$, $\mathbf{k}_\mathrm{R}$, and $\mathbf{k}_\mathrm{T}$ are the wave vectors of incident, reflected, and transmitted neutron beams, respectively.

$$n_1 \cos \alpha_1 = n_2 \cos \alpha_2 \tag{54}$$

and the Fresnel equation for reflectivity

$$r = \frac{\sin \alpha - n \sin \beta}{\sin \alpha + n \sin \beta} \tag{55}$$

As a result, the neutron reflectivity R is given by

$$R \equiv rr^* = \left| \frac{\sin \alpha - n \sin \beta}{\sin \alpha + n \sin \beta} \right|^2. \tag{56}$$

Usually, R is measured as a function of Q_z, and a depth profile $\rho(z)$ is obtained by analyzing $R(Q_z)$. In more general, the reflected amplitude from a medium $n = n_1$ to one with $n = n_2$, r_{12}, is given by

$$r_{12} = \frac{n_1 \sin \alpha_1 - n_2 \sin \alpha_2}{n_1 \sin \alpha_1 + n_2 \sin \alpha_2}. \tag{57}$$

In the case of neutron reflection from a smooth flat surface as shown in Figure 10, R is given by assuming single scattering (Born approximation),

$$R \equiv R_\mathrm{F} = \frac{16\pi^2 \rho^2}{Q_z^4} \tag{58}$$

Here, R_F is called Fresnel reflectivity. If a thin film has a finite thickness, one needs to integrate the reflectivity along the z-direction,

$$R = R_\mathrm{F} \left| \int \frac{d\rho(z)}{dz} e^{iQ_z z} dz \right|^2 \tag{59}$$

This is a modulation of Fresnel reflectivity and the structural information along z is obtained from $d\rho(z)/dz$. For example, the neutron reflectivity from a rough surface with variance σ^2 is given by

$$R = R_\mathrm{F} (\Delta\rho)^2 \exp(-Q_z^2 \sigma^2) \tag{60}$$

Another example is the neutron reflectivity of a thin film with thickness d is obtained by

$$R = \left| \frac{r_{01} + r_{12} \exp(2i\beta)}{1 + r_{01} r_{12} \exp(2i\beta)} \right|^2 \tag{61}$$

$$\beta = \frac{2\pi}{\lambda} n_1 d \sin\theta \tag{62}$$

As described in Section 8.6.5, NR has been extensively used in soft condensed matter science since the late 1980s. A brief review is given in Section 8.8.3.5 together with a future direction of NR. For more details in the theory and applications, see references [4, 7].

8.3.3 Inelastic Neutron Scattering (INS)

As discussed in Section 8.1.2, the energy of thermal and cold neutrons is of the same order of those of molecular motions and excitations. Hence, neutron inelastic scattering (INS), more often called "quasielastic neutron scattering (QENS)," is used for investigations of the dynamics of soft condensed matter, such as local segment motions of polymer chains, glass transition, oscillation of vesicles, local motions of protein, etc. (Sec. 8.1.4).

8.3.3.1 Quasielastic Neutron Scattering (QENS)

In order to investigate the dynamics, one observes the changes of neutron energy $\hbar\omega$ (or neutron velocity v) before and after scattering. The energy change of neutrons by scattering is given by

$$\hbar\omega = \frac{mv'^2}{2} - \frac{mv^2}{2} \cong mv(v' - v) \tag{63}$$

The neutron velocity changes by collision of a neutron with a scattering element ($v \to v'$). For inelastic scattering measurements, an energy change is analyzed by a triple axis spectrometer or a various types of time-of-flight (TOF) spectrometers [14]. A comprehensive review on QENS is given by Sakai [5]. However, since the observable variable in inelastic scattering is a second derivative of the cross section with respect to the scattering angle and wavenumber, i.e., $\partial^2\sigma/\partial\Omega\partial\omega$, the intensity is much smaller than that in elastic scattering. Hence, it takes much longer time for obtaining data with suitable counting statistics than light scattering or X-ray scattering. In order to circumvent this disadvantage, neutron spin echo (NSE) spectroscopy is often used for soft condensed matter science [22].

8.3.3.2 Neutron Spin Echo (NSE)

NSE, developed by Mezei [23, 24], has the following advantages; (1) a high energy resolution is achieved by separating wavelength distribution and energy dispersion, (2) energy resolution is better than nano eV, and (3) it allows one to observe the intermediate correlation function $S(Q, t)$ directly. Here, let us describe the basic concept of NSE in the following [7]. The details of NSE are described in Ref. [25].

The probability that an energy change given by Equation (63) is described by the intermediate correlation function $S(\mathbf{Q}, \omega)$. Here, $\mathbf{Q}$ is the momentum transfer (the scattering vector),

$$\hbar\mathbf{Q} = m\mathbf{v}' - m\mathbf{v} \tag{64}$$

As shown in Figure 12, NSE allows one to observe neutron energy change by scattering as a phase shift of Larmor precession motion [7].

(1) Incident neutron spin $(0, 0, p_z)$: Align the incident neutron spin parallel to z-axis.

(2) $\pi/2$ flip: Turn the neutron spin by 90° to $(0, -p_z, 0)$ with respect to z-axis using a $\pi/2$ flipper.

(3) The first presession $(p_z \sin\phi, -p_z \cos\phi, 0)$: The neutron spin undergoes a Larmor precession during propagating in the 1st precession coil. The phase shift ϕ_1 is given by

$$\phi_1 = \frac{\gamma H l}{v} \tag{65}$$

This is proportional to the magnetic field H, the Larmor frequency γ, and the traveling distance in the precession coil l, but is inversely proportional to the neutron velocity v.

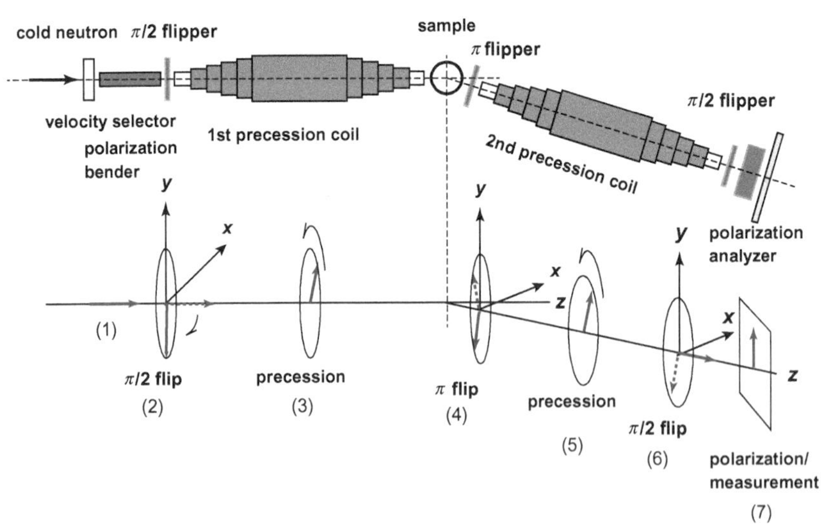

FIGURE 12 Schematic of neutron spin echo spectrometer (upper) and the spin directions at each step (lower).

(4) π flip $(p_z \sin\phi_1, p_z \cos\phi_1, 0)$: Invert one of the axes x or y by using π flipper, and flip the phase; $\phi_1 \rightarrow \phi_2$

(5) The 2nd precession $(p_z \sin(\phi_1 - \phi_2), p_z \cos(\phi_1 - \phi_2), 0)$: Similar to the case of 1st precession, the neutron undergoes a Larmor precession. The resultant phase shift is a function of ω given by

$$\phi_1 - \phi_2 = \frac{\gamma Hl}{v} - \frac{\gamma Hl}{v'} \cong \frac{\gamma Hl}{v^2}(v' - v) = \frac{\gamma Hl}{v^2}\hbar\omega = \frac{\gamma Hl m^2 \lambda^3}{2\pi h^2}\omega \tag{66}$$

(6) $\pi/2$ flip $(p_z \sin(\phi_1 - \phi_2), 0, p_z \cos(\phi_1 - \phi_2))$: Turn one of the components of the spin to the z-direction.

(7) Polarization $(0, 0, p_z \cos(\phi_1 - \phi_2))$: Measure the polarization P of the neutron spin.

$$P = \langle \cos(\phi_1 - \phi_2) \rangle = \left\langle \frac{\gamma Hl m^2 \lambda^3}{2\pi h^2}\omega \right\rangle \tag{67}$$

$$= \frac{\int S(\mathbf{Q}, \omega) \cos(\omega t) d\omega}{\int S(\mathbf{Q}, \omega) d\omega} \tag{68}$$

$$= \frac{S(\mathbf{Q}, t_\mathrm{F})}{S(\mathbf{Q})} \tag{69}$$

Here, $S(\mathbf{Q})$ in the denominator is the static structure factor (proportional to the scattering intensity) and t_F is the relaxation time called Fourier time and is given by

$$t_\mathrm{F} = \frac{\gamma Hl m^2 \lambda^3}{2\pi h^2} \tag{70}$$

On the other hand, the numerator $S(\mathbf{Q}, t_\mathrm{F})$ is a time correlation function called the intermediate correlation function and is corresponding to the time correlation function $g^{(1)}(Q, t)$ in dynamic light scattering (DLS) [26, 27]. The difference between NSE and DLS lies in its spatial range, i.e., the Q range. That is, DLS covers the dynamic range of colloidal objects (from a few tens nm to μm order), while NSE covers from a few to a few tens nm. The characteristic feature of NSE is to use neutron Larmor precession just like a stopwatch and to measure the energy and momentum transfer change with a high precision. Note that the scattering intensity obtained by NSE $I(Q, t_\mathrm{F})$ is related to the correlation function $S(Q, \omega)$ in the frequency space via Wiener-Khinchin theorem,

$$I_\mathrm{NSE}(Q, t_\mathrm{F}) \sim \int_{-\infty}^{\infty} S(Q, \omega) \frac{1 + \cos(\omega t_\mathrm{F})}{2} d\omega \tag{71}$$

NSE spectroscopy is extensively used for investigations of soft condensed matter, particularly in glass transition of polymer [28], melt-polymer dynamics [29, 30], polymer solutions [31], microemulsions [32], and protein solutions [33, 34], and the technology related to NSE spectroscopy is still on the stage of developing [14]. A breakthrough work and a summary of a comprehensive review are given in Sections 8.6.4 and 8.8.3.6, respectively.

8.4 POLYMERIC SYSTEMS

8.4.1 Polymer and Neutron Scattering

Polymer is a generic name for a large molecule consisting of a number of chemically bonded small monomeric units called "monomers" or "segments." Most polymers are chain-like large molecules and do not have a given shape and take a random conformation. Hence, polymer chains are usually treated statistically. The principal properties of polymers were described by Flory [35]. The framework of the theory of polymers was established in the 1950s as symbolized by the famous textbook "*Principles of Polymer Chemistry*," and the theory of polymer solutions was established in the 1970s [36, 37]. The mean-field picture of the theories, however, had various limitations when applying to semidilute, concentrated solutions and bulk, because of a many-body problem.

de Gennes circumvented this limitation and developed a scaling theory of polymer physics. de Gennes pointed out in his book that "a second stage in the physics of polymers has evolved, because of the availability of new experimental and theoretical tools" [38]. Those are neutron diffraction, (laser dynamic) light scattering, and theoretical methods including functional integrals, Feynman diagrams, many-body theories, and Monte Carlo simulations. *It should be noted here that neutron diffraction (neutron scattering) was one of the breakthrough methodology in development of polymer science.*

The development of polymer physics was followed by Doi-Edwards theory on the dynamics of polymer melt and concentrated solutions in their book entitled "*The Theory of Polymer Dynamics*" [39]. They generalized a new concept for polymer dynamics, "reptation" proposed by de Gennes [40] and constructed a systematic theory on polymer dynamics. Reptation is the thermal motion of long polymer chains in melt or concentrated polymer solutions. Figure 13 schematically shows reptation of a polymer chain. A long polymer chain is allowed only a sliding motion along its length direction because of topological constraints of neighboring chains. Since such motion is analogous to a motion of reptile, this thermal motion is called reptation. The hypothesis of reptation was experimentally verified by neutron spin echo measurements, which will be described later (see Sections 8.6.4 and 8.8.3.6) and references cited in Section 8.3.3 (melt-polymer dynamics [29, 30] and polymer solutions [31]).

The most significant characteristic feature of polymer is that the space occupied by a polymer chain is much larger than its net volume, Nv (N; the degree of polymerization, v; the monomer molecular volume), just like a "cotton floss." If N is large enough, a polymer solution is space-filled by

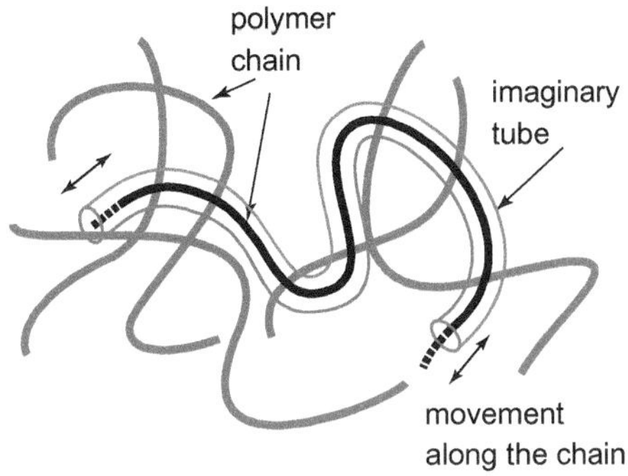

FIGURE 13 Relaxation of polymer chain. A polymer chain are allowed to move only along its length direction as if it is in a tube.

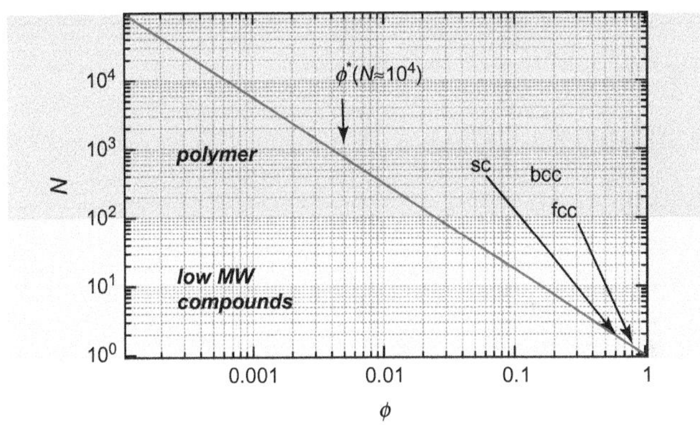

FIGURE 14 Comparison of space-filling ratio between polymer ($N \gg 1$) and low-molecular weight compounds ($N \approx 1$).

the polymers even the polymer concentration is very low. Figure 14 schematically shows the relationship between the polymer volume fraction ϕ and N. It is well known that the space-filling fractions for simple cubic (SC), body-centered cubic (bc), and face-entered cubic (fcc) are 0.52, 0.68, and 0.74, respectively. On the other hand, the polymer volume fraction for space filling could be as low as 5×10^{-4} for polystyrene with $N = 10^4$. This is one of the reason why a super-adsorbent polymer made of cross-linked polymer networks can absorb a few hundred times as large amount of water as its weight (see Section 8.5).

As shown in Figure 15, polymer chains in a dilute solution are localized ($\phi \ll \phi^*$; regime I), and the local concentration is finite in the "envelope" of the polymer chain and none outside. Thus a dilute polymer solution is not a mean-field system. When the polymer concentration is increased, there appears a concentration at which polymer chain envelopes overlap (semidilute polymer solutions; $\phi \approx \phi^*$; regime II). By further increasing concentration,

FIGURE 15 Concentration dependence of polymers: I, dilute solution; II, semidilute solution; and III, concentrated solutions.

polymer chains are heavily overlapped with each other (concentrated or molten state, $\phi \gg \phi^*$; regime III). This concentration is called the chain-overlap concentration (or C^* concentration) [38]. Let us discuss the scattering functions for polymer solutions in each regime.

8.4.2 Scattering Functions of Polymeric Systems

8.4.2.1 Radius of Gyration and Debye Function

Generally, a polymer chain takes a random coil conformation. Such a polymer chain is described statistically as follows,

$$P(R,N) = \left(\frac{3}{2\pi Na^2}\right)^{3/2} \exp\left(-\frac{3R^2}{2Na^2}\right) \tag{72}$$

Here, $P(R, N)$ is the probability function of finding chain ends at distance R (the end-to-end distance) for a chain with the segment length of a and the degree of polymerization (the number of segments in a chain) of N.

A polymer chain is not necessary to be a linear chain, but there are various types, such as branched, star shaped, and comb-like chains. In order to describe such chains, the end-to-end distance R is not relevant, and the radius of gyration R_g is a more common and universal variable to describe the polymer size.

Take a look at Figure 16. The radius of gyration is defined by

$$\mathbf{R}_G = \frac{1}{N}\sum_{n=1}^{N}\mathbf{R}_n \tag{73}$$

$$R_g^2 = \frac{1}{N}\sum_{n=1}^{N}\left\langle(\mathbf{R}_n - \mathbf{R}_G)^2\right\rangle \tag{74}$$

$$= \frac{1}{2N^2}\sum_{n=1}^{N}\sum_{m=1}^{N}\left\langle(\mathbf{R}_m - \mathbf{R}_n)^2\right\rangle \tag{75}$$

The radius of gyration square R_g^2 for a Gaussian chain with the segment length a and the segment number N is given by

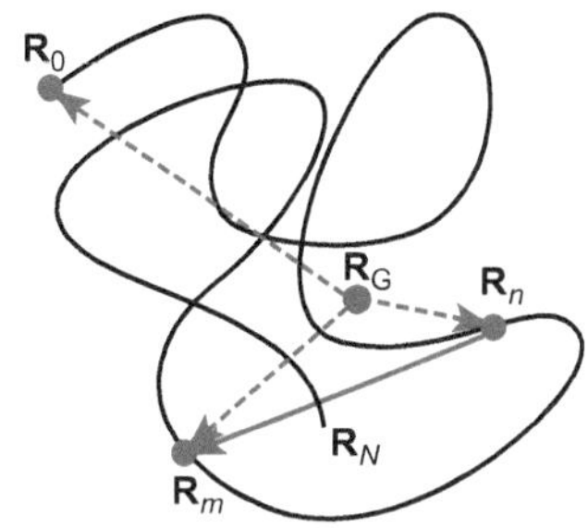

FIGURE 16 A polymer chain consisting of N segments. $\mathbf{R}_m$, the position vector of pointing segment m; $\mathbf{R}_G$, the position vector of the center of gravity.

$$R_g^2 = \frac{Na^2}{6} \tag{76}$$

The corresponding scattering function is obtained by,

$$g_D(Q) = \frac{1}{N}\sum_{n=1}^{N}\sum_{m=1}^{N}\langle \exp[i\mathbf{Q}\cdot(\mathbf{R}_m - \mathbf{R}_n)].\rangle \tag{77}$$

$$= \frac{1}{N}\sum_{n=1}^{N}\sum_{m=1}^{N}\exp\left[-\frac{|m-n|}{6}a^2Q^2\right] \tag{78}$$

After double summation, it reads,

$$g_D(u) = \frac{2N}{u^2}(e^{-u} - 1 + u) \tag{79}$$

$$u \equiv \frac{Na^2}{6}Q^2 = R_g^2Q^2 \tag{80}$$

Equation (79) is called *Debye function* and this is the basic scattering function appearing in discussions of various polymeric systems. Now, let us discuss the scattering functions of polymeric systems in different concentration regimes.

8.4.2.2 Regime I: Dilute Solutions

In this concentration regime, individual polymer chains are not overlapped and undergo Brownian motion. Hence, the scattering function is given as a sum of Debye function as given by the following function,

$$\frac{d\Sigma}{d\Omega} = n_p V^2 (\Delta\rho)^2 g_D(u) \tag{81}$$

$$\approx n_p V^2 (\Delta\rho)^2 e^{-R_g^2Q^2/3} \tag{82}$$

for small $Q(\ll R_g)$ region. Here, n_p is the number density of the polymer chains and $\Delta\rho$ is the scattering length density difference between polymer and solvent. Note that Equation (82) is the same as the Guinier function

(Equation (38)). When the polymer-solvent and/or polymer-polymer interaction have to be taken into account, Equation (82) needs to be modified to

$$\frac{HC}{\mathrm{d}\Sigma/\mathrm{d}\Omega}=\frac{1}{M_\mathrm{w}}\left[1+\frac{R_\mathrm{g}^2}{3}Q^2+\ldots\right]+2A_2C+\ldots \tag{83}$$

This equation is called *Zimm equation*, and the weight-average molecular weight M_w and the second virial coefficient A_2 can be obtained from a Zimm plot. Here, H and C are the contrast factor and the polymer concentration, respectively. Note that the radius of gyration R_g in Equation (83) is the z-averaged radius of gyration.

If polymer chains are in a good solvent, A_2 is positive, and individual polymer chains are expanded longer than their unperturbed dimension. The internal polymer concentration in the envelope of the polymer chain is $\phi_\mathrm{int} \simeq Na^3/R_F^3$. Here, $R_\mathrm{F} = N^{3/5}a$ is the so-called Flory radius.

8.4.2.3 Regime II: Semidilute Solutions

When the polymer concentration becomes high enough to touch each other ($\approx \phi^* \simeq Na^3/R_F^3$), a screening effect by neighboring chains becomes significant and individual polymer chains are no more visible by scattering. Such a polymer solution is regarded as an assembly of "blobs" consisting of subpolymer chains as shown in Figure 17. In this case, even a long polymer chain can be regarded as an "ideal" polymer chain consisting of blobs. Here, the blob size is characterized by the so-called correlation length ξ. In this case, the correlation function $\gamma(r)$ becomes the Yukawa potential (Debye-Hückel potential) type function and is given by

$$\gamma(r)=\frac{\xi}{r}e^{-r/\xi} \tag{84}$$

The corresponding scattering function is obtained by its Fourier transform,

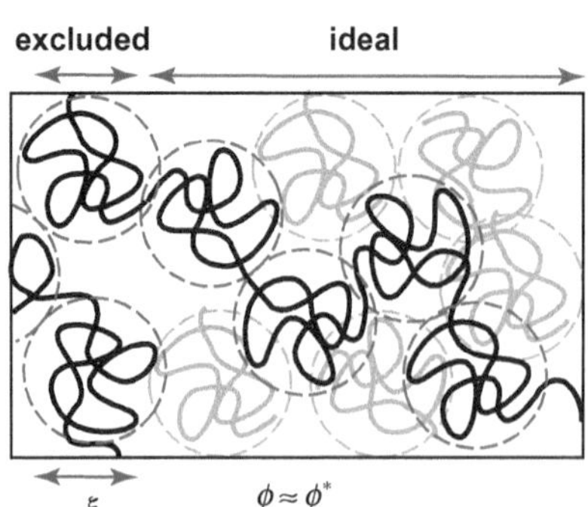

FIGURE 17 Blob model in polymer solution. In a semidilute solution, $\phi \approx \phi^*$, the range where excluded volume effect reaches is limited to the screening length, i.e., the correlation length, ξ, due to screening effect of other chains. Hence, a long polymer chain (highlighted with black curve) can be regarded as an "ideal" polymer chain consisting of blobs with size ξ.

$$\frac{d\Sigma}{d\Omega} \propto \frac{1}{1+\xi^2 Q^2} \tag{85}$$

This is well-known *Ornstein-Zernike function* as discussed in Section 8.3.1.3 and in Appendix A.5. Hence, such a scattering behavior of semidilute polymer solution is analogous to a gaseous system with density fluctuations or a binary system with concentration (or composition) fluctuations [41]. In the case of gaseous systems and binary systems, the scattering intensity and the correlation length diverge by approaching the critical point, while a smooth crossover occurs in polymer solutions by increasing ϕ at ϕ^*.

8.4.2.4 Regime III: Concentrated Solutions and Polymer Blends

In concentrated polymer solutions and melt (Figure 15 (right)), concentration fluctuations are suppressed to the segment level and the scattering intensity becomes weak and the scattering function does not carry any information about the polymer chains (such as the molecular mass N or size R_g). However, in the case of polymer blends having different scattering length densities, e.g., hydrogenous and deuterated polymer mixtures, the scattering function of such blend systems provides significant information. The scattering function for polymer blends is given below,

$$\frac{1}{S(Q)} = \frac{1}{\phi_A N_A g_D(Q;N_A)} + \frac{1}{\phi_B N_B g_D(Q;N_B)} - 2\chi \tag{86}$$

This is called *de Gennes scattering function* [38]. Here, $S(Q)$ is the response function (proportional to the scattering intensity), and ϕ_i and N_i are the volume fraction and the degree of polymerization of the component i, respectively. χ is the interaction parameter. $g_D(Q;N_i)$ is the Debye function for the component i and is given by

$$g_D(Q;N_i) = \frac{2}{u_i^2}\left(e^{-u_i} - 1 + u_i\right) \tag{87}$$

For asymmetric polymer blends having different segment volumes, Equation (86) needs to be modified to [42]

$$\frac{1}{v_0 S(Q)} = \frac{1}{v_A \phi_A N_A g_D(Q;N_A)} + \frac{1}{v_B \phi_B N_B g_D(Q;N_B)} - 2\frac{\chi}{v_0} \tag{88}$$

Here, v_A and v_B are the segment volumes of component A and B, respectively, and v_0 is the geometric mean of v_A and v_B, i.e., $v_0 = \sqrt{v_A v_B}$. This de Gennes function is often used to study the miscibility or critical phenomena of binary polymer blends (see Section 8.6.2).

8.4.2.5 Block Copolymers in Bulk or Concentrated Systems

Block copolymers are a class of polymers in which different kinds of polymer chains are chemically bonded. There are A-B type diblock copolymers,

A-B-A type triblock copolymers, A-B-C type triblock copolymers, and so on. Block copolymers have been used for polymer surfactant (see, for example, Section 8.7.1), high-impact plastics, and other applications. Recently, block copolymers are applied to more advanced materials, such as nano-fabrication and quantum dots. Because of the chemical junction, block copolymer cannot undergo ("macroscopic") phase separation, while it undergoes "microphase" separation. The scattering function for molten block copolymers was derived by Leibler [43].

In the case of A-B type block copolymers with the segment numbers of fN and $(1 - f)N$, the scattering function is given by

$$S(Q) = \frac{1}{H(x) - 2\chi} \tag{89}$$

$$H(x) = \frac{f_D(1,x)/N}{f_D(f,x)f_D(1-f,x) - \frac{1}{4}[f_D(1,x) - f_D(f,x) - f_D(1-f,x)]^2} \tag{90}$$

$$f_D(f,x) = \frac{2}{x^2}[e^{-fx} - 1 + fx] \tag{91}$$

$$x \equiv \frac{Q^2Na^2}{6} \tag{92}$$

Here, χ is the interaction parameter of A and B.

Figure 18 shows the scattering functions of (a) a polymer blend with $N_A = N_B = 100$ and $\chi = 0$ and of (b) a block copolymer with $N = 100, f = 0.5$, and

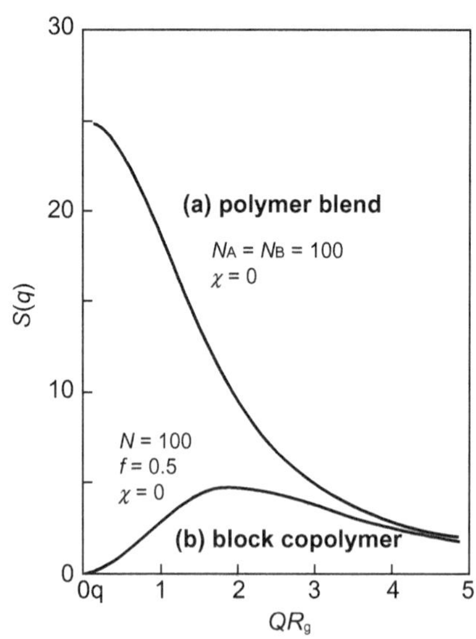

FIGURE 18 Comparison of scattering functions for (a) polymer blend with $N_A = N_B = 100$ and $\chi = 0$ and (b) block copolymer with $N = 100, f = 0.5$, and $\chi = 0$.

$\chi = 0$. One of the marked difference is that $S(Q = 0) = 0$ for the block copolymer. This is because the concentration fluctuations of block copolymer vanishes in the long scale, i.e., the thermodynamic limit ($Q \rightarrow 0$) due to the chemical bonding between A and B chains.

Equation (92) has a scattering maximum around $Q \approx 1/a\sqrt{N}$. When the product of χ and N, χN, becomes larger than the critical value, $\chi N(> 10.5)$, the block copolymer undergoes microphase separation and its scattering function cannot be given by Equation (92). After this function was proposed, investigations of order-disorder transition, evaluation of miscibility have been extensively developed and are hot topic even today (see Section 8.7).

8.5 RUBBERS AND GELS

Rubbers and gels are soft condensed matter consisting of long polymer chains interconnected to each other via chemical or physical cross-linkers. Rubbers have been used in daily life for more than a century, as tires, rubber rings, shock absorbers/dumpers, and so on. Gels, on the other hand, have even longer history in human life as food and glues. Recently, gels are also used in cosmetics and in medical cares, etc. Gels are classified, depending on the medium, to hydrogels (water), organogels (oil), or aerogels (air) [44, 45]. Rubbers and gels have been extensively investigated by scattering methods since 1980s. A highlighted event was the workshop on the "Molecular Basis of Polymer Networks" held in 1988 in Jülich. The preface of the proceeding says "*During the last decade the advent of Small-Angle Neutron Scattering (SANS) has allowed the molecular structure of polymer networks to be studied and thus the reliability of the theories to be tested directly at the secular level...*" [46].

One of the reasons why polymer networks had been studied by SANS in early days was the need to elucidate the structure and deformation/reinforcing mechanism of tire, i.e., vulcanized and carbon-reinforced polymer networks. Recently, SANS studies on gels are a hot topic because of several novel gels having extraordinary toughness have been developed, such as slide-ring gels, double network gels, nanocomposite gels, and tetra-PEG gels. In this section, a brief outline of SANS studies on polymer networks is given. Several references are listed for further reading [47–49].

8.5.1 Scattering Functions of Polymer Gels: Effects of Cross-Linking

Because of introduction of cross-links, the structure of polymer networks is markedly different from that of polymeric systems. In most cases, the scattering intensity increases exclusively in a low Q-region. Such an increase has been observed since the late 1960s and called "cross-linking inhomogeneities" [50, 51]. In the 1990s, it was recognized that the concentration fluctuations in polymer gels consist of thermal fluctuations and static (spatial)

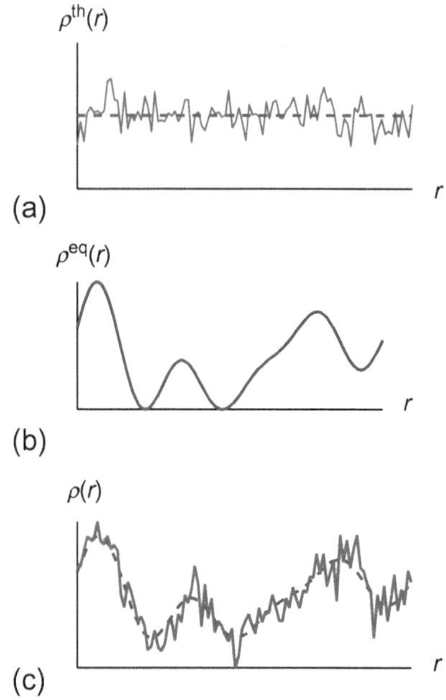

FIGURE 19 Schematic representation of the density profiles of (a) the thermal density fluctuations $\rho^{th}(r)$, (b) the inhomogeneous equilibrium density fluctuations introduced by cross-linking $\rho^{eq}(r)$, and (c) the concentration profile containing both dynamic and frozen fluctuations $\rho(r)$. *Reproduced from Ref. [57].*

inhomogeneities [46, 52–56]. The former are the dynamic fluctuations similar to those present in semidilute polymer solutions. However, the latter are characteristic of gels and are originated from cross-linking.

As shown in Figure 19, polymer gels have two types of concentration fluctuations. This concept comes from the theory of "Statistical Physics of Polymer Gels" by Panyukov and Rabin [55]. Before going into the sophisticated theory, let us discuss phenomenological theories of polymer gels. Polymer solutions have only thermal density fluctuations, $\rho^{th}(r)$ (a). These fluctuations have zero-mean average both in time and ensemble averages (ergodic medium). On the other hand, by introducing cross-links, the concentration fluctuations are fixed to some extent, at least topological points of view and inhomogeneous equilibrium density fluctuations, $\rho^{eq}(r)$, are built up (b). As a result, polymer gels have two types of concentration fluctuations as schematically shown in Figure 19c and the concentration profile, $\rho(r)$, is given by,

$$\rho(r) = \rho^{th}(r) + \rho^{eq}(r) \tag{93}$$

Accordingly, the scattering intensity function of polymer gels, $I(Q)$, is described by a sum of the scattering intensities of the corresponding polymer solution, $I_{soln}(Q)$, and of an excess scattering, $I_{ex}(Q)$, as follows,

$$I(Q) = I_{soln}(Q) + I_{ex}(Q) \tag{94}$$

Here, $I_{soln}(Q)$ is the scattering function for semidilute polymer solutions. It is known that $I_{soln}(Q)$ is described by the OZ function (Equation (43)) as discussed in Section 8.3,

$$I_{soln}(Q) = \frac{I_{soln}(0)}{1 + \xi^2 Q^2} \tag{95}$$

where ξ is the correlation length, i.e., a length measure of screening by neighboring polymer chains.

For the excess scattering, on the other hand, several phenomenological scattering functions are proposed. For example,

(i) a Lorentz function [46, 58]

$$I_{ex}(Q) = \frac{I_{ex}(0)}{1 + \Xi^2 Q^2} \tag{96}$$

(ii) a stretched exponential function [59]

$$I_{ex}(Q) = I_{ex}(0) \exp\{-(\Xi Q)^\alpha\} \tag{97}$$

(iii) a Debye-Bueche function [60]

$$I_{ex}(Q) = \frac{I_{ex}(0)}{(1 + \Xi^2 Q^2)^2} \tag{98}$$

where Ξ is a length scale characterizing the inhomogeneities in the gel, and α is an exponent in the range of 0.7 to 2 [61, 62]. These functions for $I_{ex}(Q)$ are introduced in order to describe additional fluctuations and/or solid-like inhomogeneities. Equation (96) assumes that the gel has another correlation length, Ξ ($\gg \xi$) in addition to ξ. Equation (97) is an extended form of the Guinier equation ($\alpha = 2$) [63], where noninteracting domains of higher or lower monomer densities are assumed to be randomly distributed in the network. For example, Geissler et al. examined the value of α and concluded $\alpha = 2$ for end-linked poly(dimethylsiloxane) swollen in toluene [64], and 0.7 for poly(vinylacetate) gels [61]. Equation (98) is the Debye-Bueche function (Equation (44)) [61], which represents a two-phase structure with a sharp boundary as discussed in Section 8.3. In the case of gels, it corresponds to highly cross-linked domains and loosely cross-linked matrix. The validity of these functions and their physical meaning will be discussed later.

It should be noted that the contribution of the static inhomogeneities is enhanced by swelling or stretching the gel. This is due to the fact that by deformation (e.g., swelling or stretching) a coarsely cross-linked region deforms larger than a densely cross-linked region, resulting in enhancement of the concentration difference between the two regions [65, 66]. Figure 20 schematically shows network structures of (a) a reactor-batch gel well above

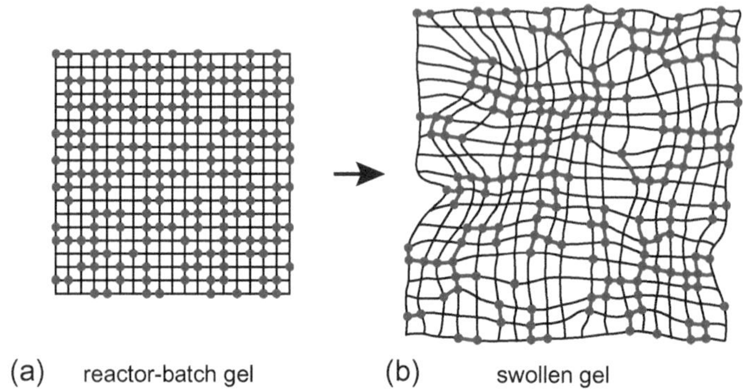

FIGURE 20 Schematic representation showing the structures of (a) a reactor-batch gel well above the chemical gelation threshold and (b) a gel in swollen state. The thin line, red circles, and thick red lines represent polymer chains, cross-linkers, and clusters of first chemical neighbor junctions (frozen blobs). Each lattice represents a blob characterized by the screening length of the excluded volume interaction ξ, while thick red clusters are characterized by Ξ. Modified from the original figure in [65]. *Reproduced from Ref. [65].*

its gelation threshold and (b) a gel in overswollen state by solvent [65]. In a reactor batch, blobs are closely packed. In addition, cross-links are not "visible" by scattering experiments although cross-links are intensionally highlighted with red circles in the figure. Hence, the scattering intensity profiles are not so different from that of the system before introducing cross-links. On the other hand, by swelling the gel with a solvent, swelling inhomogeneities are created due to nonuniform cross-linking as shown in Figure 20, resulting in large concentration fluctuations and increase in the scattering intensity, particularly in low-Q region ($Q \ll 1/\xi$). Such increase in the scattering intensity also occurs by stretching a gel as studied by Rouf et al. [53].

Figure 21 shows series of SANS profiles of (a) monomer-cross-linked PNIPA gels and (b) polymer-cross-linked PNIPA gels, where PNIPA denotes poly(*N*-isopropylacrylamide) [67]. Figure 21 (a) is a typical example of scattering behavior of polymer gels. That is, a polymer solution (no cross-links) gives the lowest scattering intensity, $I(Q)$, and $I(Q)$ increases with increasing cross-linker concentration C_{BIS}, particularly at low Q region. Here, C_{BIS} is the concentration of cross-linker (*N*, *N*′-methylenebisacrylamide). However, in the case of polymer-cross-linked gels (b), $I(Q)$ is rather insensitive to gamma ray "dose." This is due to the fact that the introduction of cross-links is just a pinning of concentration fluctuations in the system and the concentration fluctuations in the parent polymer solutions are preserved without significant disturbance. On the other hand, the monomer-cross-linked systems are prepared by simultaneous polymerization and cross-linking, which build up significant frozen-concentration fluctuations during the chemical reaction. This is one of the important notes in discussing cross-linking inhomogeneities of polymer gels.

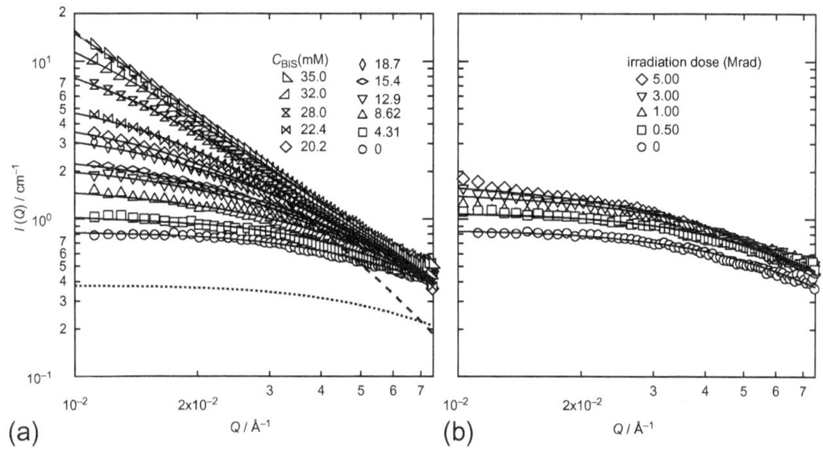

FIGURE 21 SANS profiles of polymer gels (a) made by radical polymerization of *N*-isopropylacrylamide (NIPA) monomers in the presence of different cross-linker concentrations of cross-linker C_{BIS} and (b) those prepared from poly(*N*-isopropylacrylamide) (PNIPA) solutions by gamma-ray irradiation with different doses. The lines are fits with theoretical functions described later. *Reproduced from Ref. [67].*

8.5.2 Charged Gels

Typical applications of polymer gels are water-absorbing materials, such as diapers and sanitary goods, which were realized by introduction of charge groups (dissociable groups in water), e.g., carboxyl anion ($-COO^-$) and ammonium cation ($-NH_3^+$). These charged groups accompany their counter ions, which lead to a large increase in the osmotic pressure and suppression of concentration fluctuations in a gel. As a result, an introduction of chargeable groups in a gel results in a suppression of the scattering intensity.

The scattering intensity for randomly cross-linked weakly charged gels was developed by Rabin and Panyukov [68]. The structure factor, $S(Q)$, is given by

$$S(Q) = G(Q) + C(Q) \tag{99}$$

Here, $G(Q)$ and $C(Q)$ are the correlators representing thermal density fluctuations and static (frozen) density inhomogeneities, respectively. They are given by

$$G(Q) = \frac{a^{-3}\phi N}{1 + w(Q)g(Q)}, \tag{100}$$

$$C(Q) = \frac{a^{-3}\phi N}{[1 + w(Q)g(Q)]^2(1+Q^2)^2} \times \left\{6 + \frac{9}{w_0(Q) - 1 + (1/2)Q^2(\phi_0/\phi)^{2/3}\phi_0^{-1/4}}\right\} \tag{101}$$

where a and N denote the segment length and the average degree of polymerization between cross-links, respectively. The parameters ϕ_0 and ϕ are the volume fractions of polymer at preparation and at observation, respectively. By introducing the dimensionless wave vector, $\hat{Q} \equiv aN^{1/2}Q$, $g(Q)$ is given by

$$g(Q) = \frac{1}{\hat{Q}^2/2 + (4\hat{Q}^2)^{-1} + 1} + \frac{2(\phi/\phi_0)^{2/3}\phi_0^{1/4}}{(1+\hat{Q}^2)^2} \tag{102}$$

The dimensionless quantities $w(Q)$ and $w_0(Q)$ in Eqs. (100) and (101) represent the effective second virial coefficients in the final and in the preparation states, respectively, and are given by

$$w(Q) = (1 - 2\chi + \phi)\phi N + \frac{l_B f^2 \phi N^2}{\hat{Q}^2 + l_B f \phi N}, \tag{103}$$

$$w_0(Q) = \phi^{5/4} N + \frac{l_B f_0^2 \phi_0 N^2}{\hat{Q}^2 (\phi_0/\phi)^{2/3} + l_B f_0 \phi_0^{54} N} \tag{104}$$

where χ is the Flory interaction parameter at the state of measurement. The parameters f_0 and f denote the degrees of ionization at preparation and at measurement, respectively, and l_B is the dimensionless Bjerrum length given by $l_B \equiv (4\pi/a)L_B$ with Bjerrum length, L_B. Thus, $S(Q)$ is expressed as a function of a set of parameters in the preparation state [$w_0(Q)$, ϕ_0, and f_0] and those under conditions of observation [$w(Q)$, ϕ, and f].

Figure 22 shows examples of theoretical structure factors, $S(Q)$, with different N's for (a) $\chi_c = 0.70$, (b) $\chi_c = 0.83$ [69]. Here, the other values are fixed to $\phi = \phi_0 = 0.07, f = f_0 = 0.0457$, i.e., the same condition for observation as for preparation. Note that the charge density "CD" is inversely proportional to N. The presence of the scattering peak in weakly charged gels has been reported experimentally [70–73]. In Figure 22 (a), the peak intensity, $S(Q^*)$, increases with increasing CD (decreasing N). On the other hand, in Figure 22 (b), $S(Q^*)$ increases with decreasing CD (increasing N). These figures (a) and (b) indicate that an inflection occurs in $S(Q)$ with respect to χ. In general, it has been reported that an introduction of cross-links into a gel (i.e., increasing in CD) leads to an increase in the inhomogeneities in the gels [62, 74, 75] as well as discussed in Section 8.5.1. Therefore, the observation of the inflection in $S(Q)$ is quite interesting.

The criterion for the inflection in $S(Q^*)$ is approximately given by [69]

$$\chi_c = \frac{1}{2}\left\{\phi + 1 + \frac{f}{\phi}\right\} \tag{105}$$

This analytical expression is plotted in Figure 23 with a thin dotted line. This is in good agreement with the numerical results (solid lines). It should be noted that Equation (105) defines the line along $w(0) \equiv w(Q = 0) = 0$.

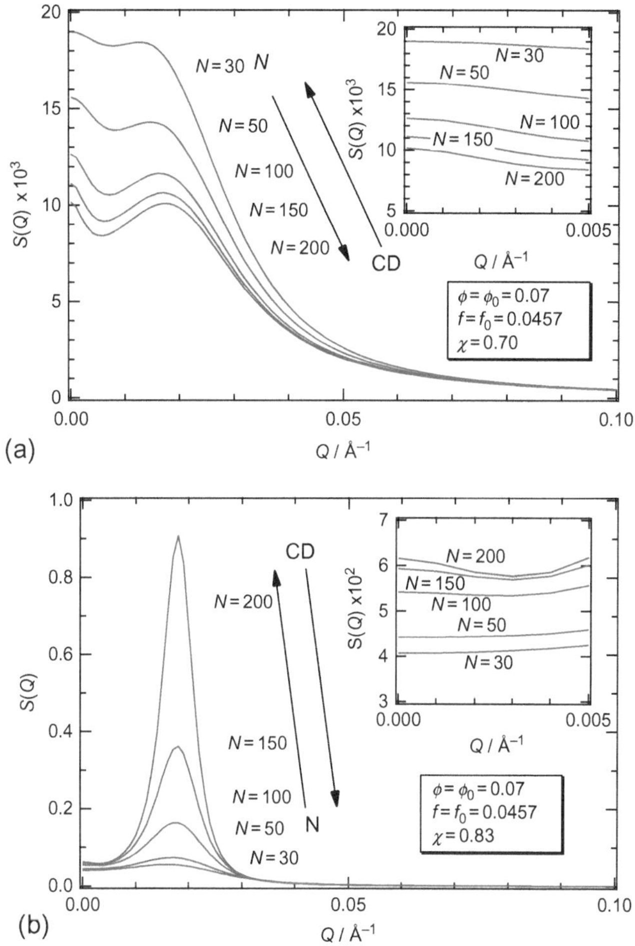

FIGURE 22 *CD* ($\propto 1/N$) dependence of the theoretical structure factor, *S*(*Q*), with $\phi = \phi_0 = 0.07, f = f_0 = 0.0457$, and (a) $\chi_c = 0.70$, (b) $\chi_c = 0.83$. The lines are fits with theoretical functions described later. *Reproduced from Ref. [69].*

Therefore, the inflection is governed by the sign of $w(Q = 0)$, which is given by

$$w(0) = N\{(1 - 2\chi + \phi)\phi + f\} \tag{106}$$

When f is large enough, there appears a region where electrostatic effects dominate ($w(0) > 0$). In this region, increasing N (decreasing CD) leads to a larger $w(0)$ and suppression of $S(Q)$ as shown in Figure 22 (a). On the other hand, for small f's, electrostatic interaction is not important and there appears a region with $w(0) < 0$. In this region, increasing N (decreasing CD) leads to a more negative $w(0)$ and increase of $S(Q)$ as shown in Figure 22 (b). This

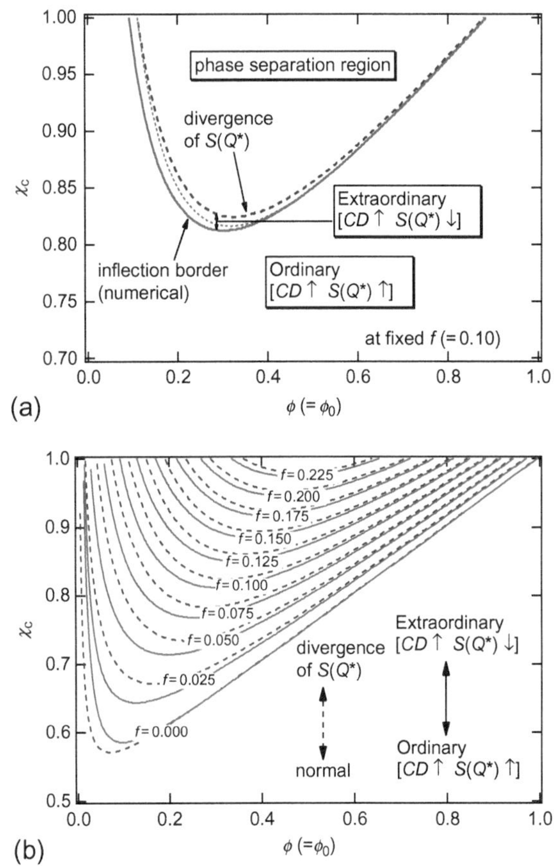

FIGURE 23 (a) ϕ dependence of the critical χ_c for a weakly charged gel ($f = 0.1$). The thick solid and broken curves represent the border at which the inflection takes place and the phase stability boundary, respectively. The analytically obtained inflection border is also shown with a thick dotted line. Below the thick solid curve, an ordinary behavior is observed where the inhomogeneity increases with CD. On the other hand, above this curve, an extraordinary behavior (opposite to the ordinary behavior) appears. (b) The same as (a) but with various fs. The lines are fits with theoretical functions described later. *Reproduced from Ref. [69].*

demonstration shows that pinning, i.e., an increase in CD, and introduction of charges f have competing effects on the structure and stability of polymer gels.

8.5.3 Physical Gels

Physical gels are defined as gels cross-linked by "physical cross-links," such as helices, crystallites, or microdomains [76]. Various types of polysaccharides, gelatin gels, oil gelators [77–79], and slime (i.e., polyvinyl alcohol-borate complex gels) [80], are also classified to physical gels.

8.5.3.1 Amorphous Physical Gels and Gelators

Guenet et al. carried out pioneering scattering studies on physical gels of poly(vinylchloride) (PVC) in organic solvent and reported that both pregel and gel are formed with fibrillar strands of stacked PVC chains [81]. Interestingly, they reported that there is no marked difference in the scattering patterns after normalization with polymer concentration as shown in Figure 24. Note that this is one of marked differences between chemical and physical gels and the structure of physical gels are similar to that of polymer-cross-linked gels (see Figure 21). A comprehensive review on physical gels composed of low-molecular mass gelators is reported by Terech [77].

8.5.3.2 Physical Gels by Crystallization

Kobayashi et al. reported physical gelation of syndiotactic polystyrene dispersed in chloroform studied by time-resolved SANS and infrared spectroscopy [82]. In this case, the scattering intensity slowly grew with time in the scale of 10 h or longer. This is due to local ordering of polymer chains with TTGG-type helical chain conformation, followed by cluster formation, where T and G denotes trans and gauche conformation, respectively.

8.5.3.3 Physical Gels by Thawing

Kanaya et al. studied poly(vinyl alcohol) (PVA) gels made by quenching PVA solutions in 60/40 (DMSO/H_2O) mixture homogenized at 100 °C to 25 or −20 °C [83–85]. They reported that SANS intensity $I(Q)$ of the PVA gels is well described by the Ornstein-Zernike (OZ) function and Porod's law,

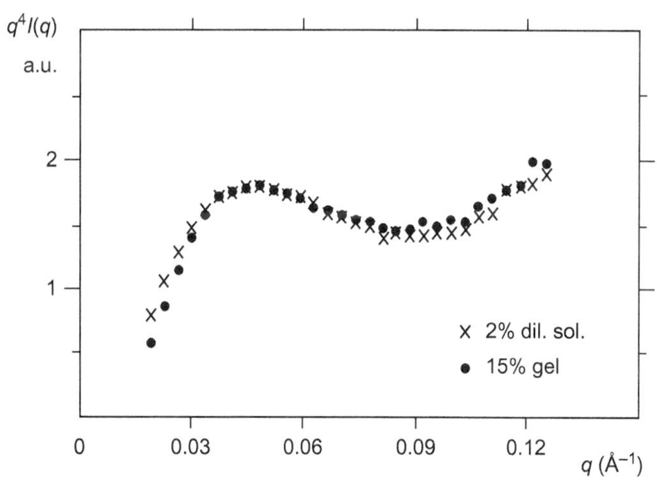

FIGURE 24 $q^4I(q)$ vs q for a 15%-gel (●) and a 2%-solution (X) of HTPVC in diethyl oxalate, where HTPVC is an abbreviation of high-temperature PVC. The scattering curve of the dilute solution has been multiplied by 7.5 to compensate the concentration effect on the scattering intensity. *Reproduced from Ref. [81].*

respectively, for low and large Q-regions. The low-Q upturn was assigned to a phase-separated structure by spinodal decomposition. In order to investigate the gelation process, they also carried out ultra-small-angle neutron scattering (USANS) together with time-resolved light scattering measurements.

8.5.3.4 Physical Gels in Block Copolymer/Solvent Systems

Amphiphilic block copolymer solutions are known to undergo sol-gel transition in water due to aggregation of hydrophobic component or microphase separation. The most familiar one may be poloxamer, consisting of poly(ethylene oxide)-poly(propylene oxide)-poly(ethyelene oxide) triblock copolymer (PEO-PPO-PEO), as will be described in Section 8.7.1. It is more commonly called Pluronic® (BASF). Here, let us introduce other examples having interesting behaviors.

Sugihara et al. investigated heat-induced sol-gel transition in aqueous solutions of pHOVE-b-pMOVE block copolymer [86], while Fuse et al. reported a water-induced gelation in acetone solution of pPhOVE-b-pMOVE block copolymer, where pHOVE, pMOVE, and pPhOVE are abbreviations of poly(hydroxyethyl vinyl ether), poly(2-methoxyethyl vinyl ether), and poly(2-phenoxyethyl vinyl ether), respectively [87]. Because of the difference in the affinity to water, these block copolymers are sensitive to temperature or solvent, resulting in aggregation or microdomain formation of more hydrophobic components. In particular, in the case of the latter [87], acetone solutions of pPhOVE-pMOVE undergo tremendous viscosity thickening by adding small amount of water as a result of formation of spherical domains highly ordered in a body-centered cubic (bcc) lattice with interdomain entanglements of tethered chains. The SANS curves also changed markedly by addition of water.

8.6 BREAKTHROUGH WORKS IN POLYMER SCIENCES

In this section, breakthrough works in polymer science investigated by neutron scattering are highlighted.

8.6.1 Size of Polymer Chains

It is well known that amorphous flexible polymer chains obey Gaussian statistics (Equation (72)) and the end-to-end distance of a polymer chain R is proportional to the square root of the number of steps consisting of the chain, N. This is the starting point of the statistical physics of polymers. However, it was just an assumption until the beginning of the 1970s. Although Kuhn predicted Gaussian chain behavior for molten polymer in his theory of rubber elasticity [88], no decisive evidence was given for polymers in glassy state. In the beginning of the 1970s, German [2], British [3], and French groups [89] independently carried out SANS experiments on amorphous polymer chains in bulk and confirmed the Gaussian nature of polymer chains. Cotton

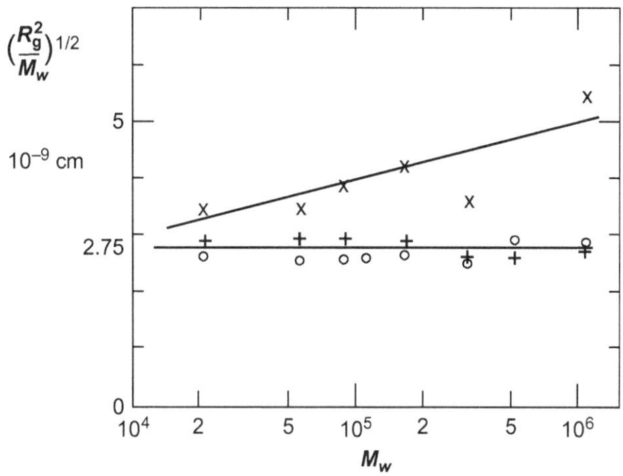

FIGURE 25 Relationship between the radius of gyration of polystyrene (PS), R_g, and the weight-average molecular weight, M_w. ×, in hydrogen sulfide (HS); +, in Θ solvent; ○, in bulk. *Reproduced from Ref. [89].*

et al. studied a mixture of hydrogenous polystyrene (PS) and deuterated polystyrene (DPS) and obtained a scaling rule as a function of molecular weight shown in Figure 25 [89].

Figure 25 shows that the radius of gyration, R_g, for amorphous PS chains is the same as that in a θ solvent (i.e., a solvent in which polymer coils act like ideal chains) for a wide range of molecular weights $21{,}000 \leq M_w \leq 1{,}100{,}000$, and is given by

$$\left(R_g^2\right)^{1/2} \propto M_w^{1/2} \tag{107}$$

Here, the proportional constant K is 2.75×10^{-9} cm. This is the experimental confirmation of the prediction (Equation (76)). In addition, it was confirmed that the scattering function is described by the Debye function (Equation (79)), discussed in the previous section. On the other hand, the size of PS chains in hydrogen sulfide (HS) was larger than that for unperturbed state due to excluded volume effect. The segment length estimated from the proportional constant was $a \equiv \left(6R_g^2/M_w\right)^{1/2} m_{PS}^{1/2} = 0.688$ nm. This study together with those by Kirste et al. [2] and Ballard et al. [3] was the starting point of polymer physics with neutron scattering, and has significance as the proof of Gaussian nature of polymer chains.

8.6.2 Critical Phenomena in Polymer Blends

In order to determine the size of labeled polymer chains in a polymer matrix, the concentration of the labeled chains was necessary to be low enough to neglect interference effect between the labeled chains (dilute assumption).

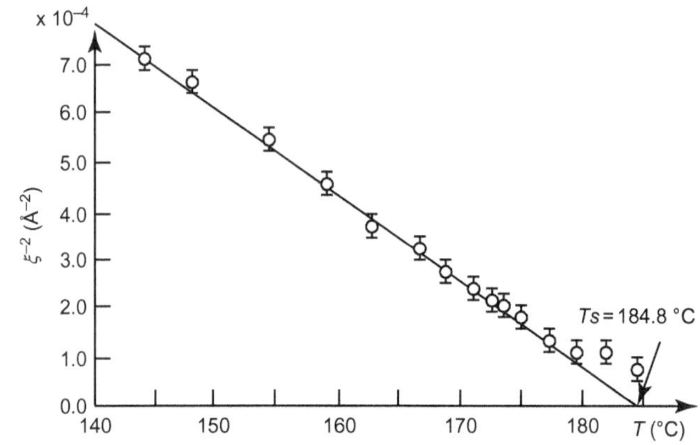

FIGURE 26 Inverse squared correlation length, ξ^{-2}, vs T plot for deuterated polystyrene (DPS)-poly(vinyl methyl ether) blend ($\phi_{PVME} = 0.408$). *Reproduced from Ref. [91].*

However, de Gennes derived a scattering function (Equation (86)) for binary polymer blends [38] available for any compositions based on the random phase approximation [90]. This function had accelerated studies on critical phenomena and miscibility of polymer blends. For example, Herkt-Maetzky et al. carried out SANS experiments on a series of polymer blends of deuterated polystyrene (DPS)($M_w = 47$k; k $\equiv 10^3$) and poly(vinyl methyl ether) (PVME) ($M_w = 99$k), and observed a critical phenomenon by increasing temperature [91]. They fitted the SANS curves with the Ornstein-Zernike function and confirmed that the correlation length diverges by approaching the critical point. Figure 26 shows the temperature dependence of the correlation length ξ for DPS/PVME blend with $\phi_{PDMS} = 0.408$. The inverse square of ξ is linear with T, and it vanishes at 184°C, meaning that ξ diverged at this temperature. They concluded that the critical phenomena of the blend are classified to the mean-field because the exponents for the correlation length ξ and the susceptibility (proportional to the scattering intensity) were 0.5 and 1.0, respectively.

Shibayama et al. obtained a phase diagram shown in Figure 27 for DPS ($M_w = 255$k)/PVME ($M_w = 99$k) blends, where the binodal line and spinodal line were obtained with light scattering and SANS, respectively [42]. They also reported that the neutron cloud point curve (denoted by + in the figure) was in good agreement with the binodal line. Here, the neutron cloud point is the temperature where the transmission or the scattering intensity changes drastically. Furthermore, they discussed the composition and temperature dependence of the correlation length and the interaction parameter based on a modified de Gennes function (Equation (88)) by taking into account the asymmetry of the monomeric volumes of the components.

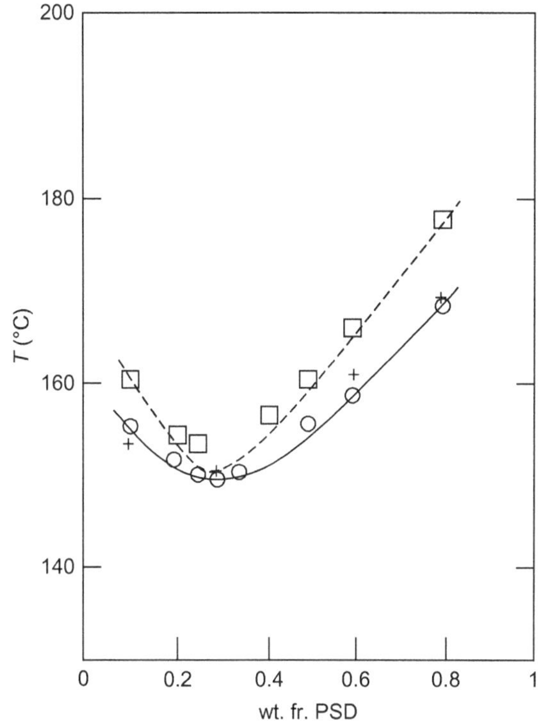

FIGURE 27 Phase diagram for DPS (M_w = 255k)/PVME (M_w = 99k) blends. Solid line and circles denote binodal line, dashed line and squares denote spinodal line, plus sign (+) denotes; neutron cloud point. *Reproduced from Ref. [42].*

8.6.3 Quantum-Phase Separation of Isotope Blends

Isotope blends undergo phase separation in liquid state at $T\Delta S_m \to 0$. Here, ΔS_m is the difference in the mixing entropy S_m between phase-separated state and mixed state. Such quantum-phase separation has been known exclusively in ^{3}He and ^{4}He mixture at $T \to 0$. In the case of molten polymer, on the other hand, it was anticipated that a quantum-phase separation might occur at $N \to \infty$ because of $\Delta S_m \sim N^{-1}$. As a matter of fact, Bates and Wignall carried out SANS experiments, and observed a quantum-phase separation in polymeric systems for the first time [92]. This work is regarded as an example demonstrating an analogy between polymeric systems and quantum fluids.

8.6.4 Direct Observation of Reptation Motion of Molten Polymer by Spin Echo Spectroscopy

The dynamics of polymer chains in the molten state are suppressed because of entanglements. de Gennes predicted that a polymer chain moves like a reptile

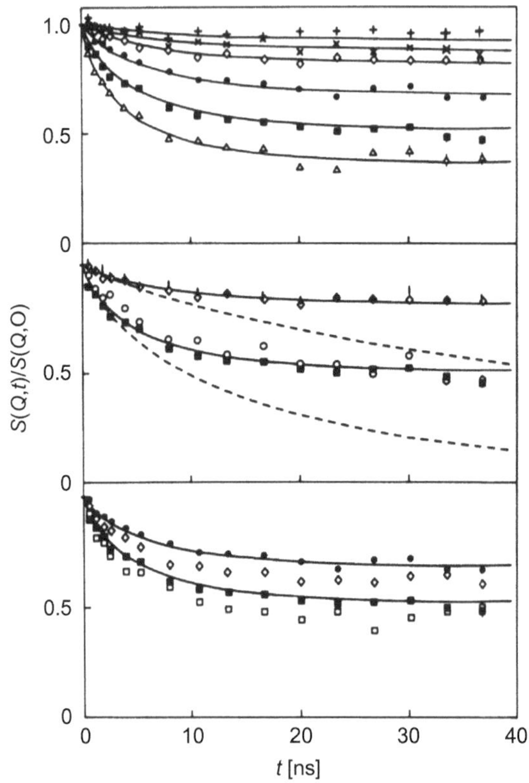

FIGURE 28 Neutron spin echo results. Upper: Homopolymer at 492 K (+; $Q = 0.058$ Å^{-1}, ×; $Q = 0.068$ Å^{-1}, ◊; $Q = 0.078$ Å^{-1}, ●; $Q = 0.097$ Å^{-1}, ■; $Q = 0.116$ Å^{-1}, $Q = 0.135$ Å^{-1}). Middle: comparison of homopolymer and copolymer at $Q = 0.078$ Å^{-1}(| and diamond) and $Q = 0.116$ Å^{-1} (○ and ■). The broken line shows the intermediate scattering function predicted for Rouss relaxation. Bottom: Homopolymers at 492 K and 523 K. ◊; $Q = 0.097$ Å^{-1}, □; $Q = 0.116$ Å^{-1}. The solid lines in all of the figures denote a fitting result with Ronca theory. *Reproduced from Ref. [30].*

(a snake) in a tube formed its neighboring polymer chains [40]. This type of motion is called "reptation" (see Figure 13). Richter et al. observed this reptation for the first time by neutron spin echo spectroscopy [30] (see Sec. 8.3.3.2). Figure 28 shows the result for alternating AB type copolymer consisting of poly(ethylene-propylene) in molten state. The upper figure is the intermediate scattering functions at various Qs for homopolymer. The middle is the comparison of homopolymer and the copolymer, and the bottom is the comparison at 492 K and 523 K. The solid line is the fitting results with a theory by Ronca et al. [93], which took into account of entanglement effects. This figure shows a Rouse relaxation at short-time region, while a plateau at longer time originates from entanglement effects. From this work, it was confirmed that the reptating length was 4.7 nm and tube diameter was ca. 5.0 nm. This work confirmed the validity of de Gennes' theory and contributed to the Nobel Prize in Physics 1991 to him.

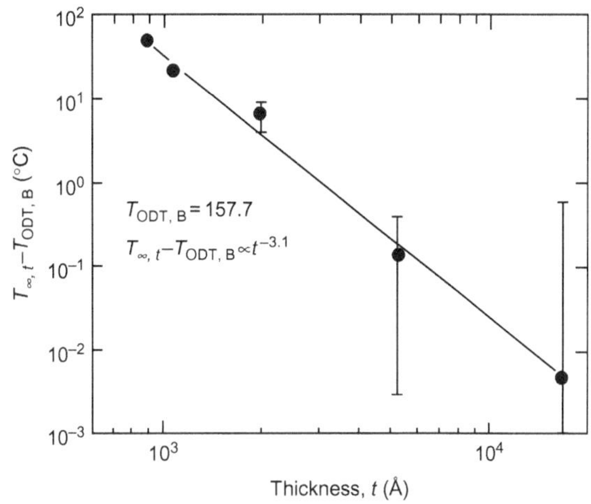

FIGURE 29 Film thickness dependence of order-disorder transition temperature, T_{ODT}, of poly(styrene-deuterated methyl methacrylate) thin films determined with NR. *Reproduced from Ref. [94].*

8.6.5 Order-Disorder Transition of Block Copolymer Thin Films

It is known that surface-induced ordering takes place in binary liquids, liquid crystals, polymer blends, and so on. Russell and coworkers applied NR (see Section 8.3.2) to soft matter for the first time, and carried out order-disorder transition in block copolymer thin films [94]. As described in Section 8.4.2.5, block copolymers having a symmetric composition undergo microphase separation to a lamellar microdomain structure whose periodicity is of the order of the component polymer chain. The microdomain structure undergoes order-to-disorder transition by lowering χN less than 10.5 [43]. Here, χ and N are the Flory-Huggins' interaction parameter and the number of segment of the polymer chain, respectively. However, this transition temperature was shifted upward significantly for poly(styrene-deuterated methyl methacrylate) thin films. As shown in Figure 29, they also observed film thickness t dependence of the order-disorder transition temperature, T_{ODT}. This work made NR one of typical measuring tools for soft-matter thin films.

8.6.6 Volume Phase Transition of Polymer Gels

If a condition is satisfied, a piece of gel undergoes a discrete and drastic change in its volume upon a slight change of thermodynamic variable, such as temperature, solvent composition, pH, ionic strength, etc. Such a discrete volume change was discovered by Tanaka and named "volume phase transition" [95]. Figure 30 shows temperature dependence of SANS curves of poly(*N*-isopropylacrylamide-*co*-acrylic acid) (PNIPA/AAc) copolymer gels. Poly(*N*-isopropylacrylamide) (PNIPA) is temperature responsive polymer in

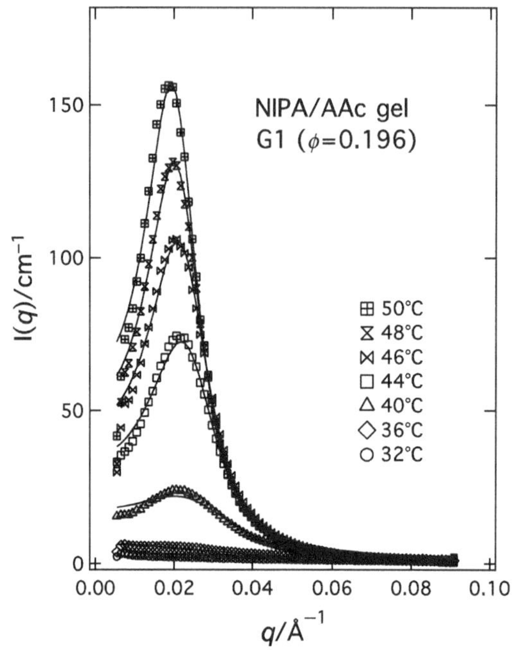

FIGURE 30 Temperature dependence of SANS curves for weakly charged poly(*N*-isopropylacrylamide-*co*-acrlyic acid) copolymer gels (the polymer volume fraction $\phi = 0.196$). By increasing temperature, a characteristic peak corresponding to a periodic distance ≈40 nm grows, indicating presence of a frustrated state in the gel. The solid lines are fits with Borue-Erukhimovich scattering function [96]. *Reproduced from Ref. [72].*

water, and it undergoes collapse transition around 32°C. However, by copolymerizing with acrylic acid (AAc) monomer together with cross-linker, a thermosensitive gel is formed, and it undergoes a discrete volume transition at a temperature higher than that for PNIPA homopolymers as a result of antagonistic interactions between swelling and collapsing. Shibayama et al. studied this volume phase transition by SANS [72]. At a low temperature, swelling is dominant and polymer gels are uniform, resulting in a weak scattering represented by OZ function. However, by increasing temperature, SANS intensity increases accompanying a scattering peak. This peak is characteristic of polyelectrolytes. The curves are the fits with Borue-Erukhimovich theory [96]. This SANS work showed that polymer gels near volume phase transition are in highly frustrated state and have a nm-order periodic structure.

8.6.7 Contrast Variation SANS on Surfactant Effects of Block Copolymers

Jakobs et al. reported that an addition of small amount of amphiphilic block copolymer to a ternary system of water, oil, and nonionic surfactant boosts

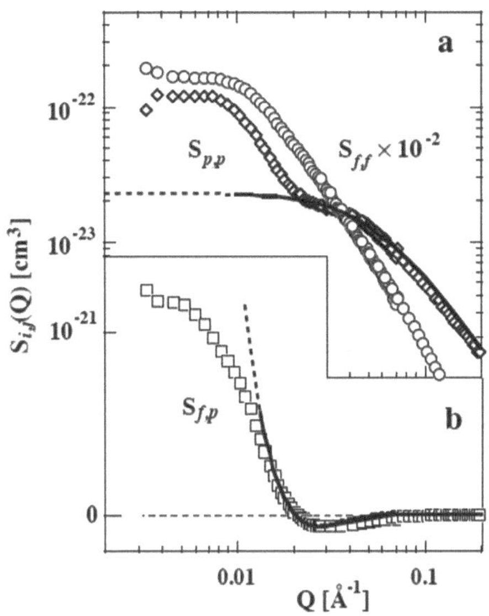

FIGURE 31 Experimental partial scattering functions. (a) $S_{p,p}$ (diamond) in comparison with $S_{f,f}$ (circles) and calculated polymer coil scattering (solid line). (b) $S_{f,p}$ (square) contains the interference between surfactant film and extended polymer layer and fit to the calculated intensity (solid line). *Reproduced from Ref. [98].*

up the surfactant effect more than a few tens times [97]. This means that the amount of surfactant can be reduced to a few tenths. Endo et al. elucidated this surfactant effect by contrast variation SANS in which the scattering length density of solvent was changed stepwise [98]. They reported that poly(ethylenepropylene)-*co*-poly(ethylene oxide) (PEP-PEO) added to a ternary mixture is localized at the water-oil interface and dispersed by forming a mushroom structure and is swollen. Figure 31 shows the experimental partial scattering functions, $S_{i,j}(Q)$. Here, the subscripts i, j are combinations of $= w, o, f, p$ for water, oil, film (or surfactant), and polymer, respectively. The resemblance of $S_{p,p}$ and $S_{f,f}$ for small $Q < 0.02$ Å^{-1} indicates that the polymer decorates the surfactant film. Figure 31 (a) shows a fit to the Zimm formula [8] indicating interchain interaction at the interface. Figure 31 (b) shows that the film-polymer scattering $S_{f,p}$ has a minimum at $Q \simeq 0.03$ Å^{-1}, indicating a presence of maximum in the polymer density. These experimental results show that the polymer is tethered to the interface, and a mushroom structure formed by the PEP chains in the oil phase and by the PEO chain in water is not influenced by the presence of neighboring membrane [98]. This work demonstrated that contrast variation SANS (CV-SANS) is a powerful tool for structural characterization of multicomponent soft condensed matter. The details of CV-SANS are described in Appendix B.

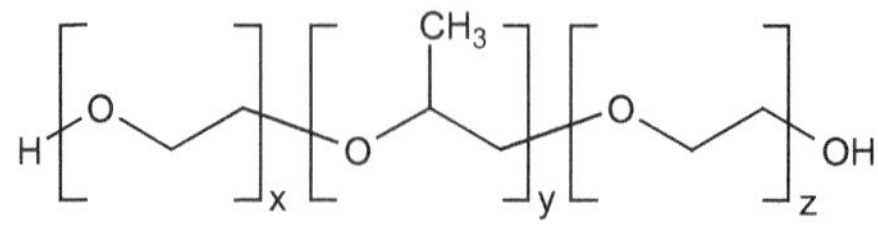

FIGURE 32 Chemical structure of poloxamer. *x*, *y*, *z* are the degrees of polymerization (number of the repeating unit) of each block chains.

8.7 NEW DIRECTIONS

In this section, among a huge number of studies in the literature, a few topics shedding new directions of neutron science, and/or a large number of citations are introduced.

8.7.1 Poloxamer

Poloxamer is an amphiphilic block copolymer, consisting of poly(ethylene oxide)-poly(propylene oxide)-poly(ethyelene oxide) triblock copolymer (PEO-PPO-PEO) as shown in Figure 32.

It is more commonly called Pluronic® (BASF). Since the middle block is hydrophobic and the two end blocks are hydrophilic, the poloxamer behaves as polymer surfactant. It is used as nonionic polymer surfactant. They can function as antifoaming agents, wetting agents, dispersants, thickeners, and emulsifiers.

Mortensen and coworkers studied micelle formation and crystallization [99], and phase behavior of melt and aqueous solutions of Pluronic® [100, 101]. Figure 33 shows the phase diagram. It has a variety of structures depending on concentration and temperature, e.g., unimers, random networks, micelles, micellar networks. Mortensen and Talman carried out cryo-electron microscopy (Cryo-TEM) together with SANS on Pluronic® solutions and succeeded in complete microstructural characterization of aqueous solutions [102]. Cryo-TEM provided direct information about the building blocks of these systems, i.e., spheroidal micelles. They also demonstrated the potential of selective electron beam radiolysis to enhance contrast in radiation-sensitive, inherently low-contrast systems.

Goldmints et al. studied unimer-to-micelle transition [103]. At low concentration (1%), where no intermicellar interactions were evident, aggregation numbers were obtained using the contrast variation method. At higher concentrations a three-parameter model accounting for the presence of water in the core and corona of the micelle, and for micelle-micelle interactions, was proposed. Furthermore, Alexandridis et al. studied temperature-dependent aggregation behavior [104] and micelle structures in mixed solvents of water and formamide, ethanol, or glycerol [105]. Holzwarth and coworkers investigated temperature-dependent aggregation behavior [104]. Pedersen et al. combined SANS and Monte Carlo simulations, and proposed a model with spherical

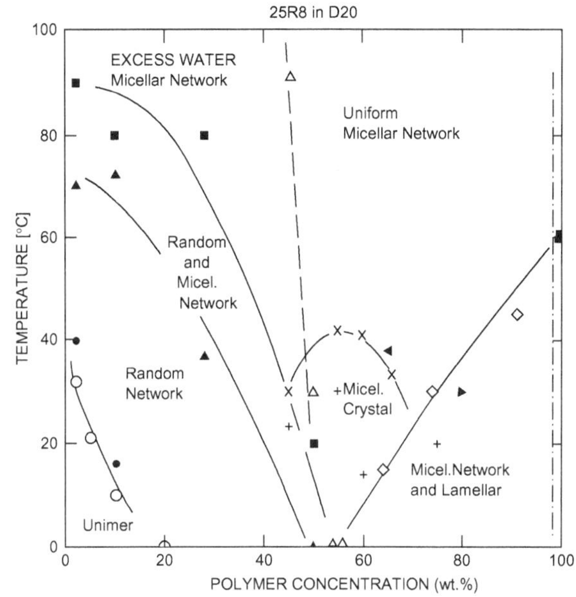

FIGURE 33 Phase diagram of the Pluronic system. Closed symbols refer to neutron-scattering results, open symbols to light scattering, and crosses to rheological data. *Reproduced from Ref. [100].*

core of PPO with some water surrounded by a corona of the PEO block [106]. The corona chains are noninteracting and obey Gaussian statistics, but are expelled from the core region. Bahadur et al. [107] studied salt effect on the structure of Pluronic.®

These works introduced here are just a small fraction of a large number of studies on poloxamers. One of the reasons why poloxamers have been investigated by SANS is its variety of phase behavior, applications, particularly to bioengineering field. Since there are many variations in poloxamer with different numbers of x, y, and z in spite of its simple structure (Figure 32), there still remain a large number of studies on poloxamer with SANS.

8.7.2 Kinetics of Amphiphilic Molecules in Lipid Vesicles

Spherical-bilayer-shell structures consisting of amphiphilic molecules are called vesicles. Vesicles are intensively investigated as a model system for studying the structure and functions of cells. Vesicles are also focused in medical and pharmaceutical sciences as drug-delivery systems (DDS). In vivo, material transfer/exchange occurs across vesicles.

Nakano et al. carried out time-resolved SANS (TR-SANS) experiments on large unilamellar vesicles (LUV) consisting of phospholic lipid, dimyristoylphosphatidylcholine (DMPC). They prepared deuterated and hydrogeneous

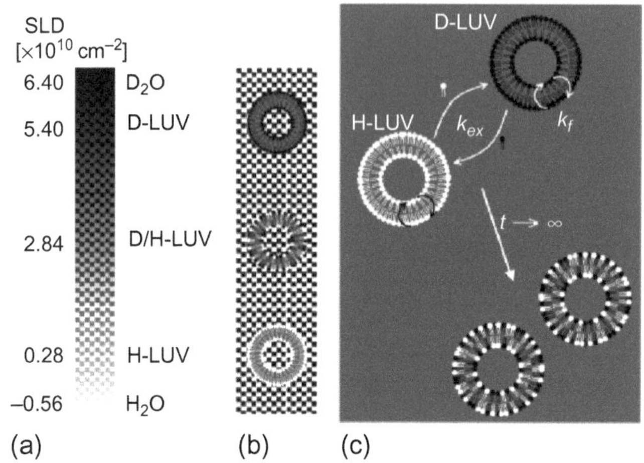

FIGURE 34 Principle of the detection of lipid dynamics by TR-SANS. (a) Calculated SLD's of LUV's and solvents are graded, where high and low SLD are represented by black and white, respectively. SLD of D-H-LUV corresponds to that of water with 50 vol% D_2O (contrast matching). (b) In 50 vol% D_2O, D-LUV (top) and H-LUV (bottom) are "visible," while D-H-LUV (middle) is invisible by neutrons under contrast-matching conditions. (c) Lipid exchange between D- and H-LUV and flip-flop in the membranes reduces the contrast of LUV's with solvent and LUV's become less visible, which can be detected as a decrease in the scattering intensity by TR-SANS experiment. *Reproduced from Ref. [108].*

DMPC (D-LUV, H-LUV) in 1:1 ratio, and observed mixing process of DMPC in a mixed solvent of D_2O/H_2O [108]. Figure 34 shows schematically the mixing process of H-LUV and D-LUV in D_2O/H_2O mixture. The observed kinetics was explicitly represented by a simple model that includes two independent kinetic parameters, i.e., the rates of transbilayer and interbilayer exchange. The observed rates k_f (transbilayer) and k_{ex} (interbilayer) were $k_f = 4.6 \times 10^{-1}$ and $k_{ex} = 1.35 \times 10^{-1}$ at 37.0°C. Carrying out similar experiments at different temperatures, they successfully obtained the enthalpy and entropy of these exchange reactions. This work demonstrates that this technique is perfectly suited for the determination of lipid exchange kinetics in equilibrium because it is not necessary to introduce fluorescence or spin labels into the lipids which may change their physical and chemical properties. This method could also be applicable to evaluation of the activity of the factors relevant to lipid migration, such as translocase and lipid transfer proteins.

8.7.3 Shish-Kebab Structure

Kimata et al. studied the shish-kebab structure of polypropylene (PP) extruded from a mold [109]. Mixtures of labeled PP (low, medium, and high-MW deuterated PPs) and hydrogeneous PP were prepared and extruded with an extruder. SANS results on thus deformed PP films showed that the labeled PP chains participate in shish formation irrespective of their molecular weights. Since this observation denies the common sense of the past (shish parts are formed exclusively by high-molecular weight polymer), studies on polymer crystallization under shear or extrusion have been reactivated.

8.7.4 Structure Formation of Ion-Pair and Salt

A mixture of water and 3-methylpyridine (3MP) is classified to a lower-critical solution temperature (LCST) system and has the critical point at 310 K ($\phi_w \approx 0.7$). Here, ϕ_w is the water fraction. Sadakane et al. observed that addition of small amount of sodium tetraphenylborate ($NaBPh_4$) increases LCST and widens the miscibility window [110]. Interestingly, the system exhibited a critical opalescent color in yellow to green. They explained this interesting phenomenon, i.e., the appearance of a periodic structure in such a simple mixture, with a coupling model of critical concentration fluctuations and solvation effect. That is, a very strong hydrophobic salt, i.e., $NaBPh_4$, is localized at the water-oil (3MP) interface, and it behaves as surfactant. As a result, this "surfactant" leads to an increase in solubility as well as structure formation in visible light range. Furthermore, they observed that it forms even spherical domains of the order of 10 μm for small fractions of 3MP ($\phi_{3MP} \simeq 0.1$) and the domains exhibit optical anisotropy. Based on these observation, they concluded onion-like multivesicle formation. Corresponding SANS patterns are shown in Figure 35. A lamellar structure with a marked periodicity was observed at low temperatures, but its periodicity decreased and finally disappeared by increasing temperature [110]. This is an ultimate "surfactant" system with thinnest interface. Hence, the physics behind these phenomena may open a new direction of application of surfactants.

8.7.5 Tough Polymer Gels

Polymer gels are known to be soft and fragile material. The fragility limits the range of applications of gels exclusively to food, glue, paste, etc. However, novel gels with large swellability with 1000 folds in volume, large deformability with a few tens times in length, strong gels tougher than cartilage, and so on, have been developed one to another after entering the 21st century [111–113]. As shown in Figure 36 [114], the structures of these novel gels are rather unique. As discussed in Section 8.5, typical gels (chemical gels) have inhomogeneous structure as shown in Figure 36 (a). On the other hand, slide ring (SR) gels (b) have movable cross-links which are able to slide along polymer chains so as to minimize local strain in the network just like "pulley" [111]. Nanocomposite (NC) gels (c), comprising long polymer chains,

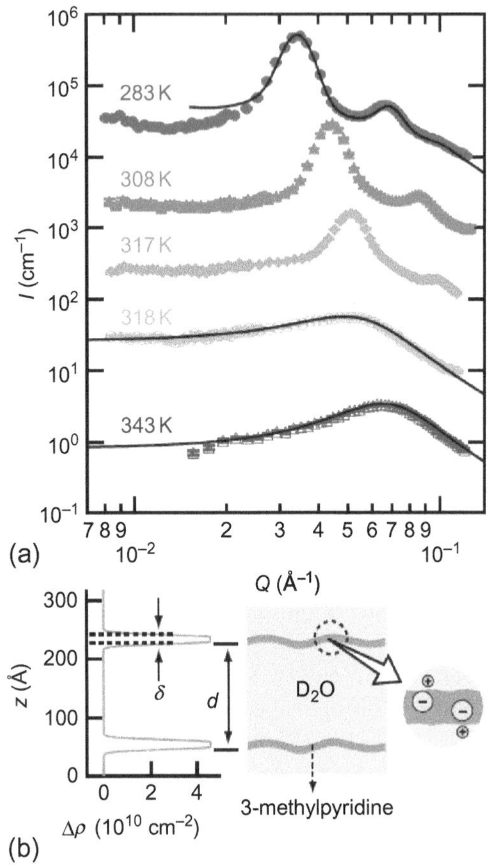

FIGURE 35 (a) Temperature dependence of SANS intensity from 3MP containing a small amount of $NaBPh_4$. The first-order transition occurs slightly below $T = 318$ K. The data at lower temperatures are shifted by multiplying 10 for better visualization. (b) Schematic representations of the scattering length density variation (left) and the distribution of the ingredients (right). *Reproduced from Ref. [110].*

inorganic clay, and water, exhibit extraordinarily toughness against deformation [112]. Tetra-PEG gels, fabricated by cross-end-coupling of tetra-arm poly(ethylene glycol) (PEG) having complimentary end functional groups, have a nearly ideal network structure free from dangling chains, loops, and entanglements (see more in detail, Section 8.7.5.2) [113].

8.7.5.1 Contrast Variation of Nanocomposite Gels

Nanocomposite (NC) gels developed by Haraguchi et al. [112] (Figure 36 (c)) consist of poly(*N*-isopropylacrylamide) (PNIPA) and synthetic clay, Laponite. NC gels have very large deformability (more than 10 times), extraordinary

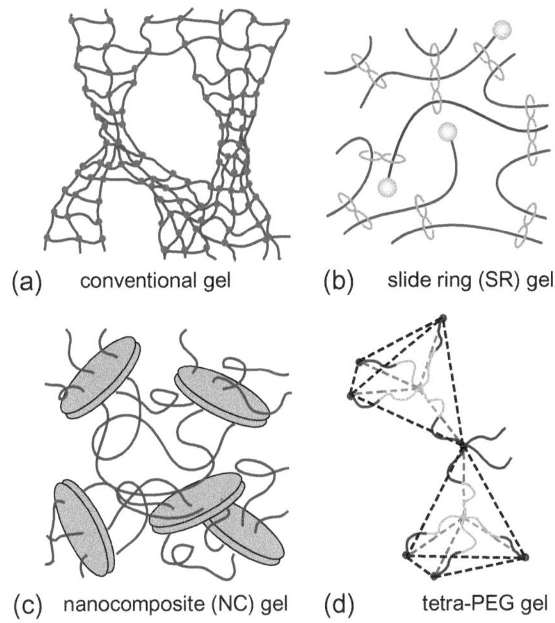

FIGURE 36 Comparison of conventional gels and novel gels developed recently. (a) Conventional gels, (b) slide-ring gels, (c) nanocomposite gels, and (d) tetra-PEG gels. *Reproduced from Ref. [115].*

toughness (of the order of hundred MPa in elastic modulus and 100 MPa in strength at break), and high optical transparency. A CV-SANS method was applied to elucidate the structure and the relationship between polymer and clay [116]. The details of CV-SANS are described in Appendix B.

Figure 37 shows the scattering length densities ρ_i of polymer (P), clay (C), and solvent (S). By tuning the scattering density of the solvent (mixtures of H_2O and D_2O), Miyazaki et al. prepared NC gels in solvents with five different scattering densities as shown in Figure 37 and carried out SANS experiments. Figure 38 shows (a) observed SANS results (symbols) and reconstructed SANS functions with partial scattering functions (lines), and (b) partial scattering functions for clay-clay ($S_{CC}(Q)$), polymer-polymer ($S_{PP}(Q)$), and polymer-clay ($S_{CP}(Q)$) correlations obtained by decompositions. By analyzing these functions, it was concluded that clay platelets with 30 nm long and 1 nm thick are randomly oriented on which high density of polymer chains are adsorbed. The clay platelets play as cross-linker (plane cross-linker) and the adsorbed polymer chains play as a reservoir for large deformation. Hence, CV-SANS revealed that such a unique structure provides the toughness and large deformability of NC gels.

Nishida et al. investigated the deformation mechanism of NC gels by CV-SANS [117]. It was a challenging experiment because both scattering contrast and stretching ratios had to been precisely controlled. They obtained

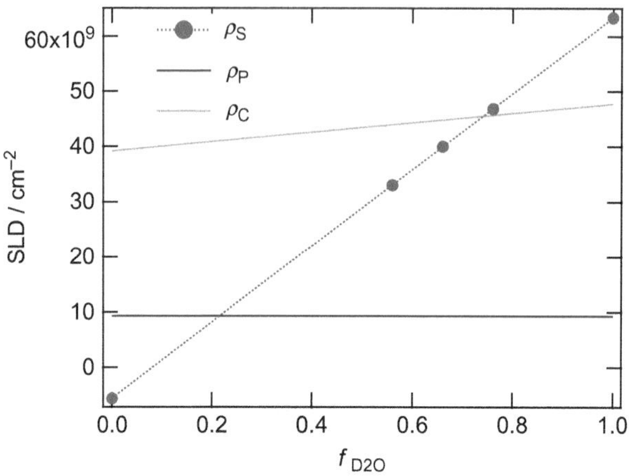

FIGURE 37 Scattering length densities, ρ_i (i = W (water), C (clay), and P (PNIPA)) as a function of D_2O fraction, ϕ_{D2O}. *Reproduced from Ref. [116].*

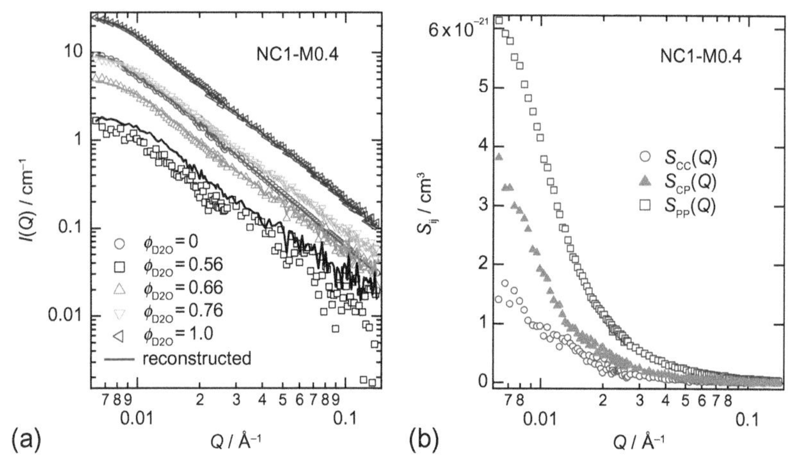

FIGURE 38 (a) Results of CV-SANS for NC gel. The solid lines are the reconstructed scattering intensity functions from decomposed partial scattering functions. (b) Partial scattering functions of the clay-clay $S_{CC}(Q)$, the clay-polymer cross term $S_{CP}(Q)$, and the polymer-polymer $S_{PP}(Q)$ decomposed from $I(q)$'s in (a). *Reproduced from Ref. [116].*

a series of SANS functions and decomposed to the respective partial scattering functions (see Appendix B). Figure 39 shows the results after decomposition. At $\lambda = 1$, the scattering intensities are weak and the patterns are isotropic, suggesting that both clay platelets and polymer chains are oriented randomly in the NC gel. Here, λ is the stretching ratio. By increasing λ, SANS curves became anisotropic. At $\lambda = 9$, a strong streak pattern appears in the perpendicular direction, indicating that polymer chains are highly stretched

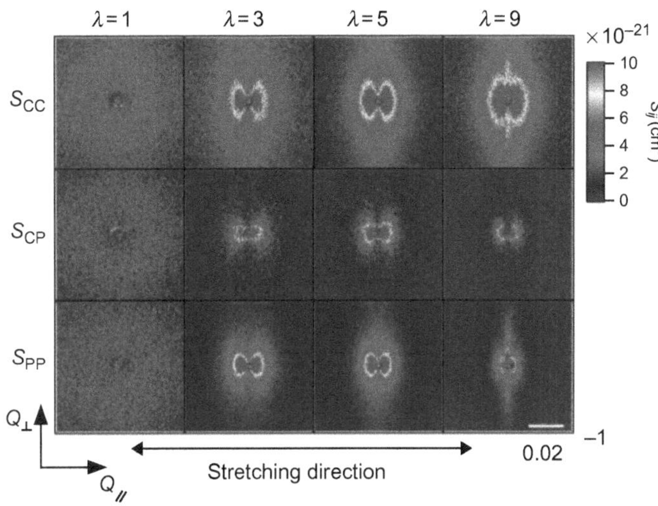

FIGURE 39 Two-dimensional SANS partial scattering functions for NC15 obtained by CV-SANS method; clay scattering S_{CC}, clay polymer cross term S_{CP}, polymer scattering S_{PP}. λ is the stretching ratio, $Q_{/}$ and $Q_{\perp}$ are the Q vectors along the stretching and perpendicular direction, respectively. *Reproduced from Ref. [117]*

in the parallel direction. It is quite important to note that $S_{CP}(Q_{//}, Q_{\perp})$ is positive and looks similar to $S_{CP}(Q_{//}, Q_{\perp})$ at $\lambda = 3$. This indicates that there exists strong correlation between clay and polymer and the clay-polymer cross term behaves similar to the clay-clay self-term. They concluded that clay platelets orient parallel to the stretching direction accompanying stretching of polymer chains along the stretching direction as well as peeling off of polymer chains from the adsorbed layer. This peeling-off phenomenon is characteristic and is one of the main reasons of the high extensibility and high toughness of NC gels.

8.7.5.2 Near Ideal Network Structure

Sakai et al. succeeded in novel polymer gels consisting of poly(polyethylene glycol) (PEG) called Tetra-PEG gels [113]. It was verified by SANS together with mechanical measurements and swelling measurements that Tetra-PEG gels comprise "near ideal polymer networks" which do not have defects, dangling chains, or entanglements [118]. SANS functions of tetra-PEG gels are described simply by Ornstein-Zernike function (Equation (43)) without excess scattering term discussed in Section 8.5.1 not only in as-prepared state [118], but also in swollen state [119]. Figure 40 shows reduced SANS functions of Tetra-PEG gels. As shown in the figure, all of SANS curves of Tetra-PEG gels with different polymer volume fractions ϕs and molecular weights $M_w = 10k$, 20k, and 40k are reduced to a single master curve. On the other hand, SANS curves for linear PEG ($M_w = 100k$) observed by Hammouda

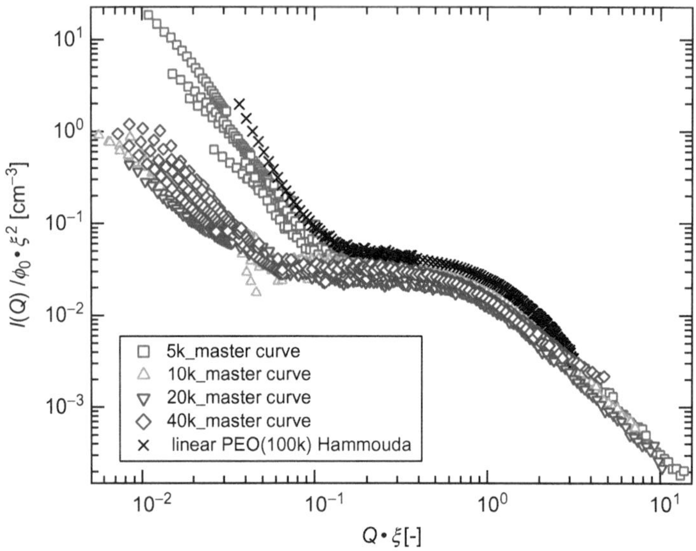

FIGURE 40 Reduced SANS plots, $I(Q)/\phi_0\xi^2$ vs $Q\xi$, for Tetra-PEG gel-5k (red square), -10k (green triangle), -20k (blue inverted triangle), and -40k (purple diamond), where ϕ_0 is the polymer volume fraction. Crosses indicate the reduced scattering intensity function for linear PEG chains (M_w = 100k) in water observed by Hammouda et al. [120]. *Reproduced from Ref. [119].*

[120] and Tetra-PEG (M_w = 5k) have nontrivial upturn at low Q region. This is due to clustering of polymer chains at chain ends (linear PEG) and defects in the network (5k-Tetra-PEG). The smaller upturn in the low Q regions for the other Tetra-PEG gels is due to imperfectness of the network structure arising in a very long distance much larger than the mesh size ξ.

Tetra-PEG gels with nearly ideal networks are also fabricated in an ionic liquid [121]. The structure of Tetra-PEG gels in ionic liquids was found to be similar or even more uniform than Tetra-PEG hydrogels.

Figure 41 shows photographs of Tetra-PEG gels (40k) films mounted on a stretching device at the stretching ratio of (a) $\lambda = 1$ and (b) $\lambda = 5$, and (c) the stress-stretching ratio curves of Tetra-PEG gels (20k and 40k). The dashed curves in Figure 41 (b) show the shape of the gel for the eye. As shown in this figure, Tetra-PEG gels are highly deformable up to $\lambda \approx 8$ (20k-gel) and $\lambda \approx 12$ (40k-gel).

Figure 42 shows two-dimensional SANS patterns of Tetra-PEG (20k) gel films at λ = 1.0, 3.0, and 5.0 [122]. Interestingly, the SANS curves show no anisotropy even at $\lambda = 5$ as is a strong contrast against the case of NC gels (see Figure 39). These experimental observations strongly support the ideal network structure of Tetra-PEG gels. That is, due to the absence of heterogeneities, neither up-turn in scattering intensity nor anisotropy appears by deformation. This is a marked difference from conventional gels.

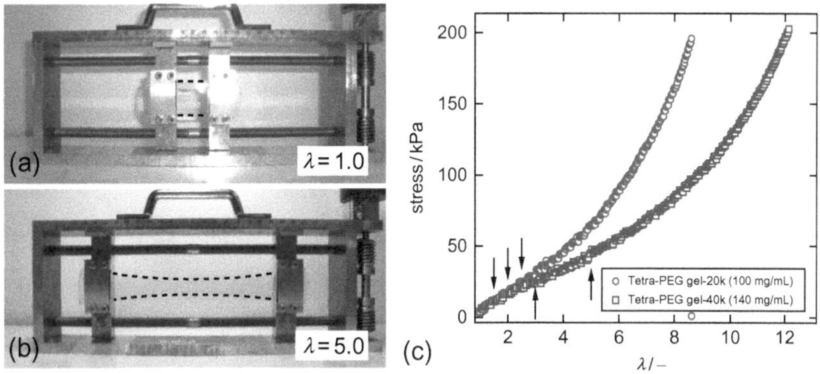

FIGURE 41 Photographs of Tetra-PEG gel-40k in (a) nonstretched state ($\lambda = 1.0$) and (b) stretched state ($\lambda = 5.0$). (c) Stretching stress-strain curves of Tetra-PEG gel-20k and -40k. The stretching ratios at which SANS experiments were carried out are marked by arrows. *Reproduced from Ref. [122].*

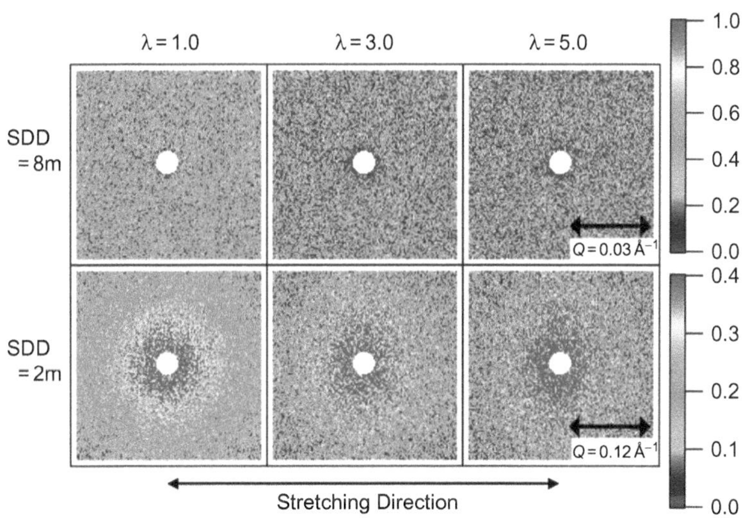

FIGURE 42 2D SANS patterns of Tetra-PEG gel-20k (100 mg/mL) at SDD = 8 m (top) and 2 m (bottom), where SDD is the sample-to-detector distance. Stretching ratios are $\lambda = 1.0$, 3.0, and 5.0. *Reproduced from Ref. [122].*

SANS is a very powerful tool for investigations of the structure and deformation mechanism of polymer gels as was reported by Shibayama [114, 115]. This is because (i) scattering contrast can be easily introduced simply by using deuterated solvent and (ii) a variety of sample environments, such as deformation, stretching, and swelling, is easily set up owing to strong material-penetration power of neutron beam. Hence, not only SANS but also various types of neutron-scattering techniques will be applied to polymer gels for further elucidation of their structure and for development of novel soft condensed matter having unique properties.

8.8 FUTURE DIRECTIONS OF NEUTRON SCATTERING IN SOFT CONDENSED MATTER

In this section, future directions of neutron-scattering research in soft condensed matter physics are predicted. The predictions are made by considering the advantages of neutron scattering and by extrapolating the past and present research streams together with recent advancement of instrumentation. Here, we would like to point out (i) high-resolution/high-brilliance SANS, (ii) complementary SANS and SAXS, (iii) computational-science-aided neutron scattering, (iv) high-pressure neutron-scattering experiments, and (v) rheo-SANS.

8.8.1 High-Resolution/High-Brilliance SANS

Iwase et al. upgraded an SANS system by installing a high-resolution area detector and focusing collimation system at SANS-U, Univ. Tokyo (see Figure 9) [15]. Furthermore, the area detector was optimized by tuning the thickness of the scintillator ($ZnS/^6LiF$) [123]. These kinds of upgrade/optimization improve neutron instruments, which are one of the future directions of neutron scattering in soft condensed matter because structural characterization with a lower-Q limit and a more brilliant neutron beam are always demanded in soft condensed matter science, particularly for *time-resolved SANS* experiments to monitor time-dependent systems such as chemical reactions and transient phenomena.

8.8.2 Complementary SANS and SAXS

It is always useful to combine SANS with small-angle X-ray scattering (SAXS) experiments. SAXS is superior to SANS in signal-to-noise ratio, spatial resolution, and time resolution. On the other hand, SANS is advantageous in nonradiation-damage measurements, low parasitic scattering, and so on. In addition to these complementarity, difference in the scattering contrast between SANS and SAXS is useful to precise determination of the structure particularly for multicomponent systems.

Ballauff et al. demonstrated complementary use of SAXS and SANS for studying polymer colloids [124]. Hiroi et al. also utilized both SAXS and SANS for structural studies on amphiphilic conetwork gels [125]. More recently, Inoue et al. demonstrated the usefulness of SANS and SAXS combination in elucidation of a mechanism of subunit dynamics in homo-oligomers of $\alpha\beta$-crystallin [126]. They observed a decrease of SANS intensity with time upon mixing of isotope mixtures and confirmed the existence of subunit exchange with deuteration-assisted SANS (DA-SANS). Here, SAXS was used to confirm that degradation upon mixing did not occur. They proposed that transiently liberated subunits, namely, "traveling subunits," play a role in subunit exchange, and that protein function is regulated by these traveling subunits.

It would be ideal to place an SAXS instrument next to an SANS instrument. This allows users to work on both instruments without time loss or without sample deterioration. The latter point is of particular importance for biological samples. In the future, simultaneous measurements of SAXS and SANS would be available by installing X-ray generator next to an SANS sample stage.

8.8.3 Computational-Science-Aided Neutron Scattering

It is natural that the more developed soft condensed matter the more complicated and sophisticated the structure, architecture, dynamics, and properties have. As a result, simple and primitive scattering approach would not be applicable. One of solutions to overcome such difficulty is to combine scattering experiments and computational science. Here, let us pick up a few but representative reviews.

8.8.3.1 Ab Initio Shape Determination by Bead Models

Volkov and Svergun [127] demonstrated scattering functions for monodisperse systems of noninteracting particles (e.g., from dilute protein solutions). They showed that a comprehensive description is achieved in the bead modeling methods. Scattering patterns from geometrical bodies with different shapes and anisometry (solid and hollow spheres, cylinders, prisms) are computed and the shapes are reconstructed ab initio using envelope function and bead modeling methods. The procedure is described to analyze multiple solutions provided by bead-modeling methods and to estimate stability and reliability of the shape reconstruction. They also demonstrated that flat shapes are more difficult to restore than elongated ones and types of shapes are indicated, which require additional information for reliable shape reconstruction from the scattering data. This type of analysis has been intensively used in SAXS studies on protein solutions. However, it will be also familiar in SANS.

8.8.3.2 Reverse Monte Carlo Method

Pedersen and Schurtenberger proposed off-lattice Monte Carlo simulations on semiflexible polymer chains with and without excluded volume interactions [128]. The model used in the simulations was a discrete representation of the wormlike chain model of Kratky and Porod applied in the pseudocontinuous limit. The scattering functions of the semiflexible chains were determined with a good precision for a wide range of L/b ratios, where b is the statistical segment length and L is the contour length of the chain. Numerical approximations to these functions were determined which interpolate between the simulated functions, and these could be used in the analysis of experimental scattering data. The approximations were used in least-squares fitting of experimental SANS data from polystyrene in a good solvent.

Arleth et al. [129] studied sodium dodecyl sulfate (SDS) wormlike micelles in water with NaBr using SANS. The scattering data were analyzed using a novel approach based on polymer theory and the results of Monte Carlo simulations. The method made it possible to give a full interpretation of the scattering data, even for the entangled micellar solutions occurring at high concentrations and high salinities. Analysis of the scattering data at zero scattering angle demonstrated that the length of the micelles increases according to a power law as a function of concentration in the studied interval. The analysis furthermore showed that the length of the micelles increases exponentially with increasing salinity. The scattering data in the full range of scattering angles were analyzed using a model for polydisperse wormlike micelles where excluded volume effects are taken into account via an expression based on the polymer reference interaction site model (PRISM) [130].

8.8.3.3 SANS

Licos [131] regarded soft condensed matter as "complex fluids" whose spatial extent lies in the domain between 1 nm and 1 μm and presented a comprehensive review on recent progress of "soft particles"; star-shaped polymers and starburst molecules known as dendrimers. He showed that a combination of scattering techniques together with computer simulations and analytical methods from the realm of theoretical physics can shed light on the unusual properties of such systems. Starting from star-polymer conformation, he discussed effective interactions, structure and phase behavior, gelation and melting of dense star-polymer solutions, and glass states of star-polymer mixtures from both scattering experiments and theory/simulation.

A similar treatment was done on dendrimers. It was concluded that a deeper understanding of the rheological and dynamical properties of complex fluids can be achieved only by a combination of experiment, theory, and simulation.

8.8.3.4 QENS

The general trend in soft condensed matter is to study systems of increasing complexity. This is facilitated by the capability of quasielastic neutron scattering (QENS) to selectively probe spatially resolved dynamical modes at a molecular level (Section 8.3.3.1). The large number of recent publications using QENS for investigating complex and multicomponent soft matter systems serves as recognition of the suitability of this technique by the scientific community (see, for example, Ref. [5]). Exploiting its complementarity with molecular dynamics (MD) simulations and other experimental techniques is the basis of a successful methodology for this scientific challenge.

Sakai and Arbe [5] illustrate the potential of QENS with three kinds of soft materials whose structural units increase in size/complexity: lipids, polymers, and biomolecules. Not only for pure lipid dynamics, e.g., dipalmitoylphosphatidylcholide (DPPC) hydrated bilayers, but also for stabilization of membranes, proteins with sugars, hydration of membrane proteins, glass transition of synthetic polymers, polymer blends, polymers under confinement, and multicomponent systems, combination of QENS (and other experimental techniques), and MD simulations are becoming a common methodology.

8.8.3.5 NR

Elliott et al. [132] discussed synergies between simulations and neutron-scattering experiments on confined polymer systems, i.e., polymer thin films, and showed advantages of these two techniques. They emphasized that molecular simulations and neutron reflectivity are both extremely valuable tools for determining the structure of soft matter at interfaces and under confinement, and high-resolution structural information provided by these techniques allows us to obtain a thorough understanding at the molecular level. Computer simulations of fully detailed atomistic models are limited in their applicability up to the order of a few nanometers. So larger-scale coarser grained simulations are normally necessary.

They showed an example of short polystyrene chains grafted to a graphite surface in dry (vacuum) conditions and in toluene, a good solvent. The grafting density and the solvent conditions not only influence the shape and extent of individual PS chains but also their mutual interactions. This type of information can be assessed using radial distribution functions (RDF) $g(r)$. Owing to the two-dimensional nature of the systems under investigation, however, conventional spherically symmetric RDFs are not well suited. Instead, they were able to calculate 2D RDFs, which measure the spatial correlations parallel to the surface (in-plane) and the correlations perpendicular to the surface (out-of-plane) separately.

8.8.3.6 NSE

Ring polymers have very different dynamics from other polymers because of lack of chain ends. Richter et al. [133] discussed that the dynamics of ring melts is uniquely addressed by neutron spin echo spectroscopy (NSE; Section 8.3.3.2). The subdiffusive center of mass motion at short times agrees well with simulation as well as theoretical concepts. In the internal dynamics, the basic length scale of the ring molecule, such as the loop size, manifests itself clearly. They showed that the experiments reveal strong evidence for loop motions and call for further theoretical work describing them. NSE is the best means to study the dynamics from local segment motion to center of mass diffusion and/or reptation.

8.8.4 High-Pressure Experiments

Although a large number of studies of soft condensed matter have been reported, pressure P, which is one of the fundamental thermodynamic variables (together with temperature T, volume V, and chemical potential μ), has been excluded in most cases. This is mainly due to difficulty of instrumentation of high-pressure environment. However, soft condensed matter is very sensitive to pressure. For example, protein denaturation occurs around a few hundred MPa [134, 135]. Neutron scattering is suitable for pressure-dependent studies of soft condensed matter because of the following reasons: (i) Since neutron has high penetration power to materials, cell design is relatively easier than the case of X-ray scattering. (ii) Soft condensed matter is more pressure sensitive than hard condensed matter and it exhibits interesting/intriguing behaviors with respect to pressure, most of which have not been intensively investigated yet. Particularly, it is known that proteins and water-soluble polymers exhibit complicated phase behavior in aqueous system due to various types of molecular interactions, including van der Waals interaction, hydrophobic interaction, electrostatic interaction [134, 136–138]. Hence, there are numerous numbers of research topics related to high-pressure neutron experiments. Let us begin with an instrumentation.

8.8.4.1 High-Pressure Cell

Figure 43 shows an inner cell type high-pressure cell, PCI-400-SANS (Teramex, Co. Ltd., Kyoto, Japan; now Syn Corporation, Kyoto, Japan) developed for SANS experiments. The PCI-400-SANS is an inner cell type

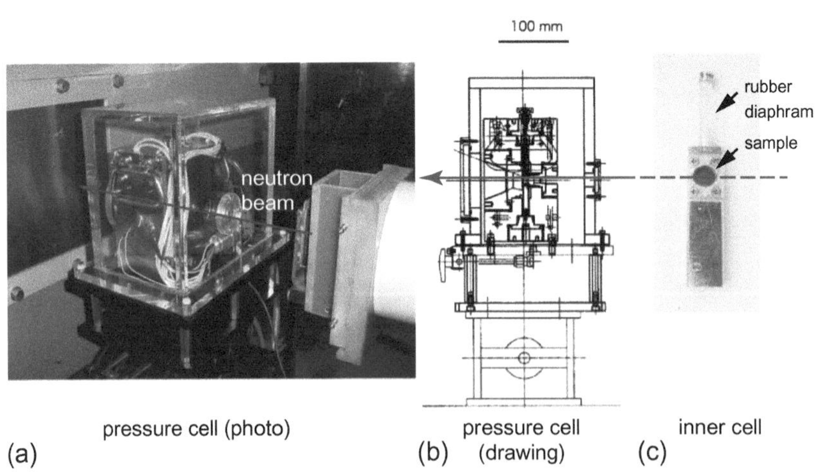

FIGURE 43 Inner cell type high-pressure cell. (a) Photograph of high-pressure cell mounted on SANS instrument, SANS-U [144], (b) drawing, (c) photograph of inner cell. Pressure is transmitted via rubber diaphragm.

pressure cell, which was designed to isolate the sample from the pressurizing medium. The basic concept of the pressure cell (for light scattering) is described elsewhere [139]. The inner cell (Figure 43 (c)) has a pair of quartz windows 2 mm thick. The applied pressure is transmitted via a rubber diaphragm connected to the inner cell. The outer chamber was filled with D_2O, and the pressure was controlled by pressurizing D_2O by a double-cylinder hand pump. Since the sample is isolated from pressurizing medium, samples can be changed without disturbing the pressure cell. The achievable pressure is 400 MPa. The temperature of the cell is controlled with a water-circulating bath. The cell was used for high-pressure SANS experiments on pressure-induced phase separation of poly(*N*-isoproptylacylamide) aqueous solutions and gels [140], weakly charged poly(*N*-isopropyl acrylamide) solutions and gels [141, 142], and hydrophobically solvated block copolymer solutions [143].

8.8.4.2 Pressure/Temperature-Induced Phase Separation of Polymer Solutions

Osaka et al. reported that block copolymers of poly(2-(2-ethoxy)ethoxyethyl vinyl ether)-*b*-poly(2-methoxyethyl vinyl ether) (EOEOVE-*b*-MOVE block copolymer) in heavy water undergo sophisticated phase behavior depending on T and pressure P, a convex-upward phase diagram in the $P - T$ [143]. This type of convex-upward phase behavior is often observed not only in polymer-aqueous systems [145–147] but also in protein solutions [138] and various types of organic compounds in water [148].

Figure 44 shows SANS profiles of 15% EOEOVE-MOVE block copolymer heavy water solution [143]. First of all, the SANS curve at room temperature and at atmospheric pressure ($P = 0.1$ MPa) indicates presence of microphase separated structure in the solution. The solid lines in the figure indicate the fitting results with Hosemann's paracrystal theory [149]. From the fitting analysis, it was found that EOEOVE forms spherical domains with the radius of 18 nm accompanying MOVE corona chains in bcc packing structure with the lattice spacing $a = 75$ nm and its fluctuations (the so-called Hosemann's g-parameter) of $\Delta a/a = 0.15$. However, by increasing P, this microdomain structure disappeared at $P = 150$ MPa, followed by reformation of microdomain structure by further increasing in P as indicated by reappearance of a peak.

Based on a series of SANS experiments, it was concluded that EOEOVE-MOVE heavy water solutions exhibit a $P - T$ phase behavior as shown in Figure 45. First of all, there are three phases, i.e, one phase (I and III), microphase separated phase (II and IV), and macrophase separated phase, divided by the temperature curves T_A and T_B. Furthermore, these T_A and T_B curves have a maximum at $P = P_0$. Regarding the T dependence, EOEOVE chains become insoluble against water at T_A and form microdomains. This is followed by macrophase separation due to insolubilization of MOVE at T_B. On the other hand, there are two regions, strongly hydrophobic solvated (SHS)

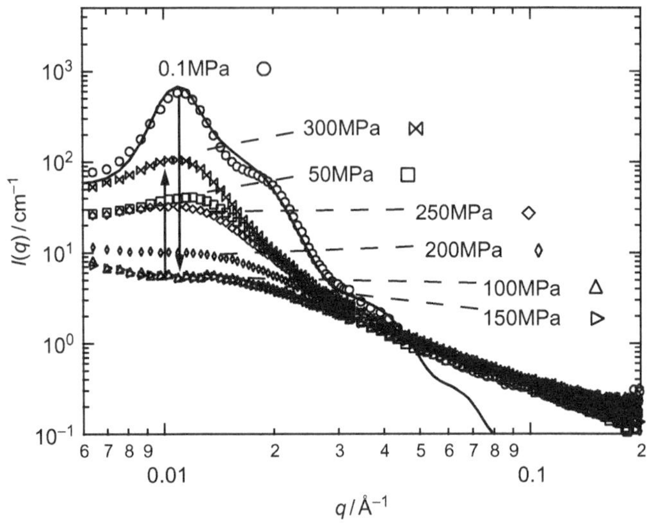

FIGURE 44 Pressure dependence of SANS curves for 15% EOEOVE-MOVE heavy water solution at $T = 45°C$. By increasing P, scattering peaks disappeared. However, by further increase in P results in reappearance. The solid lines show the fitting results with paracrystal theory [149]. *Reproduced from Ref. [143].*

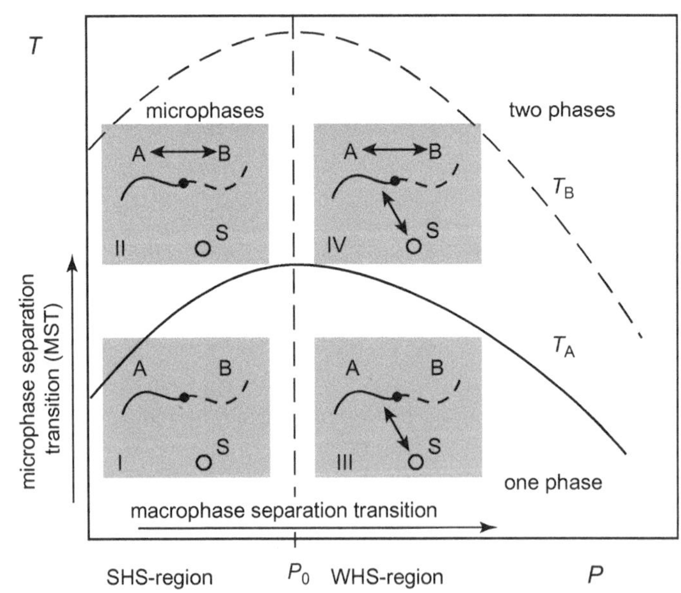

FIGURE 45 Pressure phase diagram of EOEOVE-MOVE heavy water solutions. A, B, and S denote EOEOVE and MOVE block chains, and solvent, respectively. Curves T_A and T_B divide the phases to one phase, microphase separated phase, and macrophase separated phases. P_0 is the pressure at maximum miscibility. *Reproduced from Ref. [143].*

region in low-pressure side ($P < P_0$) and weakly hydrophobic solvated (WHS) regions in high-pressure side ($P > P_0$). In general, phases in a $P - T$ phase diagram obey Clausius-Clapeyron law, i.e.,

$$\frac{dP}{dT} = \frac{\Delta H_m}{T \Delta V_m} \tag{108}$$

Here, ΔH_m and ΔV_m are the enthalpy and entropy changes on mixing. It is known that not only EOEOVE-MOVE aqueous solutions but also many water-soluble polymers including protein solutions exhibit negative change in mixing enthalpy, i.e., $\Delta H_m < 0$. Owing to the observed value of $\Delta H_m = -2.1$ kJ/mol for EOEOVE-MOVE heavy water solutions, the value ΔV_m was estimated to be $\Delta V_m = -2.4$ cm^3/mol. These values support the convex-upward behavior of Figure 45, i.e., decrease in $|\Delta V_m|$ with increasing P, and changes of its sign at P_0. This experimental finding indicates that $(P \leqslant P_0)$ is a necessary condition for protein folding and its higher-order structure formation. As a matter of fact, a similar behavior was obtained for β-lactoglobulin (β-LG) aqueous solutions [150].

8.8.4.3 Future Directions of High-Pressure Neutron Scattering

Hydrophobic interaction has been one of the intriguing problems in physical chemistry, theoretical chemistry, and biology [151]. The concept of "hydrophobic bond" proposed by Kauzmann [152] is now recognized as the "hydrophobic interaction" [153, 154] and as one of the fundamental molecular interactions ruling the structures and functions of proteins, nucleic acids, and so on [136]. As Kauzmann addressed, studies on its pressure effects are highly anticipated for better understanding of the hydrophobic interaction in addition to its temperature effects [134].

It is well known that pressure is an important variable to determine physical properties of hydrocarbons [148], polymers [155], biomacromolecules [138], and proteins in aqueous solutions [137]. β-LG molecules unfold above some pressure threshold [156, 157]. Panick et al. investigated differences between P-induced and T-induced denaturation and aggregation of β-LG by infrared spectroscopy and small-angle X-ray scattering [157]. Recently, a comprehensive review about functional improvement of β-LG by high hydrostatic pressure was given by Lopez-Fandino [135]. The behavior of denaturation and unfolding/refolding of β-LG was extensively discussed as a complex function of P, T, pH, and ionic strength.

As discussed above, although soft condensed matter is very sensitive to pressure, intensive studies on the phase behaviors and molecular interactions of soft condensed matter are still in immature. Hence, it is quite promising that neutron scattering will shed a brilliant light in further understanding of soft condensed matter with respect to pressure.

8.8.5 Rheo-SANS

As introduced in the introductory part of this chapter (Section 8.1.1), soft condensed matter can be easily deformed by weak mechanical force. Hence, the study of flow behaviors of soft condensed matter is a multidisciplinary and active research area that encompasses but is not limited to colloids, polymers, foams, gels, liquids, and biological systems and has direct relevance and application in academia and industry [158, 159]. Scientific interest lies in both fundamental and practical applications and can, in general, be categorized as an attempt to understand: (i) the fluid structural reorganization as a result of flow and/or (ii) the relation between flow and stress governing the bulk rheological properties of a system. Hence, there is a clear advantage to directly correlating the rheological response of a fluid to the internal structure on all relevant length scales.

In polymer-filler systems, polymer chains and anisotropic particles orient in a flow field so as to minimize hydrostatic friction. Amphiphilic molecules forming wormlike chains change their structures, from spherical micelles to wormlike micelles, and finally break to fragments by increasing shear rate [160]. Some multicomponent systems exhibit thixotropy (viscosity thinning by agitation) or dilatancy (viscosity thickening by agitation) by applying a flow field. In order to study flow behaviors of these complex fluids, simultaneous measurements of rheology and structure are highly demanded. Such demands are satisfied by flow-SANS (SANS measurements under flow) and Rheo-SANS (simultaneous measurements of Rheology and SANS).

The most common Couette cell consists of the vertical geometry in which there are two configurations with incident beam normal to the 1-3 plane (radial) and 2-3 plane (tangential) [158]. Here, the 1-, 2-, and 3-directions are defined as the velocity (v), velocity gradient (Δv), and vorticity (ω) directions, respectively. Many devices have been built in house and simply impose a steady-shear flow on the sample during SANS measurements (Ref. [158] and references therein). A Couette cell fixture optimized for neutron scattering is now available together with a Paar Physica rheometer®, which is stress controlled and optimized in torque, temperature, and sample loading for low viscosity fluids in the temperature range $T \simeq -20°\text{C}$ to $\simeq 200°\text{C}$. Figure 46 shows (A) illustration of the three planes of flow under investigation in Rheo-SANS instruments, (B) 1-2 plane shear cell, and (C) schematic of the 1-2 plane of shear together with 2D SANS patterns [159]. Hence, now, it is available of Rheo-SANS investigations with any configurations.

A comprehensive review on flow-SANS and Rheo-SANS applied to soft condensed matter is given by Eberle and Porcar [158]. They showed examples on hydrogels, polymer/nanoclay systems [161], polymer melts and blends, emulsions, membrane phases of amphiphilic molecules, wormlike micelles [160], soft micelles (block copolymer micelles), and biological systems.

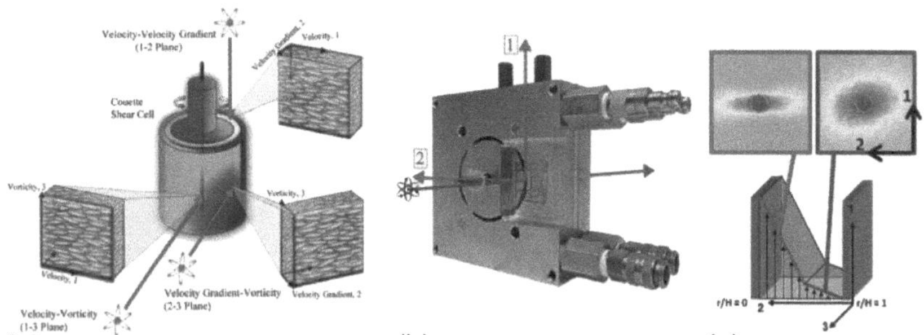
(a) (b) (c)

FIGURE 46 (a) Illustration of the three planes of flow under investigation in Rheo-SANS and flow-SANS experiments. (b) 1-2 plane shear cell sample environment labeled with (1) velocity, (2) velocity gradient, and (3) vorticity shear flow directions. (c) Schematic of the 1-2 plane of shear accompanied by examples of 2D SANS scattering patterns of oriented and isotropic structures in the high (red outline, $r/H = 0.2$) and low (blue outline, $r/H = 0.8$) shear bands for a steady-shear banding condition. Here, H and r are the radius of the rotor and the distance from the edge to the center of vortex, respectively. *Reproduced from Ref. [159].*

8.8.5.1 Viscosity-Thickening of Wormlike Micelle Aqueous System

Shear-induced thickening/thinning phenomena of aqueous rodlike micellar solutions of cetyltrimethylammonium bromide (CTAB) and sodium *p*-toluene sulfonate (NapTS) were investigated by Rheo-SANS [160]. The aqueous CTAB/NapTS solutions were classified into five different categories dependent on their flow behavior and micellar structure. By increasing salt concentration and/or shear rates, the micelles underwent morphological transition from (i) spherical or short-rodlike micelles to (ii) long-rodlike micelles without entanglements, followed by (iii) those with entanglements. These transitions were recognized as changes in flow behavior from Newtonian to shear-thickening (an increase in η_{flat}) and shear-thinning flow (a decrease in η_{flat}), respectively, as shown in Figure 47. In the latter two cases, anisotropic SANS patterns were observed around these critical shear rates $\dot{\gamma}$.

8.8.5.2 Rheo-Contrast Variation-SANS

Takeda et al. carried out Rheo-SANS measurements on polymer-clay system consisting of poly(ethylene oxide) (PEO) and synthetic clay laponite [161]. In order to extract the polymer-clay partial scattering function, S_{PC}, in addition to polymer-polymer, S_{PP}, and clay-clay partial scattering functions, S_{CC}, contrast variation SANS was also employed. This is the first example of Rheo-CV-SANS experiments. Figure 48 shows 2D SANS patterns of (a) clay dispersion, (b) PEO solution, and (c) a mixture of clay and PEO solution. In the stationary state, all of the patterns are isotropic, but only the mixture becomes anisotropic in a shear flow (Figure 48 (c) right). This

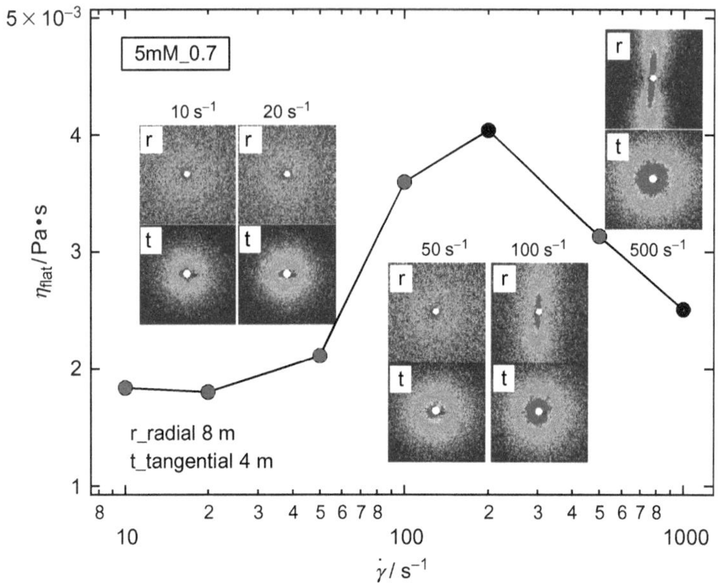

FIGURE 47 Two-dimensional SANS patterns for aqueous CTAB/NapTS solutions at C_D = 5 mM and C_S/C_D = 0.7 together with flow curves: η_{flat} versus $\dot{\gamma}$. Here, C_D and C_S/C_D denote the CTAB (detergent) concentration and the ratio of the NapTS (salt) to CTAB concentrations. r and t indicate radial (1-3 plane) (SDD = 8 m) and tangential (SDD = 4 m) (2-3 plane) configurations, respectively. *Reproduced from Ref. [160].*

change in the pattern is due to orientation of clay along the flow direction as a result of bridging of clays by stretched polymer chains as depicted in Figure 49.

8.8.5.3 Future Directions of Rheo-SANS

These examples shown above deal with ensemble average and time average structures with respect to the steady-state rheology. Since soft condensed matter is a "complex fluid," its rheological behavior is rich, particularly in start up of flow, flow reversal, stress relaxation, creep, etc. Time-resolved small- and large-amplitude oscillatory strain rheology will be available provided by intense neutron beam, which will provide new information about nonequilibrium dynamics and structural evolution.

McLeish and coworkers investigated the fluid behavior of entangled polymer melt in polymer processing macroscopically and molecular scale by SANS, rheological measurements, and theoretical prediction based on tube theory [163]. Such direct comparison of real- and reciprocal-space structures with molecular theory/simulation and SANS would be very powerful tools in further understanding of soft condensed matter.

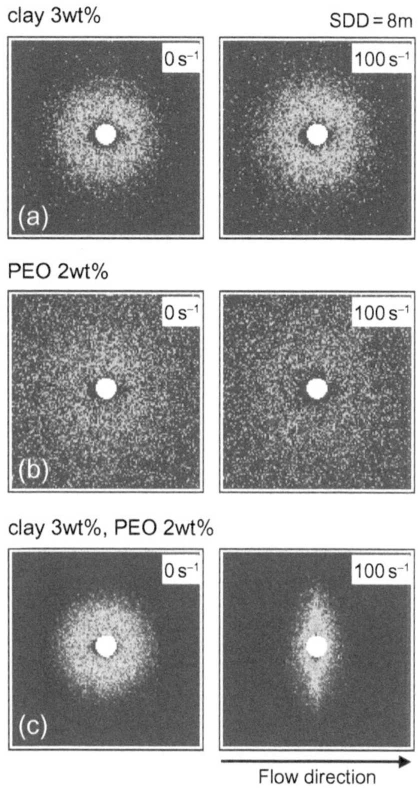

FIGURE 48 SANS profiles of (a) clay dispersion, (b) PEO solution, and (c) the mixed solutions in a stationary (0 s^{-1}) and flow (100 s^{-1}). *Reproduced from Ref. [162].*

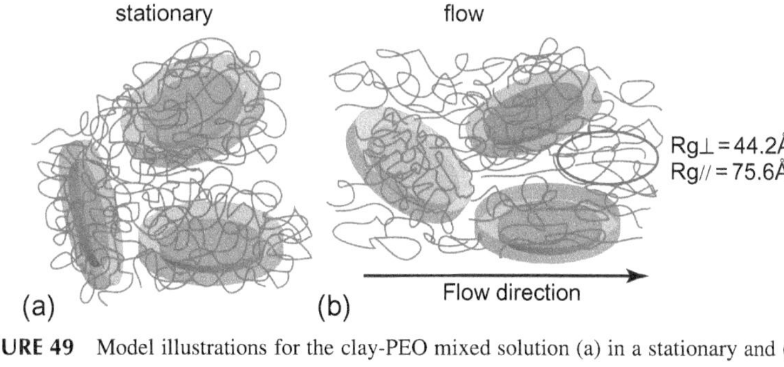

FIGURE 49 Model illustrations for the clay-PEO mixed solution (a) in a stationary and (b) in a flow. Dark green platelet; clay, light green platelet; adsorbed polymer chain layer, line; polymer chain. *Reproduced from Ref. [162].*

APPENDIX A. SCATTERING FUNCTIONS OF SPHERES WITH INTERPARTICLE INTERACTIONS

One often encounters an analysis of scattering functions for spherical particles dispersed in a medium, where the volume fraction of the particles is not negligibly small but there are noticeable interparticle interactions, e.g., hard-sphere interaction, electrostatic interactions, and excluded volume effect (resulting from a core-corona structure in block copolymer systems). Here, we discuss scattering functions for such systems along with the discussion given by Guiner and Fournet [63], Percus and Yevick [164], and Hayter and Penfold [165].

A.1 Scattering Function of Isolated Spheres

The scattering function for spherical objects with radius R is given by [63]

$$I(Q) = Kn_{\mathrm{p}}V_{\mathrm{p}}^2F(Q) \tag{A.1}$$

$$F(Q) = \Phi^2(QR) \tag{A.2}$$

$$\Phi(QR) = \frac{3}{(QR)^3}[\sin(QR) - (QR)\cos(QR)] = \sqrt{\frac{9\pi}{2}}\left\{\frac{J_{3/2}(QR)}{(QR)^{3/2}}\right\} \tag{A.3}$$

Here, K is a contrast factor ($K = (\Delta\rho)^2$ for neutron scattering), n_{p} is the number density of spheres, V_{p} is the volume of single sphere $= (4\pi/3)R^3$, $F(Q)$ is the form factor, and $J_\nu(x)$ is the Bessel function of the first kind (cylindrical Bessel function). When the number density increases, the interparticle distance becomes small and interference between particles becomes significant. As a result, the scattering function is modulated by the interparticle interference effect.

A.2 Interparticle Interference

In order to take into account the interparticle interference effect, Zernike and Prins derived the following equation [166],

$$I(Q) = Kn_{\mathrm{p}}V_{\mathrm{p}}^2F(Q)S(Q) \tag{A.4}$$

where $S(Q)$ is the interference factor (or dimensionless structure factor),

$$S(Q) = 1 + 4\pi n_{\mathrm{p}}\int_0^\infty [g(r) - 1]\frac{\sin(Qr)}{Qr}r^2\mathrm{d}r \tag{A.5}$$

Here, $g(r)$ is the radial distribution function.

A.3 Debye Equation for Spherical Systems

Debye solved Equation (A5) for dilute "hard-sphere gas" spheres [167],

$$g(r) = \begin{cases} 0 & r < 2R \\ 1 & r \geq 2R \end{cases} \tag{A.6}$$

$$I(Q) = Kn_pV_p^2\Phi^2(QR)S_D(Q) \tag{A.7}$$

$$S_D(Q) \equiv [1 - 8n_pV_p\Phi(2QR)] \tag{A.8}$$

Here, $S_D(Q)$ is the Debye interference factor.

A.4 Fournet Equation

However, the Debye's spherical function (Equation (A8)) becomes nonphysical (negative at low Q) when the volume fraction of spheres $\phi = n_pV_p$ becomes equal to or larger than 1/8. Fournet modified the Debye's function by taking account of the three-body interaction [168], and derived the following scattering intensity equation,

$$I(Q) = Kn_pV_p^2F(QR)\frac{1}{1 - n_p(2\pi)^{3/2}\varepsilon\beta(QR)} \tag{A.9}$$

$$\beta(QR) = \frac{2}{(2\pi)^{1/2}}\int_0^\infty [\exp^{-\psi(r)/kT} - 1]\frac{\sin(qr)}{qr}r^2\mathrm{d}r \tag{A.10}$$

where $\varepsilon \approx 1$ and $\psi(r)$ is the interparticle potential. This Fournet's equation can be solved for $\psi(r) = \infty(0 < r \leq 2R)$, $\psi(r) = 0(0 > 2R)$,

$$I(Q) = Kn_pV_p^2\Phi^2(QR)S_F(Q) \tag{A.11}$$

$$S_F(Q) \equiv \frac{1}{1 + 8n_p\varepsilon V_p\Phi(2QR)} \tag{A.12}$$

Here, $S_F(Q)$ is the Fournet interference factor. At small volume fractions, the Debye and Fournet equations are equivalent but the latter is more realistic at high packing densities because Fournet equation takes into account of the three-body interaction. However, since Equation (A12) still contains an unknown variable, it is required to derive an explicit form of scattering intensity. Before addressing it, let us discuss Ornstein-Zernike formalism which relates direct correlation function and the density-density correlation function.

A.5 Ornstein-Zernike Function

Ornstein and Zernike argued that the two-body correlation function $h(\mathbf{r}_{12}) \equiv g(\mathbf{r}) - 1$ can be regarded as caused by (i) a direct influence of 1 on 2 ("direct correlation function") $c(\mathbf{r}_{12})$ which should be short-ranged, and (ii) an indirect influence propagated directly from 1 to a third particle at $\mathbf{r}_3$ which in turn exert its total influence on particle 2 as shown in Figure A1 [17]. Integrating over $\mathbf{r}_3$, they thus derived [17],

$$h(\mathbf{r}_{12}) = c(\mathbf{r}_{12}) + n_p\int c(\mathbf{r}_{13})h(\mathbf{r}_{32})\mathrm{d}r_3 \tag{A.13}$$

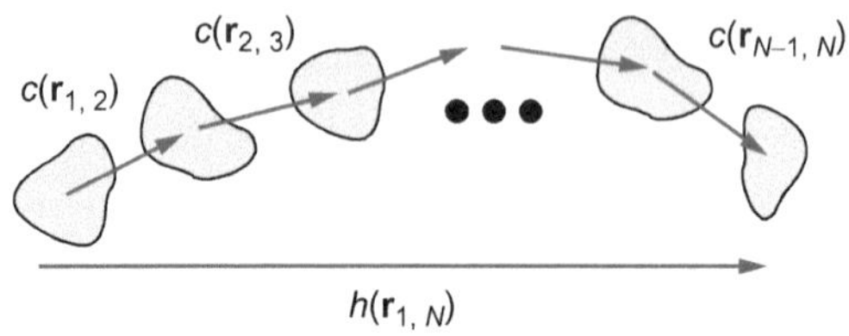

FIGURE A1 Ornstein-Zernike model. The correlation $h(\mathbf{r}_{1,\ N})$ is caused by (i) a direct correlation function $c(\mathbf{r}_{i,\ i+1})$ and (ii) an indirect influence propagated directly from 1 to a third particle at $\mathbf{r}_i$. The cartoon shows the part (ii).

This equation is called the Ornstein-Zernike integral equation and is extensively discussed by many workers including Fisher [169], Barker and Henderson [3], Thiele [170], and Hayter and Penfold [165].

By solving Eq. (A13) in Fourier space (Q space); $H(Q) = \int h(\mathbf{r})\exp(-i\mathbf{Q}\cdot\mathbf{r})d\mathbf{r}$ and $C(Q) = \int c(\mathbf{r})\exp(-i\mathbf{Q}\cdot\mathbf{r})d\mathbf{r}$, i.e., the Fourier transform of $H(Q)$ is given by

$$H(Q) = \frac{C(Q)}{1 - n_\mathrm{p}C(Q)} \tag{A.14}$$

where $C(Q)$ is the Fourier transform of the direct correlation function. On the other hand, the structure factor $S(Q) = \int g(\mathbf{r})\exp(-i\mathbf{Q}\cdot\mathbf{r})d\mathbf{r}$ is given by

$$S(Q) = 1 + n_\mathrm{p}H(Q) \tag{A.15}$$

Here, the second term of Equation (A15) means the modulation of the structure factor due to interparticle interference. By substituting Equation (A14) into Equation (A15), one obtains

$$S(Q) = \frac{1}{1 - n_\mathrm{p}C(Q)} \tag{A.16}$$

In the case of Ornstein-Zernike formalism, $C(Q)$ is expanded in Q and the following equation is obtained,

$$C(Q) = C(0) + \sum_{k=1} \frac{1}{k!}\left\{\frac{\partial^k C(Q)}{\partial Q^k}\right\}_{Q=0} Q^k \tag{A.17}$$

$$= C(0) + \frac{1}{2}C''(0)Q^2 + \mathcal{O}(Q^4) \tag{A.18}$$

Note that $C(Q)$ is a direct correlation function for a neighboring two bodies so that it does not diverge even near critical point. By evaluating the coefficient and defining a length parameter "correlation length" ξ as

$$\xi^2 \equiv \frac{C''(0)/2}{1 - n_\mathrm{p}C(0)} \tag{A.19}$$

one obtains the so-called Ornstein-Zernike function,

$$I(Q) = I(0)S(Q) = \frac{I(0)}{1 + \xi^2 Q^2} \tag{A.20}$$

where ξ is the correlation length. Equation (A20) is widely used to describe critical phenomena, screening effect, etc. Hence,

$$S_{OZ}(Q) = \frac{1}{1 + \xi^2 Q^2} \tag{A.21}$$

Note that the Fourier transform of Equation (A21) gives the correlation function for a density-fluctuating system,

$$\gamma(r) = \frac{\gamma}{r} \exp\left(-\frac{r}{\xi}\right) \tag{A.22}$$

A.6 Percus-Yevick Equation

The direct correlation function appeared in Equation (A13) is given by Percus and Yevick [164],

$$c(r) \simeq \left\{e^{-\psi(r)/kT} - 1\right\} e^{\psi(r)/kT} g(r) \tag{A.23}$$

As an approximate equation of the direct correlation function $C(Q)$, the following equation is given [170, 171]

$$c(r) = -(\alpha + \beta s + \gamma s^3) \tag{A.24}$$

$$\alpha = \frac{(1 + 2\phi)^2}{(1 - \phi)^4} \tag{A.25}$$

$$\beta = -\frac{6\phi(1 + \phi/2)^2}{(1 - \phi)^4} \tag{A.26}$$

$$\gamma = \frac{1}{2}\frac{\phi(1 + 2\phi)^2}{(1 - \phi)^4} \tag{A.27}$$

$$s = \frac{r}{2R}, \quad \phi = \frac{4\pi R^3 n_p}{3} \tag{A.28}$$

Here, ϕ is the volume fraction of hard sphere. In terms of these equations, the direct-correlation function is given by

$$C(Q,R,\phi) = -4\pi \int_0^{2R} (\alpha + \beta s + \gamma s^3) \frac{\sin(qr)}{qr} r^2 \mathrm{d}r \tag{A.29}$$

and the density-density-correlation function is obtained as,

$$S(Q) \equiv S_{\mathrm{PY}}(Q) = S(Q,R,\phi) = \frac{1}{1+24\phi[H_{\mathrm{PY}}(A)/A]} \tag{A.30}$$

Here, $A \equiv 2QR$ and $H_{\mathrm{PY}}(A)$ is given by

$$H_{\mathrm{PY}}(A) = \frac{\alpha}{A^2}(\sin A - A\cos A) \tag{A.31}$$

$$+\frac{\beta}{A^3}[2A\sin A + (2-A)\cos A - 2] \tag{A.32}$$

$$+\frac{\gamma}{A^5}\{-A^4\cos A + 4[(3A^2-6)\cos A + (A^3-6A)\sin A + 6]\} \tag{A.33}$$

A.7 Modified Percus-Yevick Equation

Fournet generalized his scattering equation (Equation (A12)) to account for the case where the spherical particles are surrounded by a "coat" of other molecules (e.g., protein molecules surrounded by a layer of water molecules) so that the minimum distance between hard-sphere centers is increased from $2R_{HS} = 2R$ to $2R_{HS}(> 2R)$ as shown in Figure A2. Yarusso and Cooper applied this concept to the analysis of sulfonated polystyrene ionomers [172], which has a core-shell structure (core; ionic domain, shell; hydrophobic shell). They modified the interference term of the Fournet equation (Equation (A12)) as written by,

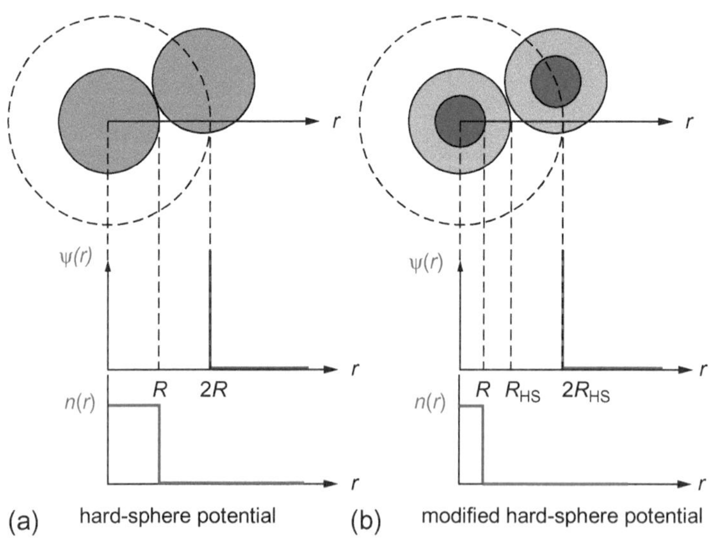

FIGURE A2 (a) Hard-sphere potential and (b) modified hard-sphere potential. $n(r)$ is the number density of the particle around $r = 0$. The lightly shaded region in (b) shows an additional excluded volume.

$$I(Q) = Kn_{\rm p}V_{\rm p}^2\Phi^2(QR)\frac{1}{1+8n_{\rm p}V_{\rm p}'\Phi(2QR_{HS})} \tag{A.34}$$

Here, $V_{\rm p}' = (4\pi/3)R_{HS}^3$, $\phi' = (R_{HS}/R)^3\phi$, and $A = 2QR_{HS}$. Kinning et al. modified the Percus-Yevick equation and took into account the corona thickness in addition to the core radius for a block copolymer system with a spherical microdomain structure [173]. They successfully reproduced SAXS curves with this modification, i.e., from $2R_{HS} = 2R$ to $2R_{HS}(> 2R)$. It is needless to mention that the excess thickness in the radius $\Delta R \equiv R_{HS} - R$ corresponds the corona thickness and proposed the modified Percus-Yevick equation,

$$I(Q) = Kn_{\rm p}V_{\rm p}^2\Phi^2(QR)\frac{1}{1+24\phi'(H_{\rm PY}/A)} \tag{A.35}$$

Here, $A = 2QR_{\rm PY}$. These equation allows to evaluate the scattering functions with weak interparticle interference.

Hence, the normalized scattering function $I_{\rm N}(Q, R, \phi)$ is given in terms of $\Phi(0) = 1$ and $S(0) = 1/\alpha$ in Equation (A25),

$$I_{\rm N}(Q,R,\phi) \equiv \frac{I(Q)}{I(0)} = \frac{S(Q)F(Q)}{S(0)F(0)} = S(Q,R,\phi)\Phi^2(QR)\alpha \tag{A.36}$$

Equation (A36) is an improvement of the Debye (A8) and Fournet equations (A10), and it is useful for analyzing high particle densities.

A.8 Hayter-Penfold Equation

Hayter and Penfold derived a structure factor for a macroion solution [165]. They assumed that the time-averaged structure of a macroion solution to be determined only by the repulsive screened Coulomb pair potential between finite macroions and solved the Ornstein-Zernike equation in the mean spherical approximation to obtain closed analytic forms for the direct-correlation function $c(r)$ and the structure factor $S(Q)$. Here, the counter ions and solvent are treated as a uniform neutralizing background, which determines the screening of the potential. It was confirmed that the Percus-Yevick hard sphere solution is recovered in the zero-charge limit. As charge is added to the macroions at given volume fraction, the isothermal compressibility decreases and $S(Q)$ shows increasing structure, eventually exhibiting solid-like behavior. The results provide a useful model basis for studying not only interacting colloidal systems but also finite ion screened one component plasmas in general. The repulsive potential between two identical spherical macroions of radius R is given by

$$\psi(r) = 4\pi\varepsilon_0\varepsilon R^2\psi_{\rm sur,0}^2\frac{\exp\{-\kappa(r-2R)\}}{r}, (r > 2R) \tag{A.37}$$

where r is the interionic center-to-center distance, $\psi_{sur,\,0}$ is the surface potential, ε is the dielectric constant of the solvent, ε_0 is the permittivity of the vacuum, and κ is the Debye-Hückel inverse screening length.

$$\psi_{sur,0} = \frac{z_m}{4\pi\varepsilon_0\varepsilon R(1+\kappa R)} \tag{A.38}$$

where z_m is the electronic charge on the macroion. The intermacroion potential is given by

$$\frac{\psi(r)}{k_B T} = \frac{\gamma_{ion}\exp(-\kappa x)}{x} \tag{A.39}$$

where $x = r/2R$, $k = 2\kappa R$, $K = 2QR$ and

$$\gamma_{ion} = \frac{4\pi\varepsilon_0\varepsilon R^2\psi_{sur,0}^2}{k_B T}e^k \tag{A.40}$$

After heavy mathematical treatment, Hayter and Penfold obtained the following equation,

$$S(Q) = S_{HP}(Q) = S(Q,R,\phi,\psi_{sur,0},\kappa,\varepsilon) = \frac{1}{1+24\phi\{H_{HP}(A,\psi_{sur,0},\kappa,\varepsilon)/A\}} \tag{A.41}$$

where

$$H_{HP}(A,\psi_{sur,0},\kappa,\varepsilon) = H_{PY}(A) + H'(A,\psi_{sur,0},\kappa,\varepsilon) \tag{A.42}$$

Here is a sophisticated function of $\psi_{sur,0}$, κ, and ε, which needs to be referred to the original paper [165]. Note that for the point charge, $S(Q)$ is reduced to the Debye-Hückel result,

$$S(Q) \rightarrow 1 - \frac{n_p z_m}{k_B T\varepsilon_0\varepsilon}\frac{1}{Q^2+\kappa^2} \tag{A.43}$$

A.9 Freltoft-Kjems-Sinha Equation

Freltoft, Kjems, and Sinha proposed the structure factor for fractal aggregates of silica particle, having power-law correlations [174]. The structure factor is given by,

$$S(Q) = 1 + \frac{C(d_f-1)\Gamma(d_f-1)\xi^{d_f}}{(1+Q^2\xi^2)^{d_f/2}}\frac{(1+Q^2\xi^2)^{1/2}}{Q\xi} \times \frac{\sin\{(d_f-1)\arctan(Q\xi)\}}{d_f-1} \tag{A.44}$$

where C is a constant and d_f is the fractal dimension of the aggregates and $\Gamma(x)$ is the gamma function of argument x, and ξ is the correlation length of the aggregates.

APPENDIX B. CONTRAST VARIATION

Because of large difference in the scattering lengths between H and D, a mixture of H- and D-isotopes provides a wide range of scattering length densities. The SANS contrast variation method has been widely used for structural investigations of complex systems, e.g., colloidal spheres [175, 176], polymer aggregates or micelles [177, 178], microemulsions [98], biological macromolecules [179, 180].

B.1 Two-Component Systems

First, let us discuss a two-component system consisting of components "1" and "2." There are four correlations, i.e., "1-1," "1-2," "2-1," and "2-2." The corresponding (partial) structure factors are $S_{11}(Q)$, $S_{12}(Q)$, $S_{21}(Q)$, and $S_{22}(Q)$, respectively. However, because of the Babinet's principle,

$$S_{11}(Q) = S_{22}(Q) = -S_{12}(Q) = -S_{21}(Q) \equiv S(Q). \tag{B.1}$$

Hence, only one structure factor $S(Q)$ is observed. Hence, the scattering function is simply given by

$$I(Q) = (\rho_2 - \rho_1)^2 S(Q) \tag{B.2}$$

B.2 Multicomponent Systems

If the number of components becomes larger than 2, every two-pair correlation has to be considered including self-correlations. Hence, the scattering function for an n-component system can be written by summing up all correlations by considering the scattering contrast,

$$I(Q) = \sum_{i=1}^{n} \rho_i^2 S_{ii}(Q) + 2\sum_{1<i<j}^{n} \rho_i \rho_j S_{ij}(Q) \tag{B.3}$$

Here, ρ_i is the scattering length density of the ith component. Since incompressibility assumption can be applied to most condensed systems, the scattering function can be rewritten to

$$I(Q) = \sum_{i=2}^{n} (\rho_i - \rho_1)^2 S_{ii}(Q) + 2\sum_{1<i<j}^{n} (\rho_i - \rho_1)(\rho_j - \rho_1) S_{ij}(Q) \tag{B.4}$$

Here, the "1" is the reference component, such as solvent. Note that the number of correlation functions is reduced to $\{n + n(n-1)/2\}$ to $\{(n-1) + (n-1)(n-2)/2\}$. For example, in the case of a three-component system, the number is reduced from 6 to 3.

B.3 Contrast Matching Method

Let us consider a three-component system more explicitly. See Figure B1. From Equation (B4), the scattering intensity function is written as follows,

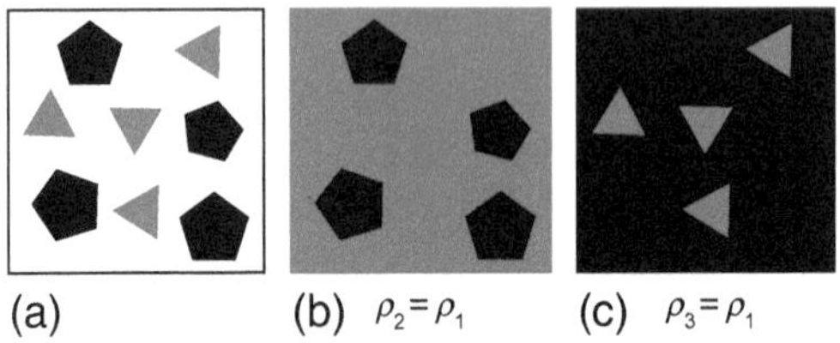

FIGURE B1 Contrast matching of three-component system. (a) $\rho_2 \neq \rho_3 \neq \rho_1$, (b) $\rho_2 = \rho_1$, and (c) $\rho_3 = \rho_1$.

$$I(Q) = (\rho_2 - \rho_1)^2 S_{22}(Q) + (\rho_3 - \rho_1)^2 S_{33}(Q) + (\rho_2 - \rho_1)(\rho_3 - \rho_1) S_{23}(Q) \tag{B.5}$$

If one can tune the scattering length density of the "solvent" (=1) to be the same as that of the component "2," i.e., $\rho_2 = \rho_1$, the scattering function is simplified to

$$I(Q) = (\rho_3 - \rho_1)^2 S_{33}(Q) \tag{B.6}$$

Similarly, for $\rho_3 = \rho_1$

$$I(Q) = (\rho_2 - \rho_1)^2 S_{22}(Q) \tag{B.7}$$

Hence, the information about the correlations between "3-3" and "2-2," i.e., the partial scattering functions $S_{33}(Q)$ and $S_{22}(Q)$ can be obtained, respectively. This is called "contrast matching method" and has been widely used in neutron scattering. Particularly, in the case of three-component systems consisting of solutes "2" and "3" and solvent "1," the scattering length density of the solvent "1" can be tuned by using a mixture of hydrogeneted-"1" and deuterated-"1." An example is a nano-composite gels consisting of polymer P (= "2"), inorganic clay C (= "3"), and a solvent S (= "1").

However, the contrast matching method has a disadvantage. It does not provide any information about the cross-terms, e.g., $S_{23}(Q)$. The contrast variation method can allow us to obtain it. Let us discuss it in the following.

B.4 Contrast Variation Method

Different from the contrast matching method, the scattering length density of the matrix is varied continuously in the contrast variation (CV) method, e.g., $\rho_1^\alpha, \rho_1^\beta, \ldots, \rho_1^\nu$, and a series of scattering measurements are conducted at ν different conditions. Here, the number of ν is usually taken larger than the number of the component n so as to improve the reliability of the analysis from experimental uncertainties, such as counting statistics and evaluation of the scattering length densities.

$$\begin{pmatrix} {}^{\alpha}I(Q) \\ {}^{\beta}I(Q) \\ \vdots \\ {}^{\nu}I(Q) \end{pmatrix} = \mathbf{M} \begin{pmatrix} S_{22}(Q) \\ S_{23}(Q) \\ S_{33}(Q) \end{pmatrix} \tag{B.8}$$

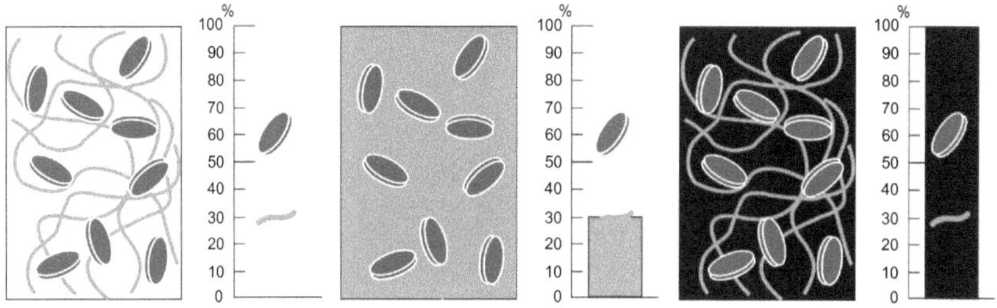

FIGURE B2 Contrast variation method for a three-component system of polymer (solid line), clay (disc), and solvent. The relative scattering length density of the solvent was varied from 0% (H_2O) to 30% and 100% (D_2O).

$$\mathbf{M} = \begin{pmatrix} {}^{\alpha}\Delta\rho_{21}^2 & {}^{\alpha}\Delta\rho_{21}\cdot{}^{\alpha}\Delta\rho_{31} & {}^{\alpha}\Delta\rho_{31}^2 \\ {}^{\beta}\Delta\rho_{21}^2 & {}^{\beta}\Delta\rho_{21}\cdot{}^{\beta}\Delta\rho_{31} & {}^{\beta}\Delta\rho_{31}^2 \\ \vdots & \vdots & \vdots \\ {}^{\nu}\Delta\rho_{21}^2 & {}^{\nu}\Delta\rho_{21}\cdot{}^{\nu}\Delta\rho_{31} & {}^{\nu}\Delta\rho_{31}^2 \end{pmatrix} \tag{B.9}$$

The partial structure factors $S_{ij}(Q)$ can be obtained by multiplying the inverse SLD matrix $\mathbf{M}^{-1}$ from the left-hand side of Equation (B8). Since the matrix is not a square but a ($\nu \times n$) matrix, $S_{ij}(Q)$ is obtained via the singular value decomposition. Figure B2 shows an example of the contrast variation method. The scattering contrast of a three component system consisting of polymer, clay, and water, can be tuned by varying the composition of water (H_2O) and heavy water (D_2O).

B.5 Physical Meaning of the Partial Structure Factor

The partial structure factors thus obtained by the CV method are subject to be analyzed similarly to a two-component system. The physical meaning of the self-terms, such as S_{22} and S_{33}, is trivial. On the other hand, the cross-term, such as S_{23}, is dependent on the geometry of the component in the system. Let us discuss the physical meaning of the cross-term. The self-term and the cross-term are defined by

$$S_{ii}(Q) = A_i(Q)\cdot A_i^*(Q) = |A_i(Q)|^2 \tag{B.10}$$

and

$$S_{ij}(Q) = \frac{A_i(Q)\cdot A_j^*(Q) + A_i^*(Q)\cdot A_j(Q)}{2}, \tag{B.11}$$

respectively. Hence, a self-term is always positive, while a cross-term is dependent on how the components are partitioned in the sample. By using incompressibility assumption, the cross-terms are obtained,

$$\begin{aligned} 2S_{23}(Q) &= -S_{22}(Q) - S_{33}(Q) + S_{11}(Q), \\ 2S_{31}(Q) &= S_{22}(Q) - S_{33}(Q) - S_{11}(Q), \\ 2S_{12}(Q) &= -S_{22}(Q) + S_{33}(Q) - S_{11}(Q) \end{aligned} \tag{B.12}$$

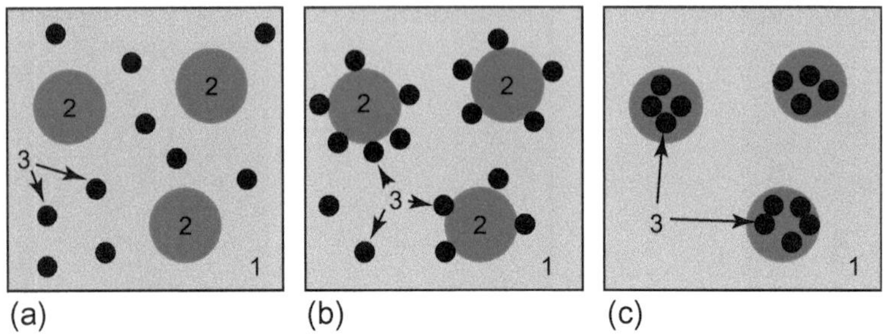

FIGURE B3 Physical meaning of the cross-terms. (a) Random distribution of "2" and "3," (b) particles "3" are preferentially adsorbed on "2," (c) particles "3" are included in "2."

By knowing the above-mentioned nature of the partial structure factors, let us discuss typical three cases of the arrangement of the components, (a) random distribution of components "2" and "3," (b) localization of component "3" on the outer-surface of "2," and (c) inclusion of "3" in "2" as schematically shown in Figure B3. Since "1" is the solvent (or matrix), the most interesting term is $S_{23}(Q)$. In the case of (a), $S_{23}(Q) = 0$ irrespective of Q because there is no correlation between "2" and "3." In the case of (b), there must be a positive correlation between "2" and "3," and $S_{23}(Q) > 0$ at large Q (Porod's law). However, in the case of (c), the Babinet's principle readily leads to $S_{23}(Q) < 0$. This example demonstrates the usefulness of the CV method for structural analysis of multicomponent system.

REFERENCES

[1] P.-G. de Gennes, Nobel Lecture, "Soft Matter" Nobel Foundation, 1991 Report.
[2] R.G. Kirste, W.A. Kruse, J. Schelten, Makromol. Chem. 162 (1973) 299–303.
[3] D.G.H. Ballard, J. Schelten, G.D. Wignall, Eur. Polym. J. 9 (1973) 965.
[4] J. Penfold, I.M. Tucker, Large-scale structures, in: F. Fernandez-Alonso, D.L. Price (Eds.), Experimental Methods in the Physical Sciences, Neutron Scattering: Fundamentals, vol. 44, Academic Press, London, 2013, pp. 353–413.
[5] V.G. Sakai, A. Arbe, Curr. Opin. Colloid Interf. Sci. 14 (2009) 381–390.
[6] F. Fernandez-Alonso, D.L. Price, Neutron Scattering: Fundamentals, Experimental Methods in the Physical Sciences, vol. 44, Academic Press, London, 2013.
[7] T. Brückel, G. Heger, D. Richter, R. Zorn, Neutron Scattering, Forschungszentrum Jülich GmbH, Jülich, 2005.
[8] J.S. Higgins, H.C. Benoit, Polymers and Neutron Scattering, Clarendon Press, Oxford, 1994.
[9] J. Dawidowski, J.R. Granada, J.R. Santisteban, F. Cantargi, L.A.R. Palomino, Neutron scattering lengths and cross sections, in: F. Fernandez-Alonso, D.L. Price (Eds.), Experimental Methods in the Physical Science, Neutron Scattering: Fundamentals, vol. 44, Academic Press, London, 2013, pp. 471–528.
[10] M. Shibayama, M. Nagao, S. Okabe, T. Karino, J. Phys. Soc. Jpn 74 (2005) 2728–2736.
[11] M. Shibayama, T. Matsunaga, M. Nagao, J. Appl. Crystallogr. 42 (2009) 621–628.
[12] B. Jacrot, Rep. Prog. Phys. 39 (1976) 911–953.

[13] D.L. Price, F. Fernandez-Alonso, An introduction to neutron scattering, in: F. Fernandez-Alonso, D.L. Price (Eds.), Neutron Scattering: Fundamentals, Academic Press, London, 2013, pp. 1–137.
[14] M. Arai, Experimental techniques, in: F. Fernandez-Alonso, D.L. Price (Eds.), Experimental Methods in the Physical Science, Neutron Scattering: Fundamentals, vol. 44, Academic Press, London, 2013, pp. 245–320.
[15] H. Iwase, H. Endo, M. Katagiri, M. Shibayama, J. Appl. Cryst. 44 (2011) 558–568.
[16] G. Porod, Kolloid Z. 125 (1952) 51–57.
[17] L.S. Ornstein, F. Zernike, Proc. Acad. Sci. 17 (1914) 793–806.
[18] P. Debye, A.M. Bueche, J. Appl. Phys. 20 (20) (1949) 518–525.
[19] G. Beaucage, J. Appl. Cryst. 28 (1995) 717–728.
[20] M.J. Grundy, R.M. Richardson, S.J. Roser, J. Penfold, R.C. Ward, Thin Solid Films 159 (1988) 43–52.
[21] S.H. Anastasiadis, T.P. Russell, S.K. Satija, C.F. Majkrzak, Phys. Rev. Lett. 62 (1989) 1852–1855.
[22] M.R. Johnson, G.J. Kearley, Dynamics of atoms and molecules, in: F. Fernandez-Alonso, D.L. Price (Eds.), Experimental Methods in the Physical Sciences, Neutron Scattering: Fundamentals, vol. 44, Academic Press, London, 2013, pp. 415–470.
[23] F. Mezei, Z. Physik 255 (1972) 146–160.
[24] F. Mezei, Neutron Spin Echo, Lecture Note in Physics, 128 Springer-Verlag, Berlin, 1980.
[25] J. Major, B. Farago, F. Mezei, Neutron optics and spin labeling methods, in: F. Fernandez-Alonso, D.L. Price (Eds.), Experimental Methods in the Physical Science, Neutron Scattering: Magnetic and Quantum Phenomena, vol. 48, Academic Press, London, 2015, pp. 1–43.
[26] B.J. Berne, R. Pecora, Dynamic Light Scattering With Application to Chemistry, Biology, and Physics, Robert E. Krieger Publishing, Malabar, 1990.
[27] B. Chu, Laser Light Scattering, second ed., Academic Press, London, 1991.
[28] B. Frick, D. Richter, Science 267 (1995) 1939–1945.
[29] D. Richter, B. Ewen, B. Farago, T. Wagner, Phys. Rev. Lett. 62 (18) (1989) 2140–2143.
[30] D. Richter, B. Farago, L.J. Fetters, J.S. Huang, B. Ewen, C. Lartigue, Phys. Rev. Lett. 64 (1990) 1389–1392.
[31] D. Richter, J.B. Hayter, F. Mezei, B. Ewen, Phys. Rev. Lett. 41 (1978) 1484–1487.
[32] B. Farago, D. Richter, J.S. Huang, S.A. Safran, S.T. Milner, Phys. Rev. Lett. 65 (1990) 3348–3351.
[33] Z. Bu, R. Biehl, M. Monkenbusch, D. Richter, D.J.E. Callaway, PNAS 102 (2005) 17646–17651.
[34] R. Inoue, R. Biehl, T. Rosenkranz, J. Fitter, M. Monkenbusch, A. Radulescu, B. Farago, D. Richter, Biophys. J. 99 (2010) 2309–2317.
[35] P.J. Flory, Principles of Polymer Chemistry, Cornell Univ., Ithaca, NY, 1953
[36] P.J. Flory, Statistical Mechanics of Chain Molecules, Wiley-Interscience, New York, NY, 1969.
[37] H. Yamakawa, Modern Theory of Polymer Solutions, Harper and Row, New York, NY, 1971.
[38] P.G. de Gennes, Scaling Concepts in Polymer Physics, Cornell University, Ithaca, NY, 1979.
[39] M. Doi, S.F. Edwards, The Theory of Polymer Dynamics, Oxford Univ. Press, Oxford, 1986.
[40] P.G. de Gennes, J. Chem. Phys. 55 (1971) 572–579.

[41] H.E. Stanley, Introduction to Phase Transition and Critical Phenomena, Oxford University Press, New York, NY, 1971.
[42] M. Shibayama, H. Yang, R.S. Stein, C.C. Han, Macromolecules 18 (1985) 2179–2187.
[43] L. Leibler, Macromolecules 13 (1980) 1602–1617.
[44] D. de Rossi, K. Kajiwara, Y. Osada, A. Yamauchi, Polymer Gels, Plenum Press, New York, NY, 1991.
[45] Y. Osada, K. Kajiwara, Gels Handbook, Academic Press, New York, NY, 2000.
[46] A. Baumgartner, C.E. Picot, Molecular Basis of Polymer Networks, Springer Proceedings in Physics, 42 Springer, Berlin, 1989.
[47] M. Shibayama, Small-angle neutron scattering on gels, in: P. Pecora, R. Borsali (Eds.), Soft Matter Characterization, vol. 2, Springer-Verlag, New York, NY, 2008, pp. 783–832.
[48] M. Shibayama, Inhomogeneous structure and dynamics of condensed soft matter, in: T. Imae, T. Kanaya, M. Furusaka, N. Torikai (Eds.), Neutrons in Soft Matter, John Wiley & Sons, Hoboken, NJ, 2011, pp. 493–516.
[49] M. Shibayama, Polym. J. 43 (2011) 18–34.
[50] R.S. Stein, J. Polym. Sci. B7 (1969) 657–660.
[51] K. Dusek, W. Prins, Adv. Polym. Sci. 6 (1969) 1–102.
[52] A. Onuki, J. Phys. II France 2 (1992) 45–61.
[53] C. Rouf, J. Bastide, J.M. Pujol, F. Schosseler, J.P. Munch, Phys. Rev. Lett. 73 (1994) 830–833.
[54] J. Bastide, S.J. Candau, Structure of gels as investigated by means of static scattering techniques, in: J.P. Cohen Addad (Ed.), The Physical Properties of Polymer Gels, John Wiley, New York, NY, 1996, pp. 143–295.
[55] S. Panyukov, Y. Rabin, Phys. Rep. 269 (1996) 1–132.
[56] M. Shibayama, Macromol. Chem. Phys. 199 (1998) 1–30.
[57] M. Shibayama, Bull. Chem. Soc. Jpn 79 (2006) 1799–1819.
[58] S. Candau, J. Bastide, M. Delsanti, Adv. Polym. Sci. 44 (1982) 27–71.
[59] S. Mallam, F. Horkay, A.M. Hecht, E. Geissler, Macromolecules 22 (1989) 3356–3361.
[60] W. Wu, M. Shibayama, S. Roy, H. Kurokawa, L.D. Coyen, S. Nomura, R.S. Stein, Macromolecules 23 (1990) 2245–2251.
[61] F. Horkay, A.M. Hecht, S. Mallam, E. Geissler, A.R. Rennie, Macromolecules 24 (1991) 2896–2902.
[62] Y. Cohen, O. Ramon, I.J. Kopelman, S. Mizrahi, J. Polym. Sci. Polym. Phys. Ed. 30 (1992) 1055–1067.
[63] A. Guinier, G. Fournet, Small-Angle Scattering of X-Rays, John Wiley & Sons, Inc., New York, NY, 1955
[64] S. Mallam, F. Horkay, A.M. Hecht, A.R. Rennie, E. Geissler, Macromolecules 24 (1991) 543–548.
[65] J. Bastide, L. Leibler, Macromolecules 21 (1988) 2647–2649.
[66] E. Mendes Jr, P. Lindner, M. Buzier, F. Boue, J. Bastide, Phys. Rev. Lett. 66 (1991) 1595–1598.
[67] T. Norisuye, N. Masui, Y. Kida, M. Shibayama, D. Ikuta, E. Kokufuta, S. Ito, S. Panyukov, Polymer 43 (2002) 5289–5297.
[68] Y. Rabin, S. Panyukov, Macromolecules 30 (2) (1997) 301–312.
[69] M. Shibayama, F. Ikkai, Y. Shiwa, Y. Rabin, J. Chem. Phys. 107 (13) (1997) 5227–5235.
[70] F. Schosseler, F. Ilmain, S.J. Candau, Macromolecules 24 (1991) 225–234.
[71] F. Schosseler, A. Moussaid, J.P. Munch, S.J. Candau, J. Phys. II 1 (10) (1991) 1197–1219.

[72] M. Shibayama, T. Tanaka, C.C. Han, J. Chem. Phys. 97 (1992) 6842–6854.
[73] M. Shibayama, F. Ikkai, S. Inamoto, S. Nomura, C.C. Han, J. Chem. Phys. 105 (1996) 4358–4366.
[74] A.M. Hecht, F. Horkay, E. Geissler, J.P. Benoit, Macromolecules 24 (1991) 4183–4187.
[75] A. Moussaid, S.J. Candau, J.G.H. Joosten, Macromolecules 27 (8) (1994) 2102–2110.
[76] J.M. Guenet, Thermoreversible Gelation of Polymers and Biopolymers, Academic Press, New York, NY, 1992.
[77] P. Terech, R.G. Weiss, Chem. Rev. 97 (1997) 3133–3159.
[78] S. Okabe, K. Ando, K. Hanabusa, M. Shibayama, J. Polym. Sci. B Polym. Phys. Ed. 42 (2004) 1841–1848.
[79] P. Dastidar, S. Okabe, K. Nakano, K. Iida, M. Miyata, N. Thonai, M. Shibayama, Chem. Mater. 17 (4) (2005) 741–748.
[80] M. Shibayama, H. Kurokawa, S. Nomura, M. Muthukumar, R.S. Stein, S. Roy, Polymer 33 (1992) 2883–2890.
[81] H. Abied, A. Bnilet, J.M. Guenet, Colloid Polym. Sci. 268 (1990) 403–413.
[82] M. Kobayashi, T. Yoshioka, M. Imai, Y. Itoh, Macromolecules 28 (1995) 7376–7385.
[83] T. Kanaya, M. Ohkura, H. Takeshita, K. Kaji, M. Furusaka, H. Yamaoka, G.D. Wignall, Macromolecules 28 (9) (1995) 3168–3174.
[84] H. Takeshita, T. Kanaya, K. Nishida, K. Kaji, Phys. Rev. E 61 (2000) 2125–2128.
[85] H. Takeshita, T. Kanaya, K. Nishida, K. Kaji, Phys. B 311 (2002) 78–83.
[86] S. Sugihara, K. Hashimoto, S. Okabe, M. Shibayama, S. Kanaoka, S. Aoshima, Macromolecules 37 (2004) 336–343.
[87] C. Fuse, S. Okabe, S. Sugihara, S. Aoshima, M. Shibayama, Macromolecules 37 (20) (2004) 7791–7798.
[88] W. Kuhn, Kolloid Z. 68 (1934) 2–15.
[89] J.P. Cotton, D. Decker, H. Benoit, B. Farnoux, J.S. Higgins, G. Jannink, R. Ober, C. Picot, J. des Cloizeaux, Macromolecules 7 (1974) 863–872.
[90] D. Bohm, D. Pines, Phys. Rev. 92 (1953) 609–625.
[91] C. Herkt-Maetzky, J. Schelten, Phys. Rev. Lett. 51 (1983) 896–899.
[92] F.S. Bates, G.D. Wignall, Phys. Rev. Lett. 57 (1986) 1429–1432.
[93] G. Ronca, J. Chem. Phys. 79 (1983) 1031–1043.
[94] A. Menelle, T.P. Russell, S.H. Anastasiadis, S.K. Satija, C.F. Majkrzak, Phys. Rev. Lett. 68 (1992) 67–70.
[95] T. Tanaka, Phys. Rev. Lett. 40 (1978) 820–823.
[96] V. Borue, I. Erukhimovich, Macromolecules 21 (11) (1988) 3240–3249.
[97] B. Jakobs, T. Sottmann, R. Strey, J. Allgaier, L. Willner, D. Richter, Langmuir 15 (1999) 6707–6711.
[98] H. Endo, J. Allgaier, G. Gompper, B. Jakobs, M. Monkenbusch, D. Richter, T. Sottmann, R. Strey, Phys. Rev. Lett. 85 (2000) 102–105.
[99] K. Mortensen, Prog. Colloid Polym. Sci. 91 (1993) 69–71.
[100] K. Mortensen, W. Brown, E. Jargensen, Macromolecules 27 (1994) 5654–5666.
[101] K. Mortensen, J. Phys. Condens. Matter 8 (1996) A103–A124.
[102] K. Mortensen, Y. Talmon, Macromolecules 28 (1995) 8829–8834.
[103] I. Goldmints, F.K. von Gottberg, K.A. Smith, T. Alan Hatton, Langmuir 13 (1997) 3659–3664.
[104] L. Yang, P. Alexandridis, B. Lindman, D.C. Steytler, M.J. Kositza, J.F. Holzwarth, Langmuir 16 (2000) 8555–8561.
[105] P. Alexandridis, L. Yang, Macromolecules 33 (2000) 5574–5587.

[106] J.S. Pedersen, M.C. Gerstenberg, Colloids Surf. A Physicochem. Eng. Asp. 213 (2003) 175–187.
[107] N.J. Jain, V.K. Aswal, P.S. Goyal, P. Bahadur, J. Phys. Chem. B 102 (1998) 8452–8458.
[108] M. Nakano, M. Fukuda, T. Kudo, H. Endo, T. Handa, Phys. Rev. Lett. 98 (2007) 238101.
[109] S. Kimata, T. Sakurai, Y. Nozue, T. Kasahara, N. Yamaguchi, M. Shibayama, J.A. Kornfield, Science 316 (2007) 1014–1017.
[110] K. Sadakane, A. Onuki, K. Nishida, S. Koizumi, H. Seto, Phys. Rev. Lett. 103 (2009) 167803.
[111] Y. Okumura, K. Ito, Adv. Mater. 13 (7) (2001) 485–487.
[112] K. Haraguchi, T. Takehisa, Adv. Mater. 14 (2002) 1120–1124.
[113] T. Sakai, T. Matsunaga, Y. Yamamoto, C. Ito, R. Yoshida, S. Suzuki, N. Sasaki, M. Shibayama, U. Chung, Macromolecules 41 (14) (2008) 5379–5384.
[114] M. Shibayama, J. Phys. Soc. Jpn 78 (4) (2009) 041008(1)–041008(7).
[115] M. Shibayama, Soft Matter 8 (2012) 8030–8038.
[116] S. Miyazaki, H. Endo, T. Karino, K. Haraguchi, M. Shibayama, Macromolecules 40 (2007) 4287–4295.
[117] T. Nishida, H. Endo, N. Osaka, H. Li, K. Haraguchi, M. Shibayama, Phys. Rev. E 80 (2009) 030801(1)–030801(4).
[118] T. Matsunaga, T. Sakai, Y. Akagi, U. Chung, M. Shibayama, Macromolecules 42 (2009) 1344–1351.
[119] T. Matsunaga, T. Sakai, Y. Akagi, U. Chung, M. Shibayama, Macromolecules 42 (16) (2009) 6245–6252.
[120] B. Hammouda, D. Ho, S. Kline, Macromolecules 35 (2002) 8578–8585.
[121] H. Asai, K. Fujii, T. Ueki, T. Sakai, U. Chung, M. Watanabe, Y.S. Han, T.H. Kim, M. Shibayama, Macromolecules 45 (2012) 3902–3909.
[122] T. Matsunaga, H. Asai, Y. Akagi, T. Sakai, U. Chung, M. Shibayama, Macromolecules 44 (2011) 1203–1210.
[123] H. Iwase, M. Katagiri, M. Shibayama, J. Appl. Cryst. 45 (2012) 507–512.
[124] M. Ballauff, Adv. Eng. Mater. 13 (2011) 793–802.
[125] T. Hiroi, S. Kondo, T. Sakai, E.P. Gilbert, Y.-S. Han, T.-H. Kim, M. Shibayama, Macromolecules 49 (2016) 4940–4947.
[126] R. Inoue, T. Takata, N. Fujii, K. Ishii, S. Uchiyama, N. Sato, Y. Oba, K. Wood, K. Kato, N. Fujii, M. Sugiyama, Sci. Rep. 6 (2016) 29208-1–29208-7.
[127] V.V. Volkov, D.I. Svergun, J. Appl. Cryst. 36 (2003) 860–864.
[128] J.S. Pedersen, P. Schurtenberger, Macromolecules 29 (1996) 7602–7612.
[129] L. Arleth, M. Bergstrom, J.S. Pedersen, Langmuir 18 (2002) 5343–5353.
[130] K.S. Schweizer, A. Yethiraj, J. Chem. Phys. 98 (1993) 9053–9079.
[131] C.N. Licos, Soft Matter 2 (2006) 478–498.
[132] I.G. Elliott, D.E. Mulder, P.T. Träskelin, J.R. Ell, T.E. Patten, T.L. Kuhl, R. Faller, Soft Matter 9 (2009) 4612–4622.
[133] D. Richter, S. Goossen, A. Wischnewski, Soft Matter 11 (2015) 8535–8549.
[134] W. Kauzmann, Nature 325 (1987) 763–764.
[135] R. Lopez-Fnadino, Crit. Rev. Food. Sci. Nutr. 46 (2006) 351–363.
[136] B. Alberts, D. Bray, J. Lewis, M. Raff, K. Roberts, J.D. Watson, Molecular Biology of the Cell, Garland Publishing Inc., New York, NY, 1994
[137] K. Heremans, L. Smeller, Biochim. Biophys. Acta 18 (1998) 353–370.
[138] L. Smeller, Biochim. Biophys. Acta 1595 (2002) 11–29.
[139] M. Matsumoto, K. Murakoshi, Y. Wada, S. Yanagida, Chem. Lett. 29 (2000) 938–939.

[140] M. Shibayama, K. Isono, S. Okabe, T. Karino, M. Nagao, Macromolecules 37 (2004) 2909–2918.
[141] I.R. Nasimova, T. Karino, S. Okabe, M. Nagao, M. Shibayama, Macromolecules 37 (23) (2004) 8721–8729.
[142] I.R. Nasimova, T. Karino, S. Okabe, M. Nagao, M. Shibayama, J. Chem. Phys. 121 (19) (2004) 9708–9715.
[143] N. Osaka, M. Shibayama, Phys. Rev. Lett. 96 (4) (2006) 048303(1)–048303(4).
[144] S. Okabe, T. Karino, M. Nagao, S. Watanabe, M. Shibayama, Nucl. Instrum. Methods Phys. Res. A 572 (2007) 853–858.
[145] K. Otake, R. Karaki, T. Ebina, C. Yokoyama, S. Takahashi, Macromolecules 26 (9) (1993) 2194–2197.
[146] E. Kato, J. Chem. Phys. 106 (1997) 3792–3797.
[147] L.P.N. Rebelo, Z.P. Visak, H.C. de Sousa, J. Szydlowski, R. Gomes de Azevedo, A.M. Ramos, V. Najdanovic-Visak, M. Nunes da Ponte, J. Klein, Macromolecules 35 (2002) 1887–1895.
[148] S. Sawamura, K. Kitamura, Y. Taniguchi, J. Phys. Chem. 93 (1989) 4931–4935.
[149] R. Hosemann, S.N. Bagchi, Direct Analysis of Diffraction by Matter, North-Holland, Amsterdam, 1962.
[150] N. Osaka, S. Takata, T. Suzuki, H. Endo, M. Shibayama, Polymer 49 (12) (2008) 2957–2963.
[151] H.S. Frank, M.W. Evans, J. Chem. Phys. 13 (1945) 507–532.
[152] W. Kauzmann, Adv. Protein Chem. 14 (1959) 1–63.
[153] J.H. Hildebrand, J. Phys. Chem. 72 (1968) 1841–1842.
[154] G. Nemethy, H.A. Scheraga, W. Kauzmann, J. Phys. Chem. 72 (1968) 1842 1842.
[155] S. Kunugi, K. Takano, N. Tanaka, Macromolecules 30 (1997) 4499–4501.
[156] E.M. Dumay, M.T. Kalichevsky, J.C. Cheftel, J. Agric. Food Chem. 42 (1994) 1861–1868.
[157] G. Panick, R. Malessa, R. Winter, Biochemistry 38 (1999) 6512–6519.
[158] A.P.R. Eberle, L. Porcar, Curr. Opin. Colloid Interf. Sci. 17 (2012) 33–43.
[159] A.K. Gurnon, C.R. Lopez-Barron, A.P.R. Eberle, L. Porcar, N.J. Wagner, Soft Matter 10 (2014) 2889–2898.
[160] M. Takeda, T. Kusano, T. Matsunaga, H. Endo, T. Shikata, M. Shibayama, Langmuir 27 (5) (2011) 1731–1738.
[161] M. Takeda, T. Matsunaga, T. Nishida, H. Endo, T. Takahashi, M. Shibayama, Macromolecules 43 (18) (2010) 7793–7799.
[162] T. Matsunaga, H. Endo, M. Takeda, M. Shibayama, Macromolecules 43 (11) (2010) 5075–5082.
[163] J. Bent, L.R. Hutchings, R.W. Richards, T. Gough, R. Spares, P.D. Coates, I. Grillo, O.G. Harlen, D.J. Read, R.S. Graham, A.E. Likhtman, D.J. Groves, T.M. Nicholson, T.C.B. McLeish, Science 301 (2003) 1691–1695.
[164] J.K. Percus, G.J. Yevick, Phys. Rev. 110 (1958) 1–12.
[165] J.B. Hayter, J. Penfold, Mol. Phys. 42 (1) (1981) 109–118.
[166] F. Zernike, J.A. Prins, Z. Physik 41 (1927) 184–194.
[167] P. Debye, Phys. Z. 28 (1927) 135–141.
[168] P.G. Fournet, Acta Crystallogr. 4 (1951) 293–301.
[169] M.E. Fisher, J. Math. Phys. 5 (7) (1964) 944–962.
[170] E. Thiele, J. Chem. Phys. 39 (1963) 474.
[171] M.S. Wertheim, Phys. Rev. Lett. 10 (1963) 321–323.
[172] D.J. Yarusso, S.L. Cooper, Macromolecules 16 (1983) 1871–1880.

[173] D. Kinning, E.L. Thomas, Macromolecules 17 (1984) 1712–1718.
[174] T. Freltoft, J.K. Kjems, S.K. Sinha, Phys. Rev. B. 33 (1986) 269–275.
[175] M.H.G. Duits, R.P. May, A. Vrij, C.G. de Kruif, J. Chem. Phys. 94 (1991) 4521–4531.
[176] P. Bartlett, R.H. Ottewill, J. Chem. Phys. 96 (1992) 3306–3318.
[177] D. Richter, D. Schneiders, M. Monkenbusch, L. Willner, L.J. Fetters, J.S. Huang, M. Lin, K. Mortensen, B. Farago, Macromolecules 30 (1997) 1053–1068.
[178] A. Poppe, L. Willner, J. Allgaier, J. Stellbrink, D. Richter, Macromolecules 30 (1997) 7462–7471.
[179] R. Willumeit, N. Burkhardt, G. Diedrich, J. Zhao, K.H. Nierhaus, H.B. Stuhrmann, J. Mol. Struct. 383 (1996) 201–211.
[180] N.N. Niimura, Y. Minezaki, I. Tanaka, S. Fujiwara, M. Ataka, J. Cryst. Growth 200 (1999) 265–270.

Chapter 9

Ionic Conductors and Protonics

Maths Karlsson* and Adrien Perrichon
Department of Chemistry and Chemical Engineering, Chalmers University of Technology, Göteborg, Sweden
Corresponding author: e-mail: maths.karlsson@chalmers.se

Chapter Outline

9.1 SOLID-STATE IONICS

Over the past 10-20 years, solid-state ionics, *i.e.*, the study of solid ion conducting materials, or so-called electrolytes, and their usage in various technological applications such as batteries, thermoelectric converters, chemical sensors, and fuel cells, has developed into an important and fast-growing field within materials science. This development is driven by the need for cleaner and more sustainable sources of energy, which is one of the grand challenges in the 21st century. In this context, the hydrogen fuel cell, in combination with efficient and functional storage of hydrogen (the fuel), is a promising yet challenging solution for the creation of a more sustainable and environment friendly society [1, 2].

Experimental Methods in the Physical Sciences, Vol. 49. http://dx.doi.org/10.1016/B978-0-12-805324-9.00009-1

At the heart of any hydrogen fuel cell is the high conductivity of either oxide-ions or protons in a solid electrolyte, which separates the anode and cathode, to produce electricity and water (Figure 1). Solid electrolytes are traditionally classified according to the temperature window in which they are highly conducting, which then determines the operating temperature window of the fuel cell. High-temperature materials (above ≈500 °C) are used in solid oxide fuel cells (SOFCs) and typically comprise oxide-ion conducting oxides, such as yttria-stabilized zirconia (YSZ), whereas low-temperature (below 100 °C) materials typically encompass water-filled sulfonated polymers such as Nafion. The high-temperature materials are nowadays being commercialized and adopted in fuel-cell driven combined heat and power devices or efficiently coupled with gas turbines, albeit the high operating temperature does impose significant challenges on materials choices to avoid interconnect oxidation, sealing failure, and thermal delamination of the membrane electrode assembly. The low-temperature materials are used in polymer electrolyte membrane fuel cells (PEMFCs) in the automotive industry, *e.g.*, in cars and buses. These materials are, however, naturally limited to low temperatures due to their need of being humidified, exhibit low tolerance toward carbon and sulfur impurities, and require expensive electrocatalysts such as platinum.

With the aim to avoid the problems associated with high- and low-temperature operation, much research is now directed toward reducing the operating temperature of SOFCs, and increasing the operating temperature of PEMFCs, respectively, to the so-called intermediate-temperature range (100–500 °C), which would partly solve these issues. Currently available intermediate temperature proton conductors include (i) solid acids such as cesium dihydrogen phosphate (CsH_2PO_4) and cesium hydrogen sulfate ($CsHSO_4$), which are, however, limited by their thermal stability window,

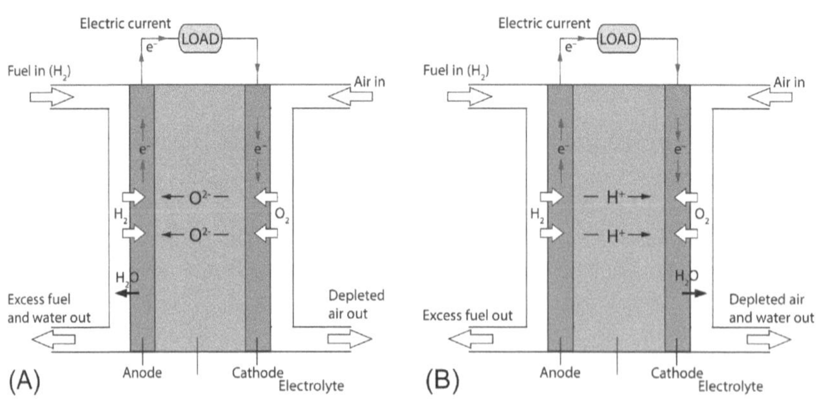

FIGURE 1 Schematics of the operation of an SOFC utilizing (A) oxide-ion and (B) proton conducting electrolytes, respectively. *Adapted with permission from Ref. [3].*

with the majority of compounds melting around 250 °C, and (ii) proton conducting oxides, yet restricted by a strongly temperature-dependent proton conductivity, leading to a limited proton conductivity in the intermediate/low-temperature range. To ensure that the total internal resistance (electrolyte + electrodes) of a fuel cell is sufficiently small, the targeted value for the area-specific resistivity of the electrolyte is set at 0.15 Ω cm^2. Because most ceramic films can be reliably produced using conventional ceramic fabrication routes at thicknesses down to ~15 μm, it follows that the conductivity of the electrolyte must exceed 0.1 S cm^{-1} [1]. This requirement is, however, not met by any currently available material (Figure 2). Thus, the discovery of new oxide-ion and/or proton conducting materials, which combine high chemical and mechanical stability with high conductivity in the intermediate-temperature range, is critical to future breakthroughs in the development of devices driven by intermediate-temperature fuel-cell technology. Such an improvement depends on increasing the understanding of the fundamental science of key materials properties such as crystal structures and ion conduction mechanisms, in the most promising classes of materials, and the exploration of completely new materials. For this purpose, neutron scattering is a powerful tool that has been applied successfully to studies of both oxide-ion and proton conducting materials, and its role in this area of research is vast.

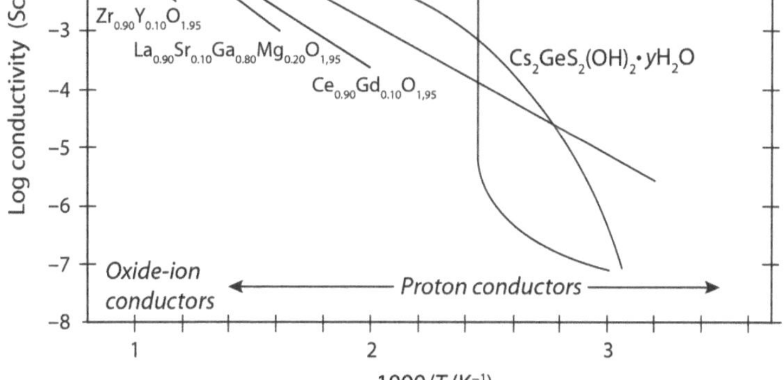

FIGURE 2 Conductivity of state-of-the-art electrolytes over a wide temperature range. Note the gap of highly conducting materials in the intermediate-temperature range ≈100–500 °C. The conductivity of $Cs_2GeS_2(OH)_2 \cdot yH_2O$ is taken from Ref. [13]. *Adapted with permission from Ref. [12].*

The aim of this chapter is not to give an exhaustive account for all the neutron scattering studies done in this field, nor is it aimed at providing a review of the fuel-cell technology, as excellent reviews can be found elsewhere [1, 4–11]. Rather, this chapter focuses on proton and oxide-ion conducting oxide materials, partly biased by the interest of the authors, and aims to give a flavor of how the variety of neutron methods can be used to unravel important information about the structure and dynamics of these technologically important materials. Properties of key materials that are addressed include crystal structure, proton and oxide-ion sites, proton concentration depth profiles in thin films, vibrational dynamics, localized and long-range diffusion of atoms, and macroscopic conductivity. We also demonstrate the important role/capacity of computer modeling as a complementary tool to neutron scattering. The chapter finishes with our personal thoughts on prospects for future work in this field.

9.2 PROTON CONDUCTORS

The proton is the only ion, which has no electron shell of its own, therefore it strongly polarizes and interacts with its environment. In terms of proton transport, this implies a strong coupling between the short- and long-range proton motions with the static structure and dynamics of its surrounding.

Perhaps in the most simple case, the proton does not migrate as H^+ but rather diffuses as part of a protonated species (*e.g.*, H_3O^+), coupled with a simultaneous counter-diffusion of un-protonated species (*e.g.*, H_2O) in the material. This "vehicle-type" mechanism is most frequently encountered in liquids or in materials with liquid-like properties, or in solids with open structures that may allow for the passage of large ions or molecules. In many materials, however, the proton does not act as a passenger of such a "vehicle," but rather jumps from one "vehicle" (or proton site) to another one. The rate of such a proton motion is generally dependent not only on the proton's local static structural environment but also on its dynamics, which is often a crucial part of the proton diffusion. In the case of proton diffusion in hydrogen-bonded liquids, such as water, imidazole, or phosphoric acid [14], such structural dynamics is usually the structural rearrangement of the hydrogen-bond network, which requires thermally activated hydrogen bond-breaking and bond-forming processes.

While proton diffusion in hydrogen-bonded networks comprises the interaction of many species, proton conduction mechanisms in oxides or other ceramic materials are usually less complex. This has to do with their generally rigid and often periodic structure, and that protons are usually present only at relatively low concentration, implying that proton-proton interactions are negligible. However, the protons form strong hydrogen-bonding interactions with neighboring oxide-ions and it is the structural and dynamical properties of these surrounding lattice ions (the host lattice) that ultimately determine the material's proton conducting properties. In the following section, we discuss in detail how protons may be incorporated into oxide materials, where they find their energetically most favorable sites, and how they move between these sites.

9.2.1 Proton Conducting Oxides/Perovskites

The first report on proton conduction in an oxide material dates back to the 1980s, when Iwahara and co-workers [15] observed this phenomenon in perovskite structured acceptor-doped strontium and barium cerates, $SrCeO_3$ and $BaCeO_3$. These ABO_3 type materials are assembly of corner-sharing BO_6 octahedra and 12-fold coordinated A site ions, where A represents a relatively large cation with oxidation state +1, +2, or +3, and B is a cation with oxidation state between +1 and +7, *cf.* Figure 3 (a). After more than three decades of research, $SrCeO_3$ and $BaCeO_3$-based compounds have become well-established proton conducting oxide systems. The role of the acceptor-doping, such as Y substituted for Ce in $BaCeO_3$, is to create an oxygen deficient structure ($BaCe_{1-x}Y_xO_{3-x/2}$, where x refers to the doping level), which can be hydrated by heat treatment in a humid atmosphere [16]. During this process, water molecules from the gaseous phase dissociate on the surface of the sample into hydroxyl groups ($-OH^-$) and protons (H^+). The hydroxyl groups occupy then nearby oxygen vacancies, while the remaining protons bind to oxygen atoms of the perovskite host lattice. The protons can diffuse (jump/hop) from one oxygen atom to another and, with time, they will therefore diffuse into the bulk of the material. At the same time as protons diffuse into the bulk, the counter diffusion of oxygen vacancies from the bulk to the surface allows the dissociation of other water molecules. This leads to an increase of the proton concentration in the material and, ideally, this process continues until the (bulk) oxygen vacancies are filled, leading to a fully hydrated structure of the form $BaCe_{1-x}Y_xO_3H_x$.

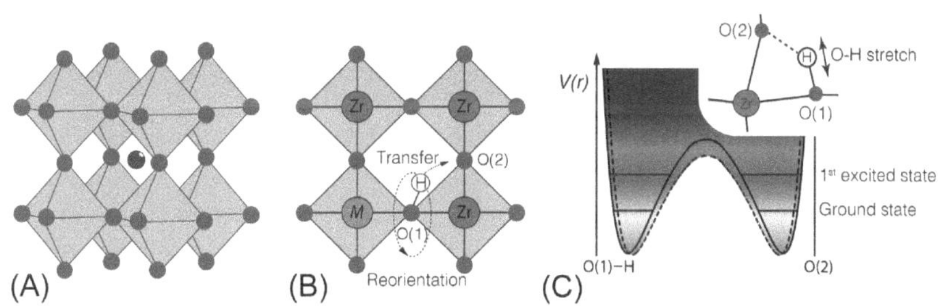

FIGURE 3 Schematics of (A) the cubic ABO_3 type perovskite structure, with corner-sharing BO_6 octahedra and "central" A type ions, and (B) the two elementary local diffusional processes in a cubic-structured $AB_{1-x}M_xO_3H_x$ type perovskite, cut in the (100) plane. M is a dopant atom. (C) The proton occupies a double Morse type potential, $V(r)$, with a barrier height dependent on the separation distance between the neighboring oxygens, O(1) and O(2). The horizontal lines depict vibrational levels for the O-H stretch mode. The dashed curve indicates the change in potential energy due to lattice phonons. Inset shows the direction of the vibration.

9.2.2 Mechanisms of Proton Diffusion

The current understanding of proton conduction in oxides/perovskite materials is based primarily on results obtained from computer simulations and quasielastic neutron scattering (QENS) on bulk powder samples of acceptor doped $BaZrO_3$ and $BaCeO_3$, as recently summarized [17]. In a simplified picture, the conduction mechanism may be divided into two elementary diffusional processes, namely (i) proton transfer between neighboring oxygen atoms, here O(1) and O(2), and (ii) reorientational motion of the -OH group between such transfers, whereas the long-range proton diffusion occurs as a series of transfers and reorientations [17]. A schematic illustration of these two elementary processes is shown in Figure 3 (b).

In more detail, the local energy environment of the proton along the O(1)–H···O(2) transfer direction may be modeled by a double Morse type potential, with a barrier height dependent on the separation distance between the two oxygen atoms [Figure 3 (c)]. From this, one can see that the O-H vibrational dynamics are very closely related to proton transport. For example, if the first excited state of the O-H stretch mode is populated, due to thermal energy or optical excitation, the barrier height is effectively lowered. Such a lowering increases the probability for proton transfer, either *directly* by the "immediate" O(1)-H bond-breaking and relocation of the proton to form H-O (2), or *indirectly* by the excited-state O-H mode decaying into lower-energy lattice modes that may alter the potential energy surface so that the transfer of protons from O(1) to O(2) becomes phonon-assisted. The importance of phonons, or vibrational motions of the perovskite host lattice, is illustrated in Figure 4 which shows proton traces sampled by quantum molecular dynamics (MD) simulations involving intra-octahedron transfer (a) and inter-octahedra proton transfer (b). As can be seen, the transition state configurations for proton transfer are shown for both cases. In the case of intra-octahedron transfer, the saddle state is characterized by B-O bond elongations and strong contraction of the O(1)H-O(2) separation, whereas in the case of inter-octahedra transfer, strong tilting of the participating octahedra is involved. Thus, the transfer process, which is considered to be rate-limiting for long-range proton diffusion, is also intimately linked to the O(1)-O(2) separation distance, which is changing with time because of the vibrational motions of the perovskite host lattice. The idea of phonon-assisted proton transfer in oxides has been supported both by theoretical studies [19–21] and by pump-probe infrared spectroscopy experiments [22, 23]. However, the local chemistry and structure complicates the description of proton transport mechanisms in oxides, that are still poorly understood even for the most simple materials/structure types [17].

9.3 OXIDE-ION CONDUCTORS

Oxide-ion conducting materials are used nowadays as oxygen sensors, oxygen separation membranes, for catalysis, and as electrolytes for SOFCs. All of

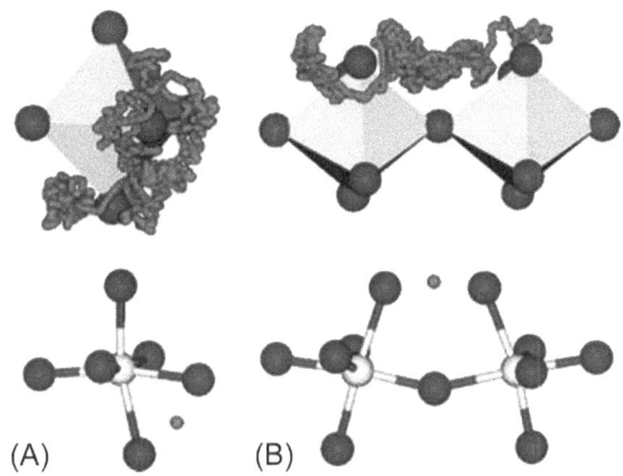

FIGURE 4 Proton traces sampled by quantum molecular dynamics simulations involving intra-octahedron transfer (A) and inter-octahedra proton transfer (B). The transition state configurations for proton transfer are shown for both cases: in the case of intra-octahedron transfer, this is characterized by B-O bond elongations and strong contraction of the O(1)H-O(2) separation; in the case of inter-octahedra transfer, severe tilting of the participating octahedra is involved. *Adapted with permission from Ref. [18].*

these devices offer the potential of enormous commercial and economic benefits provided suitable high performance materials can be developed. The diffusion of oxide-ions is generally the result of thermally activated hopping of oxide-ions between sites of a crystalline oxide material, with a superimposed drift in the direction of the electric field. The oxide-ion conductivity is consequently strongly temperature dependent. At high temperatures, >800 °C, it approaches values close to 1 S cm^{-1} for the best materials (Figure 2). This is higher than the maximum achievable conductivity in the best proton conducting oxides, comparable to the proton conductivity of Nafion, and thus even to the conductivity of liquid electrolytes. This is clearly a remarkable property of these solids and to understand its origins we must make several important observations. The first observation is that the crystal must contain connected and partially occupied oxygen sites, equivalent to those occupied by the lattice oxygen ions. Second, the energy involved in the process of migration from one site to the unoccupied equivalent site must be small, less than 1 eV. For an oxide-ion conductor, this small energy barrier to migration would seem at first sight difficult to attain, oxide-ions being normally the largest components of the lattice with an ionic radius of approximately 1.4 Å. Intuitively, it would be expected that the smaller metal ions would be more likely to have an appreciable mobility in the lattice. However, in certain crystal types, this is not the case and it is the oxide-ions that are the main charge carriers. These restrictions limit the pool of materials to a few families of structures, described in the following.

9.3.1 Oxide-Ion Conducting Oxides

Traditionally, oxide-ion conductors are fluorite-type materials, of general formula AO_2, where A is a large tetravalent cation and O is oxygen [Figure 5 (a)]. Oxide-ion conductivity is achieved by acceptor-doping to the A site, to create oxygen vacancies, *e.g.*, $A_{1-x}M_xO_{2-x/2}$, where M is a trivalent ion. Specific examples of materials are yttria-stabilized zirconia (YSZ) $Zr_{1-x}Y_xO_{2-d}$, zirconia-scandia $Zr_{1-x}Sc_xO_{2-d}$ [24–26], and ceria-gadolinia $Ce_{1-x}Gd_xO_{2-d}$ [27–30] materials.

Oxide-ion conductors are also found within the versatile perovskite family, with in particular the well-studied strontium-doped $LaGaO_3$ materials, *i.e.*, $La_{1-x}Sr_xGa_{1-y}Mg_yO_{3-d}$ (LSGM) [31–35], that operate in the temperature range $T = 650$–$800\ °C$ (Figure 2). The formation of oxygen vacancies is analogous to proton conducting equivalents, such as Y-doped $BaZrO_3$, which also are oxide-ion conductors at high temperatures. Other important classes of oxide-ion conductors include the brownmillerite-type oxides, $A_2B_2O_5$, *e.g.*, $Ba_2In_2O_5$, and A_2BO_4-type materials, *e.g.*, La_2NiO_4, which are schematized in Figure 5 (b and c). The brownmillerite structure may be viewed as an inherently oxygen deficient variant of the perovskite structure with alternating octahedral and tetrahedral layers, such as InO_6 and InO_4 layers in $Ba_2In_2O_5$, with the oxygen vacancies arranged in 1-dimensional channels in the tetrahedral layers. The La_2NiO_4-type oxide may be viewed as an intergrowth of alternating AO and ABO_3 layers. Both structures are thus highly anisotropic and may accommodate oxygen excess, in the vacancy channel and in interstitial sites, respectively, through oxidation of the transition metal element. Specific examples include

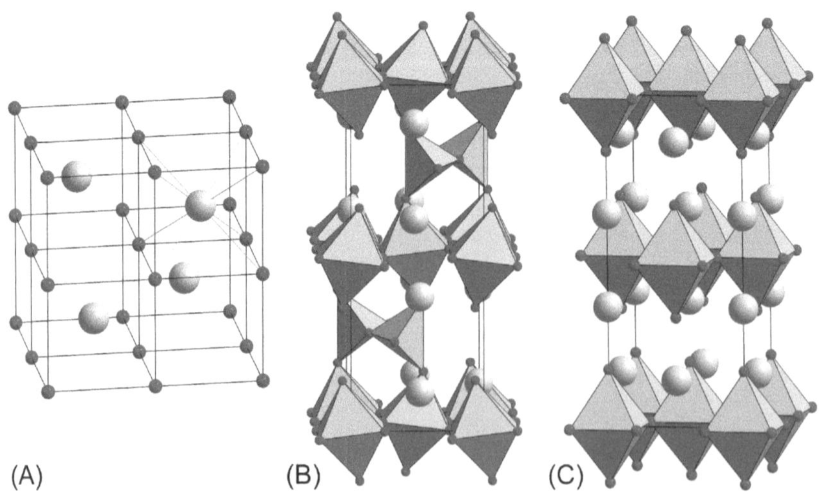

FIGURE 5 Schematic illustration of (A) the fluorite AO_2, (B) the brownmillerite $A_2B_2O_5$, and (C) the A_2BO_4-type structures.

$BaInO_{2.5+d}$ [36, 37] and Sr(Fe/Co)$O_{2.5+d}$ [38–40], and RE_2MO_{4+d} with RE = La, Nd, Sm, Pr and M = Ni, Co, Cu [41–47]. More recently discovered structure types include the LAMOX family, based on the La_2MoO_9 composition [48–51], the apatite family with lanthanum-germanate $La_{10-x}Ge_6O_{26+d}$ and silicate $La_{10-x}Si_6O_{26+d}$ [52–55], and the BIMEVOX family with typical composition of $Bi_2Mg_xV_{1-x}O_{5.5-d}$ [56, 57]. It can be mentioned that while these materials are all predominantly oxide-ion conductors, some show mixed ionic-electronic conductivity, *e.g.*, strontium-doped lanthanum-cobaltate $La_{1-x}Sr_xCoO_{3-d}$ [58–61], with in particular $La_{1-x}Sr_xCo_{1-y}Fe_yO_{3-d}$, or $Sm_{1-x}Sr_xCoO_{3-d}$, that can operate down to T = 600 and 500 °C, respectively, *i.e.*, in to the upper end of the intermediate temperature window.

9.3.2 Mechanisms of Oxide-Ion Diffusion

On a microscopic level, there are three generally accepted mechanisms associated with the diffusion of oxide-ions in an oxide material, known as vacancy mechanism, interstitial mechanism, and interstitialcy mechanism. In the vacancy mechanism, an host oxide-ion diffuses by jumping to a nearby oxygen vacancy. In the interstitial mechanism, an interstitial oxide-ion moves from the interstitial site to another one, whereas in the interstitialcy mechanism, the interstitial oxide-ion displaces an oxide-ion from its nominal host-lattice position to a neighboring vacant interstitial site, then diffuses to the newly found vacant site. Therefore, the difference between the interstitial and interstitialcy mechanisms is that the former one does not require the presence of other defects except from the interstitial ions themselves, whereas interstitialcy diffusion implies a permanent displacements of both defects and host ions, *cf.* Figure 6.

Regardless of the mechanism, the diffusion of oxide-ions is dependent on thermally activated hopping, which is, as mentioned above, a strongly activated process that requires high temperatures for high oxide-ion mobility to be observed. Generally, the conductivity shows an Arrhenius relationship, *i.e.*, $D = D_0 \exp(-E_a/k_B T)$, where D is the diffusion constant, D_0 a prefactor representing the maximum diffusion at infinite temperature, E_a the activation energy, and k_B the Boltzmann constant, which is expected in the case of a classical behavior of thermally activated stochastic oxide-ion jumping. However, at low temperatures, the diffusion might not be purely classical and can therefore deviate from an Arrhenius-like behavior. Examples of phenomenological approaches to describe ion diffusion mechanisms deviating from an Arrhenius dependency are given by the Vogel-Tam-Fucher (VTF) [62–64] and Williams-Lander-Ferry (WLF) [65] equations, which have been applied successfully for, *e.g.*, polymer electrolytes [66], silver conducting AgI composite glasses [67], and crystallized chalcogenides [68], as well as for oxide-ion conducting $La_2Mo_{2-y}W_yO_9$ [69] and lithium ion conducting $La_{2/3-x}Li_{3x}TiO_3$ [70]. However, these approaches

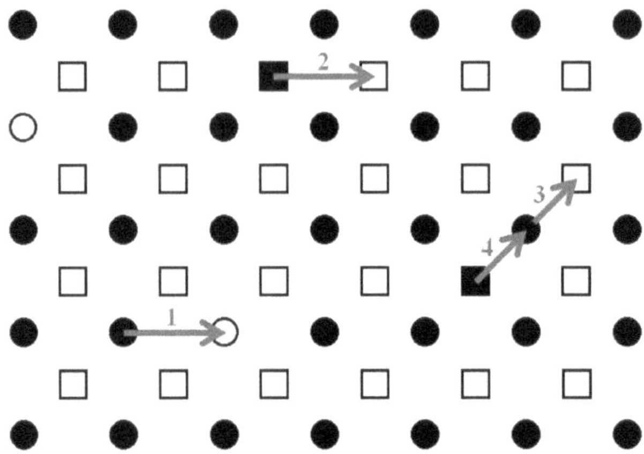

FIGURE 6 Scheme of mechanisms of diffusion on a 2D grid of "host" oxygen sites (circle), and "interstitial" sites (square). Occupied sites are filled and vacancies empty. The vacancy mechanism (1) consists of a direct hopping, from an oxygen atom of the host lattice to a vacant oxygen site of the host lattice. The interstitial mechanism (2) is also a direct hopping, from an oxygen atom on an interstitial site to another vacant interstitial site. The interstitialcy mechanism (3 then 4) is a two-step process, where the interstitial oxygen atom pushes its neighbor on the host lattice to a vacant interstitial site, then recombine in the newly vacant site of the host lattice.

do not provide information about the microscopic nature of the diffusion. To the best of our knowledge, only in the case of oxide-ion conducting brownmillerite-type oxides and related K_2NiF_4-type oxides, a microscopic model for the non-Arrhenius dependence of the moderate-temperature oxide-ion mobility has been proposed, associated with a phonon-assisted transport behavior [71–73]. For any scenario, the diffusion of oxide-ions in oxide materials is, exactly as for proton conducting materials, determined primarily by the static structure and dynamics of the host material.

9.4 NEUTRON SCATTERING

9.4.1 Role/Capacity of Neutron Scattering

The discovery of new ionically conducting materials for future technological high-performance devices depends on increasing the understanding of the coupling of currently known materials' macroscopic properties (like ionic conductivity) to its atomic structure and dynamics, as well as the exploration of entirely new classes of materials. To achieve this goal, neutron scattering is one of the most powerful tools because it provides a unique combination of methods capable of probing a material's equilibrium structure and atom dynamics as well as nonequilibrium transport properties. Notably, neutron scattering offers the potential to access information in both space and time simultaneously, through the momentum ($\hbar\mathbf{Q}$, $\mathrm{\AA}^{-1}$) and energy ($\hbar\omega$, meV) transferred in the scattering event. This capability makes neutron scattering a unique tool for exploring the atomic-scale structure and dynamics of materials. Additionally, neutrons are particularly sensitive to the hydrogen element (H), and the strength of the neutron-nucleus interaction varies considerably between isotopes of the same element. The extremely large incoherent scattering cross section of hydrogen (80 barns), much larger than all other nuclei (<10 barns), means one can observe hydrogen virtually free from constructive interference, implying that information about individual hydrogen atoms can be obtained. This characteristic is especially useful in studies of hydrogen dynamics, since information about single-particle (self) dynamics is usually most interesting. Deuterium on the other hand is a strong coherent scatterer and thus useful for obtaining structural information, such as hydrogen/deuterium positions. Furthermore, while the interaction between neutrons and matter is for most materials weak and hence requires a relatively large amount of the sample (typically several grams), the neutrons may be collimated and reflected on the surface of thin-film samples which will give information about the structure of the film. In summary, neutron scattering is an excellent tool for investigating structures (neutron diffraction (ND) and neutron reflectivity (NR)), vibrational dynamics and local structural properties (inelastic neutron scattering (INS)), diffusional dynamics (quasielastic neutron scattering (QENS)), and proton concentration (prompt-gamma activation analysis (PGAA) and polarized neutron diffraction), and hence offers good opportunities to advance the state of the art of the understanding of ion conducting solids.

9.4.2 Neutron Scattering and Computer Simulations

While neutron scattering techniques alone will always be an indispensable tool for materials analysis, it cannot provide us will all the information we need to know about materials. It is therefore important, and in many cases necessary, to draw in complementary information about materials from the usage of other techniques. Although a large number of complementary techniques could be mentioned, it is beyond the scope of this chapter to discuss them all here. Rather we focus on the role of computer modeling techniques, which are highly complementary to neutron scattering because, from a model of atomic positions, neutron scattering spectra can be straightforwardly calculated on an absolute scale and compared with experimental data. In particular, modeling methods are useful prior to experiments to predict the outcome and thus optimize the use of neutron beam time, as well as to guide future experiments. Similarly, neutron scattering data can be used to test and benchmark computational methods and results. In sum, this means that the value of neutron data can often be considerably enhanced by the use of complementary data obtained from other methods, and in the same way data obtained with other techniques can be enhanced by the use of neutron data. The following sub-sections provide a short description of the different computational methodologies commonly used in the study of structure and dynamics of materials. Emphasis is laid on the complementary with neutron scattering.

9.4.2.1 Structural Modeling Techniques

Atomistic simulations, nowadays mainly computer simulations, determine the lowest energy configuration of the crystal structure by employing energy minimization procedures. The calculations rely upon the specification of an interatomic potential, which describes the total energy of the system as a function of atomic coordinates, and are generally performed within the framework of density functional theory (DFT), on small structural models of a few hundred atoms and at $T = 0$ K. While this method is well suited when the system is showing a singular ground-state, complications arise for structures showing several iso-energetic minima of the potential energy surface, or for structures featuring pronounced anharmonic dynamics. Furthermore, a fundamental difference between crystalline structures, as described in crystallography, and the structural models used in computer modeling, is that the computed models must be exact. It implies that there must be no partial site occupancy—there is an atom, or there is not. As a consequence, and in particular for materials with crystallographic "defects," such as many of the oxide-ion and proton conducting materials, defects must be localized and the symmetry applied to the calculations, if any, is in general lower than the symmetry of the crystallographic average description from experiments. Information about the energetic ground-state, hence on the structure, can be discussed in relation with experimental results as obtained from, *e.g.*, neutron diffraction.

The simulations can be based on either classical or quantum-mechanical (*ab initio*) approaches. While classical simulations, generally based on parameterized core-shell potentials, allow the description of structures up to thousands of atoms, a proper representation of the crystal potential is often lacking when considering complex structures with strong electronic correlations, or the description of diffusion events—where bonds are breaking and forming. In the case of quantum-mechanical simulations, due to limited computing power, only a few inequivalent atoms can be simulated—of the order of 200 atoms. Considering these two major limitations—the full occupancy of sites and the limited number of inequivalent atoms—the play in atomistic calculations is to find a relevant approximation of the real material structure to use as model structure. Hence, all the results of a simulation are discussed in the limit where the structural model is accurate enough to properly describe the real material.

Of specific concern for ion conducting materials is the calculation of properties such as potential energy barriers and saddle state structures, which may give an "average picture" about the ion conduction mechanisms. While these results are not directly comparable with those from neutron scattering experiments, they can be compared with, for instance, the activation energy determined experimentally, and provide diffusion models to interpret QENS data.

Furthermore, atomistic calculations can also be used to calculate phonon dispersion relations using, for example, the direct method through the calculations of Hellmann-Feynman forces. This method is only exact if a singular ground-state is found, and is limited to harmonic systems. Away from these conditions, if the restoring forces are lacking or not symmetric, the dynamical matrix cannot be properly diagonalized and phonon modes with imaginary frequencies may arise from the calculation. Using the one-phonon approximation and the set of calculated phonon eigenvectors, it is straightforward to calculate the neutron scattering function, which can be compared with neutron scattering data.

9.4.2.2 Modeling of Dynamics

Molecular dynamics (MD) simulations, *i.e.*, calculations of the evolution of a structure with time, are employed in the study of, *e.g.*, ion-diffusion pathways, and to estimate diffusion coefficients and mean-square displacements in materials. Another strength of MD simulations is the possibility to compute vibrational density of states (DOS) from power spectral density analysis, or using the velocity auto-correlation function. When weighted with the neutron scattering lengths, calculated densities can be compared with experimental DOS as can be obtained from INS experiments. These calculations are generally performed with classical mechanics using core-shell potential models to describe the electronic interactions between atoms, and involve extended structural models of several thousand atoms, at a set temperature. MD simulations can also be performed within the DFT framework. For the latter, while the limited number of atoms and simulation length do not provide enough statistics to get accurate diffusion coefficients, it allows the modeling of structures at finite temperature, and give access to the diffusion pathway.

9.5 CASE STUDIES

Examples of recent neutron work on ionically conducting materials, which is here briefly reviewed, include studies with ND (Sec. 9.5.1), NR (Sec. 9.5.2), INS (Sec. 9.5.3), and QENS (Sec. 9.5.4). No attempt has been made to be complete in this description, but rather to highlight the different types of information that can be obtained. Likewise, the description of the various neutron methods is limited to the key characteristics of each method, as a comprehensive and contemporary review of the fundamentals of neutron scattering has been published recently [74].

9.5.1 Neutron Diffraction

ND is an extremely powerful and widely used tool for structural studies of solid-state materials generally. While conventional ND is a powerful tool to access information on the average structures of materials and, to a lower extent, information at the local level through specifically designed models such as split sites, there is often a need for a more precise characterization of defect localization and associated defective clusters—the region surrounding the defect and, usually, strongly strained by it. To address this need, neutron total scattering analysis is nowadays an important, complementary, approach to the structural studies of crystalline materials. The technique consists of measuring both the coherent and incoherent part of the scattering, and is usually processed with pair distribution function analysis (PDF) or analysis of the radial distribution function (RDF). Structural models proposed to describe the data can then be fitted using a least square minimization procedure, maximum entropy method, or even using reverse Monte Carlo (RMC) modeling. While the information obtained from Rietveld analyses are limited to an average representation of the structure, PDF and RDF, through the incoherent part of the scattering, provide information on deviations from the average structure of materials. In the following, we review a handful of examples of recent ND studies, aimed at the determination of proton positions, proton concentration, and nonaverage structural properties such as local ordering of oxygen vacancies, to illustrate the breath of information that can be obtained.

9.5.1.1 Determination of Proton Sites and Concentration

Information about proton sites is an essential piece of information we need to know about proton conducting oxides. Examples of previous ND studies aimed at elucidating the nature of proton sites in such materials include the investigations by Alonso and Turrillas [75], who reported on proton sites in the $(H_3O)SbTeO_6$ pyrochlore structure; Jayaraman *et al.* [76], who reported on the proton positions in the deuterated phase of the brownmillerite type oxide $Ba_2In_2O_5$; Knight [77], who reported on the double perovskite $BaCe_{0.9}Y_{0.1}O_{2.95}$; Sata *et al.* [78], who reported on Sc-doped $SrTiO_3$; Sosnowska *et al.* [79], who reported on $Ba_3Ca_{1.18}Nb_{1.82}O_{9-\delta}$; Kendrick *et al.* [80], who reported on $La_{0.6}Ba_{0.4}ScO_{2.8}$; Ito *et al.* [81], who reported on the location of deuterons in $BaSn_{0.5}In_{0.5}O_{2.75}$ using a combined Rietveld and maximum entropy study; Ahmed *et al.* [82] and Nagasaki *et al.* [83], who reported on the deuterium positions in $BaZr_{0.5}In_{0.5}O_{2.75}$; Nagasaki *et al.* [84], who reported on a deuterated sample of $BaSn_{0.5}In_{0.5}O_{2.75}$ by using a difference maximum entropy method; and Norberg *et al.* [85], who reported on an investigation of a deuterated sample of $BaTi_{0.5}In_{0.5}O_{2.75}$, using both Rietveld analysis and RMC modeling techniques. An important result from several of these studies is the observation of (on average) eight equivalent proton/deuterium sites around each oxygen, which correspond to the tilting and hydrogen-bond formation of the proton to a neighboring oxygen and are caused by local structural distortions induced by the net difference in charge and/or ionic radius between the dopant atom and host-site atoms in the material. The presence of local structural distortions implies, however, a variety of different types of proton sites, manifested by, *e.g.*, large atomic displacement parameters (ADPs) of the oxygen atoms [82].

Whereas ND data also give information about the concentration of protons or deuterium atoms, as well as the other constituent atoms in the studied materials, the actual proton concentration is more readily and routinely determined by measuring the weight change of the sample during hydration or dehydration using thermogravimetric analysis (TGA). However, this approach is not suitable for all types of materials. For example, hydrated perovskites containing elements that may change oxidation state upon heating may decrease in mass because of oxygen loss in addition to the evaporation of water molecules. Alternative, complementary, techniques for the determination of proton concentrations are then needed. Two examples are neutron prompt-gamma activation analysis (PGAA), which is based on the detection of element-specific gamma rays emitted due to the absorption of neutrons by these elements, and neutron diffraction with polarization analysis. The latter technique, which will be here highlighted, is based on measuring the spin-flip and nonspin-flip cross sections for three orthogonal orientations of the neutron beam polarization, then extracting from linear combinations of these the nuclear coherent + isotope incoherent, spin incoherent, and magnetic contribution [86, 87]. The hydrogen concentration is obtained from the incoherent part after subtraction of the contribution of all other atoms in the material according to

$$\left(\frac{d\sigma}{d\Omega}\right)^{\mathrm{H}}_{\mathrm{inc}} = \left(\frac{d\sigma}{d\Omega}\right)_{\mathrm{inc}} - \sum_{i \neq \mathrm{H}} \frac{\sigma^{i}_{\mathrm{inc}} n^{i}}{4\pi}, \qquad (1)$$

where $(d\sigma/d\Omega)^{\mathrm{H}}_{\mathrm{inc}}$ is referred to as the incoherent scattering cross section of hydrogen and $(d\sigma/d\Omega)_{\mathrm{inc}}$ that over the sum of all atoms in the material. $\sigma^{i}_{\mathrm{inc}}$ refers to the incoherent scattering cross section of the ith element in the compound. Dividing $(d\sigma/d\Omega)^{\mathrm{H}}_{\mathrm{inc}}$ by $\sigma^{\mathrm{H}}_{\mathrm{inc}}/4\pi$ yields the hydrogen molar content per formula unit. As an example, Figure 7 shows the coherent, incoherent, and magnetic scattering responses of the two proton conducting perovskites $BaZr_{0.9}M_{0.1}O_{2.95}$ (M = Y and Sc) [88]. Using the formulae above, and taking into account only the **Q**-region free from Bragg peaks and small-angle scattering due to the powder nature of the samples, the proton concentration is found to be approximately 340% in the Y-doped sample and 180% in the Sc-doped one, *i.e.*, much higher than the expected 100% if taking into account only the nominal concentration of unfilled oxygen vacancies. The reason for the higher-than-expected proton concentration is yet unclear but may possibly be due to the presence of hydroxide secondary phases. Such phases have been observed in similar materials [89], but would not be distinguishable from protonic defects that are part of the perovskite lattice.

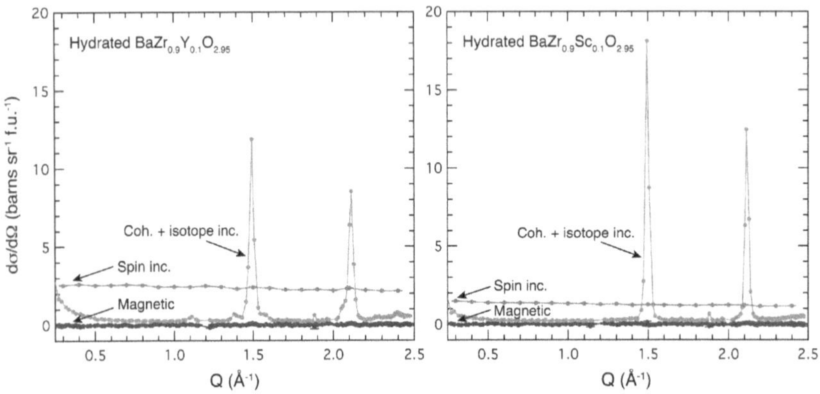

FIGURE 7 Coherent, incoherent, and magnetic scattering signals of $BaZr_{0.9}M_{0.1}O_{2.95}$ (M = Y and Sc) as measured on the D7 spectrometer at the ILL. *Adapted with permission from Ref. [88].*

9.5.1.2 Determination of Local Structural Properties

Since proton and oxide-ion conducting oxides are highly defective materials, it is easy to understand that their performance are ultimately determined by the local structural properties. This implies also that conventional ND experiments—which will only give information about the average structural properties of the investigated samples—will give essential but not sufficient information about the structure of materials. In this context, the usefulness of ND coupled not only to Rietveld refinement but also to PDF analysis and/or RMC modeling has already been mentioned and will be here highlighted by recent experiments. A specific example is the work by Malavasi *et al.* [90] on Ga-based melilite $La_{1+x}Sr_{1-x}Ga_3O_{7+0.5x}$ oxide-ion conductors. The melilite structure is an intergrowth of cationic $(La/Sr)_2$ and anionic corner-sharing tetrahedral Ga_3O_7 layers, that shows, for the oxygen enriched composition, pure oxide-ion mobility through an interstitial mechanism—the excess oxygen atoms being adapted in interstitial sites in the pentagonal rings of the Ga_3O_7 layers. ND measurements with subsequent Rietveld analysis of $La_{1.5}Sr_{0.5}Ga_3O_{7.25}$ confirm the average structure and suggest that excess oxygen atoms are located in high-symmetry sites at the center of the pentagonal rings, yet while showing large ADPs. Further analyses using PDF analysis of total scattering data reveal an extra correlation peak that cannot be reproduced using as model the average Rietveld description and, when taken into account, suggest that interstitial oxygen atoms are shifted from the center of the pentagon rings, as illustrated in Figure 8.

Another specific example is the work of Norberg *et al.* [91, 92] on the oxide-ion conductors Bi_2O_3 [93], Bi_3YO_6 [91], and $Bi_3Nb_{0.5}Y_{0.5}O_{6.5}$ [92]. The fluorite Bi_2O_3 material, in its high-temperature defect phase δ-Bi_2O_3 above 720 °C, exhibits the highest reported oxide-ion conductivity of

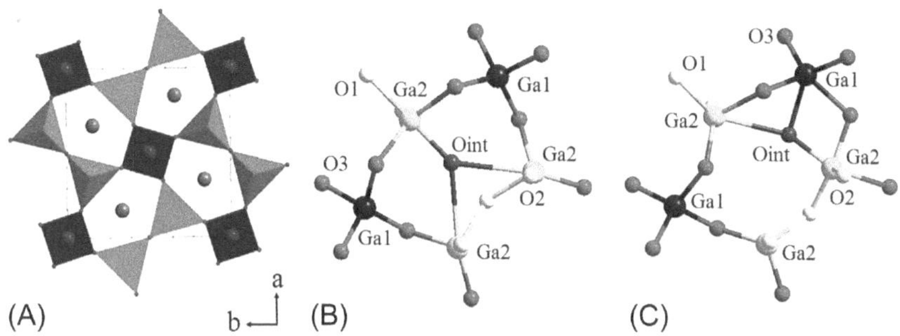

FIGURE 8 (A) Melilite structure of $LaSrGa_3O_7$ projected along the *c*-axis, showing the Ga tetrahedral units connected through bridging oxygen atoms to form distorted pentagonal rings. (B) Local view of one of the two structural models of $La_{1.50}Sr_{0.50}Ga_3O_{7.25}$, obtained from Rietveld refinement of neutron powder diffraction data, with the interstitial oxygen located in the high symmetry site at the center of the pentagon ring. (C) Local view of one of the two structural models of $La_{1.50}Sr_{0.50}Ga_3O_{7.25}$, obtained from fitting of the PDF (total scattering data), showing that the interstitial oxygen site is off-centered in the pentagonal ring. *Adapted with permission from Ref. [90].*

about 1 S cm^{-1}. Tremendous efforts have been undertaken to extend this excellent conducting property to the intermediate temperature regime, mostly through cation substitution with the aim of stabilizing the δ polymorph. In the case of, for example, double substitution of Bi^{3+} by Y^{3+} and Nb^{5+}, good conductivities in the region 400–830 °C have been reported. However, phases exhibit varying degrees of structural ordering, from short-range ordering to incommensurate superlattice ordering. The short-range defect structure of $Bi_3Nb_{0.5}Y_{0.5}O_{6.5}$ has been investigated by neutron total scattering coupled with RMC modeling, and has evidenced, using the angular distribution function, that oxygen vacancies are predominantly ordered in the [110] direction. Such information is key to develop a detailed understanding of the oxide-ion conduction mechanism in these materials.

9.5.2 Neutron Reflectivity

Neutron reflectivity (NR) involves shining a highly collimated beam of neutrons on a flat surface—usually a thin film or layers of thin films deposited on a substrate material—and measuring the intensity of the reflected neutrons as a function of the reflected angle (θ) or scattering vector perpendicular to the sample surface, $\mathbf{Q}_z = (4\pi/\lambda)\sin\theta$, where θ is the incident and reflected angle and λ is the wavelength of the incident neutrons. The interference of waves reflected from the different interfaces of a film gives rise to fringes in the reflectivity, $R(\mathbf{Q}_z)$, and through the modeling of the reflectivity a wealth of information about the film, such as its thickness, density, surface roughness, and atomic concentration depth profiles, may be obtained. Notably, NR measurements probe the neutron scattering length density (*nSLD*) profile along the normal to the surface of the sample. The *nSLD* is defined as $nSLD(z) = \sum_{i=1}^{m} b_i \rho_i(z)$, where b_i and $\rho_i(z)$ are the coherent neutron scattering length and the number density of the isotope i, respectively, as a function of depth. NR measurements are thus highly complementary to X-ray reflectivity (XRR) measurements, which probe the electronic scattering length density profile, $eSLD(z) = \sum_{i=1}^{m} f_i \rho_i(z)$, where f_i is the atomic scattering factor of atomic species i. While reflectivity techniques have been extensively used for studies of magnetic and biological systems, NR (and XRR) are both severely underused techniques for studies of proton and oxide-ion conductors.

9.5.2.1 Determination of the Structure and Proton Concentration Depth Profile in Thin Films

The initial NR study of a proton conducting oxide was done very recently and focused on the investigation of the structure and proton concentration depth profile in epitaxial thin films of $BaZr_{0.53}In_{0.47}O_{3-\delta}H_{0.47-2\delta}$ [94]. The NR data measured at four different conditions of the film, hydrated, deuterated, and dehydrated, this last measured at both room temperature and 400 °C, are shown in Figure 9 (a). The corresponding *nSLD*s of the thicker layer are shown in Figure 9 (b). The difference in *nSLD* for the three samples reflects the difference in coherent scattering length of H (-3.74 fm) and D (6.67 fm), thus confirming the insertion of H/D in the film. A key result is that the reflectivity curves can be adequately modeled by a two-layer model characterized by a 3–4 nm thick, proton-rich, near-surface layer, and a thicker layer beneath; a schematic depiction of this two-layer system is shown in Figure 10 (a). While this near-surface layer either retains or reabsorbs protons after desorption and cooling to room temperature, the bulk of the film absorbs and desorbs protons in the expected manner. Besides being of fundamental interest, these results are of relevance for the development of next-generation proton conducting oxides with optimized proton conductivity and hence for the development of miniaturized SOFCs in actual operations that need to be periodically taken into and out of service (*i.e.*, hydrated and dehydrated). The pioneering results also demonstrate the applicability and potential of reflectivity techniques for gaining unique insight into the structural properties of proton conducting oxide materials.

9.5.3 Inelastic Neutron Scattering

In inelastic neutron scattering (INS), also known as neutron vibrational spectroscopy, the intensity of the scattered neutrons is measured as a function of both momentum transfer, **Q**, and transferred energy, $\hbar\omega$. In comparison to inelastic scattering by X-rays, neutrons have a much lower incident energy, implying that the energy resolution is much higher, on the order of meV to neV. These are exactly the energies needed to study vibrational dynamics in

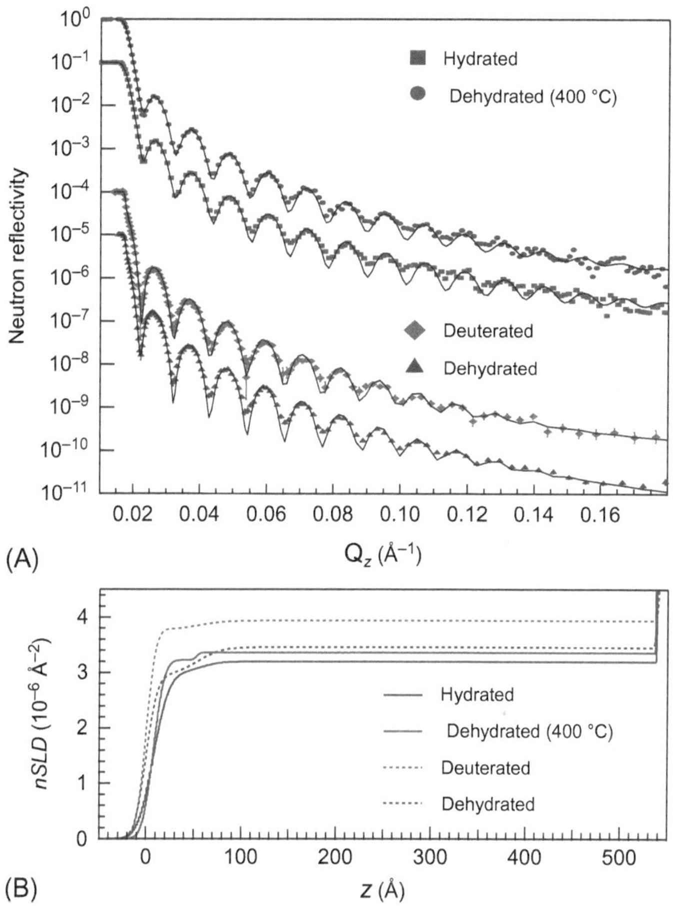

FIGURE 9 (A) NR data (points) and best fits (lines) for the $BaZr_{0.53}In_{0.47}O_{3-\delta}H_{0.47-2\delta}$ film in the four different experimental conditions. The reflectivity curves for the dehydrated (at 400 °C), deuterated, and dehydrated conditions have been scaled by one, four, and five orders of magnitude, respectively. (B) *nSLD* profiles as obtained from the fits based on the two-layer model. *Adapted with permission from Ref. [94].*

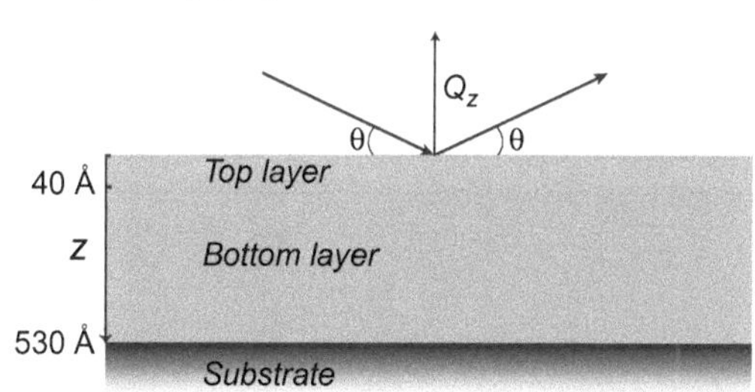

FIGURE 10 Schematic depiction of the specular reflectivity geometry and the double-layer model used to analyze the reflectivity data. The reflection angle is equal to the incident angle and the scattering vector, $\mathbf{Q}_z$, is perpendicular to the sample surface. *Adapted with permission from Ref. [94].*

solid-state materials. Furthermore, in comparison to Raman spectroscopy, INS does not rely on any selection rule and all modes are in principle measurable [95, 96]. Moreover, Raman spectroscopy and infrared absorbance spectroscopy (the other optical technique for measuring vibrational spectra) can only be used to measure near the Brillouin zone center and theoretical calculations of band intensities are in general difficult, hence information is usually only taken from the positions of the observed modes. On the contrary, INS can be used to measure the dispersion curves of phonons and is a highly quantitative method, where band intensities scale with the neutron scattering cross sections of the vibrating species and are relatively easily modeled and interpreted from first-principles phonon calculations.

9.5.3.1 *Investigation of Proton Local Environments*

The first INS study on proton conducting perovskites was done by Karmonik *et al.* [97], who investigated the vibrational proton dynamics in $SrCe_{0.95}M_{0.05}O_{3-\delta}$ (M = Sc, Ho, Nd) and found large differences between the three samples, particularly in relation to the O-H wag modes. Later, Karlsson *et al.* [98] reported on a systematic investigation of the O-H wag mode in $BaIn_xZr_{1-x}O_{3-x/2}$ (x = 0.20, 0.50, and 0.75). More recent examples include studies of the brownmillerite $Ba_2In_2O_5$, which may be described as an oxygen-deficient variant of the perovskite structure with alternating layers of InO_4 tetrahedra and InO_6 octahedra [99], *cf.* Figure 11 (a).

Upon hydration, *i.e.*, while filling oxygen vacancies with water molecules, a structural phase transition occurs to the layered, proton conducting $BaInO_3H$ material [Figure 11 (b)]. Using a combination of INS and DFT calculations, Bielecki *et al.* [99] investigated the nature of proton sites present. Figure 12 shows the INS spectra at T = 10 K of $BaInO_3H$ recorded at two different

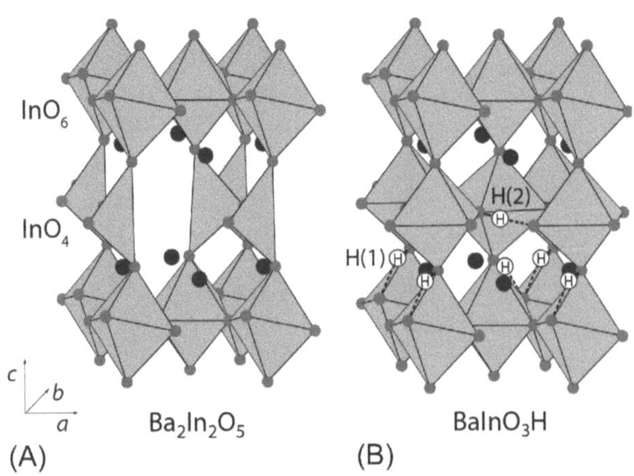

FIGURE 11 Schematic structures of (A) $Ba_2In_2O_5$ and (B) $BaInO_3H$. Covalent bonds are indicated by solid lines, and hydrogen bonds are indicated by dashed lines.

instruments, MAPS and TOSCA at the ISIS neutron source. Figure 12 (a) clearly shows an O-H stretch band peaking at around 3410 cm^{-1}, with shoulders at 3280 cm^{-1} and 3540 cm^{-1}, and an intense band at 860 cm^{-1}, assigned as the O-H wag mode fundamental (0 → 1 transition). The weaker and broader bands at 1662 and 2414 cm^{-1} are assigned as the first (0 → 2 transition) and second overtones (0 → 3 transition), respectively. The assignment to higher-order transitions is supported by the momentum transfer, **Q**, dependence of the modes as seen in Figure 12 (c). The maximum intensity of the mode occurs at larger **Q** as the transition energy increases, which is expected for an overtone progression [100]. The broad feature at 4270 cm^{-1}, which is also observed in the IR spectra of the similar materials $BaZr_{1-x}In_xO_{3-x/2}$ ($x = 0.25$–0.75) [101], is most likely associated with a combination of O-H stretch and bend modes. This combination is commonly seen in inorganic hydroxyl containing materials, *e.g.*, $LiOH{\cdot}H_2O$ [102] and hydroxyapatite [103]. Figure 12 (b) shows that the spectrum recorded on TOSCA is similar to that obtained on MAPS but with higher spectral resolution in the low-frequency region. The O-H wag mode at 869 cm^{-1} is asymmetric with a pronounced shoulder at 761 cm^{-1} and has several weaker peaks at 633, 1008, 1163, and 1227 cm^{-1} associated with it. The spread in O-H wag frequencies shows that there are several different O-H distances present.

9.5.3.2 *Investigation of Oxide-Ion Local Environments and Lattice Dynamics*

Here, we follow a recent investigation of the local structural properties of the nonstoichiometric, mixed electronic/oxide-ion conductor Nd_2NiO_{4+d} [72, 73]. While the high-temperature regime of ionic diffusion is well characterized by an Arrhenius law and the diffusion pathway predicted to be the interstitialcy mechanism [104–107], the room-temperature mobility cannot be explained by sole thermally activated stochastic oxygen hopping. For this reason, INS experiments coupled to *ab initio* MD simulations have been performed on three phases, $Nd_2NiO_{4.0}$, $Nd_2NiO_{4.10}$, and $Nd_2NiO_{4.25}$, with the aim to determine the impact of excess oxygen on the lattice dynamics, and evaluate the presence of possible lattice activated diffusion processes.

A good agreement between experimental and calculated DOS for the three stoichiometries is found, as shown in Figure 13. Features *A* and *B* in the DOS correspond, respectively, to modes polarized along the [100] and [110] crystallographic directions. The clear decrease of *A* and increase of *B*, along with the evolution of the apical oxygen atom nuclear density, suggest that the presence of excess oxygen atoms activates the [110] dynamics of the apical oxygen atoms. The diffusion pathway is found to be similar to the one reported at high-temperature, as shown in Figure 14 and described as interstitialcy. The activation of the [110] dynamics effectively shorten the distance between the apical oxygen atom and the next-vacant interstitial site, hence favoring the occurrence of diffusion events.

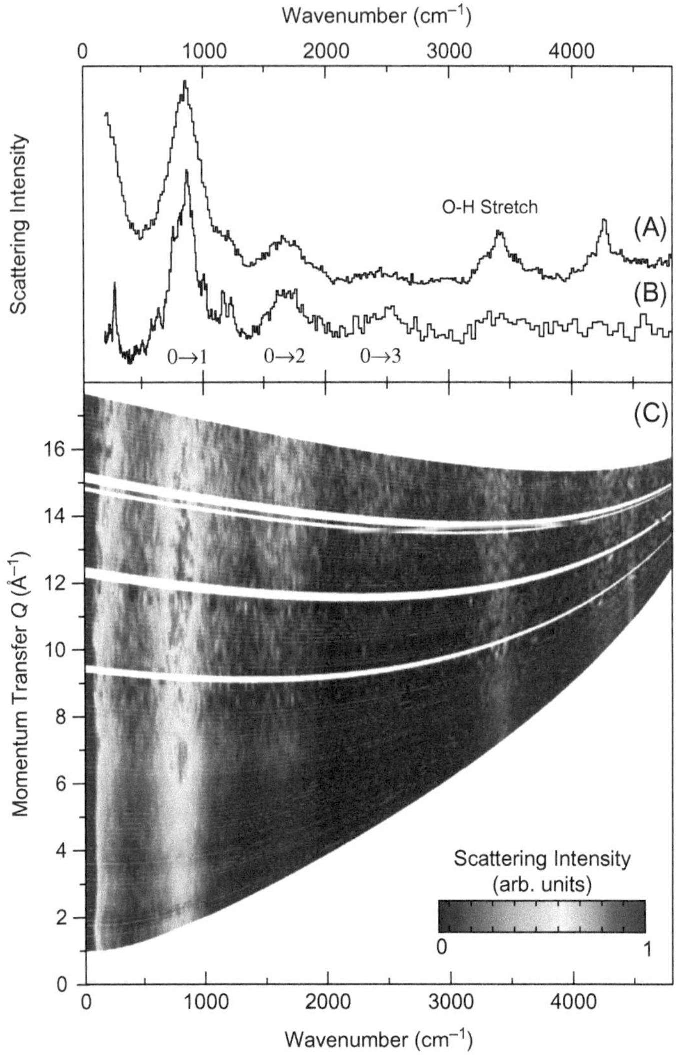

FIGURE 12 INS spectrum at $T = 10$ K of $BaInO_3H$ recorded with (a) MAPS ($E_i = 650$ meV) and (b) TOSCA. (c) $S(\mathbf{Q}, \omega)$ map of $BaInO_3H$ recorded on MAPS ($E_i = 650$ meV). *Adapted with permission from Ref. [99].*

9.5.4 Quasielastic Neutron Scattering

QENS has played a central role in studies of ionic motion in materials but is, according to us, underused for studies of proton and oxide-ion conducting materials. The usefulness of QENS comes from the fact that it gives access to the relevant time and length scales on which the atomic-scale dynamics of protons occur. Furthermore, in studies of proton conductors, the strong scattering signal from protons provides a good contrast in experiments and

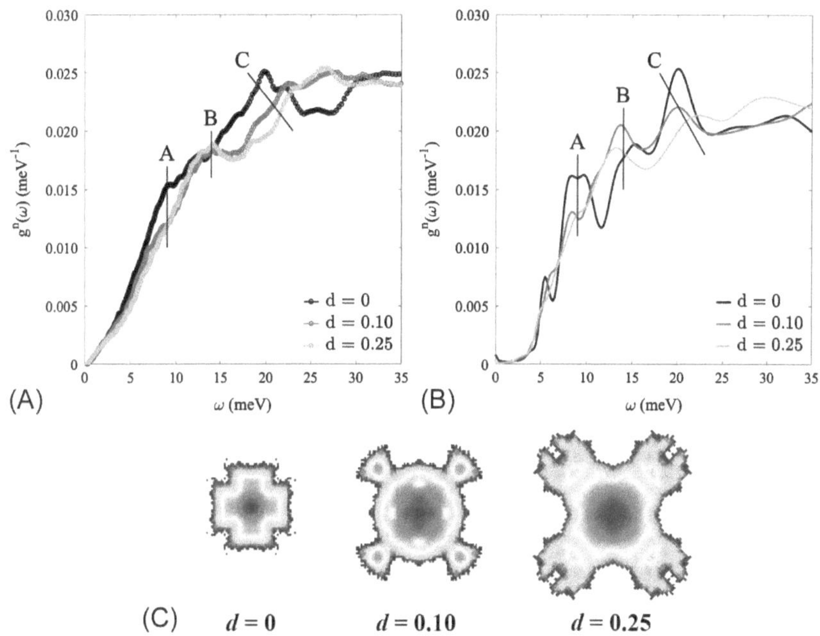

FIGURE 13 (a) Experimental DOS from INS measurements and (b) computed DOS from *ab initio* MD trajectories of $Nd_2NiO_{4.0}$ (black), $Nd_2NiO_{4.10}$ (dark gray), and $Nd_2NiO_{4.25}$ (light gray) at $T = 310$ K. (c) Nuclear densities of apical oxygen atoms, projected along the stacking c-axis direction, in atomic displacement space. *Adapted with permission from Ref. [72].*

FIGURE 14 Detail of the rocksalt layer of the $Nd_2NiO_{4.10}$ supercell in the $T = 310$ K molecular dynamics. Wired spheres represent Nd^{3+} ions and O^{2-} ions on apical sites, and plain sphere O^{2-} ion in interstitial site. Atoms are drawn at their initial positions. Clouds represent the ensemble of positions occupied by the atoms over 40 ps of the molecular dynamics. Connected clouds from apical to interstitial sites are marked with a line and evidence oxygen mobility.

enables studies of proton self-diffusion in systems containing only relatively small amounts of protons. It follows that the total dynamical structure factor, $S(\mathbf{Q},\omega)$, can, in most cases, be approximated with the incoherent dynamical structure factor, *i.e.*, $S(\mathbf{Q},\omega) \approx S_{inc}(\mathbf{Q},\omega)$. In practice, the measured structure factor is a convolution of $S_{inc}(\mathbf{Q},\omega)$ and the resolution function of the instrument, $R(\mathbf{Q},\omega)$, *i.e.*,

$$S_{meas}(\mathbf{Q},\omega) = S_{inc}(\mathbf{Q},\omega) \otimes R(\mathbf{Q},\omega). \quad (2)$$

In the case of quasielastic scattering, $S_{inc}(\mathbf{Q},\omega)$ may be expressed as

$$S_{inc}(\mathbf{Q},\omega) = A_0(\mathbf{Q})\delta(\omega) + \sum_i A_i(\mathbf{Q})L_i(\mathbf{Q},\omega), \quad (3)$$

where the $L_i(\mathbf{Q},\omega)$s are Lorentzian functions of the form

$$L_i(\mathbf{Q},\omega) = \frac{2}{\pi}\frac{\Gamma_i(\mathbf{Q})}{(2\hbar\omega)^2 + \Gamma_i(\mathbf{Q})^2}, \quad (4)$$

in which $\hbar\omega$ is the neutron energy transfer and $\Gamma_i(\mathbf{Q})$ the *i*th Lorentzian's full width at half maximum. $\Gamma_i(\mathbf{Q})$ relates to a characteristic relaxation, also known as correlation, time, τ_i, according to $\tau_i = 2\hbar/\Gamma_i$. The Lorentzian functions thus represent the quasielastic scattering, which appears as a broadening of the elastic line and contains information about both the time-scale and **Q** dependence of the observed dynamics. Studying their dependence on **Q** and temperature will bring a wealth of information about the diffusive process(es) in the system under investigation.

9.5.4.1 Investigations of Proton Dynamics in Proton Conducting Oxides

The majority of previous QENS studies on proton conducting oxides have been performed using time-of-flight (TOF) and backscattering (BS) methods, which give access to the time-scale of picoseconds, extended up to $\approx$1 ns in some cases, and on perovskite structured materials, such as doped variants of $BaZrO_3$ and $BaCeO_3$-based compounds. Figure 15 shows, as an example from previous work, the QENS spectra of the proton conducting perovskite $SrCe_{0.95}Yb_{0.05}O_{2.975}$, as reported by Hempelmann *et al.* [108, 109]. As can be observed, there is a clear quasielastic broadening, which becomes increasingly more and more pronounced with increasing temperature. Detailed analysis of the QENS spectra showed that the quasielastic broadening could be adequately reproduced by two components, corresponding to two different localized proton motions with different activation energies, suggesting the existence of two different proton sites. Even more detailed analysis showed that the QENS data were in agreement with a two-state model superimposed on the Chudley-Elliott model. The Chudley-Elliott model suggests in this case that the proton conduction occurs as a series of proton jumps (transfers) between neighboring oxygens, O(1) and O(2), and rotational diffusion of the -OH group between such transfers, as described above, whereas the two-state model indicates that the proton diffusion consists of some sort of trapping and release events.

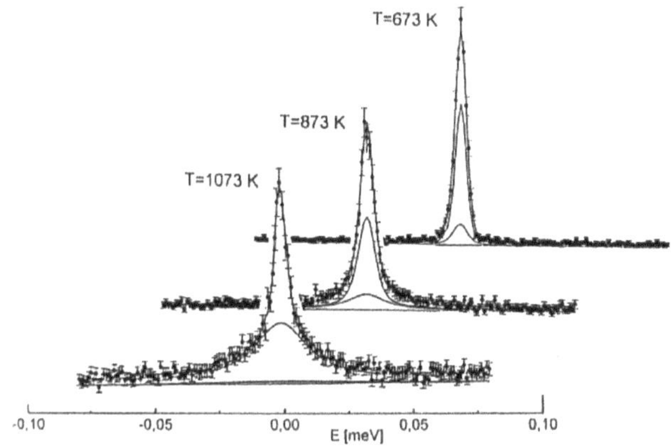

FIGURE 15 Neutron scattering spectra $S(\mathbf{Q}, \omega)$ of $SrCe_{0.95}Yb_{0.05}H_{0.02}O_{2.985}$ for different temperatures at momentum transfer $\mathbf{Q} = 0.29$ Å^{-1}. The solid lines show the total scattering function as well as the two components of the two-state model. *Adapted with permission from Ref. [108].*

Further examples include the work of Karmonik *et al.* [110], who investigated the nature of proton dynamics in the double perovskite structured material $Ba[Ca_{0.39}Nb_{0.61}]O_{2.91}$ which, similarly to the previous studies on $SrCe_{0.95}Yb_{0.05}O_{2.975}$, gave support for a proton conduction mechanism divided into proton jumps and reorientational motions. Subsequently, Pionke *et al.* [111] reported on the proton self-diffusion constant for protons in the same material, whereas Groß *et al.* [112] and Beck *et al.* [113] investigated the influence of particle size on the atomic-scale proton diffusion in $BaZr_{0.85}M_{0.15}O_{2.925}$ (M = Y, In, and Ga). An important result was the observation of a clear quasielastic signal in measurements of a microcrystalline sample, whereas measurements on a nanocrystalline equivalent did not manifest any dynamics accessible on the picosecond time-scale as probed by the spectrometer. It was speculated that in the nanocrystalline samples, the protons are concentrated mainly to the grain boundary (GB) regions or to the surface of the crystallites. More recently, Wilmer *et al.* [114] investigated the nature of proton dynamics in $BaZr_{0.90}Y_{0.10}O_{2.95}$, whereas Malikova *et al.* [115] performed a simultaneous structural and dynamic study on $BaCe_{0.90}Y_{0.10}O_{2.95}$, by combining ND with QENS. A key result was that the dehydration occurs over a large temperature range (from 500 °C to 800 °C), providing evidence for the presence of proton sites with a variable degree of hydrogen bonding strengths to the perovskite lattice. Slodczyk *et al.* [116] investigated proton dynamics in Yb-doped $BaZrO_3$, and they observed the onset of proton diffusion at a temperature of approximately 600 °C, while Karlsson *et al.* [117] reported on a QENS study of $BaZr_{0.90}M_{0.10}O_{2.95}$ (M = Y and Sc). The authors revealed a localized proton motion on the picosecond time-scale and with a small activation energy of 10–30 meV, for both materials. Braun *et al.* [118] reported on the Y-doped material, $BaZr_{0.90}Y_{0.10}O_{2.95}$, and found two different activation energies for proton diffusion at different ranges of temperature. Colomban *et al.* [119] reported on a change in local proton dynamics across a structural phase transition in (Ba/Sr)$Zr_{1-x}Ln_xO_{3-\delta}$, while Slodczyk *et al.* [116] investigated proton dynamics in $SrZr_{0.90}Yb_{0.10}O_{2.95}$ and $BaZr_{0.90}Yb_{0.10}O_{2.95}$. Measurements taken in the temperature range from $T = 400$ to 700 °C showed on a peak in the mean square displacement of the protons at around $T = 500$ °C, suggesting a maximum in proton diffusivity around this temperature. Most recently, Chen *et al.* [120] measured simultaneously macroscopic proton conductivity and atomic-scale proton diffusivity in $BaCe_{0.8}Y_{0.2}O_{2.9}$, by means of an *in situ* impedance spectroscopy cell in combination with the QENS measurements. In agreement with some of the previous works, the QENS data could be interpreted in terms of interoctahedral proton transfer.

Most recently, Noferini *et al.* [88] investigated the nature of proton dynamics in hydrated samples of the two proton conducting perovskites $BaZr_{0.9}M_{0.1}O_{2.95}$ (M = Y and Sc), she referred to as 10Y:BZO and 10Sc:BZO, respectively, using NSE spectroscopy at the spin-echo instrument IN11 at the ILL. In comparison to TOF and BS techniques, NSE probes the so-called intermediate scattering function, $I(\mathbf{Q}, t)$, rather than $S(\mathbf{Q}, \omega)$. Figure 16 shows, as an example, the $I(\mathbf{Q}, t)$, as extracted from the two samples, $BaZr_{0.9}M_{0.1}O_{2.95}$ (M = Y and Sc). The quite monotonic decrease of $I(\mathbf{Q}, t)$ suggests that the recorded dynamics do not follow a basic microscopic process characterized by a unique activation energy, but rather point toward the contribution of varying relaxation processes governed by a wide distribution of activation energies due to a range of different local structures present. Further information about the proton dynamics can be obtained from the offset in $I(\mathbf{Q}, t)$ at the

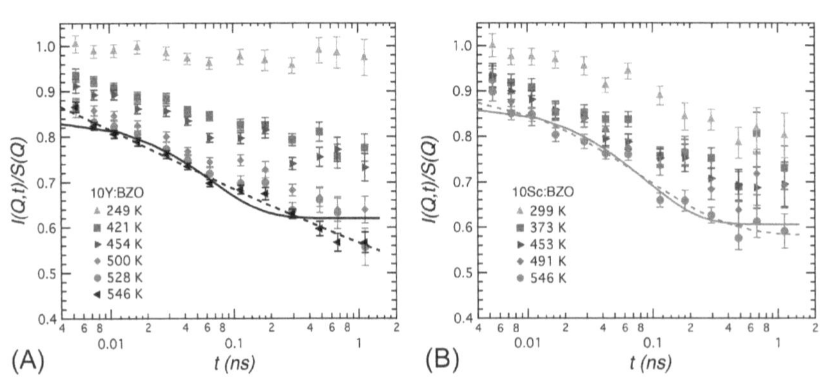

FIGURE 16 Temperature dependence of the normalized intermediate scattering function, $I(\mathbf{Q}, t)/S(\mathbf{Q})$, averaged between 0.7 and 1.2 Å^{-1} for (a) 10Y:BZO and (b) 10Sc:BZO. Solid lines are free fits to the 546 K data using single exponential functions, $I(\mathbf{Q},t)/S(\mathbf{Q}) = 0.62 + 0.23 \cdot \exp(-t/0.07)$ for 10Y:BZO and $I(\mathbf{Q},t)/(\mathbf{Q}) = 0.61 + 0.27 \cdot \exp(-t/0.09)$ for 10Sc:BZO. Dashed lines are free fits to the 546 K data using stretched exponential functions, $I(\mathbf{Q},t)/S(\mathbf{Q}) = 0.4 + 1 \cdot \exp[-(t/0.03)^{0.2}]$ for 10Y:BZO and $I(\mathbf{Q},t)/S(\mathbf{Q}) = 0.58 + 0.34 \cdot \exp[-(t/0.09)^{0.6}]$ for 10Sc:BZO. *Adapted with permission from Ref. [88].*

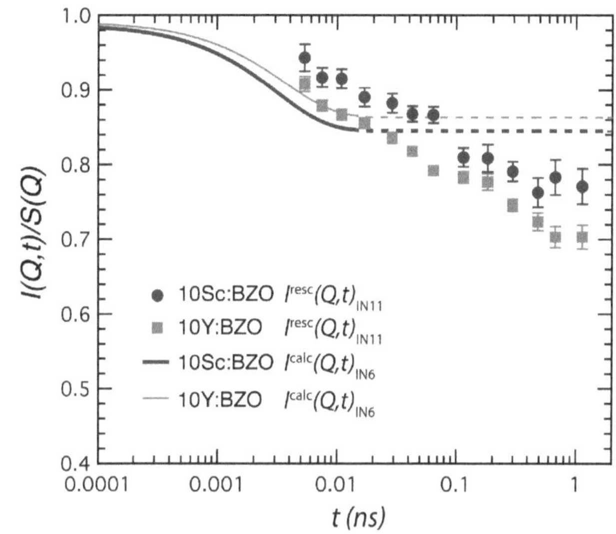

FIGURE 17 $I(\mathbf{Q}, t)$ of 10Sc:BZO and 10Y:BZO computed from the present data and approximated from data measured with the TOF spectrometer IN6 at Institut Laue-Langevin on a shorter time-scale. Dashed line indicates the extension of the TOF calculated $I(\mathbf{Q}, t)$ beyond the resolution limit of IN6. *Adapted with permission from Ref. [88].*

shortest Fourier times investigated here (~5 ps). This offset evidences the presence of proton dynamics at a shorter time-scale than the NSE time-window accessible on IN11. Figure 17 compares the $I(\mathbf{Q}, t)$ as obtained from the present experiment and Fourier-transformed from a previous TOF experiment where this shorter time-scale was examined. An important result is that the $I(\mathbf{Q}, t)$s obtained from the two different instruments overlap in the time range of 5-10 picoseconds, albeit an offset in the data for the Sc-doped sample is obvious. The resolution of the TOF spectrometer limits the validity of the response and reduces the significance of its line shape to a curve adapting the stretched exponential characteristics followed by the IN11 response. This is in agreement with several other works, pointing toward the presence of different proton sites and activation energies for localized proton dynamics in the materials. In this context, one should note that following the monotonic decay of the $I(\mathbf{Q}, t)$ should extend the footprint of the proton dynamics into the time domain of oxygen diffusion. However, we cannot exclude that a prominent portion of the hydrogen atoms remain immobile on longer time-scales forming a plateau in $I(\mathbf{Q}, t)$. This remains to be verified by future experiments.

9.5.5 Case Studies on Other Structure Types

Clearly, research on proton and oxide-ion conducting ceramics continues to be focused primarily on ABO_3-type perovskites, $A_2B_2O_5$-type brownmillerites, and AO_2-type fluorite materials, due to their high promise. However, a variety of other classes of proton and oxide-ion conducting ceramics, which show appreciable ionic conductivities at intermediate to high temperatures, have been developed and are also subject to both fundamental and applied research. Examples of proton conducting materials include the more complex perovskites of the form $Ba_3Ca_{1.18}Nb_{1.82}O_{8.72}$ and $Sr_3CaZr_{0.5}Ta_{1.5}O_{8.75}$ [121–123], which possess cation ordering leading to a doubling of the unit-cell, gallium-based oxides (*e.g.*, $LaBaGaO_4$ [124]), pyrochlores (*e.g.*, $La_2Zr_2O_7$ [125, 126]), phosphates (*e.g.*, $LaPO_4$ [127, 128]), niobates and tantalates (*e.g.*, $LaNbO_4$ [129]), tungstates (*e.g.*, La_6WO_{12} [130]), solid acids (*e.g.*, $CsHSO_4$ [131]), hydrated alkali thio-hydroxogermanates (*e.g.*, $Cs_2GeS_2(OH)_2 \cdot yH_2O$ [13, 132]), tungsten-bronze titanate/niobate systems (*e.g.*, $Ba_{0.6}Mg_{0.067}Nb_{0.933}O_3$ [133]), and so-called cupsidine systems (*e.g.*, $La_4\ (Ga_{2-x}Ti_xO_{7+x/2})O_2$, $x = 0–2$ [134]). Examples of oxide-ion conducting materials include, *e.g.*, $NdBaInO_4$-based materials [135, 136] and $LaSrGa_3O_7$-based melilite oxides [90]. Other examples are BIMEVOX [137], $Na_{0.5}Bi_{0.5}TiO_3$ [138], and $CeNbO_{4.25}$-based [139] compounds.

In addition to materials displaying pure proton or oxide-ion conduction, materials in which hydride-ions (H^-) rather than protons (H^+) are the conducting species are now accumulating increased attention. Bridges *et al.* [140] investigated the hydride-ion dynamics in the transition-metal oxide hydride $LaSrCoO_3H_{0.7}$, using QENS. The $LaSrCoO_3H_{0.7}$ structure is related to the $A_2B_nO_{3n+1}$ family of oxides, containing rocksalt (La/Sr)O layers separated by rectangular perovskite $(La/Sr)CoO_2H_{0.7}$ layers, *cf.* Figure 18 (a). A schematic diagram of the proposed hopping mechanism for hydride-ion diffusion within the perovskite-type layer, as extracted from the QENS data, is shown in Figure 18 (b). The hydride anions hop from an occupied (gray) to a vacant

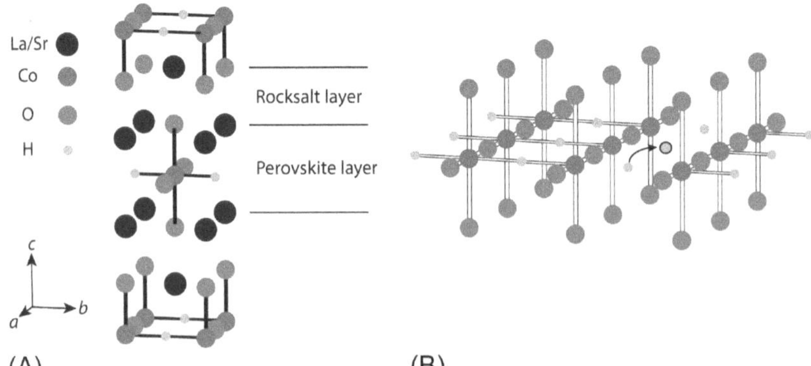

FIGURE 18 (a) The structure of $LaSrCoO_3H_{0.7}$ consists of rocksalt (La/Sr)O layers separated by perovskite $(La/Sr)CoO_2H_{0.7}$ layers. (b) Schematic diagram of the proposed hopping mechanism for hydride-ion diffusion within the perovskite-type layer of $LaSrCoO_3H_{0.7}$. *Adapted with permission from Ref. [140].*

(hollow sphere) site along the *a*-axis of the orthorhombic unit cell. The hopping was found to initiate in the temperature range used for synthesis of the material and occurs in conjunction with loss of hydrogen from the sample. The obtained activation energy and hydride-ion diffusion coefficient derived from QENS are *ca.* 200–230 meV and 6×10^{-5}–10×10^{-5} cm^2 s^{-1} in the temperature range of $T = 700$–750 K, which is similar to what has been found for proton conducting perovskites, *ca.* 400 meV and 1×10^{-5} cm^2 s^{-1}.

Recently, Kobayashi *et al.* [141] discovered a new class of perovskite-type oxyhydrides, $ATiO_{3-x}H_x$ (A = Ba, Sr, Ca, x <0.6). The oxyhydrides $ATiO_{3-x}H_x$ are the first ones to be based on the perovskite structure, where the O^{2-} and H^- ions form the octahedral environment around Ti, which is in a mixed IV/III oxidation state. In contrast with most other oxyhydrides the systems $ATiO_{3-x}H_x$ are stable in air all the way up to *ca.* $T = 400$ °C, above which hydrogen is released. This observation leads to the conclusion that the hydride species in $ATiO_{3-x}H_x$ have to be mobile. This finding is highly exciting because it implies that the perovskite structural framework can accommodate both, cation (proton) and anion (hydride) conduction. This opens up for a great scope of new basic science to determine and understand both the structural and dynamical properties related to hydrogen, now stretching from protic to hydridic species, in materials.

9.6 PERSPECTIVES

In this chapter dedicated to neutron scattering studies of ionic conductors and protonics, our goal was to give the reader an idea of the basics, the scope and diversity of this research field. Partly biased by our own interest we have chosen to focus on proton and oxide-ion conducting oxides, given a snapshot of recent neutron studies on these materials and tried to point out the most important topics, especially the ones still needing major advancements or breakthroughs.

With a view to the future, we envisage an increasing use of neutron scattering techniques for studies of structure and dynamics of proton and oxide-ion conducting materials. In particular, we see a large opportunity for studies of thin film phenomena by neutron reflectivity, which is a clearly underused technique in this field. Future studies should also take advantage of the development of new neutron sources, instruments, and sample environments, which will allow for investigations *in-operando* and/or that before may have been limited by intensity and/or resolution and hence perhaps not within reach. For the application of proton and oxide-ion conductors in miniaturized solid oxide fuel cells, which are currently accumulating increased attention, one would indeed use a film (<1 μm) of the electrolyte sandwiched between the anode and cathode [1, 142]. Differently from powder samples, such films represent a confined state and furthermore can be textured, strained, and exhibit varying degrees of crystallinity, which may affect significantly the performance of the electrolyte. This opens up for a great scope of new basic science, which is central to the field of solid-state ionics generally, and crucial for the development/discovery of new materials with higher ionic conductivity. This is in turn critical to future breakthroughs in the development of next-generation, environment friendly, devices utilizing ion conduction materials, such as batteries, fuel cells, hydrogen sensors, and thermoelectric converters, to name a few. Even small or gradual advancements of the current understanding of ion conduction mechanisms may therefore have considerable environmental and economic impact.

ACKNOWLEDGMENTS

Maths Karlsson is grateful for financial support from the Swedish Research Council (Grant No. 2010-3519 and 2011-4887) and the Swedish Foundation for Strategic Research (Grant No. ICA10-0001).

ABBREVIATIONS OF COMMONLY USED WORDS IN ALPHABETICAL ORDER

ADP	Atomic Displacement Parameters
DFT	Density Functional Theory
DOS	Density of States
INS	Inelastic Neutron Scattering
QENS	Quasielastic Neutron Scattering
MD	Molecular Dynamics
ND	Neutron Diffraction
NR	Neutron Reflectivity
NSE	Neutron Spin Echo
PDF	Pair Distribution Function
PEMFC	Polymer Electrolyte Membrane Fuel Cell
PGAA	Prompt-Gamma Activation Analysis
RDF	Radial Distribution Function
RMC	Reverse Monte Carlo
SOFC	Solid Oxide Fuel Cell
TGA	Thermogravimetric Analysis
TOF	Time-of-Flight
XRR	X-Ray Reflectivity

REFERENCES

[1] B.C.H. Steele, A. Heinzel, Nature 414 (2001) 345–352.
[2] L. Malavasi, C.A.J. Fisher, M.S. Islam, Chem. Soc. Rev. 39 (2010) 4370.

[3] M. Karlsson, Neutron scattering of proton-conducting ceramics, in: G.J. Kearley, V.K. Peterson (Eds.), Neutron Applications in Materials for Energy, Springer, 2015 Chapter 9.
[4] D.J.L. Brett, A. Atkinson, N.P. Brandon, S.J. Skinner, Chem. Soc. Rev. 37 (2008) 1568.
[5] A. Orera, P.R. Slater, Chem. Mater. 22 (2010) 675.
[6] A.J. Jacobson, Chem. Mater. 22 (2010) 660.
[7] J.A. Kilner, Faraday Discuss. 134 (2007) 9.
[8] V.V. Kharton, F.M.B. Marques, A. Atkinson, Solid State Ionics 174 (2004) 135.
[9] H.L. Tuller, Phys. Chem. Chem. Phys. 11 (2009) 3023.
[10] N.Q. Minh, Solid State Ionics 174 (2004) 271.
[11] J.B. Goodenough, Annu. Rev. Mater. Res. 33 (2003) 91.
[12] K.D. Kreuer, ChemPhysChem 3 (2002) 771–775.
[13] S.R. Poling, C.R. Nelson, S.W. Martin, Chem. Mater. 17 (2005) 1728–1735.
[14] K.D. Kreuer, S.J. Paddison, E. Spohr, M. Schuster, Chem. Rev. 104 (2004) 4637–4678.
[15] H. Iwahara, T. Esaka, H. Uchida, N. Maeda, Solid State Ionics 3–4 (1981) 359–363.
[16] K.D. Kreuer, Annu. Rev. Mater. Res. 33 (2003) 333–359.
[17] M. Karlsson, Phys. Chem. Chem. Phys. 17 (2015) 26–38.
[18] K.D. Kreuer, Mechanisms of proton conduction in perovskite-type oxide, in: T. Ishihara (Ed.), Perovskite Oxide for Solid Oxide Fuel Cells, Springer Science + Business Media, Berlin, Germany, 2009.
[19] A.L. Samgin, Solid State Ionics 136–137 (2000) 291.
[20] T. Tomoyose, N. Shimoji, K. Wakamura, J. Phys. Soc. Jpn 74 (2005) 3011–3015.
[21] A.L. Samgin, J. Phys. Chem. Solids 74 (2013) 1661–1668.
[22] E.J. Spahr, L. Wen, M. Stavola, L.A. Boatner, L.C. Feldman, N.H. Tolk, G. Lüpke, Phys. Rev. Lett. 102 (2009) 075506.
[23] E.J. Spahr, L. Wen, M. Stavola, L.A. Boatner, L.C. Feldman, H.H. Tolk, G. Lüpke, Phys. Rev. Lett. 104 (2010) 205901.
[24] S.P.S. Badwal, F.T. Ciacchi, D. Milosevic, Solid State Ionics 136–137 (2000) 91–99.
[25] S. Sarat, N. Sammes, A. Smirnova, J. Power Sources 160 (2006) 892–896.
[26] R. Devanathan, S. Thevuthasan, J.D. Gale, Phys. Chem. Chem. Phys. 11 (2009) 5506–5511.
[27] M. Sahibzada, B.C.H. Steele, K. Zheng, R.A. Rudkin, I.S. Metcalfe, Catal. Today 38 (1997) 459–466.
[28] K. Zheng, B.C.H. Steele, M. Sahibzada, I.S. Metcalfe, Solid State Ionics 86–88 (1996) 1241–1244.
[29] B.C.H. Steele, Solid State Ionics 129 (2000) 95–110.
[30] O.A. Marina, C. Bagger, S. Primdahl, M. Mogensen, Solid State Ionics 123 (1999) 199–208.
[31] T. Ishihara, H. Matsuda, Y. Takita, J. Am. Chem. Soc. 116 (1994) 3801–3803.
[32] M. Feng, J.B. Goodenough, Eur. J. Solid State Inorg. Chem. 31 (1994) 663–672.
[33] K. Huang, M. Feng, J.B. Goodenough, C. Milliken, J. Am. Ceram. Soc. 81 (1998) 2581–2585.
[34] P.R. Slater, J.T.S. Irvine, T. Ishihara, Y. Takita, J. Solid State Chem. 139 (1998) 135–143.
[35] S. Tao, J.T.S. Irvine, J.A. Kilner, Adv. Mater. 17 (2005) 1734–1737.
[36] C.E. Mohn, N.L. Allan, C.L. Freeman, P. Ravindran, S. Stølen, J. Solid State Chem 178 (2005) 346–355.
[37] S.A. Speakman, J.W. Richardson, B.J. Mitchell, S.T. Misture, Solid State Ionics 149 (2002) 247–259.

[38] E. Bakken, N.L. Allan, T.H.K. Barron, C.E. Mohn, I.T. Todorov, S. Stølen, Phys. Chem. Chem. Phys. 5 (2003) 2237–2243.

[39] A. Nemudry, M. Weiss, I. Gainutdinov, V. Boldyrev, R. Schöllhorn, Chem. Mater. 10 (1998) 2403–2411.

[40] R. Le Toquin, W. Paulus, A. Cousson, C. Prestipino, C. Lamberti, J. Am. Chem. Soc. 128 (2006) 13161–13174.

[41] S.J. Skinner, Solid State Sci. 5 (2003) 419–426.

[42] V.V. Kharton, A.P. Viskup, A.V. Kovalevsky, E.N. Naumovich, Solid State Ionics 143 (2001) 337–353.

[43] J.A. Kilner, C.K.M. Shaw, Solid State Ionics 154/155 (2002) 523–527.

[44] S.J. Skinner, J.A. Kilner, Ionics 5 (1999) 171–174.

[45] S.J. Skinner, J.A. Kilner, Solid State Ionics 135 (2000) 709–712.

[46] V.V. Kharton, A.P. Viskup, E.N. Naumovich, F.M.B. Marques, J. Mater. Chem. 9 (1999) 2623–2629.

[47] V.V. Vashook, S.P. Tolochko, I.I. Yushkevich, L.V. Makhnach, I.F. Kononyuk, H. Altenburg, J. Hauck, H. Ullmann, Solid State Ionics 110 (1998) 245–253.

[48] P. Lacorre, F. Goutenoire, O. Bohnke, R. Retoux, Y. Laligant, Nature 404 (2000) 856–858.

[49] F. Goutenoire, O. Isnard, R. Retoux, P. Lacorre, Chem. Mater. 12 (2000) 2575–2580.

[50] I.R. Evans, J.A.K. Howard, J.S.O. Evans, Chem. Mater. 17 (2005) 4074–4077.

[51] F. Goutenoire, O. Isnard, E. Suard, O. Bohnke, Y. Laligant, R. Retoux, P. Lacorre, J. Mater. Chem. 11 (2001) 119–124.

[52] S. Nakayama, Y. Higushi, Y. Kondo, M. Sakamoto, J. Mater. Chem. 5 (1995) 1801–1805.

[53] E. Kendrick, M.S. Islam, P.R. Slater, J. Mater. Chem. 17 (2007) 3104–3111.

[54] S. Nakayama, Y. Higushi, Y. Kondo, M. Sakamoto, Solid State Ionics 170 (2004) 219–223.

[55] H. Zhang, F. Li, J. Jin, Q. Wang, Y. Sun, Solid State Ionics 179 (2008) 1024–1028.

[56] F. Abraham, M.F. Debreuille-Gresse, G. Mairesse, G. Nowogrocki, Solid State Ionics 28 (1988) 529–532.

[57] I. Abrahams, F. Krok, J. Mater. Chem. 12 (2002) 3351–3362.

[58] Y. Teraoka, H.M. Zhang, K. Okamoto, N. Yamazoe, Mat. Res. Bull. 23 (1988) 51–58.

[59] S.B. Adler, Solid State Ionics 111 (1998) 125–134.

[60] R.H.E. van Doorn, A.J. Burggraaf, Solid State Ionics 128 (2000) 65–78.

[61] Z. Wang, R. Peng, W. Zhang, X. Wu, C. Xia, Y. Lu, J. Mater. Chem. A. 1 (2013) 12932–12940.

[62] H. Vogel, Phys. Z. 22 (1921) 645–646.

[63] G.S. Fulcher, J. Am. Ceram. Soc. 8 (1925) 339–355.

[64] G. Tammann, W. Hesse, Z. Anorg. Allg. Chem. 156 (1926) 245–257.

[65] M.L. Williams, R.F. Landel, J.D. Ferry, J. Am. Chem. Soc. 77 (1955) 3701–3707.

[66] F. Gray, M. Armaud, Polymer electrolytes, in: J.O. Besenhard (Ed.), Handbook of Battery Materials, Wiley-VCH Verlag GmbH., Weinheim, Germany, 1999.

[67] N. Kuwata, T. Saito, M. Tatsumisago, T. Minami, J. Kawamura, Solid State Ionics 175 (2004) 679–682.

[68] M. Ribes, G. Taillades, A. Pradel, Solid State Ionics 105 (1998) 159–165.

[69] S. Georges, O. Bohnké, F. Goutenoire, Y. Laligant, J. Fouletier, P. Lacorre, Solid State Ionics 177 (2006) 1715–1720.

[70] O. Bohnke, C. Bohnke, J.L. Fourquet, Solid State Ionics 91 (1996) 21–31.

[71] W. Paulus, H. Schober, S. Eibl, M. Johnson, T. Berthier, O. Hernandez, M. Ceretti, M. Plazanet, K. Conder, C. Lamberti, J. Am. Chem. Soc. 130 (2008) 16080–16085.

[72] A. Perrichon, A. Piovano, M. Boehm, M. Zbiri, M.R. Johnson, H. Schober, M. Ceretti, W. Paulus, J. Phys. Chem. C 119 (2015) 1557–1564.

[73] A. Piovano, A. Perrichon, M. Boehm, M.R. Johnson, W. Paulus, Phys. Chem. Chem. Phys. 18 (2016) 17398–17403.

[74] M. Arai, Experimental techniques, in: F. Fernandez-Alonso, D.L. Price (Eds.), Experimental Methods in the Physical Sciences, Neutron Scattering, vol. 44, Academic Press, Amsterdam, 2013, pp. 245–316.

[75] J.A. Alonso, X. Turrillas, Dalton Transactions, The Royal Society of Chemistry, 2005, pp. 865.

[76] V. Jayaramann, A. Magrez, M. Caldes, O. Joubert, F. Taulelle, J. Rodriguez-Carvajal, Y. Piffard, L. Brohan, Solid State Ionics 170 (2004) 25.

[77] K.S. Knight, Solid State Ionics 127 (2000) 43.

[78] N. Sata, K. Hiramotom, M. Ishigame, S. Hosoya, N. Niimura, S. Shin, Phys. Rev. B 54 (1996) 15795.

[79] I. Sosnowska, R. Prezenioslo, W. Schafer, W. Kockelmann, R. Hempelmann, K. Wysocki, J. Alloys Compd. 328 (2001) 226.

[80] E. Kendrick, K.S. Knight, M.S. Islam, P.R. Slater, Solid State Ionics 178 (2007) 943–949.

[81] T. Ito, T. Nagasaki, K. Iwasaki, M. Yoshino, T. Matsui, N. Igawa, Y. Ishii, Solid State Ionics 178 (2007) 13–17.

[82] I. Ahmed, C.S. Knee, M. Karlsson, S.G. Eriksson, P.F. Henry, A. Matic, D. Engberg, L. Börjesson, J. Alloys Compd. 450 (2008) 103–110.

[83] T. Nagasaki, T. Ito, M. Yoshino, K. Iwasaki, S. Shiotani, H. Fukazawa, N. Igawa, Y. Ishii, J. Nucl. Sci. Technol. 6 (Suppl.) (2008) 122–127.

[84] T. Nagasaki, S. Shiotani, N. Igawa, M. Yoshino, K. Iwasaki, H. Fukazawa, W. Utsumi, J. Solid State Chem. 182 (2009) 2632–2639.

[85] S.T. Norberg, S.M.H. Rahman, S. Hull, C.S. Knee, S.G. Eriksson, J. Phys. Condens. Matter. 25 (2013) 454214.

[86] J.R. Stewart, P.P. Deen, K.H. Andersen, H. Schober, J.-F. Barhélémy, J.M. Hillier, A.P. Murani, T. Hayes, B. Lindenau, J. Appl. Cryst. 42 (2009) 69–84.

[87] O. Moze, T.J. Hicks, A.C. McLaren, Phys. Chem. Miner. 5 (1980) 309–314.

[88] D. Noferini, M.M. Koza, P. Fouquet, G.J. Nilsen, M.C. Kemei, S.M.H. Rahman, M. Maccarini, S. Eriksson, M. Karlsson, J. Phys. Chem. C 120 (26) (2016) 13963–13969.

[89] N. Jalarvo, C. Haavik, C. Kongshaug, P. Norby, T. Norby, Solid State Ionics 180 (2009) 1151–1156.

[90] A. Mancini, C. Tealdi, L. Malavasi, Int. J. Hydrog. Energy 37 (2012) 8073–8080.

[91] I. Abrahams, X. Liu, S. Hull, S.T. Norberg, F. Krok, A. Kozanecka-Szmigiel, M.S. Islam, S.J. Stokes, Chem. Mater. 22 (2010) 4435–4445.

[92] X. Liu, I. Abrahams, S. Hull, S.T. Norberg, M. Holdynski, F. Krok, Solid State Ionics 192 (2011) 176–180.

[93] S. Hull, S.T. Norberg, M.G. Tucker, S.G. Eriksson, C.E. Mohn, S. Stølen, Dalton Trans. 40 (2009) 8737–8745.

[94] L. Mazzei, M. Wolff, D. Pergolesi, J.A. Dura, L. Börjesson, P. Gutfreund, M. Bettinelli, T. Lippert, M. Karlsson, J. Phys. Chem. C 120 (2016) 28415–28422.

[95] Deleted during review.

[96] B. Schrader, Infrared and Raman Spectroscopy, VCH Verlagsgesellschaft, Weinheim, 1995.

[97] Ch. Karmonik, T.J. Udovic, R.L. Paul, J.J. Rush, K. Lind, R. Hempelmann, Solid State Ionics 109 (1998) 207–211.
[98] M. Karlsson, A. Matic, S.F. Parker, I. Ahmed, L. Börjesson, S.G. Eriksson, Phys. Rev. B 77 (2008) 104302.
[99] J. Bielecki, S.F. Parker, D. Ekanayake, S.M.H. Rahman, L. Börjesson, M. Karlsson, J. Mater. Chem. A 2 (2014) 16915–16924.
[100] S.F. Parker, S.M. Bennington, A.J. Ramirez-Cuesta, G. Auffermann, W. Bronger, H. Nerman, J. Am. Chem. Soc. 125 (2003) 11656–11661.
[101] M. Karlsson, M.E. Björketun, P.G. Sundell, A. Matic, G. Wahnström, D. Engberg, L. Börjesson, I. Ahmed, S.G. Eriksson, P. Berastegui, Phys. Rev. B 72 (2005) 1–7, 094303.
[102] S.F. Parker, K. Refson, R.I. Bewley, G. Dent, J. Chem. Phys. 134 (2011) 084503.
[103] G.M. Taylor, S.F. Parker, P.C.H. Mitchell, J. Mol. Struct. 251–253 (2003) 123–126.
[104] A. Chroneos, D. Parfitt, J.A. Kilner, R.W. Grimes, J. Mater. Chem. 20 (2010) 266–270.
[105] D. Parfitt, A. Chroneos, J.A. Kilner, R.W. Grimes, Phys. Chem. Chem. Phys. 12 (2010) 6834–6836.
[106] A. Kushima, D. Parfitt, A. Chroneos, B. Yildiz, J.A. Kilner, R.W. Grimes, Phys. Chem. Chem. Phys. 13 (2011) 2242–2249.
[107] M. Yashima, N. Sirikanda, T. Ishihara, J. Am. Chem. Soc. 132 (2010) 2385–2392.
[108] R. Hempelmann, Ch. Karmonik, Th. Matzke, M. Cappadonia, U. Stimming, T. Springer, M.A. Adams, Solid State Ionics 77 (1995) 152–156.
[109] T. Matzke, U. Stimming, C. Karmonik, M. Soetramo, R. Hempelmann, F. Güthoff, Solid State Ionics 86–88 (1996) 621–628.
[110] Ch. Karmonik, R. Hempelmann, J. Cook, F. Güthoff, Ionics 2 (1996) 69.
[111] M. Pionke, T. Mono, W. Schweika, T. Springer, H. Schober, Solid State Ionics 97 (1997) 497–504.
[112] B. Groß, Ch. Beck, F. Meyer, Th. Krajewski, R. Hempelmann, H. Altgeld, Solid State Ionics 145 (2001) 325–331.
[113] C. Beck, S. Janssen, B. Gross, R. Hempelmann, Scr. Mater. 44 (2001) 2309–2313.
[114] D. Wilmer, T. Seydel, K.D. Kreuer, Mater. Res. Soc. Proc. 972 (2007) 15.
[115] N. Malikova, C.K. Loong, J.M. Zanotti, F. Fernandez-Alonso, J. Phys. Chem. C 111 (2007) 6574–6580.
[116] A. Slodczyk, P. Colomban, N. Malikova, O. Zaafrani, S. Longeville, J.-M. Zanotti, O. Lacroix, B. Sala, Solid State Ionics 252 (2013) 7–11.
[117] M. Karlsson, I. Ahmed, A. Matic, S.G. Eriksson, Solid State Ionics 181 (2010) 126.
[118] A. Braun, S. Duval, P. Ried, J. Embs, F. Juranyi, T. Strässle, U. Stimming, R. Hempelmann, P. Holtappels, T. Graule, J. Appl. Electrochem. 39 (2009) 262103.
[119] P. Colomban, A. Slodczyk, D. Lamago, G. Andre, O. Zaafrani, O. Lacroix, S. Willemin, B. Sala, J. Phys. Soc. Jpn 79 (2010) 1.
[120] Q. Chen, J. Banyte, X. Zhang, J.P. Embs, A. Braun, Solid State Ionics 252 (2013) 2–6.
[121] K.C. Liang, A.S. Nowick, Solid State Ionics 61 (1993) 77–81.
[122] K.C. Liang, Y. Du, A.S. Nowick, Solid State Ionics 69 (1994) 117–120.
[123] H.G. Bohn, T. Schober, T. Mono, W. Shilling, Solid State Ionics 117 (1999) 219–228.
[124] F. Schönberger, E. Kendrick, M.S. Islam, P.R. Slater, Solid State Ionics 176 (2005) 2951.
[125] T. Shimura, M. Kumori, H. Iwahara, Solid State Ionics 86–88 (1996) 685–689.
[126] T. Omata, K. Okuda, S. Tsugimoto, S. Otsuka-Yao-Matsuo, Solid State Ionics 104 (1997) 249–258.
[127] K. Amezawa, H. Maekawa, Y. Tomii, N. Yamamoto, Solid State Ionics 145 (2001) 233.
[128] N. Kitamura, K. Amezawa, Y. Tomii, N. Yakamoto, Solid State Ionics 162–163 (2003) 161.

[129] R. Haugsrud, T. Norby, Nat. Mater. 5 (2006) 193.

[130] A. Magraso, C. Frontera, D. Marrero-López, P. Núñez, Dalton Trans. 46 (2009) 10273–10283.

[131] S.M. Haile, D.A. Boysen, C.R.I. Chisholm, R.B. Merle, Nature 410 (2001) 910–913.

[132] S.A. Poling, C.R. Nelson, S.W. Martin, Mater. Lett. 60 (2006) 23–27.

[133] E. Kendrick, M.S. Islam, P.R. Slater, Solid State Ionics 176 (2005) 2975.

[134] M.C. Martín-Sedeño, D. Marrero-López, E.R. Losilla, L. León-Reina, S. Bruque, P. Núñez, M.A.G. Aranda, Chem. Mater. 17 (2005) 5989.

[135] K. Fujii, M. Shiraiwa, Y. Esaki, M. Yashima, S.J. Kim, S. Lee, J. Mater. Chem. A 3 (2015) 11985–11990.

[136] K. Fujii, Y. Esaki, K. Omoto, M. Yashima, A. Hoshikawa, T. Ishigaki, J.R. Hester, Chem. Mater. 26 (2014) 2488–2491.

[137] D.S. Khaerudini, G. Guan, P. Zhang, A. Abudula, Ionics 22 (2016) 93–97.

[138] M. Li, M.J. Pietrowski, R.A. De Souza, H. Zhang, I.M. Reaney, S.N. Cook, J.A. Kilner, D.C. Sinclair, Nat. Mater. 13 (2014) 31–35.

[139] S.S. Pramana, T. Baikie, T. An, M.G. Tucker, J. Wu, M.K. Schreyer, F. Wei, R.D. Bayliss, C.L. Kloc, T.J. White, A.P. Horsfield, S.J. Skinner, J. Am. Chem. Soc. 138 (2016) 1273–1279.

[140] C.A. Bridges, F. Fernandez-Alonso, J.P. Goff, M.J. Rosseinsky, Adv. Mater. 18 (2006) 3304–3308.

[141] Y. Kobayashi, O.J. Hernandez, T. Sakaguchi, T. Yakima, T. Roisnel, Y. Tsujimoto, M. Morita, Y. Noda, Y. Mogami, A. Kitada, M. Ohkura, S. Hosokawa, Z. Li, K. Hayashi, Y. Kurano, J. Kim, N. Tsuji, A. Fujiwara, Y. Matsushita, K. Yoshimura, K. Takegoshi, M. Inoue, M. Takano, H. Kageyama, Nat. Mater. 11 (2012) 507–511.

[142] E. Traversa, Electrochem. Soc. Interf. 18 (2009) 49.

Chapter 10

High-Temperature Levitated Materials

Louis Hennet*,[1], Dirk Holland Moritz[†], Richard Weber[‡] and Andreas Meyer[†]

*Conditions Extrêmes et Matériaux: Haute Température et Irradiation, Orléans, France; †Institut für Materialphysik im Weltraum, Deutsches Zentrum für Luft- und Raumfahrt, Köln, Germany; ‡Materials Development, Inc., Arlington Heights, IL, United States

[1]Corresponding author: e-mail: louis.hennet@cnrs-orleans.fr

Chapter Outline

10.1 INTRODUCTION

Neutron scattering techniques are important tools for studying the structure and dynamics of high-temperature melts. On the one hand, wide-angle neutron scattering (WANS) is a source of structural information through the determination of the static structure factor $S(Q)$ and the calculation of the corresponding pair-distribution function $g(r)$ allowing the study of the short- and medium-range order (SRO and MRO) in the melt. On the other hand, quasielastic neutron scattering (QENS) is one of the best techniques for probing the structural relaxation of solids and liquids. In particular, the technique of time-of-flight (TOF) QENS is well adapted to the study of atomic diffusion in liquid samples. In this technique, relaxation times and diffusion coefficients can be directly derived from the measured dynamic structure factor $S(Q,E)$.

However, the study of high-temperature melts is often hampered by high-temperature limits of available furnaces, which also use containers that bear the risk of chemical reactions between the melt and the container material, resulting in sample contamination. Contact with container walls also tends to induce the heterogeneous nucleation of crystals and impede investigation of deeply undercooled liquids and formation of some glasses.

For WANS experiments, standard furnaces are usually composed of a cylindrical vanadium heater, which gives no diffraction peaks because of the negligible coherent neutron scattering cross section. Nevertheless, the accessible maximum temperatures are limited to about 1100°C. For QENS experiments, it is possible to use Nb or Ta heaters enabling to work at higher temperatures that can reach 2000°C. Although ceramic refractory sample crucibles (SiC or Al_2O_3) can be used [1,2], QENS experiments have been limited so far to low melting and/or chemically fairly inert metallic systems and to temperatures close above the liquidus temperature.

These limitations encouraged the development of containerless techniques that are powerful tools enabling the study of high-temperature melts. By eliminating completely any contact between sample and container, it is possible to study the sample with a very high degree of control and to access very high temperatures. An additional advantage of levitation methods is the possibility to undercool hot liquids down to several hundred of degrees below their equilibrium freezing point, since heterogeneous nucleation processes at crucible walls are suppressed.

Various containerless techniques have been applied to determine physical properties of levitated liquids. A review of all techniques has been published by D. L. Price [3]. Some of them have been combined with x-ray and neutron scattering techniques to study the structure and the dynamics of molten materials [4,5].

This chapter is divided into five main sections. In the first two, we recall rapidly the formalism used in neutron scattering techniques and the data analysis procedure, and we give a short description of the instruments at which levitation environments are developed and routinely used. The third section is a description of all levitation techniques that have been combined with neutron techniques to study high-temperature melts. In particular, it describes various developments made at the DLR in Cologne (Germany), at the CEMHTI in Orleans (France), and by MDI in Arlington Heights (IL, the United States).

The two last parts present, respectively, various studies of the structure (using WANS) and dynamics (using QENS) of some metallic and oxide melts carried out using all levitation techniques.

10.2 THEORETICAL BACKGROUND/DATA ANALYSIS

For a summary of the technique and theoretical background for neutron scattering from liquids and glasses, the reader can find a full description of the formalism in Refs. [6,7]. Regarding the dynamics, an interesting review can be found in the first volume [8]. We will just recall here some theoretical backgrounds on neutron scattering that will be useful in this chapter.

10.2.1 Structure (WANS)

10.2.1.1 Formalism

As opposed to crystals, the structure of liquids is characterized by the absence of long-range order, and the structural information is then limited to the description of the mean arrangement of the atoms around each atomic species. This structural information is extracted from the total pair-distribution function $g(\mathrm{r})$, which is related to the probability of finding an atom at a distance r from another taken at the origin. From this function, it is then possible to determine useful information on interatomic distances and coordination numbers.

The $g(\mathrm{r})$ can be determined by neutron or x-ray diffraction. In particular, in a neutron diffraction experiment from an n-component material, the mean differential scattering cross section per atom can be written as a function of the total interference function $F(\mathrm{Q})$ or the total structure factor $S(\mathrm{Q})$ as

$$\frac{\mathrm{d}\sigma}{\mathrm{d}\Omega} = \overline{b^2} + F(\mathrm{Q}) = \overline{b^2} + \bar{b}^2(S(\mathrm{Q}) - 1) \tag{1}$$

with $\bar{b} = \sum_{\alpha=1}^{n} c_\alpha b_\alpha$ and $\overline{b^2} = \sum_{\alpha=1}^{n} c_\alpha b_\alpha^2$

where c_α and b_α are, respectively, the atomic concentration and the coherent scattering length of species α present in the sample. The b_α values for all elements can be found, for instance, in the appendix of the first volume [9].

Using the Faber-Ziman (FZ) formalism [10], $F(\mathrm{Q})$ and $S(\mathrm{Q})$ can be expressed as a weighted sum of the different partial structure factors $S_{\alpha\beta}(\mathrm{Q})$ for atomic species α and β:

$$F(\mathrm{Q}) = \sum_{\alpha,\beta=1}^{n} c_\alpha c_\beta b_\alpha b_\beta \left[S_{\alpha\beta}(\mathrm{Q}) - 1\right] \tag{2}$$

$$S(\mathrm{Q}) = \sum_{\alpha,\beta=1}^{n} \frac{c_\alpha c_\beta b_\alpha b_\beta}{\bar{b}^2} S_{\alpha\beta}(\mathrm{Q}) = \sum_{\alpha,\beta=1}^{n} w_{\alpha\beta} S_{\alpha\beta}(\mathrm{Q}) \tag{3}$$

The total pair-distribution function $g(\mathrm{r})$ is then directly obtained with a Fourier transform of $S(\mathrm{Q})$:

$$g(\mathrm{r}) = 1 + \frac{1}{2\pi^2 \rho_0} \int_0^{\mathrm{Q_{max}}} \mathrm{Q}(S(\mathrm{Q}) - 1) \frac{\sin \mathrm{Q}r}{r} \mathrm{dQ} \tag{4}$$

where ρ_0 is the number of atoms per unit volume, r is a distance in real space, and $\mathrm{Q_{max}}$ is the maximal Q-value of the measurement. The radial distribution function is then defined as

$$RDF(\mathrm{r}) = 4\pi\rho_0 \mathrm{r}^2 g(\mathrm{r}) \tag{5}$$

Like $S(\mathrm{Q})$, $g(\mathrm{r})$ is also a weighted sum of partial pair-distribution functions $g_{\alpha\beta}(\mathrm{r})$:

$$g(\mathrm{r}) = \sum_{\alpha,\beta=1}^{n} w_{\alpha\beta} g_{\alpha\beta}(\mathrm{r}). \tag{6}$$

The mean coordination number $CN_{\alpha\beta}$ gives the average number of β atoms in a spherical coordination shell of radius $\mathrm{r_1} \le \mathrm{r} \le \mathrm{r_2}$ centered on an atom of type α and is obtained by integrating in real space over a peak of the partial radial distribution function arising from a specific $g_{\alpha\beta}(\mathrm{r})$ function:

$$CN_{\alpha\beta} = 4\pi\rho_0 c_j \int_{\mathrm{r_1}}^{\mathrm{r_2}} \mathrm{r}^2 g_{\alpha\beta}(\mathrm{r}) \mathrm{dr} \tag{7}$$

In addition to the previous description of the partial structure factors in terms of atomic pairs, the Bhatia-Thornton (BT) formalism [11] defines, in the case of a binary system, an alternative set of three partial structure factors: $S_{\mathrm{NN}}(\mathrm{Q})$ describes the distributions of atomic number density and is related to the topological short-range order of the system. $S_{\mathrm{CC}}(\mathrm{Q})$ describes the distribution of concentration and is related to the chemical short-range order and finally $S_{\mathrm{NC}}(\mathrm{Q})$, which describes the correlation between the two.

These two descriptions of the partial structure are connected by simple linear combinations. More details can be found in [7].

For an n-component material, there are $\mathrm{n(n+1)/2}$ partial functions, and it can be difficult to obtain accurate value of distances and coordination numbers due to the large overlaps of each contribution. It is then useful to have the possibility to use more selective techniques such as neutron diffraction with isotopic substitution (NDIS). See Ref. [7] for more details. It can be seen from Eq. (2) that it is possible to modify the weighting factors $w_{\alpha\beta}$ of each element by using another isotope. By making measurements on an appropriate number of isotopes, it is possible to extract the partial structure factors and pair-distribution functions.

10.2.1.2 Data Analysis

This section describes the procedure typically used to derive the structure factor from the differential scattering cross section measured on levitated samples. The procedure is usually done using available software such as the program CORRECT [12].

The first performed correction is the background subtraction using a measurement of the levitator without sample. The self-absorption of the specimen is corrected according to the method described by Paalman and Pings [13], developed for a specimen of cylindrical geometry, but giving a relatively good approximation for spherical samples. The multiple scattering contributions are corrected using the method of Blech and Averbach [14]. The effect of inelastic scattering is corrected according to the procedure described by Yarnell et al. [15], which is based on the method of Placzek [16].

The signal is then corrected for geometric effects and detector efficiency using a measurement performed on a standard vanadium sample (nearly perfect incoherent scatterer) of the same size as that of the samples in which should give a flat diffractogram.

The next step consists of an autonormalization of the scattered intensity. Specifically, the high-Q asymptote of the corrected intensity (after accounting for the Placzek inelasticity effect) is scaled so as to be equal to the theoretical self-scattering $\overline{b^2}$ that depends only on the known composition of the sample. Finally, $S(Q)$ is obtained using Eq. (1).

For paramagnetic materials, a contribution of the incoherent magnetic scattering contribution has to be also subtracted [17].

10.2.2 Dynamics (QENS)

As described in the first volume [6,8], the dynamics of solids and liquids can be studied using inelastic neutron scattering (INS) and QENS, which are two complementary techniques allowing to investigate the collective excitations and structural relaxations, respectively. So far, levitated liquids have been mainly studied by QENS for obtaining diffusion coefficients, and the rest of the chapter will be focused on this technique.

10.2.2.1 Formalism

The dynamic properties of a system are described by the *van Hove* function $G(\mathrm{r},t)$ that is the time-dependent generalization of $g(\mathrm{r})$:

$$G(\mathrm{r},t) = (1/N)\sum_{\alpha=1}^{N}\sum_{\beta=1}^{N}\left\langle \delta\left[\mathrm{r}+\mathrm{r}_\alpha(0)-\mathrm{r}_\beta(t)\right]\right\rangle \tag{8}$$

In this case, $G(\mathrm{r},t)$ is related to the probability of finding an atom at r at time t if there was an atom at $\mathrm{r}=0$ at time 0. If the particle is the same, this gives the self-term $G_s(r,t)$, and a different particle leads to the distinct part $G_d(r,t)$:

$$G(r,t) = G_s(r,t) + G_d(r,t) \tag{9}$$

The self and distinct parts give rise to incoherent and coherent scattering, respectively.

$G(r,t)$ is obtained from a space Fourier transform of the intermediate scattering function $I(Q,t)$:

$$I(Q,t) = (1/N)\sum_{\alpha=1}^{N}\sum_{\beta=1}^{N}\langle \exp\{iQ.[r_\alpha(0) - r_\beta(t)]\}\rangle \tag{10}$$

In liquids, $I(Q,t)$ describes the decay in time of the density fluctuations at fixed wave vector Q. It is obtained itself with an energy Fourier transform of the dynamic structure factor $S(Q,E)$, which is the measured quantity.

As used by Johnson and Kearley in a previous volume [8], in the rest of this chapter, we will also refer to the frequency ω and use $S(Q,\omega)$ that is equivalent to $S(Q,E)$ since the neutron energy transfer $E = \hbar\omega$.

In a neutron scattering experiment, the measured intensity is proportional to the double-differential cross section:

$$\frac{d^2\sigma}{d\Omega d\omega} = \frac{k_f}{k_i}\frac{\sigma}{4\pi}S(Q,\omega) \tag{11}$$

that can be written as a sum of the coherent and incoherent parts

$$\frac{d^2\sigma}{d\Omega d\omega} = \frac{d^2\sigma_{coh}}{d\Omega d\omega} + \frac{d^2\sigma_{inc}}{d\Omega d\omega} = \frac{k_f}{k_i}\left[\frac{\sigma_{coh}}{4\pi}S_{coh}(Q,\omega) + \frac{\sigma_{inc}}{4\pi}S_{inc}(Q,\omega)\right] \tag{12}$$

where $\sigma_{coh} = 4\pi\bar{b}^2$ and $\sigma_{inc} = 4\pi\left(\overline{b^2} - \bar{b}^2\right)$

The coherent dynamic structure factor reflects the collective motion of atoms, whereas the incoherent part reflects motions of single particles. For liquid dynamics involving diffusive motion, the self-part of the van Hove correlation function $G_s(r,t)$ follows Fick's law:

$$D\nabla^2 G_s(r,t) = \frac{\partial}{\partial t}G_s(r,t) \tag{13}$$

If we consider that all diffusing particles start at the origin at time zero, the solution of Eq. (13) can be written as

$$G_s(r,t) = \frac{1}{(4\pi Dt)^{3/2}}\exp\left(-\frac{r^2}{4Dt}\right) \tag{14}$$

D denotes the self-diffusion coefficient of the atoms. The space Fourier transform of $G_s(r,t)$ gives the intermediate scattering function:

$$I_s(Q,t) = \exp\left(-DQ^2 t\right) \tag{15}$$

Its time Fourier transform leads to the dynamic structure factor:

$$S_{\text{inc}}(\text{Q},\omega)=\frac{1}{\pi}\frac{\Gamma_{1/2}(\text{Q})}{\omega^2+\Gamma_{1/2}^2(\text{Q})} \tag{16}$$

The neutron spectrum is then a Lorentzian function with a Q^2-dependent half width at half maximum (HWHM):

$$\Gamma_{1/2}(\text{Q})=D\text{Q}^2 \tag{17}$$

We will see that in a limited temperature range around the melting temperature the temperature dependence of the diffusion process in the studied melts is well described by an Arrhenius law:

$$D=D_0\exp(E_A/k_BT) \tag{18}$$

with k_B the Boltzmann constant, T the temperature, and E_A the activation energy for the diffusion process.

10.2.2.2 Data Analysis

The measured scattering intensity requires some corrections in order to obtain the desired quantity $S(\text{Q},\omega)$. To perform this data-reduction stage, various software can be used, such as the FRIDA [18] or LAMP [19,20] programs.

In these software tools, the measured number of counts, saved as a function of the scattering angle and time-of-flight $I(2\theta,tof)$, is first normalized to the monitor. This makes it independent of the incident neutron flux and counting time.

A measurement with a vanadium sample is used for the detector-efficiency corrections and to determine the energy resolution $R(\text{Q},\omega)$ described by a Gaussian function.

The background subtraction is then made by using spectra recorded with the empty levitator.

After the conversion from time of flight into energy transfer, a constant-energy binning is applied, and the spectra are sorted with constant Q-values: $I(\text{Q},\omega)$.

At the end of this correction stage, the scattering function $S(\text{Q},\omega)$ is still convoluted with the instrument resolution function $R(\text{Q},\omega)$:

$$I(\text{Q},\omega)=R(\text{Q},\omega)\otimes S(\text{Q},\omega) \tag{19}$$

The signal can be directly studied in the (Q,ω) space. In order to work in the time domain, the intermediate scattering function $I(\text{Q},t)$ is obtained using a Fourier transform (FT) of Eq. (19):

$$I(\text{Q},t)=\frac{FT[S(\text{Q},\omega)]}{FT[R(\text{Q},\omega)]} \tag{20}$$

10.3 NEUTRON SCATTERING INSTRUMENTS

This section described briefly the neutron instruments where levitation systems have been used for studying high-temperature melts. These are mainly at the Institut Laue-Langevin (ILL) in Grenoble (France) [21], at Spallation Neutron Source (SNS) in Oak Ridge (TN, the United States) [22] and at Forschungsreaktor München II (FRM II) in Garching (Germany) [23].

10.3.1 For WANS Experiments

For studying the structure of disordered materials and in particular high-temperature melts, there are basically two kinds of instrument to enable the determination of the structure factor $S(Q)$.

(i) Two-axis diffractometers, usually available at neutron reactor sources, such as the instruments D20 [24] and D4c [25] at the ILL are described hereafter. In this case, a monochromatic neutron beam with an incoming wavelength λ is directed onto the sample, and the scattered intensity is detected as a function of the scattering angle 2θ. The scattering vector is then calculated from $Q = 4\pi/\lambda \sin\theta$.
(ii) TOF diffractometer is used in pulsed neutron sources, such as the NOMAD [26] instrument at SNS, in which a pulsed white beam is sent from the source onto the sample. The scattered beam is then measured as a function of the time-of-flight τ by a detector placed at a constant scattering angle. If L is the total flight path and m the neutron mass, $Q = 2\,mL/\hbar\tau \sin\theta$.

10.3.1.1 D20 (ILL)

The D20 instrument is a very-high-intensity two-axis diffractometer equipped with a large position-sensitive microstrip detector. A detailed description can be found in Hansen et al. [27]. Due to the very high incoming neutron flux, a diffraction pattern over the full angular range can be obtained in a matter of seconds allowing real-time experiments on very small samples. Various kinds of sample environments can be also easily integrated into the instrument including levitation setups.

Levitation experiments are performed using a copper monochromator Cu (200) working in transmission geometry with a fixed vertical focusing. It provides wavelengths of 0.82, 0.88, 0.94 and 1.30 Å.

A wide-angle microstrip detector composed of 1536 cells covers an angular range of 153.6° with a resolution of 0.1°/cell. The positioning of the beam stop installed after the sample results in a minimum useful diffraction angle of 5° giving a Q-range of about 0.67–14.9 $Å^{-1}$ for the shortest wavelength.

10.3.1.2 D4c (ILL)

A precise description of the D4c neutron diffractometer can be found elsewhere [28]. This is a two-axis instrument optimized for structural investigations of liquids and amorphous materials. It has basically three main components:

(i) The monochromator. CNL experiments are performed using a Cu-focusing monochromator [220] giving a working wavelength of about 0.5 Å.
(ii) The vacuum chamber that consists of a cylinder 550 mm high and 460 mm in diameter where the aerodynamic levitation environment [29], described in the next paragraph, is enclosed.
(iii) A wide-angle detector allowing to measure the diffracted beam over a 1.3–140° angular range, corresponding to a scattering vector Q-range of about 0.3–24 $Å^{-1}$. This detector is composed of nine microstrip detectors with an angular aperture of 8°. Since there is a gap of 7° between each microstrip detector, it is necessary to make measurements at several positions to produce a complete diffractogram. In practice, scans of five or six angular positions are used in order to equilibrate counting statistics.

10.3.1.3 NOMAD (SNS)

The nanoscale-ordered materials diffractometer (NOMAD) instrument is a neutron time-of-flight diffractometer used to measure pair-distribution functions. NOMAD combines the high neutron flux of the SNS with wide-angular detector coverage. A detailed description of the instrument including its optical configuration, neutron-detector layout, and chopper arrangement is given in Ref. [30].

The combination of the detector coverage and bandwidth of neutron energies provides an accessible momentum-transfer range of 0.04–100 $Å^{-1}$. Measurements are routinely made to a Q_{max} of 50 $Å^{-1}$ and with a typical resolution $\Delta Q/Q$ of 0.8%. Various sample environments can be integrated into the instrument; both aerodynamic CNL [31] and electrostatic ESL [32] levitators have been used.

10.3.2 For QENS Experiments

10.3.2.1 IN6 (ILL)

IN6 is a time-focusing time-of-flight spectrometer designed for quasielastic and inelastic scattering, enabling to study the dynamics and relaxation properties in condensed matter [33].

A monochromatic beam is provided by an array of three composite pyrolytic graphite monochromators, which uses the full height of the guide. A vertical focusing of the neutron beam leads to a beam size of $3 \times 5\ cm^2$ at the sample position. At the instrument, four incident wavelengths are available: 4.1, 4.6, 5.1, and 5.9 Å. A 12 cm beryllium filter is also used to remove the second-order reflection from the monochromator.

The monochromatic beam is then pulsed by a Fermi chopper, with a small slot length to ensure a good transmission, and placed at a distance of 38 cm from the sample. To prevent frame overlaps occurring when the chopper is rotating fast, an additional anti-overlap chopper is used. It rotates in phase with the Fermi chopper but at a lower speed.

The sample position can accept various environments placed in cylindrical chambers that can be evacuated. The vertical positioning of the sample box is motorized.

The detector bank consists of 337 elliptical ^{3}He detectors placed at a distance of about 248 cm from the sample, and the scattered beam is measured over a horizontal angular range of 10–115° with a vertical detection angle of $\pm 15°$. To minimize the background, a box filled with helium is placed between the sample and the detectors.

Levitation experiments are performed using an incident neutron wavelength of 5.1 Å providing a Q-range of 0.3–2.1 $Å^{-1}$ and a spectrometer resolution of 70 μeV.

10.3.2.2 TOFTOF (FRM II)

The TOFTOF spectrometer at FRM II [34] is described in detail in Refs. [35,36]. It is a direct-geometry disc-chopper time-of-flight spectrometer for cold neutrons. It is used for both inelastic and quasielastic neutron scattering experiments with high neutron flux and high-energy resolution, providing also an excellent signal-to-background ratio.

The neutrons coming from the reactor cold source, operating with liquid D_2 at 25 K, are directed through an s-shaped curved neutron guide, acting also as a velocity filter, to the primary spectrometer. The latter is equipped with a system of seven high-speed carbon fiber chopper discs and a focusing secondary neutron guide including supermirrors, which reduces the size of the primary beam from 44×100 to 23×47 mm^2 (width $\times$ height). The beam size can be further reduced with a slit system placed 18 cm in front of the sample. The primary beam monitor is positioned directly behind these slits. A beam stop is mounted about 16 cm behind the sample.

The scattered neutrons are detected by 1006 squeezed ^{3}He detector tubes mounted at a distance of 4 m from the sample and covering angles from -15 to 140°. Measurements can be performed in a wavelength range of 1.4–12 Å with resolutions between 3 meV and 2 μeV, depending on the wavelength and the rotation speed of the chopper system. The range of energy transfer accessible is between -30 and 50 meV.

This instrument can accept a wide range of sample environments, and the sample chamber is accessible from the top and via a small door from the side.

10.4 LEVITATION TECHNIQUES

Four main levitation methods have been combined with scattering techniques at neutron sources: electromagnetic levitation (EML) [37], electrostatic levitation (ESL) [32,38,39], conical nozzle levitation (CNL) [40,41], and single-axis acoustic levitation [42]. However, only the first three are used for studying high-temperature melts at neutron sources and will be described in the following sections.

Before starting, it is worth mentioning that it is also possible to find hybrid systems combining several methods: aerodynamic levitation with inductive heating [43,44], electrostatic-aerodynamic systems [45,46], or aeroacoustic levitators [47,48].

10.4.1 Aerodynamic Levitation (CNL)

10.4.1.1 Principle

The CNL technique is based on works reported by Winborne et al. [49] and Coutures et al. [50] and on first developments described by Weber and Nordine [51] in which specimens were stably levitated by a gas flow in a convergent-divergent nozzle and laser-heated.

The principle of a typical system combining CNL with laser heating is represented in Figure 1.

A spherical sample (1–4 mm in diameter) is placed in a levitator, which includes a convergent-divergent conical nozzle that channels a regulated gas flow upwards to counteract the gravity. The gas flow is precisely regulated using mass flow controllers. The condition for a stable levitation is then obtained when the sample weight Mg is compensated by the force exerted by the gas stream on the sample [52]:

$$F_{CNL} = \int \left[\rho u^2 + p\right] \mathrm{d}A = Mg \qquad (21)$$

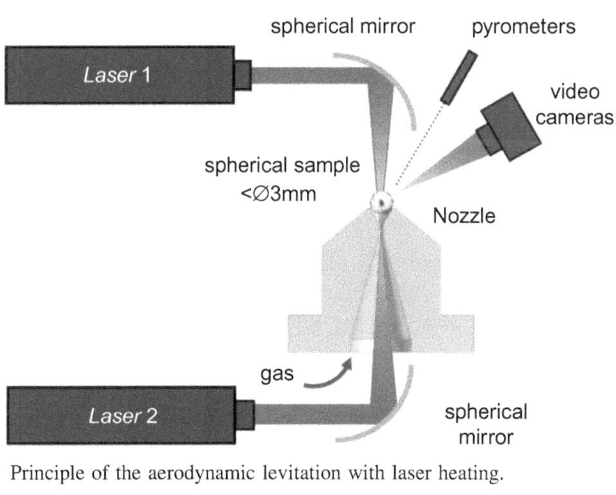

FIGURE 1 Principle of the aerodynamic levitation with laser heating.

Here, ρ, u, and p are, respectively, the gas density, the vertical gas flow velocity, and the gas pressure. The integral is performed over the specimen's surface A.

In the existing setups developed by the CEMHTI [29] and MDI [31] and described hereafter, the heating is performed using CO_2 lasers focused onto the sample from the top and from the bottom through the hole inside the nozzle. The focusing can be achieved either by using spherical mirrors (CEMHTI) or by combining a ZnSe lens with flat mirrors (MDI). The temperature is measured using optical pyrometers (see Section 10.4.4), and video cameras are used to monitor the sample during the levitation process.

The advantages of CNL are the simplicity and compactness of the levitation device, making it easy to combine with neutron techniques. It can be also used with all kinds of materials. The sample levitates inside the nozzle exposing a large part of its volume above the top of the nozzle to the incident neutron beam.

10.4.1.2 Existing Setups

The first neutron diffraction measurements on aerodynamically levitated liquids were performed by Landron et al. on liquid alumina at the SANDALS instrument at ISIS (Rutherford Appleton Laboratory, the United Kingdom) [40]. A detailed description of the setup can be found in Ref. [53]. A similar CNL environment was also developed by Weber et al. at IPNS (Intense Pulsed Neutron Source, Argonne, IL, the United States) [41]. These two groups have then improved the technique and developed new CNL environments, described below. These were utilized at the ILL and SNS neutron sources.

- **At the D4c instrument at ILL (WANS)**

Figure 2 is a schematic view of the levitation apparatus. The spherical sample with a diameter of about 2.7 mm is placed in the levitator. The upper part containing the cone that is in the neutron beam is made of vanadium, which has a very small coherent neutron scattering length, and its contribution is easy to remove. The entire levitator support is water cooled during the experiment, and the bottom part of the nozzle is made of aluminum in order to compensate the low thermal conductivity of vanadium and increase the cooling efficiency.

The position of the levitator in the vacuum chamber is vertically adjusted according to the neutron beam height. It is checked by using a piece of B_4C mounted on the levitator and taking a neutron picture of the direct beam on a neutron camera.

The samples are heated to the desired temperature by two 125 W CO_2 lasers ($\lambda = 10.6$ μm) directed from above. The primary laser beams (~5 mm in diameter) are focused on the sample by means of spherical mirrors at two

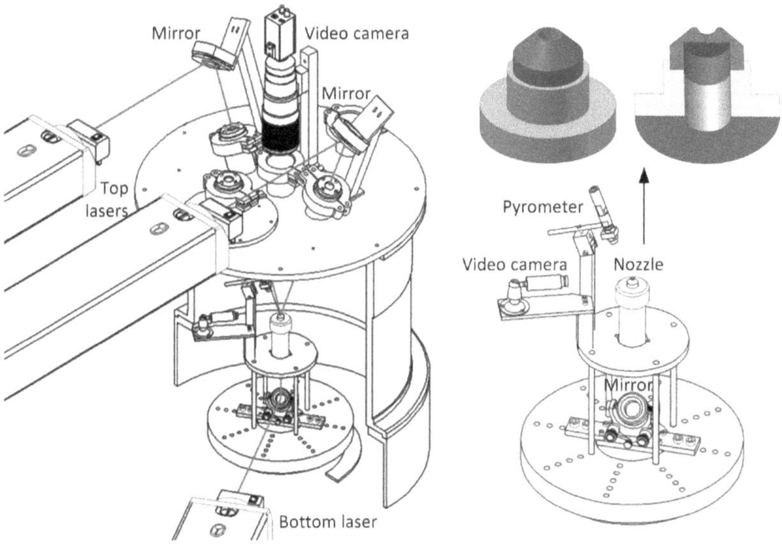

FIGURE 2 Levitation setup integrated into the D4c diffractometer at the ILL. Left: Schematic view of the experimental arrangement. Bottom right: Detailed view of the sample environment. Top right: Schematic representation of the nozzle with the Vanadium made cover.

different angles in order to obtain a homogeneous temperature distribution. The final diameters are about 1 mm at the sample position. A third laser directed at the sample through the nozzle from below is used to compensate the cooling of the sample by the gas flow.

The temperature of the sample is measured in the 800–3000°C temperature range with an optical pyrometer operating at a wavelength of 0.85 μm. The pyrometer is integrated into the vacuum chamber and focused directly on the sample to avoid window corrections.

Video images of the sample taken from above are continuously displayed in order to monitor the sample during the levitation process. Video images can be also recorded with a horizontal camera in order to determine the sample position in the levitator and to monitor its vertical stability.

Before starting the experiments, the chamber is initially pumped down to about 10^{-4} mbar and then filled with the levitation gas up to atmospheric pressure. The gas outlet is connected to an oil bubbler to avoid any backflow of air into the chamber and provide a reproducible atmosphere.

A set of vertical and horizontal B_4C slits produced a rectangular beam on the sample. The vertical slits are placed very near the nozzle. The top of the lower part is vertically adjusted in order to mask the vanadium nozzle from the incident beam and reduce the scattering from it. The final beam size at the sample position is around 5 mm vertically and 9 mm horizontally.

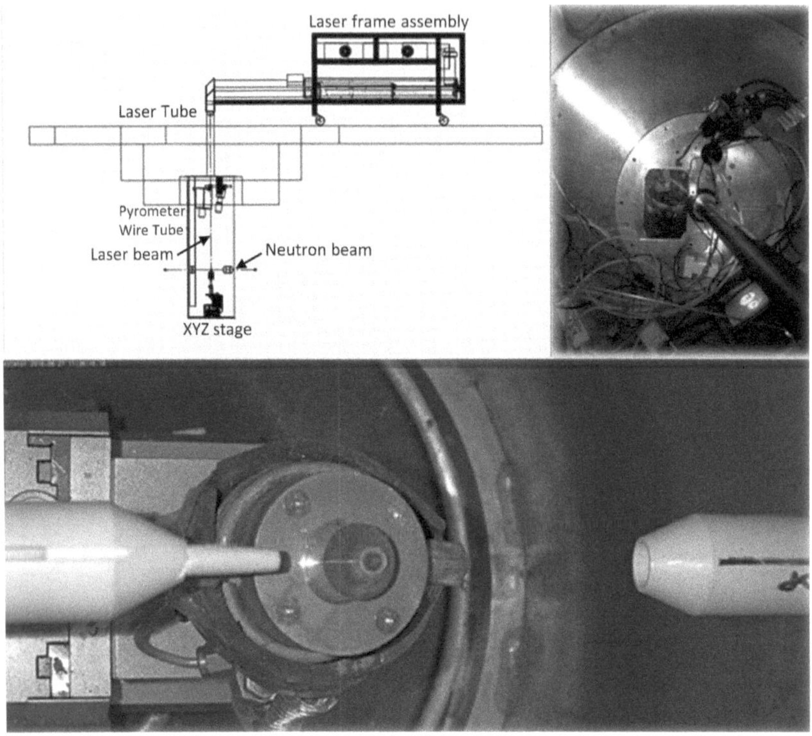

FIGURE 3 Layout of the levitation setup at the NOMAD instrument at SNS. Top left—diagram illustrating the main components. The laser head is mounted horizontally above the neutron beam and directed into the sample well. Top right—photograph showing a view down onto the top plate over the sample well. Bottom—view into the sample well. The levitation nozzle is near the center of the image and mounted on an X-Y-Z stage. The direction of the neutron beam runs from left to right and intercepts the sample. An optical pyrometer used to measure the sample temperature is visible to the right of the image.

- **At the NOMAD instrument at SNS (WANS)**

The layout of the aerodynamic levitator at the NOMAD instrument is illustrated in Figure 3. The levitator components are arranged inside a *ca.* 400 mm inside diameter sample well fabricated from 1 mm wall thickness aluminum alloy. The neutron beam enters the chamber through a vanadium window, and then, it passes through a beam-defining boron nitride aperture tube. A ^{10}B powder-filled absorber is installed in the downstream port directly opposite to the aperture tube to block the direct beam.

The levitator comprises a liquid-cooled plenum that houses a converging-diverging levitation nozzle. Various nozzles can be used including those made of solid vanadium, aluminum alloy with a vanadium insert, and aluminum alloy. Nozzles typically use a 90° cone with a 1 mm diameter orifice. Samples are placed on the levitation nozzle using a vacuum chuck. The incident neutron beam profile is modified by using an aperture so that it only enters the top part of the sample that is above the levitation nozzle. The plenum mounts onto a precision motorized X-Y-Z stage located on the base of the chamber. This arrangement is used to precisely locate the sample in the neuron beam. The heating laser beam, optical pyrometer, and video camera are then aligned with the sample. Heating is performed using a partially focused carbon dioxide laser beam from a 400 W laser. The beam is steered using a mirror with precision adjustable mounts. The sample temperature can be controlled using a PID loop with the pyrometer as the sensor. The laser setup is operated as a Class 1 laser with embedded Class 4 laser. This laser arrangement makes it easier for users to work with the instrument.

Once the system is fully aligned and before running calibrations or experiments, the chamber is purged to displace air with high-purity argon. Two flow systems are used. One delivers argon at a rate of *ca.* 2000 cm^3/min into a copper bustle pipe located on the base of the chamber. The second flow system meters gas to the levitation nozzle to levitate samples. By purging the chamber with gas of uniform and constant composition, a steady background can be maintained. It was found that operating without the purge can lead to variations in background due to the air mixing with the argon in the chamber. The system was calibrated using vanadium spheres. All functions of the levitator are controlled remotely using a computer control program.

- **At the IN6 instrument at ILL (QENS)**

A 3-D drawing of the CNL environment developed at IN6 is presented in Figure 4. This setup is described in detail in Refs. [54,55]. The entire

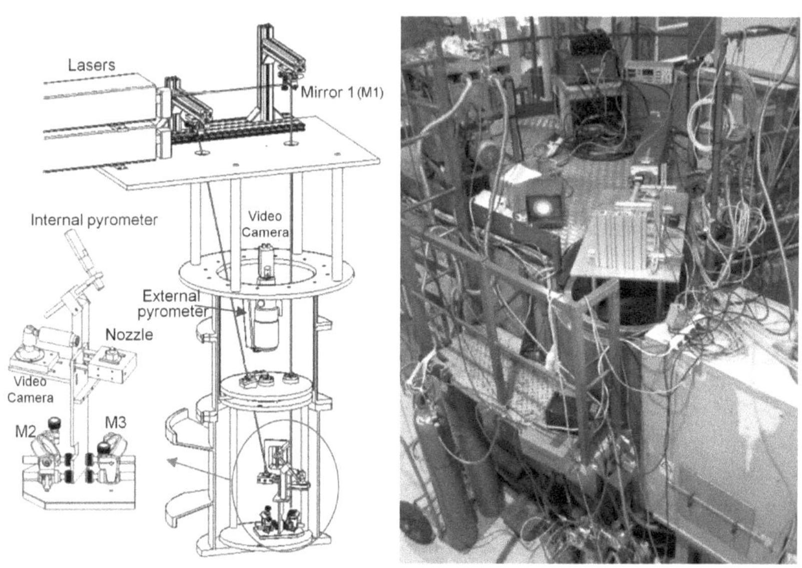

FIGURE 4 Levitation environment designed at the IN6 instrument.

apparatus is mounted on the same support structure in order to assure a stable alignment of the laser optics. A side opening in the vacuum chamber allows an easy access for sample changing.

In order to improve the statistics, an aluminum nozzle was specially designed to levitate samples with a diameter of 4 mm that corresponds to a volume about three times greater than the 2.7 mm diameter used for diffraction. Between one-third and one-half of the sample levitates inside the cone. However, aluminum is relatively transparent to neutrons. Since inevitably some of the incident beam will strike the Al nozzle, the amount of Al in the beam was minimized, keeping enough metal for the cooling efficiency.

The working wavelength of 5.1 Å is above the Bragg cutoff of aluminum. Hence, the elastic scattering from the nozzle is strongly reduced, and the contribution of the empty nozzle compared to the total signal is relatively low. Nevertheless, the nozzle was further machined in order to reduce the amount of aluminum in the beam. In particular, we made a conical nozzle head shape (see Figure 5).

The position of the sample is aligned to the beam by taking an image of cadmium piece mounted at the sample position. In addition to the upstream slits, the incident neutron beam is further reduced by a 10 mm punched cadmium (Cd) mask mounted in front of the nozzle.

The sample is heated by means of two 125 W CO_2 lasers ($\lambda = 10.6$ μm). The primary laser beam (providing most of the heat to the sample) is focused from above by a spherical mirror onto the sample with a final beam diameter of 2 mm. The second laser beam is mainly used for temperature-homogeneity purposes. It is reflected through two flat and one concave copper mirrors before hitting the sample from below, passing through the nozzle's conical interior.

The temperature of the levitated sample is measured in the two temperature ranges of 623–2073 K and 1073–3273 K by two pyrometers working at wavelengths of 1.65 and 0.85 μm, respectively. They make it possible to have a dual temperature measurement in the range of 800–1800°C. For optimizing the compactness of the setup, one is placed outside the chamber and measure the sample luminance through a silica window. The second is placed inside the chamber that has also the advantage to avoid window corrections.

Before data acquisition, the sample chamber is pumped down to low pressure and then filled with the levitation gas at atmospheric pressure.

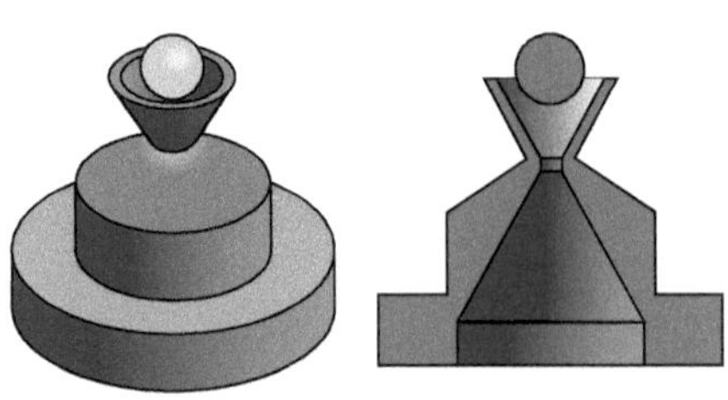

FIGURE 5 Optimized conical geometry of the nozzle with a head thickness reduced to 1 mm.

10.4.2 Electromagnetic Levitation (EML)

10.4.2.1 Principle

The electromagnetic levitation [56] is one of the oldest techniques used for containerless experiments. The method was patented by Muck in 1923 [57] and further developed in the 1950s by Okress et al. [58].

The sample is placed in a radio-frequency alternating inhomogeneous electromagnetic field B, generated by a levitation coil composed of several windings like shown in Figure 6. Eddy currents are then induced in the sample leading to an inductive heating.

The power P absorbed by the sample due to ohmic losses of the induced currents is given by [59]

$$P = -\frac{B^2 \omega 4\pi}{2\mu_0 \; 3} a^3 H(s) \quad \text{with } s = a\sqrt{\frac{\omega \sigma \mu_0}{2}} \tag{22}$$

where μ_0 is the magnetic permeability, a is the radius of the sample, σ is the electric conductivity of the sample, and ω is the frequency of the alternating field. The function $H(s)$ is given by

$$H(s) = \frac{9}{4s^2}\left(s\frac{\sinh(2s) + \sin(2s)}{\cosh(2s) - \cos(2s)} - 1 \right) \tag{23}$$

At the same time, the interaction of the Eddy currents with the magnetic field of the coil leads to Lorentz force F_L that counteracts the gravity and makes it possible to levitate the sample [59]:

$$F_L = -\frac{\nabla B^2 \, 4\pi}{2\mu_0 \; 3} a^3 L(s) \tag{24}$$

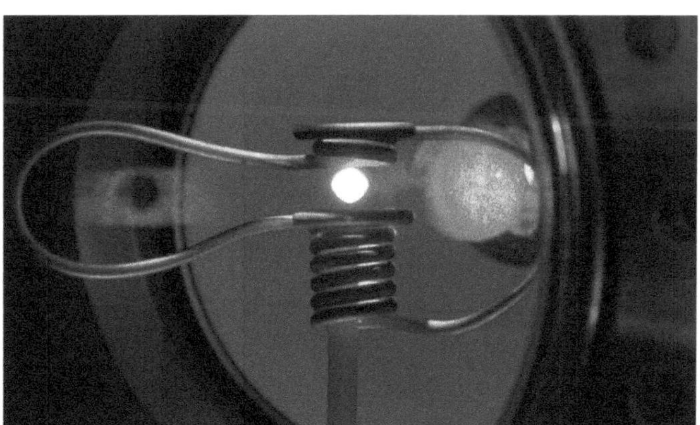

FIGURE 6 Electromagnetically levitated liquid droplet. *Image from DLR.*

where

$$L(s) = \frac{3}{4}\left(1 - \frac{2}{2s}\frac{\sinh(2s) - \sin(2s)}{\cosh(2s) - \cos(2s)}\right) \tag{25}$$

As seen previously, the levitation force and the heating power are intimately coupled. The accessible temperature range is then limited especially for low temperatures. The temperature of the sample is thus controlled by forced convection by varying a gas flow over the sample surface created by a nozzle installed close to the sample. Moreover, only electrically conducting materials can be levitated.

The levitation coil is situated in a vacuum chamber that is first evacuated and then filled with ultrahigh purity gas. Depending on the processed material, pure He, Ar, or a He/(4%)H_2 mixture is used as processing gas. If compatible with the sample, the addition of hydrogen may be favorable in order to remove contamination of the samples by oxides due to its reducing properties. Such oxides may act as heterogeneous nucleation sites and thus limit the undercooling of the melt.

As for CNL, the temperature is measured using optical pyrometers, and video images are continuously displayed for monitoring the sample during the levitation.

The combination of electromagnetic levitation with neutron scattering techniques has been mainly developed at the DLR. A first setup has been integrated into the D20 at the ILL for structural investigation and a slightly modified one at the TOF-TOF instrument of FRMII for studying the dynamics.

10.4.2.2 Existing Setups

A schematic representation of the EML arrangement integrated into the D20 (WANS) and TOFTOF (QENS) instrument is presented in Figure 7.

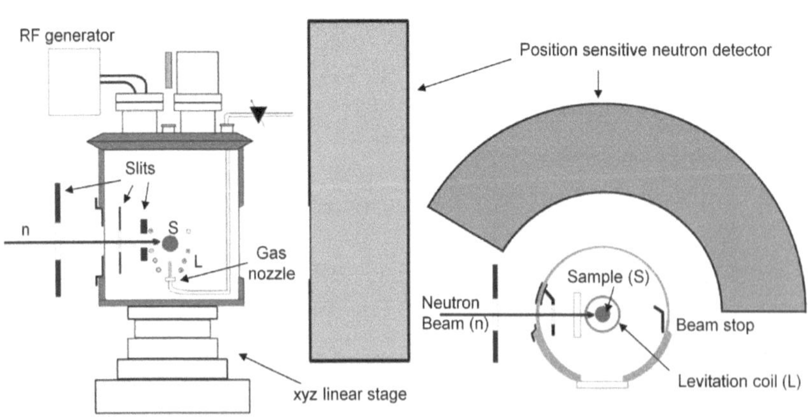

FIGURE 7 Side (right) and top (left) schematic view of the levitation chamber used in the combination with WANS at the D20 instrument. For more details, see Ref. [60].

A detailed description of the setup can be found in Holland-Moritz et al. [60], and we give here only few details.

The levitation coil, especially designed for making diffraction measurements, is placed in the vacuum chamber, which is evacuated down to a pressure of 10^{-7} mbar and then backfilled with the high-purity gas used during the levitation process.

For most of the experiments, the coil is made using five windings in the lower part and four counterwindings at the top part for sample stabilization. The gap between the upper and the lower windings of the coil is sufficiently broad to provide a clear view of the sample by the neutron beam.

The levitation coil is powered using a 5 kW RF generator and operated at a frequency of about 150 kHz.

The temperature of the sample (6–10 mm in diameter) is measured with pyrometers. It can be manually or remotely controlled by adjusting the flow of cooling gas (He or He/H_2) directed onto the sample using a nozzle made of Al_2O_3 or fused silica and placed few millimeters below the sample.

The levitation chamber is mounted on a xyz linear stage allowing a precise adjustment of the position of the levitated sample in the beam. The vertical and horizontal sizes of the neutron beam arriving onto the sample are defined using a set of Cd and B_4C slits. They are adjusted in order to optimize the irradiated sample volume, keeping the Cu windings of the levitation coil outside the beam.

The vacuum chamber is equipped with a 1.5 mm thick aluminum window covering an angle of 200° and through which the incoming neutrons enter the chamber, and the diffracted beam is measured by the detector. Appropriate Cd shields are installed inside the chamber to eliminate any scattering from the Al window due to the incoming direct beam. A Cd beam stop is also placed behind the sample to prevent the direct irradiation of the back wall of the chamber.

10.4.3 Electrostatic Levitation (ESL)

10.4.3.1 Principle

Electrostatic levitation uses a vertical electric field E produced by two capacitor plates, to which a high voltage is applied (Figure 8). The sample itself must be charged, and levitation is achieved through the Coulomb force:

$$F_{ESL} = qE \tag{26}$$

where q is the charge on the sample.

According to Earnshaw's theorem, in a static electric field without space charges, no potential minimum exists [61]. Electrostatic levitation is then intrinsically unstable, and the sample position needs to be controlled in all three directions in space. Thus, in addition to the vertical control by the top and ground electrode, the horizontal sample position is controlled by two pairs

FIGURE 8 Electrostatically levitated liquid $Ti_{39.5}Zr_{39.5}Ni_{21}$ droplet at 1150 K [38]. The sample levitates between the two vertical electrodes. Also, the two pairs of side electrodes are visible. *Image from DLR.*

of side electrodes (see Figure 8). The sample position is then monitored using a fast feedback mechanism.

Due to the high voltages applied, electrostatic levitation requires processing under UHV conditions or under high gas pressures [45] in order to avoid electric breakthrough. The sample must be charged initially, and the charge must be maintained; that is, the loss of charge by evaporation must be compensated by the photoelectric effect, when irradiating the samples with UV-light or by thermionic emission at high temperatures.

When it levitates, the sample is laser heated, and temperatures of more than 2500 K are accessible.

10.4.3.2 Existing Setups

While the ESL technique was routinely used for performing containerless diffraction experiments using synchrotron radiation [62,63], the first generations of electrostatic levitators were only able to levitate relatively small samples with a volume up to about 10 mm^3, making this method less attractive for neutron scattering studies.

The combination of ESL with neutron scattering techniques is also technically challenging since the relatively sophisticated environment has to be compatible with the geometric constraints of the instruments.

In 2003, Aoki et al. demonstrated the feasibility of neutron diffraction experiments using ESL [39]. However, only diffractograms recorded for solid samples have been shown. In 2011, Kordel et al. developed a compact ESL system dedicated for performing WANS and QENS experiments at the D20

(ILL) and TOFTOF (FRM II) instruments [38]. This device makes it possible to levitate sample of up to 6.5 mm in diameter (Figure 8). This means an increase of the typical mass of the samples (0.5–1 g) and therefore of the scattering signal by one order of magnitude.

Recently, Mauro et al. developed another ESL environment for use at the NOMAD instrument (SNS) [32].

Figure 9 shows a schematic representation of the electrostatic levitation furnace described by Kordel et al. [38] (left) and its integration into the neutron scattering environment (right).

The levitation part is composed by the top and ground electrodes placed vertically and associated with the two pairs of side electrodes (SEs). They are operated using high voltages of 40 and 3 kV, respectively.

The sample position in the three directions is monitored by using the cast shadows of the sample lighted by two diode lasers on two crossed Si position-sensitive photodetectors (PSDs). Each of them measures the movement of the sample in two perpendicular directions. The high voltages supplied to the electrodes are then adjusted accordingly every 2 ms, using a sample position control algorithm [64].

This set of electrodes is placed in a high vacuum chamber that can be evacuated down to a pressure of about 10^{-7} mbar.

The sample is heated using a 75 W infrared fiber-coupled diode laser operating at a wavelength of 808 nm and sufficient to melt metals with high melting temperatures (above 2000 K).

The temperature is measured using a fiber-coupled two-color pyrometer working in the wavelength ranges of 1650–1750 nm and 1850–2000 nm.

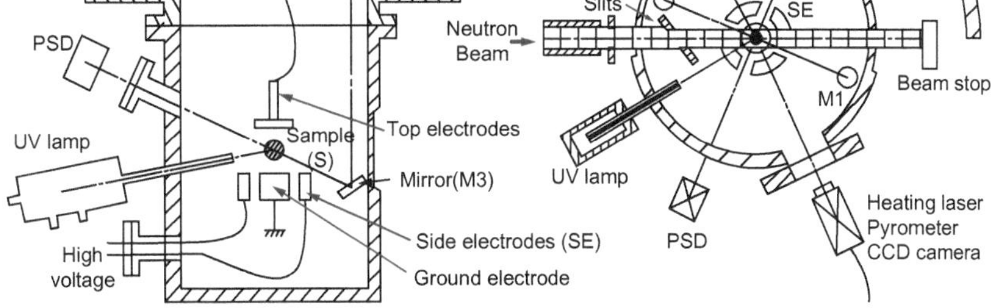

FIGURE 9 Side (right) and top (left) schematic view of the levitation chamber used in the combination with WANS at the D20 instrument. For more details, see Ref. [38].

A UV lamp is used to compensate the charge loss due to the sample evaporation during heating. At temperatures above about 1200 K, the thermionic emission becomes sufficiently efficient for charging the sample.

Evaporation of samples can also lead to changes in their composition. If significant mass loss occurs during the experiment, it is important to verify the sample composition by chemical analysis.

As for EML, the neutron beam is adjusted in size by using an adjustable aperture. The vacuum chamber is equipped with a 2 mm thick aluminum window covering an angle of 220° horizontally and ±20° vertically.

With this design of the chamber, a 130° of unshaded scattering angle is available for measurements. For the diffraction experiment at D20, the direct neutron beam is also blocked inside the chamber by a Cd beam stop to avoid secondary scattering from the exit neutron window.

10.4.4 Temperature Measurements

For temperature measurements on levitated samples, optical pyrometry is the ideal method. This contactless technique is very easy to use but, its drawback, at least for single-color pyrometers, is the need for spectral emissivity data in order to derive the true temperature T from the apparent temperature Ta measured by the pyrometer. T is related to Ta through Wien's law:

$$\frac{1}{T}-\frac{1}{Ta}=\frac{\lambda}{C_2}\ln\varepsilon_\lambda \tag{27}$$

where $C_2 = 1.4388$ cm K is Planck's second radiation constant, and ε_λ is the spectral emissivity of the material at the operating wavelength λ of the pyrometer.

As a practical matter, window reflection losses are typically included in the spectral emissivity term. It is important to verify that the pyrometer windows are clean and free of deposits in order for the data to be valid. In vacuum system, evaporation from a hot sample may cause a coating to form on the pyrometer window.

When possible, corrections are applied using values published in the literature. When no emissivity values are available, the pyrometer can be calibrated by comparing the melting or crystallization temperatures obtained during the free cooling of the sample with that expected from the phase diagram, assuming that the spectral emissivity is independent of the temperature in the studied temperature range.

10.5 STRUCTURAL INVESTIGATIONS

10.5.1 Study of Metallic Melts

In the past, it has been shown that metallic melts can be deeply undercooled below their freezing temperature if heterogeneous nucleation can be avoided.

Such large undercooling that can reach 20% of the melting temperatures T_L were first observed in 1950 by Turnbull for a wide variety of liquid metals [65].

To explain these large undercoolings of pure metals, Frank postulated an icosahedral short-range order (ISRO) in the melt [66].

EML and ESL are ideal containerless methods for studying deeply undercooled metals, and in the following, we present various results obtained using WANS on some liquid metals and alloys.

10.5.1.1 Experiments Using EML

- **Monoatomic metallic metals**

This part summarizes a work describing neutron scattering experiments on several pure transition-metal melts (Ni, Fe, and Zr) both in the stable liquid above the melting point T_L and in the undercooled state [37].

EML was applied to deeply undercool liquids for extended periods of time necessary to perform diffraction measurements.

The neutron diffraction experiments were performed at the D20 instrument (ILL), described previously, using wavelengths of 0.94 Å (Ni) and 0.88 Å (Fe and Zr).

The structure factors for the different melts are presented in Figure 10. They were measured at several temperatures above and below T_L at a maximum undercooling of about 140 K for Fe and 290 K for Ni and Zr.

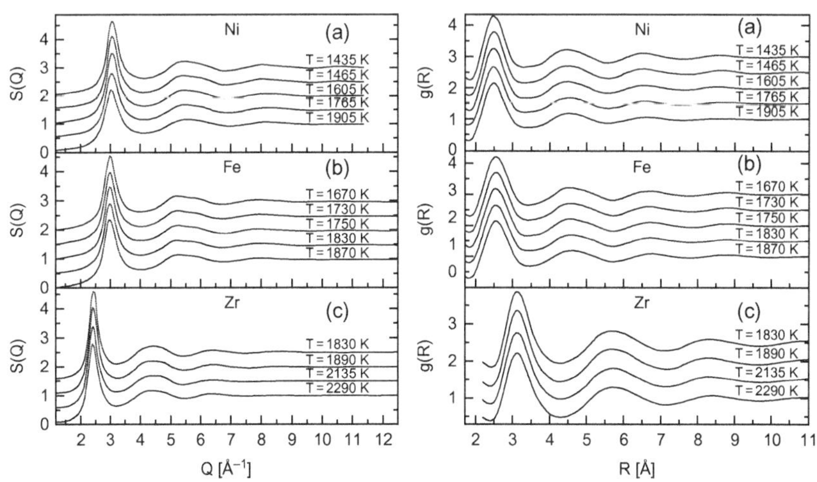

FIGURE 10 Structure factor $S(Q)$ and pair-correlation functions $g(r)$ of Ni (a), Fe (b), and Zr (c) melts at different temperatures T [$T_L(\mathrm{Ni})=1726$ K, $T_L(\mathrm{Fe})=1811$ K, and $T_L(\mathrm{Zr})=2125$ K]. *Figures reproduced with permission from Ref. [37].*

For all studied melts and temperatures, the *S*(Q)s exhibit a shoulder on the right-hand side of the second oscillation. This feature has been identified as a signature of ISRO.

The interatomic distances and coordination numbers derived from the corresponding *g*(r)*s* are given in Table 1 for the lowest and highest measured temperatures. For the three liquids, the shortest distance (r_1) remains essentially the same, while the second-neighbor distance (r_2) slightly decreases with decreasing *T*.

The integration under the first peak in *RDF*(r) leads to a nearest-neighbor coordination number (*CN*) around 12, which also slightly increases with decreasing *T* for all liquids. This value is a characteristic of several types of SRO. In this study, by comparing the measured structure factor with simulations involving different structural models, ISRO was evidenced in all of the investigated monoatomic metallic liquids, already above T_L and more pronounced in the undercooled state with decreasing *T*.

This work was the first experimental proof of the hypothesis by Frank [66] that was only based on topological ordering arguments.

- **Liquid alloys**

In the past, EML has been used for studying a large variety of alloys such as Al-Fe and Al-Co [67], Al-Fe-Co [68], or Ti-Fe [69]. This section presents a study of Ni-Zr-based melts performed by Holland-Moritz et al. [70]. In order to explain the peculiar atomic dynamics in these alloys, the authors have investigated the short-range structure of three $Ni_{36}Zr_{64}$ melts prepared with natural Ni, ^{58}Ni, and ^{60}Ni at a temperature of $T = 1375$ K by neutron diffraction with isotopic substitution.

TABLE 1 Nearest- (r_1) and Second- (r_2) Neighbor Distances and Coordination Numbers CN Derived from the Diffraction Experiments on the Different Melts for the Highest and Lowest Measured Temperature

Element	*T* (K)	r_1 (Å)	r_2 (Å)	*CN*
Ni	1905	2.48+0.02	4.53+0.05	11.2+0.5
Ni	1435	2.49+0.02	4.43+0.05	12.3+0.5
Fe	1870	2.55+0.02	4.55+0.05	12.3+0.5
Fe	1670	2.55+0.02	4.52+0.05	12.6+0.5
Zr	2290	3.12+0.02	5.74+0.05	11.9+0.5
Zr	1830	3.13+0.02	5.71+0.05	12.2+0.5

From Ref. [37].

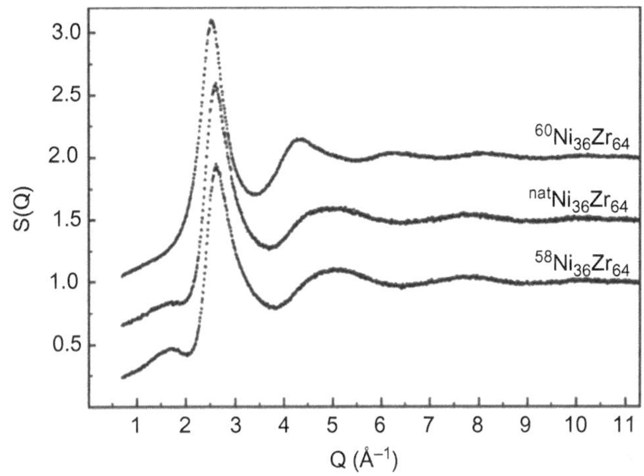

FIGURE 11 Total structure factors measured by neutron scattering for $^{58}Ni_{36}Zr_{64}$, $^{60}Ni_{36}Zr_{64}$, and $^{nat}Ni_{36}Zr_{64}$ at T = 1375 K. *Figure reproduced with permission from Ref. [70].*

The experiments were carried out at the D20 instrument at the ILL using electromagnetic levitation. The static structure factors of the liquids were measured at a neutron incident wavelength of = 0.94 Å.

Figure 11 shows the total structure factors *S*(Q) obtained for the three $Ni_{36}Zr_{64}$ samples prepared with different Ni isotopes. Marked differences between these structure factors are visible, especially with the ^{60}Ni isotope that has a much smaller neutron scattering length (2.8 fm) compared with ^{nat}Ni (10.3 fm) and ^{58}Ni (14.4 fm).

The partial structure factors were calculated from these three total structure factors by using both Faber-Ziman [10] and Bhatia-Thornton formalisms [11] that give complementary information. The results are, respectively, presented at the bottom and top of Figure 12 (left). The corresponding FZ and BT partial pair-distribution functions are shown in Figure 12 (right).

The BT concentration-concentration partial-correlation function $g_{CC}(r)$ is characterized by a remarkable minimum at r ~ 2.7 Å. This minimum is a signature of a chemical short-range order prevailing in the liquid with a preference for unlike neighbors corresponding to an affinity for the formation of Ni-Zr bonds. This observation is in accordance with the significantly larger amplitude of the first maximum in the $g_{NiZr}(r)$ partial compared with the first maxima of $g_{NiNi}(r)$ and $g_{ZrZr}(r)$.

The positions of the first maxima of the three FZ $g_{\alpha\beta}(r)$ lead to the nearest-neighbor distances $r_{NiNi} = 2.55$ Å and $r_{ZrZr} = 3.20$ Å slightly larger than in the molten pure elements (see Table 1). Conversely, r_{NiZr} found at 2.69 Å is slightly smaller than the average of the nearest-neighbor distances of both pure elements.

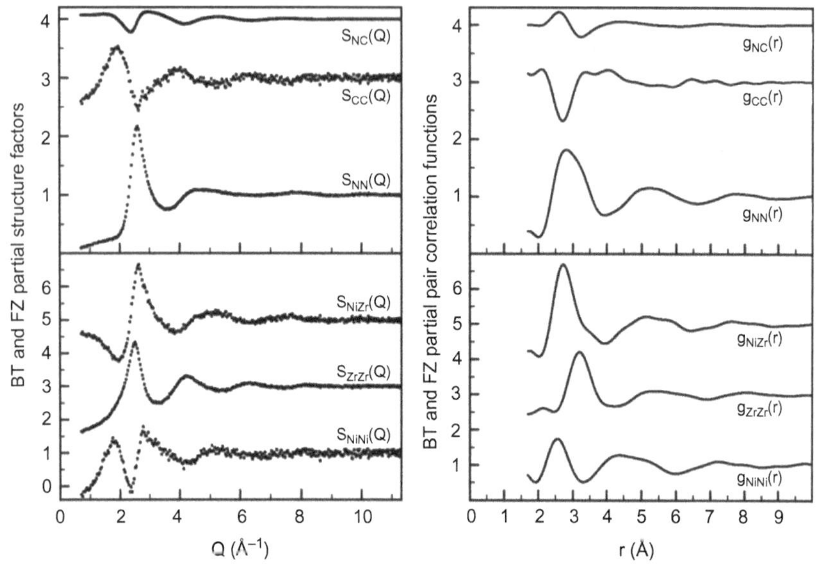

FIGURE 12 Left—static partial Bhatia-Thornton (upper part) and Faber-Ziman (lower part) structure factors of liquid $Ni_{36}Zr_{64}$ at $T=1375$ K. Right—Partial BT (upper part) and FZ (lower part) pair-correlation functions of liquid $Ni_{36}Zr_{64}$ at $T=1375$ K.

Nearest-neighbor coordination numbers $CN_{NiNi}=2.5$, $CN_{ZrZr}=10.4$, $CN_{ZrNi}=8.8$, and $CN_{NiZr}=5.0$ were calculated by integrating the FZ partial radial distribution functions $RDF_{\alpha\beta}(r)$ ($\alpha,\beta=$Zr,Ni) around their first maximum. This leads to an average coordination number CN = 13.9 in agreement with the nearest-neighbor coordination number $CN_{NN}=13.8$, calculated from the BT $RDF_{NN}(r)$. The latter describes the average number of the nearest neighbors of an arbitrary atom without distinguishing the different elements. This value is much higher than the typical values of CN = 12 reported for most metallic melts [37,71] including the melts of the pure components Ni and Zr presented in the previous section (see Table 1).

All these results (distances and coordination numbers) indicate a higher local density of packing in molten $Ni_{36}Zr_{64}$ than in pure liquid Ni and Zr.

This study allowed to conclude that $Ni_{36}Zr_{64}$ melts exhibit a pronounced chemical short-range order with preferred Ni-Zr nearest neighbors. This study shows also that, contrary to most of the other metallic melts investigated so far, no indications for the existence of an icosahedral short-range order in stable $Ni_{36}Zr_{64}$ melts were found. This has been attributed to the larger difference of the atomic radii of Ni and Zr ($r_{Zr}/r_{Ni}=1.29$). For more details, see Ref. [70].

10.5.1.2 Experiments Using ESL

ESL has been also applied to study various liquid alloys including for instance Zr-Cu [72] or Ni-Nb [73]. For comparing with the EML results shown above, this section presents also a study of $Zr_{64}Ni_{36}$ melts performed by Kordel

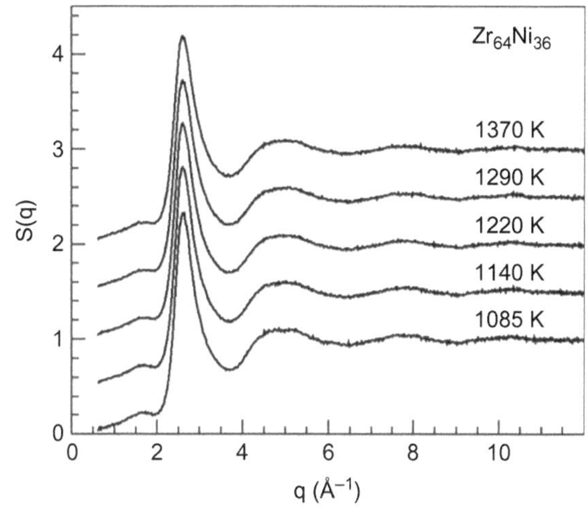

FIGURE 13 Total structure factors of liquid $Zr_{64}Ni_{36}$ for different temperatures in the stable liquid and down to about 200 K below the melting temperature ($T_M = 1283$ K). The curves are shifted by multiples of 0.5 for clarity. *Figure reproduced with permission from Ref. [38].*

et al. [38]. Neutron diffraction experiments were performed at the D20 diffractometer at the ILL using a neutron wavelength of 0.94 Å. The scattered intensity was measured over a 2θ angular range of 4–134° giving an accessible Q-range of approximately 0.6–12 Å^{-1}.

Figure 13 shows the static total structure factors of liquid $Zr_{64}Ni_{36}$ for different temperatures ranging from 1085 to 1370 K. They were determined from the measured diffractograms using the data treatment described previously (Section 10.2.1.2) and detailed in Ref. [60]. In this range of temperatures, the peak positions and the general shape of the structure factor do not change significantly, and only small changes are observed in the amplitudes of the oscillations, indicating that the short-range order remains essentially the same. As observed also for other metallic melts [37,68], the characteristic features of the structure factors become more pronounced, if the temperature is decreased.

In the previous section, we have presented a study of the short-range order of stable $Zr_{64}Ni_{36}$ melts at $T = 1375$ K performed at the same diffractometer (D20), using the EML method (Ref. [70]). The use of isotopic substitution of Ni made it possible to determine the partial structure factors.

For comparing the results obtained using the two different levitation methods, Figure 14A presents the total S(Q) of $Zr_{64}Ni_{36}$ measured with natural Ni using the EML technique and the measurement performed with ESL at essentially the same temperature ($T = 1370$ K). While both curves have the same shape, it is evident that the oscillations in the structure factor measured by ESL are slightly more pronounced.

This difference has been attributed to the contribution of secondary scattering from the coil when the sample is in place and thus not removed by

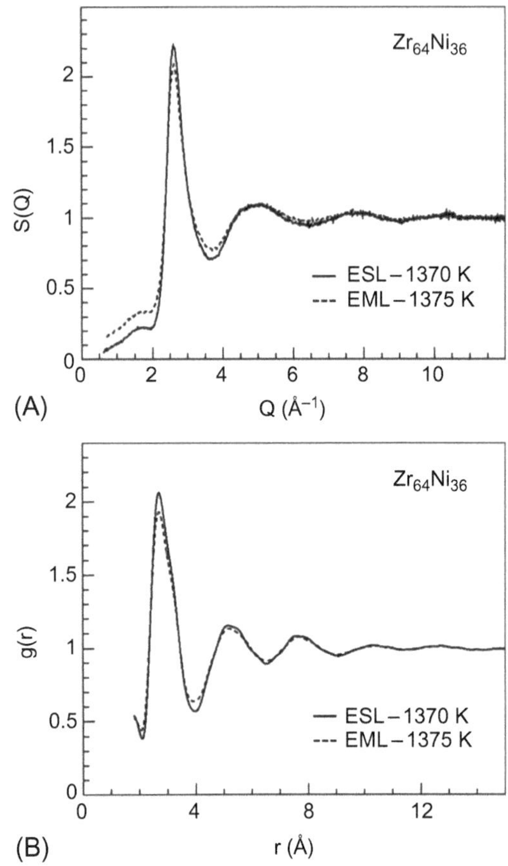

FIGURE 14 Total static structure factor (A) and total pair-correlation function (B) of $Zr_{64}Ni_{36}$ using the EML and the ESL techniques.

the background subtraction using the signal measured with the empty levitator. Taking this into account leads to a good agreement between both measurements.

The effect of the secondary scattering on the total pair-correlation functions $g(\mathrm{r})$ is shown in Figure 14B where both curves differ only in the amplitude of the oscillations. This means that interatomic distances derived from both measurements are identical. It was found that the influence on the coordination numbers calculated from $RDF(\mathrm{r})$ is also negligible.

This shows that, despite the limitations of the EML technique due to secondary scattering at the water-cooled levitation coil, the main results on the short-range order are essentially not affected by these problems and thus results obtained using ESL and EML lead to the same conclusions.

A recent study performed on the $Zr_{64}Ni_{36}$ alloys using ESL at the NOMAD instrument (SNS) gives similar results [32].

10.5.1.3 Experiments Using CNL

CNL can be also used for studying liquid metals or semiconductors. In the past, the method has been combined with synchrotron radiation to study the structure of various melts such as Si [74], Ni [52], Zr [75], or Ti-Zr-Ni alloys [76].

Even if this levitation method is not ideal for studying metallic melts using neutron diffraction compared with EML and ESL, good results can, however, be obtained. In particular, Gruner et al. applied this technique for studying the structure of some Ni-Si alloys using NDIS at the D4c instrument at the ILL [77].

In this work, spherical samples (2.5 mm in diameter) were made from NiSi and $NiSi_2$ alloys, prepared by mixing pure silicon ($b=4.15$ fm) with different nickel isotopes: (i) natural nickel ^{nat}Ni ($b=10.31$ fm), (ii) ^{58}Ni ($b=14.4$ fm), (iii) ^{60}Ni ($b=2.8$ fm), and (iv) ^{62}Ni ($b=-8.7$ fm).

All measurements were performed at about 50 K above the liquidus temperature of the two studied alloys. The incident wavelength, calibrated using a crystalline Ni reference sample, was $\lambda=0.6975$ Å leading to a scattering vector Q-range of 0.25–16.8 $Å^{-1}$.

The total structure factors obtained with both alloys are shown in Figure 15, and the corresponding total pair-correlation functions $g(r)$ calculated using Eq. (4) are presented in Figure 16. The use of the ^{62}Ni isotope with its negative scattering length results in unusual weighting factors for the two samples and enhanced amplitudes in the total structure factors. This isotope leads also to negative weighted contributions of Ni-Si pairs to the ^{62}Ni data, allowing to make various observations.

Like atoms contribute to the total $S(Q)$ at a Q-value of about 2 $Å^{-1}$ and the prepeak in the total structure factors, most pronounced in the ^{58}Ni data, also found at about 2 $Å^{-1}$, was attributed to Ni-Ni pairs. In the r-space, unlike atoms approach closer than like atoms, clearly seen in a negative contribution in the ^{62}Ni-g(r) at about 2.3 Å.

The procedure for extracting the partial structure factors is well described in Gruner et al. [77]. In order to use all the four available data sets simultaneously, Reverse Monte Carlo (RMC) simulations (see Refs. [78,79] for details) were performed. From the model configuration, the partial pair-correlation functions $g_{\alpha\beta}(r)$ are calculated (Figure 17), and by Fourier transformation, the partial structure factors $S_{\alpha\beta}(Q)$ can be derived and were in good agreement with the experimental partial structure factors. This finding encouraged the use of the model obtained from the RMC procedure for the full analysis.

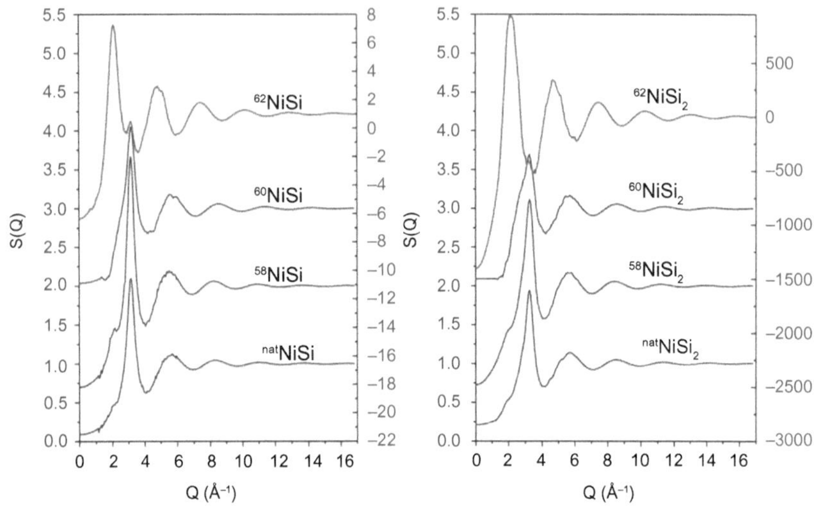

FIGURE 15 Total structure factors of the liquid NiSi (left) and $NiSi_2$ (right) alloys. The right ordinates apply only for the ^{62}Ni measurement. For clarity, the curves are shifted by one.

The bond distances and the partial coordination numbers for all atomic pairs calculated with the RMC simulations are listed in Table 2. As observed on the total *g*(r)s, unlike atoms approach each other closer than the like atoms in both alloys. The Ni-Si first coordination shell is rather sharp compared with those of the other pairs. The oscillation corresponding correlations between Ni atoms range up to 15 Å, while the correlations for the other pairs are rather limited. It can be also noticed that the first Si-Si shell is split into two subshells in the case of the NiSi alloys.

This study concludes that both alloys exhibit a strong tendency to heterocoordination in the first coordination shell and a certain charge transfer between the species is observed. It seems that the atomic structure of the liquid NiSi alloy is strongly influenced by Si's tendency to forming covalent bonds while the structure of liquid $NiSi_2$ moreover resembles that of the ordered solid alloy.

10.5.2 Study of Oxide Melts

Contrary to metals, simple oxides contain at least two elements giving three different partial pairs. In the past, some of them have been studied using neutron diffraction combined with CNL: Al_2O_3 [40,80] and Y_2O_3 [81,82]. However, in most cases, the studied samples are mixtures of several oxides leading to a large number of atomic pairs. This is the case of the CaO-Al_2O_3-SiO_2 (CAS) ternary phase diagram in which glasses and melts are of both technological and scientific interest. There are 10 different atomic pairs in CAS

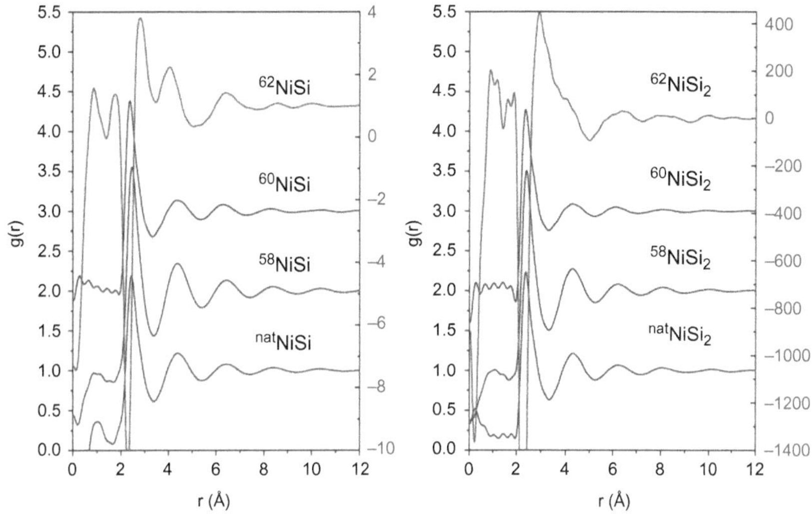

FIGURE 16 Total pair-correlation functions of the liquid NiSi (left) and $NiSi_2$ (right) alloys obtained by Fourier transformation of the $S(Q)$s. The right ordinates apply only for the 62Ni measurement. For clarity, the curves are shifted by one.

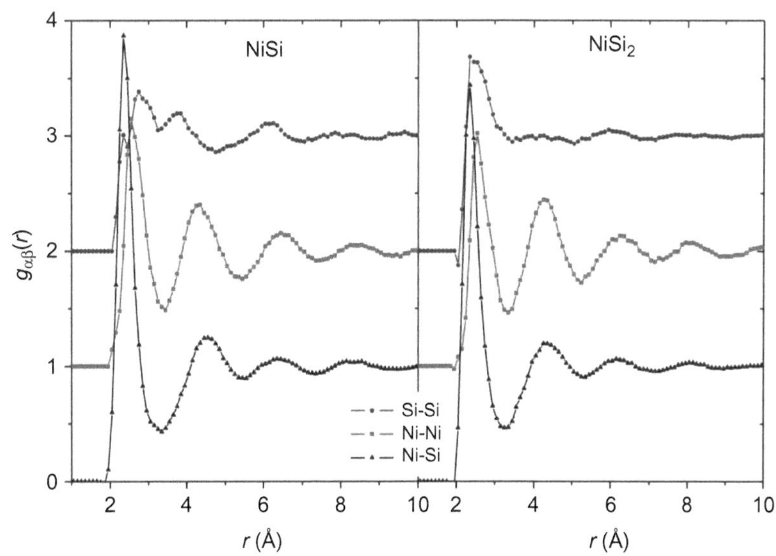

FIGURE 17 Partial pair-correlation functions of the NiSi (left) and $NiSi_2$ (right) liquid alloys obtained from the RMC model. The curves are shifted by one for clarity.

TABLE 2 Main Structural Parameters: Radii of the First Two Coordination Spheres and Partial Coordination Numbers in the First Shell

	NiSi				$NiSi_2$			
	Ni-Ni	*Ni-Si*	*Si-Ni*	*Si-Si*	*Ni-Ni*	*Ni-Si*	*Si-Ni*	*Si-Si*
r^{I} (Å)	2.59	2.37		2.77	2.54	2.35		2.53
				3.72				
r^{II} (Å)	4.35	4.54		6.16	4.28	4.38		6.03
r^{II}/r^{I}	1.68	1.92		2.22	1.69	1.86		2.38
CN^{I}	4.9	5.4	5.4	4.2	2.5	6.5	3.2	4.7

From Ref. [77].

compounds. In order to have precise structural information, it is thus important to study the binary oxides first.

This section presents results obtained for some compositions in the two joins, $CaO\text{-}Al_2O_3$ and $CaO\text{-}SiO_2$.

10.5.2.1 *Experiments at D4*

This section presents a neutron diffraction study performed by Drewitt et al. on three $(CaO)_x(Al_2O_3)_{1-x}$ samples (2.7 mm in diameter), denoted CA2 ($x=0.33$), CA ($x=0.50$), and C3A ($x=0.75$) [83]. The melting points of CA2, CA, and C3A are 2045, 1878, and 1814 K, and diffraction measurements were taken for the liquid samples at 1900 K (CA) and 2170 K (CA2 and C3A). The data treatment was performed using the procedure detailed in Section 10.2.1.2.

The total structure factors $S(Q)$ and the corresponding total pair-distribution functions $g(r)$ for all liquids are plotted in Figure 18 Molecular dynamic (MD) computer simulations from Ref. [83] are also presented and show a good overall agreement with the experimental measurements in both Q- and r-spaces.

For all samples, the first peak at $r=1.78$ Å in $g(r)$ functions can be attributed to the nearest-neighbor Al-O correlations. For the C3A sample, the peak found at 2.35 Å arises from Ca-O bonds. Decreasing the CaO concentration, this peak is found at about 2.30 Å in CA and barely resolvable from the next correlations. It is no more distinguishable in CA2.

The next peak found around 2.95 Å arises from O-O bonds.

The average Al-O coordination numbers CN_{AlO} were obtained by integrating over the first peak in $RDF(r)$. A value $CN_{AlO}=4.3\pm0.2$ was obtained

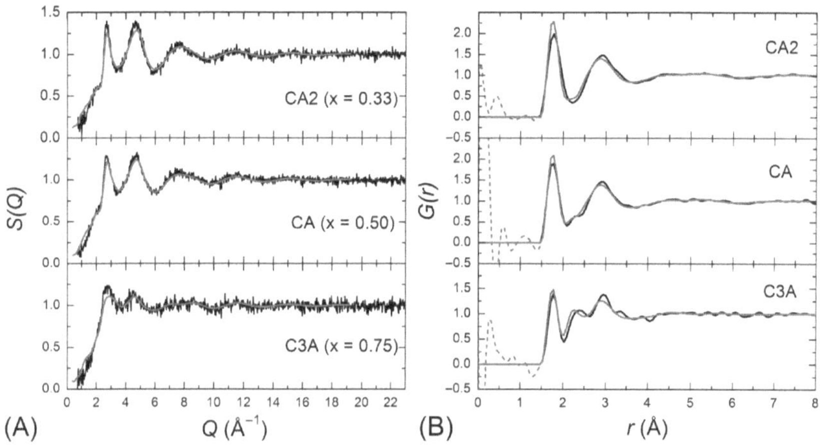

FIGURE 18 The total structure factors $S(Q)$ (A) and corresponding total pair-distribution functions $g(r)$ (B) for the $(CaO)_x\ (Al_2O_3)_{1-x}$ liquids obtained from neutron diffraction. The $S(Q)$ and $g(r)$ functions obtained from MD simulations reported in [83] are also shown. In figure (B), the broken light-gray curves show the extent of the unphysical low-r features.

for the CA2 composition indicating the presence of predominantly fourfold coordinated Al-O but with a small concentration of higher-coordinated species. However, the Al-O peak experiences a degree of overlap from the Ca-O correlation, which may lead to an overestimate of the value. For the CA composition, CN_{AlO} remains at 4.3 ± 0.2 and reduces to 4.0 ± 0.2 for C3A.

For CA and C3A, the Ca-O peak in $RDF(r)$ yields an average coordination number $CN_{CaO} = 3.9 \pm 0.5$ and 4.0 ± 0.5, respectively. However, the Ca-O peak experiences significant overlap by the O-O correlation. In particular, it is not possible to calculate CN_{CaO} for CA2.

For comparison, the MD simulation results give average coordination numbers CN_{AlO} of 4.20 (CA2), 4.13 (CA), and 4.04 (C3A) and CN_{CaO} of 6.40 (CA2), 6.20 (CA), and 5.62 (C3A).

While the experimental values for the Al-O correlations are in good agreement with the simulations, the coordination numbers are largely underestimated for CaO pairs. This is mainly due the overlapping contribution of the O-O pair that hides the asymmetry of the CaO correlations, which corresponds to higher-coordinated Ca, leading to lower coordination numbers.

A solution to go further is to use more selective techniques, and the next paragraph presents a work of Drewitt et al. who completed the previous study by a precise analysis of CA ($CaAl_2O_4$) using NDIS [84].

The experiments were performed at the D4c diffractometer with an incident wavelength of 0.4978 Å giving a maximal Q-value of 23.45 Å^{-1}. Diffraction patterns were measured in the liquid state at 1973 K. The counting times were of 24 h for each sample. Additional experimental details can be found in Ref. [84].

Three samples of $CaAl_2O_4$ were prepared using ^{nat}Ca [$b_{nat}=4.70$ fm], ^{44}Ca [$b_{44}=1.45$ fm], and a 50:50 mixture of the two ^{mix}Ca [$b_{mix}=3.08$ fm].

The samples used in this study were synthesized by melting the mixed powders in a platinum crucible. The resulting ceramics were melted using a CO_2 laser and rapidly cooled to form optically transparent and colorless spherical glass with a diameter of about 2.54 mm.

From the experiment, three different total interference functions denoted $^{nat}F(Q)$, $^{44}F(Q)$, *and* $^{mix}F(Q)$ were obtained at high temperature.

It should be noted that as the samples contain six different pairs, making only three experiments is not sufficient to obtain all partial functions as for binary alloys (see Section 10.5.1). However, the procedure described below makes it possible to derive important additional structural information on these compositions.

Contrary to $S_{\alpha\beta}$(Q) functions containing Ca that have different weighting factors, the $S_{\alpha\beta}$(Q) with $\{\alpha,\beta\}\neq$Ca have identical weights and can therefore be eliminated by taking a difference function such as

$$\begin{aligned}\Delta^{Ca}(\mathrm{Q}) &= {}^{nat}F(\mathrm{Q}) - {}^{44}F(\mathrm{Q}) \\ &= c_{Ca}^2\left(b_{nat}^2 - b_{44}^2\right)[S_{CaCa}(\mathrm{Q}) - 1] + 2c_{Ca}\left(b_{nat} - b_{44}\right)\delta^{Ca}(\mathrm{Q}) \end{aligned} \quad (28)$$

where δ^{Ca}(Q) involves only $S_{Ca\beta}$(Q) with $\beta\neq$Ca

$$\delta^{Ca}(\mathrm{Q}) = c_{Al}b_{Al}[S_{CaAl}(\mathrm{Q}) - 1] + c_O b_O[S_{CaO}(\mathrm{Q}) - 1] \quad (29)$$

$b_{Al}=3.449$ and $b_O=5.803$ fm [9]. Alternatively, the $S_{Ca\beta}$(Q) with $\beta\neq$Ca can be eliminated by taking a weighted difference function such as

$$\Delta^x(\mathrm{Q}) = \frac{b_{nat}\,{}^{44}S(\mathrm{Q}) - b_{44}\,{}^{nat}S(\mathrm{Q})}{b_{nat} - b_{44}} = \delta^x(\mathrm{Q}) - c_{Ca}^2 b_{nat} b_{44}[S_{CaCa}(\mathrm{Q}) - 1] \quad (30)$$

where δ^x(Q) involves only correlations with $\{\alpha,\beta\}\neq$Ca

$$\delta^x(\mathrm{Q}) = c_{Al}^2 b_{Al}^2[S_{AlAl}(\mathrm{Q}) - 1] + c_O^2 b_O^2[S_{OO}(\mathrm{Q}) - 1] + 2c_{Al}c_O b_{Al} b_O[S_{AlO}(\mathrm{Q}) - 1] \quad (31)$$

Moreover, δ^x(Q), δ^{Ca}(Q), and S_{CaCa}(Q) can be extracted individually by solving the matrix equation:

$$\begin{bmatrix} {}^{nat}F(\mathrm{Q}) \\ {}^{mix}F(\mathrm{Q}) \\ {}^{44}F(\mathrm{Q}) \end{bmatrix} = \begin{bmatrix} c_{Ca}^2 b_{nat}^2 & 2c_{Ca}b_{nat} & 1 \\ c_{Ca}^2 b_{mix}^2 & 2c_{Ca}b_{mix} & 1 \\ c_{Ca}^2 b_{44}^2 & 2c_{Ca}b_{44} & 1 \end{bmatrix} \cdot \begin{bmatrix} S_{CaCa}(\mathrm{Q}) - 1 \\ \delta^{Ca}(\mathrm{Q}) \\ \delta^x(\mathrm{Q}) \end{bmatrix} \quad (32)$$

The real-space difference functions, denoted by ΔG^{Ca}(r), δG^{Ca}(r), ΔG^x(r), and δG^x(r), are obtained by replacing $S_{\alpha\beta}$(Q) by its corresponding partial pair-distribution $g_{\alpha\beta}$(r) in Eqs. (28)–(31), respectively.

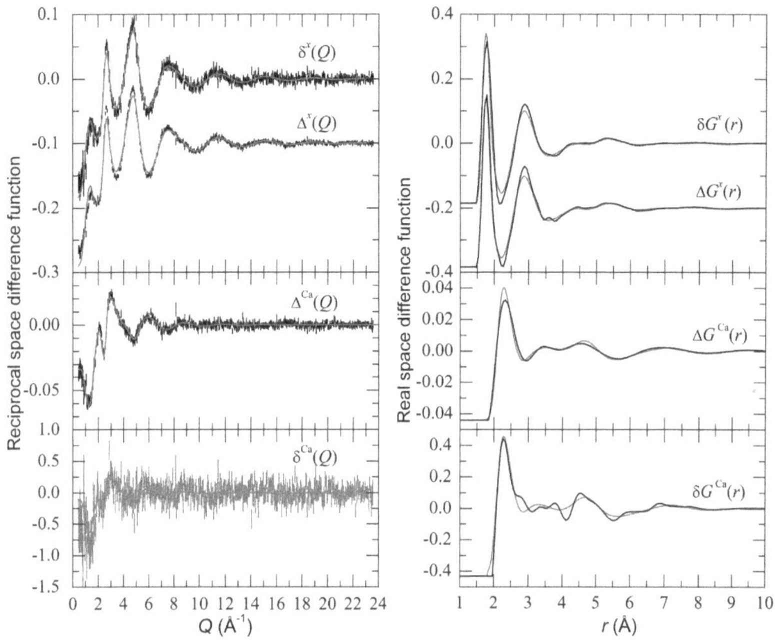

FIGURE 19 Left: The reciprocal-space difference functions for liquid $CaAl_2O_4$. The solid (red) curves show the difference functions calculated from the partial structure factors generated from the MD simulations. Right: The real-space difference functions as obtained by Fourier transforming the corresponding reciprocal-space functions (left). The red curves show the functions calculated from the partial pair-distribution functions generated directly from the MD simulations.

The difference functions obtained in the reciprocal and real spaces are shown in Figure 19 together with MD simulations of the liquid at 2500 K, previously reported in Ref. [83]. The difference functions from MD are in good overall agreement with the diffraction results.

The first and second peaks in $\Delta G^x(\mathrm{r})$ and $\delta G^x(\mathrm{r})$ arise from nearest-neighbor Al-O and O-O correlations, respectively. For the liquid, they give distances of $r_{AlO}=1.77$ Å and $r_{OO}=2.89$ Å such that the mean O-Al-O bond angle is 108.9° as compared with 109.47° for a regular tetrahedron.

As observed with the previous classical neutron diffraction, the Al-O distance is consistent with a structure made predominantly from AlO_4 and AlO_5 polyhedra [85] such that the measured coordination number for O around Al, $CN_{AlO}=4.2$, points to 80% AlO_4 and 20% AlO_5 units. A small tail on the Al-O peak at ~2 Å may, however, indicate a small fraction of AlO_6 units.

The first peak in $\Delta G^{Ca}(\mathrm{r})$, $\delta G^{Ca}(\mathrm{r})$, is attributed to nearest-neighbor Ca-O correlations and gives a distance $r_{CaO}=2.30$ Å and coordination number $CN_{CaO}=6.0$ in the liquid state. These results are in good agreement with the MD simulations also reported in Figure 19, which find $r_{CaO}=2.29$ Å and $CN_{CaO}=6.2$.

10.5.2.2 Experiments at SNS

The CNL system at NOMAD has been used for a variety of experiments on solid and liquid oxides [30,31,80,86] and metallic alloys [87]. Some experiments have used isotopic substitution to measure first-order difference functions in melts. By exploiting the different neutron scattering power of isotopes of some elements, it is possible to extract partial structure factors from the data. The first sample that was investigated was calcium silicate $CaSiO_3$ composition.

Samples were prepared using ^{43}Ca and natural Ca to provide neutron contrast. Molten samples were run at a temperature of 1650°C for a period of 3 h. Compared with the glass at ambient temperature, the Ca-O peak in the pair-distribution function is significantly broadened in the liquid and extends to longer radial distances. The difference function is in good agreement with molecular dynamics simulations. The results are consistent with the formation of CaO_6 and CaO_7 "channels" in the liquid.

Measurements on molten rare-earth oxides were used to investigate the coordination number of cations in liquids and glasses compared with the corresponding crystals [76]. The data show that in many cases there is a large decrease in average cation coordination when the crystal melts.

Ushakov et al. [86] investigated the thermal expansion and phase content of rare-earth zirconate crystals at temperatures above 1500°C where container reactions would hamper measurements using conventional methods. Recent work at the Advanced Photon Source has used a "rotating nozzle" CNL that

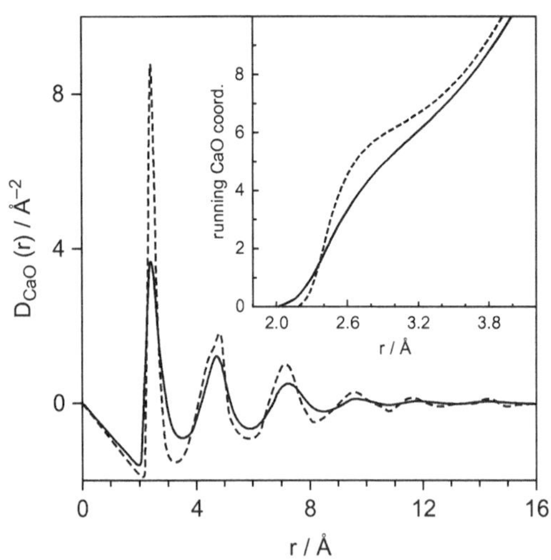

FIGURE 20 Partial distribution functions for Ca-O. The black line shows data for the liquid at 1650°C and the blue line for the glass at ambient temperature. The inset compares the running coordination number in the liquid and glass. *Reprinted with permission from (J. Phys. Chem. B, 116 (2012) 13439). Copyright (2012) American Chemical Society (Ref. [31]).*

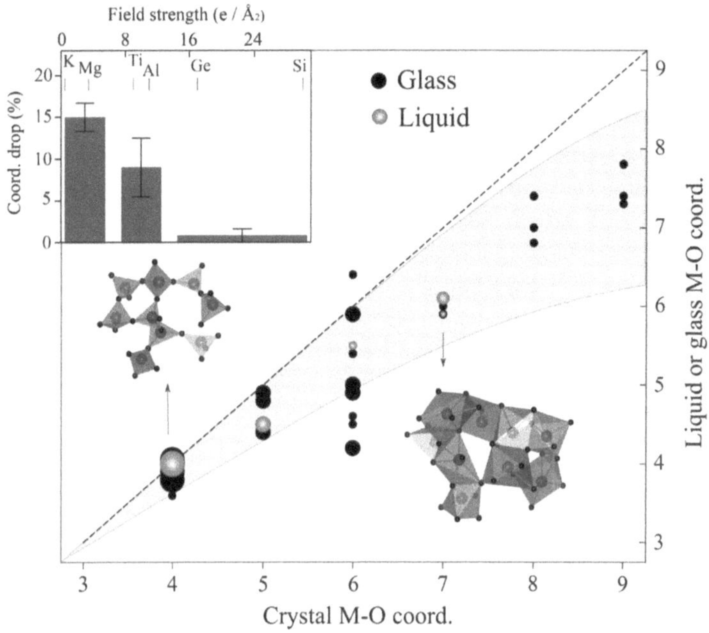

FIGURE 21 Cation coordination in the glasses and liquids *versus* the corresponding crystals. The inset shows the decrease in coordination as a function of the cation field strength. *Figure reproduced with permission from Ref. [82].*

can improve the quality of data from polycrystalline materials by rotating the sample during the measurement [87].

Santodonato et al. [88] used a combination techniques including neutron diffraction, x-ray diffraction, and transmission electron microscopy to investigate phase selection in high-entropy metallic alloys. Aerodynamic levitation was used to study samples at high temperatures above the range that can be accessed using conventional furnaces.

Overall, combining the high neutron flux available at NOMAD with aerodynamic levitation provides a powerful tool for measurements of structure of materials in extreme conditions. The levitator is now in the SNS user program, and work is continuing to investigate a variety of materials. Recent results of experiments on molten titanates are being analyzed. The use of isotopic substitution is very effective for titanium-bearing compositions because of the high contrast between ^{46}Ti and ^{48}Ti bound coherent scattering lengths. The details of the levitator design and operation are continuously being improved to obtain better signal to background and enable faster measurements on a wider variety of materials. The potential for work with redox atmospheres is being explored to provide complementary data to measurements currently being made at the Advanced Photon Source using high-energy x-ray diffraction and XANES spectroscopy.

10.6 DYNAMICAL INVESTIGATIONS

10.6.1 Study of Metallic Melts Using EML

10.6.1.1 Monoatomic Liquids

This section presents a study of the dynamics of levitated nickel melts using QENS in which A. Meyer et al. determined the diffusion coefficient at various temperatures in the liquid and undercooled states [89]. Few metals exhibit relatively large incoherent scattering cross sections, and in addition to Ni, only liquid Ti has been also studied by combining EML with QENS analysis [90].

The experiments were performed at the TOFTOF instrument using an incident neutron wavelength of 5.4 Å giving an accessible Q-range of 0.4–2.2 $Å^{-1}$ at zero energy transfer, well below the position of the first structure factor maximum of liquid Ni at about 3 $Å^{-1}$ (see Section 10.5.1). The measured intensity is then dominated by the incoherent scattering. The resolution function was measured on a solid vanadium sample at room temperature and modeled using a Gaussian function with an energy resolution of about 78 μeV full width at half maximum (FWHM).

Spherical Ni samples with a diameter of 8 mm were levitated under a He/4% H_2 gas mixture. The QENS measurements were performed at different temperatures ranging from 200 K above to more than 200 K below the melting point (1726 K). The acquisition times were 3 h for temperatures below the melting point and 1 h for temperatures above the melting point.

Figure 22 displays the dynamic structure factor $S(Q,\omega)$ for different temperatures and values of the momentum transfer Q. All spectra are well described by fits with Eq. (16).

Figure 23 shows the Q^2 dependence of the width $\Gamma_{1/2}$ (FWMH) of the quasielastic line for different temperatures ranging from 1514 to 1940 K with linear fits to the small q part of the curves. As observed in multicomponent glass-forming alloys [91,92], the linear Q^2 dependence of $\Gamma_{1/2}$ holds up to about 1.5 $Å^{-1}$. At larger Q-values, the coherent scattering contributes strongly to the signal. Using Eq. (17), the Ni self-diffusion coefficient D is derived from the slope of the linear fit for each temperature. The obtained values (Figure 24) ranging from $(2.09 \pm 0.08) \times 10^{-9}\ m^2/s$ (1514 K) to $(4.54 \pm 0.23) \times 10^{-9}\ m^2/s$ (1940 K) are in excellent agreement with the results obtained previously using Al_2O_3 containers in a limited range of temperatures (1735–1795 K) [2].

In the studied temperature range, the temperature dependence of the diffusion coefficients D is well described by an Arrhenius law (Eq. 18) with an activation energy $E_A = 0.47 \pm 0.03$ eV/at and a prefactor $D_0 = (77 \pm 8) \times 10^{-9}\ m^2\ s^{-1}$ (dashed line in Figure 24). This shows that the diffusive dynamics in liquid Ni above the melting point and in the undercooling regime are essentially the same at the atomic scale.

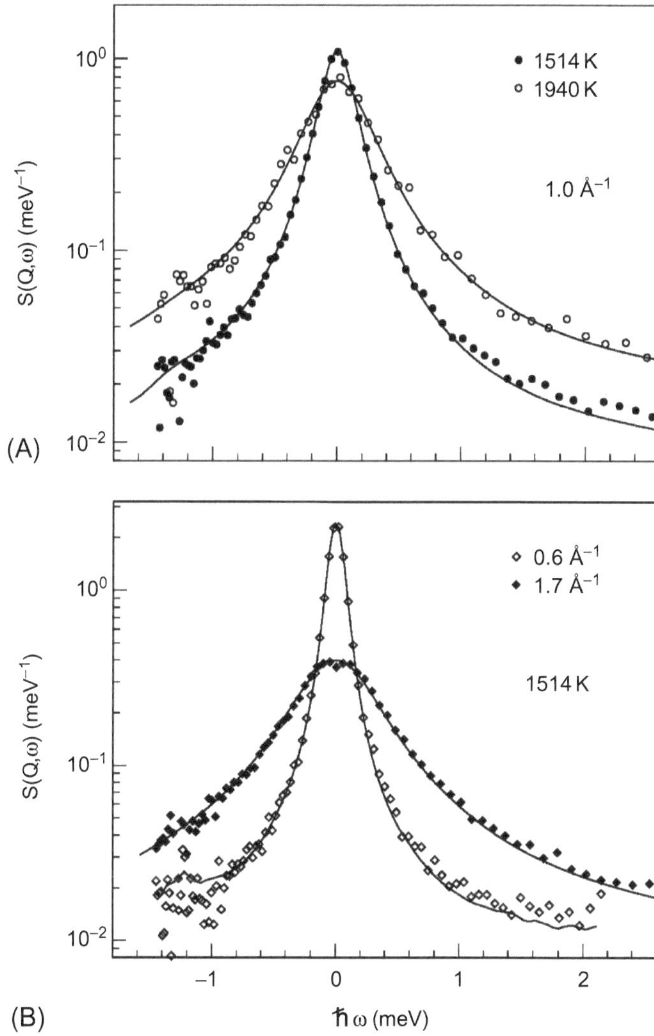

FIGURE 22 Dynamic structure factor $S(Q,\omega)$ of liquid Ni: (A) at fixed momentum transfer $Q = 1\ Å^{-1}$ for the two extreme temperatures (1514 K and 1940 K) and (B) at $T = 1514$ K (undercooled state) for two Q-values (0.6 and 1.7 $Å^{-1}$). The lines are fits with a Lorentz function convoluted by the instrumental energy resolution function. *Figure reproduced with permission from Ref. [89].*

This study shows that the combination of the EML technique with quasielastic neutron scattering makes it possible to determine self-diffusion coefficients in a wide range of temperatures, including deep undercooling regimes thanks to the absence of heterogeneous nucleation from crucible walls. More details about this study can be found in Meyer et al. [89]. As we will see in the next paragraph, the developed technique was further used for studying metallic alloy melts.

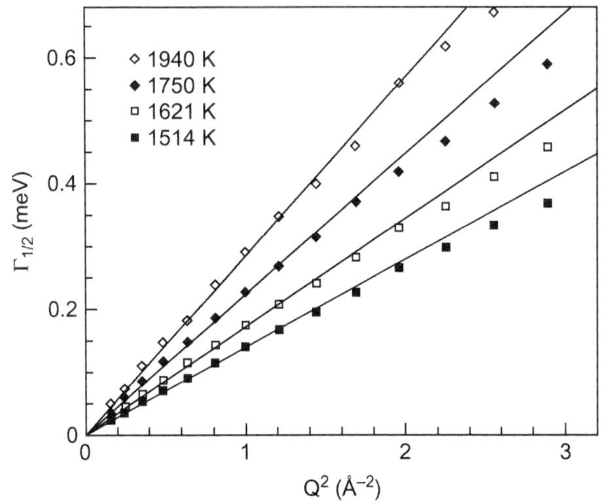

FIGURE 23 Q^2 dependence of the linewidth of the scattering function $S(Q,\omega)$ for different temperatures. Lines are fits with Eq. (**17**) in the Q-range of 0.6–1.5 $Å^{-1}$. *Figure reproduced with permission from Ref. [89].*

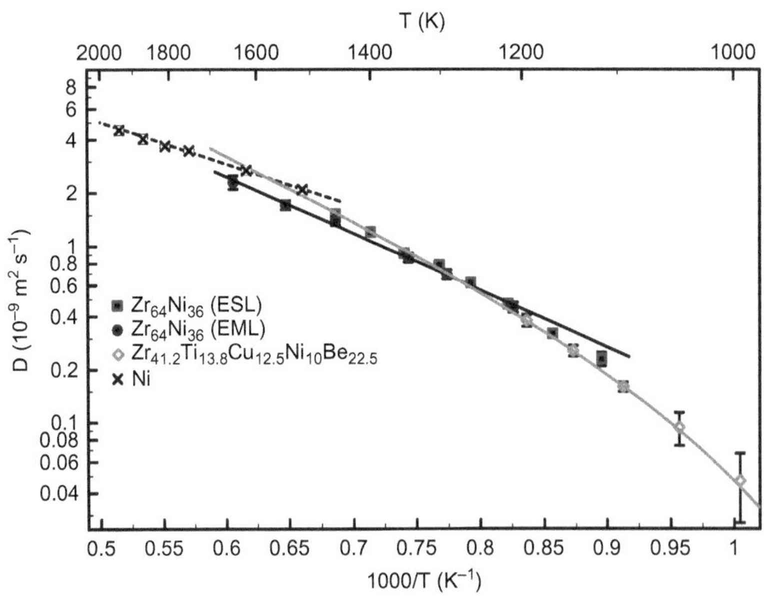

FIGURE 24 Temperature dependence of the Ni self-diffusion coefficient in liquid $Zr_{64}Ni_{36}$ processed by the EML [70] and the ESL technique, self-diffusion coefficient in pure liquid Ni and the average Ni, and Ti self-diffusion coefficient in the bulk metallic glass-forming alloy $Zr_{41.2}Ti_{13.8}Cu_{12.5}Ni_{10}Be_{22.5}$ (Vitreloy 1). The gray solid line represents a fit of a Tc scaling law to the Vit1 data with $Tc = 850$ K and $\gamma = 2.5$.

10.6.1.2 Metallic Alloys

The dynamics of various metallic alloys have been studied using EML, including, for example, Al-Ni [93] or Ni-Si [94] melts. This paragraph summarizes a study performed by Holland-Moritz et al. [70] on levitated liquid $Ni_{36}Zr_{64}$. The QENS experiments were performed at the TOFTOF instrument using an incident neutron wavelength of 5.1 Å. The energy resolution was 95 μeV (FWHM) at zero energy transfer.

The measurements were performed at different temperatures ranging from above the liquidus temperature T_L (1283 K) down to temperatures in the undercooled state. Some of these measurements are presented in Figure 25.

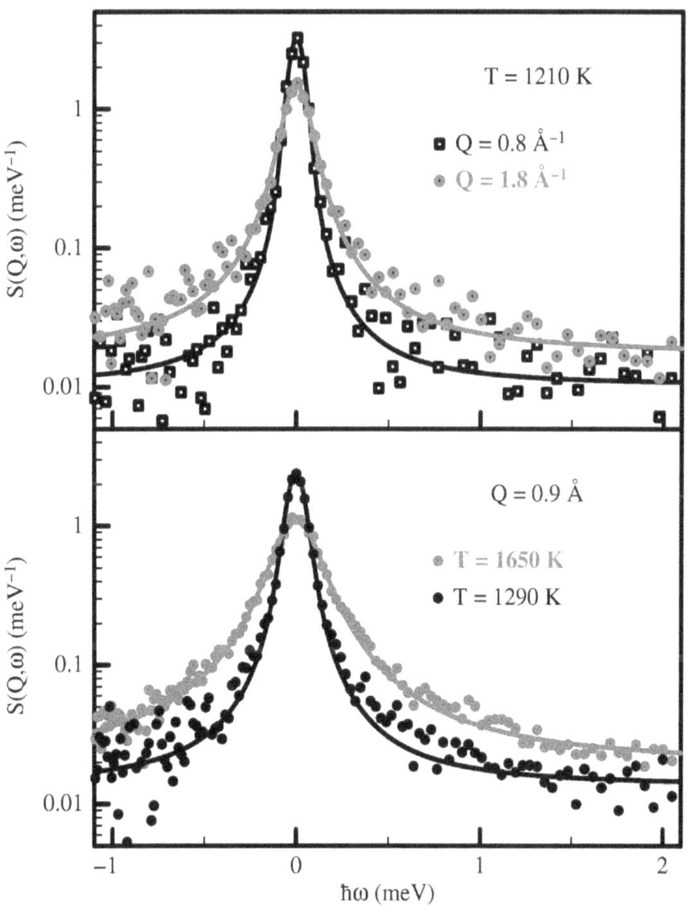

FIGURE 25 Dynamic structure factors for liquid $Ni_{36}Zr_{64}$ at $T = 1210$ K for two different scattering vectors Q = 0.8 and 1.8 $Å^{-1}$ (upper part) and at Q = 0.9 $Å^{-1}$ for two different temperatures $T = 1290$ and 1650 K (lower part). *Figure reproduced with permission from Ref. [70].*

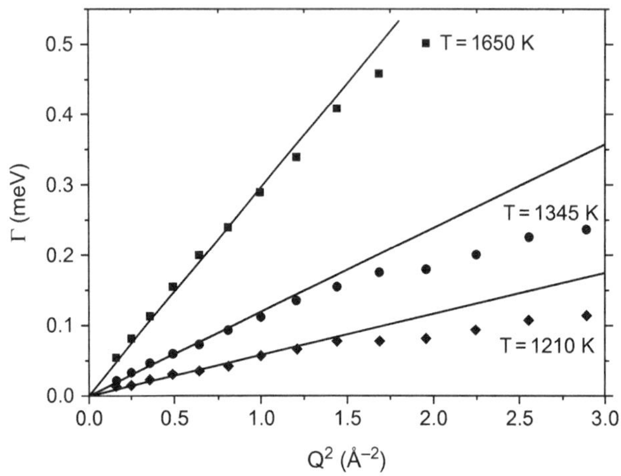

FIGURE 26 Full width at half maximum Γ of the Lorentzian curves as a function of Q^2 for liquid $Ni_{36}Zr_{64}$ at three different temperatures.

The bottom part of the graph shows the dynamic structure factors $S(Q,\omega)$ of $Ni_{36}Zr_{64}$ at two temperatures around $T = 1290$ K and well above the melting point ($T = 1650$ K) at a momentum transfer $Q = 0.9$ Å^{-1}. The upper part of Figure 25 shows the signal obtained in the undercooled state ($T = 1210$ K) at two different values of momentum transfer ($Q = 0.8$ and 1.8 Å^{-1}).

In all cases, the quasielastic signal is well described with a Lorentzian function convoluted with the instrumental energy resolution function measured with a vanadium sample (see the line plots in Figure 25).

Figure 26 shows the evolution of Γ as a function of Q^2 for different temperatures ranging from 1650 down to 1210 K. Since the incoherent scattering cross section of ^{nat}Ni (5.2 b) is much higher than for Zr (0.02 b), for momentum transfer below $Q \sim 1.2$ Å^{-1}, the dynamic structure factors are dominated by the incoherent scattering of the Ni atoms. Therefore, the slope of the $\Gamma(Q^2)$ curves at small Q (Figure 26) is directly related to the Ni self- diffusion coefficient, giving values (Figure 28) ranging from $D = (0.45 \pm 0.06) \times 10^{-9}$ m^2/s (1210 K) to $(2.3 \pm 0.2) \times 10^{-9}$ m^2/s (1650 K).

In the studied temperature range, the temperature dependence of the diffusion coefficient exhibits also an Arrhenius-type behavior (solid black curve in Figure 28) with an activation energy $E_A = 0.64 \pm 0.02$ eV and a pre-exponential factor of $D_0 = (2.1 \pm 0.3) \times 10^{-7}$ m^2/s. This value of E_A for the Ni self-diffusion is larger than the values observed in other melts. For instance, in pure Ni, $E_A(\text{Ni}) = 0.47$ eV [89] (see previous section), and a value $E_A(\text{Al-Ni}) = 0.36$ eV has been reported in liquid $Al_{80}Ni_{20}$ [95]. This behavior has been related to the different short-range order in these systems.

10.6.2 Study of Metallic Melts Using ESL

This section describes QENS experiments performed on liquid $Zr_{64}Ni_{36}$ at the TOFTOF instrument by Kordel et al. [38]. Measurements were made using ESL at various temperatures above the melting point ($T_L = 1283$ K) and in the undercooled state.

Figure 27 shows the dynamic structure factors $S(Q,\omega)$ measured at 1460 K for two values of the momentum transfer (Q = 0.5 and 1.1 Å^{-1}). All spectra were recorded using an incident neutron wavelength of 7.0 Å, and with the experimental settings of the instrument, the resolution measured on a vanadium sample was 55 μeV (FWHM).

As previously, spectra were modeled using Lorentzian functions convoluted with the instrumental resolution, and the Ni self-diffusion was deduced from the slope of the Q^2 evolution of the HWHM at small Q-values.

In Figure 24, the temperature dependence of the Ni self-diffusion coefficients in liquid $Zr_{64}Ni_{36}$, measured using both ESL and EML technique [70] (previous section), is compared and shows a very good agreement. Thanks to the high vacuum used for the sample processing with ESL; it was possible to reach lower temperatures of undercooling than with EML.

Figure 24 shows also the Arrhenius law determined previously with EML (solid black line). It is noteworthy that the temperature range in which the self-diffusion coefficient exhibits an Arrhenius behavior is for temperatures largely above the critical temperature Tc of the mode-coupling theory, which

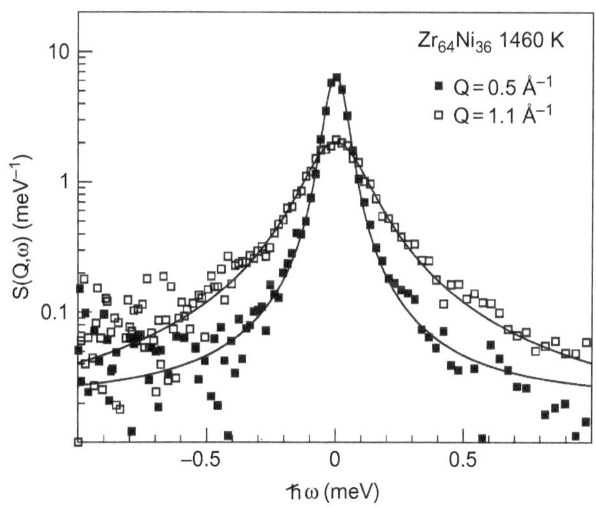

FIGURE 27 Dynamic structure factor $S(Q,\omega)$ of $Zr_{64}Ni_{36}$ at 1460 K for Q = 0.5 and 1.1 Å^{-1}. The lines are fits with a Lorentz function convoluted by the instrumental energy resolution function. *Figure reproduced with permission from Ref. [38].*

predicts a $[(T-T_c)/T_c]^\gamma$ temperature dependence of the self-diffusion coefficient when approaching a critical temperature T_c. This was observed previously with the average Ni and Ti self-diffusion coefficient of the bulk metallic glass-forming alloy $Zr_{41.2}Ti_{13.8}Cu_{12.5}Ni_{10}Be_{22.5}$ (Vitreloy 1) [92] (see gray solid line in Figure 24).

By using the ESL technique, larger undercoolings have been obtained on liquid $Zr_{64}Ni_{36}$, as compared with the EML measurements. At the lower temperatures, small deviations from the Arrhenius law become evident, indicating an approach of the critical temperature of mode-coupling theory.

10.6.3 Experiments Using CNL

10.6.3.1 *Study of Metallic Melts*

This section describes QENS measurements on $Ni_{75}Si_{25}$ melts performed at the IN6 instrument (ILL) by F. Demmel et al. [96].

The spherical samples of about 2.7 mm in diameter (75.2 mg) were made from weighted amounts of pure nickel and silicon melted in the levitator to spherical shape and homogeneous alloys. The pyrometer was calibrated at the melting point of the alloy ($T_L = 1513$ K).

The IN6 spectrometer was operated with an incoming neutron wavelength of 5.12 Å well above the Bragg cutoff of aluminum (nozzle material). The energy resolution, measured using a vanadium sample, is described by a Gaussian function with a FWHM = 0.077 meV.

QENS experiments were performed at four different temperatures: 1920, 1800, 1535, and 1420 K. In addition, a background measurement was taken with the empty levitator and an argon-gas-filled levitation chamber. All spectra were recorded for about 3–4 h per temperature.

Since the incoherent neutron cross section of Si is negligible, as for Ni-Zr alloys (previous section), the measured signal at small Q-values is dominated by the incoherent scattering of the Ni atoms.

Figure 28(A) shows the dynamic structure factor $S(Q,\omega)$ obtained at 1535 K with the $Ni_{75}Si_{25}$ sample at Q = 0.6 Å in addition of the energy resolution and the signal from the empty levitator further subtracted from the sample measurement. Despite the small sample size, the quality of the signal is relatively good.

To extract the line width of $S(Q,\omega)$, a fit with a single Lorentzian function convoluted with the resolution function was applied.

Figure 28(B) presents the spectra obtained at a scattering vector Q = 0.5 Å^{-1} for the highest and lowest measured temperatures (1920 and 1420 K). The fits using Lorentzian functions are also displayed. The derived HWHM are shown in Figure 29 for all temperatures at Q-values between 0.3 and 0.9 Å^{-1} in addition of the modeling using Eq. (16).

The temperature dependence of the Ni self-diffusion coefficients follows an Arrhenius behavior (Eq. 18) even in the undercooled state, and hence, a diffusion process activated by a single activation energy is observed.

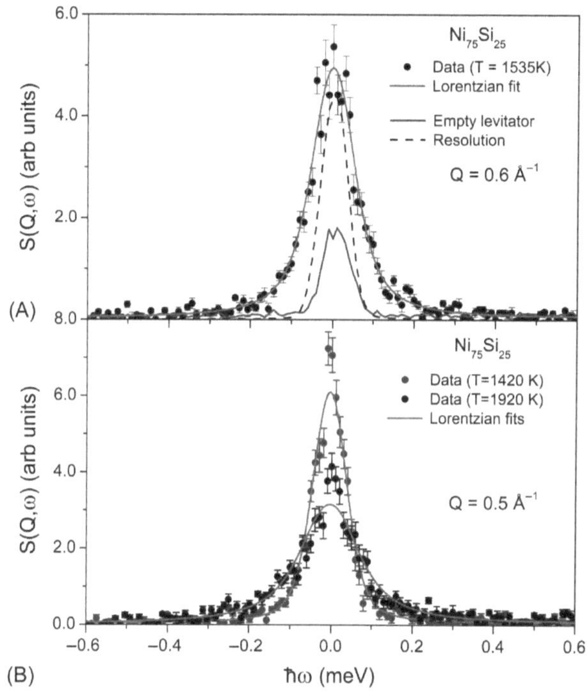

FIGURE 28 (A) Dynamic structure factor $S(Q,\omega)$ of liquid $Ni_{75}Si_{25}$ (T = 1535 K) measured at a Q-value = 0.6 $Å^{-1}$. The figure includes also with the Lorentzian fit to the data (red line), the empty levitator measurement (blue line), and the resolution measured on a vanadium sphere (dashed line). (B) Dynamic structure factors $S(Q,\omega)$ measured at Q = 0.5 $Å^{-1}$ for two temperatures T = 1420 and 1920 K (plain circles). The red lines are the Lorentzian fits.

The activation energy $E_A = 0.556 \pm 0.040$ eV obtained for $Ni_{75}Si_{25}$ is found to be 10% larger compared with pure Ni for which $E_A = 0.47 \pm 0.030$ eV [89] (see Section 10.6.1.1). That increase in activation energy might be the reason for the reduced mobility compared with pure Ni. A structural investigation on a $Ni_{50}Si_{50}$ melt found evidence for the remaining covalent bonding between silicon atoms [77], whereas at a higher Si concentration, no evidence was reported.

10.6.3.2 Study of Oxide Melts

This part presents a study of the dynamics of liquid $CaAl_2O_4$ performed at the IN6 instrument (ILL) and described in Refs. [54,55].

The coherent and incoherent cross section for Ca,Al and O are listed in Table 3. It is noteworthy that σ_{inc} is small for all elements leading to a nearly purely coherent scattering system. This has two consequences:

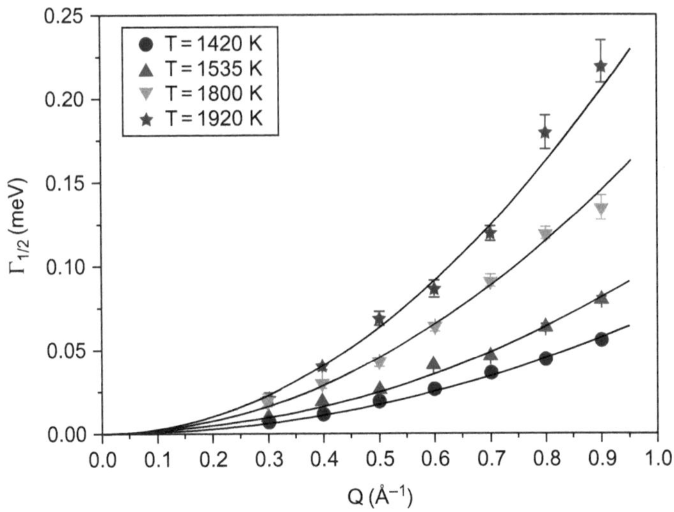

FIGURE 29 Q-dependence of the linewidth of the scattering function $S(Q,\omega)$ of liquid $Ni_{75}Si_{25}$ for different temperatures. Lines are fits with Eq. (**17**). *Figure reproduced with permission from Ref. [96].*

TABLE 3 Coherent and Incoherent Cross Sections for the Different Atomic Species in $CaAl_2O_4$

Element	b_{coh} [fm]	σ_{coh} [barns]	σ_{inc} [barns]
O	5.805(4)	4.232(6)	0.000(8)
Al	3.449(5)	1.495(4)	0.0082(6)
Ca	4.70(2)	2.78(2)	0.05(3)

From Ref. [9].

(1) To measure properly single-atom diffusion and/or density of vibrational states with QENS, it is necessary to cover a range of dynamic structure factor $S(Q,\omega)$ that approaches the self-scattering limit corresponding to single-atom diffusion.

(2) As seen previously, in a system dominated by one incoherent scatterer such as Ni, the values of relaxation time and diffusion constant extracted from the QENS data are basically those of that incoherent scatterer. In a coherent scattering system such as $CaAl_2O_4$, the derived relaxation times and diffusion constants are an average over all the atomic species, weighted by concentration and coherent neutron scattering length.

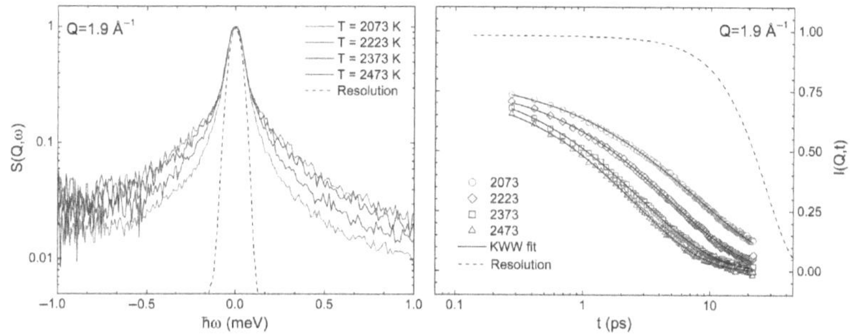

FIGURE 30 Left: Dynamic structure factors $S(Q,\omega)$ of liquid $CaAl_2O_4$ at four selected temperatures for $Q=1.9\ \text{Å}^{-1}$(near the FSDP). Right: Corresponding intermediate scattering functions $I(Q,t)$. The solid lines are stretched exponential fits.

The instrument was then operated at an incident wavelength of 5.1 Å providing an energy resolution of 69 μeV. The accessible range of momentum transfer was between 0.3 and 2.4 Å^{-1} including the position of the first peak in $S(Q)$, allowing to study relaxation processes at the length scale of the nearest neighbors since the coherent part dominates the total scattering of the samples.

The measurements were performed at 2473 K in the liquid state down to 1753 K in the undercooled state. That corresponds to an undercooling by 125 K (melting point of 1878 K).

Figure 30 (left) shows selected structure factors $S(Q,\omega)$ in a logarithmic scale. The acquisition time was about 12 h per temperature except for the two lowest temperatures where a 2 mm sample was measured for about 24 h.

The dynamic structure factors were then Fourier transformed to give the intermediate scattering functions $I(Q,t)$ displayed in Figure 30 (right). These spectra clearly show stretched functional forms for all the investigated temperatures. Such behavior is usually explained by the presence of different superposed simple exponential relaxations and is a typical feature of fragile glass-forming systems.

The decay of $I(Q,t)$ was then modeled using a stretched exponential function KWW (Kohlrausch-Williams-Watts):

$$I(Q,t) = A\exp\left[-(t/\tau)^{\beta}\right] \tag{33}$$

where A is the amplitude, τ is the relaxation time, and β (≤ 1) is the stretching exponent that quantifies the deviation from simple exponential decay ($\beta = 1$). All the three parameters A, τ, and β depend on both the momentum transfer and the temperature. τ and β are also strongly correlated; we decided to consider the mean relaxation time:

$$\langle\tau\rangle = \int_0^{\infty} e^{(-t/\tau)^{\beta}} dt = \frac{\tau}{\beta}\Gamma\left(\frac{1}{\beta}\right) \tag{34}$$

where Γ is the gamma function.

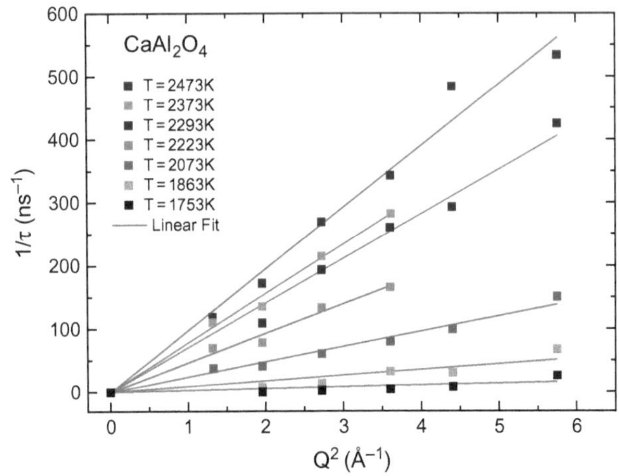

FIGURE 31 Q^2-dependence of the inverse of the mean relaxation time of liquid $CaAl_2O_4$ for all studied temperatures. The solid lines are linear fits used to estimate the diffusion coefficients.

As shown in Figure 31, the inverse of the mean relaxation time follows a Q^2 dependence for $Q > 1$ Å^{-1} even though we are dealing with coherent scatterers. Such dependence is a characteristic feature of stochastic translational diffusion of individual atoms. The diffusion coefficients $D = 1/<\tau> Q^2$ were estimated from linear fits to all the data points giving values ranging from $D = 0.30 \pm 0.08 \times 10^{-10}$ m^2/s at 1753 K to 9.80 ± 0.30 m^2/s at 2473 K.

Over the studied temperature range, the temperature dependence of the diffusion coefficient follows an Arrhenius behavior with and activation energy $E_A = 1.8 \pm 0.1$ eV/atom $= 172 \pm 9$ kJmole^{-1}. Urbain et al. [97] measured the shear viscosity of liquid $CaAl_2O_4$ and estimated an activation energy $E_A = 189 \pm 5$ kJmole^{-1} from the temperature dependence of the viscosity, in good agreement with the value obtained using QENS.

10.7 CONCLUSION AND PERSPECTIVES

In this chapter, we have shown that EML, ESL, and CNL are complementary levitation techniques allowing the study of the structure and dynamics of many different materials for temperatures ranging from above the melting point down to deep undercooling temperatures. The quality of the obtained data is quite good, allowing derivation of information in good agreement with data from other experimental techniques or simulations.

The combination of these containerless methods with neutron diffraction provides important tools for deriving structural information from metallic and oxide melts and to obtain useful information about SRO and MRO including coordination numbers and atomic distances.

With oxide samples, $S(Q)$ and $g(r)$ are dominated by all oxygen bonds with a strong contribution of O-O correlations that can lead to large overlaps with the other contributions. With metallic melts, contrary to the covalent bonds in oxide systems that lead to relatively well-defined bond lengths, there is a broad distribution of atomic pair distances.

We have shown that in both cases, the use of NDIS makes it possible to derive partials functions and get a better description of the liquid structure at the atomic scale. As performed in various studies presented here, the use of complementary x-ray measurements provides important result that help for the data analysis [38,70,82,83].

The use of modeling techniques such as molecular dynamic simulations [98] or Reverse Monte Carlo [77] makes it possible to obtain various other structural details not available experimentally. All this structural information is useful to explain the dynamic properties of the high-temperature melts.

We have also shown that the time-of-flight quasielastic neutron scattering technique is a technique very well adapted to the study of atomic diffusion in levitated liquid samples.

When incoherent scatterers such as Ti or Ni are present, the combination of all containerless processing techniques with QENS experiments allows to determine self-diffusion coefficients with high precision. In all the studied samples, the diffusion follows an Arrhenius law at temperatures close to the melting temperature. However, with the deep undercooling accessible with ESL, small deviations to the Arrhenius behavior are visible at the approach of the critical temperature as described by the mode-coupling theory. This effect is not visible when the achieved undercooling is limited.

For oxide liquids such as calcium aluminates that contain purely coherent scatterers, the determined coherent diffusion coefficient activation energies are in good agreement with published data.

Nowadays, only few groups have combined neutron experiments with levitation techniques. In addition to the three groups at the origin of this chapter (CEMHTI, MDI, and DLR) and that routinely make experiments since about more than 15 years, a team led by K. F. Kelton of Washington University in St Louis (WUSTL) (MO, the United States) just installed an ESL levitator at SNS. As mentioned in the chapter, the latter was successfully combined with WANS measurements but is also designed for studying the dynamics on several instruments [32].

For the future, works will be devoted to the optimization of the existing setups and the use of new neutron techniques.

Contrary to ESL and EML, the CNL technique is currently limited by the maximum size of the sample that can be stably levitated during enough time for making neutron experiments. In the future, this size should be increased in order to improve the statistics.

In addition to the techniques described in this chapter, small-angle neutron scattering (SANS) is also an interesting technique to apply for investigating the development of medium-range order, demixing phenomena, clustering effects, or magnetic properties. Several setups have been developed by the DLR and the CEMHTI and are currently tested at the ILL.

Taking into account the kinematic restrictions that make it hard to perform spectroscopic measurements with INS in a part of the (Q,E) domain (low-Q and high-E), INS would be also an interesting domain to explore in the future, on liquid materials with adapted sound velocities.

The perspectives of these works are also closely related to the use of new high flux sources such as SNS or the new European Spallation Source (ESS) under construction. These machines should provide high flux for QENS/INS experiments and wide Q-range for diffraction measurements.

ACKNOWLEDGMENTS

The authors are grateful to the staff members of the various instruments for their technical help and assistance during the experiments: Joerg Neuefeind (NOMAD) at SNS; Pierre Convert (deceased) and Thomas Hansen (D20), Gabriel Cuello and Henry Fischer (D4c), and Michael Marek Koza (IN6) at the ILL; and Tobias Unruh, Jürgen Peters, Giovanna Simeoni, and Wiebke Lohstroh (TOFTOF) at FRM II.
The authors also thank their colleagues from the DLR, MDI, and CEMHTI for their participation in all the studies presented in this chapter and directors of the ILL, SNS, and FRM II neutron sources for the provision of beamtime.

REFERENCES

[1] A. Meyer, J. Wuttke, W. Petry, O.G. Randl, H. Schober, Phys. Rev. Lett. 80 (1998) 4454.
[2] S. Mavila Chathoth, A. Meyer, M.M. Koza, F. Juranyi, Appl. Phys. Lett. 85 (2004) 4881.
[3] D.L. Price, High-Temperature Levitated Materials, Cambridge University Press, Cambridge, 2010.
[4] I. Egry, D. Holland-Moritz, Eur. Phys. J. Spec. Top. 196 (2011) 131–150.
[5] L. Hennet, V. Cristiglio, J. Kozaily, I. Pozdnyakova, H.E. Fischer, A. Bytchkov, J.W.E. Drewitt, M. Leydier, D. Thiaudière, S. Gruner, S. Brassamin, D. Zanghi, G.J. Cuello, M. Koza, S. Magazù, G.N. Greaves, D.L. Price, Eur. Phys. J. Spec. Top. 196 (2011) 151–165.
[6] D.L. Price, F. Fernandez-Alonso, An introduction to neutron scattering, in: F. Fernandez-Alonso, D.L. Price (Eds.), Experimental Methods in the Physical Sciences, Neutron Scattering, vol. 44, Academic Press, Amsterdam, 2013, pp. 1–136.
[7] H.E. Fischer, A.C. Barnes, P.S. Salmon, Rep. Prog. Phys. 69 (2006) 233–299.
[8] M.R. Johnson, G.J. Kearley, An introduction to neutron scattering, in: F. Fernandez-Alonso, D.L. Price (Eds.), Experimental Methods in the Physical Sciences, Neutron Scattering, vol. 44, Academic Press, Amsterdam, 2013, pp. 415–469.
[9] J. Dawidowski, J.R. Granada, J.R. Santisteban, F. Cantargi, L.A. Rodríguez Palomino, Neutron scattering lengths and cross sections, in: F. Fernandez-Alonso, D.L. Price (Eds.), Experimental Methods in the Physical Sciences, Neutron Scattering, vol. 44, Academic Press, Amsterdam, 2013, pp. 471–528.
[10] T.E. Faber, J.M. Ziman, Philos. Mag. 11 (1965) 153–173.
[11] A.B. Bhatia, D.E. Thornton, Phys. Rev. B 2 (1970) 3004–3012.

[12] M.A. Howe, R.L. McGreevy, P. Zetterström, NFL Studsvik Internal Report CORRECT: A Correction Programme for Neutron Diffraction Data, 1996.
[13] H.H. Paalman, C.J. Pings, J. Appl. Phys. 33 (1962) 2635–2639.
[14] I.A. Blech, B.E. Averbach, Phys. Rev. 137 (1965) A1113.
[15] J.L. Yarnell, M.J. Katz, R.G. Wenzel, S.H. Koenig, Phys. Rev. A 7 (1973) 2130–2144.
[16] G. Placzek, Phys. Rev. 86 (1972) 377–388.
[17] Y. Waseda, K. Suzuki, Phys. Status Solidi 39 (1970) 669–678.
[18] J. Wuttke, F. Kargl, http://sourceforge.net/projects/frida/.
[19] LAMP, The Large Array Manipulation Program, http://www.ill.eu/data_treat/lamp/the-lamp-book/.
[20] D. Richard, M. Ferrand, G.J. Kearley, J. Neutron Res. 4 (1996) 33–39.
[21] Institut Laue Langevin, https://www.ill.eu/fr/, 2017 (accessed 02.01.17).
[22] Spallation Neutron Source, https://neutrons.ornl.gov/sns, 2017 (accessed 02.01.17).
[23] Forschungsreaktors München II, https://www.frm2.tum.de/en/home/, 2017 (accessed 02.01.17).
[24] D20, https://www.ill.eu/instruments-support/instruments-groups/instruments/d20/description/instrument-layout/, 2017 (accessed 02.01.17).
[25] D4c, https://www.ill.eu/instruments-support/instruments-groups/instruments/d4/description/instrument-layout/, 2017 (accessed 02.01.17).
[26] NOMAD, https://neutrons.ornl.gov/nomad, 2017 (accessed 02.01.17).
[27] T.C. Hansen, P.F. Henry, H.E. Fischer, J. Torregrossa, P. Convert, Meas. Sci. Technol. 19 (2008) 034001.
[28] H.E. Fischer, G.J. Cuello, P. Palleau, D. Feltin, A.C. Barnes, Y.S. Badyal, J.M. Simonson, Appl. Phys. A 74 (2002) S160.
[29] L. Hennet, I. Pozdnyakova, A. Bytchkov, V. Cristiglio, P. Palleau, H. Fischer, G.J. Cuello, M. Johnson, P. Melin, D. Zanghi, S. Brassamin, J.-F. Brun, D.L. Price, M.-L. Saboungi, Rev. Sci. Instrum. 77 (2006) 053903.
[30] J. Neuefeind, M. Feygenson, J. Carruth, R. Hoffmann, K.K. Chipley, Nucl. Instrum. Methods Phys. Res. B 287 (2012) 68–75.
[31] L.B. Skinner, C.J. Benmore, J.K.R. Weber, S. Tumber, L. Lazareva, J. Neuefeind, L. Santodonato, J. Du, J.B. Parise, J. Phys. Chem. B 116 (2012) 13439–13447.
[32] N.A. Mauro, A.J. Vogt, K.S. Derendorf, M.L. Johnson, G.E. Rustan, D.G. Quirinale, A. Kreyssig, K.A. Lokshin, J.C. Neuefeind, K. An, X.-L. Wang, A.I. Goldman, T. Egami, K.F. Kelton, Rev. Sci. Instrum. 87 (2016) 013904.
[33] IN6, https://www.ill.eu/instruments-support/instruments-groups/instruments/in6/description/instrument-layout/, 2017 (accessed 02.01.17).
[34] TOFTOF, http://www.mlz-garching.de/toftof, 2017 (accessed 02.01.17).
[35] T. Unruh, J. Neuhaus, W. Petry, Nucl. Instrum. Methods Phys. Res. A 580 (2007) 1414–1422.
[36] T. Unruh, A. Meyer, J. Neuhaus, W. Petry, Neutron News 18 (2007) 22–24.
[37] T. Schenk, D. Holland-Moritz, V. Simonet, R. Bellissent, D.M. Herlach, Phys. Rev. Lett. 89 (2002) 075507.
[38] T. Kordel, D. Holland-Moritz, F. Yang, J. Peters, T. Unruh, T. Hansen, A. Meyer, Phys. Rev. B 83 (2011) 104205.
[39] H. Aoki, P.-F. Paradis, T. Ishikawa, T. Aoyama, T. Masaki, S. Yoda, Y. Ishii, T. Itami, Rev. Sci. Instrum. 74 (2003) 1147.
[40] C. Landron, L. Hennet, T. Jenkins, N. Greaves, J.P. Coutures, A.K. Soper, Phys. Rev. Lett. 86 (2001) 4839.

[41] J.K.R. Weber, C.J. Benmore, J.A. Tangeman, J. Siewenie, K.J. Hiera, J. Neutron Res. 11 (2003) 113–121.

[42] V. Cristiglio, I. Grillo, M. Fomina, F. Wien, E. Shalaev, A. Novikov, S. Brassamin, M. Réfrégiers, J. Pérez, L. Hennet, Biochim. Biophys. Acta (2016).

[43] G. Mathiak, I. Egry, L. Hennet, D. Thiaudière, I. Pozdnyakova, D.L. Price, Int. J. Thermophys. 26 (2005) 1151–1166.

[44] J. Brillo, A. Bytchkov, I. Egry, L. Hennet, G. Mathiak, I. Pozdnyakova, D.L. Price, D. Thiaudière, D. Zanghi, J. Non-Cryst. Solids 352 (2006) 4008–4012.

[45] P.-F. Paradis, T. Ishikawa, J. Yu, S. Yoda, Rev. Sci. Instrum. 72 (2001) 2811–2815.

[46] P.-F. Paradis, T. Ishikawa, J. Yu, S. Yoda, Meas. Sci. Technol. 16 (2005) 452–456.

[47] J.K.R. Weber, D.S. Hampton, D.R. Merkley, C.A. Rey, M.M. Zatarski, P.C. Nordine, Rev. Sci. Instrum. 65 (1994) 456–465.

[48] P.C. Nordine, D. Merkley, J. Sickel, S. Finkelman, R. Telle, A. Kaiser, R. Prieler, Rev. Sci. Instrum. 83 (2012) 125107.

[49] D.A. Winborne, P.C. Nordine, D.E. Rosner, N.F. Marley, Metall. Trans. B 7 (1976) 711–713.

[50] J.P. Coutures, D. Massiot, C. Bessada, P. Echegut, J.C. Rifflet, F. Taulelle, C.R. Acad. Sci. II 310 (1990) 1041–1045.

[51] J.K.R. Weber, S. Krishnan, P.C. Nordine, JOM 43 (1991) 8–14.

[52] S. Krishnan, D.L. Price, J. Phys. Condens. Matter 12 (2000) R145–R176.

[53] C. Landron, L. Hennet, J.-P. Coutures, T. Jenkins, C. Aletru, N.G. Greaves, A.K. Soper, G. Derbyshire, Rev. Sci. Instrum. 71 (2002) 1745–1751.

[54] J. Kozaily, L. Hennet, H.E. Fischer, M. Koza, S. Brassamin, S. Magazu, F. Kargl, Phys. Status Solidi C 8 (2011) 3155–3158.

[55] J. Kozaily, Structure and Dynamics of Calcium Aluminosilicate Melts (Ph.D. Thesis), University of Orleans, 2012.

[56] G. Jacobs, I. Egry, K. Maier, D. Platzek, J. Reske, R. Frahm, Rev. Sci. Instrum. 67 (1996) 3683.

[57] O. Muck, German Patent No. 42204, 1923.

[58] E.C. Okress, E.M. Wroughton, C. Comenetz, P.N. Brace, J.C.K. Kelly, J. Appl. Phys. 23 (1952) 545.

[59] P.R. Rony, The Electromagnetic Levitation of Metals, Tech. Rep. UCRL-11411, Lawrence Radiation Lab., University of California, Berkeley, USA, 1964.

[60] D. Holland-Moritz, T. Schenk, P. Convert, T. Hansen, D.M. Herlach, Meas. Sci. Technol. 16 (2005) 372–380.

[61] S. Earnshaw, Trans. Cambridge Philos. Soc. 7 (1842) 97.

[62] K.F. Kelton, G.W. Lee, A.K. Gangopadhyay, R.W. Hyers, T.J. Rathz, J.R. Rogers, M.B. Robinson, D.S. Robinson, Phys. Rev. Lett. 90 (2003) 195504.

[63] G.W. Lee, A.K. Gangopadhyay, K.F. Kelton, R.W. Hyers, T.J. Rathz, J.R. Rogers, D.S. Robinson, Phys. Rev. Lett. 93 (2004) 037802.

[64] T. Meister, H. Werner, G. Lohöfer, D. Herlach, H. Unbehauen, Control Eng. Pract. 11 (2003) 117.

[65] D. Turnbull, J. Appl. Phys. 21 (1950) 1022.

[66] F.C. Frank, Proc. R. Soc. Lond. A 215 (1952) 43–46.

[67] D. Holland-Moritz, T. Schenk, V. Simonet, R. Bellissent, P. Convert, T.C. Hansen, J. Alloys Compd. 342 (2002) 77–81.

[68] T. Schenk, V. Simonet, D. Holland-Moritz, R. Bellissent, T.C. Hansen, P. Convert, D.M. Herlach, Europhys. Lett. 65 (2004) 34–40.

[69] D. Holland-Moritz, O. Heinen, R. Bellissent, T. Schenk, D.M. Herlach, Int. J. Mater. Res. 97 (2006) 948–953.
[70] D. Holland-Moritz, S. Stüber, H. Hartmann, T. Unruh, T. Hansen, A. Meyer, Phys. Rev. B 79 (2009) 064204.
[71] Y. Waseda, The Structure of Non-Crystalline Materials, McGraw-Hill, New York, 1980.
[72] D. Holland-Moritz, F. Yang, T. Kordel, S. Klein, F. Kargl, J. Gegner, T. Hansen, J. Bednarcík, I. Kaban, O. Shuleshova, N. Mattern, A. Meyer, Europhys. Lett. 100 (2012) 56002.
[73] D. Holland-Moritz, F. Yang, J. Gegner, T. Hansen, M.D. Ruiz-Martín, A. Meyer, J. Appl. Phys. 115 (2014) 203509.
[74] S. Ansell, S. Krishnan, J.J. Felten, D.L. Price, J. Phys. Condens. Matter 10 (1998) L73–L78.
[75] L. Hennet, S. Krishnan, I. Pozdnyakova, V. Cristiglio, G. Cuello, H. Fischer, A. Bytchkov, F. Albergamo, D. Zanghi, J.F. Brun, S. Brassamin, M.L. Saboungi, D.L. Price, Pure Appl. Chem. 79 (2007) 1643–1652.
[76] G.W. Lee, A.K. Gangopadhyay, K.F. Kelton, R.W. Hyres, T.J. Rathz, M.B. Robinson, J. Rogers, L. Hennet, S. Krishnan, Mater. Res. Soc. Symp. Proc. 754 (2003) 135–140.
[77] S. Gruner, J. Marczinke, L. Hennet, W. Hoyer, G.J. Cuello, J. Phys. Condens. Matter 21 (2009) 385403.
[78] R.L. McGreevy, L. Pusztai, Mol. Simul. 1 (1988) 359–367.
[79] R.L. McGreevy, J. Phys. Condens. Matter 13 (2001) R877–R913.
[80] L.B. Skinner, A.C. Barnes, P.S. Salmon, L. Hennet, H.E. Fischer, C.J. Benmore, S. Kohara, J.K.R. Weber, A. Bytchkov, M.C. Wilding, J.B. Parise, T.O. Farmer, I. Pozdnyakova, S.K. Tumber, K. Ohara, Phys. Rev. B 87 (2013) 024201.
[81] V. Cristiglio, L. Hennet, G. Cuello, I. Pozdnyakova, A. Bytchkov, P. Palleau, H.E. Fischer, D. Zanghi, M.-L. Saboungi, D.L. Price, J. Non-Cryst. Solids 353 (2007) 993–995.
[82] L.B. Skinner, C.J. Benmore, J.K.R. Weber, J. Du, J. Neuefeind, S.K. Tumber, J.B. Parise, Phys. Rev. Lett. 112 (2014) 157801.
[83] J.W.E. Drewitt, S. Jahn, V. Cristiglio, A. Bytchkov, M. Leydier, S. Brassamin, H.E. Fischer, L. Hennet, J. Phys. Condens. Matter 23 (2011) 155101.
[84] J.W.E. Drewitt, L. Hennet, A. Zeidler, S. Jahn, P.S. Salmon, D.R. Neuville, H.E. Fischer, Phys. Rev. Lett. 109 (2012) 235501.
[85] A.C. Hannon, J.M. Parker, J. Non-Cryst. Solids 274 (2000) 102–109.
[86] S.V. Ushakov, A. Navrotsky, J.K.R. Weber, J.C. Neuefeind, J. Am. Ceram Soc. 98 (2015) 3381–3388.
[87] J.K.R. Weber, A. Tamalonis, C.J. Benmore, O.L.G. Alderman, S. Sendelbach, A. Hebden, M.A. Williamson, Rev. Sci. Instrum. 87 (2016) 073902.
[88] L.J. Santodonato, Y. Zhang, M. Feygenson, C.M. Parish, M.C. Gao, J.K.R. Weber, J.C. Neuefeind, Z. Tang, P.K. Liaw, Nat. Commun. 6 (2015) 5964.
[89] A. Meyer, S. Stüber, D. Holland-Moritz, O. Heinen, T. Unruh, Phys. Rev. B 77 (2008) 092201.
[90] A. Meyer, J. Horbach, O. Heinen, D. Holland-Moritz, T. Unruh, Diffusion in materials – Dimat2008, in: A. Agüero, J.M. Albella, M.P. Hierro, J. Phillibert, F.J. Pérez Trujillo (Eds.), Defect and Diffusion Forum Series, vols. 289–292, 2009, pp. 609–614.
[91] A. Meyer, Phys. Rev. B 66 (2002) 134205.
[92] A. Meyer, W. Petry, M. Koza, M.-P. Macht, Appl. Phys. Lett. 83 (2003) 3894.
[93] S. Stüber, D. Holland-Moritz, T. Unruh, A. Meyer, Phys. Rev. B 81 (2010) 024204.
[94] A.I. Pommrich, A. Meyer, D. Holland-Moritz, T. Unruh, Appl. Phys. Lett. 92 (2008) 241922.

[95] J. Horbach, S.K. Das, A. Griesche, M.-P. Macht, G. Frohberg, A. Meyer, Phys. Rev. B 75 (2007) 174304.
[96] F. Demmel, L. Hennet, S. Brassamin, D.R. Neuville, J. Kozaily, M.M. Koza, Phys. Rev. B 94 (2016) 014206.
[97] G. Urbain, Rev. Int. Hautes Temp. Réfract. 20 (1983) 135–139.
[98] V. Cristiglio, G.J. Cuello, L. Hennet, I. Pozdnyakova, M. Leydier, J. Kozaily, H.E. Fischer, M.R. Johnson, D.L. Price, J. Non-Cryst. Solids 356 (2010) 2492–2496.

Chapter 11

High-Pressure Neutron Science

Malcolm Guthrie[1]
European Spallation Source, Lund, Sweden
[1]*Corresponding author: e-mail: malcolm.guthrie@esss.se*

Chapter Outline

11.1 INTRODUCTION

Pressure, the simple application of force to an area, is perhaps able to affect more dramatic changes on the structure and properties of matter than any other experimental variable. One way to visualize this is to contrast the effect of both temperature and pressure on volume and, correspondingly, free energy. In 1989, Jeanloz explored this in the case of CsI [1], finding that pressures of only ~16 GPa were sufficient to exceed the typical bonding energies in the solid. Given that pressures in excess of 100 GPa are now routinely accessible in the laboratory, we are thus able to explore a wide range of structural states, chemistry, and material properties that would be inaccessible by other means.

Experimental Methods in the Physical Sciences, Vol. 49. http://dx.doi.org/10.1016/B978-0-12-805324-9.00011-X

The ability to manipulate matter so profoundly is one of the key drivers of interest in high-pressure science. Examples of dramatic changes in structure and properties abound in the literature. Pressure can control conductivity—transforming semiconductors (e.g., silicon or germanium), insulators (such as oxygen [2]; CsI [3]; hydrogen [4]), and even the inert gases Ar, Kr, Xe, and Rn [5] into fully metallic states. Working in the opposite direction, pressure can force metals to become insulating (and, in the case of sodium [6], making them transparent to visible light).

Pressure can also modify bonding state, such as tuning the bond hybridization in carbon. In pure carbon, this famously leads to the stabilization of the diamond over the graphite form with complete alteration of the material's electronic, mechanical, and thermal properties. Meanwhile, in unsaturated hydrocarbons, completely new low-dimensional polymers have been formed under pressure [7]. And the simultaneous application of pressure and high temperatures has been shown to convert simple methane to the higher order molecules ethane, propane, and butane [8] revealing a potential route for the abiogenic synthesis of petroleum.

While pressures of 5–10 GPa and beyond are typically required to modify chemical bonds and alter electronic structure, far more modest pressures can have a dramatic effect on biological systems. Pressures approaching 1 GPa can destroy living organisms through direct protein denaturing, disruption of hydrogen bonding, and secondary effects due to the room-temperature freezing of intracellular water at pressures above 0.9 GPa. Other important biological effects include the mechanical stability of cell membranes and the ultimate robustness of DNA strands to pressure. Although the ability of pressure to kill bacteria is now used commercially to sterilize food (a process known as "pascalization") [9], there is some experimental evidence that *E. coli* bacteria may remain viable to pressures as high as 2.0 GPa [10,11], with successive generations demonstrating an evolutionary response to the stress of pressure.

With such a rich tapestry of phenomena to study under high pressure, it is natural that powerful neutron techniques should be used to quantify these. Indeed, high-pressure neutron experiments have been attempted since the 1960s, with the first dedicated beamline developed by 1966 [12]. However, the historical trajectory of high-pressure science has been driven in recent decades by the substantial advances in brightness of synchrotron (and now free-electron laser) X-ray sources. These have enabled a vast expansion of accessible pressure-temperature space, with the current record for a static pressure diffraction measurement standing at 640 GPa [13], while temperatures exceeding 3000 K are routine achieved using laser heating. Moreover, a diverse range of both structural and dynamic techniques is available, currently including inelastic X-ray spectroscopy, X-ray absorption fine structure and near-edge spectroscopy, imaging and tomography, nuclear resonant inelastic scattering and synchrotron Mössbauer, and X-ray magnetic dichroism.

Despite the dominance of synchrotron techniques, the story of high-pressure neutron science has continued to evolve and, due to recent investments in new sources, this evolution has begun to accelerate. A principal driver has been the construction of the two liquid-target spallation sources, the Spallation Neutron Source (SNS) in Oak Ridge, TN, the United States, and the Material and Life Sciences Facility (MLF) in Tokai, Japan, which have each achieved new levels of neutron brightness. The intensity of these sources has been used to conduct neutron diffraction at unprecedented pressures approaching ~100 GPa [14] and is just beginning to expand the palette of techniques to include vibrational spectroscopy up to 10 GPa (see Section 11.6). The European Spallation Source (ESS), currently under construction in Lund, Sweden, will be the brightest neutron source in history and will create an opportunity to realize yet further the full potential of these recent developments.

This chapter will begin with a historical overview of the 50-year-old field of high-pressure neutron science. The adoption of synchrotron X-ray sources and the resulting rapid growth of the high-pressure scientific community provide an important context, which will also be outlined. Following this, the fundamentals of high-pressure neutron science will be described, from the basic principles of pressure generation to the implementation of high pressure at different neutron facilities. Specialized data analysis techniques, which are relevant for high pressure, will be described, and a brief outline of science in the fields of biology, chemistry, and material science will be given.

11.2 HIGH-PRESSURE INSTRUMENTATION

The story of high-pressure research begins with instrumentation: devices are needed to generate and sustain pressure. The application places stringent requirements on the materials that can be used and on the geometries that can be employed. As described below, the ability to maintain static pressures in the gigapascal range dates to the early 1900s. These techniques evolved over the following 50 years and, when early neutron sources (reactors, primarily developed to support nuclear efforts during the Second World War) began to be used for science, attempts to measure material properties under high pressure using neutrons were soon made.

11.2.1 The Beginnings of High-Pressure Instrumentation

The ability to generate pressures high enough to increase the density of gases significantly dates back to the industrial revolution. James Watt patented one of the first commercially successful engines that used high-pressure steam to generate motion as early as 1781 [15]. Soon after, these engines were adapted to drive compressors that could increase pressures still further.

However, densifying liquids and solids is vastly harder, requiring the direct compression of chemical bonds and pressures of at least several GPa. One of the simplest technologies for achieving such pressures is the piston cylinder. Here, a force applied over a large area at the back of a piston is transferred to a smaller area at the tip of the piston, thus enhancing pressure. This pressure is contained by a cylinder, which has an inner diameter just larger than the piston's outer diameter and by seals, which prevent pressurized material escaping through the small gap between the inner surface of the cylinder and the piston.

An early pioneer of piston-cylinder technology was Sir Charles Parsons, who experimented with many designs and claimed pressures as high as 1.5 GPa [16]. One of Parsons' goals was the artificial synthesis of diamond, which requires both heat and pressure. His illustrations (see Figure 1) show all of the key features of a monobloc piston-cylinder design and also indicate how a flow of electricity through the sample could be used to simultaneously apply heat. Parsons also experimented with the use of chemical energy to simultaneously apply both heat and pressure (Figure 1).

The piston-cylinder design is limited to a maximum elastic limit of 1.4 GPa (for the best available material, that is, high-tensile steel [17]). While this can be increased by several factors by using a compound cylinder, there are intrinsic limitations to the maximum pressure that can be sustained. It was party due to these limitations that Percy Bridgman developed a radically different design in the early 1900s—opposed anvils (see Figure 2). Again, the principle of operation was conceptually simple: the sample is crushed between two "anvils," formed from a very hard material, while being prevented from escaping perpendicular to the force by a seal provided by a gasket. Bridgman developed the technique for many decades, eventually reaching pressures of ~10 GPa [18]. These pressures enter deeply into the regime where matter can be fundamentally altered, and Bridgman uncovered a great many new phenomena in his career, ultimately leading to the award of the Nobel Prize in Physics in 1946.

As is the case with the piston cylinder, the available materials limit the ultimate pressure that can be achieved. Bridgman used Carboloy™, a composite tungsten carbide material, which has a very high bulk modulus and is also very hard, comparable with sapphire. In common with all very hard materials, tungsten carbide is also brittle. One of Bridgman's innovations was to support the carbide with a high-tensile steel binding ring; this approach greatly increases the maximum load that can be sustained [17].

Throughout the 1950s, there was continued innovation in the field of high pressure. Working for General Electric, H. Tracy Hall was finally the first to synthesize diamond routinely in the laboratory in 1954 [19]. Hall's device, called a belt apparatus [20], also relied on opposing Carboloy™ pistons but used a supportive "belt," itself formed from Carboloy™. Both anvils and belt

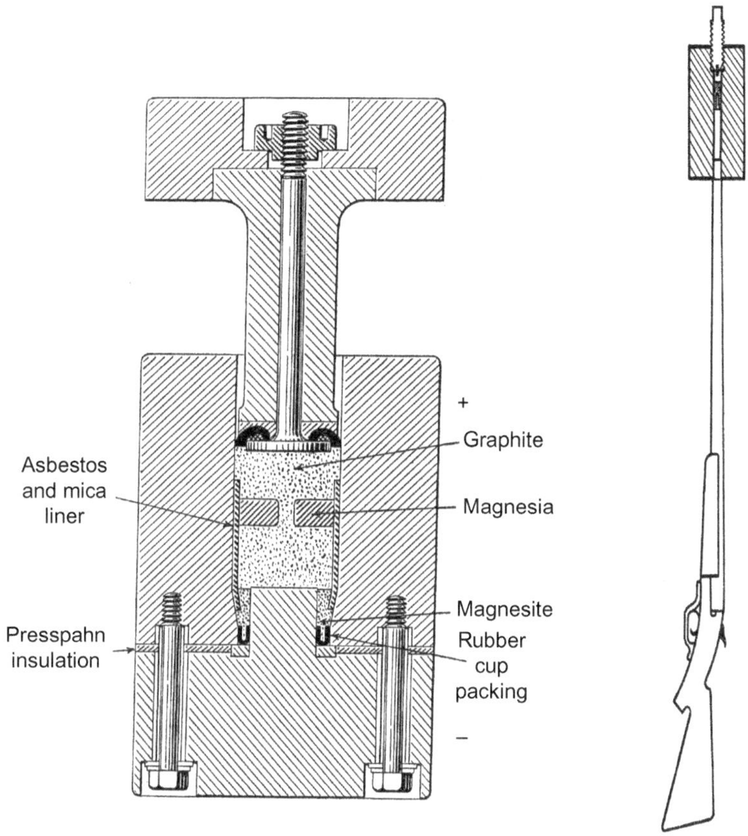

FIGURE 1 Some of Sir Charles Parsons' attempts to synthesize diamond with high pressures and temperature. (Left) A piston-cylinder design, using rubber, brass, and leather to form seals. The sample (graphite) was also simultaneously heated by passing an electric current through the arrangement. (Right) shows an alternative approach employing chemical energy. The chamber at the end of the muzzle was filled with acetylene and oxygen (with an excess of the former used as a source of carbon). Parsons believed he successfully formed a tiny piece of diamond with this approach, but this was never confirmed. *Image reprinted with permission from Ref. [16].*

were then supported by a series of steel binding rings. The extreme hardness of the Carboloy™ was critical in order to sustain the required pressures (in excess of ~5 GPa). Hall also developed an entirely different design of press, a "tetrahedral multianvil" apparatus with which he also successfully synthesized diamond [21]. In this design, four anvils were independently pushed onto an extrudable tetrahedral gasket (made of pyrophyllite) containing the carbon precursor. The four anvils, again formed from tungsten carbide, were each cylinders with precisely machined triangular faces that, when placed together, formed a tetrahedral volume for the sample. Like the opposed anvil,

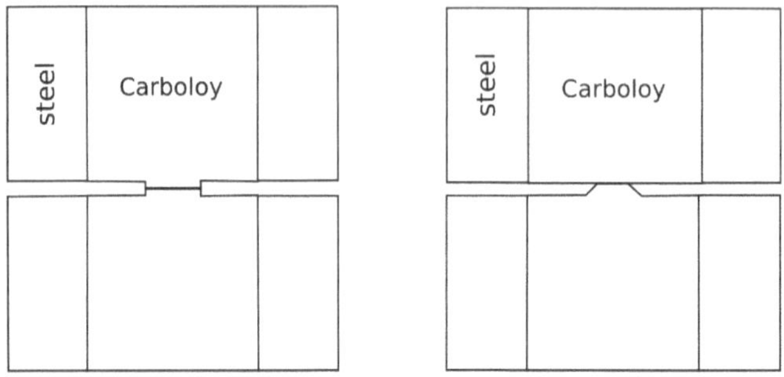

FIGURE 2 Two designs of Carboloy anvils, with outer steel binding rings, developed by P.W.Bridgman in the course of his career. The design on the right features a conical, truncated boss, which could reach higher pressures than the design on the left. *Image adapted with permission from Ref. [18].*

the multianvil approach has also seen widespread use since its inception, with multiple variations, such as six-axis cubic arrangements and double-stage anvils (see review by Liebermann et al. [22]).

By the 1960s, interest in the new field of high-pressure research had grown to the point where 150 laboratories in the United States alone had high-pressure programs [12]. In light of this momentum, the time seemed right for Robert M. Brugger (at the time head of the nuclear technology branch of the Idaho Nuclear Corporation) to propose construction of the first dedicated high-pressure beamline in the world at the Materials Testing Reactor in Idaho Falls, ID, the United States [12]. The Brugger diffractometer was also one of the first neutron beamlines to use time-of-flight diffraction techniques, which were considered ideally matched to the limited angular aperture of the pressure cells. Later, following the advent of accelerator-driven, pulsed neutron sources, time of flight would become a widely adopted technique.

As shown in Figure 3, the entire instrument was built around a dedicated sample environment consisting of two opposing hydraulic presses. Tracy Hall consulted on the project and proposed a modification of his belt apparatus, expected to achieve 8 GPa and 1000 °C. However, this (and a cubic multianvil press that was also proposed) appears to have never been constructed. The highest pressure reached looks to have been 7.3 GPa, using Bridgman style flat anvils formed from alumina [23]. With a few exceptions (described in Section 11.2.4), neutron experiments were not to exceed this pressure for another 30 years, and developments in other technologies were soon to eclipse their capabilities.

11.2.2 The Diamond Anvil Cell (DAC)

For the vast majority of the high-pressure work in the first half of the 20th century, the key material used in the anvils of the pressure cells was cemented

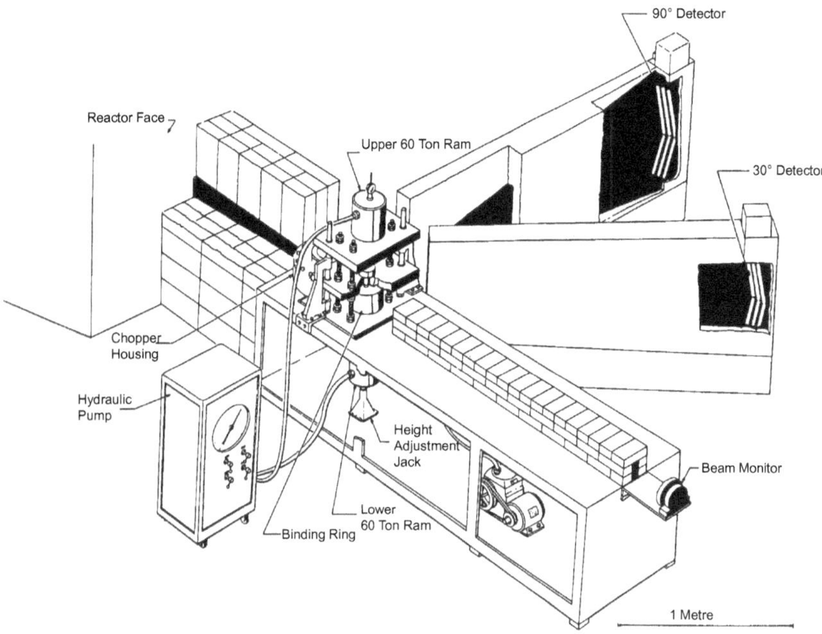

FIGURE 3 The Brugger high-pressure neutron diffractometer. *Image reprinted with permission from Ref. [95].*

tungsten carbide [18]. However, a glance at Table 1 immediately suggests that diamond, with its superlative and unsurpassed hardness, should be a far more attractive material for the role of an anvil. Of course, Bridgman was aware of this and conducted many tests of diamond anvils; however, he concluded by 1941 that "[its] superiority is too slight to justify the expense except perhaps for very special purposes" [18]. In this respect, rather uncharacteristically, history has proved Bridgman's assessment to be incorrect.

In addition to the significant expense and lack of availability, another key challenge of diamond and a possible reason why Bridgman was unsuccessful is that (unlike tungsten carbide) natural diamonds have a single-crystal microstructure. This means that their mechanical properties are highly anisotropic, yielding easily when pushed along certain directions in the crystal lattice (a fact that is exploited in faceting and polishing of diamond) while exhibiting unrivaled hardness in other directions [24,25]. From the perspective of pressure cell design, this means that alignment—in particular, parallelism of the anvil faces—is critical.

This challenge began to be overcome in the late 1950s when Weir et al. developed the first diamond anvil cell [26]. Although only capable of reaching 3 GPa, much lower than the tungsten carbide technology of the time, this device exploited yet another of the advantages of diamond—its optical transparency—to enable the first infrared optical measurements under pressure.

TABLE 1 Some Mechanical Properties of Materials Used for Anvils in High-Pressure Cells

Material	Hardness (Vickers)	Young's modulus (GPa)	Bulk modulus (GPa)	Ultimate tensile strength (MPa)
Sintered tungsten carbide	2600	530–700	350–400	344
Sapphire	~2000	386–435	240	400
Diamond	10,000 (nominal)	1050	560	Not known

Although the first diamond cells were able to generate only modest pressures, the designs continued to improve. The innovations of using a metal gasket to contain the sample [27], ruby fluorescence to measure pressure [28–30], and improvements in anvil alignment led to the achievement of static pressures approaching 50 GPa by 1975 [31]. With continued miniaturization of the anvil tip and the development of beveled anvils in 1978 [32], the highest achievable pressure had soared to 170 GPa. However, under such enormous pressures, the sample volume became truly microscopic, with diameters of less than ~50 μm and thicknesses of only ~10 μm. Such vanishingly small samples would be extremely challenging to detect with the relatively weak sources of neutrons available at the time; however, a new technology was just emerging that could not be better suited for probing matter contained within diamond anvil cells.

11.2.3 The Rise of Synchrotrons

Not only are diamonds transparent to visible light, but also, with the low atomic weight of carbon, they are easily penetrated by hard X-ray radiation. The first X-ray diffraction measurements using diamond cells date to the late 1960s and revealed abrupt changes in Bragg peak positions, as a function of pressure—a direct observations of structural phase transitions with atomic resolution. However, these experiments, which were conducted on in-house laboratory X-ray sources, were limited in terms of pressure by divergent, low-flux beams. Both key characteristics of the beam are encapsulated in the concept of beam brilliance—the total number of photons per second, per unit area, and per unit solid angle in a given wavelength band [33].

For measurements using a diamond anvil cell, brilliance is the critical parameter for the probe radiation. The total number of photons (or, as described later, neutrons) must be maximized, but the beam must also enter and exit cleanly through the cell without touching components such as the seats or gaskets. Although synchrotron radiation was first observed as early as 1947 [34], it wasn't until the early 1980s that the so-called second-generation synchrotrons provided beams that were dedicated to scientific measurements [34]. These sources, with brilliances that exceeded laboratory sources by a full seven orders of magnitude (see Figure 4), were tailor-made for science using diamond anvils cells. Almost immediately, researchers began to conduct such high-pressure diffraction and spectroscopy measurements and, by the end of the decade, were studying atomic-level structure and dynamics at unprecedented pressures, well above 100 GPa, on dedicated high-pressure synchrotron beamlines.

In fact, the output of high-profile science from DAC-based synchrotron experiments was so great that when third-generation facilities, with extremely bright undulator and wiggler sources, began to come online in the late 1990s, dedicated, high-pressure facilities became almost ubiquitous. A 10-fold reduction in beam sizes to 5–10 μm enabled further expansion of the pressures that could be reached with diffraction measurements, with the current record standing at 640 GPa [13]. Throughout the previous two decades, high-pressure research at synchrotrons also matured to the point where a great many more techniques were enabled with DACs including inelastic and X-ray absorption spectrometry, nuclear resonant inelastic scattering, X-ray magnetic circular dichroism, and both direct and transmission imaging and tomography.

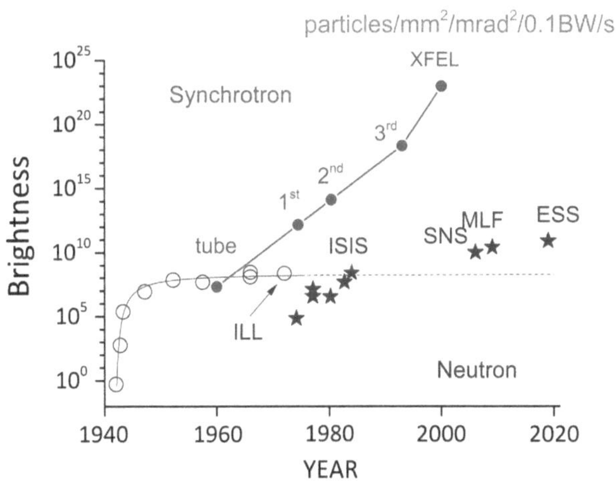

FIGURE 4 Brugger plot, where X-ray sources are shown as solid (red online) circles and neutron sources as open circles (steady state) and solid stars (pulsed), both blue online. In this comparison, the peak brightnesses of the pulsed neutron sources are given. In all cases, the time-averaged brightness of the ILL reactor exceeds that of the pulsed sources, with the exception of the ESS.

11.2.4 Neutron Instrumentation at High Pressure

Although the stratospheric growth of high-pressure synchrotron science throughout the previous 30 years was not matched at neutron sources, pressure has been a mainstay at many facilities, and innovations have continued. The main limitations are directly attributable to the limited intensity of available neutron beams. Brugger had already recognized this issue and commented on it in an article simply entitled "We need more intense thermal-neutron beams," penned in 1968 [35]. Here, Brugger showed that the evolution in neutron flux, as a function of time had all but plateaued by the 1950s. This so-called 'Brugger plot' was updated in 1986 by Carpenter and Yelon [36] to show the evolution of the newer accelerator-driven sources. Figure 4 shows an updated version of the plot, comparing both reactor and spallation sources up to and including the ESS, which will begin to generate neutrons around 2020.

Also apparent from Figure 4 is that the improvements in X-ray technologies have dramatically exceeded those of neutrons, with third-generation synchrotrons capable of brightness more than eight orders of magnitude greater than the SNS. As a result of the relatively weak sources, neutron sample volumes have remained large, many orders of magnitude greater than those used in X-ray studies. As a consequence, most of the high-pressure neutron work in the 1960s, 1970s, and 1980s was conducted below 3 GPa using a combination of gas cells, piston cylinders, and clamp devices, such as the McWhan cell [37].

Although higher pressures could be reached with DACs, the small volumes made neutron science extremely challenging, but not impossible. Instructive examples include a first attempt in the late 1980s by a Russian group who were able to see diffraction peaks from a single crystal of pure D_2 to as high as 31 GPa [38,39]. However, quantitative diffraction intensities were not obtained, and only the positions of the three strongest diffraction peaks were measured. Subsequently, this technique found a successful niche in the measurement of magnetic Bragg peaks under pressure that owes much to the work of the late Igor Goncharenko (see, e.g., [40–44]). Here, in order to be successful, the measurements used longer neutron wavelengths, where backgrounds could be more easily controlled and where detector efficiencies are high. Longer wavelengths also allowed the use of neutron focusing optics to increase flux (at the expense of increased divergence). To optimize all of these variables, Goncharenko built a dedicated high-pressure diffractometer on beamline G 6-1 at the LLB reactor in Saclay [45]. He then went on to conduct many experiments with both diamond anvil cells and, where these had insufficient volume, larger sapphire anvils [41]. While these steps enabled diffraction intensities to be measured, the use of long wavelengths led to a reduction in coverage in reciprocal space, and only the longest d-spacing reflections were accessible.

Of course, high-resolution structural information is also important, and indeed, many of the advantages of neutrons for crystallographic studies require the accurate measurement of Bragg intensities to large momentum transfers. With the sources of the day, this demanded larger sample volumes simply not accessible with diamond anvils. It was the pressing need to conduct full crystallographic studies in the 2.5–20 GPa range that drove the development of a new device in the 1990s—the Paris-Edinburgh (PE) press.

Arising from a collaboration between the Université Pierre et Marie Curie in France and the University of Edinburgh in the United Kingdom, the PE press is a compact hydraulic press with a standard capacity of 200 metric tons. In its most common configuration, the device uses two opposed anvils with a modified Bridgman design. The anvils, formed from tungsten carbide, are supported by a steel binding ring in tension [46]. The key modification from the (flat) Bridgman design was the use of a hemispheric depression for the sample and a toroidal groove for the gasket, which was in turn adapted from a design by Khvostantev [47]. By 1993, the PE press with toroidal anvils had demonstrated high-quality neutron crystallographic measurements could be conducted up to ~10 GPa [46,48]. In 1995, by replacing tungsten carbide with sintered diamond and by reducing the sample volume, pressures in excess of ~20 GPa were obtained, almost a full order of magnitude increase on the McWhan cell [49].

The success of the PE press was such that it could be used to justify the construction of a dedicated high-pressure neutron diffractometer. This instrument, PEARL, was built at the ISIS Spallation Source. As with G 6-1 at Saclay, the diffractometer itself was built around and optimized for the pressure vessel (in this case, the PE press). Unlike G 6-1, PEARL is a time-of-flight machine, which is described in more detail in Section 11.4.2.

Having been in operation for 20 years now, PEARL has firmly established the field of quantitative neutron crystallography in the 2–20 GPa regime [50]. And, in addition to conventional crystallography, has also been at the forefront of developing techniques with significantly poorer signal levels. From the diffraction perspective, these include attempts to study the structure of liquids [51], amorphous ice [52], glasses [53,54], and also pair-distribution studies of local structure in disordered crystalline materials [55] (also, see review by Salmon and Zeidler [56]). It is perhaps partly due to the success of PEARL, and also the continuing importance of high-pressure neutron diffraction techniques that both of the latest neutron sources (the MLF in Japan, and the SNS in the United States) also have dedicated high-pressure beamlines.

Continued improvements in neutron flux have led to a revisiting of DAC techniques for neutrons, almost 30 years after the first Russian measurements. Much of this work has been based at the SNAP beamline at the SNS, where there has been a focused effort on using DACs for quantitative crystallography. At the time of writing, the highest pressure at which a full Rietveld analysis of neutron diffraction intensities stands at 43 GPa [57]. Meanwhile, quantitative diffraction intensities have been measured to as high as 94 GPa [14].

Another extreme, where much remains to be explored is that of very low temperatures, combined with high pressure. As reviewed by Ridley and Kamenev [58], such work can also benefit from work using very small sample sizes, as the lower force required for a given pressure means cells can be much less massive. Goncharenko conducted many experiments using sapphire and diamond anvil cells in liquid He cryostats, reaching temperatures as low as 0.1 K and even combining this with magnetic fields up to 7.5 T (e.g., [40,41,59–61]) with the lowest temperatures achieved using dilution inserts. An alternative, more recent design of sapphire anvil device, the "Sputnik" cell [62] is driven by a gas-filled "bellows" allowing pressure change at low temperature and that incorporates a fiber-based ruby fluorescence system to measure pressure *in situ* [28].

Still another extreme is that of crystallographic complexity. Some very recent work, using DACs, has been used to determine complex neutron structures of amino acids and pharmaceuticals up to 0.5 GPa [63]. This approach uses a Laue technique first developed at VIVALDI at the ILL [64] and now implemented at the KOALA diffractometer at ANSTO in Sydney (this is discussed further in Sections 11.4.1 and 11.5.1).

In addition to diffraction techniques, spectroscopic measurements are also possible under pressure. The earliest neutron measurements of dynamics under high pressure date back to Lechner's measurements of phonons in lead up to 0.3 GPa in the 1960s [65]. The first high-pressure spectroscopic measurements at a spallation sources were conducted on TFXA at ISIS in the 1990s and examined both lattice dynamics and atomic vibrations up to 0.2 GPa (see e.g., [66–68]).

At higher pressures, there have been several measurements of phonon dispersion relationships—up to 12 GPa range using PE-type presses with 5–10 mm^3 single-crystal sample volumes [69]. Meanwhile, studies of powdered samples of H_2O have exploited the large incoherent scattering length of H to achieve sufficient signal levels up to $\sim$2.5 GPa [70]. Very recently, Klotz et al. [71] and Bove et al. [72] have demonstrated quasielastic neutron scattering (QENS) measurements in liquid H_2O up to 3 GPa yielding measurements that were sensitive to the proton diffusion in the liquid.

As the availability of large artificial diamonds made by chemical vapor deposition increases, vibrational spectroscopy measurements at pressures exceeding 10 GPa are becoming possible (see Section 11.6).

11.3 DATA REDUCTION CONSIDERATIONS FOR HIGH PRESSURE CELLS

An unavoidable reality of all high-pressure neutron experiments is that the beam must enter and exit through some form of containment in order to reach the sample. In addition, the beam scattered by the sample must itself exit through the container walls. When the container is a pressure vessel, the corrections that must be made for the interaction of container and beam are especially challenging as the intense pressures can modify the atomic structure of the vessel itself.

The two key issues that must be dealt with are the background and the attenuation due to the cell. In general, both effects are wavelength dependent and, as such, are most challenging to deal with when using an energy dispersive technique such as time-of-flight diffraction. Unique approaches to this have been developed, and these are discussed here where the most general (wavelength-dependent) corrections are outlined.

11.3.1 The Cell Background

As the neutron beam transits the pressure vessel to reach the sample, this generates an additional neutron signature due to the vessel. This effect, the cell background, can be substantially larger than that from the sample and must be mitigated. The palette of materials described in Table 2 each give their own characteristic backgrounds. Clearly, the most important factor is the scattering strength of a given material—assuming equal nuclear density, stronger scatterers will give higher backgrounds. Table 2 shows the basic neutron scattering properties of some of the atomic constituents of the most common pressure cell materials.

TABLE 2 Neutron Scattering Properties of Common Pressure Cell Construction Materials

Element	Coherent scattering length (fm)	Coherent cross section (barn)	Incoherent cross section (barn)	Absorption cross section at 2200 m s^{-1} (barn)
C	6.646	5.51	0.001	0.0035
Al	3.449	1.495	0.0082	0.231
O	5.803	4.232	0.0008	000019
Ti	−3.438	1.485	2.87	6.09
Fe	9.45	11.22	0.4	2.56
Zr	7.16	6.44	6.46	0.185

Data taken from Sears [73].

Crystalline microstructure is also an important factor to consider. The key differences between single and polycrystalline materials are illustrated by their respective Ewald constructions (see e.g., [74]). While the former has a discrete lattice of points in reciprocal space, the latter (assuming a perfectly, random powder) will consist of continuous spherical surfaces. Therefore, in the polychromatic scenario of a ToF experiment, where a continuum of wavelengths between the two limiting Ewald spheres is present, the single crystal will give rise to a background of spatially localized Bragg spots, while the powder will give rise to "sheets" of background scattering at constant *d*-spacing, but at different wavelengths, and scattering angles related by Bragg's law.

In addition to these coherent sources of background, a certain percentage of neutrons are scattered by nuclear interactions that randomize the neutron's phase. The intensity of this "incoherent" scattering is quantified by the incoherent cross section of the nucleus (fourth column, Table 2). Incoherent scattering is wavelength independent and uniformly distributed in space. Some elements, famously vanadium, have predominantly incoherent scattering and so give rise to only very weak Bragg peaks. Another way to obtain a "Bragg free" background is to identify combinations of materials that have negative and positive *coherent* scattering lengths such that their contributions cancel. Neutrons scattered by nuclei with a negative scattering length experience a phase change of π and so, when combined with a positive scattering material in a substitutional alloy, can cancel out the coherent scattering. An example of such a "null-scattering" alloy is $Ti_{0.68}Zr_{0.32}$ [75], which is in common use as a gasket material for Paris-Edinburgh cells and for some high-pressure gas cells.

Lastly, γ-ray background may also be an issue for certain detector technologies. The production of γ-rays can be substantial and typically arises from neutron-absorbing shielding or collimation. This is because the energetic nuclear states arising as a result of neutron capture typically decay via the emission of high-energy photons of electromagnetic radiation. The high energy of this radiation makes it difficult to shield, requiring significant thicknesses of dense absorbing material such as lead.

11.3.2 Cell Attenuation

A second important consideration is that of cell attenuation, which results in a wavelength-dependent modification of both the incident and diffracted beams. This effect must be corrected in order to obtain reliable diffraction intensities from the sample. In general, both the incident and diffracted beams experience attenuation that is due to both the pressure vessel and the sample itself. This attenuation has two main components: simple mass absorption (neutron capture) and structured attenuation due to Bragg scattering. The first of these effects is the exponential reduction in transmitted beam intensity that depends only on the path length traversed in the material, the nuclear density, and absorption cross section of the component nuclei. The second effect of "Bragg attenuation" is the result of losses in transmitted radiation due to Bragg diffraction as the beam passes through the material.

Simple mass absorption is the easiest correction to calculate as it depends only on the density and absorption cross sections of the nuclei in the materials along with the scattering geometry. As such, it is unaffected by the complexities of (in principle, pressure-dependent) microstructure. The general approach is to build a mathematical model describing the sample and container, and to use this to calculate the path lengths through each material involved. While this is based on simple geometry, the, sometimes complex, shape of the pressure vessels can introduce difficulties. Corrections of this kind have, however, been developed for X-ray work (with diamond anvil cells) [76] and more recently for neutron work at KOALA using diamond anvil cells [63].

In the case of a polycrystalline material, each Bragg reflection can be sampled by a continuous range of wavelengths at different scattering angles, according to Bragg's law. The highest possible scattering angle of 180° gives the maximum wavelength (λ_{max}) that a given reflection of d-spacing (d) is sampled as $\lambda_{max} = 2d$. Below this "Bragg Edge," the transmitted beam will be attenuated by radiation lost to Bragg reflections. Each successive d-spacing creates a new edge, resulting in a sawtooth-shaped attenuation that is most pronounced for strongly scattering materials (see Figure 5). The depth of each edge is dictated by the microstructure of the attenuator and is generally difficult to calculate. Consequently, generating an attenuation correction for this effect usually relies on experimental measurement of profiles such as those shown in Figure 5.

In the case of a single crystal (such as a diamond anvil), the attenuation only occurs at specific wavelengths, dependent on the orientation of the anvil relative to the incident beam. An example of the resultant dips in transmitted spectrum as a neutron beam passes through the two diamonds in a diamond anvil cell is shown in Figure 6. As with the polycrystalline case, these dips must be measured experimentally as their depth is determined by the mosaic spread in the crystal and is highly dependent on the strain condition of the anvil (i.e., it varies with pressure).

The DAC scenario is particularly challenging in that, unlike the polycrystalline case, the attenuation function is highly dependent on anvil orientation. While the total attenuation can be relatively straightforwardly determined by a simple transmission measurement, the individual effects of each anvil must be decoupled as only the upstream anvil affects the beam that reaches the sample.

One approach that has been developed [78] is to construct a model of the transmitted spectrum, including the effect of the two individual diamonds. This model can then be fitted to an experimental measurement of the

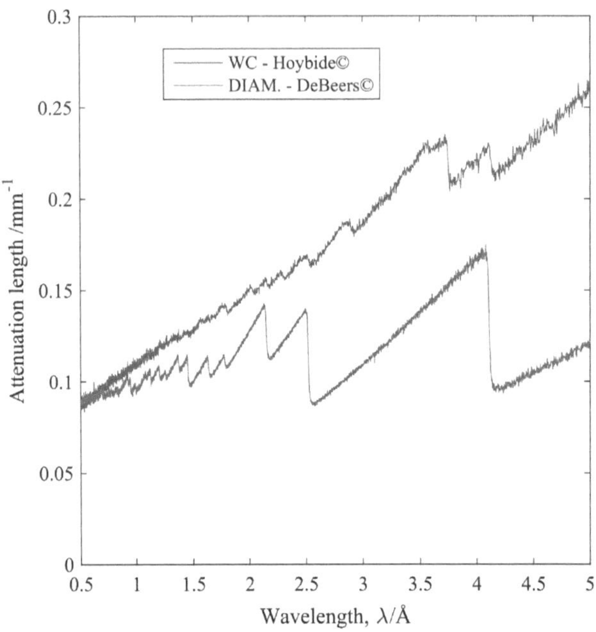

FIGURE 5 Attenuation length for tungsten carbide WC (upper trace, blue online) and sintered diamond (lower trace, red online) [77]. This nicely illustrates the difference between mass absorption and Bragg attenuation. While WC has a higher absorption, its scattering is weaker, so the sharp Bragg edges are less pronounced. Meanwhile, the diamond has a lower absorption, but much more pronounced Bragg edges as diamond is a strong coherent scatterer.

transmitted spectrum, and, from this, the individual contributions to attenuation can be extracted. The resulting attenuation profile can be used to correct the diffraction intensities for diamond attenuation.

Regardless of whether the attenuation is in the form of sharp single-crystal dips or broader Bragg edges, the effect manifests at distinct wavelengths. In diffraction, the object of the measurement is to determine the intensity of features—Bragg peaks—that do not lie at fixed wavelength, but at fixed *d*-spacing. Bragg's law describes how a given reflection, with *d*-spacing, will be measured at a different wavelength depending on scattering angle. In the case where an area detector is used, with pixels spanning a range of scattering angles, the spectra measured in each pixel must be first converted to *d*-spacing and then summed. This procedure, called diffraction focusing, ensures that all Bragg peaks line up in every pixel. The exact same process must be applied to the attenuation; however, as this is a function of wavelength (λ), it has the opposite effect of "blurring" features that are sharp in wavelength space, which will now move to different values of *d*-spacing depending on angle.

While focusing the attenuation correction, it is critical to run the summation over the same pixels that the data are measured in, taking account of

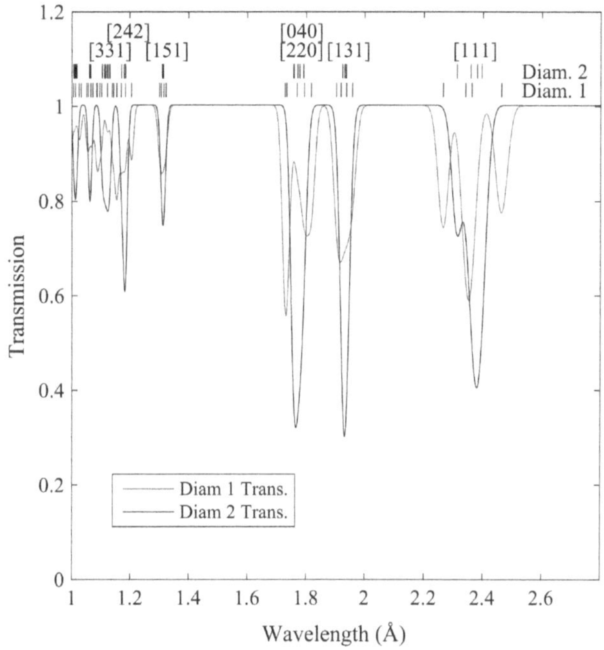

FIGURE 6 Transmission through two, approximately 3mm thick, diamond anvils: upstream anvil in light gray (red online), downstream anvil in black. Both anvils are approximately aligned with their 100 crystal axes parallel to the beam; however, slight variation in this angle can cause equivalent reflections to satisfy the Bragg condition to appear at different wavelengths. The vertical ticks above the spectra indicate predicted positions for the dips for each diamond, determined from its orientation matrix. Numbers in square brackets above the ticks give the hkl class for groups of reflections (only the first seven are shown for clarity).

those pixels that may be masked by the scattering aperture of the pressure cell. As a given sample Bragg reflection can transit into and out of a fixed wavelength dip, the correction must be weighted to reflect the number of pixels that are reflected. This can be visualized in a ToF *versus* scattering angle (2θ) plot, such as that shown in Figure 7. As described in the caption, the 200 sample Bragg reflection is only affected by the "111" anvil dip (due to the set of equivalent 111 diamond reflections) over a narrow range of 2θ angles (approx. 86–96°). This weighting must be included in the final attenuation correction.

11.4 HIGH-PRESSURE NEUTRON FACILITIES

Each neutron facility is unique and has particular expertise in specific areas of high-pressure research. The purpose of this section is to provide a snap shot of

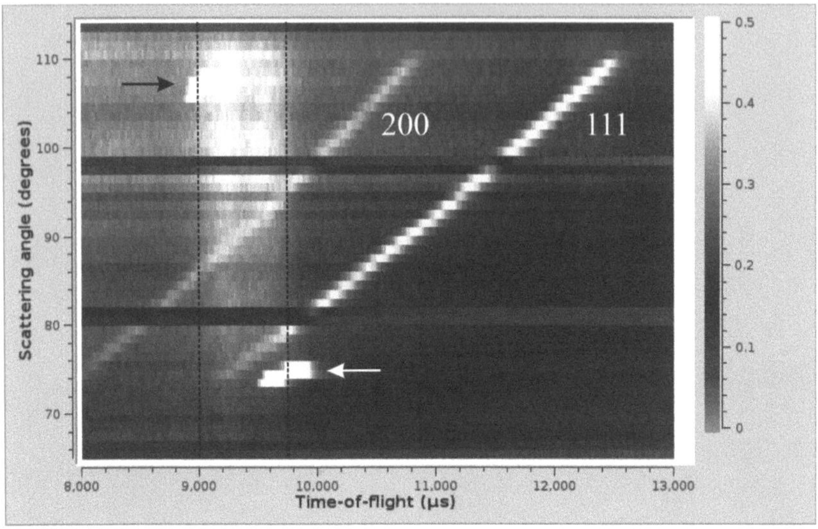

FIGURE 7 False color image of diffraction intensities of a Ni powder sample in a DAC measured in the 90° detector banks on the SNAP diffractometer at SNS. These are shown as a function of 2θ and ToF using the Mantidplot software package [79]. The ToF (or equivalently wavelength, λ) range corresponding to dips in the single-crystal diamond transmission actually appear as additional background in the image (the radiation lost in transmission rescatters to enhance the background at those particular wavelengths). So, the affected wavelengths are readily visible as a broad vertical band, delineated by black dashed lines, between 9000 and 9750 μs. The oblique lines labeled with the miller indices (200) and (111) are due to the Ni powder sample, and transit to higher ToF (or λ) as 2θ increases, according to Bragg's law. The 200 Ni peak is seen to enter the dip at approximately 87° and exit again at around 95°, outside these angles, its intensity is not attenuated.

current scope of HP neutron capabilities at a nonexhaustive subset of facilities across the world (See Figure 8). Due to its likely importance for the field, a discussion of the ESS is also included.

11.4.1 Reactor Sources

Research reactor sources have been the mainstay of neutron diffraction since their first appearance in the 1950s. Wikipedia contains an exhaustive list of 48 facilities [80], including some now defunct sources. Here, the focus is on those reactors that the author is aware have notable high-pressure activity. These are the LLB and ILL in France, the IBR-2 in Dubna, Russia, and the OPAL reactor in Sydney, Australia.

At 58 MW, the ILL [81] is one of the most powerful sources in the world. Having been operational since 1967, it now has a well-established set of high-pressure capabilities. These include "traditional" hardware, such as gas and McWhan cells that have been used since the 1970s. They also include more recent devices, such as the Paris-Edinburgh cell. Of particular note have been

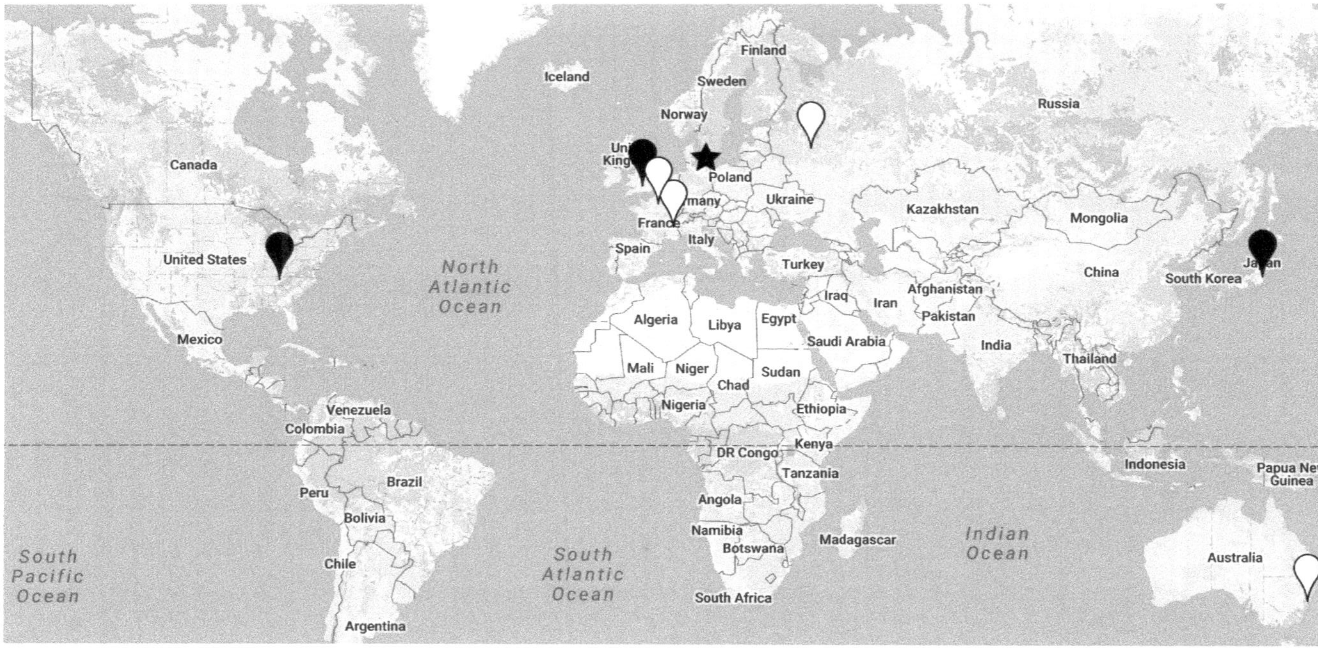

FIGURE 8 Some of the neutron sources with significant high-pressure activity that are currently in operation. Black (blue online) indicates spallation, while white (red online) indicates reactor sources (n.b. there are also reactor sources at both ORNL and J-PARC; however, the main high-pressure activities at both facilities use the spallation sources). The black (blue online) star indicates the location of the future ESS in Lund, Sweden. *Map data © 2016 Google, INEGI.*

efforts to operate the PE press at low temperature. This is challenging as the mass of the cell exceeds 70 kg; however, such measurements have been conducted as low as 3 K [82]. High-pressure single-crystal Laue diffraction was also developed at the ILL, on the VIVALDI diffractometer [64], using a novel approach of a curved image-plate detector that covered almost the entirety of reciprocal space. Recently, this program has moved to the KOALA beamline at ANSTO in Australia [63] (see below).

The LLB reactor [83] in Saclay, France, is a medium-power reactor running at 14 MW. It's most notable contributions to high pressure are on the G 6-1 diffractometer [45], although some pioneering phonon measurements were also made on the 1T1 triple-axis instrument [69]. As mentioned previously, G 6-1 was developed to have a dedicated high-pressure mode, incorporating a focusing device, which directed a relatively long-wavelength (4.749 Å) beam from a pyrolytic graphite monochromator. The use of a long wavelength was critical; in addition to enabling focusing, it also maximized detector efficiency and the effectiveness of neutron shielding, leading to very low backgrounds. A range of specially designed cells using diamond and sapphire anvils not only were able to go to record breaking (at the time) pressures of 43 GPa [84] but also in some cases were compatible with low temperatures and high magnetic fields.

Unlike the majority of reactors, which operate in steady state, the IBR-2 research reactor [85] has a unique pulsed mode of operation. A high-pressure beamline constructed at the DN-12 beamline was able to conduct both diffraction in the range of 0–10 GPa and temperatures down to 15 K and, unusually, also inelastic incoherent scattering experiments [86,87]. In the last 2 years, an upgraded high-pressure diffractometer has been established on the DN-6 beamline enabling 30 times improvement of neutron count rate. Currently, pressures up to 35 GPa have been achieved using diamond anvil cells, which can also be simultaneously cooled to 4 K [88].

Operated by the Australian nuclear science and technology organization, the OPAL reactor [89], near Sydney, Australia, has two beamlines that conduct high-pressure research. The WOMBAT high-intensity diffractometer is able to run with a PE cells up to 10 GPa. Meanwhile, the KOALA Laue diffractometer runs the continuation of the DAC work begun on VIVALDI instrument at the ILL. Recent diffraction work on KOALA has demonstrated the feasibility of determining rather complex crystallographic structures, including pharmaceuticals and amino acid structures at high pressure [63]. A unique attribute of this approach is that DAC used (a modified Merrill-Bassett design [90]) can also be used for X-ray measurements enabling dual radiation studies of the same crystal under the same conditions.

Lastly, Six Anvil Press for High pressure Radiography and diffraction (SAPHiR) [91] is under construction at the FRM II research reactor, where it will share a beamline with the POWTEX instrument. Once operational, this facility will be the only multianvil press that views a neutron beam in Europe.

11.4.2 Spallation Sources

Accelerator-driven spallation sources came into being during the 1970s through a collaboration between American and Japanese scientists. The prototype device zero-gradient synchrotron intense neutron generator-prototype (ZING-P) was built at Argonne National Laboratory between 1974 and 1975. This was followed by the construction of the first dedicated source—the intense-pulsed neutron source (IPNS) [92], which began operations in 1981.

The key difference between the spallation sources and equivalent reactors is that the possibility to extract heat from the target between pulses enables much higher instantaneous powers to be sustained. As seen in Figure 4, the instantaneous brightness of the spallation sources exceeded the most powerful reactors around 1980. In addition to flux considerations, spallation sources are politically easier to fund and build as they do not require nuclear fuel, and there is no possibility of a meltdown. At the present time, particularly in the wake of the Fukushima Daiichi reactor meltdown in 2011, governments around the world appear to be reducing investment in nuclear reactors.

A notable attribute of spallation sources is that they are naturally suited to time-of-flight (ToF) diffraction techniques, as the neutrons are generated in bunches synchronized to the accelerator pulses. Time of flight is a useful technique for high-pressure studies, as reciprocal space is explored in an energy—*versus* angle-dispersive fashion. This can be helpful as high-pressure devices often have a restricted scattering aperture. In fact, the constraints of high-pressure sample environment were one of the drivers behind the development of the ToF technique. As explained by Brugger in 1966, "A new method of neutron diffraction, the time of flight method, has been developed because it is more compatible with the geometrical restrictions of the pressurised samples" [12]. The Brugger high-pressure diffractometer at the Idaho Falls Research Reactor used a chopper to create a pulsed beam from the steady-state source. Another advantage of ToF is that volumes of reciprocal space can be accessed with stationary detectors, allowing making calibration robust and helping with easing background reduction.

Although some high-pressure research was carried out at IPNS (notably on the special environment powder diffractometer (SEPD)), the technique really demanded higher flux sources. These soon became available at the ISIS Facility [93] in Oxfordshire, the United Kingdom, and Lujan Center [94] in Los Alamos, NM, the United States, in the 1980s. This increase in available flux raised the possibility of reducing sample volumes by a factor of $\sim$10 from 1 cm^3 to $\sim$100 mm^3 while maintaining adequate signal to conduct full crystallographic analyses.

The relaxing of the volume constraint led to the development of new devices. In Europe, perhaps the most notable development was that of the Paris-Edinburgh press, mentioned previously in Section 11.2.4. Initially commissioned on the general-purpose POLARIS diffractometer, at ISIS, the cell was then moved to the TEST beamline where a purpose built diffractometer was constructed. Operational since 1996 and now called PEARL, the instrument bears notable similarities to the original beamline of Brugger 30 years earlier (see Figure 9).

In its 20 years of operation [50], PEARL has matured into an extremely well calibrated and effective diffractometer. Many years of effort invested in optimizing attenuation corrections (see Section 11.3.2) ensure that not only atomic positions but also more subtle measurements, for example, of multisite disorder and atomic motion, can be attempted [96]. It currently has a broad user base that spans the geophysical, planetary, materials, and chemical communities [50]. In addition to high pressures, up to around 23 GPa, PEARL is also capable of collecting data to temperatures as low as 80 K (e.g., [97,98]). By using an internal resistive heater, incorporated inside the gasket, temperatures of up to 2000 K [99] can be obtained, which are measured using a neutron resonance technique [100]. Currently, the high-temperature set up has a reduced maximum pressure, relative to room temperature, of around 7 GPa due to instability of the gasket [99].

In the second half of the 2000s, two new next-generation spallation sources were constructed: the SNS in Oak Ridge, TN, the United States [101], and the Material and Life Sciences Facility (MLF) in Tokai, Japan [102]. Both sources use targets consisting of circulating liquid mercury and are driven by accelerators with maximum powers exceeding 1 MW. Although operation includes some substantial technical challenges, liquid sources have the advantage of being able to be continuously cooled, which is critical at the high proton beam densities that are injected.

The large accelerator powers of the liquid sources translate to high neutron fluxes on sample. As has been amply demonstrated on the X-ray synchrotrons, these high fluxes (when concomitant with controlled background levels) are naturally suited to high-pressure research. Both the SNS and MLF have dedicated high-pressure diffraction beamlines: SNAP and PLANET, respectively, designed to optimally exploit the intense beams available. While both beamlines use the ToF diffraction techniques, interestingly, there are substantial differences between the designs of both instruments.

Constructed on the BL-11 beamline, the PLANET diffractometer has a more traditional geometry featuring two detectors both at 90°. Unlike PEARL or SNAP, it uses ^{3}He technology for its detectors, providing extremely good discrimination for background γ-radiation. The heart of the PLANET diffractometer is a custom-designed, massive six-axis multianvil press, which can apply uniaxial stress in three perpendicular directions and can simultaneously achieve up to 16 GPa and 1000 °C [103] (Figure 10). Despite weighing 30 tons, the entire press can be translated out of the beam allowing for the use of other sample environments, such as PE presses and DACs (as in Figure 11). To ensure minimal signal is observed from the sample

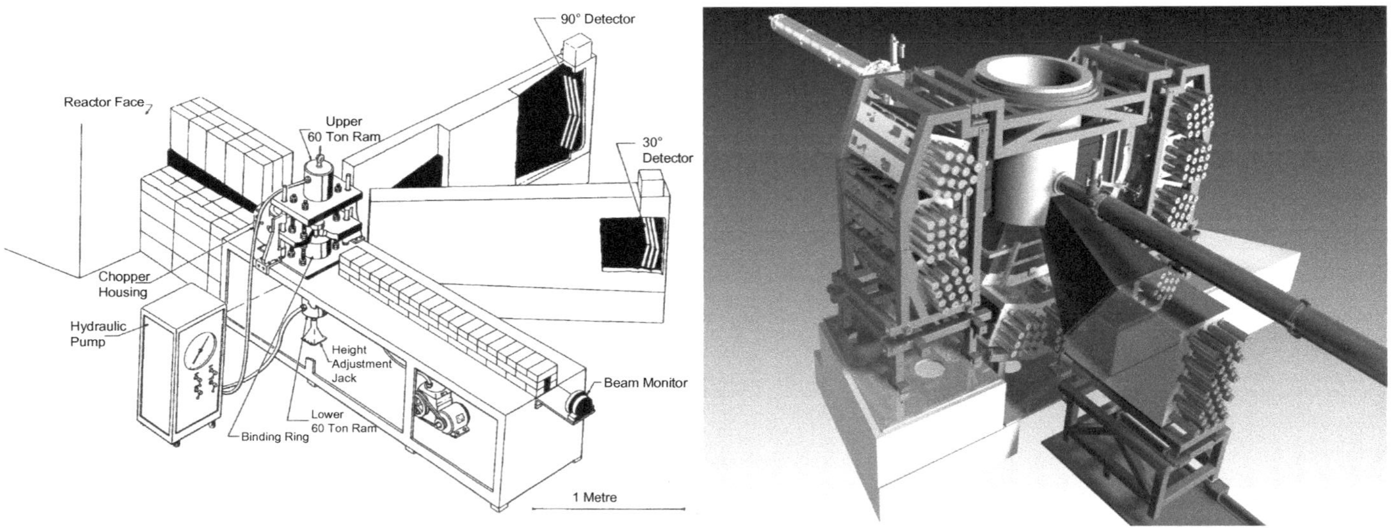

FIGURE 9 Comparison of the Brugger [95] (left) and PEARL (right) diffractometers. Both instruments use the ToF technique for diffraction. They also exploit the geometric advantages of 90° detectors, which allow the optimum situation for collimating out pressure cell scatter. PEARL improves on the Brugger design by featuring a large arc of 90° detectors, which can be summed together to maximize counting statistics. In order to extend Q range, for a given incident wavelength spectrum, both instruments feature detectors at other angles, in the case of PEARL, a second "longitudinal" bank provides access from scattering angles extending from $2\theta = 30°$ to 150°. Lastly, both instruments were constructed around, and fully optimized for, specific high-pressure sample environments.

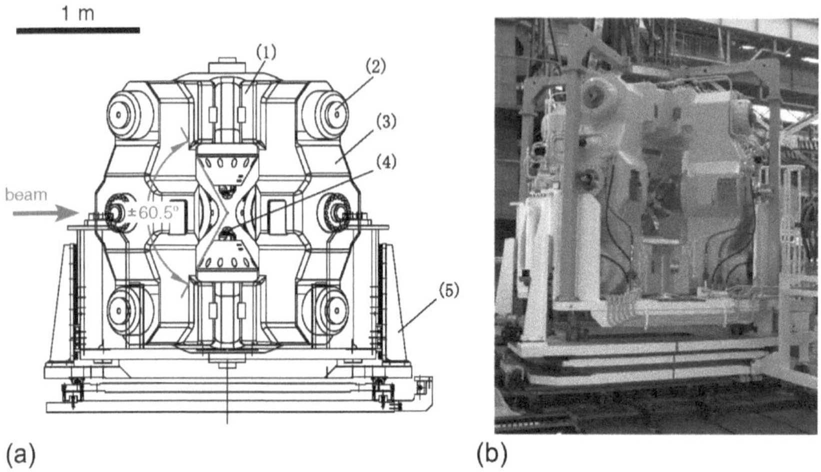

FIGURE 10 Atsuhime, the six-axis press developed especially for the PLANET diffractometer at the MLF neutron source. (a) Shows a schematic illustrating the 60.5° scattering angle that is accessible within the plane at 90° to the incident beam. (b) Shows the actual press itself, prior to its installation on the beamline. *Image reprinted with permission from Ref. [103].*

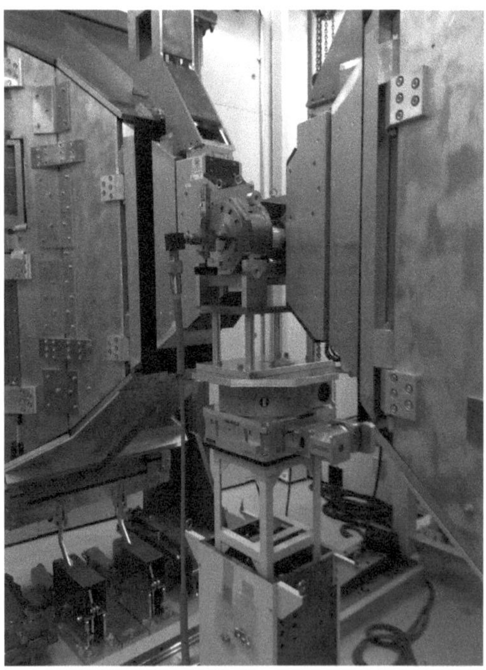

FIGURE 11 The PLANET diffractometer. A PE cell is shown installed with its load axis parallel to the incident beam, which enters from the far side of the cell. The large metal housings on both side of the cell contain shielding for the 90° diffraction detectors. Mounted to the front of these, closest to the cell, are exchangeable fine collimation elements, which can be used to define a small gauge volume within the sample, excluding background from the anvils. *Photograph by M. Guthrie.*

environment, radial collimators are used, with two optional gauge volumes (with thicknesses of 1.5–3 mm) (see Figure 11).

The SNAP diffractometer on Beamline-3 at the SNS, Oak Ridge, TN, in the United States, has adopted a distinctly different approach. Unlike PEARL and PLANET, the detectors on SNAP are movable and can be positioned over a range of angles within a horizontal plane. The sample space is also unique and bears a closer resemblance to a synchrotron than a neutron instrument being based around a six-axis sample stage (a hexapod) surrounded by an easily accessible optical bench.

The flexible design of SNAP can be an advantage in that it allows for multiple pressure cells including gas cells and PE cells, while remaining compatible with low-temperature environments via a closed-cycle refrigerator. SNAP is notable for its recent development of DACs for quantitative crystallography [14,57,104], and its highly accessible design was vital during the development phase of this technique as it permitted easy and rapid adjustments for the testing of collimation and shielding and for alignment of the pressure cell.

The implementation of DACs on SNAP required careful development of beamline components, including precision-mounted collimation, diffracted beam collimation, and the implementation of a downstream neutron monitor (the latter being vital to correct for the diamond anvil attenuation described in Section 11.3.2). In addition, and in collaboration with the Geophysical Laboratory in Washington, DC, in the United States, new diamond anvil cells were developed, which were able to increase the DAC sample volume by almost a factor of 100 (Figure 12). Although volumes are still small by neutron standards, at ~0.05 mm^3 the resulting signal was sufficient for full Rietveld refinement of the data.

FIGURE 12 Diamond anvil pressure cells used for neutron diffraction studies on SNAP *ca.* 2014. Left is the first cell used in the development work: a panoramic cell (originally designed for inelastic X-ray scattering [105]), which was used to compress specially designed, conically supported anvils. Middle is the first custom-made neutron cell, which used a sintered diamond version of the conical seats and reached 94 GPa [14]. Both of these cells are transferred to a simple load frame, driven by a gas membrane, in order to provide the required force. Right shows the current latest design, where the membrane is incorporated inside the body of the cell. *Photograph by M. Guthrie.*

A key aspect of the GL neutron DACs is the use of large, conically supported diamonds [14,106]. The conical design is advantageous as it maximizes the contact area between diamond and seat. In addition, as the axial force increases, a perpendicular, radial force is built up that supports the diamond. Typical culet diameters are currently in the 1.3–1.6 mm range, which are substantially larger than equivalent X-ray cells. Forces of order 10 metric tons are required to achieve 60–70 GPa. The current designs of cells are driven by gas-filled membrane systems, which provide the added advantage of remote *in situ* operation. As larger diamonds become available, the maximum culet size for a given pressure will likely increase.

11.4.3 The European Spallation Source

The ESS [107], which is currently under construction in Lund, Sweden, continues the historical trend towards higher power with a design specification of 5 MW. The ESS will have a solid tungsten target consisting of 36 radial sectors that must rotate in order to dissipate the enormous heat load from the powerful proton beam (that will peak at 125 MW). The rotation speed will be such that each sector will only see one pulse per revolution, allowing time for heat to be extracted, which will be achieved by a constant flow of helium coolant.

Another feature of the ESS design is that it will have a much longer pulse width than other spallation facilities ($\sim$3 ms compared with 0.1–0.2 ms, see Figure 13 [108]). An important consequence of this is that steps must be taken to mitigate the degradation of diffraction resolution that would normally arise from a long pulse. Equation (1) describes the diffraction resolution of a time-of-flight machine, $\delta d/d$, where d is the d-spacing, T the total time of flight and L the flight path as follows:

$$\frac{\delta d}{d} = \left[\delta\theta^2\cot^2\theta + \left(\frac{\delta T}{T}\right)^2 + \left(\frac{\delta L}{L}\right)^2\right]^{1/2} \tag{1}$$

The main contribution to the uncertainty in time of flight, δT, is the pulse width, as it's not possible to know whether a detected neutron originated at the beginning or end of the pulse. As seen in Eq. (1), this means that—for diffraction—the total time of flight must also be increased in order to retain a similar fractional uncertainty. As a consequence, the ESS diffraction beamlines will be much longer than those at short-pulse facilities, with some reaching lengths $\sim$160 m. With such long beamlines, the use of guides to transport neutrons from moderator to sample is essential. As these guides rapidly lose efficiency below wavelengths of $\sim$1 Å, some of the greatest advantages of the ESS will be found with cold neutrons.

The possibilities for high pressure at the ESS are enormous and multifacetted. An obvious growth area will be high-pressure diffraction studies.

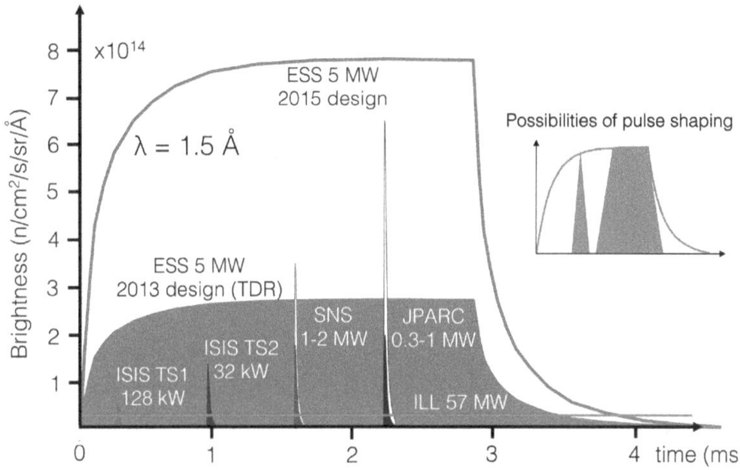

FIGURE 13 ESS pulse structure for 1.5 Å neutrons (2015 design [108]) compared with other facilities in terms of neutron brightness (neutrons per centimeter squared per second per steradian per Å bandwidth) [109]. The total number of neutrons generated per pulse is proportional to the area under the curve. Not only does the instantaneous brightness exceed any other existing spallation source, but also the time-averaged brightness is greater than even a steady-state reactor such as the ILL. Inset shows how pulse-shaping choppers can be used to select sections of the total pulse, exchanging flux for improved energy or wavelength resolution and defining specific Bragg peak shapes. This contrasts with the case of a short-pulsed source, where the Bragg peak shape is fixed by the physics of the neutron-moderator interaction.

Calculations [110] suggest that, particularly at longer wavelengths, the ESS could outperform SNAP by factors approaching 100. This should be sufficient to allow neutron DAC studies to come of age, allowing routine diffraction work at pressures well above 50 GPa.

A second growth area is likely to be in spectroscopic studies of correlated electron phenomena and quantum quasiparticles. Such studies benefit heavily from the increased flux of cold neutrons at the ESS. From a technical perspective, the main challenge of these experiments is to reach very low temperatures, sub-1 K, at relatively low <5 GPa pressures and with sample volumes sufficient for neutron spectroscopic studies. The intense beams of the ESS will enable this, as miniaturization of the sample allows much less massive cells, which are more readily cooled.

Beyond the traditional fields of high-pressure neutron diffraction and spectroscopy, it is also likely that the extreme brightness of the ESS will enable advances in high-pressure SANS, imaging and even spin echo. In order to coordinate these diverse high-pressure techniques, the ESS already has a dedicated high-pressure platform within its Scientific Activities Division. This will be responsible for developing and operating a range of sample environment and also has the broader goal of engaging the extensive and vibrant high-pressure communities in neutron science.

Lastly, there is a clear opportunity to substantially increase the pressure range available to neutron techniques by constructing state-of-the-art dedicated high-pressure neutron diffractometers. A concept for such a beamline—which would be entirely optimized around sub-mm samples and DACs—has already been proposed and garnered substantial support [110]. Such an instrument could enable neutron diffraction under pressures approaching 150 GPa and temperatures from 1 to 2000 K: a completely new regime for neutrons. However, at present, this instrument is not among the 15 beamlines funded by the initial ESS construction budget, and will have to compete in the future for one of the remaining 7 beam ports.

11.5 HIGH-PRESSURE NEUTRON SCIENCE

The remaining sections of this chapter will attempt to survey some of the high-pressure science that is conducted at existing facilities. As a result of the enormous breadth of activity, it would be impossible to cover everything, and the goal here is simply to give a flavor of what has been achieved. As is appropriate for this publication, emphasis will be given to work in biology, chemistry, and material science. In a great many fields, there is substantial room for growth, either through future technological improvements or simply by engaging new communities. For this reason, emphasis will be given to novel and emergent areas accompanied by some speculation on possible future directions.

11.5.1 Biological Systems

Neutron diffraction has much to offer in studies of biological systems, due to both the ubiquity of water in the chemistry of life and important role of hydrogen bonding in fundamental biomolecules such as proteins and larger structures such as cell membranes. Moreover, the relatively weak interaction strength of neutrons avoids the severe beam damage that remains a fundamental challenge for synchrotron X-ray studies.

As mentioned in Section 11.1, even very modest pressures—below or just exceeding 1 GPa—can induce a great diversity of interesting biophenomena. Indeed, it is not uncommon for modern high-pressure meetings to devote large fractions of the program to biological, pharmaceutical, and food science [111], and a recent survey of the field [112] runs to 730 pages.

Despite this, the role of neutrons remains relatively modest. However, there has been a recent and promising breakthrough in the field of high-pressure crystallography. A key problem for X-ray-based diffraction studies is the location of hydrogen atoms, whose positions can often only be inferred using geometric guesses, chemical intuition or *ab initio* theory. In this sense, neutron radiation, which *is* sensitive to the hydrogen location, is highly complementary to X-ray studies. However, as the structure itself is modified by high pressure; it is critical to collect both datasets under the same conditions.

As mentioned previously (Section 11.4.1), this has been achieved by Binns et al. [63] who have developed a cell suitable for *both* X-ray and neutron single-crystal diffraction allowing complementary measurements of the exact same sample under exactly the same conditions. Moreover, this cell has been successfully used to study rather complex systems including the naturally occurring amino acid L-arginine dihydrate ($C_6H_{14}N_4O_2$,2(H_2O) shown in Figure 14) [113].

Pressure-induced denaturation of proteins is another fascinating area of study. If sufficient pressure is applied (as high as 0.5–0.7 GPa in one of the earliest studied cases of egg albumin [114]), the unfolding that occurs in response to pressure is irreversible. At more moderate pressures, the unfolding is reversible and makes pressure a useful tool for exploring different conformations that may not be accessible by temperature alone (see Figure 15).

One of the most useful neutron probes of protein denaturation is small-angle neutron scattering (SANS), which is sensitive to structural details on the relevant lengthscales of 1–100 nm. And there are many studies using SANS on biological systems under high pressure, although these are currently limited to <0.4 GPa due to the large sample-size requirements.

High-pressure SANS scattering on lipid mesophases, model biomembranes, and proteins was reviewed in 2002 [115]. An example system where SANS has been used more recently is in studies of calmodulin, a protein that binds calcium and that can be *reversibly* unfolded by pressures below 0.3 GPa [116].

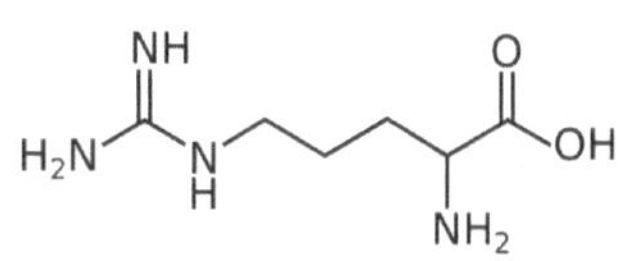

FIGURE 14 Skeletal formula of arginine.

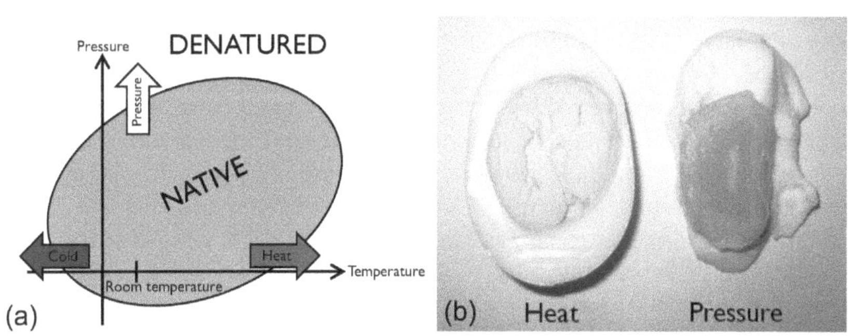

FIGURE 15 (a) shows how proteins can be denatured by either temperature of high pressure. (b) contrasts an egg denatured with heat (left) or pressure (right). As high temperatures can trigger chemical reactions, which can affect flavor, high-pressure processing is a useful alternative where raw flavors are desirable (e.g., guacamole or orange juice). *Image reprinted from Ref. [176].*

This protein exists in calcium-free and calcium-saturated forms, which unfold in different temperature regimes. The use of pressure was exploited to permit studies and comparisons of unfolding in both systems along isotherms. Another system that has been studied is casein micelles, a framework of proteins that stabilize nanocrystals of insoluble colloidal calcium in solution (and are an important milk component) have been studied up to 0.35 GPa. This study revealed evidence for irreversibility though the formation of "submicelle" structure [117].

While many biological structures are strongly affected by pressures as low as 0.1 GPa, there is also a perceived need for higher pressures. One of the goals of the Integrated Infrastructure Initiative for Neutron Scattering and Muon Spectroscopy (NMI3 [118]) was the development of a new SANS cell that has successfully reached 0.5 GPa [119]. This new capability is expected to have applications in studies of Alzheimers disease, sterilization (or "Pascalization") of food and cleaning of seawater. Over the next decade, particularly by exploiting new and upgraded neutron sources, conducting SANS measurements at pressures as high as 1.0 GPa is a clear goal.

11.5.2 Chemistry and Materials Science

One of the fascinating effects of high pressure is the ability to change the, seemingly immutable, characteristics of the elements. Traditional lines of demarcation separating, say, metals from nonmetals or covalent from ionic materials, are routinely blurred as density increases. As a consequence, high-pressure research offers unique insights to chemists and access to potentially important new materials. One way to visualize the role of pressure is to imagine it lending an additional dimension to the, otherwise flat, periodic table: each element, as a function of pressure, having many different states, each with their own bonding possibilities. This, of course, vastly increases the number of combinations of elements that can be used to form new materials. Thus, one of the key uses of high pressure in chemistry and material science is the synthesis of new materials with interesting or useful properties.

A second important use of high pressure, particularly in material science, is to enable direct and systematic measurements of structure-property relationships. Pressure can affect subtle changes in structure, such as tuning of bond lengths or octahedral tilt angles, or more dramatic changes involving fully reconstructive phase transitions. By combining structural measurements, such as diffraction, with direct *in situ* property determinations (e.g., electric transport measurements), the relation between structure and behavior can be explored in a clean and simple fashion. Conventionally, these effects are explored with doping; however, this can introduce many complications including the formation of defects and changes in electronic structure, which pressure avoids.

In the following sections, both uses of high-pressure research are explored in a number of specific cases. While far from being exhaustive, these examples have been chosen to be illustrative of key scientific challenges that may become tractable in the next decade.

11.5.2.1 Light Elements and Superconductivity

Much of traditional high-pressure research has focused on the behavior of simple systems. At the present time, essentially all of the elements of the periodic table have been studied to the multiple 100's of GPa accessible to modern synchrotron techniques. These detailed investigations have collectively and decisively disproved long-standing assumptions that, at the highest densities, all elements must eventually adopt simple, close-packed structures. In contrast, unexpected behavior appears to be the norm and surprising levels of complexity have been observed in even the simplest systems.

A good example is the behavior of the light alkali metals Li and Na: both exhibit maxima in their melting curves (a negative melting line indicating that the liquid state is denser than the solid, according to the Clausius-Clapeyron relation [120]), and both exhibit rather complex crystalline structures in the solid at high pressure. Lithium, for example, has a melting temperature of 190 K at around 45 GPa, which is lower than its ambient pressure-melting temperature by over 250 K [121] (although the exact value of the melting temperature is contested [122], there is general agreement on the existence of a melting minimum under high pressure that is colder than at ambient). Moreover, in the vicinity of the melting temperature minimum, complex crystal structures with as many as 88 atoms in the unit cell are found [121] (at ambient pressure, Li has a simple body-centered cubic structure with only two atoms in the unit cell). Even greater complexity is found in Na, which has a structural phase with 512 atoms in the unit cell at ~118 GPa [123].

In focusing on neutron studies, however, a different light element is inarguably the most prominent: hydrogen, along with its heavy isotope deuterium. Consisting of a single proton and a single electron (plus a neutron in the case of deuterium) and adopting a simple hexagonal close packed structure in the solid state at low pressure, this simplest of atoms is expected to exhibit highly exotic and complex behavior as pressure intensifies. The prediction by Wigner and Huntington that hydrogen would metallize under pressure, becoming the lightest alkali metal, dates to 1935 [4]. Later, toward the end of the 1960s, Neil Ashcroft pointed out that, were metallic hydrogen created, it would likely be a room-temperature superconductor [124]. More recently, in 2004, it was also speculated that—also at high pressure—this same metallic hydrogen would have a liquid ground state, transitioning from a superconductor to a superfluid at low temperatures [125].

Neutron diffraction would clearly be an invaluable tool for studying this fascinating behavior; however, the enormous pressures required have mostly inhibited its use. When Wigner and Huntington predicted metallization, they speculated that "even under the assumption of the most advantageous compressibility...the pressure necessary is 250,000 atm (25 GPa)" [4]. Indeed, the research conducted since has shown that Wigner and Huntington's calculation was an enormous underestimate. In the interim, there have been multiple experimental claims of the observation of metallic hydrogen [126,127]. One of the more recent claim by Eremets and Troyan put the pressure at 260–300 GPa [128]; however, this proved to be controversial (a rebuttal being published within months of Eremets' publication [129]). Very recent work extends studies of H, D, and HD mixtures up to almost 400 GPa, using optical spectroscopy and has mapped out the phase diagram under these extreme conditions, revealing several new phases [130,131] (Figure 16). The highest pressure phase V observed to be stable in H_2 up to 380 GPa appears to remain molecular, although with tantalizing indications of the onset of metallization [130].

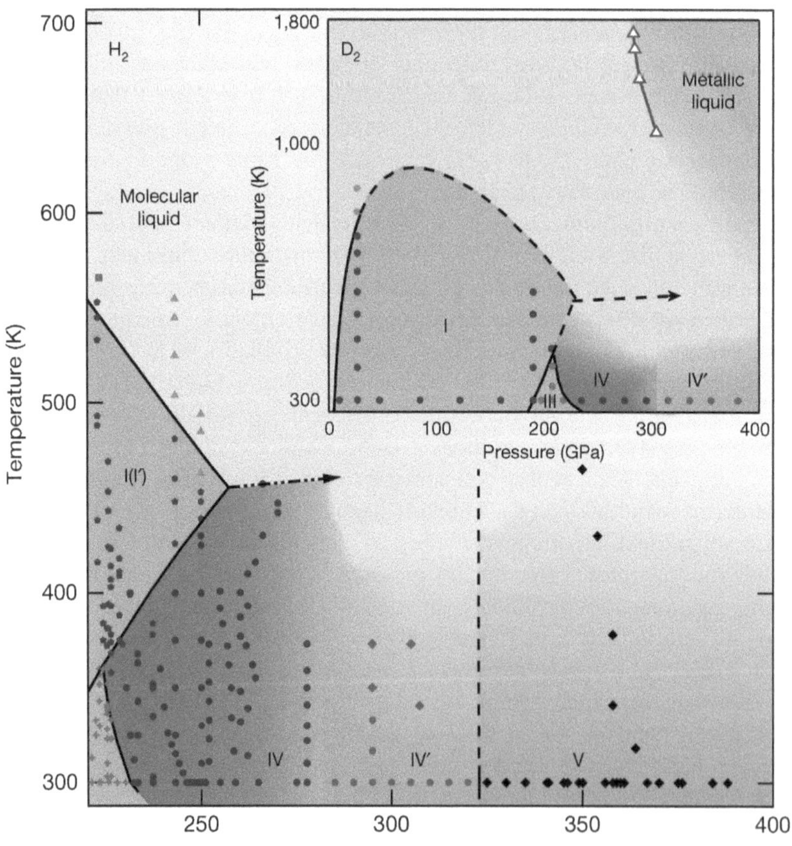

FIGURE 16 High-pressure phase diagrams of hydrogen (main) and deuterium (inset). *Image reprinted with permission from Ref. [131].*

The highly controversial nature of the high-pressure states of hydrogen would benefit enormously from accurate structural information from neutron diffraction. At present, the only information on possible structure comes from theory [132]. As such, this hydrogen/deuterium system provides a strong scientific driver for neutron diffraction capability in the 200–300 GPa pressure range.

To date, the highest pressure published neutron diffraction data on deuterium were measured by Goncharenko and Loubeyre [40]. Operating at the absolute limit of the technology of the time, they were able to measure diffraction intensities from the three strongest reflections of hexagonal close-packed phase I deuterium (the 101, 100, and 110 reflections) from a single crystal of D_2 (embedded in a solid He pressure medium) at 38 GPa. By cooling to $\sim$5 K, they crossed the transition line to phase II, the so-called "broken-symmetry" phase, where the D_2 molecules lose their perfect quantum spherical symmetry. On the basis of these (and X-ray diffraction) data, Goncharenko and Loubeyre claimed that phase II is incommensurate, with only local orientational order, a structure that remains incompatible with the many theoretical predictions [133].

As discussed previously (see Section 11.4.2), recent advances in DAC instrumentation combined with more powerful neutron sources have demonstrated the feasibility of neutron diffraction at pressures approaching or even exceeding 100 GPa. At the present time, it seems likely that the enhanced brightness of the ESS—which is optimal for small, high-pressure samples—may substantially increase this limit, potentially even allowing neutron diffraction in the regime of the phase I–IV-liquid triple point of deuterium (although experiments at these extremes would likely require an investment in a dedicated high-pressure instrument).

Another concept that has been explored is that there may be alternative routes to realize the exotic phenomena predicted to occur in high-pressure hydrogen. In 2005, Neil Ashcroft proposed that this might be achieved by using covalent bonds to pull hydrogen atoms into extremely close proximity, simulating the effect of pressure [134]. This suggestion triggered an avalanche of theoretical and experimental work [135–144] that focused particularly on the group 14 hydrides (silane, etc.), solvated electron systems like Li/NH_3 [145], and hydrides of the alkali and alkaline earth elements (KH_8 and NaH_8). The diverse structural phenomena uncovered have been studied with theory, spectroscopy, and X-rays, but structural detail on the H(D) atoms remains hidden without neutron diffraction at the required pressures.

Taking silane as a specific example—in the 30–90 GPa range, this complex system exhibits insulator-metal transitions [135,139,146,147], high-temperature superconductivity [136], P-induced amorphization, and polymerization [138,148]. Similar phenomena were also found in germane [149] and stannane [137]. Neutrons are clearly required to locate H/D atoms bonded to increasingly heavy Group 14 atoms, and such studies may also become feasible in the next decade.

Beyond hydrogen-rich materials, another system that demands neutron investigation at pressures higher than those currently within reach are alloys formed in the Li/Be/B ternary system. Again, interest here is the identification of new routes to high T_c superconductivity. For example, Li and Be, which are immiscible at ambient conditions, have been shown by calculations [150] to exhibit a range of stable stoichiometries above 20 GPa. Further calculations show more structures stabilizing at pressures up to 80 GPa [151]. As both Be and B^{11} have extremely high coherent neutron cross sections, neutrons are an ideal structural probe of stoichiometry and structure.

The preceding examples show ways in which new superconductors might be synthesized using high pressure. However, systematic studies of existing materials are equally important. Traditionally, the experimental parameter in such investigations has been composition; however, pressure provides a simple and "clean" variable to probe the behavior of interest. Indeed, high pressures can directly induce superconductivity, increasing the number of elements that superconduct by over 76% [152] (see Figure 17). Pressure can

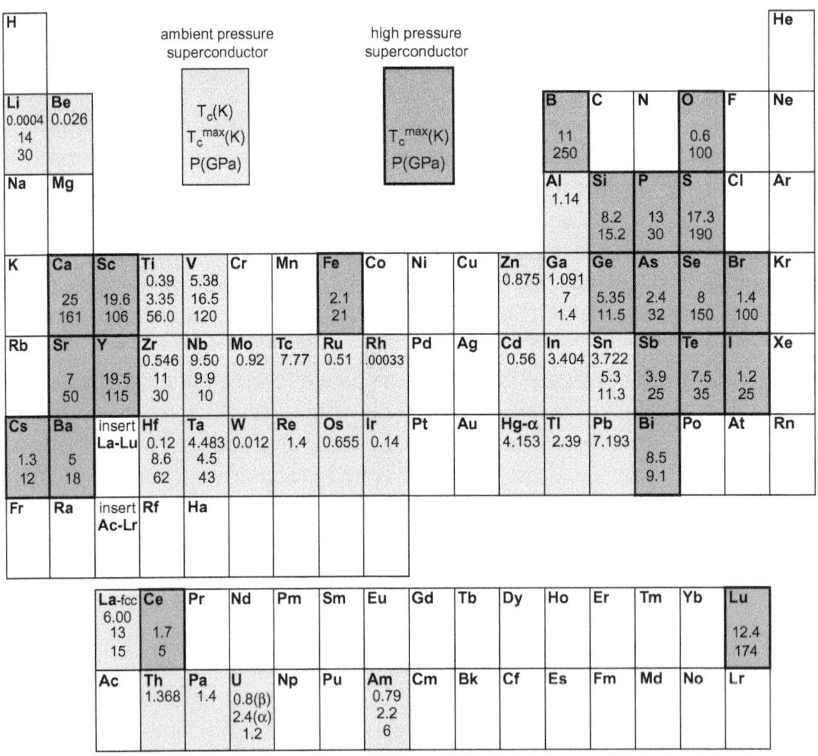

FIGURE 17 Periodic table showing ambient pressure superconducting elements (shaded, yellow online) and those additional elements that superconduct at high pressure (shaded, bold outline, green online). *Image reprinted with permission from Ref. [152].*

also tune the superconducting temperature T_c, and the highest values ever recorded have been measured under high pressure: an earlier record T_c of 164 K found at 31 GPa in a Hg-based cuprate [153,154] was recently exceeded when superconductivity was observed in the hydrogen-sulfide system at ~200 K and pressures above 150 GPa [155].

11.5.2.2 New Carbon Architectures

The most famous example of high-pressure synthesis of industrially useful material (already discussed in Section 11.2.1) is the direct conversion of graphite to diamond. The graphite-diamond phase transition is highly kinetically hindered. Indeed, simple "cold" compression of graphite does not form diamond, leading instead to a poorly understood form of carbon that, while transparent under high pressure, back converts to opaque, electrically conducting graphite on decompression (see review by Wang and Lee [156]). This intriguing phase was recently revisited in a theoretical study, which proposed it is a new carbon allotrope with Cmmm symmetry, stable above 10 GPa [157].

In order to overcome the deep kinetic barriers in the carbon system, both pressure *and* heat must be applied simultaneously. Once diamond has formed, its exceptionally strong network of covalent bonds creates a deep internal energy minimum, enabling its metastable recovery to ambient pressure. At ambient pressure, the energy barrier converting back conversion of diamond to graphite is so high that temperatures exceeding 1700 °C are needed to back convert [158] (however, if exposed to oxygen during heating, diamond will oxidize, forming carbon dioxide at the much lower temperature of 700 °C [159]).

The minimum required conditions of diamond synthesis (using a catalyst, such as iron) of ~6 GPa and ~1400 °C [160] are accessible to modern industrial techniques and, today, synthetic diamond is produced in ~20 metric ton amounts annually in the United States alone [161]. However, diamond can also be synthesized under conditions of ultrahigh vacuum using chemical vapor deposition (CVD) technology. At the time of writing, single crystals of CVD diamond are being grown at volumes exceeding 1.5 cm^3 (>26 carat) and there is no intrinsic reason, beyond cost, why they could not be substantially larger [162].

However, interest in carbon materials goes far beyond diamond. In recent years, various research groups have begun to study the response of various different carbon architectures to high pressures. For example, extensive work has been carried out on carbon nanotubes under pressure, which can be transformed into families of allotropes with important mechanical properties including superhardness and high-tensile strength [163]. Recent theoretical work has highlighted a sensitivity to hydrostaticity [164] with different stress environments leading to different materials and transition conditions.

Amorphous and mesoporous carbons have also been compressed, leading respectively to amorphous [165] and mesoporous [166] diamond-like materials, respectively, both of which can both be recovered metastably to ambient pressure. Even an amorphous carbon aerogel has been transformed into an ultralow-density nanocrystalline diamond form [167]. Neutron studies of these nano- and mesoarchitectured materials offer substantial advantages over X-ray techniques, as they can access the high values of momentum transfer needed to obtain high-resolution information on the local structural arrangements on this weak X-ray scatterer. However, few of these systems have been studied with neutrons as the large pressures involved limit the maximum sample sizes. Indeed, neutron studies of noncrystalline carbons may be a growth area as increasingly powerful sources become available.

Beyond pure carbon, high-pressure studies of benzene have also yielded entirely new nanophases materials. The ability of pressure to polymerize benzene—leading to a whitish solid after removal of pressure—has been known for over a century [168]. However, the nature of the material formed remained a mystery. As it happened, the key to uncovering the structure of polybenzene was pure serendipity, when it was accidentally discovered that, when the polymer is downloaded from high pressure at a slow rate of ~1 GPa/h, it crystallizes. Detailed studies of the recovered crystalline polymer, including neutron pair-distribution measurements, revealed the formation of extended 1-D structures, with diamond-like sp^3 bonded carbons forming a "nanothread" with a diameter of only 3.6 Å that may extend for many 100's of nm in length [7]. As with diamond, this material can be recovered to ambient pressure and, calculations show that it has the highest tensile strength of any material [169].

11.6 OTHER FUTURE APPLICATIONS OF HIGH PRESSURE

There remain a great many possibilities for growing the field of high-pressure neutron science, many of which are already being enabled for the first time by the latest generation of high-brightness sources. With the ESS on the horizon, even more may be possible within the next decade. Here, we look at some "nonmainstream" applications for high pressure that may be able to develop in the near future.

As discussed earlier in Section 11.2.4, an area that is already seeing early developments is high-pressure vibrational spectroscopy. Much of this activity is occurring at the SNS, primarily using the VISION vibrational spectrometer [170]. VISION is unique in that its design enables count rates two orders of magnitude higher than other comparable instruments.

Some of the first test measurements on VISION involved studies of high-pressure synthesized materials, which were recovered to ambient pressure before measurement. Despite only relatively tiny amounts of material being available (~7 mg), quantitative vibrational spectra have been obtained [171]. This important proof of concept then led to the design and construction of a dedicated DAC for INS. The current version has a sample volume exceeding 1.5 mm^3 and has been compressed to 5 GPa with usable vibrational spectra measured (see Figure 18). INS measurements under such conditions are unprecedented, and look set to open up a whole new field of science.

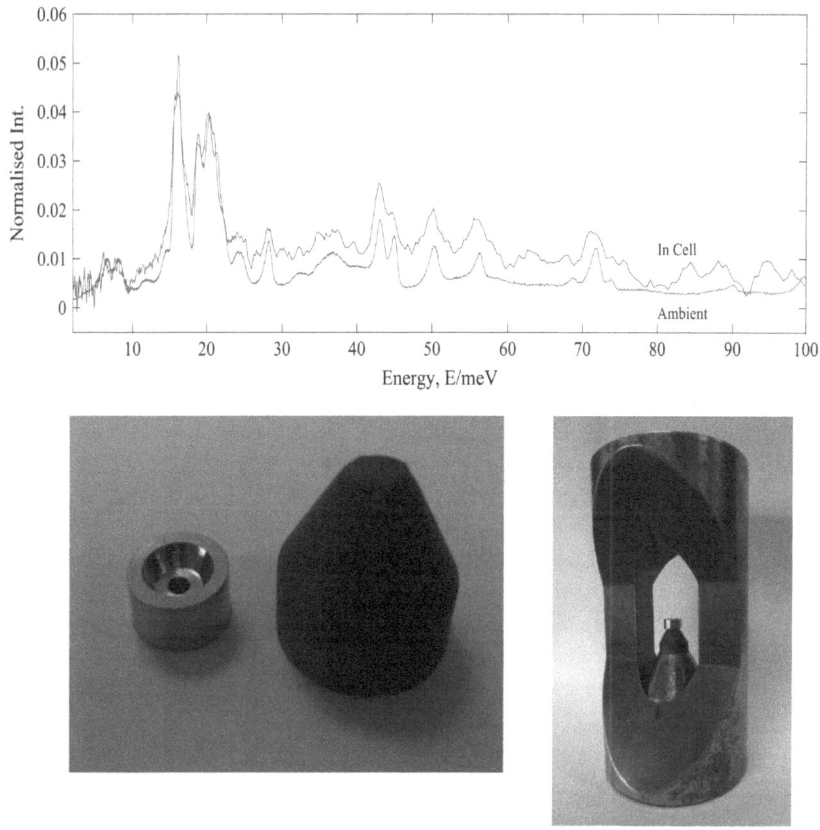

FIGURE 18 Top shows vibrational spectra from hexamethylbenzene measured inside a DAC with single-crystal anvils, compared with a reference sample of 440 mg measured at ambient. Bottom right shows a new prototype cell, which was specifically designed for VISION and uses large polycrystalline diamond composite anvils. Bottom left shows one of these anvils and a specially designed stainless steel gasket [172].

Other inelastic spectroscopies that are sensitive to the extremely weak excitations associated with quasiparticles, such as magnons, polarons would also benefit from developments in instrumentation and high-pressure sample environment. Pressure is a perfect way to tune these excitations as it can introduce strong modifications without chemically altering the material under study. Such science is a key target for the ESS spectrometer BIFROST [173], scheduled to enter user operations in 2023. The key technical challenge for such work is to develop pressure cells that not only have sufficient volume but also are able to access very low temperatures below 1 K. Combining these extreme conditions with those of intense magnetic fields is another key goal.

Further new high-pressure science will be possible at both of the high-pressure neutron beamlines that have a multianvil capability (PLANET at the MLF and SAPHiIR [91] under construction at the FRM II reactor).

Among other applications, these will enable an imaging capability. Although X-ray radiography and tomography under high pressure is well developed, very little work has been done using neutrons. The high absorption of hydrogen makes such studies particularly useful for examining water diffusion through mineral assemblies, with applications in the study of subduction zones. Another strongly absorbing element is Li, which leads to the possibility of *in operando* neutron imaging studies of lithium ion batteries.

Neutron imaging studies can also be conducted with PE presses, and some test data from one of these are shown in Figure 19. While these images illustrate the potential of the technique, improvements in the possible experimental resolution would enable much more detail to be observed. Also required would be sophisticated attenuation corrections, which can take account of the changing geometry of the cell as the pressure is increased. Lastly, incorporating a rotator, such as that developed for single-crystal measurements in the PE press [175], would enable full tomographic studies.

Also mentioned previously (e.g., in Section 11.5.1) has been SANS work using pressure cells. The vast majority of this work has previously been conducted in the 0.1–0.2 GPa range, using gas cells and sapphire-windowed devices. Although sapphire-windowed SANS cells exist, diamond windows have not yet been implemented. If diamonds of sufficiently large size become available in the next decade, implementing them in SANS devices would be likely to increase the maximum achievable pressures fivefold to exceed 1 GPa.

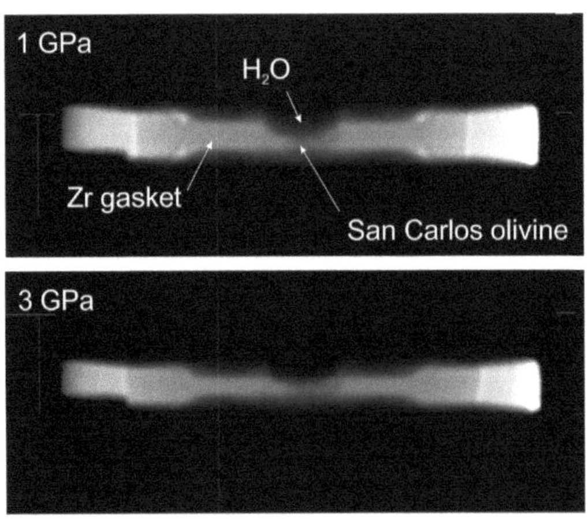

FIGURE 19 Neutron transmission data of a sintered plug of San Carlos olivine (courtesy B. O. Mysen) in contact with liquid H_2O at room temperature in a PE press. The WC anvils above and below the sample are effectively opaque at this energy; however, the transparent Zr gasket allows the water-mineral interface to be observed clearly up to at least 3 GPa. These data were collected on beamline CG-1D at the HFIR reactor at ORNL with the assistance of H. Bilheux and J.J. Molaison [174].

11.7 SUMMARY

From our current vantage point in the second half of the 2010s, dedicated high-pressure instrumentation at the newest facilities is realizing substantial breakthroughs in high-pressure neutron science. Meanwhile, ongoing improvements in detector design, neutron optics, and high-pressure sample-environment hardware continue apace. When viewed in the context of still brighter sources, such as the ESS, which will begin to come on line by the end of the decade, there is a real sense of momentum in the field. Particularly in the fields of biology, chemistry, and material science, there is substantial scope for growth, either through substantial increases in maximum pressure, minimum temperature or even the implementation of completely new experimental techniques.

APPENDIX A

Summary of Typical Devices Used for High-Pressure Neutron Work

Cell type	Description	Current max. pressure (GPa)	Sample vol (mm^3)
Gas	A thick-walled cylinder connected to a high-pressure gas supply	<1	2000
Clamp	A supported piston-cylinder-type device where an applied force is locked in by a clamping mechanism	2–3	2000
Paris-Edinburgh/TAP-98	Moderate 50–400 T capacity hydraulic presses used to force opposed anvils together. The sample is contained by a metal gasket. These devices have also been used to drive modified multianvil set-ups	23	30–100
Multianvil	Used on the PLANET beamline [103] at the MLF and the SAPHiR beamline [91] at the FRM II reactor source, a massive device that compresses an assembly of anvils. In a typical assembly, eight cubic anvils with truncated vertices are placed together such that the adjacent vertices form an octahedral sample space	15	30–100
Gem-based	Opposed diamond or sapphire anvil cells optimized for neutrons. The sample is contained in a metal gasket	100	0.05–1.5

REFERENCES

[1] R. Jeanloz, Annu. Rev. Phys. Chem. 40 (1) (1989) 237–259.

[2] S. Desgreniers, Y.K. Vohra, A.L. Ruoff, J. Phys. Chem. 94 (3) (1990) 1117–1122.

[3] M.I. Eremets, K. Shimizu, T.C. Kobayashi, K. Amaya, J. Phys. Condens. Matter 10 (49) (1998) 11519.

[4] E. Wigner, H.B. Huntington, J. Chem. Phys. 3 (12) (1935) 764–770.

[5] A. Koufos, D. Papaconstantopoulos, J. Supercond. Nov. Magn. 28 (12) (2015) 3525–3533.

[6] Y.M. Ma, M. Eremets, A.R. Oganov, Y. Xie, I. Trojan, S. Medvedev, et al., Nature 458 (7235) (2009) 182–185.

[7] T.C. Fitzgibbons, M. Guthrie, E.-S. Xu, V.H. Crespi, S.K. Davidowski, G.D. Cody, et al., Nat. Mater. 14 (1) (2015) 43–47.

[8] A. Kolesnikov, V.G. Kutcherov, A.F. Goncharov, Nat. Geosci. 2 (8) (2009) 566–570.

[9] M.E.G. Hendrickx, D. Knor (Eds.), Ultra High Pressure Treatment of Foods, Springer Science + Business Media, LLV, New York, 2012.

[10] A. Sharma, J.H. Scott, G.D. Cody, M.L. Fogel, R.M. Hazen, R.J. Hemley, et al., Science 295 (5559) (2002) 1514–1516.

[11] D. Vanlint, R. Mitchell, E. Bailey, F. Meersman, P.F. McMillan, C.W. Michiels, et al., MBio 2 (1) (2011) 1–3.

[12] R.M. Brugger, R.B. Bennion, T.G. Worlton, E.R. Peterson, Neutron Diffraction Studies of Samples at High Pressures IDO-17170, AEC Research and Development Report, 1966.

[13] L. Dubrovinsky, N. Dubrovinskaia, V.B. Prakapenka, A.M. Abakumov, Nat. Commun. 3 (2012) 1163.

[14] R. Boehler, M. Guthrie, J.J. Molaison, A.M. dos Santos, S. Sinogeikin, S. Machida, et al., High Pressure Res. 33 (3) (2013) 546–554.

[15] R.L. Hills, Power From Steam: A History of the Stationary Steam Engine, Cambridge University Press, Cambridge, 1989.

[16] C.A. Parsons, Philos. Trans. R. Soc. Lond. A Math. Phys. Eng. Sci. 220 (571–581) (1920) 67–107.

[17] S. Klotz, Techniques in High Pressure Neutron Scattering, CRC Press, Taylor & Francis Group, Boca Raton, FL, 2013.

[18] P.W. Bridgman, J. Appl. Phys. 12 (6) (1941) 461–469.

[19] Diamond synthesis was achieved one year earlier, in 1953, in Stockholm, Sweden by the ASEA (Allmänna Svenska Elektriska Aktiebolaget) Company. However, the complicated process was not commericialised and the result wasn't reported until decades after that of GE.

[20] H.T. Hall, Rev. Sci. Instrum. 31 (2) (1960) 125–131.

[21] H.T. Hall, Rev. Sci. Instrum. 29 (4) (1958) 267–275.

[22] R.C. Liebermann, High Pressure Res. 31 (4) (2011) 493–532.

[23] D.R. Mccann, L. Cartz, R.E. Schmunk, Y.D. Harker, J. Appl. Phys. 43 (4) (1972) 1432–1436.

[24] http://university.langantiques.com/index.php/A_History_Of_Diamond_Cutting (accessed January 3, 2017).

[25] G.E. Harlow (Ed.), The Nature of Diamonds, Cambridge University Press, Cambridge, United Kingdom, 1998.

[26] C.E. Weir, E.R. Lippincott, A. Vanvalkenburg, E.N. Bunting, J. Res. Natl. Bur. Stand. 63 (1) (1959) 55–62.

[27] Valkenburg A.V., Conference Internationale Sur-les-Hautes Pressions, LeCreusot, Saonne-et-Loiure, France, 1965.

[28] G.J. Piermarini, S. Block, J.D. Barnett, R.A. Forman, J. Appl. Phys. 46 (6) (1975) 2774–2780.
[29] R.A. Forman, S. Block, J.D. Barnett, G.J. Piermari, Bull. Am. Phys. Soc. 17 (3) (1972) 351–357.
[30] R.A. Forman, S. Block, J.D. Barnett, G.J. Piermarini, Science 176 (4032) (1972) 284–285.
[31] G.J. Piermarini, S. Block, Rev. Sci. Instrum. 46 (8) (1975) 973–979.
[32] H.K. Mao, P.M. Bell, Science 200 (4346) (1978) 1145–1147.
[33] J. Nielsen, Elements of Modern X-Ray Physics, John Wiley, Chichester, West Sussex, UK, 2011.
[34] A.C. Thompson, D.T. Attwood, E.M. Gullikson, M.R. Howells, J.B. Kortright, A.L. Robinson, et al., X-Ray Data Booklet, Published by Lawrence Berkeley National Laboratory, University of California, Berkely, California, 2001.
[35] R.M. Brugger, Phys. Today 21 (12) (1968) 23–30.
[36] J. Carpenter, W. Yelon, Meth. Exp. Phys. 23 (A) (1986) 99–196.
[37] D.B. Mcwhan, D. Bloch, G. Parisot, Rev. Sci. Instrum. 45 (5) (1974) 643–646.
[38] S.P. Besedin, I.N. Makarenko, S.M. Stishov, V.P. Glazkov, I.N. Goncharenko, A.V. Irodova, et al., High Pressure Res. 4 (1990) 447–449.
[39] V.P. Glazkov, S.P. Besedin, I.N. Goncharenko, A.V. Irodova, I.N. Makarenko, V.A. Somenkov, et al., JETP Lett. 47 (12) (1988) 763–767.
[40] I. Goncharenko, P. Loubeyre, Nature 435 (7046) (2005) 1206–1209.
[41] I.N. Goncharenko, High Pressure Res. 24 (1) (2004) 193–204.
[42] I.N. Goncharenko, P. Cadavez-Peres, I. Mirebeau, O.L. Makarova, T. Le Bihan, M. Mezouar, Phys. Rev. B 68 (21) (2003) 214418.
[43] I.N. Goncharenko, O.L. Makarova, L. Ulivi, Phys. Rev. Lett. 93 (5) (2004) 055502.
[44] I.N. Goncharenko, I. Mirebeau, Phys. Rev. Lett. 80 (5) (1998) 1082–1085.
[45] http://www-llb.cea.fr/spectros/pdf/g61-llb.pdf (accessed January 3, 2017).
[46] J.M. Besson, R.J. Nelmes, G. Hamel, J.S. Loveday, G. Weill, S. Hull, Phys. B 180 (1992) 907–910.
[47] L.G. Khvostantsev, V.N. Slesarev, V.V. Brazhkin, High Pressure Res. 24 (3) (2004) 371–383.
[48] J.M. Besson, G. Weill, G. Hamel, R.J. Nelmes, J.S. Loveday, S. Hull, Phys. Rev. B 45 (6) (1992) 2613–2619.
[49] S. Klotz, J.M. Besson, G. Hamel, R.J. Nelmes, J.S. Loveday, W.G. Marshall, et al., Appl. Phys. Lett. 66 (14) (1995) 1735–1737.
[50] C.L. Bull, N.P. Funnell, M.G. Tucker, S. Hull, D.J. Francis, W.G. Marshall, High Pressure Res. 36 (4) (2016) 493–511.
[51] T. Strassle, A.M. Saitta, Y. Le Godec, G. Hamel, S. Klotz, J.S. Loveday, et al., Phys. Rev. Lett. 96 (6) (2006) 067801.
[52] S. Klotz, G. Hamel, J.S. Loveday, R.J. Nelmes, M. Guthrie, A.K. Soper, Phys. Rev. Lett. 89 (28) (2002) 285502.
[53] M. Wilding, M. Guthrie, C.L. Bull, M.G. Tucker, P.F. McMillan, J. Phys. Condens. Matter 20 (24) (2008) 244122.
[54] P.S. Salmon, J.W.E. Drewitt, D.A.J. Whittaker, A. Zeidler, K. Wezka, C.L. Bull, et al., J. Phys. Condens. Matter 24 (43) (2012) 439601.
[55] M. Guthrie, C.L. Bull, R.J. Nelmes, J.S. Loveday, T. Strassle, S. Klotz, Acta Crystallogr. A61 (2005) c94.
[56] P.S. Salmon, A. Zeidler, J. Phys. Condens. Matter 27 (13) (2015) 133201.
[57] M. Guthrie, R. Boehler, C.A. Tulk, J.J. Molaison, A.M. dos Santos, K. Li, et al., Proc. Natl. Acad. Sci. U. S. A. 110 (26) (2013) 10552–10556.

[58] C.J. Ridley, K.V. Kamenev, Z. Kristallogr. 229 (3) (2014) 171–199.
[59] I.N. Goncharenko, Phys. Rev. Lett. 94 (20) (2005) 205701.
[60] I.N. Goncharenko, P. Cadavez-Peres, I. Mirebeau, O.L. Makarova, Europhys. Lett. 54 (6) (2001) 807–812.
[61] I.N. Goncharenko, I. Mirebeau, Europhys. Lett. 37 (9) (1997) 633–638.
[62] M.K. Jacobsen, C.J. Ridley, A. Bocian, O. Kirichek, P. Manuel, D. Khalyavin, et al., Rev. Sci. Instrum. 85 (4) (2014) 043904.
[63] J. Binns, K.V. Kamenev, G.J. McIntyre, S.A. Moggach, S. Parsons, IUCrJ 3 (Part 3) (2016) 168–179.
[64] G.J. McIntyre, L. Melesi, M. Guthrie, C.A. Tulk, J. Xu, J.B. Parise, J. Phys. Condens. Matter 17 (40) (2005) S3017–S3024.
[65] R. Lechner, G. Quittner, Phys. Rev. Lett. 17 (25) (1966) 1259–1261.
[66] A.V. Belushkin, M.A. Adams, S. Hull, L.A. Shuvalov, Solid State Ion. 77 (1995) 91–96.
[67] J. Rubín, J. Bartolomé, M. Adams, K. Stroessner, Solid State Commun. 97 (3) (1996) 229–233.
[68] A.V. Belushkin, M.A. Adams, Phys. B Condens. Matter 234 (1997) 37–39.
[69] S. Klotz, Z. Kristallogr. 216 (8) (2001) 420–429.
[70] T. Strassle, A.M. Saitta, S. Klotz, M. Braden, Phys. Rev. Lett. 93 (22) (2004) 225901.
[71] S. Klotz, T. Strassle, L.E. Bove, Appl. Phys. Lett. 103 (19) (2013) 193504
[72] L.E. Bove, S. Klotz, T. Strassle, M. Koza, J. Teixeira, A.M. Saitta, Phys. Rev. Lett. 111 (18) (2013) 185901.
[73] V. Sears, Neutron News 3 (3) (1992) 26–37.
[74] N.W. Ashcroft, N.D. Mermin, Solid State Physics, Holt, New York, 1976.
[75] https://www.ill.eu/instruments-support/sample-environment/equipment/high-pressures/tizr-null-matrix-alloy/ (accessed January 2, 2017).
[76] R.J. Angel, J. Appl. Cryst. 37 (2004) 486–492.
[77] C.L. Bull, Personal Comm., 2016.
[78] M. Guthrie, C.G. Pruteanu, M.-E. Donnelley, J. Molaison, A.M. dos Santos, J.S. Loveday, et al., J. Appl. Cryst. 50 (1) (2017) 76–86. in press.
[79] J. Taylor, O. Arnold, J. Bilheaux, A. Buts, S. Campbell, M. Doucet, et al., Bull. Am. Phys. Soc. 57 (2012).
[80] https://en.wikipedia.org/wiki/Neutron_research_facility (accessed January 2, 2017).
[81] https://www.ill.eu/ (accessed December 21, 2016).
[82] E. Bourgeat-Lami, J.-F. Chapuis, J. Chastagnier, S. Demas, J.-P. Gonzales, M.-P. Keay, et al., Phys. B Condens. Matter 2 (2006) 1303–1305.
[83] http://www-llb.cea.fr/en/ (accessed December 21, 2016).
[84] I.N. Goncharenko, I. Mirebeau, A. Ochiai, Hyperfine Interact. 128 (1–3) (2000) 225–244.
[85] http://flnph.jinr.ru/en/34/?group=34 (accessed December 21, 2016).
[86] V.L. Aksenov, A.M. Balagurov, V.P. Glazkov, D.P. Kozlenko, I.V. Naumov, B.N. Savenko, et al., Phys. B Condens. Matter 265 (1–4) (1999) 258–262.
[87] http://flnph.jinr.ru/en/facilities/ibr-2/instruments/dn-12 (accessed January 2, 2017).
[88] D.P. Kozlenko, Personal Comm., 2016.
[89] http://www.ansto.gov.au/AboutANSTO/OPAL/ (accessed December 21, 2016).
[90] L. Merrill, W.A. Bassett, Rev. Sci. Instrum. 45 (2) (1974) 290–294.
[91] http://www.mlz-garching.de/saphir (accessed January 2, 2017).
[92] http://www.aai.anl.gov/history/project_pages/ipns.html (accessed January 2, 2017).
[93] http://www.isis.stfc.ac.uk/ (accessed December 21, 2016).
[94] http://lansce.lanl.gov/lujan/ (accessed December 21, 2016).

[95] R.M. Brugger, R.B. Bennion, T.G. Worlton, Phys. Lett. A 24 (13) (1967) 714–717.
[96] R.J. Nelmes, J.S. Loveday, W.G. Marshall, G. Hamel, J.M. Besson, S. Klotz, Phys. Rev. Lett. 81 (13) (1998) 2719–2722.
[97] S. Klotz, T. Strassle, A.M. Saitta, G. Rousse, G. Hamel, R.J. Nelmes, et al., J. Phys. Condens. Matter 17 (11) (2005) S967–S974.
[98] R.J. Nelmes, J.S. Loveday, T. Strassle, C.L. Bull, M. Guthrie, G. Hamel, et al., Nat. Phys. 2 (6) (2006) 414–418.
[99] Y. Le Godec, M.T. Dove, S.A.T. Redfern, M.G. Tucker, W.G. Marshall, G. Syfosse, et al., High Pressure Res. 23 (3) (2003) 281–287.
[100] H.J. Stone, M.G. Tucker, F.M. Meducin, M.T. Dove, S.A.T. Redfern, Y. Le Godec, et al., J. Appl. Phys. 98 (6) (2005) 064905.
[101] https://neutrons.ornl.gov/sns (accessed January 2, 2017).
[102] http://j-parc.jp/MatLife/en/ (accessed January 2, 2017).
[103] A. Sano-Furukawa, T. Hattori, H. Arima, A. Yamada, S. Tabata, M. Kondo, et al., Rev. Sci. Instrum. 85 (11) (2014) 113905.
[104] M. Guthrie, R. Boehler, J.J. Molaison, A.M. dos Santos, C.A. Tulk, Amer. Cryst. Assoc. Trans. Ser. 44 (2013) 149–169.
[105] H.K. Mao, J. Xu, V.V. Struzhkin, J. Shu, R.J. Hemley, W. Sturhahn, et al., Science 292 (5518) (2001) 914–916.
[106] R. Boehler, K. De Hantsetters, High Pressure Res. 24 (3) (2004) 391–396.
[107] https://europeanspallationsource.se/ (accessed January 2, 2017).
[108] https://europeanspallationsource.se/unique-capabilities-ess (accessed July 26, 2016).
[109] K. Andersen, Personal Comm., 2016.
[110] http://doi.org/10.5281/zenodo.154478 (accessed).
[111] In the 2015 meeting of the European High Pressure Research Group in Madrid, some 15% of the oral presentations were on biological, food and pharmaceutical science.
[112] K. Akasaka, H. Matsuki (Eds.), High Pressure Bioscience: Basic Concepts, Applications and Frontiers, Springer, Dordrecht, The Netherlands, 2015.
[113] J. Binns, Development of High-Pressure Single-Crystal Neutron Diffraction on KOALA (Ph.D. Thesis), School of Chemistry, University of Edinburgh, 2016.
[114] P.W. Bridgman, J. Biol. Chem. 19 (4) (1914) 511–512.
[115] R. Winter, Biochim. Biophys. Acta Protein Struct. Mol. Enzymol. 1595 (1–2) (2002) 160–184.
[116] G. Gibrat, G.H.B. Hoa, C.T. Craescu, L. Assairi, Y. Blouquit, B. Annighöfer, et al., Biochim. Biophys. Acta Protein Proteomics 1844 (9) (2014) 1560–1568.
[117] A.J. Jackson, D.J. McGillivray, Chem. Commun. 47 (1) (2011) 487–489.
[118] http://nmi3.eu/about-nmi3.html (accessed January 2, 2017).
[119] http://nmi3.eu/news-and-media/scientific-highlights/a-new-pressure-cell-for-sans-experiments-up-to-5-kbar.html (accessed January 2, 2017).
[120] https://en.wikipedia.org/wiki/Clausius%E2%80%93Clapeyron_relation (accessed January 2, 2017).
[121] C.L. Guillaume, E. Gregoryanz, O. Degtyareva, M.I. McMahon, M. Hanfland, S. Evans, et al., Nat. Phys. 7 (3) (2011) 211–214.
[122] A.M.J. Schaeffer, W.B. Talmadge, S.R. Temple, S. Deemyad, Phys. Rev. Lett. 109 (18) (2012) 185702.
[123] E. Gregoryanz, L.F. Lundegaard, M.I. McMahon, C. Guillaume, R.J. Nelmes, M. Mezouar, Science 320 (5879) (2008) 1054–1057.
[124] N.W. Ashcroft, Phys. Rev. Lett. 21 (26) (1968) 1748–1749.

[125] E. Babaev, A. Sudbo, N.W. Ashcroft, Nature 431 (7009) (2004) 666–668.
[126] R.J. Hemley, H.K. Mao, Science 244 (4911) (1989) 1462–1465.
[127] P. Loubeyre, F. Occelli, R. LeToullec, Nature 416 (6881) (2002) 613–617.
[128] M.I. Eremets, I.A. Troyan, Nat. Mater. 10 (12) (2011) 927–931.
[129] W.J. Nellis, A.L. Ruoff, I.F. Silvera, Has metallic hydrogen been made in a diamond anvil cell? 2012, arXiv 1201.0407.
[130] R.T. Howie, P. Dalladay-Simpson, E. Gregoryanz, Nat. Mater. 14 (5) (2015) 495–499.
[131] P. Dalladay-Simpson, R.T. Howie, E. Gregoryanz, Nature 529 (7584) (2016) 63–67.
[132] S. Azadi, B. Monserrat, W.M.C. Foulkes, R.J. Needs, Phys. Rev. Lett. 112 (16) (2014) 165501.
[133] C.J. Pickard, R.J. Needs, Nat. Phys. 3 (7) (2007) 473–476.
[134] N.W. Ashcroft, Phys. Rev. Lett. 92 (18) (2004) 187002.
[135] X.-J. Chen, V.V. Struzhkin, Y. Song, A.F. Goncharov, M. Ahart, Z. Liu, et al., Proc. Natl. Acad. Sci. U. S. A. 105 (1) (2008) 20–23.
[136] M.I. Eremets, I.A. Trojan, S.A. Medvedev, J.S. Tse, Y. Yao, Science 319 (5869) (2008) 1506–1509.
[137] G. Gao, A.R. Oganov, P. Li, Z. Li, H. Wang, T. Cui, et al., Proc. Natl. Acad. Sci. U. S. A. 107 (4) (2010) 1317–1320.
[138] M. Hanfland, J.E. Proctor, C.L. Guillaume, O. Degtyareva, E. Gregoryanz, Phys. Rev. Lett. 106 (9) (2011) 095503.
[139] M. Martinez-Canales, A.R. Oganov, Y. Ma, Y. Yan, A.O. Lyakhov, A. Bergara, Phys. Rev. Lett. 102 (8) (2009) 087005.
[140] L. Sun, A.L. Ruoff, C.-S. Zha, G. Stupian, J. Phys. Condens. Matter 18 (37) (2006) 8573–8580.
[141] R. Szczesniak, A.P. Durajski, Solid State Commun. 172 (2013) 5–9.
[142] Y.K. Wei, N.N. Ge, G.F. Ji, X.R. Chen, L.C. Cai, D.Q. Wei, Comput. Mater. Sci. 88 (2014) 116–123.
[143] Y. Yao, D.D. Klug, Proc. Natl. Acad. Sci. U. S. A. 107 (49) (2010) 20893–20898.
[144] C. Zhang, X.-J. Chen, Y.-L. Li, V.V. Struzhkin, R.J. Hemley, H.-K. Mao, et al., J. Supercond. Nov. Magn. 23 (5) (2010) 717–719.
[145] E. Zurek, X.-D. Wen, R. Hoffmann, J. Am. Chem. Soc. 133 (10) (2011) 3535–3547.
[146] J. Feng, W. Grochala, T. Jaron, R. Hoffmann, A. Bergara, N.W. Ashcroft, Phys. Rev. Lett. 96 (1) (2006) 017006.
[147] D.Y. Kim, R.H. Scheicher, S. Lebegue, J. Prasongkit, B. Arnaud, M. Alouani, et al., Proc. Natl. Acad. Sci. U. S. A. 105 (43) (2008) 16454–16459.
[148] T.A. Strobel, A.F. Goncharov, C.T. Seagle, Z. Liu, M. Somayazulu, V.V. Struzhkin, et al., Phys. Rev. B 83 (14) (2011) 144102.
[149] G.Y. Gao, A.R. Oganov, A. Bergara, M. Martinez-Canales, T. Cui, T. Iitaka, et al., Phys. Rev. Lett. 101 (10) (2008) 107002.
[150] J. Feng, R.G. Hennig, N.W. Ashcroft, R. Hoffmann, Nature 451 (7177) (2008) 445–448.
[151] A. Hermann, B.L. Ivanov, N.W. Ashcroft, R. Hoffmann, Phys. Rev. B 86 (1) (2012) 014104.
[152] M. Debessai, J.J. Hamlin, J.S. Schilling, Phys. Rev. B 78 (6) (2008) 064519.
[153] L. Gao, Y.Y. Xue, F. Chen, Q. Xiong, R.L. Meng, D. Ramirez, et al., Phys. Rev. B 50 (6) (1994) 4260–4263.
[154] A. Yamamoto, N. Takeshita, C. Terakura, Y. Tokura, Nat. Commun. 6 (2015) 8990.
[155] A.P. Drozdov, M.I. Eremets, I.A. Troyan, V. Ksenofontov, S.I. Shylin, Nature 525 (7567) (2015) 73–76.

[156] Y. Wang, K.K.M. Lee, J. Superhard Mater. 34 (6) (2012) 360–370.
[157] M. Amsler, J.A. Flores-Livas, L. Lehtovaara, F. Balima, S.A. Ghasemi, D. Machon, et al., Phys. Rev. Lett. 108 (6) (2012) 065501.
[158] T. Evans, P.F. James, Proc. R. Soc. Lond. A Math. Phys. Eng. Sci. 277 (1369) (1964) 260–269.
[159] P. John, N. Polwart, C.E. Troupe, J.I.B. Wilson, Diamond Relat. Mater. 11 (3–6) (2002) 861–866.
[160] H.P. Bovenkerk, F.P. Bundy, H.T. Hall, H.M. Strong, R.H. Wentorf, Nature 184 (4693) (1959) 1094–1098.
[161] http://minerals.usgs.gov/minerals/pubs/commodity/diamond/mcs-2014-diamo.pdf (accessed January 2, 2017).
[162] Y. Tsach, Personal Comm., 2016.
[163] H. Niu, X.-Q. Chen, S. Wang, D. Li, W.L. Mao, Y. Li, Phys. Rev. Lett. 108 (13) (2012) 135501.
[164] M. Hu, Z. Zhao, F. Tian, A.R. Oganov, Q. Wang, M. Xiong, et al., Sci. Rep. 3 (2013) 1331.
[165] Y. Lin, L. Zhang, H.-K. Mao, P. Chow, Y. Xiao, M. Baldini, et al., Phys. Rev. Lett. 107 (17) (2011) 175504.
[166] L. Zhang, P. Mohanty, N. Coombs, Y. Fei, H.-K. Mao, K. Landskron, Proc. Natl. Acad. Sci. U. S. A. 107 (31) (2010) 13593–13596.
[167] P.J. Pauzauskie, J.C. Crowhurst, M.A. Worsley, T.A. Laurence, A.L.D. Kilcoyne, Y. Wang, et al., Proc. Natl. Acad. Sci. U. S. A. 108 (21) (2011) 8550–8553.
[168] P.W. Bridgman, Phys. Rev. 3 (3) (1914) 153–203.
[169] D. Stojkovic, P.H. Zhang, V.H. Crespi, Phys. Rev. Lett. 87 (12) (2001) 125502.
[170] https://neutrons.ornl.gov/vision (accessed January 2, 2017).
[171] X. Li, L.L. Daemen, Personal Comm., 2016.
[172] L.L. Daemen, B. Haberl, Personal Comm., 2016.
[173] https://europeanspallationsource.se/article/bifrost-prismatic-approach-neutron-spectroscopy (accessed January 2, 2017).
[174] A portion of this research used resources at the High Flux Isotope, a DOE Office of Science User Facility operated by the Oak Ridge National Laboratory.
[175] J. Fang, C.L. Bull, H. Hamidov, J.S. Loveday, M.J. Gutmann, R.J. Nelmes, et al., Rev. Sci. Instrum. 81 (11) (2010) 113901.
[176] C. Ferstl, P. Ferstl, Natl. Food Lab. White Pap. Ser. 10 (2013) 1–6.

Chapter 12

Engineering Applications

Wanchuck Woo*, **Masato Ohnuma**[†] and **Xun-Li Wang**[‡,1]
Korea Atomic Energy Research Institute, Daejeon, South Korea; †Hokkaido University, Sapporo, Japan; ‡City University of Hong Kong, Kowloon, Hong Kong
[1]Corresponding author: e-mail: xlwang@cityu.edu.hk

Chapter Outline

12.1 INTRODUCTION

We begin by defining the scope of engineering applications to be discussed in this chapter. The application of neutron scattering techniques in engineering is a broad subject. Here we are mainly interested in the application of neutron

Experimental Methods in the Physical Sciences, Vol. 49. http://dx.doi.org/10.1016/B978-0-12-805324-9.00012-1

scattering to the understanding of mechanical behaviors or mechanical properties of materials or structure components. To this end, recent developments in the following areas are noteworthy. Neutron diffraction has been effectively used to determine the stress distribution in structure components. In-situ loading studies with neutron diffraction provide insights into deformation mechanisms at the microscopic level. Small-angle scattering is a powerful probe to characterize the chemistry and morphology of nanoscale structure features, which determine the mechanical properties. These developments form the pillars of a de-facto new discipline, which is categorically referred to as Engineering Application. Sometimes it is also called engineering diffraction.

Although of relatively short history, the field of engineering applications has been developing very rapidly. There have been a number of excellent review articles and books on the subject; some examples are given in Refs. [1–5]. This book chapter is intended to provide an overview in broad context of expanding applications, and an update of exemplary studies.

Mechanical properties are perhaps the most important functionalities of materials and structure components that are used in our daily life. First, for almost any applications, the structural components need to be strong and durable. This requirement applies to structures both large and small, ranging from microelectronics in computer chips to infrastructures such as bridges or nuclear power plants. Secondly, the structure components are not static and do not work alone; they are subjected to mechanical loading, thermal environment, electric or magnetic fields, and often a combination of all of these. Lastly, there is always a budgetary constraint. For mass production, in particular, the structure components must be economical to produce. A mechanical engineer therefore needs to understand all of the above requirements, in order to design structures for functionality, aesthetics, and endurance to withstand the forces and the environment which they will be subjected to; and last but not least, to determine the best way to manufacture the structure components and to ensure that they will operate safely without failure.

There have been several high-profile cases of spectacular failures which serve to illustrate the need to quantify mechanical properties. Unfortunately, many of them ended up with grave tragedy. On June 3 1998, a high-speed intercity express (ICE) train in Germany derailed and crashed into a road bridge in the northern part of Germany, see Fig. 1. Until today, it remains one of the worst high-speed-rail disasters in the world, in which more than one hundred people died and another hundred were injured. Investigations found the cause to be a single fatigue crack in a wheel which, when it failed, caused the train to derail. Neutron diffraction based residual stress measurements played an important role to help establish the cause. Neutron diffraction measurements on a series of railway wheels with varying service life demonstrated how residual stress could build up [6], which eventually initiated fatigue cracks.

FIGURE 1 Aerial view of the crash site of a German high speed train in Eschede, Northern Germany, on June 3 1998.

On May 25, 2002, China Airlines Flight 611, a regularly scheduled passenger flight from Taiwan to Hong Kong, crashed into the Taiwan Strait 20 min after take-off, killing all 225 people on board. The recovered flight data recorder revealed a midair break-up before the crash. Subsequent investigations found that the break-up was caused by improper repairs to the aircraft many years earlier. Multiple fatigue cracks were found in areas where deep scratches had been polished out. Crash investigators estimated that a long fatigue crack of about 6 feet may have caused the plane to break up in midair.

Fatigue cracks are also a major concern for military aircrafts. On October 25, 2008, the US Navy ordered an inspection of hundreds of its Hornet fighter jets, after discovering fatigue cracks on more than a dozen aircraft deployed overseas [7]. The Hornet was designed for the Navy and Marine Corps at a cost of roughly $57 million each. Each Hornet fighter jet was to be inspected for cracks in a hinge that connects the plane's wing with the aileron, which are flaps that control a plane's banking movements and help to stabilize the aircraft in flight.

On June 17, 2013, a large containership, the MOL Comfort, broke into two while travelling from Singapore to Jeddah on the Indian Ocean. Although there were no casualties amongst the crew members, the damage to the ship was extensive, and part of the ship eventually sank. The exact cause of the fracture has not been determined, but initial investigations have pointed to residual stress in the welded structure as a possible cause.

The above examples of structure failures underscore the need to quantify residual stresses inside structure components. Application of neutron scattering in mechanical studies dates back to the early 1980s. In the early days, much of the work was strongly oriented towards the engineering aspect, involving spatial mapping of the residual stress distribution in structure components and the use of these residual stress data in design and life predictions. In the early 1990s, advances in the experimental technique reached a level where the experimental data could be used to quantitatively calibrate the calculations by 3D finite element modeling [8], which are extensively used in product design by industry.

Opportunities for fundamental research emerged when it became evident that some of the experimental data could not be understood within the framework of the continuum theory and simple thermal-mechanical simulations. For example, Holden et al. [9] found a very large difference between the axial strains measured with different reflections in bent Incoloy-800 steam generator tubing. This means that grains with different orientations responded differently to the applied bending force. In fact, at some locations, the axial strains measured with different reflections are of opposite signs. Experimental observations like these called for in situ investigations of interactions between grains in polycrystalline materials and the associated deformation mechanisms. In broader terms, these in situ studies apply to the structure response caused by mechanical load or other external stimuli, including electric or magnetic fields.

At a fundamental level, the mechanical properties of a material are governed by its microstructures. Phase transformation is an effective method to manipulate the microstructures. Here, the use of the term "microstructure" is rather liberal, which includes such information as crystalline structure, orientation relationships among the grains, grain size, and stress distribution in the different grains and phases, among others.

Phase transformation creates a two-phase material, which, if employed properly, can dramatically improve the mechanical properties. A good example is precipitation hardening, in which heat treatment produces a nanometer-sized secondary phase which can increase the yield strength of a variety of alloys, including aluminum, magnesium, nickel, titanium, and some steels and stainless steels. The precipitates, which are usually a hard phase, impede the movement of dislocations, leading to the strengthening effect. A well-known example is the so-called Guinier-Preston (GP) zones, a coherent nanometer-sized precipitate, in aluminum alloys.

Small-angle neutron scattering (SANS) is a powerful tool to characterize the nanoscale microstructures. As described in detail in Chapter 1.4.6 [10], SANS gives information of large scale structures in condensed matter. Among the different microstructural features, SANS gives the quantitative information of the hetero-structures, such as the secondary phase, precipitates, and voids. For example, SANS along with its X-ray counterpart, small-angle X-ray scattering (SAXS), have been extensively used to characterize GP zones in aluminum alloys to explain the hardening effect [11].

In the following section, the neutron scattering methodology is briefly described for residual stress determination, in situ study of deformation and phase transformation, and SANS measurements. Select examples are presented to highlight the applications.

12.2 RESIDUAL STRESS MEASUREMENTS

Residual stresses are inherently introduced into materials through a variety of thermal and/or mechanical processes such as welding, forming, and heat treatments [12,13]. Excessive residual stresses are known to degrade the structural integrity and performance of components [14–16]. Particularly in structures made by welding or joining, an abrupt failure is often caused by the combined effect of residual stresses and applied loads. For example, the midship breakup of MOL Comfort, a large container vessel, as shown in Fig. 2(a) [18], was purportedly caused by welding residual stresses. Detailed investigations are ongoing and call for precise knowledge of internal stresses in full scale.

As the use of large-scale high-strength metallic structures increases in various industries, the need to ensure safe operation of the structures has led to an emphasis on assessment of fracture and fatigue-related failures [19]. In general, the fracture toughness has been examined by impacting an instant load into a window frame-type notch tip of the specimen under the constant load and temperature gradient, as shown in Fig. 2.1(b) and (c) [17,20,21].

As the distribution of residual stresses is of practical importance for the postweld mechanical properties and fracture behavior, precise measurement of the residual stresses in welds has been a long standing task in the field of

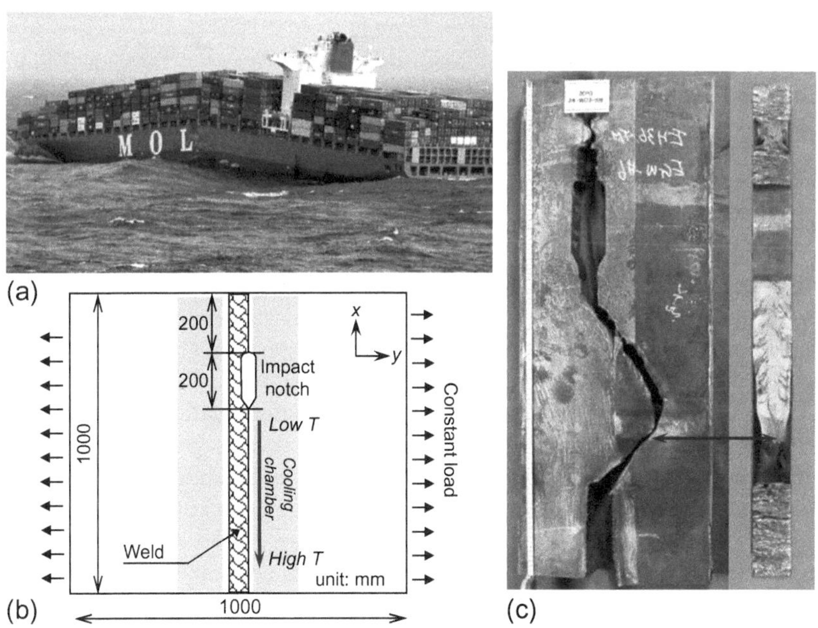

FIGURE 2 (a) Photo in news article (the loss of the container vessel on June 2013), (b) schematic of a large-scale brittle-fracture test specimen, (c) brittle crack-propagation paths and crack-arrest locations in a 80 mm thick weld. *Figure reproduced with permission from Ref. [17].*

materials joining [22,23]. It has been known that tensile residual stresses can reduce the load and cause fracture of a component and conversely local compression can confine fatigue crack growth to regions adjacent to the tearing of a crack edge [24,25]. Thus, it is of critical importance to determine the maximum magnitude and spatial distributions of the residual stress, because this information can provide an indication of the significance of crack initiation and the integrity of the final weld components. This is particularly the case when heavy-section or very thick steel plate and pipes are used.

There are numerous experimental and theoretical studies of residual stress and its influence on the mechanical properties in welds. Various destructive and nondestructive methods have been utilized to characterize the residual stresses in welds. Destructive techniques (for example, hole drilling, contour method, and indentation technique) require the specimen to be drilled or cut mechanically to measure the strain components [26]. Nondestructive methods (for example, laboratory X-ray, synchrotron X-ray, and neutron diffraction methods), based mainly on Bragg diffraction, determine residual stresses from the changes in interplanar lattice spacing (*d*-spacing), which measure the expansion or contraction of the lattice planes [27]. Thus, the lattice plane acts as a natural strain gauge inside materials.

Neutron diffraction has been well established as a means to assess residual stresses in a wide range of engineering components [28–31]. The deep penetration capability of neutrons into most metallic materials makes neutron diffraction a powerful tool for measuring residual stresses through welds. Furthermore, the unique volume-averaged bulk measurements characteristic of the scattering beam and the capability to map in three dimensions are well suited for engineering purposes. In this regard, the neutron-diffraction measurement method has been gaining importance and is being increasingly used for determination of residual stresses in welded materials [32–44].

The following milestones briefly summarize the development of residual stress measurements in welds obtained by neutron diffraction. In the early 1980s, Allen et al. suggested using high resolution neutron diffraction to measure small changes in *d*-spacing to determine the residual stress tensor averaged over a specific volume in a welded component [32]. Krawitz and Winholtz then determined that stress-free reference standards are required in welding cases in order to convert the measured changes in peak position to strain and stress tensors when considering the metallurgical variations in microstructure during welding [35]. In the 1990s, a number of neutron diffraction measurements of residual stresses were performed based on the theoretical background and example experimental results are summarized by Feng et al. [8], Wang et al. [36], Gianni et al. [37], and Stone et al. [38], and Woo et al. [45]. More recently, the time-of-flight (TOF) diffraction enabled by pulsed beams from spallation sources, has been gaining a great deal of attention [30,31,46]. A pulsed beam allows simultaneous collection of full information on all (*hkl*) reflections from a specimen, whereas the continuous monochromatic beam generated by nuclear reactors measures a representative peak. A pulsed beam source provides a variety of useful information in neutron research. For example, the simultaneous observation of multiple peaks is well suited for the studies of changes in phase and texture and in situ deformation behavior in many polycrystalline materials. By using a spallation neutron source, Carr et al. [42] and Holden et al. [43] observed that intergranular stress, which is the residual stress equilibrating over a number of grain dimensions and also known as microstress [39–41], underlies the long range macroscopic residual stress in several types of welds.

12.2.1 Principle of Strain Measurements

The principle of neutron diffraction measurement of residual strains/stresses is based on Bragg's law:

$$\lambda = 2d_{hkl} \sin\theta_{hkl} \tag{1}$$

where λ is the wavelength of the neutron, d_{hkl} is the lattice spacing of atomic planes characterized by Miller indices $\{hkl\}$, and $2\theta_{hkl}$ is the scattering angle of the peak being measured [27–30]. Equation (1) allows us to measure interplanar lattice spacings along the direction normal to the lattice planes in a scattering volume, Fig. 3 [47]. Here, the scattering vector ($\boldsymbol{Q}$) is defined by the difference between the scattered and incident wave vectors.

There are two types of neutron beams: continuous or pulsed [30,31]. For the continuous beam (the monochromatic beam generated by a nuclear reactor), Fig. 3(a), the wavelength (λ) is known and fixed. The neutron-diffraction measurement gives the *d*-spacing (d_{hkl}) changes as a function of $2\theta_{hkl}$, Eq. (1). It is similar to the well-known X-ray diffraction method. The scattering angle of $\pi/2$, however, is generally used for neutron diffraction because it gives a cubical gauge volume. As a result, this approach provides a representative (*hkl*) peak satisfying Bragg's law. For the pulsed neutron beam (wavelength-sorted white beam from a spallation source), the diffraction angle ($2\theta_{hkl}$) is fixed and the arrival time (t) of neutrons at detectors gives λ (t is proportional to λ). Fig. 3(b) shows the TOF technique, which uses the elapsed time (t) for each neutron traveling from a source to a detector. As a result, the distinct peaks can be measured in a whole range of *d*-spacings and separate response of each (*hkl*) orientation among grains with their plane normal parallel to the diffraction vector ($\boldsymbol{Q}$) can be determined.

The single (*hkl*) peak obtained by continuous beam diffractions can be analyzed using the least-squares Gaussian fitting method. For the pulsed neutron beam, the full diffraction pattern from the multiple crystallographic lattice planes in a particular specimen direction can be analyzed by the Rietveld refinement method [48]. Rietveld analysis makes use of the entire diffraction pattern by fitting all the diffraction peaks to a structural model of the material. Hence, the Rietveld approach allows accurate determination

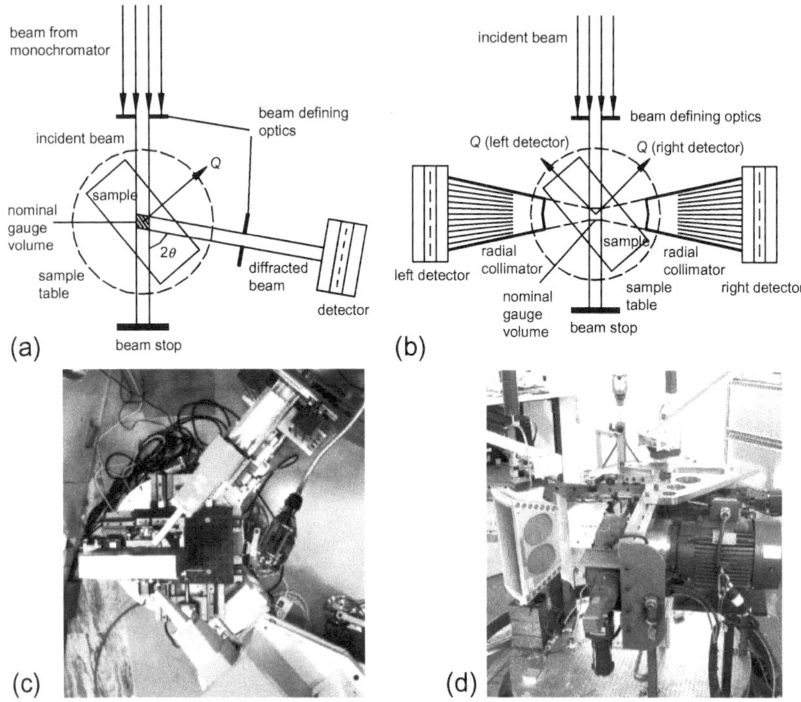

FIGURE 3 Schematic illustration of (a) reactor based diffractometer and (b) spallation neutron source based diffractometer for stress measurements. (c) An actual setup for a reactor-based experiment [47]. (d) Setup for in situ frictional stir welding experiment [45]. *Figure reproduced with permission from Refs. [45,47].*

of *d*-spacing with much shorter count times than is typically required for individual peak refinements. The general structure analysis system (GSAS) is a widely-used computer program for this type of analysis [49]. Once the *d*-spacing is determined, the elastic lattice strain (ε) in a particular direction is calculated in terms of the deviation in the lattice spacing (d) relative to the unstrained (stress-free) lattice spacing (d_o) [30]:

$$\varepsilon = (d - d_o)/d_o \tag{2}$$

Generally, Hooke's law is used to convert elastic strain to the principal residual stresses (σ_x, σ_y, σ_z):

$$\sigma_i = \frac{E_{hkl}}{1 + \nu_{hkl}}\left[\varepsilon_{ii} + \frac{\nu_{hkl}}{1 - 2\nu_{hkl}}\left(\varepsilon_{xx} + \varepsilon_{yy} + \varepsilon_{zz}\right)\right] \tag{3}$$

where $i = x$, y, or z corresponding to the three principal directions, respectively. The quantities of E_{hkl} and ν_{hkl} are the plane-specific analogues of Young's modulus and Poisson's ratio, respectively.

12.2.2 Enhanced Capability of Deep Penetration by Neutron Diffraction

Neutron diffraction is widely used for the stress measurements deep inside engineering structures and components [14,26,30]. Limitation of the penetration path length in iron is known to be about 60 mm, which corresponds to the available plate thickness of about 40 mm [50,51]. However, various heavy industries demand stress measurements in ferritic steel weld plates with thicknesses up to about 80 mm now, and even thicker plates in the near future. The maximum available path length (l_m) can be increased by using a larger sampling (gauge) volume and/or longer measurement times. However, the maximum size of the gauge volume is limited by the required spatial resolution. Increasing measurement time is not effective either because the intensity of the diffraction peak exponentially decreases as the neutron beam path length increases through the thickness of the material [30].

An effective way to increase the available penetration path length is through the selection of the neutron wavelength (λ). It has been suggested that the neutron path length can be maximized when the wavelength is located at the local minimum of the neutron total cross-section slightly off the Bragg edges [52]. Moreover, the available path length in steels can be remarkably increased with the benefit of the focusing bent perfect crystal Si monochromators [53–56]. One can compare two neutron beams having different wavelengths to show the different penetration path lengths for the low textured ferritic steel (α-Fe, bcc structure) in Fig. 4. It shows that the wavelength of 2.28 Å having a high total cross-section can reach the maximum path length (l_m) of about 65 mm with ± 100 $\mu\varepsilon$ accuracy, whereas the wavelength of 2.39 Å having a low total cross-section can increase the l_m up to 83 mm, Fig. 4(b). Therefore, it is plausible that residual stresses can be determined precisely in thick steel plates by selecting a proper wavelength.

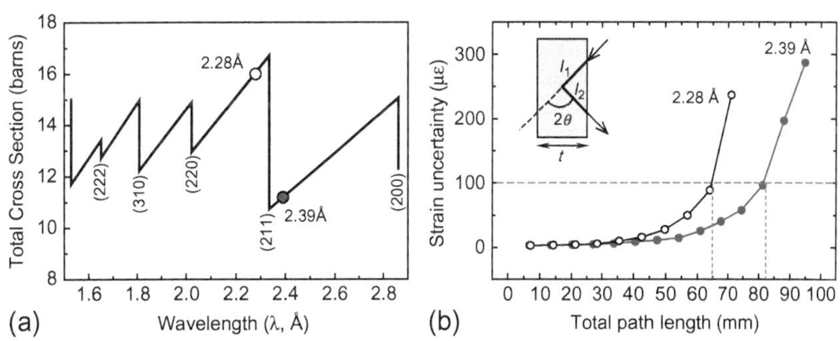

FIGURE 4 (a) Total neutron cross-section of the polycrystalline ferrite iron as a function of the wavelength. Marked wavelength near the maximum (open circle, $\lambda = 2.28$ Å) and the minimum (closed circle, $\lambda = 2.39$ Å) of the total cross-section, (b) apparent strains caused by the diffraction peak error as a function of the total path length ($l_1 + l_2$) through the thickness (t) of the steel specimen. *Figure reproduced with permission from Ref [57].*

The maximum beam-penetration capability of neutron diffraction is described as below. The criterion for the deep penetration capability of the neutron-diffraction strain measurements is mainly dependent on the decrease of the attenuation. Briefly, this is achieved by reducing the strain uncertainty (Δx), which is related to the peak height, H $(=H_o e^{-\mu l})$ and the integral peak intensity, $I\left(=I_0 e^{-\mu l}\right)$ in the determination of the peak position [50]:

$$(\Delta x)^2 = \left(u_x^2/I\right)\left[1 + 2\sqrt{2}B/H\right] \quad (4)$$

where μ_x is the standard deviation of the peak, B is the background level, μ is the attenuation coefficient, l is the path length, H_o and I_o are the peak height and integral peak intensity at the zero path length, respectively. A similar equation has been noted in the conceptual design of VULCAN, the engineering diffractometer at the Spallation Neutron Source (SNS) [58,59] Since μ is proportional to the total neutron cross-section (σ_t), the lower σ_t exponentially decreases the Δx as l increases. Thus, the properly selected λ (2.39Å) from the close vicinity of the Bragg edges can make the attenuation lower and enhance the beam penetration [52]. Furthermore, an appropriate configuration of the diffractometer can minimize the actual beam path length by changing the diffraction angle ($2\theta_s$) [60]. The peak intensity (H_o) to background (B_o) ratio needs to be higher than 1.0 for the proper statistical analysis of the neutron diffraction data [30]. In addition, it has been noted that as the neutron-beam path increases through a specimen, significant peak broadening was observed in ferritic steels, but not in austenitic steels. Such peak broadening was decreased under a magnetic field, which is applied perpendicular to the beam direction. It is suggested that peak broadening is attributed to the multiple small-angle neutron scattering on magnetic domains. Thus, sufficient magnetic fields can enhance the deep penetration capability of neutron diffraction due to less peak broadening. More details are described in ref. [61].

12.2.3 Practical Issues in Neutron-Diffraction Measurements of Residual Stresses

In the neutron diffraction measurements of residual stresses in welds, it is necessary to first measure the d-spacings with their scattering vectors parallel to the three orthogonal directions—longitudinal direction (LD) (welding direction), transverse direction (TD), and normal direction (ND)—of the welding plate [27–30]. LD, TD, and ND are generally assumed to be the three principal directions of the measured strain tensor in a given linear weld plate. If one uses the instrument in a SNS, two opposing detectors can measure two d-spacing components at the same time, Fig. 3(b). Then it is necessary to rotate the weld sample 90 degrees relative to the neutron detectors for the measurement of the third component [42]. The scattering volume is defined by a pair of slits: an incident beam square slit and a detector slit. The measurements over the scattering volume (e.g., 2 (LD) × 2 (TD) × 2 (ND) mm^3) provide volume-averaged bulk strain values of the weld. When there is no loss of spatial resolution along the LD in welds, the height of the scattering volume can be enlarged to achieve better statistics of the diffraction peak [62]. The neutron scattering volume is usually centered on the midplane of welds, but it can also focus on different locations through the thickness of the weld plate as long as the scattering volume is completely filled with the intended material [30]. It has been shown that a partially filled scattering or gauge volume can cause an artifact due to the inappropriate centroid of the sampling volume and erroneous fitting of the diffraction peaks [63].

Secondly, the unstrained lattice parameter (d_o) is required for the standard state as a reference [64]. The reference sample is used for the determination of residual stresses via neutron diffraction as shown in Eq. (2). The "stress-free" lattice spacing is often measured using a coupon taken from the initial base material or a piece of the welded specimen that is far from the process-affected zone. In welds, however, a careful determination of d_o is necessary. Due to the high temperature and the flow of molten materials during welding, there are significant microstructural changes in the weld metal and heat-affected zones. The microstructure changes (for example, redistribution of alloying elements) result in corresponding changes in the d-spacing. For example, Steuwer et al. reported that the pseudo-strain shift is about 1000 $\mu\varepsilon$ due to the solute changes in an AA7010 friction stir weld (Al-Mg-Zn system) [65]. This is equivalent to a uniaxial stress of about 70 MPa (corresponding to ~50% of residual stress). In practice, it is recommended that a few millimeter sized cubes be cut from each location of the weld and reassembled to measure the d_o at corresponding locations of welds. It is assumed that the macroscopic stresses are destroyed during cutting of the cubes but that the chemistry does not change. A long, rectangular comb-like sample is also valid for d_o measurements [64].

Considerations also should be given to intergranular anisotropy among grains in welds [43]. Intergranular strain is defined as the difference in the strain obtained from a particular (*hkl*) diffraction peak and the strain due to the plane-specific elastic modulus (known as the diffraction elastic constant) [40]. If there is a significant intergranular strain in a material, a certain (*hkl*) diffraction peak may not represent the responses for the macroscopic stress field and may only exhibit the response of the specific (*hkl*) grains. Therefore, when measuring the d and d_o from the coupons, it is necessary to select the diffraction plane that is either not or only weakly affected by intergranular strains in order to properly determine the macroscopic residual stresses. The Versailles Project on Advanced Materials and Standards (VAMAS) procedure recommends the diffraction lattice plane: (111), (311), and (422) for fcc Ni, Fe, Cu, and Al based alloys and (110), (211) for bcc Fe-based alloys [47]. In any case, the (200) plane is not recommended due to the large intergranular strains on that plane. Moreover, E_{hkl} and ν_{hkl} in Eq. (3) need to be used for the "diffraction elastic constants" for the *hkl* reflection, which can be determined by in situ measurements on a tension sample under loading [40].

12.2.4 Examples of Engineering Applications

12.2.4.1 Residual Stress Measurements of Thick Welds for Shipbuilding Applications

In general, a weld is heated locally and the temperature distributes nonuniformly by the heat source [12]. The localized heating imposed on the weld causes the formation of compressive stresses due to the constraint of localized thermal expansion by the cold material surrounding the heated region. On cooling as the heat source moves, the weld contract and cause tensile stresses in regions due to the relatively cold surrounding materials. As a result, high tensile stresses are distributed usually in regions near the weld, while balanced compressions are produced away from the weld [14].

Many manufacturing industries are aggressively pursuing new joining technologies, introducing high heat inputs into welded structures to improve manufacturing efficiency and productivity [66]. For example, instead of the conventional low heat-input methods using less than 2 kJ/mm, new extremely high heat-input joining techniques using over 50 kJ/mm have emerged recently. The high heat-input processes inherently generate detrimental tensile residual stresses and potentially degrade the fatigue strength and life of the components.

Through-thickness variations of the residual stresses are of particular importance in heavy-section and high heat-input welded structures [57]. Spatial variations of residual stresses were determined through the thickness of 70-mm thick ferritic steel welds created using low (1.7 kJ/mm) and high (56 kJ/mm) heat-inputs.

The residual stresses were measured through the thickness of the 70-mm thick welds, utilizing the two-peak combined methodology with the benefit of the wavelength dependence of neutron beam penetration [52]. First, wavelengths of 2.39 and 1.55 Å were selected for the diffraction planes (110) and (211) at scattering angles of 72.4° and 82.5°, respectively. Note that both Si (111) and Si (220) monochromators at take-off angles of 45° and 48° produced neutrons with the appropriate wavelengths. Such a configuration increased the penetration depth of neutrons significantly due to lower attenuation [52]. This was achieved, in brief, by reducing the strain uncertainty (Δx), which is related to the peak height, H ($=H_o e^{-\mu l}$) and the integral peak intensity, I ($=I_o e^{-\mu l}$) in the determination of the peak position, as described in Eq. (4) [23]. The (110) diffraction plane is advantageous for the transmission geometry and the (211) diffraction plane for the reflection geometry of the welded specimens. The combined method is available because of the similar diffraction elastic constants (225.5 GPa) of the two diffractions [30].

The scattering gauge volume of the neutron beam was defined by 4 mm wide and 8 mm (or 20 mm) high input slits and a 4 mm output slit. Then, the nominal scattering volumes of $4(x)\times 8(y)\times 4(z)\,\text{mm}^3$ for the diffraction patterns with their scattering vectors were configured to be parallel to the x direction and a volume of $20(x)\times 4(y)\times 4(z)\,\text{mm}^3$ for the y or z directions. Both volumes retained the 4-mm spatial resolution along the thickness direction (z). A total of 13 points were measured through the thickness of welded specimens starting from 5 mm from the top surfaces to 65 mm in 5 mm steps at locations from the weld centerline of 0, 30, 60, and 100 mm. Note that all strain components were measured by using the (110) diffraction plane with a wavelength of 2.39 Å except that the strain normal components (z) at the 30 to 40 depths (only 3 points) were measured by using the (211) diffraction plane with a wavelength of 1.55 Å. The measurement period was about 1 hour for each strain component achieving a strain uncertainty of about $\pm 100\ \mu\varepsilon$. However, to obtain strains in the z-direction at midthickness locations, measurement times of up to 12 hours [23] are sometimes needed.

Diffraction peaks were analyzed using a least-squares Gaussian fitting method. Once the peak position was determined, the elastic lattice strains (ε) were calculated using $\varepsilon = -\cot\theta(\theta-\theta_o) = (d-d_o)/d_o$, where the θ_o (d_o) and θ (d) are the diffraction angles (d-spacings) for the stress-free and stressed materials at each position, respectively [30]. The generalized Hooke's law was used to convert elastic strains (ε_x, ε_y, ε_z) to the residual stresses (σ_x, σ_y, σ_z) along the three orthogonal directions (x, y, and z) in a given plate. Measured diffraction elastic constants were averaged using $E_{110/211} = (2E_{110}+E_{211})/3$. This resulted in 220 GPa (at the 0 mm location) and 226 GPa (at the 30, 60, and 100 mm locations) being used. The difference between elastic constants E_{110} and E_{211} was less than ± 10 GPa. The Poisson's ratios ν_{110} and ν_{211} of 0.28 were used [30]. After strain scanning, the plate specimen, measured using the ND technique, was sectioned using electro-discharge machining to prepare a "stress free" reference sample. Comb-like reference samples were extracted along each line of strain scanning. The combs were 10 mm long (x), 4 mm wide (y), and 5 mm deep (z). The stress-free lattice spacing (d_o) was carefully measured with the assumption that there was no steep variation of d_o along the weld direction (x) due to chemical composition changes [60]. Note that the gauge volume was 8 mm^3 ($2\times 2\times 2$ mm^3) for all d_o measurements.

Significant difference in residual stresses is observed between the low heat-input (LHI) and high heat-input (HHI) welds. Fig. 5 compares the distribution of the longitudinal stress (σ_x) through the thickness of the LHI and HHI specimens. The results show that the σ_x in the LHI was mostly higher than in the HHI at the weld centerline, Fig. 5(a). This is reversed in the 30 mm location, Fig. 5(b). More importantly, the maximum σ_x was located at a depth of 5 mm below the top surface at the weld centerline in the LHI and at a depth of 20 mm below the plate surface along the 30 mm in the HHI. These maximum stress locations were similarly found by the destructive contour method as shown in Fig. 2.4(a) and (b), respectively [67].

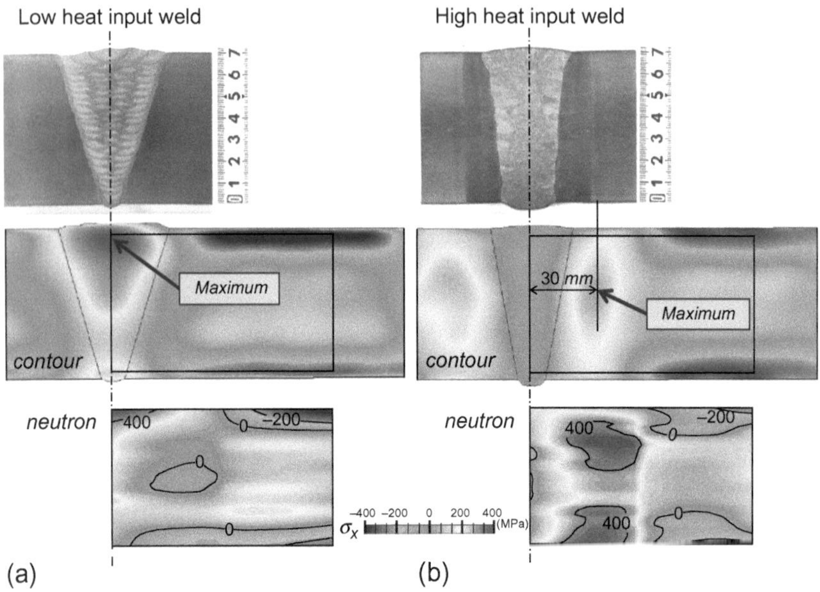

FIGURE 5 Comparison of the macrostructure and 2-D mapping of the longitudinal residual stress (σ_x) in welds: (a) low heat input weld and (b) high heat input weld, determined by the neutron diffraction and contour methods. *Figure reproduced with permission from Ref. [23].*

12.2.4.2 Residual Stress Measurements of Dissimilar Welds for Nuclear Plant Applications

Residual stresses were determined through the thickness of a dissimilar weld overlay pipe using neutron diffraction [68]. The specimen has a complex joining structure consisting of a ferritic steel (SA508), an austenitic steel (F316L), a Ni-based consumable (Alloy 182), and an overlay of Ni-base superalloy (Alloy 52M). It simulates pressurized nozzle components, which have been a critical issue in the dissimilar joining structures mainly due to primary water stress corrosion cracking (PWSCC) under the severe crack conditions of nuclear power reactors, Fig. 6(a).

In the neutron diffraction measurements of residual stresses, it is necessary to measure the interplanar spacings (d) with their scattering vectors parallel to the three orthogonal orientations of the pipe specimen, i.e., axial, hoop, and radial directions. The specific locations for the residual stress measurements are -3, -1 (overlay part), 1, 2, 3, 4, 6, 8, 10, 12, 13, and 14 mm (dissimilar weld part) marked in Fig. 6(b). Note that the outer surface of the dissimilar welded pipe was named as 0 mm. The incident neutrons have a wavelength (λ) of 1.46 Å, and the diffraction angle (2θ) was 84.6° for the (311) reflection from the fcc austenitic Alloy 182. The scattering volume was $1 \times 1 \times 20\ \text{mm}^3$ for the axial and radial components and

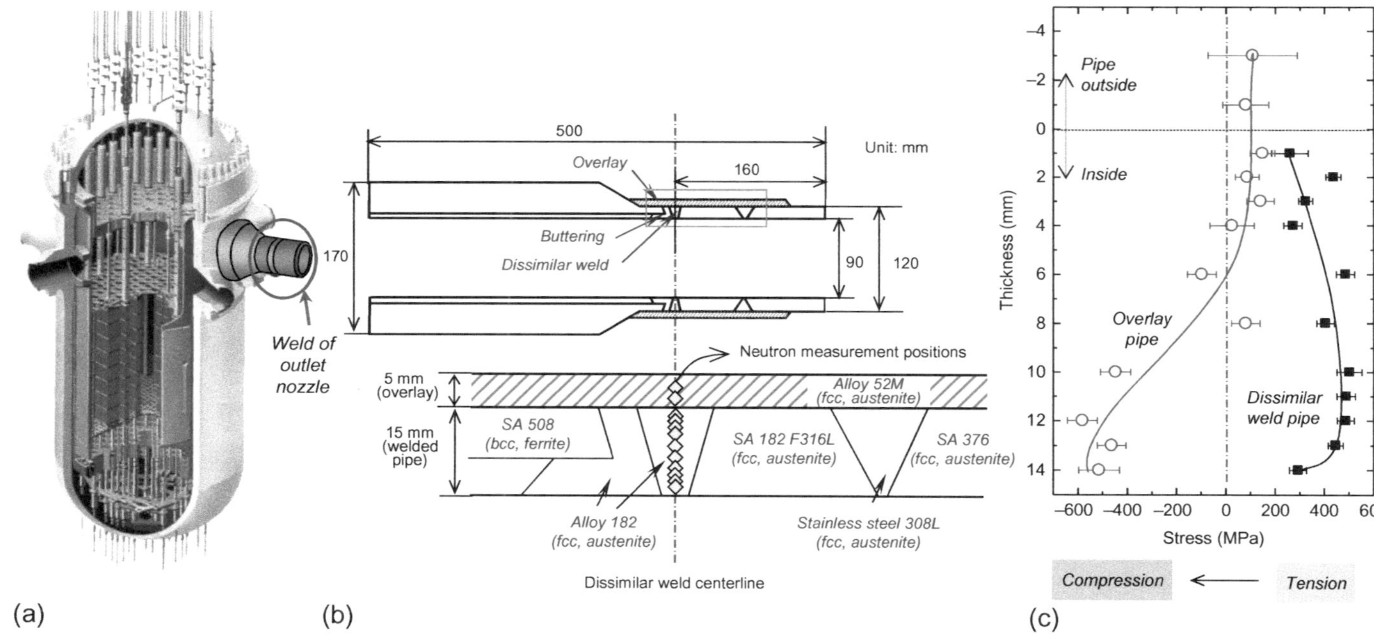

FIGURE 6 (a) Schematic of a pressurized water reactors (PWR). (b) Dimension of dissimilar weld overlay pipe specimen and measurement locations of the specimen in a neutron diffraction experiment, (c) residual stress distribution through the thickness of the specimens (hoop component). *Figure reproduced with permission from Ref [68].*

$1 \times 1 \times 8\ mm^3$ for the hoop component, which were defined by a pair of incident and detector slits. Both components retained 1-mm spatial resolution along the thickness direction (radial direction). The $\pm 3°$ oscillations of the diffraction vector (Q) can provide sufficient intensities for the peak fitting and lessen the issues of the grain size and texture in welds [30]. It should be noted that "a window" was cut in the side wall of the pipe specimens in order to decrease the total beam path length for the setup of the hoop component. The size of the window was about 30 mm in height along the axial direction and 70 mm in width along the hoop direction. It was taken at about 140 degree rotated location from the measurement location along the axis of the axial direction. The macroscopic "stress-free" lattice spacing (d_o) was also obtained from both using a 2-mm width comb-like coupon.

The distribution of residual stresses was obtained in axial, hoop, and radial components of the dissimilar welded pipe specimen [68]. Overall the three components of the residual stresses are tensile, and it increases to $300 \sim 400$ MPa near the inner wall of the dissimilar welded pipe. The tensile stress approaches the yield strength (~420 MPa) of the localized weld part in Alloy 82/182 [69]. Such high tension is plausible when considering the difference of the thermal expansion coefficient (CTE) between the dissimilar metals. Note that the CTE of F316L ($1.53 \times 10^{-5}\ K^{-1}$) is roughly 30% larger than those of SA508 ($1.15 \times 10^{-5}\ K^{-1}$) and Alloy 182 ($1.22 \times 10^{-5}\ K^{-1}$). The existence of the developed tension can be explained by the tourniquet ring load around the pipe due to the contraction of the expanded hot welding line and constraints of the surrounding cold materials during cooling [35]. Several simulations and experimental results show that high tensile residual stresses can be developed at the inner wall in girth type welded pipe structures [70].

The distribution of residual stresses was obtained in three components of the overlay pipe specimen. It is obvious that the residual stresses vary significantly along the through-thickness direction of the pipe. For example, the hoop component of the residual stress, Fig. 6(c), is mostly tension fluctuating around 100 MPa at the overlay and near the outer surface of the dissimilar welded pipe (−3 to 8 mm locations), and significantly decreases to compression (−600 MPa) from the 10 mm location toward the inner wall of the pipe. Since the hoop component can be a driving force behind the transverse crack growth through the thickness, the compressive stress is important to lessen the crack initiation and growth around the crack tip. It is likely that thermal expansion of the heating layers followed by the constraint by the existing cold dissimilar weld can cause the tension on the overlay, and in turn, it imposes compression on the inner wall of the dissimilar weld and makes a stress balancing [12]. In summary, the results show significant changes in residual stresses from tension ($300 \sim 400$ MPa) to compression (−600 MPa) through the thickness of the dissimilar weld overlay pipe specimen.

12.2.4.3 In Situ Study of Stress and Microstructure Evolution in Friction Stir Welds

Friction stir welding (FSW) is an innovative material joining process that enables the advantages of solid-state joining for fabrication of continuous linear welds without bulk melting [71]. The process utilizes a specially shaped tool that rotates and traverses along the joint line, creating heat that softens a column of material underneath the tool. The softened material flows around the tool through extensive plastic deformation and is consolidated behind the tool to form a solid-state continuous joint. The economic and technological benefits of FSW have been well established for Al alloys and other low-melting temperature materials. There have been extensive efforts to apply FSW to high-temperature materials and to extend the process as a material processing technology (friction stir processing) for property improvement.

Neutron diffraction measurements of the residual stresses in FSW have been extensively reported [45,72–75]. For the FSW Al alloys, the maximum tensile residual stresses along the welding direction are reported to be about 20 to 70% of the yield strength of the FSW joint, depending on the welding variables [73]. Wang et al. [72] showed that the strain/stress components have a characteristic peak-and-valley shape in FSW Al 6xxx alloys. Sutton et al. used neutron diffraction to construct three-dimensional maps of the residual stresses along the cross section of Al 2024-T3 FSW plate [74]. The maps showed asymmetric distribution of residual stress profiles across the weld centerline, and the highest residual stresses developed normally along the longitudinal components in the advancing side of the weld. Staron et al. investigated the influence of an enforced severe cooling on residual stresses in a FSW Al 2024-T351 alloy joint [62]. Samples were cooled with CO_2 gas along the weld seam during FSW. The results showed that tensile residual stresses were significantly decreased near the center of the weld and that the distance of the maximum residual stress locations became narrower due to the reduced development of the heat-affected zone. This work suggests that the heating generated during FSW can be a major source of the residual stresses. Woo et al. further deconvoluted the influence of frictional heating and plastic deformation on residual strains generated during FSW [75]. The comparison between a regular FSW and a modified FSW (subjected mainly to the heating effect) shows that the heat input from the tool shoulder causes a significant amount of residual strains.

Most neutron diffraction work has been limited to the so-called "static" behavior of materials (i.e., unchanging over time), primarily because of the inadequate neutron fluxes from existing facilities. Recently, an in situ time-resolved neutron diffraction method was developed and applied to determine the temperature and thermal stress inside the stir zone during FSW of Al 6061-T6 plate [45]. Measured strains by means of *d*-spacing changes as a function of distance from the tool center during FSW were decomposed into thermal and elastic strains, and then subsequently were converted to the transient temperature and stress in the bulk of materials. The temperature and thermal stress distribution along the weld centerline are shown in Fig. 7(a). The temperature decreases monotonically as the distance from the tool center increases. Because of the relatively low rotational speed (156 rpm), the maximum temperature (362°C) attained underneath the tool shoulder was at the low end of the temperature range reported in the literature.

As shown in Fig. 7(a), both the longitudinal (σ_x) and transverse (σ_y) stresses were largely in compression under the tool shoulder and were due to the thermal expansion and the forging pressure from the tool shoulder. The longitudinal stress was more compressive than the transverse stress, which can be attributed to the elongated temperature field caused by the moving heat source of friction stir tool along the longitudinal direction of the plate. Outside the tool, the thermal stresses become tensile to balance the thermal contraction of the weld region. The magnitudes of the tensile stresses (75 MPa for longitudinal stress and 60 MPa for transverse stress) are quite close to that of the

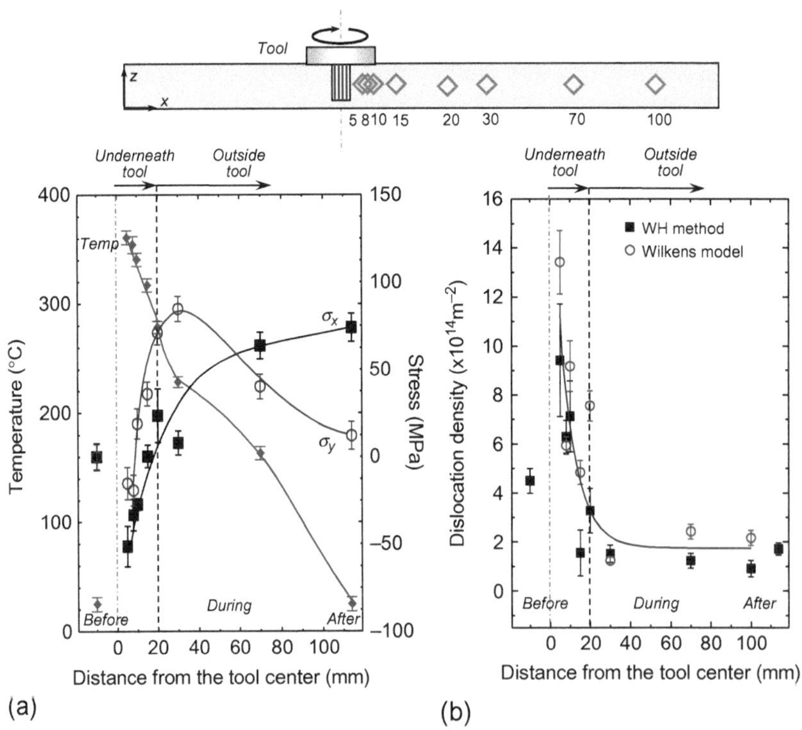

FIGURE 7 (a) Temperature and thermal stress distribution along the weld centerline behind the friction stir welding tool as function of distance from tool center (σ_x: longitudinal stress and σ_y: transverse stress), (b) dislocation density evolution as a function of distance from the tool center during friction stir welding. *Figure reproduced with permission from Refs. [45,76].*

residual stresses, which have typically been measured in other studies when the specimen completely cooled down.

It is important to quantify the grain structure because of the significant influence of the dislocation density and subgrain size on deformation behaviors [76]. Peak profile analysis has become a well-established method for the determination of microstructure in crystalline [77–80]. We applied the modified Williamson-Hall (WH) method [77,78] (see section 12.3.2.4 for details of the method) and Wilkens model [79,80] for the determination of the dislocation density (ρ) and subgrain size (D). The parameter D is the volume weighted domain size of crystallites in their direction normal to the scattering planes. The WH method deconvolutes the subgrain size effect and dislocation-induced strain effect on the peak profile broadening (ΔK, nm^{-1}) based on their different (hkl) reflection dependences [77]:

$$(\Delta K)^2 = (0.9/D)^2 + \left(\pi A b^2/2\right)\rho\left(K^2\overline{C}\right) + O\left(K^2\overline{C}\right)^2 \qquad (5)$$

where $K = 1/d$, $\Delta K = -K(\Delta d/d)$. The parameter d (nm) is the lattice-spacing and Δd (nm) is the full width at half maximum. The parameter b is the modulus of Burgers vector ($\cong a_o/\sqrt{2}$, 0.286 nm in Al) of dislocation, $\overline{C}$ is the average dislocation contrast factor, and O indicates higher-order terms which are often ignore or interpreted. Fig. 7(b) shows the dislocation density (ρ) changes as a function of distance from the tool center during FSW. Compared to the initial ρ (4.5×10^{14} m^{-2}) before FSW, significant increases of the ρ (up to 1.3×10^{15} m^{-2}) are obvious underneath the tool. It is likely due to the large straining by the severe plastic deformation with high strain rates of ~10 s^{-1} estimated by the torsion-type deformation. The high ρ decreases rapidly to $1.0 \sim 3.3 \times 10^{14}$ m^{-2} outside the tool and approaches the low ρ of 1.7×10^{14} m^{-2} after FSW. Such decreases are attributed to the recovery and/or recrystallization process at the elevated temperature (~370°C) when considering the dislocation-free equi-axed recrystallized grains in TEM observation. Zhao et al. reported that the ρ of ~9.5×10^{14} m^{-2} can decrease to 0.5×10^{14} m^{-2} as ultrafine grains were heat-treated at 300°C in an equal-channel angular pressing 7075 Al alloy [81].

12.3 IN-SITU STUDY OF DEFORMATION AND PHASE TRANSFORMATION

A major goal of in situ study is to gain fundamental knowledge of materials response to external stimuli, which could be temperature, applied load, and electric or magnetic field. For crystalline materials, it is now understood, for example, that under applied load, different crystallographic planes exhibit different strains, due to crystal elastic and plastic anisotropy [82]. The deformation behavior is also a function of the grain size, and when the grain size is reduced to nanometer scale, a very different deformation behavior emerges.

For piezoelectric materials, an electric field can generate a lattice strain. For magnetic shape memory alloys, a magnetic field introduces a mechanical strain, and vice versa. Another emerging application is in-operando study of battery, where the chemistry and structure undergo complicated changes during charging and discharge. In situ studies are key to understanding the microscopic origin of these fascinating properties. Neutron diffraction is a powerful tool for in situ studies because neutrons are highly penetrating and therefore sample environments are relatively easy to implement.

12.3.1 Experimental Considerations

Sample environments are the key to well-designed in situ studies. A standard sample environment suite for materials science or engineering diffraction includes the following:

- High temperature furnace. Conventional vacuum furnaces now can easily reach 1600°C. Atmospheric furnaces are also available. Solid and powder samples are typically loaded in a vanadium can and then mounted in a furnace. Sometimes solid samples can be hung in a furnace.
- Mechanical load-frame. This has now become a standard piece for engineering diffractometers. Two types are in routine use: screw-driven and hydraulic. The latter has the advantage of being capable of performing cyclic loading for fatigue tests. Heating has become a necessity for in situ loading studies. A variety of heating methods have been implemented, such as induction heating, lamp heating, and laser heating. A design document for a state-of-the-art load-frame can be found in Ref. [83]. A less visited, but a very important, research direction, is deformation at low temperatures. Many martensitic transformations occur below room temperature, for example. So far, only a few neutron scattering facilities are capable of studying deformation at low temperatures. The TAKUMI instrument at J-PARC has a dedicated low-temperature load-frame [84,85]. It was initially developed to address the brittleness of superconducting wires for high-field magnets.
- Magnet. Since ferritic steel is a major subject of study for structure materials, it is inevitable that a magnetic field is needed. A major application of magnetic field is in small-angle neutron scattering (see Section 12.4), to separate the scattering intensities by magnetic domains from the precipitates. For convenience, it is highly desirable to have a wide open space to accommodate the magnet, which is usually bulky. For ferritic steel, the saturating field is typically 1–2 T, so usually an electromagnet will suffice. A magnetic field is also essential for the study of magnetism related functional materials. This is illustrated in the example of ferromagnetic shape memory alloys [86–88], where the sample is able to restore its shape under a magnetic field. It has been suggested that the magnetic field

can be used to tune the phase diagram of structural steel [89]. There, a very high magnetic field (~30 T) will be needed, as early experiments demonstrated.

- Electric field. The high-voltage electric field is another desired sample environment for the study of piezoelectric materials. An example is given on determination of the lattice response to an applied electric field [90]. Here, the device is not difficult to implement, but the safety is often a primary concern which must be addressed before the experiment commences. Given the renewed interest in Pb-free ferroelectric materials, an increasing number of applications involving high voltage is anticipated [91].
- Battery charger. Energy storage materials have become a major topic of study for neutron scattering. The examples include Li-ion batteries and supercapacitors [92]. Facilities are often requested to equip the diffractometers with a charging system to enable in situ study of the chemical and structure changes during charge-discharge.
- Specialized equipment. In addition to the list of frequently requested equipment, there are also truly specialized sample environments. Some examples of unique sample equipment that enabled rather unconventional experiments include:
  - Hydrogen charging, for studying hydrogen storage materials. This was demonstrated in earlier works [93,94] at Intense Pulsed Neutron Source, Argonne National Laboratory.
  - Friction stir weld machine. A body of in situ welding work has been demonstrated by Woo et al. [45], using the SMARTS diffractometer at LANSCE, Los Alamos, see section 12.2.4.3.
  - Electrostatic levitator. This equipment is designed for the study of the structures and phase transitions in metallic liquids. An electrostatic levitator has been developed at the SNS. Feasibility of the equipment [95] was successfully demonstrated using VULCAN, and the first scientific study [96] was carried out with the NOMAD diffractometer at the SNS.

  Special sample environment of this kind are mission orientated and custom built. The typical cost for such equipment can easily reach a million USD. The electrostatic levitator at the SNS is a case in point. For this reason, the development of unique sample environment often requires a concerted effort and special grants.

From a measurements perspective, there is the typical trade-off between measurement speed and resolution. For in situ studies, the measurements are usually intensity limited. This is obviously evident for time-resolved studies. The time-resolution for a typical neutron diffraction measurement is on the order of minutes. In some cases, e.g., with large samples, a time resolution of subseconds can be achieved. Examples include phase transformation kinetics in steel [97,98].

The time-resolution can be further improved, using the stroboscopic measurement technique, if the process is repeatable. This was certainly the case for cyclic deformation, or fatigue test. The stroboscopic technique also came handy in studies of charge-discharge of energy storage materials [92]. To apply the technique, however, the process need not be cyclic in nature. The key word is repeatable. An earlier demonstration was carried out and reviewed by Izumi [99].

Event-based data acquisition at pulsed neutron source has now made stroboscopic measurements much easier, and revolutionized the way in which time-resolved neutron diffraction measurements are made, particularly for experiments with varying sample environment conditions, such as temperature, applied load, or magnetic field. In event-based data acquisition, which is natural for time-of-flight neutron instruments, the position and time-of-arrival are recorded for each captured neutron. This method of data acquisition provides an efficient and flexible way to gather and analyze neutron diffraction data. In this new paradigm of data acquisition, scattered neutrons are collected continuously and sorted later into diffraction patterns over a desired time interval, which can be correlated with the sample environment variables that are also recorded continuously as a function of time. In a time-resolved experiment, the width of this time interval defines the temporal resolution with which the scattering pattern is resolved. Event-based data acquisition is the default mode of data collection at the SNS and J-PARC, and is now widely used to study structure evolutions [92,100–102] in advanced materials.

12.3.2 Case Studies

In the following, selected examples of in situ studies on deformation and phase transformation behaviors are presented.

12.3.2.1 Deformation in Coarse Grained Materials

Polycrystalline materials consist of grains with various orientations. Under applied load, the lattice strain response within a grain, which can be readily measured by neutron diffraction, depends on the grain orientation. In the elastic regime, the lattice strain is linear and scales with the elastic anisotropy. When plastic deformation occurs, however, the lattice strain starts to deviate from the linear elastic response. The additional elastic strain that develops in the grains to accommodate inhomogeneous plastic deformation during the macroscopic elasto-plastic transition, is termed the intergranular or type II strain, ε_{II}^{hkl}, where the superscript denotes grains with (*hkl*) planes normal to the direction of measurement (specified by the scattering vector) [103].

The presence of intergranular strains has long been recognized. In 1947, Greenough [104,105] examined the residual lattice strains of polycrystalline ferritic steel after plastic tensile deformation, and pointed out the effect of plastic anisotropy. In 1987, Pitschovious et al. [106] studied the intergranular strains in their study of cold-rolled ferric steel. By using a triple-axis spectrometer to eliminate the effect of centering [107], the authors were able to determine the intergranular strains with high precision. Meanwhile, in a study of steam generator tubing, Holden et al. [9] found at some locations, the axial strains measured with {111} and {200} reflections are of opposite signs, which cannot be explained at all by elastic anisotropy alone.

Examinations of intergranular strains have led to the development of a new subfield, i.e., fundamental understanding of the deformation mechanisms in polycrystalline materials at the microscopic level. For this purpose, uniaxial loading devices were built to fit onto neutron diffractometers and the elastic lattice strains for different reflections were recorded as a function of applied stress. Fig. 8 illustrates the development of lattice strain in a coarse-grained *fcc* metals [108]. Under tensile deformation, tensile and compressive intergranular strains develop in ‹200›/LD and ‹220›/LD grains, respectively [103], where LD designates the loading direction. This is a direct result of the crystal elastic and plastic anisotropy [109–111]. The ‹111›/LD grains, on the other hand, show a minimal intergranular strain. Meanwhile, a characteristic texture also develops during deformation as grains rotate toward ‹111›/LD and to a lesser extent toward ‹200›/LD while the population of ‹220›/LD grains is depleted [103,112].

A major breakthrough in understanding the intergranular strains came from the advances in elastic-plastic self-consistent (EPSC) modeling [110,113]. However, the essential physics can be explained within the framework of the Taylor model. By considering an ensemble of grains, Holden et al. [114] demonstrated that the ‹111›/LD grains (i.e., all grains with (111) plane parallel to LD) yielded first, despite the fact that the ‹111› direction has the highest yield strength. This unusual behavior is a result of a combination of elastic and plastic anisotropy. For stainless steel, the ‹111› direction is the stiffest

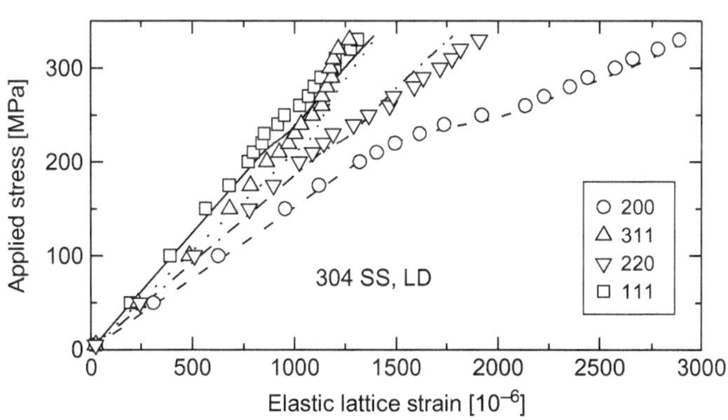

FIGURE 8 Development of lattice strain in coarse grained stainless steel. The deviation from the linear relationship indicates the presence of intergranualar strain. The lines are calculations from elasto-plastic self consistent model. *Figure reproduced with permission from Ref. [108].*

elastically. As a result, in a polycrystalline environment, ‹111›//LD grains carry the highest load, causing ‹111›//LD grains to reach the yield strength faster than other grains. Similarly, ‹100›//LD grains yield last (see Fig. 8), despite having the lowest yield strength, because the ‹100› is the softest direction. Unloading from the maximum tensile load thus leaves a tensile residual strain for ‹100›.

Signature intergranular strains develop due to the activation of select slip planes. For this reason, intergranular strains have been used as a fingerprint to gain insight in to elastic and plastic anisotropy in polycrystalline metals. For example, by measuring the intergranular strains and using polycrystal plasticity modeling to interpret the results, Wollmershauser et al. [115] demonstrated that the intergranular strain evolution of CeAg is nearly identical to that of NiAl, indicating that they share a common primary plastic deformation mechanism, in this case the ‹100›{011} "cube" slip.

12.3.2.2 Deformation in Nanostructured Materials

In coarse grained materials (grain size $D \sim$ μm), the deformation mechanism is usually dominated by dislocation activities. It has been widely anticipated that when the grain size is reduced to the nanometer regime, the deformation will change, as the critical shear stress needed to initiate the dislocation is inversely proportional to the grain size [116,117]. Shear transformation (e.g., deformation twinning) is another way by which some metals deform. However, the nucleation of twinning usually requires a very high local stress concentration or special heterogeneities [118].

It is important to recognize that in nanocrystalline materials, the grain boundary becomes an important factor. As a simple estimate, the relative fraction of grain boundary vs. grain is given by $\frac{4\pi r^2 \Delta r}{\frac{4}{3}\pi r^3} \sim 3\Delta r/r$. Assuming a grain boundary thickness of $\Delta r \sim 1$ nm, we can estimate the fraction of the grain boundary phase. For $D \sim 1$ μm, the volume fraction of the grain boundary phase is $\sim 3 \times 10^{-3}$ or 0.3%. As D decreases to anemometer regime, say $D \sim 10$ nm, the fraction of the grain boundary phase increases to 0.1, or 10%. The fraction increases even more as D approaches zero, further underscoring the importance of the grain boundary phase in nanocrystalline materials.

Alternate deformation mechanisms, particularly grain-boundary mediated plasticity must be considered. Early on, molecular dynamics simulations suggested that grain-boundary mediated deformation (e.g., grain-boundary sliding) becomes prominent in Ni with a grain size $D \leq 10$ nm [119]. Stacking fault via partial dislocation emission was also predicted in Ni with $D \sim 20$ nm [120]. However, subsequent in situ synchrotron X-ray measurements showed no evidence of stacking fault; instead, a reversible peak broadening was reported during loading/unloading, which was attributed to the lack of build-up of a residual dislocation network after the plastic deformation [121]. Thus, the nature of plastic deformation in nanocrystalline materials had been a long debate, and the fundamental question as to whether a cross-over exists from the traditional dislocation slip to grain-boundary mediated deformation lingered.

In a study by Sheng et al. [122], the deformation in nanocrystalline Ni was examined by a combination of in situ synchrotron X-ray and neutron diffraction. The salient results are summarized in Fig. 9. Fig. 9(a) shows the macroscopic stress-strain curves for a Ni sample with 20 nm grain size. The sample was loaded to ~4% plastic strain before failure. The lattice strains are plotted in Fig. 9(b). Compared to conventional *fcc* alloys in Fig. 8, the lattice strains continue to increase linearly, through the regime of plastic deformation all the way to failure. This linear behavior of the lattice strain showed that the Ni grains remained elastic, even though the nanocrystalline sample suffered ~4% plastic deformation. This means that something else in the nanocrystalline sample is taking the brunt of the plastic deformation. That leaves the grain boundaries. In other words, it is the gain boundary phase that carried the plastic deformation.

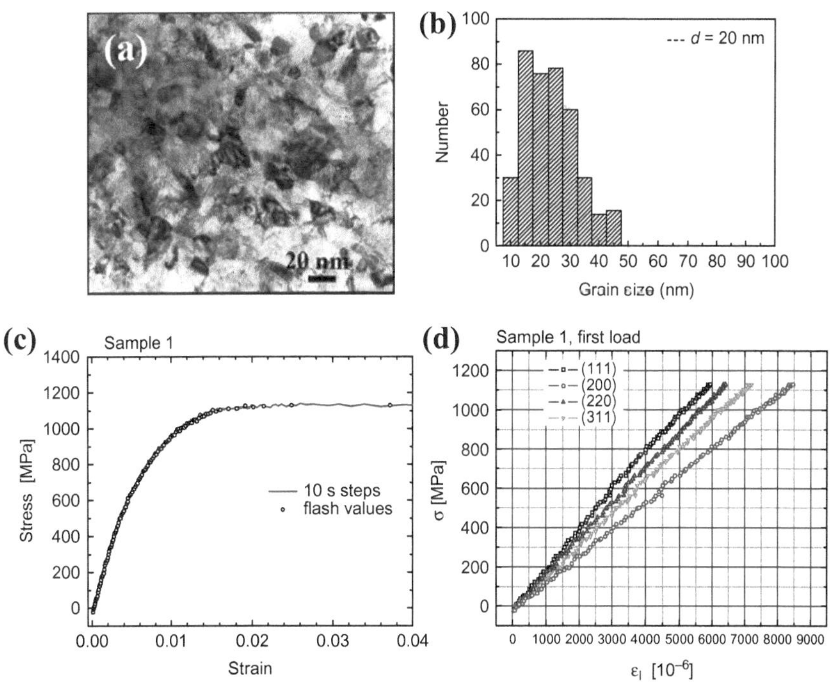

FIGURE 9 Deformation behaviors of nanocrystalline nickel. (a) TEM image of the nanocrystalline grains. (b) Grain size distribution. (c) Loading curve showing ~4% plasticity. (d) Development of lattice strain. The linear relationship indicates that Ni grains remained elastic, even when the sample has gone plastic. *Figure reproduced with permission from Refs. [122].*

More sophisticated analysis is presented in Ref. [122], in which an intergranular strain index and a texture index were used to quantify the transition as the grain size was reduced from μm to nm regime. Significant development was observed in both indices for coarse grain samples, consistent with the kind of changes caused by dislocation mediated plasticity. When the grain size is reduced to 15 nm, however, neither index showed significant changes, indicating that the Ni grains remained plastic during the entire deformation process. Plastic deformation in nanocrystalline samples is therefore mediated by the grain boundary phase. Questions remain though, on how the grain boundary carried the plastic deformation. This is a difficult question, as the structure of the grain boundary phase—its chemical make-up or the atomic arrangement—is presently unknown. It will be a topic of future study.

12.3.2.3 Deformation in Amorphous Alloys

Engineering diffraction started with common engineering materials, such as steel or aluminum and later on magnesium alloys, which are all in crystalline form. However, with the success in crystalline materials, there has been growing interest in applying the technique to noncrystalline materials, such as amorphous alloys. The interest in amorphous materials intensified when bulk metallic glasses [124,125] became available. A critical issue in the application of bulk metallic glasses is the lack of ductility. As with other glasses, bulk metallic glasses are also intrinsically brittle. However, strain measurements in amorphous materials are challenging, as discussed below. As a result, fundamental understanding of deformation in metallic glasses has been lacking, even in the elastic regime.

There are two dilemmas in diffraction measurements of amorphous materials. (1) Compared to crystalline materials, there are no sharp diffraction peaks from amorphous materials, only broad maxima. It was not clear whether small strains due to elastic or plastic deformation, which are on the order of ~1–2%, are even measurable from the shift in the broad maxima. (2) In crystalline materials, each diffraction peak has a meaning. Specifically, the position of peak (*hkl*) represents the lattice spacing of (*hkl*) planes in the direction of the scattering vector. In amorphous materials, a diffraction maxima do not represent a certain lattice spacing. The question becomes how to analyze the diffraction data to extract microscopic insights.

The first question was answered by pioneering measurements with high energy synchrotron diffraction [126]. Clear changes in diffraction pattern were demonstrated when a metallic glass was under load. The second question was answered by detailed analysis of the first sharp diffraction peak, obtained with X-ray and neutron diffraction for a variety of metallic glass alloys [127].

Despite the lack of long-range crystalline order, metallic glasses are known to exhibit some degrees of short-range order (SRO) [123,128] and medium-range order (MRO) [127]. Previous studies have revealed that the

SROs are characterized by solute-centered clusters, consisting of a minority solute atom surrounded by a majority of solvent atoms, and the MRO is constructed by packing of the clusters beyond the SRO [127]. Analysis of the pair-distribution function shows that the first sharp diffraction peak in $S(Q)$, the diffraction structure factor, describes the medium-range atomic correlation in real space, while the high-Q portion of $S(Q)$ reflects the short-range order. This knowledge about the structure provides a basis to link the change in diffraction patterns with atomic configurations in metallic glasses.

The strain analysis began by determining the strain values in reciprocal space. The ε was determined by implementing a pattern matching technique, where for a given Q value, the $S_0(Q)$ curve in the vicinity of Q was displaced by $Q \cdot \varepsilon$. A χ^2-type functional $\Delta[\varepsilon,Q]$ accounting for the difference between the measured $S_\sigma(Q_\sigma)$ and $S_{\sigma=0}[(1+\varepsilon)Q]$ was evaluated for each trial value of ε, and ε was determined when $\Delta[\varepsilon,Q]$ reaches a minimum. Details of this χ^2-method are described in Ref. [129]. Fig. 10(a) depicts the $\Delta[\varepsilon,Q]$ as a contour plot of ε and Q, for which $\sigma = 1500$ MPa. Each of the six letters (A-F) points to a local minimum, which represents the most likely value of the microscopic strain at the corresponding Q scale. The strain at location A is the largest among all (-1.3%), while the strains at B-F are vanishingly small. Thus Fig. 10(a) demonstrates, in a quantitative way, that the strain in BMGs is highly spatially inhomogeneous. The low-Q portion of the $S(Q)$, or medium-range order, responds much more sensitively to stress than the high-Q portion, or the short-range order, does.

Based on this analysis, the $S(Q)$ can be decomposed into two parts: the strain sensitive part, and strain-insensitive part. For detailed analysis, the readers are referred to Ref. [129]. The results are shown in Fig. 10(b). As might be expected, the strain-sensitive part is mostly located in the low-Q region, in the vicinity of the first sharp diffraction peak. The strain-insensitive part is located in the high-Q region, corresponding to the short-range order. A physical picture for the deformation can be visualized by imagining uniaxial compression of a flexible sandbox filled with hard bicycle wheels, see Fig. 10(c). When the sandbox is compressed, the bicycle wheels themselves do not deform, but the bicycle wheels can shuffle in space. It is thus the configuration changes in the relative positions of the bicycle wheels that characterize the deformation of the sandbox.

12.3.2.4 *Insights From Peak Width Analysis*

In addition to the peak position, the width of the diffraction peaks also carries significant insights about the deformation, especially in the plastic regime. Peak width analysis has a long history. The Williamson-Hall plot [130] is commonly used but not very precise for evaluating peak width broadening due to plastic deformation. In this section, we discuss the application of CMWP (convolutional multiple whole profile) fitting program [131,132], to the analysis of dislocation-induced peak broadening in stainless steel.

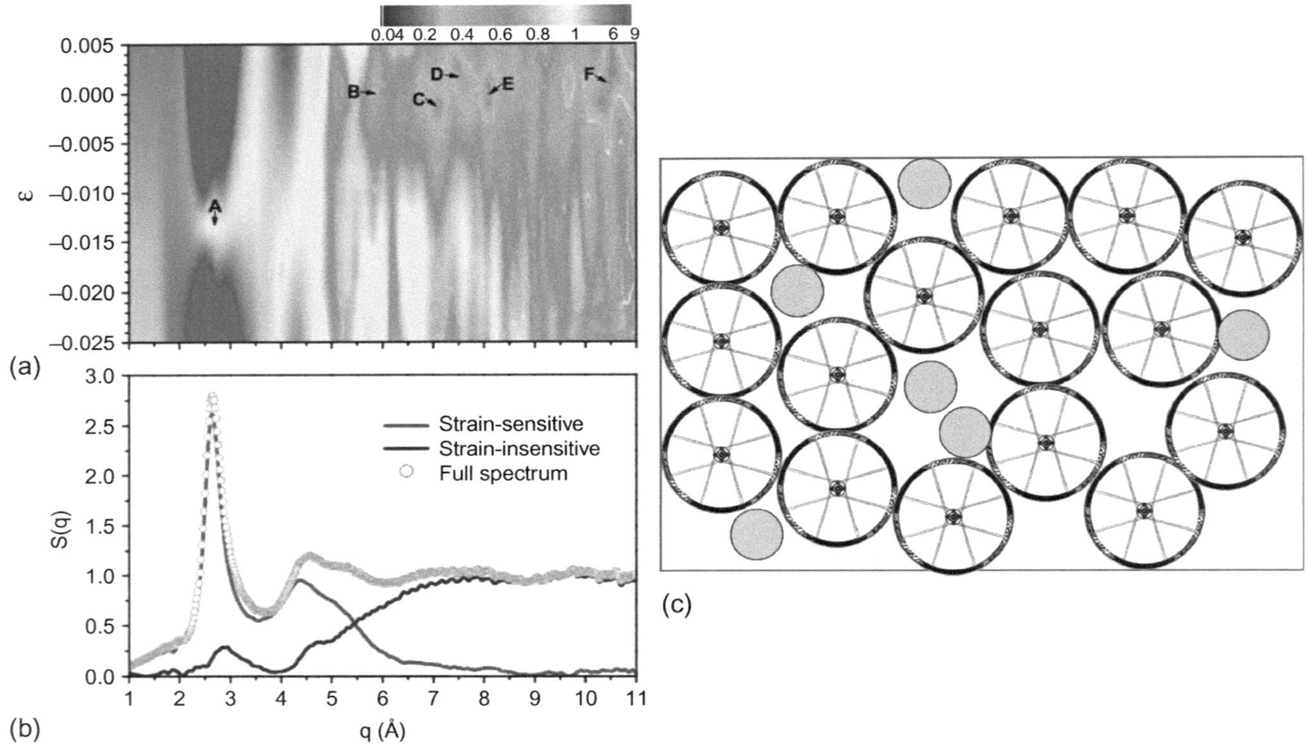

FIGURE 10 Deformation in metallic glass. (a) χ^2 functional showing the inhomogenous strain distribution during elastic deformation. (b) Separation of strain-sensitive and strain-insensitive parts in $S(Q)$. See text for details. (c) Schematic illustration of deformation in metallic glasses. Under a compressive stress, the short-range order represented by the bicylce wheels does not change. The deformation is carried out by medium-range order or configuration of the non-deformable wheels. *Figure reproduced with permission from Ref. [129].*

It is important to recognize that dislocation-induced peak broadening is strongly (*hkl*) dependent, since the different (*hkl*)-oriented grain populations deform differently [133–138]. The strain broadening is characterized by the mean square strain,

$$\left\langle \varepsilon^2_{Q,L} \right\rangle = \frac{\rho_{hkl} c_{hkl} b^2}{4\pi} f(L/R_e), \tag{6}$$

where ρ, R_e, and b are the density, the effective outer cut-off radius, and the Burgers vector of dislocations. $f(L/R_e)$ is the Wilkens function [139], where R_e is the outer cut-off radius of the dislocation strain field, and L is the Fourier variable. The subscripts Q and L in $\langle \varepsilon^2_{Q,L} \rangle$ indicate that the mean square strain depends on the diffraction vector Q and on L. In principle, both ρ and C are (*hkl*) dependent. The analysis of grain-orientation dependent dislocation density has been addressed by Ungar et al. [140]. Here we discuss a simplified case, i.e., ignoring the (*hkl*) dependence of ρ. Thus, the only (*hkl*) dependence is in the contrast factor C_{hkl}.

$$\left\langle \varepsilon^2_{Q,L} \right\rangle = \frac{\rho c_{hkl} b^2}{4\pi} f(L/R_e). \tag{7}$$

The theoretical values for C_{hkl} can be calculated numerically using the ANIZC software package [141], assuming a dislocation type.

Fig. 11 demonstrates the difference between a regular Williamson-Hall plot and a modified Williamson-Hall plot based on Eq. (7) for stainless steel subjected to tensile deformation. The modified Williamson-Hall plot takes into account the strain anisotropy, and thus a linear relationship is obtained. For the *fcc* structure stainless steel, a total of 12 slip systems have to be accounted for. The single crystal elastic constants of stainless steel with $C_{11} = 206$ GPa, $C_{12} = 133$ GPa, and $C_{44} = 119$ GPa [142] were used for the calculation of C_{hkl}. The slope of the linear relationship in the modified Williamson-Hall plot gives the average dislocation density ρ. In the case examined, ρ is on the order of 10^{14} m^{-2}, and increases with plastic strain.

A more refined analysis can provide further useful information about dislocation types. For example, for cubic materials, C_{hkl}, is given by,

$$C_{hkl} = C_0\left(1 - qH^2_{hkl}\right) \tag{8a}$$

$$H^2_{hkl} = \frac{h^2k^2 + k^2l^2 + l^2h^2}{\left(h^2 + k^2 + l^2\right)^2} \tag{8b}$$

The values of C_0 and q depend on the dislocation type. Calculations with ANIZC software have found C_0 and q to be 0.3001 and 1.7182 for edge dislocations, and 0.3237 and 2.4695 for screw dislocations.

Substituting Eqs. (8a) and (8b) into Eq. (6), it can be seen that the normalized slope, q, of $\langle \varepsilon^2_{Q,L} \rangle$ with respect to H^2_{hkl} depends on the dislocation types. Thus, by plotting $\langle \varepsilon^2_{Q,L} \rangle$ vs. H^2_{hkl}, an assessment can be made of the main

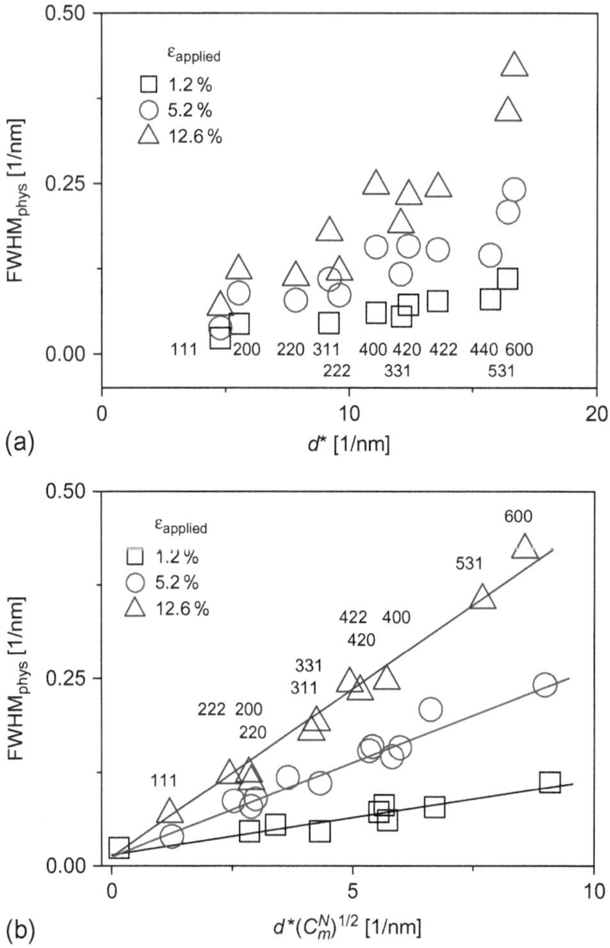

FIGURE 11 Scaling of FWHM for different reflections. (a) William-Hall plot. $d^* = Q/2\pi$. (b) Modified William-Hall plot. C_{hkl} is the contrast factor. *Figure reproduced with permission from Ref. [140].*

dislocation type. This analysis has been applied to stainless steel subjected to cyclic loading, where the slope q was found to be ~1.7, indicating an edge-type of dislocation [143]. The same analysis was applied to high entropy alloys, where a mixture of edge and screw type of dislocations were found, and subsequently confirmed by transmission electron microscopy [144].

12.3.2.5 Functional Materials

There has been increasing interest in "smart materials," where the mechanical properties and functionality are interdependent. In piezoelectric materials, for example, a large elastic strain is induced by applying an electric field, and vice versa. Because of the coupling between their mechanical, electric, and magnetic responses, smart materials are being developed for use as sensors and actuators, and, potentially, for energy harvesting.

An early application of neutron diffraction to smart materials was to experimentally determine the bulk longitudinal and transverse electrostrictive coefficients for polycrystalline samples of PMN–PT–BT piezoelectric [90]. When an electric field is applied, an electrical response and a mechanical strain are generated, which are related to the electric field by

$$P = P_s \cdot \tan\mathrm{h}\left(\frac{E}{E_0}\right) \tag{9}$$

and

$$\varepsilon = QP_S^2 \cdot \tan\mathrm{h}\left(\frac{E}{E_0}\right) \tag{10}$$

where P is the polarization vector, E is the applied electric field. E_0 is a material parameter for the polycrystalline material. Formally, the electrostriction coefficient Q is a fourth rank tensor

$$\varepsilon_{ij} = Q_{ijkl} \cdot P_k \cdot P_l \tag{11}$$

relating second order strain ε_{ij} with the first order polarization tensor P_k and P_l. Measurements of lattice strains under an applied electric field allow direct determination of the Q_{ijkl} vector. An example is shown Fig. 12. The experiment was carried out using a residual stress diffractometer, and the sample was subjected to high electric field of 1250 kV/m. A complete determination of the Q_{ijkl} tensor enables the development of predictive models for electric and mechanical responses in piezoelectric materials.

Ferromagnetic shape memory alloys (FSMA) are intermetallic ferromagnetic martensites that exhibit large magnetic-field-induced strains and stress-induced magnetization upon application of magnetic fields and/or mechanical stress. Recent development showed that off-stoichiometric $Ni_{2-x}Mn_xGa$ based FSMAs alloys can achieve a remarkable maximum strain of $\sim$10%. Moreover, in contrast to conventional shape memory alloys, FSMAs can react at a very high response frequency, up to 1 kHz [145,146]. These unique properties can potentially open up new areas of applications and, as a result, FSMAs have attracted much attention in recent years.

Unlike in conventional shape memory alloys, in which a martensitic phase transformation occurs, in FSMAs, the shape memory effect is driven by the twin boundary motion of the martensitic phase, leading to reorientation of the crystallographic twin variants [147]. Understanding the underlying mechanism for the twin boundary motion is therefore central to the development of new FSMAs and related devices. Twin reorientation occurs when the applied

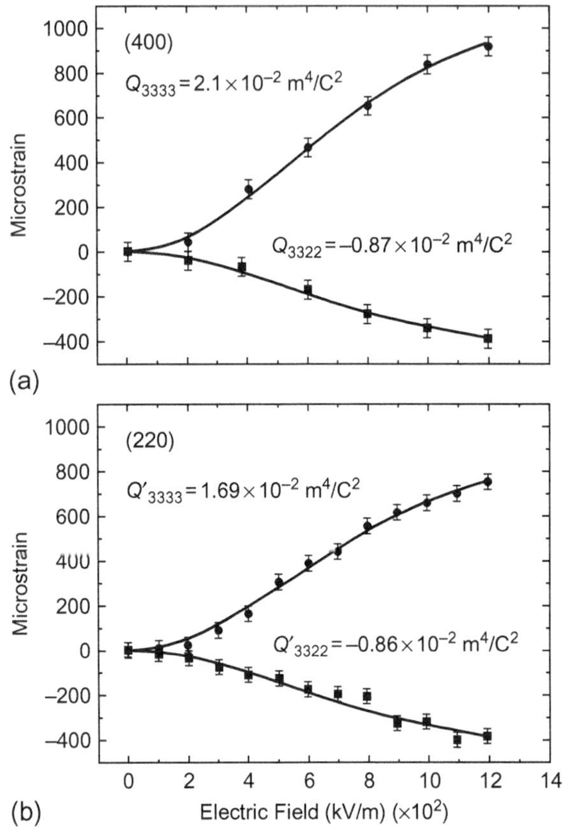

FIGURE 12 Longitudinal and transverse strain vs. field measured using neutron diffraction and the constitutive model fits used to determine the electrostrictive coefficients for the (400) and (220) reflections. Q' values are related to Q_{ijkl} through a linear combination. *Figure reproduced with permission from Ref. [90].*

stress exceeds a critical value, causing the atoms to shear and hence reorientation of twins. In the presence of a magnetic field, twinning or detwinning is achieved via magnetoelastic coupling, which can be phenomenologically described as an effective stress acting on the twin boundaries. Thus, for FSAMs, the relationship between functionality and mechanics can be understood through an atomistic level insight into the phenomenon of magnetoelastic coupling.

Using in situ neutron diffraction, twin variant reorientation in single-crystal Ni–Mn–Ga during quasi-static mechanical compression was investigated [88]. The experiments were carried out on VULCAN, the engineering diffractometer at the SNS. The experimental setup and key results from the experiments are presented in Fig. 13. The volume fraction of reoriented twin variants for different stress amplitudes were obtained from the changes in

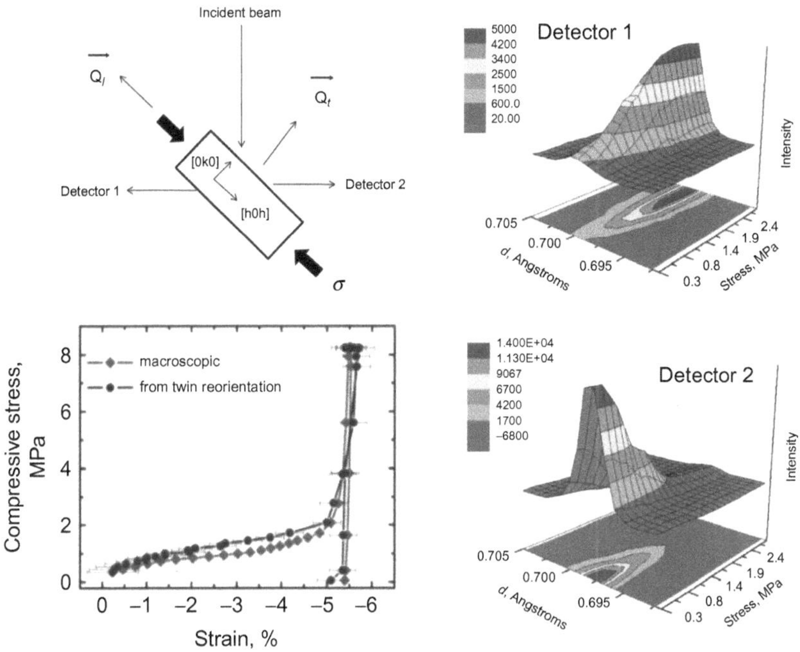

FIGURE 13 (a) Schematic setup of an in situ neutron diffraction experiment to investigate twin reorientation during compressive mechanical loading. The crystallographic orientation of the majority twin variant with respect to the compressive stress axis are shown; (b) measured intensities of $(080)_m$ diffraction peaks with scattering vectors parallel (top panel) and perpendicular (bottom panel) to the loading direction, respectively; (c) the macroscopically measured compressive strain is consistent with twin reorientation strain calculated from diffraction measurements. *Figure reproduced with permission from Ref. [88].*

integrated intensities of high-order neutron diffraction peaks, where magnetic scattering intensities vanish. It is shown that, during compressive loading, ~85% of the twins were reoriented parallel to the loading direction resulting in a maximum pseudoplastic strain of ~5.5%, which is in agreement with measured macroscopic strain.

12.3.2.6 Phase Transformation Kinetics

Phase transformation is an effective method to manipulate the microstructure of materials. Phase transformation studied on an engineering diffractometer typically requires a special sample environment that is difficult or impossible to implement on conventional powder diffractometers which usually come with a confined space area. Fig. 14 illustrates the thermomechanical processing of a low-alloy steel [88,148]. After austenitization at 1173 K, the steel specimen was cooled down to a certain temperature for isothermal transformation. Selected diffraction profiles obtained in 120 s intervals at various stages of annealing are summarized in Fig. 14. The starting microstructure at room temperature is martensite, which fully transformed to austenite at

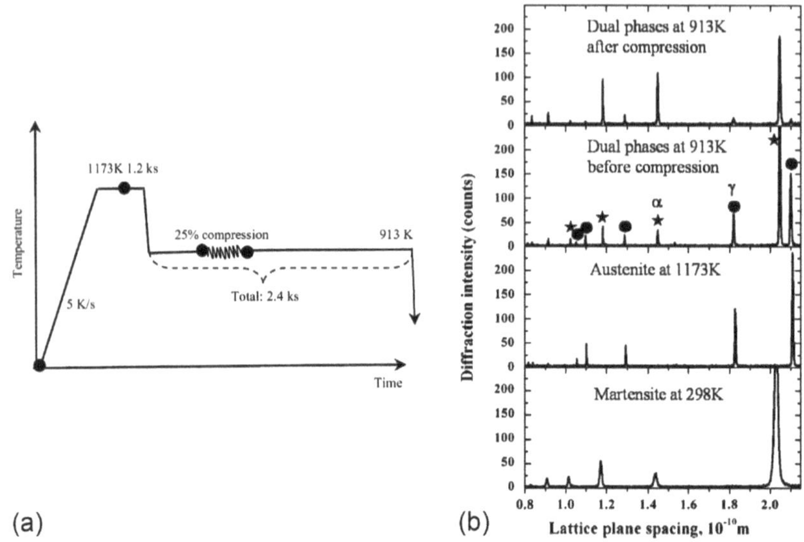

FIGURE 14 (a) Schematic drawing of a thermo-mechanical controlled progress (TMCP). (b) Neutron diffraction patterns collected at different stages during processing. The symbols identify the characteristic phases formed at different stages. *Figure reproduced with permission from Ref. [148].*

1173 K. Upon cooling, ferrite formed. The specimen was subsequently plastically deformed in compression. Significant changes in the diffraction profile can be observed: first, an increase in ferrite volume fraction; second, the change in (*hkl*) intensity ratio as evidenced by comparing the diffraction profiles before and after compression in Fig. 14. The change in ferrite volume fraction indicates accelerated transformation by plastic deformation. Finally, the change in relative peak intensities indicates development of texture in both ferrite and retained austenite.

The above example also highlights the experimental challenges in kinetic studies with neutron diffraction. Ideally, one would like to conduct a time-resolved study. Placing 10–20 points over the span of 1000–2000 seconds in thermo-mechanical processing means a measurement time of about 50–100 seconds for each data point. The instrument used at the time, ENGIN-X was barely sufficient for this purpose, as the data profile over 120 seconds demonstrated. This kind of experiment called for a new generation of engineering diffractometers. With the VULCAN at the SNS [46,59] and TAKUMI at the J-PARC [149], time-resolved measurements have become a routine. In fact, with sufficiently large samples, a time-resolution in subsecond regime has been demonstrated in a recent study of the ferrite-to-austenite phase transformation in steel [97,98]. In this context, in situ, time-resolved studies of thermo-mechanical processing, under load and at temperature, have finally become a reality.

Energy storage materials, batteries, have attracted much attention lately. High capacity batteries are an important component in the overall strategy for a secured energy future [150]. High power Li-ion batteries require an operating lifetime of over 5000 cycles and 10–15 years or more in order to achieve economic viability in battery-powered vehicles and in the future electricity infrastructure. These requirements are about one order of magnitude greater than what current technology provides in small format cells (as in, for example, laptop computers and cell phones). Unfortunately, our ability to predict battery life is extremely limited because many seemingly unrelated degradation mechanisms have been identified.

Neutron diffraction is an excellent tool in the chemistry and structure evolution of Li-ion batteries. Specific advantages, particularly with time-of-flight diffractometers at a pulsed neutron source, were nicely summarized in a recent paper by Taminato [151]. (i) Neutron diffraction allows the simultaneous collection of diffraction data over a wide range of momentum transfers. Acquiring diffraction data with high-intensity neutrons makes accurate structural analysis possible under nonequilibrium conditions in the electrode matrix at each charge or discharge reaction step. (ii) Neutrons penetrate the stainless-steel cans used for commercial cells, which enables the collection of diffraction data with sufficient accuracy to determine the structures of the electrode materials under high C-rate conditions. (iii) Neutrons can detect light elements and enable the determination of the position and concentration of lithium in the cell. (iv) Large-scale cells contain large amounts of electrode materials, which provide diffraction patterns of high quality and high intensity. For these reasons, a special powder diffractometer, SPECA, has been constructed at J-PARC to specifically address the challenges in battery materials [152].

A demonstration experiment [92] was carried out with a large format pouch battery cell, using the VULCAN diffractometer at the SNS. A 40 Ah commercial pouch cell, with the dimension $190 \times 206 \times 7\ mm^3$, was used in this study. The battery has a $Li_xNi_yMn_zCo_{(1-y-z)}O_2$ (NMC) composite cathode and a graphite composite anode. Cu and Al foils served as the cathode and anode current collectors, respectively. VULCAN is a time-of-flight neutron diffractometer, with which a diffraction pattern over a large range of d-spacings is collected simultaneously. A sampling volume of $5 \times 5 \times 20\ mm^3$, defined by an incident beam slit and a radial collimator, was located at middepth in the middle of the battery cell.

The succession of Li-Graphite intercalation phases was fully captured under a rapid charge-discharge condition (1C-rate, i.e., charge to full capacity in 1 hour), see Fig. 15. However, here, the lithiation and dilithiation pathways are distinctively different and, unlike in slowing charging experiments with which the Li-Graphite phase diagram was established, no LiC_{24} phase was found during rapid charge. Approximately 75 mol.% of the graphite converts to LiC_6 at full charge, and a lattice dilation as large as 4% was observed during a charge-discharge cycle. This work demonstrated the potential of in situ, time and spatially resolved neutron diffraction study of the dynamic chemical

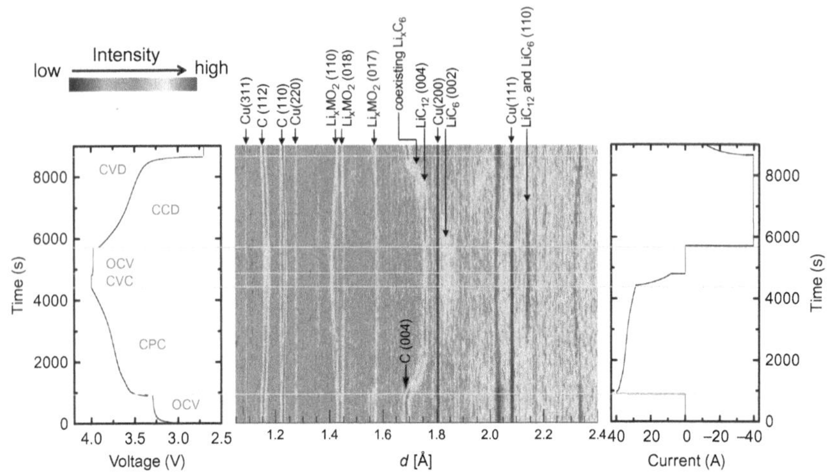

FIGURE 15 In situ neutron diffraction pattern of a large pouch battery taken during a charge and discharge cycle. The in situ diffraction data, averaged over 7 cycles, are binned into 2.5 minutes histograms. The voltage (current) is plotted in the side panel to the left (right) of the diffraction data. Voltage and current curves are shown on the left and right hand side, respectively. The acronyms used in the figure are as follows: CVD-constant voltage discharge, CCD-constant current discharge, OCV open circuit voltage, CVC-constant voltage charge, and CPC-constant power charge. *Figure reproduced with permission from Ref. [92].*

and structural changes in "real-world" batteries under realistic cycling conditions, providing insights on degradation at the microscopic level and elucidating the important role of diffusion kinetics in energy storage materials.

12.4 SMALL-ANGLE NEUTRON SCATTERING

12.4.1 Measurement Theory

The intensity (or peak) of SANS, $I(Q)$, is given by the square of the summation of waves $F_i(Q)$ scattered by all atoms, where the subscript i indicates the ith atom. At $Q=0$, the scattered waves by atoms irradiated by neutron have exactly the same phases and therefore give the intensity maximum. The difference between diffraction and SANS is simply found in the condition difference in which the phases of all scattered waves are the same, i.e., the path length difference of each scattered beam is $n\lambda$ (here, n is an integer number) in the diffraction case while the path length difference is simply zero in the SANS case. Therefore, SANS does not include any information about crystalline structures except for crystals with large lattice parameters such as oxide.

Evaluating grain size may sound suitable for SANS, but usually it is difficult to apply the method to the major (or matrix) phase. This is because the SANS sensitivity is simply proportional to the square of $\Delta\rho$, the contrast of scattering length density, which is defined as the difference between the scattering length density of the hetero-structure ρ_{hetero} and ρ_{matrix} as follows:

$$\Delta\rho = \rho_{\text{hetero}} - \rho_{\text{matrix}}$$
$$\rho = \frac{\sum n_i b_i}{V_{\text{cell}}} \tag{12}$$

Here, the n_i is the number of atoms of i in the unit volume V_{cell} and b_i is the scattering length density.

The diffraction technique is much easier for characterizing the major phase than for the minor phase. In contrast, SANS is easier for the minor phase, especially when the volume fraction is less than 10%. This is the advantage for characterizing the hetero-structure, such as precipitates, or a minor phase in a multiphase system in engineering materials. Another important feature of SANS is that the scattering intensity can be determined on an absolute scale. Thanks to the simple optics and small-angle area, one only needs to know the source intensity and detector efficiency of the instruments in order to transform the scattering intensity to absolute scattering cross section. Using a secondary standard sample, this is achieved quite easily compared to the other techniques [153]. Absolute intensity measurements allow the determination of the secondary phase in a quantitative manner. For example, the volume fraction of the secondary phase with a particle size of a few nanometers can be evaluated quantitatively, once the $\Delta\rho$ is known.

Steels are one of the most important alloys in engineering fields. In this case, the concentration of iron can be discussed using the intensity ratio between nuclear and magnetic components of SANS. This is achieved by the use of scattering length difference for iron in nuclear and magnetic scattering components [154]. Similarly, it is also possible to obtain the information of the composition using scattering length difference between X-ray and neutron [155]. The detail of the analysis of SANS has been nicely covered in chapter 1.4.6., 5.2.2, and 5.3.4 to 5.4 in Vol. 1 [10,156]. Comprehensive reviews are also available about the application of SANS and SAXS to alloys and ceramics [157,158]. Therefore, only practical aspects of SANS analysis for engineering materials, especially metals and alloys, are overviewed here. All discussions are based on the absolute intensity measurements obtained by using pinhole collimations or profiles after desmearing in the case of slit collimations.

12.4.2 Practical Aspects of Analysis in Metals and Alloys

The basic equations and concept for analysis summarized in this section is related to the equations explained in section 5 in Vol. 1. We can start the analysis using a method similar to Eq. 5.13 and 5.23 in vol. 1 as follows:

$$I(q) = \Delta\rho^2 N \int_0^\infty V^2(r) F^2(Q,r) f(r) dr \tag{13}$$

Here N is the number density, and $V(r)$ and $f(r)$ are, respectively, the volume of the particles with radius r, and the normalized probability distribution function for the particle size.

$F(Q, r)$, is the form factor and the one for sphere is given in Eq. (5.21) in ref. [10] as follows:

$$F(Q,r) = \frac{3[\sin(Qr) - Qr\cos(Qr)]}{(Qr)^3} \tag{14}$$

For most cases in engineering materials, using form factor of sphere is an acceptable approximation (more detail about the shape will be discussed in later section). $f(r)$ can be any function, such as normal or log-normal distribution function. One can also use maximum entropy analysis method without fixing any size distribution functions [159]. If we assume that it has log-normal distribution of probability at radius $R=5$ and 10 nm and the width (sigma) is 0.3, the profile gives the shape shown in Fig. 16. In a log-log scale plot, we can find important features of the SANS profile. The first one is in the low-Q region where the profile shows a relatively flat area. One can realize that this area corresponds to the range of half width of half maxima in the regular peak analysis by comparing the intensity at the minimum Q. Therefore,

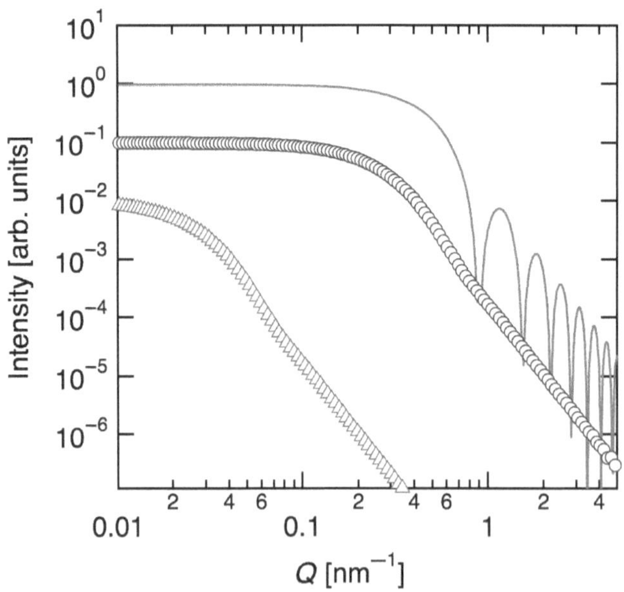

FIGURE 16 SANS profiles for spherical particles with radius $R=5$ nm (circles) and 50 nm (triangles) at the peak probablitiy in log-normal size distribution with $\sigma=0.3$, together with the scattering by single size particle with 5 nm (solid line). The profiles are normalized by $N\Delta\rho^2V^2$ and then shifted by 0.1 for circles and 0.01 for triangles. Dotted ellipsoids shows the "flat" region where the Gunier approximation is applicable. Straight solid line shows the eye-guide for Q^{-4} dependence.

observation of this Q-range is a key requirement for size analysis. In this region, the scattering profile is represented well by the Guinier approximation [160]. Then, the profile is followed by the region which shows straight dependence in the log-log plot. The intensity decreases by Q^{-4} as shown in Fig. 16, and it can also be derived by taking Qr to a large value in Eqs. (13) and (14). A Q^{-4} dependence always appears when the particles have a "flat" (or smoothed) surface and constant scattering length density inside of the object [161]. This condition is valid for most secondary phase and precipitates in the case of metals and alloys in the engineering fields. Therefore, the first step for SANS analysis is to find the flat and Q^{-4} regions in a log-log plot.

In reality, engineering alloys always include much heterogeneity, such as inclusions which form during the solidification, and precipitates which form in different processing stages, e.g., cooling or the subsequent tempering. Consider the case where two different phases are embedded in the matrix, one with the size of a micron scale and the other a nanometer scale. Assuming precipitates are located far apart, then we can ignore the interference between them. In this case, the total intensity $I_{total}(Q)$ can be described as follows:

$$I_{total}(q) = \Delta\rho_1^2 N_1 \int_0^\infty V_1^2(r_1) F_1^2(Q, r_1) f_1(r_1) dr_1 + \Delta\rho_2^2 N_2 \int_0^\infty V_2^2(r_2) F_2^2(Q, r_2) f_2(r_2) dr_2 \quad (15)$$

The calculated profile is shown in Fig. 17. If the whole Q-range is covered, we can obtain good accuracy about all parameters. However, all instruments have a certain limitation in the coverage of Q. If our observation range only covers box A, then we cannot obtain any information about the size. In the case of box B, we can analyze the precipitates with a smaller size assuming the background with Q^{-4} dependence corresponding to the tail of large-size precipitates. Therefore, the intensity can be described as

$$I_{total}(q) = AQ^{-4} + \sum_p \Delta\rho_p^2 N_p \int_0^\infty V_p^2(r_p) F_p^2(Q, r_p) f_p(r_p) dr_p + B \quad (16)$$

Here, p is the number of kinds of precipitates with different $\Delta\rho_p$ and $f_p(r_p)$, and A and B are the coefficients of the Q^{-4} term and the Q independent background caused by the incoherent scattering or some other effects in the high-Q region, respectively.

Some of the observed SANS profiles, however, may start from other types of Q dependence, often around Q^{-3} to Q^{-4}. This gives uncertainty to the analysis in the higher Q region. Perhaps the best way to remove this uncertainty is through the use of a reference sample. Large size heterogeneities are often introduced by solidification or preprocessing before the final microstructure control. As an example, in the precipitation case, the sample before tempering is a good reference. Simple subtraction from the profiles after tempering gives the information formed after tempering [162]. In addition, a sample without

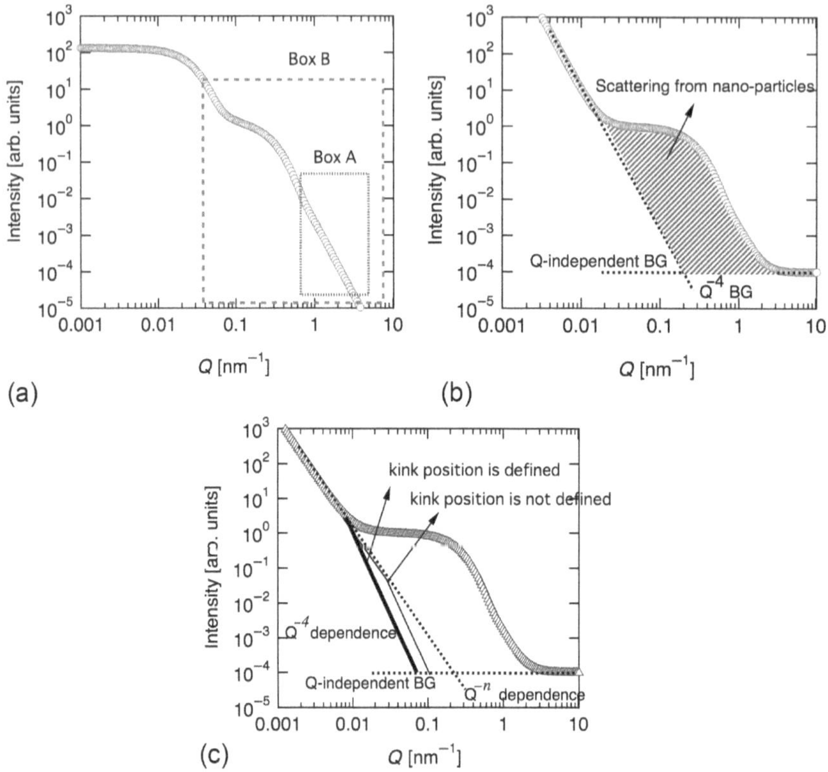

FIGURE 17 (a) SANS profiles for the sample include two different precipitates. (b) nano-precipitates with Q^{-4} and Q-independent BG's. (c) nano-precipitates with Q^{-n} and Q-independent BG's.

additive elements can be a good reference for discussing the microstructure formed by micro alloying. If no good reference can be prepared, one might assume a general Q^{-n} background instead of a fixed Q^{-4} term. A general Q^{-n} relationship can be attributed to a wide size distribution, overlapping from many different precipitates or fractal structures, which can be fitted by a certain value of n. However, most cases should show the Q^{-4} dependence if we can reach sufficiently large Q values compared to the size distribution or the fractal scale limit, except in some special arrangements, like a rod array parallel to the beam direction. For example, if the systems show Q^{-3} dependence in low-Q region, this contribution of the intensity should show a transition to Q^{-4} somewhere in the hidden area which is shown in dotted line in Fig. 17. Thus, "assumption of other exponent" does introduce uncertainty into our models. Instead, assuming Q^{-4} in the low-Q part has a well-defined meaning, i.e., the minimum contribution of Q^{-4} from large particles.

The easiest way to take size distribution into account is to assume the frequency factor as a certain function, typically using normal or log-normal size distributions. In those fittings, one should note the prefactors of both Eqs. (13) and (14).

At $Q=0$, the prefactor is $N\Delta\rho^2V^2$, meaning that the contribution of large particles is much stronger than those of small particles. For example, if the size difference is 2 with same N and $\Delta\rho$, then contribution in low-Q can be 64 times higher for the large size particles than those of smaller sizes. This brings some complexity in the interpretation of the analysis results, especially when a log-normal size distribution is used. The distribution sometimes expands to decays. In such a case, the contribution of the highest frequency in size is negligibly small to the scattering intensity. On the other hand, in the high Q region, the prefactor of Q^{-4} is proportional to total area of interface, which can be proportional to NR^2. Therefore, for the above example, the intensity difference is only a factor of 4 in Porod region.

As a good approximation, in practical analysis, a log-normal distribution is assigned to the volume fraction frequency, not to the size frequency itself for the case of engineering materials [163]. It is worth mentioning that the peak value of the highest frequency in the volume fraction often gives a good agreement with TEM observations [164].

12.4.3 Which Form Factor to Use?

As listed in Fig. 18, many form factors can be used in data analysis [166]. In the actual case of precipitates, the disc (or plate) shape, cylindrical (or needle)

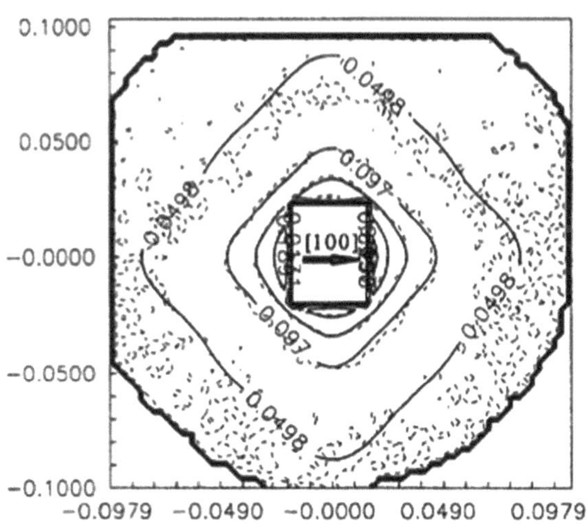

FIGURE 18 Two dimensional SANS profiles of cuboidal γ' precipitates in Ni based single crystal. *Figure reproduced with permission from Ref. [165].*

shape, and even polyhedra are often observed, because of the structural coherence between the matrix and precipitates. These coherent precipitates have certain orientations in one grain. In this case, two-dimensional scattering patterns will show an anisotropy due to the shape. In fact, single crystals, such as Ni based superalloys for high temperature applications, give a distinct anisotropy in the scattering profiles, as shown in Fig. 18 [165]. In contrast, in polycrystalline materials, there are many crystalline grains with different orientations in the area irradiated by neutrons. In such a case, the anisotropic shape is averaged out. Only particles with strong anisotropy in shape show some special features, i.e., cylinder shows Q^{-1} in the intermediate range and Q^{-2} for disc shape precipitates, while the assumption of sphere is still good enough for cubic or other polyhedron with a small anisotropy. For the fitting, the distribution of the aspect ratio c/a does have similar effect with the size distribution of sphere. In the case of c/a of $0.5 \sim 2$, the averaged results still give a good approximate solution. One can judge the shape only when the profiles show clear intermediate range of Q^{-1} or Q^{-2} followed by Q^{-4}. Otherwise, it is better to fix c/a or sigma determined by the direct observation using TEM. In any case, it is recommended to observe basic features of the morphology by TEM prior to SANS measurements and use SANS for quantitative evaluations for average size, number density, and/or volume fraction. Analysis of the region Q^{-1} or Q^{-2} are described in Ref. [167].

12.4.4 Quantitative Evaluation of Volume Fraction—Importance of Scattering Length Density

Since Eq. (14) reaches unity by taking the limit of $Q \rightarrow 0$, the SANS intensity becomes $N\Delta\rho^2V^2$ at $Q \rightarrow 0$. If we know the composition of matrix, for example, from phase equilibrium, the ρ_{matrix} and $\rho_{precipitates}$ can be calculated from Eq. (12). This works reasonably well for a relatively simple system. In the case of multicomponent system, $\Delta\rho$ can be given as a good approximation by using the ρ_{matrix} calculated from the alloy composition when the volume fraction of the precipitates or secondary phase is limited to a few %. The ρ of the precipitates or secondary phases can be calculated from the lattice constant and number of atoms in unit cell. Once we obtain these values, and hence the $\Delta\rho$, the number density (or volume fraction) can be determined from the prefactor $N\Delta\rho^2V^2$.

Note that the unit of parameters, i.e., $\Delta\rho$ and N is usually expressed in cm^{-2} and cm^{-3}, respectively. In contrast, the volume of particles is in nm^3. The $\Delta\rho^2$ is also in the prefactor of Porod region. Therefore, if we know the volume fraction, NV, of large precipitates from some other techniques, a rough evaluation of size can be made from Porod region alone.

$\Delta\rho$ is determined by compositions of both matrix and second phase. Thus it is very useful to calculate the expected value of $\Delta\rho$ before measurements to determine whether the secondary phase is visible or not. Details of scattering lengths are given in a previous volume in this series [156,168], and the list can be obtained at the NIST website [169].

12.4.5 Magnetic Scattering and "A" Value in Steel

Steel is one of the most important engineering alloys. Ferritic steels are ferromagnetic and, therefore, SANS from a ferritic steel sample always includes strong scattering caused by magnetic domain structures. To separate magnetic and nuclear component, it is necessary to apply a magnetic field. A field more than 0.3 T can be applied by a permanent or electric magnet [170–173] for ferrous alloys. If the applied field is strong enough to saturate the local magnetic moment in all regions of the microstructure, the SANS intensity can be separated into magnetic and nuclear components. For nonpolarized neutron beams, the contribution can be written as follows:

$$I(Q) = I_{\text{nuclear}}(Q) + \langle \sin^2\alpha \rangle I_{\text{magnetic}}(Q) \tag{17}$$

Here α is the angle between the local magnetization (the same with the magnetic field direction in a saturating field) and $\boldsymbol{Q}$ vector. ‹ › means the average. Fig. 19 shows a typical SANS pattern for a steel sample under 1T magnetic field. Therefore, the SANS intensity perpendicular to the magnetic field includes both nuclear and magnetic contributions while the one parallel to the

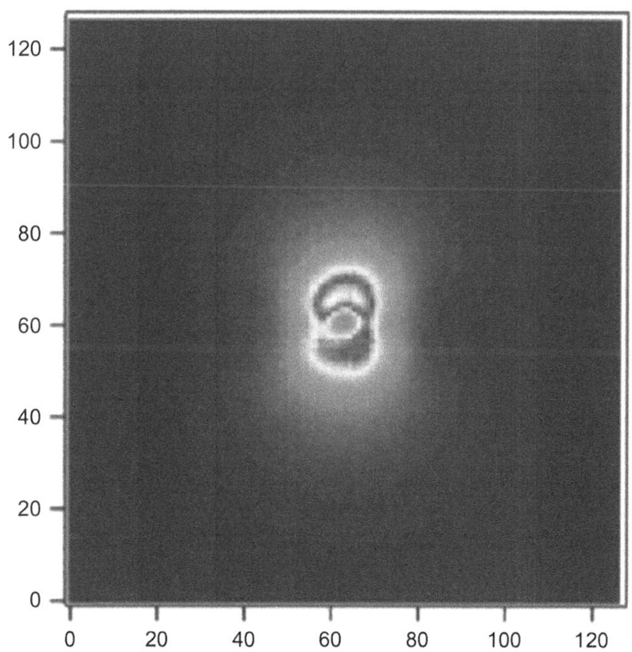

FIGURE 19 Example of 2D SANS patterns obtained under the field of 1 T.

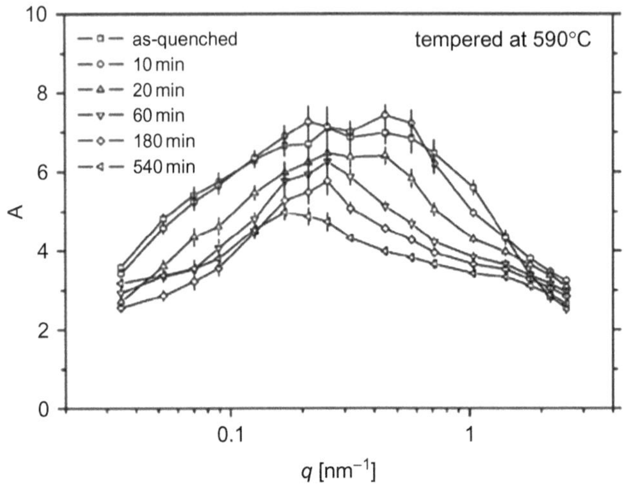

FIGURE 20 Dependence of A ratio to the tempering temperature in high-speed steel; the change indicate the concentration change of the precipitates during tempering. *Figure reproduced with permission from Ref. [154].*

Q vector includes only nuclear component. The scattering length density for the magnetic component is simply determined by the Fe concentration. The ratio of the perpendicular to the parallel parts, the so-called A ratio, given by the following equation, gives the information of Fe concentration [154,170].

$$A = \frac{I_{\perp}(Q)}{I_{\parallel}(Q)} = 1 + \frac{\Delta\rho_{mag^2}}{\Delta\rho_{nuc^2}} \tag{18}$$

Fig. 20 is one example on how to use the A ratio. Usually, other analysis techniques for compositional information of the secondary phase, which is embedded in a matrix, have some effects from the matrix. Therefore, accurate compositions, especially of main constituent elements, are quite difficult to determine when the size reaches a few nanometers, even using the most advanced techniques, i.e., TEM and Atom probe tomography (APT). In this regard, the A ratio can give rather unique qualitative information on ferromagnetic alloys.

12.4.6 Another Way to Use Different Contrast: Combination of SANS and SAXS

Similar to the A ratio, we can compare the intensity between neutron and X-ray scattering. Because neutron scattering lengths vary independently with the atomic number while those of X-ray are proportional to the atomic number, we can use two different contrasts for our analysis. This can apply not only to the ferromagnetic alloys but also to nonmagnetic or paramagnetic materials as well. This technique was initially developed in the 1970s, but very little follow-up work was done thereafter [174]. Since a good standard sample for absolute unit measurements has been proposed [153] and advanced alloys with nanometer scale precipitates are commercially realized [175], demands for characterizing nanometer second phase become high, especially in the steel industry. As a result, several studies are currently in progress. Fig. 21 is one of the examples, for an advanced steel called oxide-dispersion-strengthened (ODS) steel. In this case, the chemical composition of nanometer sized oxide precipitates was carefully analyzed and concluded that the precipitates are $Y_2Ti_2O_7$ and almost free of Fe (<6%) [155]. The authors further expanded the application to precipitates of VC and reported that in the early stage, the precipitates have off-stoichiometric composition with large amount of Fe inside [176].

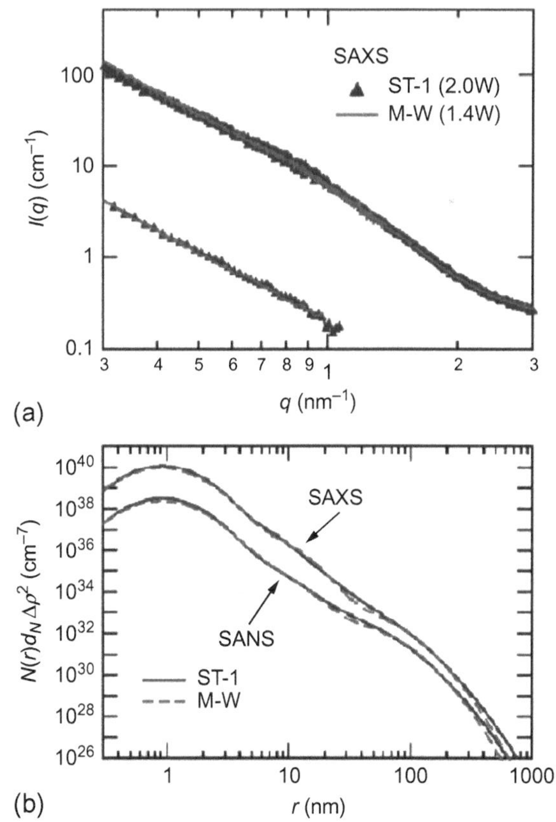

FIGURE 21 Comparison of SAXS and SANS and the frequency distribution multiply by $\Delta\rho^2$. *Figure reproduced with permission from Ref. [155].*

12.4.7 Further Expansion of SANS as a Daily Tool With a Compact Neutron Source

Due to rapid advances of nanostructured engineering materials, easier and quicker characterization tools are needed. For the characterization of microstructures inside a material, TEM and APT are now almost de-facto standard tools. However, in both cases, the sample preparation takes time. More importantly, they are destructive methods. TEM requires thin films of about 10 nm in thickness and APT needs a needle-shaped specimen with a diameter of 10 nm. Obviously, nondestructive and easy sample preparation tools are highly desired. SANS satisfy these requirements; however, the number of available instruments is too limited for SANS to be considered as a standard tool. To overcome this handicap, a new concept of SANS, installed on a compact neutron source, is now under development. The technique is based on a spallation source, using the time-of-flight (TOF) analysis, which compensates for the low flux of neutrons in the compact source by the efficient use of neutron detection. In addition, it is also important to use a cold neutron source to prevent double diffraction effect and to reach smaller Q region. The advantage of compact source is that we can optimize the entire set of instruments from source to spectrometer for one focused purpose. One example is nanoscale characterization. As described in the previous section, the competitive length scale, compared to TEM or APT, is on the order of a few nanometers. For this purpose, one can design the SANS for nanometer scale with a relaxed $\Delta Q/Q$. The first such instrument, intermediate-angle neutron scattering (iANS), was developed in Japan, covering a length of 0.2 to 10 nm^{-1} [177]. The performance of the iANS has been tested by comparing iANS data with the SANS data measured in large facilities. The results are shown in Fig. 22. Of course, the covered Q range by iANS is much narrower compared to those obtained by a SANS on a reactor source, but the data obtained by iANS is enough for nanoprecipitates. It has also been confirmed that the data obtained by iANS agreed well with those by 18m-SANS in absolute unit scale, indicating that quantitative measurements are achieved. The iANS is now under upgradation on a new 3 kW electron accelerator. Typical measurement time can be expected to reduce to a few hours. Application of a compact neutron source is also being promoted at RIKEN in Japan [178,179]. When compact neutron sources become available in many countries, perhaps SANS can be used as a first tool to characterize nanoscale precipitates, due to the easy sample preparation in engineering materials. After initial screening by SANS, the samples can be further analyzed by TEM and/or APT. This may shorten the time for nanostructure characterization and materials development. Laboratory-based iANS can also be expected for "long term in-operando" measurements, such as creep, fatigue, or oxidation that require occupying the instrument for several months by one group. In addition to SANS, other applications such as neutron imaging and even diffraction are

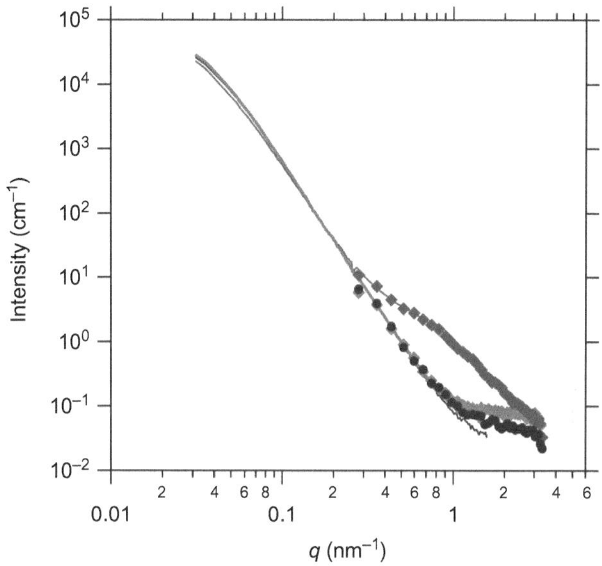

FIGURE 22 SANS profiles of high-nitrogen steels with different tempering condition measured by 18 m SANS in HANARO (solid lines) and iANS in Hokkaido University (markers). *Figure reproduced with permission from Ref. [177].*

now being regularly discussed and explored at international meetings on compact neutron sources [180].

12.5 SUMMARY AND OUTLOOK

A survey has been provided on engineering application of neutron scattering. Mechanical behaviors are the focus, and the main techniques employed are neutron diffraction and small-angle scattering. Examples included residual stress measurements in industrial components, deformation behaviors in polycrystalline metals at different length scales as well as in metallic glasses, and kinetic study of phase transformation which controls microstructure and hence the properties. These neutron scattering studies have played an important role in advancing our understanding of mechanical behaviors at the microscopic level. The fundamental knowledge gained from neutron scattering studies provides guidance on the development of new materials with superior mechanical properties. A good example is the use of SANS and iANS data in the design of new advanced steel. More importantly, however, the greater impact of engineering application goes beyond the fundamental knowledge level. The experimental stress data, for example, have been used in assessment of the mechanical integrity of structure components in a wide-range of industries, from transportation to manufacturing to nuclear energy.

The field of Engineering Application is evolving very rapidly. The scope of applications is expanding, driven both by continuous demands from the user community and the advances in instrumentation. In situ loading studies, which were a novelty in the late 1990s, have now become routine experiments, even for measurements at elevated temperatures. On the other hand, low-temperature deformation experiments are far less explored. This is a regime where deformation mechanisms beyond dislocation, e.g., twinning or martensitic phase transformation, become important. Deformation study in amorphous material is just a beginning, where even the structure response in the elastic regime has not been well understood. New capabilities, such as event-based data acquisition, enable new types of experiments with unprecedented time-resolution. This will open up new opportunities to examine fast dynamic processes that were hitherto inaccessible, e.g., rapid quenching of steel or cycling of batteries. For industrial applications, there will be increasing demand for in situ measurements of real components under realistic operating conditions. One example could be the measurements of residual strain or temperature in operating jet engines. Engineering diffractometers are ideal platforms to explore new types of measurements. From these perspectives, the field of engineering application looks very bright. As past examples have shown us, with innovative ideas, difficult experiments can be realized. In this regard, Engineering Application will continue to grow and expand into areas beyond what can be imagined today.

ACKNOWLEDGMENTS

X.L.W. acknowledges the support by the CAS-Croucher Foundation Joint Laboratory on Neutron Scattering (Project No. CityU 9500020), and Ms. Haiyan He for helping with manuscript preparation.

REFERENCES

[1] A.D. Krawitz, T.M. Holden, MRS Bull. 15 (1990) 57–64.
[2] M.T. Hutchings, A.D. Krawitz, The Measurement of Residual and Applied Stress Using Neutron Diffraction, Kluwer Academic Publishers, Dordrecht, The Netherlands, 1992.
[3] P.J. Withers, T.M. Holden, MRS Bull. 24 (1999) 17–23.
[4] X.-L. Wang, JOM 58 (2006) 52–57.
[5] I.M. Robertson, C.A. Schuh, J.S. Vetrano, N.D. Browning, D.P. Field, D.J. Jensen, M.K. Miller, I. Baker, D.C. Dunand, R. Dunin-Borkowski, B. Kabius, T. Kelly, S. Lozano-Perez, A. Misra, G.S. Rohrer, A.D. Rollett, M.L. Taheri, G.B. Thompson, M. Uchic, X.-L. Wang, G. Was, J. Mater. Res. 26 (2011) 1341–1383.
[6] M. Grosse, P. Ottlinger, Mater. Sci. Eng. A 437 (2006) 88–92.
[7] http://articles.latimes.com/2008/oct/25/business/fi-briefs25.S3 (Accessed 15 February 2017).
[8] Z. Feng, X. Wang, S. Spooner, G. Goodwin, P. Maziasz, C. Hubbard, T. Zacharia, ASME-PVP 327 (1996) 119–126.
[9] T. Holden, G. Dolling, B. Powell, J. Winegar, R. Holt, Metall. Trans. A 19 (1988) 2207–2214.

[10] J. Penfold, I.M.Tucker, Large-scale structures, in: F. Fernandez-Alonso, D.L. Price (Eds.), Experimental Methods in the Physical Sciences, Academic Press, London, UK, 2013, pp. 353–409.
[11] E. Bubeck, V. Gerold, G. Kostorz, Cryst. Res. Technol. 20 (1985) 97–103.
[12] K. Masubuchi, Analysis of Welded Structures: Residual Stresses, Distortion, and Their Consequences, Pergamon Press, New York; Oxford, 1980.
[13] P.J. Withers, H.K.D.H. Bhadeshia, Mater. Sci. Technol. 17 (2001) 366–375.
[14] Z. Feng, Processes and Mechanisms of Welding Residual Stress and Distortion, CRC Press, Boca Raton, USA, 2005.
[15] G.A. Webster, A.N. Ezeilo, Residual stress distributions and their influence on fatigue lifetimes, Int. J. Fatigue 23 (2001) S375–S383.
[16] P.J. Withers, Residual stress and its role in failure, Rep. Prog. Phys. 70 (2007) 2211.
[17] G.B. An, W. Woo, J.U. Park, V. Em, Int. J. Fract. 194 (2015) 197–203.
[18] ClassNK, World Maritime News, http://worldmaritimenews.com/, 2013 (Accessed 15 February 2017).
[19] C. Ouchi, ISIJ Int. 41 (2001) 542–553.
[20] T. Inoue, T. Ishikawa, S. Imai, T. Koseki, K. Hirota, M. Tada, Long crack arrestability of heavy-thick shipbuilding steels, in: Isope, 2007, pp. 3322–3326.
[21] International Association of Classification Societies (IACS), Rule Requirements for Use of Extremely Thick Steel Plates, 2013, p. s33.
[22] A.H. Mahmoudi, C.E. Truman, D.J. Smith, Eng. Fract. Mech. 75 (2008) 1516–1534.
[23] W. Woo, G.B. An, E.J. Kingston, A.T. DeWald, D.J. Smith, M.R. Hill, Acta Mater. 61 (2013) 3564–3574.
[24] S.Y. Lee, P.K. Liaw, H. Choo, R.B. Rogge, Acta Mater. 59 (2011) 485–494.
[25] C.J. Lammi, D.A. Lados, Metall. Mater. Trans. A 43 (2012) 87–107.
[26] G.S. Schajer, Practical Residual Stress Measurement Methods, Wiley, Chichester, UK, 2013. pp. 259–277.
[27] A.D. Krawitz, Introduction to Diffraction in Materials Science and Engineering, Wiley, Toronto, Canada, 2001. pp. 424–424.
[28] C.G. Windsor, Pulsed Neutron Scattering, Taylor & Francis, Oxford, UK, 1981.
[29] I.C. Noyan, J.B. Cohen, Residual Stress: Measurement by Diffraction and Interpretation, Springer-Verlag, New York, 1987.
[30] M.T. Hutchings, P.J. Withers, T.M. Holden, T. Lorentzen, Introduction to the Characterization of Residual Stress by Neutron Diffraction, CRC Press, Boca Raton, USA, 2005.
[31] P.J. Withers, C. R. Phys. 8 (2007) 806–820.
[32] A. Allen, C. Andreani, M.T. Hutchings, C.G. Windsor, NDT Int. 14 (1981) 249–254.
[33] D.J. Smith, R.H. Leggatt, G.A. Webster, H.J. Macgillivray, P.J. Webster, G. Mills, J. Strain Anal. Eng. Des. 23 (1988) 201–211.
[34] J.H. Root, T.M. Holden, J. Schroder, C.R. Hubbard, S. Spooner, T.A. Dodson, S.A. David, Mater. Sci. Technol. 9 (1993) 754–759.
[35] A.D. Krawitz, R.A. Winholtz, Mater. Sci. Eng. A 185 (1994) 123–130.
[36] X.L. Wang, E.A. Payzant, B. Taljat, C.R. Hubbard, J.R. Keiser, M.J. Jirinec, Mater. Sci. Eng. A 232 (1997) 31–38.
[37] A. Gianni, B. Giovanni, C. Adele, F. Fabrizio, R. Massimo, R. Franco, Meas. Sci. Technol. 10 (1999) R56.
[38] H.J. Stone, P.J. Withers, S.M. Roberts, R.C. Reed, T.M. Holden, Metall. Mater. Trans. A 30 (1999) 1797–1808.

[39] M.R. Daymond, M.A.M. Bourke, R.B. Von Dreele, B. Clausen, T. Lorentzen, J. Appl. Phys. 82 (1997) 1554–1562.
[40] J.W.L. Pang, T.M. Holden, T.E. Mason, Acta Mater. 46 (1998) 1503–1518.
[41] B. Clausen, T. Lorentzen, M.A.M. Bourke, M.R. Daymond, Mater. Sci. Eng. A 259 (1999) 17–24.
[42] D.G. Carr, M.I. Ripley, T.M. Holden, D.W. Brown, S.C. Vogel, Acta Mater. 52 (2004) 4083–4091.
[43] T.M. Holden, H. Suzuki, D.G. Carr, M.I. Ripley, B. Clausen, Mater. Sci. Eng. A 437 (2006) 33–37.
[44] A.M. Paradowska, J.W.H. Price, T.R. Finlayson, U. Lienert, P. Walls, R. Ibrahim, J. Phys. Condens. Matter 21 (2009) 124213.
[45] W. Woo, Z. Feng, X.L. Wang, D.W. Brown, B. Clausen, K. An, H. Choo, C.R. Hubbard, S.A. David, Sci. Technol. Weld. Joining 12 (2007) 298–303.
[46] X.L. Wang, T.M. Holden, G.Q. Rennich, A.D. Stoica, P.K. Liaw, H. Choo, C.R. Hubbard, Phys. B Condens. Matter 385–386 (Part 1) (2006) 673–675.
[47] G.A. Webster, R.W. Wimpory, Polycrystalline Materials Determination of Residual Stresses by Neutron Diffraction, ISO/TTA3, Technology Trends Assessment, Geneva, Switzerland, 2001.
[48] H. Rietveld, J. Appl. Crystallogr. 2 (1969) 65–71.
[49] A.C. Larson, R.B. Von Dreele, Structure 748 (2004) 86–748.
[50] P. Withers, J. Appl. Crystallogr. 37 (2004) 596–606.
[51] J.R. Santisteban, M.R. Daymond, J.A. James, L. Edwards, J. Appl. Crystallogr. 39 (2006) 812–825.
[52] W. Woo, V. Em, B.-S. Seong, E. Shin, P. Mikula, J. Joo, M.-H. Kang, J. Appl. Crystallogr. 44 (2011) 747–754.
[53] P. Mikula, J. Kulda, P. Lukáš, M. Vrána, V. Wagner, Nucl. Instrum. Methods Phys. Res. Sect. A Accel. Spectrom. Detect. Assoc. Equip. 338 (1994) 18–26.
[54] M. Popovici, W.B. Yelon, R. Berliner, A.D. Stoica, I. Ioniţă, R. Law, Nucl. Instrum. Methods Phys. Res. Sect. A Accel. Spectrom. Detect. Assoc. Equip. 338 (1994) 99–110.
[55] M.-K. Moon, V.T. Em, C.-H. Lee, P. Mikula, K.-P. Hong, Y.-H. Choi, J.-K. Cheon, U.-W. Nam, K.-N. Kong, K.-C. Jin, Phys. B Condens. Matter 368 (2005) 70–75.
[56] B.-S. Seong, V. Em, P. Mikula, J. Saroun, M.-H. Kang, J. Appl. Crystallogr. 43 (2010) 654–658.
[57] W. Woo, V. Em, P. Mikula, G.-B. An, B.-S. Seong, Mater. Sci. Eng. A 528 (2011) 4120–4124.
[58] X. L. Wang, Conceptual Design of the SNS Engineering Diffractometer, SNS Report No. IS-1.1.8.2-6035-RE-A-00, 2000.
[59] https://neutrons.ornl.gov/vulcan (Accessed 15 February 2017).
[60] V. Em, W. Woo, B.S. Seong, P. Mikula, J. Joo, M.H. Kang, K.H. Lee, J. Phys. Conf. Ser. 340 (2012) 012025.
[61] W. Woo, V. Em, E. Shin, P. Mikula, V. Ryukhtin, J. Appl. Crystallogr. 48 (2015) 350–356.
[62] P. Staron, M. Koçak, S. Williams, Appl. Phys. A 74 (2002) s1161–s1162.
[63] X.-L. Wang, Y.D. Wang, J.W. Richardson, J. Appl. Crystallogr. 35 (2002) 533–537.
[64] A. Paradowska, T.R. Finlayson, J.W.H. Price, R. Ibrahim, A. Steuwer, M. Ripley, Phys. B Condens. Matter 385–386 (Part 2) (2006) 904–907.
[65] A. Steuwer, M. Dumont, M. Peel, M. Preuss, P.J. Withers, Acta Mater. 55 (2007) 4111–4120.
[66] D. Deng, S. Kiyoshima, Comput. Mater. Sci. 62 (2012) 23–34.

[67] D. Smith, P. Bouchard, D. George, J. Strain Anal. Eng. Des. 35 (2000) 287.
[68] W. Woo, V. Em, C.R. Hubbard, H.-J. Lee, K.S. Park, Mater. Sci. Eng. A 528 (2011) 8021–8027.
[69] J.W. Kim, K. Lee, J.S. Kim, T.S. Byun, J. Nucl. Mater. 384 (2009) 212–221.
[70] M. Law, H. Prask, V. Luzin, T. Gnaeupel-Herold, Mater. Sci. Eng. A 437 (2006) 60–63.
[71] R.S. Mishra, Z.Y. Ma, Mater. Sci. Eng. R Rep. 50 (2005) 1–78.
[72] X.L. Wang, S.A. David, Z. Feng, S. Spooner, C.R. Hubbard, Neutron Diffraction Study of Residual Stresses in Friction Stir Welds, in: Proceedings of the 6th International Conference on Residual Stresses, IOM Communications, Oxford, UK, 2000, pp. 1408–1414.
[73] W. Woo, Z. Feng, X.L. Wang, S.A. David, Sci. Technol. Weld. Join. 16 (2011) 23–32.
[74] M.A. Sutton, A.P. Reynolds, D.Q. Wang, C.R. Hubbard, J. Eng. Mater. Technol. 124 (2002) 215–221.
[75] W. Woo, H. Choo, D.W. Brown, M.A.M. Bourke, Z. Feng, S.A. David, C.R. Hubbard, P.K. Liaw, Appl. Phys. Lett. 86 (2005) 231902.
[76] W. Woo, Z. Feng, X.L. Wang, K. An, B. Clausen, T.A. Sisneros, J.S. Jeong, Mater. Lett. 85 (2012) 29–32.
[77] T. Ungár, A. Borbély, Appl. Phys. Lett. 69 (1996) 3173–3175.
[78] T. Ungar, I. Dragomir, A. Revesz, A. Borbely, J. Appl. Crystallogr. 32 (1999) 992–1002.
[79] B.E. Warren, B.L. Averbach, J. Appl. Phys. 23 (1952) 497.
[80] J.S. Jeong, W. Woo, K.H. Oh, S.K. Kwon, Y.M. Koo, Acta Mater. 60 (2012) 2290–2299.
[81] Y.H. Zhao, X.Z. Liao, Z. Jin, R.Z. Valiev, Y.T. Zhu, Acta Mater. 52 (2004) 4589–4599.
[82] B. Clausen, Characterisation of Polycrystal Deformation by Numerical Modelling and Neutron Diffraction Measurements, PhD Thesis, Riso National Laboratory, Roskilde, Denmark, 1997.
[83] P.K. Liaw, H. Choo, R.A. Buchanan, C.R. Hubbard, X.L. Wang, Mater. Sci. Eng. A 437 (2006) 126–133.
[84] X. Jin, T. Nakamoto, S. Harjo, T. Hemmi, T. Umeno, T. Ogitsu, A. Yamamoto, M. Sugano, K. Aizawa, J. Abe, W. Gong, T. Iwahashi, Rev. Sci. Instrum. 84 (2013) 063106.
[85] S. Harjo, K. Aizawa, T. Kawasaki, W. Gong, T. Iwahashi, T. Ito, J. Abe, T. Nakatani, Present status of BL19 TAKUMI at J-PARC, in: 21st International Collaboration on Advanced Neutron Source (ICANS-21), 2015.
[86] A. Pramanick, X.P. Wang, K. An, A.D. Stoica, J. Yi, Z. Gai, C. Hoffmann, X.L. Wang, Phys. Rev. B 85 (2012) 144412.
[87] A. Pramanick, X.L. Wang, A.D. Stoica, C. Yu, Y. Ren, S. Tang, Z. Gai, Phys. Rev. Lett. 112 (2014) 217205.
[88] A. Pramanick, K. An, A.D. Stoica, X.L. Wang, Scr. Mater. 65 (2011) 540–543.
[89] G.M. Ludtka, F.R. Klose, R.A. Kisner, J.A. Fernandez-Baca, G.M. Ludtka, J.B. Wilgen, R.A. Jaramillo, L.J. Santodonato, X.-L. Wang, C.R. Hubbard, F. Tang, Time-resolved analyses of microstructure in advanced materials under magnetic fields at elevated temperature using neutrons, in: Proceedings of 2007 TMS Annual Meeting and Exhibition, 2007.
[90] S. Misture, S. Pilgrim, J. Hicks, C. Blue, E. Payzant, C. Hubbard, Appl. Phys. Lett. 72 (1998) 1042–1044.
[91] A. Pramanick, M.R.V. Jørgensen, S.O. Diallo, A.D. Christianson, J.A. Fernandez-Baca, C. Hoffmann, X. Wang, S. Lan, X.-L. Wang, Adv. Mater. 27 (2015) 4330–4335.
[92] X.-L. Wang, K. An, L. Cai, Z. Feng, S.E. Nagler, C. Daniel, K.J. Rhodes, A.D. Stoica, H.D. Skorpenske, C. Liang, Sci. Rep. 2 (2012) 747.

[93] A. Huq, J.W. Richardson, E.R. Maxey, D. Chandra, W.-M. Chien, J. Alloys Compd. 436 (2007) 256–260.
[94] W.-M. Chien, J. Lamb, D. Chandra, A. Huq, J. Richardson Jr., E. Maxey, J. Alloys Compd. 446–447 (2007) 363–367.
[95] N. Mauro, A. Vogt, K. Derendorf, M. Johnson, G. Rustan, D. Quirinale, A. Kreyssig, K. Lokshin, J. Neuefeind, K. An, Rev. Sci. Instrum. 87 (2016) 013904.
[96] M.L. Johnson, M.E. Blodgett, K.A. Lokshin, N.A. Mauro, J. Neuefeind, C. Pueblo, D.G. Quirinale, A.J. Vogt, T. Egami, A.I. Goldman, K.F. Kelton, Phys. Rev. B 93 (2016) 054203.
[97] Z.Z. Yu, Z.L. Feng, K. An, W. Zhang, E.D. Specht, J. Chen, X.-L. Wang, S. David, Z.Z. Yu, Z.L. Feng, Measurement of steel phase transformation kinetics by dilatometry and in-situ neutron diffraction – a comparative study, in: Trends in Welding Research 2012: Proceedings of the 9th International Conference, A Complete Proceedings, ASM International, 2013, pp. 510–513.
[98] Z. Yu, Z. Feng, K. An, W. Zhang, E.D. Specht, J. Chen, X.-L. Wang, S. David, In-situ neutron diffraction study of non-equilibrium phase transformation in advanced high-strength steels, in: Trends in Welding Research 2012: Proceedings of the 9th International Conference, A Complete Proceedings, ASM International, 2013, pp. 505–509.
[99] M. Iizumi, Physica B+C 136 (1986) 36–41.
[100] D. Ma, A.D. Stoica, K. An, L. Yang, H. Bei, R.A. Mills, H. Skorpenske, X.-L. Wang, Metall. Mater. Trans. A 42 (2011) 1444–1448.
[101] H. Nojiri, S. Yoshii, M. Yasui, K. Okada, M. Matsuda, J.S. Jung, T. Kimura, L. Santodonato, G.E. Granroth, K.A. Ross, J.P. Carlo, B.D. Gaulin, Phys. Rev. Lett. 106 (2011) 237202.
[102] X.-L. Wang, K. An, L. Cai, Z. Feng, S.E. Nagler, C. Daniel, K.J. Rhodes, A.D. Stoica, H.D. Skorpenske, C. Liang, W. Zhang, J. Kim, Y. Qi, S.J. Harris, Sci. Rep. 2 (2012) 747.
[103] J.W.L. Pang, T.M. Holden, J.S. Wright, T.E. Mason, Acta Mater. 48 (2000) 1131–1140.
[104] G. Greenough, Nature 160 (1947) 258–260.
[105] G.B. Greenough, Nature 166 (1950) 509–510.
[106] L. Pintschovius, V. Hauk, W.K. Krug, Mater. Sci. Eng. 92 (1987) 1–12.
[107] S. Spooner, X.-L. Wang, J. Appl. Crystallogr. 30 (1997) 449–455.
[108] M. Bourke, D. Brown, B. Clausen, H. Choo, P. Rancaswamy, R. Vaidyanathan, Neutron News 10 (1999) 24–30.
[109] B. Clausen, Characteristics of Polycrystalline Deformation by Numerical Modeling and Neutron Diffraction Measurements, Resø National Laboratory Report, 1997, ISBN: 87-550-2304-2305.
[110] B. Clausen, T. Lorentzen, T. Leffers, Acta Mater. 46 (1998) 3087–3098.
[111] Y.-D. Wang, H. Tian, A.D. Stoica, X.-L. Wang, P.K. Liaw, J.W. Richardson, Nat. Mater. 2 (2003) 101–106.
[112] J.W.L. Pang, R.B. Rogge, R.L. Donaberger, Mater. Sci. Eng. A 437 (2006) 21–25.
[113] T. Holden, R. Holt, C. Tomé, Phys. B Condens. Matter 241 (1997) 1246–1248.
[114] T. Holden, R. Holt, A. Clarke, J. Neutron Res. 5 (1997) 241–264.
[115] J.A. Wollmershauser, S. Kabra, S.R. Agnew, Acta Mater. 57 (2009) 213–223.
[116] S. Cheng, J.A. Spencer, W.W. Milligan, Acta Mater. 51 (2003) 4505–4518.
[117] J.R. Trelewicz, C.A. Schuh, Acta Mater. 55 (2007) 5948–5958.
[118] A.S. Argon, Strengthening Mechanisms in Crystal Plasticity, Oxford University Press, New York, USA, 2008, p. 64.
[119] H. Van Swygenhoven, P.M. Derlet, Phys. Rev. B 64 (2001) 224105.
[120] H. Van Swygenhoven, P.M. Derlet, A.G. Froseth, Nat. Mater. 3 (2004) 399–403.

[121] Z. Budrovic, H. Van Swygenhoven, P.M. Derlet, S. Van Petegem, B. Schmitt, Science 304 (2004) 273.

[122] S. Cheng, A.D. Stoica, X.L. Wang, Y. Ren, J. Almer, J.A. Horton, C.T. Liu, B. Clausen, D.W. Brown, P.K. Liaw, L. Zuo, Phys. Rev. Lett. 103 (2009) 035502.

[123] H.W. Sheng, W.K. Luo, F.M. Alamgir, J.M. Bai, E. Ma, Nature 439 (2006) 419–425.

[124] W.L. Johnson, MRS Bull. 24 (1999) 42–56.

[125] A. Inoue, Acta Mater. 48 (2000) 279–306.

[126] H.F. Poulsen, J.A. Wert, J. Neuefeind, V. Honkimaki, M. Daymond, Nat. Mater. 4 (2005) 33–36.

[127] D. Ma, A.D. Stoica, X.L. Wang, Nat. Mater. 8 (2009) 30–34.

[128] D.B. Miracle, Nat. Mater. 3 (2004) 697–702.

[129] D. Ma, A.D. Stoica, X.L. Wang, Z.P. Lu, B. Clausen, D.W. Brown, Phys. Rev. Lett. 108 (2012) 085501.

[130] G.K. Williamson, W.H. Hall, Acta Metall. 1 (1953) 22–31.

[131] T. Ungar, J. Gubicza, G. Ribarik, A. Borbely, J. Appl. Crystallogr. 34 (2001) 298–310.

[132] G. Ribárik, J. Gubicza, T. Ungár, Mater. Sci. Eng. A 387–389 (2004) 343–347.

[133] D.W. Brown, M.A.M. Bourke, R.D. Field, W.L. Hults, D.F. Teter, D.J. Thoma, S.C. Vogel, Mater. Sci. Eng. A 421 (2006) 15–21.

[134] J. Diehl, Z. Metallkd. 47 (1956) 331–343.

[135] J. Diehl, Z. Metallkd. 47 (1956) 411–416.

[136] X. Huang, A. Borrego, W. Pantleon, Mater. Sci. Eng. A 319–321 (2001) 237–241.

[137] X. Huang, G. Winther, Philos. Mag. 87 (2007) 5189–5214.

[138] G. Winther, X. Huang, Philos. Mag. 87 (2007) 5215–5235.

[139] M. Wilkens, Theoretical aspects of kinematical X-ray diffraction profiles from crystals containing dislocation distributions (Fourier transform of X ray diffraction line profiles from crystals with dislocations), in: NBS Fundamental Aspects of Dislocation Theory, vol. 2, 1970, pp. 1195–1221.

[140] T. Ungár, A.D. Stoica, G. Tichy, X.-L. Wang, Acta Mater. 66 (2014) 251–261.

[141] A. Borbely, J. Dragomir-Cernatescu, G. Ribarik, T. Ungar, J. Appl. Crystallogr. 36 (2003) 160–162.

[142] W.F. Hosford, The Mechanics of Crystals and Textured Polycrystals, Oxford University Press, USA, 1993, p. 248.

[143] X.L. Wang, Y.D. Wang, A.D. Stoica, D.J. Horton, H. Tian, P.K. Liaw, H. Choo, J.W. Richardson, E. Maxey, Mater. Sci. Eng. A 399 (2005) 114–119.

[144] Y. Wu, W. Liu, X. Wang, D. Ma, A.D. Stoica, T. Nieh, Z. He, Z. Lu, Appl. Phys. Lett. 104 (2014) 051910.

[145] Y.-W. Lai, R. Schäfer, L. Schultz, J. McCord, Acta Mater. 56 (2008) 5130–5137.

[146] M.A. Marioni, R.C. O'Handley, S.M. Allen, Appl. Phys. Lett. 83 (2003) 3966–3968.

[147] P. Molnar, P. Sittner, V. Novak, J. Prokleska, V. Sechovsky, B. Ouladdiaf, S.P. Hanulla, O. Heczko, J. Phys. Condens. Matter 20 (2008) 104224.

[148] Y. Tomota, P.G. Xu, T. Kamiyama, E.C. Oliver, Nucl. Instrum. Methods Phys. Res. Sect. A Accel. Spectrom. Detect. Assoc. Equip. 600 (2009) 313–315.

[149] https://j-parc.jp/researcher/MatLife/en/instrumentation/images/takumi.gif (Accessed 15 February 2017).

[150] M. Armand, J.M. Tarascon, Nature 451 (2008) 652–657.

[151] S. Taminato, M. Yonemura, S. Shiotani, T. Kamiyama, S. Torii, M. Nagao, Y. Ishikawa, K. Mori, T. Fukunaga, Y. Onodera, T. Naka, M. Morishima, Y. Ukyo, D.S. Adipranoto, H. Arai, Y. Uchimoto, Z. Ogumi, K. Suzuki, M. Hirayama, R. Kanno, Sci. Rep. 6 (2016) 28843.

[152] M. Yonemura, K. Mori, T. Kamiyama, T. Fukunaga, S. Torii, M. Nagao, Y. Ishikawa, Y. Onodera, D.S. Adipranoto, H. Arai, Y. Uchimoto, Z. Ogumi, J. Phys. Conf. Ser. 502 (2014) 012053.
[153] F. Zhang, J. Ilavsky, G.G. Long, J.P.G. Quintana, A.J. Allen, P.R. Jemian, Metall. Mater. Trans. A 41 (2010) 1151–1158.
[154] H. Leitner, P. Staron, H. Clemens, S. Marsoner, P. Warbichler, Mater. Sci. Eng. A 398 (2005) 323–331.
[155] M. Ohnuma, J. Suzuki, S. Ohtsuka, S.W. Kim, T. Kaito, M. Inoue, H. Kitazawa, Acta Mater. 57 (2009) 5571–5581.
[156] D.L. Price, F. Fernandez-Alonso, An introduction to neutron scattering, in: F. Fernandez-Alonso, D.L. Price (Eds.), Neutron Scattering-Fundamentals, Academic Press, London, UK, 2013, pp. 1–136.
[157] G. Kostorz, J. Appl. Crystallogr. 24 (1991) 444–456.
[158] P. Fratzl, J. Appl. Crystallogr. 36 (2003) 397–404.
[159] P.R. Jemian, J.R. Weertman, G.G. Long, R.D. Spal, Acta Metall. Mater. 39 (1991) 2477–2487.
[160] R.J. Roe, Methods of X-Ray and Neutron Scattering in Polymer Science, Oxford University Press, New York, USA, 2000, pp.150–209.
[161] A. Guinier, G. fournet, Small Angle Scattering of X-rays, John Wiley and Sons, New York, USA, 1955, p. 127.
[162] M. Ojima, M. Ohnuma, J. Suzuki, S. Ueta, S. Narita, T. Shimizu, Y. Tomota, Scr. Mater. 59 (2008) 313–316.
[163] J. Ilavsky, P.R. Jemian, J. Appl. Crystallogr. 42 (2009) 347–353.
[164] P. Kozikowski, M. Ohnuma, M. Ohta, T. Terakado, Y. Yoshizawa, S. Koppoju, M. Lewandowska, Materials Transactions, 58, 2017, in print.
[165] R. Gilles, D. Mukherji, P. Strunz, S. Lieske, A. Wiedenmann, R.P. Wahi, Scr. Mater. 39 (1998) 715–721.
[166] J.S. Pedersen, Adv. Colloid Interface Sci. 70 (1997) 171–210.
[167] L.A. Feigin, D.I. Svergun, G.W. Taylor, Determination of the integral parameters of particles, in: G.W. Taylor (Ed.), Structure Analysis by Small-Angle X-Ray and Neutron Scattering, Springer US, Boston, MA, 1987, pp. 83–91.
[168] J. Dawidowski, J.R. Granada, J.R. Santisteban, F. Cantargi, L.A. Rodríguez Palomino, Appendix: Neutron Scattering Lengths and Cross Sections, in: F. Fernandez-Alonso, D.L. Price (Eds.), Neutron Scattering - Fundamentals, Academic Press, 2013, pp. 471–528.
[169] http://www.ncnr.nist.gov/resources/n-lengths/ (Accessed 15 February 2017).
[170] K.S. Cho, S.S. Park, S.W. Baek, E.J. Shin, D.H. Choi, H.R. Yang, H. Kwon, J. Alloys Compd. 655 (2016) 395–405.
[171] M. Ohnuma, K. Hono, S. Linderoth, J.S. Pedersen, Y. Yoshizawa, H. Onodera, Acta Mater. 48 (2000) 4783–4790.
[172] A. Grob, S. Saranu, U. Herr, A. Michels, R.N. Viswanath, J. Weissmüller, Phys. Status Solidi A Appl. Res. 201 (2004) 3354–3360.
[173] H. Nishijima, Y. Tomota, Y. Su, W. Gong, J.-I. Suzuki, Metals 6 (2016) 16.
[174] V. Gerold, J.E. Epperson, G. Kostorz, J. Appl. Crystallogr. 10 (1977) 28–31.
[175] Y. Funakawa, K. Seto, Tetsu to Hagane 93 (2007) 49–56.
[176] Y. Oba, S. Koppoju, M. Ohnuma, T. Murakami, H. Hatano, K. Sasakawa, A. Kitahara, J.-i. Suzuki, ISIJ Int. 51 (2011) 1852–1858.

[177] M. Ohnuma, T. Ishida, M. Furuska, B.S. Seong, Nanosize precipitate in steels characterized by Small-Angle Neutron Scattering, in: Proceedings of Asia Steel International Conference, 2015, pp. 274–275.

[178] Y. Yamagata, K. Hirota, J. Ju, S. Wang, S.-y. Morita, J.-i. Kato, Y. Otake, A. Taketani, Y. Seki, M. Yamada, H. Ota, U. Bautista, Q. Jia, J. Radioanal. Nucl. Chem. 305 (2015) 787–794.

[179] M. Yamada, Y. Otake, A. Taketani, H. Sunaga, Y. Yamagata, T. Wakabayashi, K. Kono, T. Nakayama, Tetsu to Hagane 100 (2014) 99–101.

[180] http://www.ucans.org/index.html (accessed 8 May 2017).

Index

Note: Page numbers followed by "*f*" indicate figures and "*t*" indicate tables.

Q

R

T

X

Y

Z